42 Springer Series in Solid-State Sciences
Edited by Hans-Joachim Queisser

Springer Series in Solid-State Sciences

Editors: M. Cardona P. Fulde H.-J. Queisser

Volume 40 **Semiconductor Physics** - An Introduction By K. Seeger

Volume 41 **The LMTO Method** Muffin-Tin Orbitals and Electronic Structure
By H.L. Skriver

Volume 42 **Crystal Optics with Spatial Dispersion, and Excitons**
By V. M. Agranovich and V. L. Ginzburg

Volume 43 **Resonant Nonlinear Interactions of Light with Matter**
By V.S. Butylkin, A.E. Kaplan, Yu.G. Khronopulo, and E.I. Yakubovich

Volume 44 **Elastic Media with Microstructure II** Three-Dimensional Models
By I.A. Kunin

Volume 45 **Electronic Properties of Doped Semiconductors**
By B. I. Shklovskii and A. L. Efros

Volume 46 **Topological Disorder in Condensed Matter**
Editors: F. Yonezawa and T. Ninomiya

Volume 47 **Statics and Dynamics of Nonlinear Systems**
Editors: G. Benedek, H. Bilz, and R. Zeyher

Volume 48 **Magnetic Phase Transitions**
Editors: M. Ausloos and R. J. Elliott

Volume 49 **Organic Molecular Aggregates,** Electronic Excitation and Interaction Processes
Editors: P. Reineker, H. Haken, and H. C. Wolf

Volume 50 **Multiple Diffraction of X-Rays in Crystals**
By Shih-Lin Chang

Volume 51 **Phonon Scattering in Condensed Matter**
Editor: W. Eisenmenger, K. Laßmann, and S. Döttinger

Volume 52 **Magnetic Superconductors and Their Related Problems**
Editors: T. Matsubara and A. Kotani

Volumes 1 - 39 are listed on the back inside cover

V. M. Agranovich V. L. Ginzburg

Crystal Optics with Spatial Dispersion, and Excitons

Second Corrected and Updated Edition

With 46 Figures

Springer-Verlag Berlin Heidelberg GmbH 1984

Professor Dr. Vladimir M. Agranovich
Academy of Sciences of the USSR, Institute of Spectroscopy, Troitsk,
Moscow, r-n, 142092, USSR

Professor Dr. Vitaly Ginzburg
P.N. Lebedev Physical Institute, Academy of Sciences of the USSR
Moscow, USSR

Series Editors:
Professor Dr. Manuel Cardona
Professor Dr. Peter Fulde
Professor Dr. Hans-Joachim Queisser
Max-Planck-Institut für Festkörperforschung, Heisenbergstrasse 1
D-7000 Stuttgart 80, Fed. Rep. of Germany

Title of the 1. English Edition:
Spatial Dispersion in Crystal Optics and the Theory of Excitons
© by Interscience, London, New York, Sidney 1966
Title of the 2nd corrected and enlarged original Russian edition:
Kristallooptika s uchetom prostranstvennoi dispersii i teoriya eksitonov
© by "Nauka" Publishing House Moscow 1979

ISBN 978-3-662-02408-9 ISBN 978-3-662-02406-5 (eBook)
DOI 10.1007/978-3-662-02406-5

Library of Congress Cataloging in Publication Data. Agranovich, V.M. (Vladimir Moiseevich), 1929-. Crystal optics with spatial dispersion, and excitons. (Springer series in solid-state sciences ; v. 42). Translation of: Kristallooptika s uchetom prostranstvennoi dispersii i teorii éksitonov. 1. Crystal optics. 2. Exciton theory. I. Ginzburg, V.L. (Vitalii Lazarevich), 1916-. II. Title. III. Series.
QD941.A413 1983 548'.9 83-16717

Originally published by Springer-Verlag Berlin Heidelberg New York in 1984.
Softcover reprint of the hardcover 2nd edition 1984

Typesetting: K+V Fotosatz, 6124 Beerfelden

2153/3020-543210

Preface

Spatial dispersion, namely, the dependence of the dielectric-constant tensor on the wave vector (i.e., on the wavelength) at a fixed frequency, is receiving increased attention in electrodynamics and condensed-matter optics, particularly in crystal optics.

In contrast to frequency dispersion, namely, the frequency dependence of the dielectric constant, spatial dispersion is of interest in optics mainly when it leads to qualitatively new phenomena. One such phenomenon has been well known for many years; it is the natural optical activity (gyrotropy). But there are other interesting effects due to spatial dispersion, namely, new normal waves near absorption lines, optical anisotropy of cubic crystals, and many others.

Crystal optics that takes spatial dispersion into account includes classical crystal optics with frequency dispersion only, as a special case. In our opinion, this fact alone justifies efforts to develop crystal optics with spatial dispersion taken into account, although admittedly its influence is small in some cases and it is observable only under rather special conditions.

Furthermore, spatial dispersion in crystal optics deserves attention from another point as well, namely, the investigation of excitons that can be excited by light. We contend that crystal optics with spatial dispersion and the theory of excitons are fields that overlap to a great extent, and that it is sometimes quite impossible to separate them. It is our aim to show the true interplay between these interrelations and to combine the macroscopic and microscopic approaches to crystal optics with spatial dispersion and exciton theory.

The book covers a wide range of the crystal-optics and condensed-matter electrodynamics problems. The selection reflects, to some extent, the authors, scientific interests. Thus, this monograph should not be considered a review of all the available literature. It was our aim to deal with a broad range of problems by applying the above-mentioned approach, as in our previous publications.

To our knowledge, no other books have been devoted to spatial dispersion in crystal optics. The lack of such a treatment becomes evident in almost any textbook on crystal optics. These books usually lag behind the current literature. Many experimental and theoretical problems of linear and nonlinear optics await solution. Their investigation can be accelerated by a sufficiently comprehensive and detailed presentation of the state-of-the art.

We know that most readers of such monographs are interested only in specific problems; others find it difficult to read chapters with cumbersome calculations, etc. Therefore, we made no attempt to avoid certain repetitions and elementary explanations. Experience shows that it is not the best practice to save space in such a text by not repeating fundamentals, even several times. In our opinion, on the other hand, it proves more expedient in becoming acquainted with details and subtleties of special problems in scientific journals and papers.

The first English edition of this book was published in 1966 (first and second Russian editions – in 1965 and 1979). In addition to the quite numerous changes and additions of new material, many sections have been substantially altered or even rewritten. We are here referring to the treatments of boundary conditions, surface-polariton spectroscopy, Raman scattering of light by polaritons, the calculation of the dielectric-constant tensor by the local-field method, and experimental investigations of the effects due to spatial dispersion. These are all problems that have attracted much attention in recent years, and have been extensively discussed in the literature.

This is also true of nonlinear effects in crystal optics. But we do not deal with nonlinear crystal optics per se in the present book, except for a discussion of certain closely allied processes, such as light and x-ray scattering in crystals, the influence of external electric and magnetic fields on optical processes, and boundary conditions. In the preface to the original Russian edition, we stressed the fact that we had not made any attempt to completely cover the subject, and had especially not reviewed all the existing literature. Today, when the volume of available data has increased considerably, the aforesaid is all the more true. Nevertheless, we are confident that we have elucidated all the most important problems of linear crystal optics with spatial dispersion. This, we feel, justifies the publication of this new revised and supplemented edition.

In conclusion, the authors wish to thank T. A. Leskova, A. G. Malshukov, A. G. Molchanov and V. I. Yudson for their assistance in preparing the manuscript.

The authors also sincerely thank the translator N. Weinstein for a number of useful suggestions that have undoubtedly improved the presentation of the subject. No less gratitude is due to Dr. H. Lotsch for his extensive help in the editing of our book.

Moscow, December 1983 *V.M. Agranovich* and *V.L. Ginzburg*

Contents

1. Introduction 1
1.1 Basic Problems of Crystal Optics 1
1.2 Excitons and Polaritons 10

2. The Complex Dielectric-Constant Tensor $\varepsilon_{ij}(\omega, \boldsymbol{k})$ and Normal Waves in a Medium 18
2.1 The Tensor $\varepsilon_{ij}(\omega, \boldsymbol{k})$ and Its Properties 18
2.1.1 The Electromagnetic Field Equations and the Introduction of the Tensor $\varepsilon_{ij}(\omega, \boldsymbol{k})$ 18
2.1.2 General Properties of the Tensor $\varepsilon_{ij}(\omega, \boldsymbol{k})$ 27
2.1.3 The Approximation of Classical Crystal Optics. The Tensor $\varepsilon_{ij}(\omega, \boldsymbol{k})$ in an Isotropic Medium 37
2.2 Normal Electromagnetic Waves in a Medium 42
2.2.1 Wave Equation and Dispersion Equation 42
2.2.2 Transverse and Longitudinal Waves, "Fictitious" Longitudinal Waves and "Polarization Waves". Real, Coulomb and Mechanical Excitons 47
a) Coulomb Excitons: Longitudinal and "Fictitious" Longitudinal Waves 51
b) Coulomb Excitons: "Polarization Waves" 54
2.2.3 Multiple Roots of the Dispersion Equation 57
2.2.4 Separating the Transverse Field $\boldsymbol{E}_\perp$ and the Tensor $\varepsilon_{\perp,ij}$ .. 62
2.2.5 The Dispersion Relations for a Complex Refractive Index. Inequalities for the Region of Transparency 66
2.3 Energy Relations and Other Equations for Waves in an Anisotropic Medium 72
2.3.1 The Law of Energy Conservation in the Electrodynamics of Media Displaying Spatial Dispersion 72
2.3.2 Quadratic Functions of the Normal Wave Amplitudes ... 81
a) The Conservation of Energy for a Field 83
b) The Vector of Group Velocity 84
c) The Conservation of Momentum for a Field 93
d) Some Applications of the Poynting Theorem 94
e) The Boundary Conditions for a Gyrotropic Medium .. 95

f) The Boundary Conditions for the Optically Nonlinear Medium with the Center of Inversion ... 97
g) Spatial Dispersion and Orthogonality of Normal Waves ... 99
h) The Reciprocity Theorem ... 100
2.3.3 Certain Theorems for Ray and Wave Propagation in a Medium ... 103

3. The Tensor $\varepsilon_{ij}(\omega, \boldsymbol{k})$ in Crystals ... 108
3.1 The Tensor $\varepsilon_{ij}(\omega, \boldsymbol{k})$ for Crystals ... 108
3.1.1 Introduction of the Tensor $\varepsilon_{ij}(\omega, \boldsymbol{k})$ ... 108
3.1.2 Weak Spatial Dispersion ... 117
3.2 The Tensor $\varepsilon_{ij}(\omega, \boldsymbol{k})$ for Crystals of Various Classes ... 122
3.2.1 Gyrotropic Crystals ... 122
3.2.2 Nongyrotropic Crystals ... 129

4. Spatial Dispersion in Crystal Optics ... 136
4.1 Gyrotropic Crystals ... 136
4.1.1 Normal Waves in Gyrotropic Crystals ... 137
4.1.2 A New Wave Near the Absorption Lines in a Gyrotropic Medium ... 145
4.2 Nongyrotropic Crystals ... 149
4.2.1 Normal Waves ... 149
4.2.2 Isotropic Medium and New Waves Near Dipole Absorption Lines ... 150
a) Group-Velocity Vector ... 165
4.3 Cubic Nongyrotropic Crystals ... 167
4.3.1 Optical Anisotropy: Dipole Transitions ... 167
4.3.2 Optical Anisotropy: Quadrupole Transitions ... 171
4.3.3 Classification of the States of 'Mechanical' Excitons with $k = 0$ and Selection Rules for Quadrupole Transitions ... 174
4.3.4 New Waves Near Quadrupole Absorption Lines. Longitudinal Waves ... 185
4.4 Influence of Mechanical Stresses and External Electric and Magnetic Fields ... 188
4.4.1 Anisotropy of the Optical Properties and Selection Rules in the Presence of External Influences ... 188
4.4.2 The Explicit Dependence of the Tensor $\varepsilon_{ij}(\omega, \boldsymbol{k})$ on the Strength of Weak External Fields ... 192
a) The Magnetic-Field Inversion Effect ... 196
4.4.3 Influence of Magnetic and Electric Fields on Cadmium Sulfide (CdS) Crystals ... 198
a) Quadratic Magneto-Stark Effect ... 202
4.5 Boundary Conditions in the Case of Spatial Dispersion Near a Separate Resonance (Absorption Line) ... 205

4.5.1 The Tensor $\varepsilon_{ij}(\omega, \boldsymbol{k})$ Near an Isolated Resonance 208
4.5.2 Examples 211
a) Dielectric Tensor for Isotropic Nongyrotropic Media . 216
4.5.3 Boundary Conditions 218
a) The Transition ("Dead") Layer Problem 224
b) ABC for Molecular Crystals and the Surface Current Appearance 228
4.5.4 Reflected and Refracted Waves Near Dipole and Quadrupole Transition Frequencies in a Nongyrotropic Crystal 233
a) The Law of Energy Conservation 236
4.5.5 Reflected and Refracted Waves Near a Dipole Transition Frequency in a Gyrotropic Crystal 238
4.5.6 The Influence of a Nonhomogeneous Subsurface Layer on Light Reflection Near Exciton-Absorption Bands 242
4.5.7 Transmission of Light Through a Plane-Parallel Plate (Nongyrotropic Medium) 244
4.5.8 Transmission of Light Through a Plane-Parallel Plate (Gyrotropic Medium) 248
a) The Integral Absorption 247
4.6 Experimental Investigations of Spatial Dispersion in Crystals .. 253
4.6.1 Gyrotropic Crystals 253
4.6.2 Nongyrotropic Crystals 257

5. Surface Excitons and Polaritons 271
5.1 Polaritons at the Interface of Isotropic Media 271
5.1.1 Dispersion of Surface Waves for Lossless Media 274
5.1.2 General Case 274
5.1.3 Surface Polaritons for Layered Structures 275
a) The Brewster Waves 278
b) Propagation of Surface Waves with Damping 279
c) Reflection, Diffraction and Refraction of Surface Waves at Surface Boundaries 280
5.2 Spectra of Surface Polaritons in Anisotropic Crystals 283
5.3 Transition-Layer Effects in Surface Polariton Spectra 289
5.3.1 Transition Layers with High Electric Conductivity 290
5.3.2 Transition Layers in the Presence of a Resonance with the Surface Polariton 294
a) The Effective Boundary Conditions 294
b) Polariton Spectrum Splitting for TH Surface Waves .. 295
c) TE Surface Waves in the Transition Layers' Resonance Region 298
5.3.3 Effect of Surface Roughness (Irregularities) on the Path Length of Surface Polaritons 301

a) Surface Polariton Scattering in the Vicinity of Phase-Transition Points ... 302
5.4 Effect of Spatial Dispersion on the Spectra of Surface Polaritons, and Additional ("New") Surface Waves ... 303
5.4.1 Surface Electromagnetic Waves at the Right-Left Gyrotropic Cystal Interface ... 311
5.5 Experimental Investigations of Surface Polaritons ... 314
5.5.1 The Attenuated Total Reflection (ATR) Method ... 315
5.5.2 The Ruled Diffraction-Grating Method ... 319
5.5.3 Raman Scattering of Light by Surface Polaritons ... 319
5.5.4 Surface Polariton Propagation Over Long Distances. Crystal Optics of Surfaces ... 323
5.5.5 The Inelastic Low-Energy Electron Scattering (ILES) Method ... 324
5.5.6 Nonlinear Surface Electromagnetic Waves ... 325

6. Microscopic Theory. Calculation of the Tensor $\varepsilon_{ij}(\omega, \boldsymbol{k})$ ... 328
6.1 General Expressions for $\varepsilon_{ij}(\omega, \boldsymbol{k})$ ... 328
6.1.1 Quantum-Mechanical Derivation of $\varepsilon_{ij}(\omega, \boldsymbol{k})$ ... 328
6.1.2 Contribution of the Exciton States to the Tensor $\varepsilon_{ij}(\omega, \boldsymbol{k})$ 337
a) Refractive Index Near Exciton Resonances of Anisotropic Crystals ... 340
6.2 Mechanical Excitons and the Tensor $\varepsilon_{ij}(\omega, \boldsymbol{k})$ in Molecular Crystals and in the Classical Oscillator Model ... 342
6.2.1 Molecular Crystals. Mechanical Excitons ... 342
6.2.2 Calculating the Dielectric-Constant Tensor of Molecular Crystals by the Local-Field Method ... 348
a) Davydov Splitting ... 351
6.2.3 The Oscillator Model ... 353
6.3 Absorption ... 356
6.3.1 The Absorption Mechanism. Absorption in a First Approximation ... 356
6.3.2 Absorption of Normal Electromagnetic Waves in the Vicinity of an Exciton Transition Frequency ... 361
6.3.3 The Long-Wavelength Edge of the Exciton Absorption Bands: Raman and Brillouin-Mandelstam Scattering of Polaritons ... 363
6.4 Raman Scattering of Light and X Rays Accompanied by Exciton Production. Influence of Spatial Dispersion on Energy Losses, and on Cherenkov and Transition Radiation of Charged Particles ... 367
6.4.1 Raman Scattering of X Rays Accompanied by Exciton Production ... 368
6.4.2 Raman Scattering of Light by Polaritons ... 373
a) Raman Scattering of Light by Bulk Polaritons ... 373

b) General Expression for the Scattering Intensity 378
c) The Scattering Cross Section by Polariton. Linewidth . 380
d) Raman Scattering of Light by Surface Polaritons 383
e) Compensation Effect 387
f) Broken Symmetry Method 389
g) Coherent Anti-Stokes Raman Spectroscopy (CARS) .. 393
6.4.3 Energy Losses and Cherenkov Radiation of a Charge Travelling with Uniform Motion Through a Medium Displaying Spatial Dispersion. Transition Radiation 396

7. Conclusion 402

Appendix 406
A.1 Crystal-Symmetry Notation 406
A.2 Information from Space Group Theory 409
A.2.1 Classification of the States of Mechanical Excitons 414

Notation 417

References 419

Subject Index 437

1. Introduction

In this introductory chapter the reader is acquainted with the range of topics covered. The concept of "normal" electromagnetic waves is introduced in connection with basic problems of crystal optics. Methods are discussed, within the framework of macroscopic electrodynamics, for a theoretical investigation of these waves, making use of the dielectric-constant tensor as a characteristic of the medium. This chapter illustrates the frequency and spatial dispersion of the tensor; it indicates the main effects of spatial dispersion in crystals (gyrotropy, optical anisotropy of cubic crystals, and additional light waves). The concept of excitons and polaritons is introduced. It is shown that crystal optics with spatial dispersion, which includes all of ordinary crystal optics as a special case, is closely associated with the theory of excitons.

1.1 Basic Problems of Crystal Optics

The fundamental problem of crystal optics consists of the investigation of the propagation of plane monochromatic waves in crystals, these waves being characterized by certain values of the frequency ω and of the wave vector $\boldsymbol{k}$. Such waves, when satisfying a homogeneous wave equation, are said to be "normal" electromagnetic waves. These normal waves may be subdivided into several types, but we consider, for the moment, only uniform waves whose electric field is given by

$$\boldsymbol{E} = \boldsymbol{E}_0 \mathrm{e}^{\mathrm{i}(\boldsymbol{k}\cdot\boldsymbol{r} - \omega t)}, \quad \boldsymbol{k} = \frac{\omega}{c}\tilde{n}(\omega, \boldsymbol{s})\boldsymbol{s}, \tag{1.1.1}$$

where $\boldsymbol{E}_0$ is a complex vector (field amplitude), independent of coordinates and time (i.e., of $\boldsymbol{r}$ and t), $\tilde{n} = n + \mathrm{i}\varkappa$ is the complex index of refraction and $\boldsymbol{s} = \boldsymbol{k}/k$ is a real unit vector. The electric and magnetic induction vectors $\boldsymbol{D}$ and $\boldsymbol{B}$, respectively, can also be represented in the form of (1.1.1) if we consider normal waves.

For a given frequency ω and a given direction $\boldsymbol{s}$ of propagation several types of normal waves (denoted by the subscript l) may exist, which differ from one another by their polarization [i.e., the vector $\boldsymbol{E}_{0l}(\omega, \boldsymbol{s})$] and their index of refraction $\tilde{n}_l(\omega, s)$. The propagation of normal waves in the medium considered fully characterizes the electromagnetic properties of this medium. Generally speaking, however, it is more reasonable to choose other quantities than E_{0l} and $\tilde{n}_l$ as the fundamental (primary) characteristics in crystal optics. These characteristics of the medium are provided by the components of the complex dielectric-constant tensor

$$\varepsilon_{ij} = \varepsilon'_{ij} + \mathrm{i}\varepsilon''_{ij}, \tag{1.1.2}$$

which, by definition, link the vectors $\boldsymbol{E}$ and $\boldsymbol{D}$ in linear electrodynamics[1]

$$D_i = \varepsilon_{ij} E_j. \tag{1.1.3}$$

In the simplest case, encountered in practice only in relatively narrow ranges of transparency, the components of the tensor ε_{ij} may be considered as constants (independent of ω and $\boldsymbol{k}$). In the earlier literature (particularly when considering energy relations in which the constancy of ε_{ij} is essential) crystal optics was usually based upon this assumption. To a much better approximation frequency dispersion is taken into account, i.e., it is assumed that

$$\begin{aligned} \varepsilon_{ij}(\omega) &= \varepsilon'_{ij}(\omega) + \mathrm{i}\varepsilon''_{ij}(\omega), \\ D_i(\omega) &= \varepsilon_{ij}(\omega) E_j(\omega), \end{aligned} \tag{1.1.4}$$

where $D_i(\omega)$ and $E_i(\omega)$ are the respective Fourier components of the fields $\boldsymbol{E}$ and $\boldsymbol{D}$

$$E_i(\boldsymbol{r}, t) = \int_{-\infty}^{+\infty} E_i(\boldsymbol{r}, \omega) \mathrm{e}^{-\mathrm{i}\omega t} d\omega,$$

$$E_i(\boldsymbol{r}, \omega) = \frac{1}{2\pi} \int_{-\infty}^{+\infty} E_i(\boldsymbol{r}, t) \mathrm{e}^{\mathrm{i}\omega t} dt;$$

(the argument $\boldsymbol{r}$ was omitted in (1.1.4) as it is inessential for the case being considered).

The physical nature of the frequency dispersion is very easily understood. If the natural frequencies ω_s (or the reciprocals of the relaxation time) of the medium are smaller than or comparable to the frequency ω of the field, then electrical polarization may not occur immediately. In other words, the polarization $\boldsymbol{P}$ (and, consequently, the induction $\boldsymbol{D} = \boldsymbol{E} + 4\pi\boldsymbol{P}$) at a given instant is determined not only by the field magnitudes at that instant but also by these

---

[1] Here and throughout the book summation is to be carried out over all tensorial subscripts which occur twice. In (1.1.2) and in what follows, the tensors ε'_{ij} and ε''_{ij} are, by definition, Hermitian [the formulation of an arbitrary tensor ε_{ij} in the form (1.1.2) is unique as regards the determination of the Hermitian tensors ε'_{ij} and ε''_{ij} in terms of ε_{ij}, and vice versa].

magnitudes at previous times. When making use of the Fourier components this leads to an ω dependence of ε_{ij}. Since for most media the natural frequencies lie within the optical range, it is obvious that frequency dispersion in crystal optics is often important; the parameter characterizing the amount of frequency dispersion is the ratio ω/ω_s, ω_s being the natural frequency or inverse relaxation time essential for the spectral range considered. Only if $\omega/\omega_s \ll 1$ is the frequency dispersion small.

We shall call crystal optics that is based upon relations such as (1.1.4), which are also known as material equations, classical crystal optics.

By virtue of the symmetry principle governing the kinetic coefficients, the tensor $\varepsilon_{ij}(\omega)$ is symmetrical when there is no magnetic field (Sect. 2.1.2). For this reason the Hermitian tensors $\varepsilon'_{ij}(\omega)$ and $\varepsilon''_{ij}(\omega)$ are real and the natural optical activity (gyrotropy) cannot be explained. The phenomenon of gyrotropy, although well known for a long time, thus lies beyond the scope of classical crystal optics. For various reasons (e.g., because it is possible to introduce a nonsymmetrical tensor $\varepsilon_{ij}(\omega)$ formally and thus to describe gyrotropy) in the earlier literature this fact was frequently ignored. It is clear, however, that in crystal optics even the investigation of gyrotropy already requires that we go beyond the bounds of the approximation (1.1.4), and take spatial dispersion as well as frequency dispersion into account, within the limits of linear theory. This implies the validity of the equations

$$\begin{aligned} D_i(\omega,\boldsymbol{k}) &= \varepsilon_{ij}(\omega,\boldsymbol{k})E_j(\omega,\boldsymbol{k})\,, \\ \varepsilon_{ij}(\omega,\boldsymbol{k}) &= \varepsilon'_{ij}(\omega,\boldsymbol{k}) + \mathrm{i}\varepsilon''_{ij}(\omega,\boldsymbol{k})\,, \end{aligned} \tag{1.1.5}$$

where, obviously,

$$\begin{aligned} E_i(\boldsymbol{r},t) &= \int E_i(\omega,\boldsymbol{k})\mathrm{e}^{\mathrm{i}(\boldsymbol{k}\cdot\boldsymbol{r}-\omega t)}d\omega\, d\boldsymbol{k}\,, \\ E_i(\omega,\boldsymbol{k}) &= \frac{1}{(2\pi)^4}\int E_i(\boldsymbol{r},t)\mathrm{e}^{-\mathrm{i}(\boldsymbol{k}\cdot\boldsymbol{r}-\omega t)}d\boldsymbol{r}\, dt\,, \end{aligned} \tag{1.1.6}$$

and analogous equations apply to $\boldsymbol{D}$.

In this introductory chapter we have endeavored to outline all the problems met with in crystal optics when spatial dispersion is taken into account in its connection with the theory of excitons. It should be noted that all topics dealt with only briefly in this chapter will be subject to a detailed discussion later on. We should therefore like to recommend that readers who have not yet become sufficiently familiar with the type of problem dealt with in the book should first look through this introductory chapter and then return to it again when rereading the book. A reader who is basically interested, for instance, in what is contained in Chaps. 4 and 5 may omit Sects. 2.2.3, 4 and 5, as well as the greater part of Sect. 2.3.

The relations (1.1.5) apply to crystal optics where spatial dispersion is taken into account (the term "spatial dispersion" indicates the existence of a $\boldsymbol{k}$

dependence of ε_{ij}). Neglecting spatial dispersion is equivalent to the assumption that the electrical polarization at a given point in the medium was determined by the magnitude of the electric field at this point. This is often a satisfactory approximation, but obviously the polarization at a given point is determined, in general, by the field in a certain neighborhood of this point. In terms of Fourier components this indicates that ε_{ij} depends on the wavelength, or, equivalently, on the wave vector of the field. The amount of spatial dispersion is determined by the parameter ak or by the somewhat more descriptive parameter a/λ, where a is the characteristic dimension (the radius of the "region of influence", the radius of molecular action, etc.) and $\lambda = 2\pi/k$ is the characteristic distance over which the field in the medium changes: the wavelength in a transparent medium (though we will not observe this distinction and will always refer to λ as the wavelength, regardless of the medium). In a condensed nonmetallic medium (crystals or fluids) the radius a is of the order of the lattice constant or of the molecular dimensions, i.e., $a \sim 10^{-7}$ or 10^{-8} cm. The parameter a/λ is therefore very small even in the optical range, not to speak of the radio frequency range. As a matter of fact, $\lambda = \lambda_0/n$, with $\lambda_0 = 2\pi c/\omega_0$ being the wavelength of the radiation in vacuum. Even at $\lambda_0 = 4000$ Å $= 4 \times 10^{-5}$ cm and $n \sim 10$, the parameters $a/\lambda_0 \sim 10^{-3}$ and $a/\lambda \sim 10^{-2}$.

Spatial dispersion in crystal optics can thus be regarded as being weak in the sense that the parameter a/λ is small

$$\frac{a}{\lambda} \ll 1 . \tag{1.1.7}$$

This is the reason why it is common in crystal optics either to ignore spatial dispersion entirely or to take it into account in implicit form, as it is usually considered sufficient in investigating gyrotropy. However, the smallness of a characteristic parameter (the ratio a/λ in this case) certainly does not enable us to neglect the effects due to this parameter. This becomes especially clear when considering new phenomena which could not exist at all if the above-mentioned effect were neglected. In the case of spatial dispersion, gyrotropy may serve as an example for such a new phenomenon. In this case we may take advantage of the weakness of spatial dispersion in the following way: it is frequently sufficient for this purpose to replace the tensor's general form $\varepsilon_{ij}(\omega,\boldsymbol{k})$ by an expression which contains only terms linear in $\boldsymbol{k}$, i.e., to put

$$\varepsilon_{ij}(\omega,\boldsymbol{k}) = \varepsilon_{ij}(\omega) + \mathrm{i}\,\gamma_{ijl}(\omega)k_l \qquad \text{or} \tag{1.1.8}$$

$$\varepsilon_{ij}^{-1}(\omega,\boldsymbol{k}) = \varepsilon_{ij}^{-1}(\omega) + \mathrm{i}\,\delta_{ijl}(\omega)k_l\,, \tag{1.1.9}$$

where $\varepsilon_{ij}^{-1}(\omega,\boldsymbol{k})$ is the reciprocal of the tensor $\varepsilon_{ij}(\omega,\boldsymbol{k})$ [this tensor enters, e.g., the relation $E_i(\omega,\boldsymbol{k}) = \varepsilon_{ij}^{-1}(\omega,\boldsymbol{k})D_j(\omega,\boldsymbol{k})$], and $\gamma_{ijl}(\omega)$ and $\delta_{ijl}(\omega)$ are tensors of rank three.

Gyrotropy is a phenomenon of the order of a/λ and it can be observed only in media which possess no center of symmetry (this well-known fact will be proved in Sect. 2.1.2). In books dealing with crystal optics (see, e.g., [1.1 – 4]), the phenomenon of gyrotropy is considered along with classical crystal optics, but with insufficient generality. Gyrotropy is introduced and discussed as an isolated phenomenon and not as a special, simple case of taking spatial dispersion into account. This treatment does not hinder the analysis of a great number of properties of gyrotropic media (see, especially [1.3, 4]), but hardly any doubts may arise as to its limitations. We mention only the fact that certain results in connection with gyrotropic media are more simply and conveniently obtained on the basis of an investigation of the general properties of the tensor $\varepsilon_{ij}(\omega, \boldsymbol{k})$; this can be illustrated by way of the following example. Making use of (1.1.9) it is sufficient to consider the problem of wave propagation in a gyrotropic medium near the absorption line in order to satisfy oneself of the possibility of having three transverse normal waves (where the frequency ω and the direction $\boldsymbol{s}$ of propagation are given) instead of the two usually encountered. As far as we know, this possibility arising with gyrotropic media was pointed out not long ago [1.5] (see also Sect. 4.1.3), whereas the simple fact that new normal waves may arise when spatial dispersion is taken into account is obvious and has been known for a long time. Thus the investigation of gyrotropy can be carried out within the framework of general crystal optics when spatial dispersion is taken into account. It stands to reason, however, that the necessity of investigating other effects of spatial dispersion besides gyrotropy furnishes the main stimulus for the development of such crystal optics.

When longitudinal waves are considered it is essential to take spatial dispersion into account (we need only to recall the fact that without spatial dispersion longitudinal waves would display zero group velocity [1.6 – 9])[2]. Longitudinal waves which are especially well known in the case of a plasma (plasma waves) may also occur in crystals though in this case the waves are considerably damped. The problem of longitudinal (plasma) waves in a solid is related to the problem of discrete energy losses occurring on the passage of charged particles through thin films [1.5, 9]. We shall not deal here with this problem nor with the significance of spatial dispersion and the application of the tensor $\varepsilon_{ij}(\omega, \boldsymbol{k})$ in plasma physics [1.6 – 9], in metal optics [1.9, 10] and other fields [1.9, 11]; we shall concentrate only on crystal optics.

It is natural to ask: What part is played by spatial dispersion in a nongyrotropic medium (e.g., in crystals possessing a center of symmetry)? First of all it is clear that in this case terms of the order of a/λ vanish and only effects of the order of $(a/\lambda)^2$ can be observed. The parameter $(a/\lambda)^2$ is, of course, very small in the optical region but this fact is not decisive in considering qualita-

---

2 In all cases in which problems are dealt with that are either sufficiently general in nature or not considered in detail in the present book, reference is made particularly to monographs and reviews, and not to the original papers.

tively new phenomena. The most simple and important effect of this type is the optical anisotropy of nongyrotropic cubic crystals. In the case of cubic crystals the tensor $\varepsilon_{ij}(\omega)$ reduces to a scalar, i.e., $\varepsilon_{ij}(\omega) = \varepsilon(\omega)\delta_{ij}$, and such a medium is optically isotropic. As soon as terms of the order of $(a/\lambda)^2 \sim a^2k^2$ are taken into account, however, anisotropy arises since the tensors

$$\varepsilon_{ij}(\omega, \boldsymbol{k}) = \varepsilon_{ij}(\omega) + \alpha_{ijlm}(\omega)k_l k_m\,, \tag{1.1.10}$$

$$\varepsilon_{ij}^{-1}(\omega, \boldsymbol{k}) = \varepsilon_{ij}^{-1}(\omega) + \beta_{ijlm}(\omega)k_l k_m \tag{1.1.11}$$

do not reduce to scalars, even in the case of cubic symmetry (for further detail see Sects. 3.2.2 and 4.2).

It was *Lorentz* who indicated the appearance of an anisotropy of the order of $(a/\lambda)^2$ in cubic crystals as early as 1878 [1.12]. This conclusion was arrived at again in [1.13] on the basis of a microscopic investigation of quadrupolar transitions in crystals, and in [1.5] on the basis of (1.1.10, 11). Optical anisotropy in the absorption of nongyrotropic cubic crystals was observed for the first time in 1960 [1.14] with cuprous oxide (Cu_2O) in the range of the quadrupolar absorption line. (For observations in the transparent region see Sect. 4.7.2.) When spatial dispersion is taken into account, the cubic crystals of Cu_2O possess seven optical axes (three fourth-order axes and the four body diagonals of the cube). Of course, spatial dispersion also has an influence on the optical properties of crystals of lower symmetry (e.g., uniaxial crystals then become multiaxial). It is likewise essential when the influence of external electric and magnetic fields, and mechanical stresses, are investigated.

As is well known, the direction of a ray in a crystal (in a transparent medium this direction coincide with the direction of the group velocity vector $\boldsymbol{v}_{gr} = d\omega/d\boldsymbol{k}$) does not, in general, coincides with the direction of the wave vector $\boldsymbol{k}$. Without spatial dispersion, the angle between $\boldsymbol{k}$ and $\boldsymbol{v}_{gr}$ is always smaller than $\pi/2$ [1.15]. In the case of nonzero spatial dispersion, however, the angle between $\boldsymbol{k}$ and $\boldsymbol{v}_{gr}$ may assume any value (in particular, it can be equal to π). If in crystal optics spatial dispersion is taken into account one should therefore expect the appearance of various peculiarities (as compared with classical crystal optics) when considering the path of rays.

In favorable cases the role played by spatial dispersion increases near the absorption lines (resonances) since the refractive index n increases here, and so does the parameter $a/\lambda = an/\lambda_0$. This is particularly well known for a magnetoactive plasma [Ref. 1.6, § 12], in which case not only do the dispersion curves suffer a quantitative change but new or additional normal waves appear, too. Without spatial dispersion only two normal waves propagate with a given frequency in a given direction through an anisotropic medium; besides these, a longitudinal wave with definite frequency and zero group velocity may appear in special cases. The appearance of new waves is possible in a condensed medium as well. These include the above-mentioned longitudinal waves (at frequencies where they do not appear when spatial dispersion is neglected) and a third wave in a gyrotropic medium [1.5]. In principle, new waves (apart from the longitudinal wave) may also appear in a nongyrotropic medium. This followed, as a matter of fact, from the theory of normal electromagnetic waves in crystals, developed by *Born* in 1915 [Ref. 1.16, pp. 108 – 122] and was stated in concrete form in [1.17] where it was applied to the region of exciton lines [1.20, 21]. In this work, however, absorption was not taken into account. Moreover, near the dipole lines, the only ones dealt with in [1.17], absorption may be so intense in many cases that the influence of spatial dispersion is practically masked [1.5, 18].

Consequently, it is difficult to observe new waves near intense dipole lines and, as yet, successful experiments at low temperature only have been reported (Sect. 4.7).

Absorption near the quadrupole lines may be weak. Moreover, if the difference in the refractive indices of the new and the ordinary wave is small, this will give rise to relatively slow intensity oscillations when the crystal thickness is varied. It is therefore possible, in principle [1.19], that new waves appear near the quadrupole lines and that they can be discovered by means of intensity oscillations. According to [1.20] the intensity oscillations occurring when the thickness of Cu_2O films is varied in the region of the quadrupole line $\lambda_0 = 6125$ Å is attributed to the part played by the new (anomalous) wave. We are, however, of the opinion that the problem of the oscillations observed in [1.20] is still unsolved, as a number of experimental difficulties still exists (Sect. 4.7.2). Further investigations will be necessary.

Spatial dispersion also influences the reflection coefficient of waves from the crystal surface. In this case special attention should be paid to the investigation of the frequency dependence of the reflection coefficient [1.21] (see also Sect. 4.6). Unfortunately a study of the influence of spatial dispersion by light reflection meets with difficulties, both experimental and theoretical. The crystal surface is usually far from being perfect (roughness, surface impurities); moreover, even if we have an ideal surface, the boundary conditions must be altered or supplemented when solving a reflection problem and taking spatial dispersion into account. We thus have to analyze not only the $\boldsymbol{k}$ dependence of ε_{ij} but also the nature of the boundary conditions. It should be noted at the same time that the fact that the structure of the surface layer influences the spectra of light reflection in the region of exciton resonances is an exceptionally valuable phenomenon because it opens up new possibilities for investigations (see Chap. 5 for an investigation of the properties of the surface layer of the crystal by means of surface waves within the scope of crystal optics and spectroscopy of surfaces [1.58, 59]).

In connection with the aforesaid it should be stressed that spatial dispersion and, particularly, the appearance of new waves can, in principle, be discovered and studied not only by investigating the transmission of light through plates or reflection from crystals and other methods of linear optics. These new waves can also be excited when x rays [1.22] or charged particles [1.23] pass through crystals, or when various methods are applied which belong, basically, to nonlinear optics. The later includes Raman scattering (sometimes also called combination scattering) of light (both resonant and nonresonant) [1.24, 25], two-photon absorption [1.26], polariton-polariton resonant scattering [1.27] and Mandelstam-Brillouin scattering ([1.28] and [Ref. 1.29, Chaps. 2 and 3]) (Sects. 4.6 and 6.4). Several spectroscopic methods of linear optics have also been applied to determine the polarition dispersion. They include: reflection and interference in a cyrstal [1.30], time of flight measurements [1.31] and refraction by a prism [1.32]. These are precisely the methods employed today to obtain the maximum amount of data on spatial dispersion in crystals, i.e., the determination of the dependence of the energy of excitons (or polaritons, see also Sect. 1.2) on the wave vector. It also includes the excitons that correspond to the appearance of new waves when the dependence of the tensor $\varepsilon_{ij}(\omega,\boldsymbol{k})$ on $\boldsymbol{k}$ in the resonance region is taken into account (Sect. 4.7).

We thus know of a series of phenomena, the investigation which falls into the area of crystal optics with spatial dispersion. The main problem of the theory is to determine the relation linking $\tilde{n}_l(\omega,\boldsymbol{s})$ and $\varepsilon_{ij}(\omega,\boldsymbol{k})$, and to apply the formulas to the evaluation of experimental data. Experiments yield the complex index of refraction $\tilde{n}_l(\omega,\boldsymbol{s}) = n + \mathrm{i}\varkappa$ and, if we do not make use of crystal optics, we would be required to make measurements for a great many directions $\boldsymbol{s}$. But as it is possible to express $n_l(\omega,\boldsymbol{s})$ in terms of $\varepsilon_{ij}(\omega,\boldsymbol{k})$, it is sufficient to measure $\tilde{n}$ for only a few directions. When spatial dispersion is neglected this is immediately understood since, for a given frequency, the symmetric complex tensor $\varepsilon_{ij}(\omega)$ is characterized by at most six components (the tensors ε'_{ij} and ε''_{ij} are assumed to be diagonalized). If we take weak spatial dispersion into account, the relations are somewhat complicated but, as before, we need measurements only for relatively few directions, due to the fact that $\varepsilon_{ij}(\omega,\boldsymbol{k})$ is a simple function of $\boldsymbol{k}$.

If the tensor $\varepsilon_{ij}(\omega,\boldsymbol{k})$ is known in some approximation, all normal waves in the crystal to which this approximation applies can be assumed to be known [in particular, in establishing the dispersion law $\omega_l = \omega_1(\boldsymbol{k})$ it is equivalent to know the function $\tilde{n}_l(\omega,\boldsymbol{s}) = ck/\omega_l(\boldsymbol{k})$]. Moreover, the tensor $\varepsilon_{ij}(\omega,\boldsymbol{k})$ determines the energy loss a particle is subject to when moving through matter, the molecular forces between bodies, and the fluctuations of the electromagnetic field. Thus, it characterizes the medium (crystal) very completely [1.1, 6–11, 33].

The main task of this book is to present the fundamentals of crystal optics with spatial dispersion, within the scope of a macroscopic approach. This means that the tensor $\varepsilon_{ij}(\omega,\boldsymbol{k})$ is introduced in a general way, then certain of its properties (symmetry, dispersion relations, series expansions) are investigated and finally, with the help of the field equations, the normal waves are all determined. In addition, energy relations, ray problems, boundary conditions and the like are discussed.

There is no doubt that in many cases it is expedient just to carry out investigations by means of this macroscopic procedure and not to refer in any way to microscopic theories. However, it is necessary to develop and apply a microscopic theory in order to calculate the tensor $\varepsilon_{ij}(\omega, \boldsymbol{k})$ itself in one or another approximation, or for various crystal models. In other words, in crystal optics, as in other fields, it is a good idea to combine macroscopic and microscopic methods appropriately. Although this is now obvious, it is stressed once more, since in the previous literature on crystal optics one may find relatively cumbersome microscopic calculations carried out in order to obtain obvious results, which could be derived in a sufficiently clear way with the aid of macroscopic considerations.

Because of this fact and also due to other considerations (particularly in order to obtain greater completeness and to establish correspondence between certain experimental data and theoretical calculations), we also deal in our book with the microscopic theory as regards its relationship to macroscopic crystal optics. In this case, as is readily seen, the microscopic theory of optical effects in crystals is closely related to the branch of solid-state physics that is often called the theory of excitons. Since there is no universally adopted terminology we shall agree upon attributing the term excitons to "elementary excitations" in crystals, and also to magnons (spin waves) and electronic excitations (excitons) in superconductors [1.38–40], all of which conform to Bose statistics. Excitons also cover acoustic waves, for which, however, it is convenient to retain the standard term "phonons". With this definition it is obvious that the concept of excitons includes all normal electromagnetic waves in a crystal which, in the language of quantum mechanics and in the case of sufficiently weak absorption, are nothing else but "photons in a medium" (including here also longitudinal photons in a medium, i.e., plasmons).[3] Hence, it can be said that the general theory of excitons comprises, on the one hand, crystal optics with spatial dispersion and, on the other hand, all the theoretical models having the aim of calculating the tensor $\varepsilon_{ij}(\omega, \boldsymbol{k})$ in nonmetallic crystals. This follows, of course, mainly from the definitions of the concepts and terms.

An essential part is the discussion of ways to calculate $\varepsilon_{ij}(\omega, \boldsymbol{k})$ and to study the nature of the various approximations. The choice of the latter mainly depends on the type of crystal and the nature of the excitations under consideration. The optical branch of lattice excitations [1.41] is especially important for ionic crystals in the infrared region. In the same ionic crystals, but in the region of higher frequencies, and especially in molecular crystals and certain semiconductors, the main role is played by electronic excitations ([1.42–46] and [Ref. 1.29, Chap. 1]). These excitations can be visually represented as a moving coupled electron-hole pair (Wannier-Mott exciton) or as transitions from site to site of an excited state of the molecular concerned (Frenkel's exciton, which can also be regarded as an electron-hole pair of small radius). Either of these may be described by a wave packet, and in some

[3] The concept of photons in a medium with energy $\hbar\omega$ and momentum $(\hbar\omega n/c)\boldsymbol{s}$ has been applied to other problems in Sect. 4.4 and [1.8, 34–37].

cases this is completely justified and may be applied in quantitative calculations. More usually, however, due to the translational symmetry of the crystal, the eigenfunctions used to describe the excitations apply to the whole crystal and are of the character of modulated plane waves with the wave vector $\boldsymbol{k}$. If, for simplicity, we restrict ourselves to the case of an ideal lattice at rest, the wave function of the excitation can be written in the form [1.42]

$$\Psi_{k,m} = \mathrm{e}^{\mathrm{i}\boldsymbol{k}\cdot\boldsymbol{R}} u_{k,m}(\boldsymbol{R}, \boldsymbol{r}_i - \boldsymbol{r}_j) , \qquad (1.1.12)$$

where $\boldsymbol{R} = (\sum_i \boldsymbol{r}_i)/NV$ is the radius vector of the center of gravity of all the NV electrons (each electron having the radius vector $\boldsymbol{r}_i$), and the function $u_{k,m}$ is periodic (with the period of the lattice) with respect to $\boldsymbol{R}$. The subscript m refers to quantum numbers which cannot be expressed in terms of $\boldsymbol{k}$. If we take only the particle coordinates into account in (1.1.12) we have a problem of mechanics in which only the Coulomb interaction is considered. Therefore, the question immediately arising is how the excitations (excitons) obtained in this way are related to real excitons, and what role they play in the calculation of $\varepsilon_{ij}(\omega, \boldsymbol{k})$.

Before considering this problem, let us specify the terminology. "Real excitons" (or simply "excitons") are understood to be normal waves, i.e., exact solutions of the homogeneous problem obtained when both Coulomb and the other electromagnetic interactions are taken into account. In recent years, real excitons are frequently called polaritons [1.36, 43, 46, 47]. "Coulomb exciton" is the name we shall give to all the corresponding exact solutions of the homogeneous Coulomb problem, obtained when the transverse electromagnetic field is ignored. It should be emphasized that it is precisely the solution to the *homogeneous* field equations that concerns us here, i.e., we assume the absence of external field sources (the given currents and charges are located in the medium itself; Sect. 2.1). Finally, the solutions to the Coulomb problem are called "mechanical excitons" if there is no long-wavelength electric field (Coulomb field) $\boldsymbol{E}_{\parallel}$ or if it is ignored.

1.2 Excitons and Polaritons

Real, Coulomb, and mechanical excitons will be discussed later (see in particular Sect. 2.2.2). Some remarks follow to explain the concepts of the various types of excitons introduced here.

Longitudinal normal waves (plasmons) in which there is no magnetic field and the electric field is free from vortices represent real and Coulomb excitons at the same time. Besides this, the homogeneous Coulomb problem also has other solutions (particularly solutions which in the following are called "fictitious" longitudinal waves) that do not satisfy the complete system of field equations. A nonzero vortex-free electric field $\boldsymbol{E}_{\parallel}$ exists in "fictitious" longitudinal waves even when the wave vectors $\boldsymbol{k}$ are small. At the same time this macroscopic Coulomb field $\boldsymbol{E}_{\|}$, despite the longitudinal polarization (rot $\boldsymbol{E}_{\|} = 0$), differs in no way from an arbitrary macroscopic field (keeping in mind the equal magnitudes of ω and $\boldsymbol{k}$). Moreover, the decomposition of the field into a longitudinal and a transverse component in the general case of an anisotropic medium and an arbitrarily directed wave vector is by no means a natural procedure. This is because the field $\boldsymbol{E}$ is neither transverse nor longitudinal in the corresponding normal waves. Finally, if we consider a long-wavelength field this is done in a unified manner for the whole field (even if the normal waves are decomposed into longitudinal and normal waves) on the basis of the equations of macroscopic electrodynamics. This is one of the reasons for the introduction, side by side with Coulomb excitons, of mechanical excitons which, without external sources, can be propagated only if the actions of both the long-wavelength transverse electromagnetic field and the vortex-free macroscopic (long-wavelength) electric field are ignored. From the point of view of a solution of the mechanical problem this means that in the dynamic equations of the particles of the medium, provided there are no external sources, the vortex-free macroscopic field $\boldsymbol{E}_{\|}$ (if it is different from zero) cancels out. Hence, the point is that the Coulomb interaction in mechanical excitons is taken into account approximately, rather than completely. There is, however, another possibility of interpreting mechanical excitons, which is clearer from a mathematical standpoint: consider the inhomogeneous Coulomb problem, i.e., let us introduce certain external sources. These sources can then be chosen so that for the exact solutions of the wave equation of the Coulomb problem considered, the long-wavelength electric field $\boldsymbol{E}_{\|}$ vanishes.

We thus see that the only difference between mechanical excitons and Coulomb excitons is that the long-wavelength electric field vanishes for the former, or is in any case neglected when calculating frequencies (energy levels) and wave functions.

Neglecting the action of the long-wavelength field ($\lambda \gg a$, where a is the lattice constant) is an essential fact here. The restriction to Coulomb interaction alone in the short-wavelength (microscopic) field, as assumed in the application of (1.1.12), has no fundamental significance. Moreover, it is correct to assume that in calculations of the energy or wave functions of mechanical excitons, all short-wavelength interactions that can exist under the given conditions are taken into account (side by side with the Coulomb interaction where exchange forces are accounted for so that we may speak of magnetic interactions). As in practice all considerations are usually restricted to the Coulomb interaction alone, we shall consider only this interaction, but we do so only in order not to overload the description with details.

In an anisotropic medium the electrical polarization vector $\boldsymbol{P}$ and the induction vector $\boldsymbol{D} = \boldsymbol{E} + 4\pi\boldsymbol{P}$ are not parallel, for most directions, to the electric field-strength vector $\boldsymbol{E}$. As can easily be verified (Sect. 2.2.2), the field $\boldsymbol{E}_{\|} = -4\pi\boldsymbol{s}(\boldsymbol{s}\cdot\boldsymbol{P})$ for Coulomb excitons with the wave vector $\boldsymbol{k} = k\boldsymbol{s}$ depends on the direction s, however only for long wavelengths (i.e., for $k \to 0$). For this reason frequency of the Coulomb excitons with $E_{\|} \neq 0$ in an anisotropy medium depends on s even if $\boldsymbol{k} = k\boldsymbol{s} \to 0$. If we consider mechanical excitons, however, there is no effect of the long-wave lengthfield (i.e., in particular, of a field where $k \to 0$) and therefore the frequencies of mechanical excitons $\omega_m(\boldsymbol{k})$ do not depend on s if $k \to 0$.

It is the mechanical exciton and not the Coulomb one that plays the part of the states in the unperturbed problem of calculating $\varepsilon_{ij}(\omega, \boldsymbol{k})$ as a first approximation of perturbation theory. In other words, the wave functions $\Psi_{k,m}$ of the type (1.1.12) and the frequencies $\omega_m(\boldsymbol{k}) = W_m(\boldsymbol{k})/\hbar$ of the mechanical excitons play the same role in a crystal, when calculating ε_{ij}, as the wave functions Ψ_s and the frequencies $\omega_s = W_s/\hbar$ of the levels of separate atoms in the case of a perfect gas. For such a gas, near a dipole line, we can write $\varepsilon_{ij}(\omega)$ in the well-known approximation

$$\begin{aligned} \varepsilon_{ij}(\omega) &= \varepsilon(\omega)\,\delta_{ij}\,, \\ \varepsilon(\omega) &= \varepsilon_0 - \frac{\Omega_0^2}{\omega^2 - \omega_{0s}^2}\,, \end{aligned} \tag{1.2.1}$$

where $\Omega_0^2 = 4\pi e^2 N f_{0s}/m$, e and m are the charge and the mass of the electron, N is the atomic concentration, $\omega_{0s} = \omega_s - \omega_0$ is the frequency of the transition from state s to state 0, and $f_{0s} = \text{const}\,|\int \Psi_s^* \boldsymbol{r}\, \Psi_0 d\boldsymbol{r}\,|^2$ is the oscillator strength ([1.46] and Sect. 5.1). The frequency ω_{0s} is complex when absorption is taken into account.

In a more general treatment the expression $\int \Psi_s^* \boldsymbol{r}\, \Psi_0 d\boldsymbol{r}$ is replaced by $\int \Psi_s^* \boldsymbol{r} \exp(\mathrm{i}\boldsymbol{k} \cdot \boldsymbol{r})\, \Psi_0 d\boldsymbol{r}$; in the case of a quadrupolar transition the latter quantity reduces to $\int \Psi_s^* \boldsymbol{r} (\boldsymbol{k} \cdot \boldsymbol{r})\, \Psi_0 d\boldsymbol{r}$. For a quadrupole line we may therefore write the approximation

$$\varepsilon(\omega, \boldsymbol{k}) = \varepsilon_0 - \frac{\text{const} \cdot k^2}{\omega^2 - \omega_{0s}^2}\,. \tag{1.2.2}$$

Thus, as is also obvious from general considerations, taking quadrupole and higher multipole transitions into account means accounting for spatial dispersion of the atom (in the case of a molecule without a center of symmetry, the numerator of the expression for ε may contain first-order terms in k).

Expressions of the type (1.2.1) are obtained in the classical theory if we analyze, for example, the motion of an oscillator with charge e_*, damping constant γ and mass M subject to the action of the field $\boldsymbol{E} = \boldsymbol{E}_0 \exp(-\mathrm{i}\,\omega t)$. Thus

$$\ddot{\boldsymbol{r}} + \gamma \dot{\boldsymbol{r}} + \omega_{0s}^2 \boldsymbol{r} = \frac{e_*}{M} \boldsymbol{E}\,. \tag{1.2.3}$$

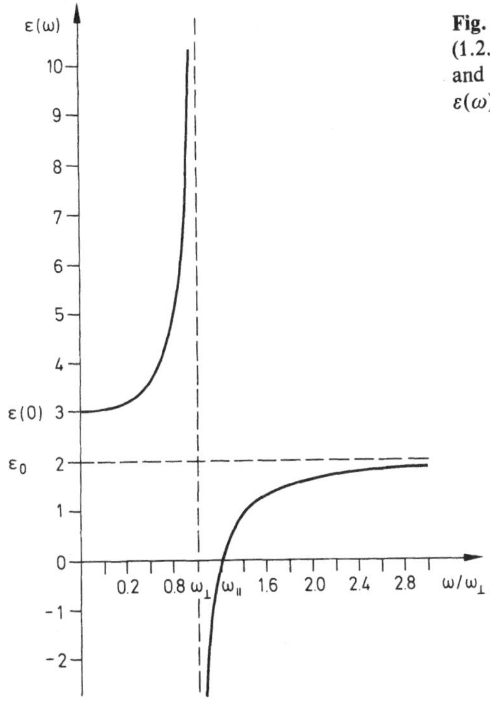

Fig. 1.1. The dielectric constant according to (1.2.1). The curves plotted assume that $\varepsilon_0 = 2$ and $\Omega_0/\omega_{0s} \equiv \Omega_0/\omega_\perp = 1$, so that $\varepsilon(\omega) = 2-(\omega^2/\omega_\perp^2-1)^{-1}$

Here the forced solution and the corresponding polarization are of the form

$$
\begin{aligned}
\boldsymbol{r} &= -\frac{e_*/M}{\omega^2-\mathrm{i}\gamma\omega-\omega_{0s}^2}\boldsymbol{E}\,,\\
\boldsymbol{P} = e_*N\boldsymbol{r} &= -\frac{e_*^2N/M}{\omega^2-\mathrm{i}\gamma\omega-\omega_{0s}^2}\boldsymbol{E}\,,
\end{aligned}
\qquad (1.2.4)
$$

where N is the concentration of oscillators. No distinction is made between the acting field and the average macroscopic field $\boldsymbol{E}$ (see Sect. 5.2.2, where this difference is taken into account). In the case of a cubic lattice consisting of two sublattices oscillating with respect to each other, the role of $\boldsymbol{r}$ is played by the relative displacement of the charges in the sublattices, and e_* and M are the corresponding effective charge and reduced mass, respectively. By definition, $\boldsymbol{P} = (\varepsilon-1)\boldsymbol{E}/4\pi$. Thus, from (1.2.4) we obtain (1.2.1) with $\varepsilon_0 = 1$ and $\Omega_0^2 = 4\pi e^2Nf_{0s}/m = 4\pi e_*^2N/M$, $\gamma = 0$.

The form of the function $\varepsilon(\omega)$ which satisfies (1.2.1) is clear from Fig. 1.1. As is wellknown and will be shown below, only transverse waves in which the wave vector $\boldsymbol{k}$ is related to the frequency by the dispersion relation

$$
c^2k^2/\omega^2 \equiv n^2(\omega) = \varepsilon(\omega) \qquad (1.2.5)
$$

can be propagated in an optically isotropic medium characterized by the dielectric constant $\varepsilon(\omega)$. For the dielectric constant (1.2.1), evidently, $\varepsilon(\omega)=\infty$ when $\omega=\omega_{0s}\equiv\omega_\perp$ and $\varepsilon(\omega)=0$ when $\omega=\omega_\| \equiv\sqrt{\omega_\perp^2+\Omega_0^2/\varepsilon_0}$, while in the frequency range $\omega_\perp<\omega<\omega_\|$, the function $\varepsilon(\omega)<0$. This and Fig. 1.1. immediately clarify the form of the dependence $\omega(k)$, which is shown in Fig. 1.2a. Longitudinal waves may also exist at the frequency $\omega_\|$. We should add that the upper branch of curve $\omega(k)$ is asymptotic to the straight line $\omega=ck/\sqrt{\varepsilon_0}$, since as $\omega\to\infty$, the dielectric constant $\varepsilon(\omega)\to\varepsilon_0$.

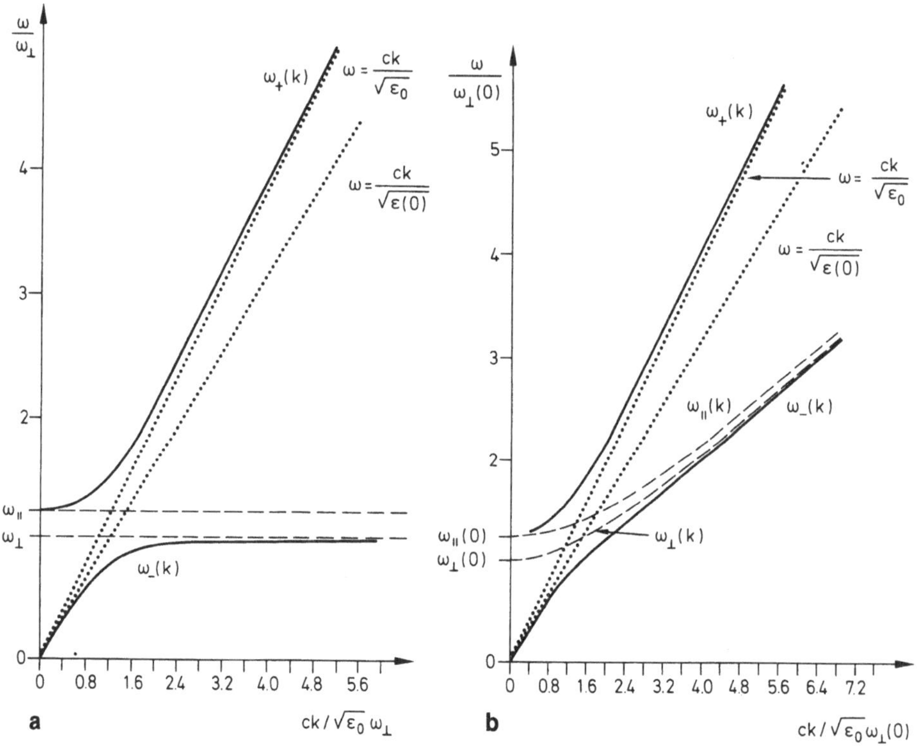

Fig. 1.2a, b. Dependence of the frequency ω on the wave vector k for normal waves (polaritons) that can propagate in an isotropic medium with the dielectric constant $\varepsilon(\omega)$ of the type (1.2.1). **(a)** No spatial dispersion [the frequency $\omega_{0s}\equiv\omega_\perp$ in (1.2.1) is independent of k]. In the calculations it was assumed that $\varepsilon(\omega)=2-(\omega^2/\omega_\perp^2-1)^{-1}$ (Fig. 1.1); **(b)** spatial dispersion is taken into account (in an equation of the type of (1.2.1) the frequency $\omega_{0s}=\omega_\perp$ is considered to depend on k for this case). It was assumed for the calculations that $\varepsilon(\omega,k)=2-[\omega^2/\omega_\perp^2(0)-\omega_\perp^2(k)/\omega_\perp^2(0)]^{-1}$, i.e., $\varepsilon_0=2$ and $\Omega_0^2/\omega_\perp^2(0)=1$, where $\omega_\perp^2(k)=\omega_\perp^2(0)+a^2k^2$ and $a^2=c^2/5$. The branches $\omega_\pm(k)$ correspond to the solution of $c^2k^2/\omega^2=\varepsilon(\omega,k)$, and the branch for the longitudinal wave $\omega_\|(k)$ is the solution of $\varepsilon(\omega,k)=0$. Actually, as will be explained in Sect. 2.2, the functions $\varepsilon(\omega,k)$ for transverse waves [branches $\omega_\pm(k)$] and for longitudinal waves [branch $\omega_\|(k)$] differ in the presence of spatial dispersion, and it is necessary to introduce two functions, $\varepsilon_{tr}(\omega,k)$ and $\varepsilon_{lo}(\omega,k)$, with $\varepsilon_{tr}(\omega,0)=\varepsilon_{lo}(\omega,0)$. In plotting the branch $\omega_\|(k)$ in Fig. 1.2b it was assumed, as indicated, that $\varepsilon_{tr}(\omega,k)=\varepsilon_{lo}(\omega,k)=\varepsilon(\omega,k)$

As $k \to 0$, the lower branch of the curve $\omega(k)$ approaches the straight line $\omega = ck/\sqrt{\varepsilon(0)}$, where $\varepsilon(0) = \varepsilon_0 + \Omega_0^2/\omega_\perp^2$. When waves are propagated in a solid (crystal), these waves, corresponding to both branches of the function $\omega(k)$ or, in the language of quantum mechanics, "photons in a medium" with energy $\hbar\omega(k)$ are now called polaritons. It is obvious, however, that only the name is new and that polaritons, within the scope of the classic pattern, are in fact identical to the normal electromagnetic waves in a solid, already investigated in the last century. More precisely, this concerns the approximation of classical crystal optics now being discussed in which spatial dispersion is neglected. If spatial dispersion is taken into consideration, then the role of the frequencies ω_{0s} in (1.2.1, 2) is played by the frequencies $\omega_m(\boldsymbol{k})$ of the mechanical excitons. Naturally, the appearance of a k dependence in the denominator of (1.2.1) leads to a change in the shape of the dispersion curves $\omega(k)$. One of the corresponding possibilities is clear from Fig. 1.2b. The qualitatively new element here, within the scope of crystal optics, is the appearance of two solutions (in a certain frequency range), i.e., two transverse normal waves (polaritons) with different k values at a given value ω of the frequency and a given polarization. The second of these solutions, having the larger value of k, is the "new wave" mentioned above. It can also be called a new polariton. (For polaritons in magnetic media see, e.g., [1.36, 48].)

By enclosing the term "new wave" in quotation marks, we wish to stress that actually no new branches appear in the spectrum of elementary excitations in the crystal. But even neglecting retardation (when investigating excitons), the spectrum has a light branch $\omega = ck$ and branches of excitons and phonons $\Omega_l = \Omega_l(k)$, where $l = 1, 2, \ldots$. Since these branches intersect, taking retardation into account (which is just what is accomplished by applying Maxwell's equations) leads only to the transformation of the branches without changing their total number. Actually, this circumstance was already realized within the scope of the classical dynamic theory of vibrations of crystal lattices [1.41, 49, 50]. Later [1.51–53], a consistent microscopic quantum theory of polaritons was advanced also for the electronic exciton region of the spectrum.

In terms of the microscopic theory, polaritons appear as a result of the diagonalization of the Hamiltonian of the crystal, which is equal to the sum of the Hamiltonians of the excitons and photons, and their interaction Hamiltonian. The state with a polariton turns out to be a superposition of the states of "bare" photons and excitons, with the relative contribution of each of these depending upon the value of the polariton wave vector. Thus, when $k \gg \Omega_l/c$ the polariton does not differ significantly from the "bare" exciton. In the region of low values of $k \lesssim \Omega_l/c$, however, the difference between the polariton and either the "bare" exciton or the "bare" photon is, in general, quite pronounced. Even though such a polariton is propagated at a velocity of the order of that of light, both electrons and nuclei are set in motion. This leads to the intensification of the scattering of polaritons by phonons, to nonlinear optical processes, and also, for example, to a huge increase, compared to the vacuum, of the inelastic scattering cross section of neutrons by laser photons propagated in crystals [1.54]. The scattering cross section turns out to be proportional to that of neutrons by nuclei. A similar problem has been treated in [1.55]. In all the cases mentioned, the intensity of the process is found to be proportional to the so-called strength function, determining the fraction of mechanical energy in the polariton (Sect. 5.3.2).

Employing one or another model of the exciton, microscopic polariton theory enables us to obtain not only the results of the phenomenological theory, based on the application of Maxwell's equations and the dielectric constant tensor, but to go beyond the limits of this theory [1.46]. Among other things, the scope of such a theory can include the investigation of the polariton interaction with short-wavelength excitations, as well as long-wavelength ones. Also, the "non-Boson" behavior of excitons at high concentration can be taken into consideration, and anharmonic members of the polariton system, leading to their scattering by one another, can be found [1.46, 56, 57]. It is precisely in investigating polariton scattering processes that dispersion is most fully manifested. As a result [1.29, 47] it has been possible to obtain the most abundant information concerning the dependence of the polariton frequency on the wave vector, including the dependence due to taking spatial dispersion into account (i.e., accounting for the finite effective mass of the exciton, as discussed in Sect. 4.6).

Due to the weakness of spatial dispersion in the optical range, we can usually restrict ourselves to terms of no higher order than k^2, in both the numerator and denominator of equations of the type (1.2.1, 2) in which the frequency ω_{0s} has been replaced by $\omega_m(\boldsymbol{k})$. In most cases we can employ even simpler equations of the type (1.1.10, 11). All this will be discussed in detail and more exactly in Chap. 6.

Here we wish only to establish the nature of the connection between the microscopic theory of excitons and crystal optics with spatial dispersion. As already mentioned, mechanical excitons play the same role in crystal optics as the atomic states in the theory of light dispersion in gases. For real excitons or polaritons (generalized to include longitudinal waves, or plasmons), this concept is in essence identical to the concept of normal electromagnetic waves (this is valid only for excitons of electromagnetic type having an appreciable macroscopic field $\boldsymbol{E}$, as considered in this book). Coulomb excitons also have a definite physical significance and it sometimes proves expedient to consider them apart from the real and mechanical excitons, since still another, important difference between a gas and a crystal exists. This difference consists in the optical anisotropy of the crystals. Owing to this anisotropy the frequencies of the absorption lines, generally speaking, do not coincide with the frequencies or the differences in the frequencies of mechanical excitons, i.e., with the poles of ε_{ij}. But, as we shall see, the frequencies of the absorption lines (poles of the refractive index $\tilde{n}$) are, in general, equal to the frequencies or the differences in the frequencies of the Coulomb excitons. The frequencies of these and real excitons (polaritons) can be found from the field equations if the tensor $\varepsilon_{ij}(\omega, \boldsymbol{k})$ is known. In turn, this tensor can be expressed in terms of the characteristics $\Psi_{k,m}$ and $\omega_m(\boldsymbol{k})$ of the mechanical or Coulomb excitons.

However, in this text we shall not go into the details of the theory of mechanical excitons, i.e., calculations aimed at finding the functions $\Psi_{k,m}$ and the frequencies $\omega_m(\boldsymbol{k})$ of mechanical or Coulomb excitons in various approximations and models. As already stressed, however, attention is paid to the microscopic theory and a quantum-mechanical expression is obtained for $\varepsilon_{ij}(\omega, \boldsymbol{k})$, a theory of excitons in molecular crystals is developed, and use is made of the symmetry of the wave functions of the exciton states, etc.

2. The Complex Dielectric-Constant Tensor $\varepsilon_{ij}(\omega, \boldsymbol{k})$ and Normal Waves in a Medium

The general properties of the dielectric-constant tensor are discussed for a spatially homogeneous anisotropic medium with spatial dispersion. Dispersion relations are analyzed, as well as relations following from the symmetry principle of kinetic coefficients. An expression is derived for the energy flux in media with spatial dispersion. A number of examples are given to demonstrate how accounting for spatial dispersion leads to corrections in Maxwell's boundary conditions. The longitudinal field is excluded from the relationship between the induction vector and electric field strength in a monochromatic plane wave. The transverse dielectric-constant tensor $\varepsilon_{\perp,ij}$, i.e., $D_i(\omega, \boldsymbol{k}) = \varepsilon_{\perp,ij}(\omega, \boldsymbol{k}) E_j(\omega, \boldsymbol{k})$, is introduced. It is shown that the resonances of the tensor $\varepsilon_{\perp,ij}(\omega, \boldsymbol{k})$ correspond to the so-called Coulomb excitons (i.e., states determined without taking retarded interaction into account), whereas the resonances of the tensor $\varepsilon_{ij}(\omega, \boldsymbol{k})$ correspond to mechanical excitons (i.e., states found without accounting for the total electric macrofield).

2.1 The Tensor $\varepsilon_{ij}(\omega, \boldsymbol{k})$ and Its Properties

2.1.1 The Electromagnetic Field Equations and the Introduction of the Tensor $\varepsilon_{ij}(\omega, \boldsymbol{k})$

The equations of an electromagnetic field in a medium can be written as

$$\left.\begin{aligned} &\operatorname{rot} \boldsymbol{B} = \frac{1}{c}\frac{\partial \boldsymbol{D}}{\partial t} + \frac{4\pi}{c}\boldsymbol{j}_{\text{ext}}, \\ &\operatorname{div} \boldsymbol{D} = 4\pi \rho_{\text{ext}}, \\ &\operatorname{rot} \boldsymbol{E} = -\frac{1}{c}\frac{\partial \boldsymbol{B}}{\partial t}, \\ &\operatorname{div} \boldsymbol{B} = 0 . \end{aligned}\right\} \tag{2.1.1}$$

Here $\boldsymbol{E}$ is the electric field strength, $\boldsymbol{D}$ and $\boldsymbol{B}$ are the electric- and magnetic-field inductions, $\boldsymbol{j}_{\text{ext}}$ and ρ_{ext} are the current and charge densities of the external sources (indicating that the quantities $\boldsymbol{j}_{\text{ext}}$ and ρ_{ext} are assumed to be given and independent of $\boldsymbol{E}$, $\boldsymbol{D}$ and $\boldsymbol{B}$). The induction $\boldsymbol{D}$ is determined by the relation

$$\frac{\partial \boldsymbol{D}}{\partial t} = \frac{\partial \boldsymbol{E}}{\partial t} + 4\pi \boldsymbol{j},$$

where $\boldsymbol{j}$ is the current density induced by the fields $\boldsymbol{E}$ and $\boldsymbol{B}$; in addition, it is sometimes convenient to introduce the polarization vector $\boldsymbol{P}$, where $\boldsymbol{D} = \boldsymbol{E} + 4\pi \boldsymbol{P}$.

All the properties of the medium are expressed in the expressions for $\boldsymbol{D}$ or $\boldsymbol{j}$ or, more exactly, in the relations linking $\boldsymbol{D}$ with the vectors $\boldsymbol{E}$ and $\boldsymbol{B}$.

Equations (2.1.1) are obtained from the microscopic equations of the field by quantum-mechanical and statistical averaging of these microscopic equations. The expression relating the current density $\boldsymbol{j}$ to the microscopic characteristics of the system will be presented in Sect. 5.1. It is important, for the present, to point out that the vectors $\boldsymbol{E}$, $\boldsymbol{D}$ and $\boldsymbol{B}$ being discussed here are statistically averaged ones. Such average vectors have a definite meaning for a medium in a state of thermodynamic equilibrium. Averaging can be carried out in the more general case as well (for instance, for metastable states corresponding to superheating, supercooling, etc.). Due to the averaging, fluctuations are not taken into consideration in (2.1.1). At the same time, the fields $\boldsymbol{E}$, $\boldsymbol{D}$ and $\boldsymbol{B}$ may vary in anyway in space and time without requiring any kind of additional averaging (apart from the quantum-mechanical and statistical kinds) of the fields with respect to $\boldsymbol{r}$. Such averaging is not only unnecessary, but, generally speaking, is unfeasible in the electrodynamics of media if spatial dispersion is properly taken into account. In a similar way, averaging over the time t is, in general, impossible when frequency dispersion is taken into consideration.

When spatial dispersion is not taken into account, the fields $\boldsymbol{E}$, $\boldsymbol{D}$ and $\boldsymbol{B}$ can be measured directly by determining the force $\boldsymbol{F} = e[\boldsymbol{E} + (\boldsymbol{v} \times \boldsymbol{B})/c]$ acting on a "point" particle having charge e and velocity $\boldsymbol{v}$. Here the particle should be located in a properly positioned, sufficiently narrow, vacant slit in the substance (if the slit is parallel, for instance, to the field $\boldsymbol{E}$, this field is the same both in the slit and in the medium adjoining the slit because of the continuity of the components of the field $\boldsymbol{E}$ tangential to the interface between the media). The situation becomes somewhat more complicated when spatial dispersion is taken into account because the conditions for a sharp interface between the media (and, particularly, between the medium and a vacuum) include additional terms, see (2.1.2). On the other hand, a direct measurement of the fields in a medium is not usually of any interest, and the field equations are most often applied either to find the dispersion relations characterizing normal waves in a medium, or to determine the coefficients of reflection and transmission for electromagnetic waves in the medium. In one way or another, we can in each concrete case arrive at relations between quantities that are measured experimentally.

Equations (2.1.1) in which all quantities are functions of $\boldsymbol{r}$ and t are the ordinary equations of macroscopic electrodynamics. It is true that somewhat different quantities and notation are often used, but, since this fact has been discussed, it should not give rise to any misunderstandings; moreover, in Sect. 2.1.3, we shall give the relations linking the notation adopted here with other widely employed notation.

If abrupt interfaces exist between the media it is necessary to use boundary conditions which can be obtained from (2.1.1) as the result of a passage to the limit. These conditions have the form

$$\begin{aligned} &\boldsymbol{E}_{1\mathrm{t}} = \boldsymbol{E}_{2\mathrm{t}}, \quad [\boldsymbol{n}, \boldsymbol{B}_2 - \boldsymbol{B}_1] = \frac{4\pi}{c}(\boldsymbol{i} + \boldsymbol{i}_{\mathrm{ext}}), \\ &B_{1\mathrm{n}} = B_{2\mathrm{n}}, \quad D_{2\mathrm{n}} - D_{1\mathrm{n}} = 4\pi(\sigma + \sigma_{\mathrm{ext}}), \end{aligned} \tag{2.1.2}$$

where $\boldsymbol{n}$ is the normal to the interface, directed from medium 1 to medium 2; the subscripts n and t refer to the normal and tangential components; $\boldsymbol{i}_{\mathrm{ext}}$ and σ_{ext} are the surface densities of the external currents and charges, respectively; and the densities $\boldsymbol{i}$ and σ are expressed in terms of $\boldsymbol{D}$ by integration over the depth of the surface layer [Ref. 2.1, § 1]

$$\boldsymbol{i} = -\frac{1}{4\pi}\int_1^2 \frac{\partial \boldsymbol{D}}{\partial t} dl \quad \text{and} \quad \sigma = \frac{1}{4\pi}\int_1^2 \operatorname{div}[\boldsymbol{n}(\boldsymbol{D} \times \boldsymbol{n})]\, dl\,.$$

The densities $\boldsymbol{i}$ and σ are nonzero, in general, when spatial dispersion is taken into consideration, in which case the relation between $\boldsymbol{D}$ and $\boldsymbol{E}$ contains derivatives with respect to the coordinates (Sect. 2.3). In this connection, while it is usually assumed that $\boldsymbol{i} = 0$ and $\sigma = 0$ when spatial dispersion is neglected, there is no particular reason for such an assumption when such dispersion is taken into account. We note also that in deriving the boundary conditions (2.1.2), it was assumed that the physical fields $\boldsymbol{E}$ and $\boldsymbol{B}$ at the interface cannot become infinite, while the induction $\boldsymbol{D}$ can tend to infinity (this is possible, for instance, when the value of $\boldsymbol{D}$ is determined by derivatives of $\boldsymbol{E}$ with respect to the normal to the interface). If the density $\boldsymbol{i}$ is proportional to a delta function, the fields $\boldsymbol{E}$ and $\boldsymbol{B}$ also cannot be assumed to be finite, but we shall not concern ourselves with this possibility. On the whole, in the following we shall deal mainly with unbounded media so that we need not refer to boundary conditions, with the exception of Sect. 4.5 which is specially devoted to this matter.

We underline, however, that in deriving (2.1.2) it was assumed, in fact, that the surface density $\boldsymbol{i}$ of the current has no normal component. But such a component exists when, for example, the surface has a thin film or, in more general terms, a certain transition layer (see also Sect. 5.6.3). It can readily be seen that if the surface current has a normal component, this leads to discontinuities of the tangential components of the field $\boldsymbol{E}$. As a matter of fact, if we integrate, for example, the y component of the equation curl $\boldsymbol{E} = -\dot{\boldsymbol{B}}/c$ with respect to the transition layer, we obtain

$$\int_1^2 \left(\frac{\partial E_z}{\partial x} - \frac{\partial E_x}{\partial z} \right) dz = -\frac{1}{c} \int_1^2 \dot{B}_y dz \, .$$

Since B_y has no singularities at the boundary, the integral at the right-hand side of the preceding equation can be omitted in the limit $1 \to 2$. Hence, in this limit

$$E_x(2) - E_x(1) = \lim_{1 \to 2} \int_1^2 \frac{\partial E_z}{\partial x} dz \, .$$

Since $E_z = (D_z - 4\pi P_z)/\varepsilon(z)$ and since the quantity P_z includes a term of the form $i_n \delta(z)$ if the surface current has a normal component, we arrive at the conclusion that the quantity $E_x(z)$ actually does have a discontinuity at $z = 0$. The magnitude of this discontinuity cannot be calculated consistently within the scope of an abrupt-boundary model, because, in this case, it is impossible uniquely to determine abruptly the value of the integral

$$\int_1^2 \frac{\delta(z)}{\varepsilon(z)} dz \, ,$$

where $\varepsilon(z) = \varepsilon_1$ at $z < 0$ and $\varepsilon(z) = \varepsilon_2$ at $z > 0$.

The set of equations (2.1.1) is a complete system, as we know, only when the relation (material equation) is indicated that enables us to express $\boldsymbol{D}$ in terms of $\boldsymbol{E}$, and if necessary, in terms of $\boldsymbol{B}$ as well. In a condensed medium this relation can usually be assumed to be linear as the fields considered are by far weaker than the atomic-scale fields $E_a \sim e/a^2 \sim 10^8$ V/cm. As a result nonlinear effects (which can be observed, e.g., in a plasma under rather easily achievable conditions [Ref. 2.2., Chap. 8]) in the optics of condensed media have only recently attracted some interest in connection with the application of lasers. Nonlinear optics [2.3 – 5] remains, however, a somewhat isolated field without any direct contact to the problem we are interested in [except for the problem of Raman scattering of light and x rays with the production of polaritons, phonons and excitons, to be dealt with in Sect. 6.4 and the problem of the boundary conditions for the optically nonlinear medium with center inversion (Sect. 2.3.2f.)].

Thus assuming the material equation to be linear, we can write it in the following general form:

$$D_i(\boldsymbol{r}, t) = \int_{-\infty}^{t} dt' \int d\boldsymbol{r}' \, \hat{\varepsilon}_{ij}(t, t', \boldsymbol{r}, \boldsymbol{r}') E_j(\boldsymbol{r}', t') \, . \qquad (2.1.3)$$

Here we merely refer to the principle of causality owing to which the induction at the instant t is determined only by the present field and the field at previous times $t' \leqslant t$. If the properties of the medium do not change with time (homogeneity in time), then the kernel $\hat{\varepsilon}_{ij}$ can depend only on the difference $t-t'$. Finally, if the medium is uniform in space, then all its points (assuming fluctuations to be ignored) will be equivalent and $\hat{\varepsilon}_{ij}$ depends only on the difference $\boldsymbol{r}-\boldsymbol{r}'$. Under such conditions

$$D_i(\boldsymbol{r},t) = \int_{-\infty}^{t} dt' \int d\boldsymbol{r}' \, \hat{\varepsilon}_{ij}(t-t', \boldsymbol{r}-\boldsymbol{r}') E_j(\boldsymbol{r}', t') \, . \tag{2.1.4}$$

In vacuo $\boldsymbol{D} = \boldsymbol{E}$ and, in the general case, $\boldsymbol{D}$ is the sum of $\boldsymbol{E}$ and $4\pi\boldsymbol{P}$. Therefore, obviously, the kernel $\hat{\varepsilon}_{ij}$ contains a delta-function type term and it is sometimes convenient to write (2.1.4) in the form

$$D_i(\boldsymbol{r},t) = E_i(\boldsymbol{r},t) + 4\pi \int_{-\infty}^{t} dt' \int d\boldsymbol{r}' \, \hat{\chi}_{ij}(t-t', \boldsymbol{r}-\boldsymbol{r}') E_j(\boldsymbol{r}', t') \, . \tag{2.1.5}$$

The same is true, of course, for (2.1.3). In contrast to the kernel $\hat{\varepsilon}_{ij}$ which contains the above-mentioned singularity, the kernel $\hat{\chi}_{ij}$ has no singularities at all, as may be expected from physical considerations.

It should be noted that the kernel $\hat{\varepsilon}_{ij}$ can be interpreted as an induced field appearing as a "response" to the delta-function field $\boldsymbol{E}$; we have in fact $D_i(\boldsymbol{r},t) = \hat{\varepsilon}_{ij}(t, \boldsymbol{r}) e_j$ for $\boldsymbol{E}(\boldsymbol{r}', t') = \boldsymbol{e}\,\delta(t')\,\delta(\boldsymbol{r}')$, $e = 1$ and $t > 0$ [see (2.1.4)]; in the case of (2.1.3) $D_i(\boldsymbol{r},t) = \hat{\varepsilon}_{ij}(t, 0, \boldsymbol{r}) e_j$.

The system of equations (2.1.1), together with the material equation (2.1.3) and the boundary conditions (2.1.2) is complete, i.e., they uniquely determine an electromagnetic field at any point in space and at any instant of time (in which case, of course, definite initial or boundary values of the fields, depending upon the nature of the problem, must be known).

Let us go over to Fourier transforms in (2.1.4, 5), assuming that

$$E_i(\boldsymbol{r},t) = \int E_i(\omega, \boldsymbol{k}) \, \mathrm{e}^{\mathrm{i}(\boldsymbol{k}\cdot\boldsymbol{r} - \omega t)} d\omega \, d\boldsymbol{k} \, ;$$

the fact that we use the same notation $\boldsymbol{E}$ for both the transforms $E_i(\omega, \boldsymbol{k})$ and the originals $E_i(\boldsymbol{r}, t)$ should not lead to any confusion because the arguments are indicated. Analogous notation is used for other quantities as well. It is not even necessary to carry out a Fourier transformation to obtain $D_i(\omega, \boldsymbol{k})$; it is sufficient to put

$$E_j(\boldsymbol{r}', t') = E_j(\omega, \boldsymbol{k}) \, \mathrm{e}^{\mathrm{i}(\boldsymbol{k}\cdot\boldsymbol{r}' - \omega t')}$$

in (2.1.4, 5), and to write a similar expression for D_i. As a result we obtain

$$D_i(\omega,\boldsymbol{k}) = \varepsilon_{ij}(\omega,\boldsymbol{k})E_j(\omega,\boldsymbol{k}) ,$$

$$\varepsilon_{ij}(\omega,\boldsymbol{k}) = \int_0^\infty d\tau \int d\boldsymbol{R}\, \mathrm{e}^{-i(\boldsymbol{k}\cdot\boldsymbol{R}-\omega\tau)}\, \hat{\varepsilon}_{ij}(\tau,\boldsymbol{R}) = \delta_{ij} + 4\pi\chi_{ij}(\omega,\boldsymbol{k}) , \qquad (2.1.6)$$

$$\chi_{ij}(\omega,\boldsymbol{k}) = \int_0^\infty d\tau \int d\boldsymbol{R}\, \mathrm{e}^{-i(\boldsymbol{k}\cdot\boldsymbol{R}-\omega\tau)}\, \hat{\chi}_{ij}(\tau,\boldsymbol{R}) ,$$

where $\tau = t - t'$ and $\boldsymbol{R} = \boldsymbol{r} - \boldsymbol{r}'$. It should be stressed that the components of $D_i(\omega,\boldsymbol{k})$ are related only to the components of $E_j(\omega,\boldsymbol{k})$ for the same values of ω and $\boldsymbol{k}$; this fact is due to the homogeneity of the medium with respect to time and space, i.e., because $\hat{\varepsilon}_{ij}$ depends only on the differences $\tau = t - t'$ and $\boldsymbol{R} = \boldsymbol{r} - \boldsymbol{r}'$. The quantity $\hat{\varepsilon}_{ij}(\omega,\boldsymbol{k})$ is called the complex dielectric constant tensor; the tensor $\chi_{ij}(\omega,\boldsymbol{k})$ the dielectric susceptibility.

The frequency dependence of the tensor $\varepsilon_{ij}(\omega,\boldsymbol{k})$ corresponds to the frequency dispersion, and its dependence on the wave vector to the spatial dispersion. The region in which the kernel $\hat{\varepsilon}_{ij}$ has significant magnitude is determined by the characteristic frequencies ω_s of the medium, by the reciprocals of the relaxation times and the characteristic dimensions a_s. The frequency ω_s usually lies within relatively wide limits. The dimensions a_s ("radius of molecular action" and the like), on the contrary, can be assumed small in many cases. In liquids and solids the molecular dimensions, the atomic distances or the lattice constants usually play the part of a_s; all these quantities are generally of the same order of magnitude and are very small compared with the wavelengths in the optical frequency range. In an isotropic plasma the characteristic dimension a is given by the Debye radius $D = \sqrt{kT/8\pi e^2 N}$ [2.2], and in a conducting medium when collisions are taken into account it is the mean-free path l. The parameters D and l can both be large compared with the wavelength or the depth of penetration of the field. Thus, it should not be thought that spatial dispersion in optics is always weak, although it is evident that it plays a smaller role in optics than does frequency dispersion[1].

The tensor $\varepsilon_{ij}(\omega,\boldsymbol{k})$ completely describes both the electrical and the magnetic properties of the medium, i.e., it takes into account the influence of

---

[1] When passing from microscopic to macroscopic electrodynamics [2.6], averaging over a "physically infinitely small" volume element and time interval is frequently carried out. This operation means that fluctuations with respect to time are neglected; it makes it impossible, however, to account for sufficiently rapid variations of the fields in space and time. Within the scope of the approach discussed here we have only a statistical averaging in conjunction with which fluctuations are neglected. As a result of this the kernel $\hat{\varepsilon}_{ij}$ in (2.1.4) can be assumed to depend on the differences $t - t'$ and $\boldsymbol{r} - \boldsymbol{r}'$. Moreover, the dependence of $\hat{\varepsilon}_{ij}$ on $\tau = t - t'$ can, in principle, be an arbitrarily "rapid" function. In other words, if frequency dispersion is taken into account and it is necessary to go over to the Fourier components $\boldsymbol{E}(\omega,\boldsymbol{r})$, etc., the fields are automatically separated with respect to the frequency and there is no necessity for another averaging of the fields over time intervals $\Delta t \ll 1/\omega$. The same can be proved for the averaging over an infinitesimally small volume if spatial dispersion is consistently taken into account and use is made of the quantities $\boldsymbol{E}(\omega,\boldsymbol{k})$, $\boldsymbol{D}(\omega,\boldsymbol{k})$, $\varepsilon_{ij}(\omega,\boldsymbol{k})$, etc.

the induction $\boldsymbol{B}$ on $\boldsymbol{D}$ [or, equivalently, on the induced current $\boldsymbol{j}=1/4\pi \times(\partial/\partial t)(\boldsymbol{D}-\boldsymbol{E})$]. In fact, from the field equation $\operatorname{rot}\boldsymbol{E}=(-1/c)\partial\boldsymbol{B}/\partial t$, we obtain the following relation for the Fourier components: $\boldsymbol{B}(\omega,\boldsymbol{k})=(c/\omega)\boldsymbol{k}\times\boldsymbol{E}(\omega,\boldsymbol{k})$. Therefore, the relation between $\boldsymbol{D}(\omega,\boldsymbol{k})$ and $\boldsymbol{B}(\omega,\boldsymbol{k})$ for $\omega \neq 0$ can always be replaced, without difficulty, by the corresponding relation between $\boldsymbol{D}(\omega,\boldsymbol{k})$ and $\boldsymbol{E}(\omega,\boldsymbol{k})$. This is valid, of course, within the scope of the more general relation, (2.1.3). If spatial dispersion is neglected, we cannot in general set $\varepsilon_{ij}(\omega,0)=\varepsilon_{ij}(\omega)$ but should also introduce the magnetic-permeability tensor $\mu_{ij}(\omega)$. We shall discuss this problem again in Sect. 2.1.3, but it should be mentioned here that in the optics of nonferromagnetic media we can and should always set $\mu_{ij}(\omega)=\delta_{ij}$ [Ref. 2.7, § 60]. Therefore, we shall henceforth assume that

$$\varepsilon_{ij}(\omega,\boldsymbol{k}\rightarrow 0)=\varepsilon_{ij}(\omega,0)=\varepsilon_{ij}(\omega)\,, \tag{2.1.7}$$

where $\varepsilon_{ij}(\omega)$ is the tensor of the complex dielectric constant if spatial dispersion is ignored. (In the case of ferromagnetic substances, the tensor $\mu_{ij}(\omega)$ cannot always be reduced to δ_{ij} even in the optical frequency range [2.8].)

In addition to the tensor $\varepsilon_{ij}(\omega,\boldsymbol{k})$, it also proves convenient to use the inverse tensor $\varepsilon_{ij}^{-1}(\omega,\boldsymbol{k})$. Obviously

$$E_i(\omega,\boldsymbol{k})=\varepsilon_{ij}^{-1}(\omega,\boldsymbol{k})D_j(\omega,\boldsymbol{k})\,, \tag{2.1.8}$$

and

$$\varepsilon_{ij}^{-1}(\omega,\boldsymbol{k})=\frac{A_{ji}(\omega,\boldsymbol{k})}{|\varepsilon_{ij}(\omega,\boldsymbol{k})|}\,,\qquad \varepsilon_{il}\varepsilon_{lj}^{-1}=\delta_{ij}\,, \tag{2.1.9}$$

where $|\varepsilon_{ij}|$ is the determinant of the tensor (matrix) ε_{ij}, and $A_{ij}=(-1)^{i+j}\Delta_{ij}$ and Δ_{ij} are the algebraic complement and the minor, respectively, referring to the element ε_{ij} of the determinant $|\varepsilon_{ij}|$(recall that

$$A_{il}\varepsilon_{jl}=A_{li}\varepsilon_{lj}=|\varepsilon_{l'm'}|\,\delta_{ij})\,.$$

In those cases for which $|\varepsilon_{ij}|\neq 0$ and the tensor ε_{ij}^{-1} exists, the use of both tensor ε_{ij} and ε_{ij}^{-1} is equally justified. If, however, $|\varepsilon_{ij}(\omega,\boldsymbol{k})|=0$ we could introduce the singular tensor ε_{ij}^{-1} which has to be handled with care. In the present text we shall take the tensor $\varepsilon_{ij}(\omega,\boldsymbol{k})$ as the fundamental one and the tensors $\varepsilon_{ij}^{-1}(\omega,\boldsymbol{k})$ will be introduced in accordance with the definition (2.1.9); in the cases of interest we shall not be concerned with the singularity of ε_{ij}^{-1}.

The variables ω and $\boldsymbol{k}$ on which the tensor $\varepsilon_{ij}(\omega,\boldsymbol{k})$ depends are, generally speaking, independent variables. This fact follows from the definition (2.1.6), but sometimes this does not seem to be as clear as it might be. The point is that in optics we often deal with the propagation of waves in the absence of external sources in the medium itself. In this case the wave vector depends on ω, for example, and in the case of uniform normal plane waves $\boldsymbol{k} = (\omega/c)\tilde{n}(\omega, \boldsymbol{s})\boldsymbol{s}$. If, however, $\boldsymbol{k} = \boldsymbol{k}(\omega)$, spatial dispersion may seem to be equivalent to frequency dispersion. The answer to the question this poses is in the following. The tensor $\varepsilon_{ij}(\omega, \boldsymbol{k})$ is introduced for fields of the general form when the medium contains sources $\boldsymbol{j}_{\text{ext}}$ and ρ_{ext}, related as it follows from (2.1.1) by the continuity equation:

$$\operatorname{div} \boldsymbol{j}_{\text{ext}} + \partial \rho_{\text{ext}}/\partial t = 0 \qquad \text{or}$$

$$\boldsymbol{k} \cdot \boldsymbol{j}_{\text{ext}}(\omega, \boldsymbol{k}) = \omega \rho_{\text{ext}}(\omega, \boldsymbol{k}) \, .$$

Under these conditions a field $\boldsymbol{E}$ can be produced with any independent values of ω and $\boldsymbol{k}$ [the component $\boldsymbol{E}(\omega, \boldsymbol{k})$ is expressed, in the final analysis, in terms of $\boldsymbol{j}_{\text{ext}}(\omega, \boldsymbol{k})$ and $\rho_{\text{ext}}(\omega, \boldsymbol{k})$; for details see Sect. 2.2.1]. Hence it is clear that all problems connected with the investigation of wave propagation can be considered if the tensor $\varepsilon_{ij}(\omega, \boldsymbol{k})$ is known. As far as we know, the converse is not true; it is in general impossible to determine the tensor $\varepsilon_{ij}(\omega, \boldsymbol{k})$ for all values of ω and $\boldsymbol{k}$ on the basis of investigating only the propagation of waves.

Since a Fourier transformation is connected with the transition from $\boldsymbol{E}(\boldsymbol{r}, t)$ and $\boldsymbol{D}(\boldsymbol{r}, t)$ to $\boldsymbol{E}(\omega, \boldsymbol{k})$ and $\boldsymbol{D}(\omega, \boldsymbol{k})$, the tensor $\varepsilon_{ij}(\omega, \boldsymbol{k})$ can be directly determined – see (2.1.6) – only if ω and $\boldsymbol{k}$ are real. It is, however, impossible to restrict oneself to real values in the electrodynamics of continuous media. Of course, the usual statement of optical problems requires that the frequency ω be real (an external source of frequency ω is given). But then the vector $\boldsymbol{k}$ automatically becomes complex since waves propagating in a medium are always damped, at least if we consider a medium in equilibrium (i.e., a medium in the state of thermodynamic equilibrium). In the range of transparency the imaginary part of $\boldsymbol{k}$ may be small or even negligible, but, as is well known, when the frequency is changed one comes inevitably into a range where damping is appreciable.

Independent physical considerations also lead to the result that the tensor $\varepsilon_{ij}(\omega, \boldsymbol{k})$ should, in general, have a meaning even when ω and $\boldsymbol{k}$ are complex. The tensor $\varepsilon_{ij}(\omega, \boldsymbol{k})$ thus relates fields of the type const $\cdot \exp[\mathrm{i}(\boldsymbol{k} \cdot \boldsymbol{r} - \omega t)]$:

$$D_{0i} \mathrm{e}^{\mathrm{i}(\boldsymbol{k} \cdot \boldsymbol{r} - \omega t)} = \varepsilon_{ij}(\omega, \boldsymbol{k}) E_{0j} \mathrm{e}^{\mathrm{i}(\boldsymbol{k} \cdot \boldsymbol{r} - \omega t)} \, . \tag{2.1.10}$$

Even if ω and $\boldsymbol{k}$ are real, the production of such a field does not comply with reality since, in fact, a wave train is always bounded in space and time. However, if the train is long enough so that the main part of its Fourier expansion is determined by the given values of ω and $\boldsymbol{k}$, the relation between $\boldsymbol{D}$ and $\boldsymbol{E}$ for the trains may be replaced by the relation (2.1.10) for monochromatic waves. Of course, this must be treated carefully since a long train and a monochromatic wave are not always equivalent, as is known from the example of the introduction of the group velocity or the expression for the energy density (Sect. 2.3.1). But, with this limitation, (2.1.10) does have a meaning even if ω and $\boldsymbol{k}$ are complex. In this case we have to deal with finite pulses (packets) of appropriate form (e.g., of the form $\boldsymbol{E}_0 \exp(\omega'' t) \exp(-\mathrm{i}\omega' t)$, where $\boldsymbol{E}_0 =$ const when $0 \leqslant t \leqslant T$ and $\boldsymbol{E}_0 = 0$ outside this interval) which are approximated by a monochromatic wave of the type (2.1.10).

The problem concerning the values of $\varepsilon_{ij}(\omega, \boldsymbol{k})$ in the complex region should be analyzed on the basis of the initial expression (2.1.6). If the medium considered is in thermodynamic equilibrium or if it is at least stable, then the kernel $\hat{\varepsilon}_{ij}(\tau, \boldsymbol{R})$ does not increase with τ. Moreover, it vanishes in the case of a nonconducting medium, i.e., $\hat{\varepsilon}_{ij}(\tau \to \infty, \boldsymbol{R}) \to 0$ (in a conducting medium $\hat{\varepsilon}_{ij}(\tau \to \infty, \boldsymbol{R}) \to$ const [2.7]). In this case the function $\varepsilon_{ij}(\omega, \boldsymbol{k})$ has no poles in the upper half ω plane (i.e., when $\omega'' \geqslant 0$, with $\omega = \omega' + \mathrm{i}\omega''$); this follows directly from (2.1.6) and the well-known theory of Fourier integrals.[2] This property of $\varepsilon_{ij}(\omega, \boldsymbol{k})$ is used when deriving the dispersion relations (Sect. 2.1.2). In the lower half ω plane, the function $\varepsilon_{ij}(\omega, \boldsymbol{k})$ has singularities, but it is well defined in a large region and may be used. In exactly the same way, the function $\varepsilon_{ij}(\omega, \boldsymbol{k})$ is analytic in the whole region of the complex variable $\boldsymbol{k}$, provided the kernel $\hat{\varepsilon}_{ij}(\tau, \boldsymbol{R})$ in (2.1.6) decreases sufficiently quickly as R increases; this is actually the case (we shall return to discuss this problem in Sect. 6.3). In other words, the function $\varepsilon_{ij}(\omega, \boldsymbol{k})$, like other similar expressions encountered in physics, is an analytic, i.e., regular (without poles) function of its arguments in a certain range of their variation. As we know from the theory of analytic functions, these are, when given in a certain finite interval (e.g., on a section of the real axis), also uniquely determined in a certain vicinity outside this interval, including the complex region. The existence of the function $\varepsilon_{ij}(\omega, \boldsymbol{k})$ for real values of ω and $\boldsymbol{k}$, and the supposition of its analyticity in a certain region are sufficient to lead to the determination of ε_{ij} in the whole complex region.

Thus, the tensor $\varepsilon_{ij}(\omega, \boldsymbol{k})$ has a definite meaning not only in the case of real values of ω and $\boldsymbol{k}$, but also in a certain region of complex values of these variables. The determination of this region represents a special problem which is to be solved, for instance, when calculating $\varepsilon_{ij}(\omega, \boldsymbol{k})$ for a definite model. Any further use of the tensor $\varepsilon_{ij}(\omega, \boldsymbol{k})$ with complex values of ω and $\boldsymbol{k}$ is based, of course, on the supposition that this tensor exists in the region of values being considered (i.e., that it is a regular function of ω and $\boldsymbol{k}$).

---

[2] Essential here is the use of the causality principle, by virtue of which (2.1.6) is integrated only over the region $t' \leqslant t$. Such a seemingly obvious requirement at $\boldsymbol{k} \neq 0$ is clearly valid only with respect to the tensor $\varepsilon_{ij}^{-1}(\omega, \boldsymbol{k})$ relating the field $\boldsymbol{E}$ to the induction $\boldsymbol{D}$, which sets up the field and can be varied by correspondingly changing the density of the external charges ρ_{ext}. The requirement of causality for $\varepsilon_{ij}(\omega, \boldsymbol{k})$ "performs" reliably only at $\boldsymbol{k} = 0$ [2.9], and, owing to analyticity, in a certain region of $\boldsymbol{k}$ values close to $\boldsymbol{k} = 0$. For this reason it is impossible to guarantee the validity of dispersion relations for the quantities $\varepsilon_{ij}(\omega, \boldsymbol{k})$ when $\boldsymbol{k} \neq 0$ (Sect. 2.1.2).

The vector $\boldsymbol{k}$ is determined by three scalars (components) and when passing over to the complex region it is thus necessary, in general, to consider three complex variables, e.g., k_x, k_y and k_z. In other words, the vector $\boldsymbol{k}$ has the form of $\boldsymbol{k} = \boldsymbol{k}' + \mathrm{i}\boldsymbol{k}''$, $\boldsymbol{k}'$ and $\boldsymbol{k}''$ being real vectors. If these vectors are not parallel the waves are said to be nonuniform. An important role is played, however, by uniform waves for which $\boldsymbol{k} = (k' + \mathrm{i}k'')\boldsymbol{s} = k\boldsymbol{s}$, where $\boldsymbol{s}$ is a real unit vector. In this case, if $\boldsymbol{s}$ is given, we have to deal with only one complex variable: $k = k' + \mathrm{i}k''$. For normal waves, $\boldsymbol{k} = \boldsymbol{k}(\omega)$ and if the waves are uniform, $\boldsymbol{k} = (\omega/c)\tilde{n}(\omega,\boldsymbol{s})\boldsymbol{s} = (\omega/c)(n + \mathrm{i}\varkappa)\boldsymbol{s}$, see (1.1.1), i.e., in this case $k' = (\omega/c)n(\omega,\boldsymbol{s})$ and $k'' = (\omega/c)\varkappa(\omega,\boldsymbol{s})$. Although in what follows we shall consider almost exclusively uniform waves the wave vector will in most cases be denoted generally by $\boldsymbol{k}$. To simplify the discussion this fact will not usually be mentioned.

As already indicated, the tensor $\varepsilon_{ij}(\omega,\boldsymbol{k})$ was introduced solely for homogeneous media. It is therefore impossible to apply (2.1.6) directly to such spatially inhomogeneous media as crystals. If, however, long wavelengths are considered ($\lambda \gg a$, a being the lattice constant), a definite meaning can be imparted to the application of a tensor of the type of $\varepsilon_{ij}(\omega,\boldsymbol{k})$ in the case of crystals as well.

Moreover, the difference between crystals and homogeneous media in this respect [as regards the application of the tensor $\varepsilon_{ij}(\omega,\boldsymbol{k})$ in optics] is of little significance. This problem will be discussed in Sect. 3.1.1., but in Sects. 2.1 – 3 we shall often speak of crystals, though, owing to the necessity for special treatment in the case of crystals, the term "medium" would be more exact if we consider the tensor $\varepsilon_{ij}(\omega,\boldsymbol{k})$ as a characteristic of such a homogeneous medium.

2.1.2 General Properties of the Tensor $\varepsilon_{ij}(\omega,\boldsymbol{k})$

The tensor $\varepsilon_{ij}(\omega,\boldsymbol{k})$ is, in general, complex even if ω and $\boldsymbol{k}$ are real. Moreover, this tensor is neither Hermitian nor symmetrical. It is therefore convenient to decompose ε_{ij} into a real (Re) and an imaginary (Im) part as well as into two Hermitian tensors ε'_{ij} and ε''_{ij}

$$\varepsilon_{ij} = \mathrm{Re}\{\varepsilon_{ij}\} + \mathrm{i}\,\mathrm{Im}\{\varepsilon_{ij}\}, \tag{2.1.11}$$

$$\varepsilon_{ij} = \varepsilon'_{ij} + \mathrm{i}\varepsilon''_{ij}\,. \tag{2.1.12}$$

It likewise proves convenient to decompose ε'_{ij} and ε''_{ij} into real and imaginary parts:

$$\begin{aligned}
\varepsilon'_{ij} &= \varepsilon'_{ij,c} + \mathrm{i}\varepsilon'_{ij,a}, & \varepsilon''_{ij} &= \varepsilon''_{ij,c} + \mathrm{i}\varepsilon''_{ij,a};\\
\varepsilon'_{ij,c} &= \mathrm{Re}\{\varepsilon'_{ij}\} = \varepsilon'_{ji,c}, & \varepsilon'_{ij,a} &= \mathrm{Im}\{\varepsilon'_{ij}\} = -\varepsilon'_{ji,a};\\
\varepsilon''_{ij,c} &= \mathrm{Re}\{\varepsilon''_{ij}\} = \varepsilon''_{ji,c}, & \varepsilon''_{ij,a} &= \mathrm{Im}\{\varepsilon''_{ij}\} = -\varepsilon''_{ji,a}.
\end{aligned} \tag{2.1.13}$$

The symmetry of the tensors $\varepsilon'_{ij,c}$ and $\varepsilon''_{ij,c}$ and the antisymmetry of $\varepsilon'_{ij,a}$ and $\varepsilon''_{ij,a}$ result from the fact that the tensors ε'_{ij} and ε''_{ij} are Hermitian. Furthermore,

$$\begin{aligned} \mathrm{Re}\{\varepsilon_{ij}\} &= \varepsilon'_{ij,c} - \varepsilon''_{ij,a}\,, \\ \mathrm{Im}\{\varepsilon_{ij}\} &= \varepsilon'_{ij,a} + \varepsilon''_{ij,c}\,. \end{aligned} \tag{2.1.14}$$

If spatial dispersion is ignored and if there is no constant magnetic field, the tensor ε_{ij} is symmetrical, see (2.1.24) below. Consequently, the tensor ε'_{ij} and ε''_{ij} are real and coincide with $\mathrm{Re}\{\varepsilon_{ij}\}$ and $\mathrm{Im}\{\varepsilon_{ij}\}$. It is therefore not necessary to decompose ε_{ij} in two ways (see also Sect. 2.1.3). Note that the dielectric constant ε_{ij} is often replaced by the conductivity σ_{ij}. Most widely used is the notation

$$\varepsilon''_{ij} = 4\pi\sigma_{ij}/\omega\,. \tag{2.1.15}$$

The complex non-Hermitian conductivity

$$\sigma_{ij} = \sigma'_{ij} + \mathrm{i}\sigma''_{ij} = -\frac{\mathrm{i}\omega}{4\pi}[\varepsilon_{ij}(\omega,\boldsymbol{k}) - \delta_{ij}] = -\mathrm{i}\omega\chi_{ij}(\omega,\boldsymbol{k}) \tag{2.1.16}$$

is also frequently adopted. Note that here and elsewhere in this book the convention $\exp(\mathrm{i}\boldsymbol{k}\cdot\boldsymbol{r} - \mathrm{i}\omega t)$ has been used. It links the induced current

$$\boldsymbol{j} = \frac{1}{4\pi}\,\frac{\partial}{\partial t}(\boldsymbol{D} - \boldsymbol{E})$$

with the field $\boldsymbol{E}$, see (2.1.6),

$$\begin{aligned} j_i(\omega,\boldsymbol{k}) &= \sigma_{ij}(\omega,\boldsymbol{k})E_j(\omega,\boldsymbol{k}) = -\frac{\mathrm{i}\omega}{4\pi}[D_i(\omega,\boldsymbol{k}) - E_i(\omega,\boldsymbol{k})] \\ &= -\frac{\mathrm{i}\omega}{4\pi}[\varepsilon_{ij}(\omega,\boldsymbol{k}) - \delta_{ij}]E_j(\omega,\boldsymbol{k}) = -\mathrm{i}\omega\chi_{ij}(\omega,\boldsymbol{k})E_j(\omega,\boldsymbol{k})\,. \end{aligned} \tag{2.1.17}$$

Equation (2.1.16) cannot be used if $\omega = 0$; we shall not dwell on this here in detail [Ref. 2.1, § 2]. We observe that the σ_{ij} in (2.1.15) and σ_{ij} in (2.1.16) denote different quantities. As a rule, σ_{ij} will not be used in the following; however in Sect. 5.1, a conductivity is introduced in accordance with the definition (2.1.6).

Let us now discuss general properties of the tensor $\varepsilon_{ij}(\omega,\boldsymbol{k})$. A real field $\boldsymbol{E}$ leads, of course, to a real induction $\boldsymbol{D}$. Consequently, the kernel $\hat{\varepsilon}_{ij}(\tau,\boldsymbol{R})$ in (2.1.16) is real and therefore

$$\varepsilon_{ij}(\omega,\boldsymbol{k}) = \varepsilon^*_{ij}(-\omega^*, -\boldsymbol{k}^*) \tag{2.1.18}$$

or, analogously,

$$\varepsilon_{ij}(\omega^*, \boldsymbol{k}^*) = \varepsilon_{ij}^*(-\omega, -\boldsymbol{k}) . \tag{2.1.18a}$$

If it is necessary or desirable to use real fields, it is convenient to write the field equation in the form

$$\boldsymbol{E}(\boldsymbol{r}, t) = \tfrac{1}{2}(\boldsymbol{E}_0 \mathrm{e}^{\mathrm{i}(\boldsymbol{k}\cdot\boldsymbol{r} - \omega t)} + \boldsymbol{E}_0^* \mathrm{e}^{-\mathrm{i}(\boldsymbol{k}^*\cdot\boldsymbol{r} - \omega^* t)}) , \tag{2.1.19}$$

and, by definition,

$$D_i(\boldsymbol{r}, t) = \tfrac{1}{2}[\varepsilon_{ij}(\omega, \boldsymbol{k}) E_{0j} \mathrm{e}^{\mathrm{i}(\boldsymbol{k}\cdot\boldsymbol{r} - \omega t)} + \varepsilon_{ij}(-\omega^*, -\boldsymbol{k}^*) E_{0j}^* \mathrm{e}^{-\mathrm{i}(\boldsymbol{k}^*\cdot\boldsymbol{r} - \omega^* t)}] . \tag{2.1.20}$$

This expression is real if (2.1.18) holds. Thus we arrive at this condition in an equivalent but, perhaps, somewhat clearer manner.

The principle of symmetry of the kinetic coefficients yields the relation

$$\varepsilon_{ij}(\omega, \boldsymbol{k}, \boldsymbol{B}_{\text{ext}}) = \varepsilon_{ji}(\omega, -\boldsymbol{k}, -\boldsymbol{B}_{\text{ext}}) , \tag{2.1.21}$$

where $\boldsymbol{B}_{\text{ext}}$ denotes the magnetic field induction which is constant with respect to time and nonzero if there is an external magnetic field or magnetic structure (a ferromagnetic or antiferromagnetic substance). For the sake of simplicity we shall usually call $\boldsymbol{B}_{\text{ext}}$ the external magnetic induction, as indicated by the subscript.

The validity of (2.1.21) for the tensors $\varepsilon_{ij}(\omega)$ and $\varepsilon_{ij}(\omega, \boldsymbol{B}_{\text{ext}})$ has been proved in [2.7]. A generalization for the case of spatial dispersion [Ref. 2.1, § 9] does not bring to light any fundamentally new aspects [since the behavior of the vectors $\boldsymbol{k}$ and $\boldsymbol{B}_{\text{ext}}$ is the same on time reversal, the transition from the relation $\varepsilon_{ij}(\omega, \boldsymbol{B}_{\text{ext}}) = \varepsilon_{ij}(\omega, -\boldsymbol{B}_{\text{ext}})$ to (2.1.21) may be considered as almost obvious][3]. Note that (2.1.21) can be proved [2.1, 7] only for real ω and $\boldsymbol{k}$. In the region of analyticity these relations are automatically maintained even if ω and $\boldsymbol{k}$ are complex.

A medium is called nongyrotropic if

$$\varepsilon_{ij}(\omega, \boldsymbol{k}) = \varepsilon_{ji}(\omega, \boldsymbol{k}) \tag{2.1.22}$$

[3] Let us briefly explain why the vector $\boldsymbol{k}$ changes its sign if t is replaced by $-t$. This is readily understood if we take into consideration that in quantizing $\hbar\boldsymbol{k} = \boldsymbol{p}$ the momentum $\boldsymbol{p}$ changes its sign, according to the laws of mechanics, as soon as t is replaced by $-t$. In the case of moving media, the tensor ε_{ij} also depends on the velocity $\boldsymbol{u}$ of the medium (we assume that within the limits of accuracy $\boldsymbol{u} = \text{const}$, and thus is independent of the coordinates and the time). Obviously, the velocity $\boldsymbol{u}$ changes its sign with the substitution $t \to -t$. Consequently, (2.1.21) can be generalized as

$$\varepsilon_{ij}(\omega, \boldsymbol{k}, \boldsymbol{B}_{\text{ext}}, \boldsymbol{u}) = \varepsilon_{ji}(\omega, -\boldsymbol{k}, -\boldsymbol{B}_{\text{ext}}, -\boldsymbol{u}) .$$

for any arbitrary ω and $\boldsymbol{k}$. If (2.1.22) is satisfied, we readily see that

$$\begin{aligned}\mathrm{Re}\{\varepsilon_{ij}(\omega,\boldsymbol{k})\} &= \varepsilon'_{ij}(\omega,\boldsymbol{k})\,, \\ \mathrm{Im}\{\varepsilon_{ij}(\omega,\boldsymbol{k})\} &= \varepsilon''_{ij}(\omega,\boldsymbol{k})\,,\end{aligned} \qquad (2.1.22\,\mathrm{a})$$

i.e., the Hermitian tensors ε'_{ij} and ε''_{ij} become real (and symmetrical).

The argument $\boldsymbol{B}_{\mathrm{ext}}$ was omitted in (2.1.22) because, strictly speaking, the tensor ε_{ij} is always asymmetric in an external magnetic field since it is equal to ε_{ij} only if $\boldsymbol{B}_{\mathrm{ext}}$ is replaced by $-\boldsymbol{B}_{\mathrm{ext}}$ (here we consider merely the external field and ignore the case of antiferromagnetic materials in the absence of an external field where the tensor ε_{ij} may remain unchanged when replacing $\boldsymbol{B}_{\mathrm{ext}}$ by $-\boldsymbol{B}_{\mathrm{ext}}$). It is thus possible to speak of gyrotropy caused by a magnetic field. But, in order not to give rise to confusion, this kind of gyrotropy will be called magnetic activity and the corresponding medium will be referred to as being magnetoactive. Unless otherwise indicated, we shall assume in the following that $\boldsymbol{B}_{\mathrm{ext}} = 0$ and only a naturally gyrotropic (active) medium will be called gyrotropic.

As readily seen from (2.1.21, 22), for a nongyrotropic medium

$$\varepsilon_{ij}(\omega,\boldsymbol{k}) = \varepsilon_{ij}(\omega, -\boldsymbol{k})\,. \qquad (2.1.23)$$

In a gyrotropic medium there must exist at least one direction which is not equivalent to its opposite. In other words, only a medium possessing no center of symmetry can be gyrotropic. The reverse statement, however, is not true: a medium having no center of symmetry can be nongyrotropic since (2.1.23) can remain valid by virtue of other elements of symmetry as well (Sect. 3.2.1).

In the absence of spatial dispersion, (2.1.23) is automatically fulfilled and the medium is always nongyrotropic, as repeatedly mentioned above. But, of course, magnetic activity exists in this case as well if $\boldsymbol{B}_{\mathrm{ext}} \neq 0$ since

$$\varepsilon_{ij}(\omega,\boldsymbol{B}_{\mathrm{ext}}) = \varepsilon_{ji}(\omega, -\boldsymbol{B}_{\mathrm{ext}}) \qquad (2.1.24)$$

and the tensor ε_{ij} is asymmetric for given ω and $\boldsymbol{B}_{\mathrm{ext}}$. [As before, we ignore the possibility that $\varepsilon_{ij}(\omega,\boldsymbol{B}_{\mathrm{ext}}) = \varepsilon_{ij}(\omega, -\boldsymbol{B}_{\mathrm{ext}})$ as may be the case in antiferromagnetic materials in the absence of an external magnetic field. In antiferromagnetic materials $\boldsymbol{B}_{\mathrm{ext}}/4\pi$ is the statistical mean of the magnetization in a given microvolume of the crystal, vanishing only after averaging over an entire elementary cell of the magnetic structure of the crystal.]

The definition of a gyrotropic medium as a medium whose tensor $\varepsilon_{ij}(\omega,\boldsymbol{k})$ is asymmetric (with $\boldsymbol{B}_{\mathrm{ext}} = 0$) is, of course, a somewhat formal one. But in Sect. 4.1 we shall see that it is precisely the asymmetry of the tensor ε_{ij} that gives rise to the physical peculiarities (rotation of the plane of polarization for normal waves in the absence of absorption) that constitute the difference between gyrotropic and nongyrotropic media.

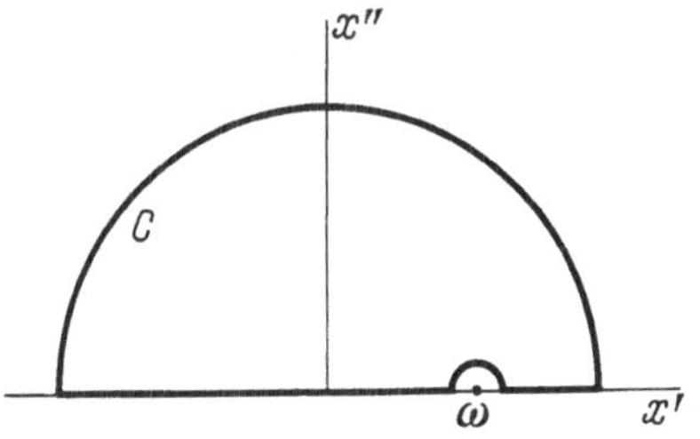

Fig. 2.1. Integration contour C

As already mentioned in Sect. 2.1.1, at $\boldsymbol{k}=0$ and at least in a certain region of $\boldsymbol{k}$ values close to this point the function $\varepsilon_{ij}(\omega,\boldsymbol{k})$ has no singularities in the upper half-plane and on the real axis of the complex variable ω. This important fact results primarily from the principle of causality due to which integration is carried out in (2.1.6) only over the interval $0 \leqslant \tau \leqslant \infty$. Hence it follows that the real and imaginary parts of $\varepsilon_{ij}(\omega,\boldsymbol{k})$ are mutually connected on the real axis ω by integral relations, namely, the so-called dispersion relations. In the case of normal conductors the tensor $\varepsilon_{ij}(\omega,\boldsymbol{k})$ has a simple pole at $\omega=0$, see, e.g., (2.1.15); in the case of superconductors there is a double pole of the form $1/\omega^2$ at $\omega=0$ [2.10]. Moreover, the kernel $\hat{\varepsilon}_{ij}(\tau,\boldsymbol{R})$ in (2.1.6) can be considered as nonincreasing with increasing τ only in the case of a stable medium (particularly a medium in equilibrium). Unless otherwise specified, we are dealing only with a nonconducting medium in equilibrium. Note that the difference between the lower and the upper half-planes of the variable ω is due to the nonequivalence of fields of the type $\exp(\omega'' t)$ when $\omega''>0$ and $\omega''<0$ (in the latter case the field was previously assumed to be infinite, [Ref. 2.7, § 62]).

In order to obtain the dispersion relations, we consider the integral

$$\int_C \frac{\varepsilon_{ij}(x,\boldsymbol{k})-\delta_{ij}}{x-\omega}dx = -\mathrm{i}\pi[\varepsilon_{ij}(\omega,\boldsymbol{k})-\delta_{ij}] + \fint_{-\infty}^{+\infty} \frac{\varepsilon_{ij}(x,\boldsymbol{k})-\delta_{ij}}{x-\omega}dx = 0\,, \tag{2.1.25}$$

where the first integral is taken over the contour C, as shown in Fig. 2.1, and the second integral along the real axis as the principal value (such integrals are denoted by the symbol $\fint$). In (2.1.25) we took into account that in the upper half-plane

$$\lim_{\omega\to\infty} \varepsilon_{ij}(\omega,\boldsymbol{k}) = \delta_{ij}\,. \tag{2.1.26}$$

If the function $\varepsilon_{ij}(\omega,\boldsymbol{k})$ is unique in the lower half-plane of ω and has a finite number of poles, then (2.1.26) applies to this region as well.

For real frequencies (i.e., if $\omega=\omega'$), (2.1.26) results from the theory of dispersion. The medium "does not have time" to react to external influences at very high frequencies and, in this respect, no longer differs from a vacuum. In the complex region, when $\omega = \omega' + \mathrm{i}\omega''$, $\omega'' > 0$ and $|\omega| \to \infty$, the field displays an extremely rapid variation with time, and the medium is thus not able to react to it. Moreover, as follows from the theory of analytic functions, if $\varepsilon_{ij}(\omega,\boldsymbol{k}) \to \delta_{ij}$ with $\omega = \omega' \to \pm\infty$, then, for $\omega = \omega' + \mathrm{i}\omega'' \to \infty$ and $\omega'' > 0$ (i.e., in the upper half-plane), $\varepsilon_{ij}(\omega,\boldsymbol{k})$ either tends to δ_{ij} or increases more rapidly than a certain exponent. From physical considerations, the latter is quite improbable and contradicts our previous statement that $\varepsilon_{ij}(\omega,\boldsymbol{k})$ has no singularities in the upper half-plane [if $\varepsilon_{ij}(\omega,\boldsymbol{k})$ grew without limits, a singular point would exist in the upper half-plane as $\omega'' \to \infty$]. This confirms the validity of (2.1.26) [2.11].

Separating real and imaginary parts in (2.1.25) we arrive at the dispersion relations in the form

$$\begin{aligned}
\mathrm{Re}\{\varepsilon_{ij}(\omega,\boldsymbol{k})\} - \delta_{ij} &= \frac{1}{\pi} \fint_{-\infty}^{+\infty} \frac{\mathrm{Im}\{\varepsilon_{ij}(x,\boldsymbol{k})\}}{x-\omega}\,dx\,, \\
\mathrm{Im}\{\varepsilon_{ij}(\omega,\boldsymbol{k})\} &= -\frac{1}{\pi} \fint_{-\infty}^{+\infty} \frac{\mathrm{Re}\{\varepsilon_{ij}(x,\boldsymbol{k})\} - \delta_{ij}}{x-\omega}\,dx\,.
\end{aligned} \tag{2.1.27}$$

We have assumed here, and in what follows, that $\varepsilon_{ij}(\omega,\boldsymbol{k})$ has no pole at $\omega = 0$. These equations evidently apply separately to each of the components of the tensor, i.e., in the general case there are 18 relations ($\mathrm{Re}\{\varepsilon_{ij}\}$ and $\mathrm{Im}\{\varepsilon_{ij}\}$ have nine components each). Introducing the symmetric tensors $\varepsilon'_{ij,s}$, $\varepsilon''_{ij,s}$ and the antisymmetric tensors $\varepsilon'_{ij,a}$, $\varepsilon''_{ij,a}$, see (2.1.13, 14), we can rewrite (2.1.27) in the form

$$\left.\begin{aligned}
\varepsilon'_{ij,s}(\omega,\boldsymbol{k}) - \delta_{ij} &= \frac{1}{\pi} \fint_{-\infty}^{+\infty} \frac{\varepsilon''_{ij,s}(x,\boldsymbol{k})}{x-\omega}\,dx\,, \\
\varepsilon''_{ij,s}(\omega,\boldsymbol{k}) &= -\frac{1}{\pi} \fint_{-\infty}^{+\infty} \frac{\varepsilon'_{ij,s}(x,\boldsymbol{k}) - \delta_{ij}}{x-\omega}\,dx\,, \\
\varepsilon'_{ij,a}(\omega,\boldsymbol{k}) &= \frac{1}{\pi} \fint_{-\infty}^{+\infty} \frac{\varepsilon''_{ij,a}(x,\boldsymbol{k})}{x-\omega}\,dx\,, \\
\varepsilon''_{ij,a}(\omega,\boldsymbol{k}) &= -\frac{1}{\pi} \fint_{-\infty}^{+\infty} \frac{\varepsilon'_{ij,a}(x,\boldsymbol{k})}{x-\omega}\,dx\,.
\end{aligned}\right\} \tag{2.1.28}$$

If the medium is nongyrotropic, i.e., if (2.1.23) is satisfied, then according to (2.1.18), for real ω and $\boldsymbol{k}$,

$$\varepsilon_{ij}(\omega,\boldsymbol{k}) = \varepsilon^*_{ij}(-\omega,\boldsymbol{k})\,,$$

$$\mathrm{Re}\{\varepsilon_{ij}(\omega,\boldsymbol{k})\} = \mathrm{Re}\{\varepsilon_{ij}(-\omega,\boldsymbol{k})\}\,,$$

$$\mathrm{Im}\{\varepsilon_{ij}(\omega,\boldsymbol{k})\} = -\mathrm{Im}\{\varepsilon_{ij}(-\omega,\boldsymbol{k})\}.$$

Under these conditions it is easy to reduce (2.1.27) on account of (2.1.22a) to

$$\varepsilon'_{ij}(\omega,\boldsymbol{k}) - \delta_{ij} = \frac{2}{\pi}\fint_0^\infty \frac{x\varepsilon''_{ij}(x,\boldsymbol{k})}{x^2-\omega^2}dx\,,$$
$$\varepsilon''_{ij}(\omega,\boldsymbol{k}) = -\frac{2\omega}{\pi}\fint_0^\infty \frac{\varepsilon'_{ij}(x,\boldsymbol{k})-\delta_{ij}}{x^2-\omega^2}dx\,. \tag{2.1.27a}$$

The wave vector $\boldsymbol{k}$ appears in (2.1.27, 28) as a parameter and remains arbitrary [it is assumed that the tensor $\varepsilon_{ij}(\omega,\boldsymbol{k})$ exists for the $\boldsymbol{k}$ values under consideration, and has no singularities in the upper half-plane and on the real ω axis]. The most important property used above is due to causality. If, in addition, the fact that interactions are propagated at finite velocities is taken into account, we obtain some additional restrictions on (2.1.3–5) for $\boldsymbol{D}(\boldsymbol{r},t)$. As a result we can arrive at dispersion relations [2.9, 11] which can be written in the form of (2.1.27, 28) for special cases only.

All the symmetry properties of the tensor $\varepsilon_{ij}(\omega,\boldsymbol{k})$ are common to the inverse tensor $\varepsilon_{ij}^{-1}(\omega,\boldsymbol{k})$, as follows directly from (2.1.9). In general, the tensors ε_{ij} and ε_{ij}^{-1} can be equally well applied over a wide range, but, in general, the use of ε_{ij} is somewhat more convenient and more widespread. At the same time, as we have already indicated above, when $\boldsymbol{k} \neq 0$, $\varepsilon_{ij}(\omega,\boldsymbol{k})$ can have singularities, in principle, when $\omega'' > 0$, while the tensor $\varepsilon_{ij}^{-1}(\omega,\boldsymbol{k})$ is an analytic function in the upper half ω plane (i.e., when $\omega'' > 0$) for all values of $\boldsymbol{k}$ [2.9]. Therefore, dispersion relations of the type (2.1.27, 28) for the tensor $\varepsilon_{ij}^{-1}(\omega,\boldsymbol{k})$ are also valid for all values of $\boldsymbol{k}$. In the case of $\varepsilon_{ij}(\omega,\boldsymbol{k})$, these dispersion relations are clearly valid only at $\boldsymbol{k} = 0$ and, owing to the analyticity with respect to $\boldsymbol{k}$, in a certain neighborhood of the point $\boldsymbol{k} = 0$. It is, however, precisely the small values of k which are dealt with in optics (or, to be more exact, small values of the dimensionless parameter ka, a being the lattice constant or some other characteristic dimension). Therefore, in optics the possible case when $\boldsymbol{k} \neq 0$ for singularities of the tensor $\varepsilon_{ij}(\omega,\boldsymbol{k})$ in the region $\omega'' > 0$ is hardly significant (such a singularity most likely appears only when the condition $ka \ll 1$ does not hold). For this reason, both with respect to the dispersion relations and within the scope of the following analysis of certain properties of the tensor $\varepsilon_{ij}(\omega,\boldsymbol{k},\boldsymbol{B}_{\mathrm{ext}})$, we shall assume that this tensor is equivalent to ε_{ij}^{-1} and, specifically, that it has no singularities when $\omega'' > 0$ for all the $\boldsymbol{k}$ values under consideration.

The conclusion as to the finiteness of $\varepsilon_{ij}(\omega,\boldsymbol{k})$ for $\omega'' \geqslant 0$ (with the exclusion, perhaps, of the point $\omega = 0$ where ε_{ij} has a pole in the case of a statically conducting medium) is, apart from causality, based on the assumption that the kernel $\hat{\varepsilon}_{ij}(\tau,\boldsymbol{R})$ in (2.1.6) does not increase with τ. If the medium is in thermodynamic equilibrium, this statement is obvious from physical arguments[4]. For a large class of nonequilibrium states the same result can be expected, and only for unstable states (media) or in the case of external sources can the kernel $\hat{\varepsilon}_{ij}(\tau, \boldsymbol{R})$ increase as $\tau \to \infty$. Moreover, at least for a medium in equilibrium, we can also approach the problem from another standpoint, namely, from the principle of growth of entropy. According to this principle, if a medium in equilibrium is subjected to any action whatsoever, heat must be given off; it cannot be absorbed or not be present at all. For an isotropic medium without spatial dispersion this requirement results in the well-known condition $(\omega/4\pi)\varepsilon''(\omega) = \sigma(\omega) > 0$, i.e., to $\varepsilon''(\omega) > 0$ if $\omega \neq 0$. It will be shown in Sect. 2.3.1 that if ω and $\boldsymbol{k}$ are real the heat liberated per unit volume is equal to

$$q = \frac{i\omega}{16\pi}[\varepsilon_{ij}^*(\omega, \boldsymbol{k}, \boldsymbol{B}_{\text{ext}}) - \varepsilon_{ji}(\omega, \boldsymbol{k}, \boldsymbol{B}_{\text{ext}})]E_{0i}E_{0j}^*$$

$$= \frac{\omega}{8\pi}\varepsilon_{ij}''(\omega, \boldsymbol{k}, \boldsymbol{B}_{\text{ext}})E_{0j}E_{0i}^*, \tag{2.1.29}$$

where $\boldsymbol{E}_0$ is the amplitude of the electric field $\boldsymbol{E} = \boldsymbol{E}_0 \exp[i(\boldsymbol{kr} - \omega t)]$. In general (i.e., if external sources exist), this field can be assumed to be completely arbitrary. As a result of (2.1.29) and the principle of growth of entropy, the condition $q(\omega) > 0$ can be written in the form

$$\varepsilon_{ij}''(\omega, \boldsymbol{k}, \boldsymbol{B}_{\text{ext}})a_i a_j^* > 0, \qquad \omega = \omega' > 0, \tag{2.1.30}$$

where $\boldsymbol{a}$ is an arbitrary complex vector; in particular, we may set $\boldsymbol{a} = \boldsymbol{s}$, where $\boldsymbol{s}$ is a real unit vector.

Let us now consider the function

$$u(\omega, \boldsymbol{k}, \boldsymbol{B}_{\text{ext}}) = \varepsilon_{ij}(\omega, \boldsymbol{k}, \boldsymbol{B}_{\text{ext}})a_i a_j^*. \tag{2.1.31}$$

Since the tensors ε_{ij}' and ε_{ij}'' are Hermitian we find from (2.1.30)

$$\text{Im}\{u\} = \varepsilon_{ij}'' a_i a_j^* > 0, \qquad \omega = \omega' > 0. \tag{2.1.32}$$

It can readily be shown [2.11] that

$$\text{Im}\{u(-\omega, \boldsymbol{k}, \boldsymbol{B}_{\text{ext}})\} < 0, \qquad \omega = \omega' < 0, \tag{2.1.33}$$

---

[4] Recall that with $\boldsymbol{E} = \boldsymbol{E}_0(\boldsymbol{r}')\delta(t')$ the induction is

$$D_i(\boldsymbol{r}, t) = \int d\boldsymbol{r}'\, \hat{\varepsilon}_{ij}(t, \boldsymbol{r} - \boldsymbol{r}')E_{0j}(\boldsymbol{r}').$$

In the case of equilibrium the "response" $\boldsymbol{D}$ cannot increase with time [this means that the induced current $\boldsymbol{j}(\boldsymbol{r}, t) = (1/4\pi)(\partial/\partial t)(\boldsymbol{D} - \boldsymbol{E})$ should tend to zero as $t \to \infty$].

making use of the properties of $\varepsilon_{ij}(\omega,\boldsymbol{k},\boldsymbol{B}_{\text{ext}})$ already indicated in the present subsection. Moreover, with or without spatial dispersion, but if $\boldsymbol{B}_{\text{ext}}=0$ (or ε_{ij} is independent of the sign of $\boldsymbol{B}_{\text{ext}}$), then

$$\operatorname{Im}\{u(-\omega,\boldsymbol{k},\boldsymbol{b}_{\text{ext}})\} = -\operatorname{Im}\{u(\omega,\boldsymbol{k},\boldsymbol{B}_{\text{ext}})\}\,. \tag{2.1.34}$$

Finally, u always tends to unity, as ω tends to infinity, since, in this limiting case $\varepsilon_{ij}\to\delta_{ij}$, see (2.1.26); here and in the following we assume that $\boldsymbol{a}=\boldsymbol{s}$, $s=1$).

The properties of $u(\omega,\boldsymbol{k},\boldsymbol{B}_{\text{ext}})$ mentioned above coincide with the properties of the function $\varepsilon(\omega)$ in an isotropic medium in equilibrium. Likewise in this last case [Ref. 2.7, § 62], we may state that under the above conditions, (2.1.32–34), the function $u(\omega)$ does not assume real values at any point in the upper half-plane, except along the imaginary axis. On this axis, $u(\omega)$ decreases monotonically from some value $u>1$ (dielectric medium) or $u=\infty$ (metal) at $\omega=\mathrm{i}\omega''=\mathrm{i}0$ to $u=1$ at $\omega=\mathrm{i}\infty$. Hence, it follows, in particular, that the function u has no zeros in the upper half-plane. This is true also for the real axis (see the initial inequalities (2.1.32, 33); for dielectrics $\varepsilon''_{ij}(\omega)=0$ at the point $\omega=0$, but in this case $u=\varepsilon'_{ij}(\omega=0)a_i a_j^*>1$ as follows from thermodynamics [Ref. 2.7, § 14]).
We thus have for the upper half-plane and the real axis[5]

$$u(\omega,\boldsymbol{k},\boldsymbol{B}_{\text{ext}})=\varepsilon_{ij}(\omega,\boldsymbol{k},\boldsymbol{B}_{\text{ext}})a_i a_j^* \neq 0\,,\qquad \omega=\omega'+\mathrm{i}\omega''\,,\quad \omega''\geqslant 0\,. \tag{2.1.35}$$

This inequality involves another one, namely

$$|\varepsilon_{ij}(\omega,\boldsymbol{k},\boldsymbol{B}_{\text{ext}})|\neq 0\,,\qquad \omega=\omega'+\mathrm{i}\omega''\,,\qquad \omega''>0\,. \tag{2.1.36}$$

In fact, if $|\varepsilon_{ij}|=0$ the set of equations $\varepsilon_{ij}a_j=0$ always has a nonzero solution $a_j\neq 0$ for which the quadratic form (2.1.35) equals zero. Under the conditions (2.1.36) the inverse tensor $\varepsilon_{ij}^{-1}(\omega,\boldsymbol{k},\boldsymbol{B}_{\text{ext}})$ exists and has no singularity. Also

$$|\varepsilon_{ij}^{-1}(\omega,\boldsymbol{k},\boldsymbol{B}_{\text{ext}})|\neq 0\,,\qquad \omega=\omega'+\mathrm{i}\omega''\,,\qquad \omega''\geqslant 0\,, \tag{2.1.37}$$

where use was made again of the fact that $\varepsilon_{ij}(\omega,\boldsymbol{k},\boldsymbol{B}_{\text{ext}})$ has no singularity at the $\boldsymbol{k}$ values under consideration.

These conclusions are valid only if q in (2.1.29) can be interpreted as a heat quantity. In this case, at least for a medium in equilibrium, we know for certain that $q>0$ and we make use of this fact. In the case of a real frequency

---

[5] The symmetry condition (2.1.34) need not be fulfilled to obtain (2.1.35). If the inequalities (2.1.32, 33) apply but condition (2.1.34) is not met, then the formulated properties of $u(\omega)$ are changed only in the respect that the role of the imaginary axis is played by a certain unique curve in the upper half ω plane, emerging from the point $\omega=0$ [2.11].

ω and no spatial dispersion, or the presence of spatial dispersion but real $\boldsymbol{k}$, q is in fact a heat quantity. But for complex $\boldsymbol{k}$ and in the presence of spatial dispersion, this is no longer true (Sect. 2.3.1). The problem of the form of the heat equation and of the application of the principle of growth of entropy has not yet been elucidated for the general case. It is moreover likely that if $\boldsymbol{k}$ is complex, heat cannot be expressed in terms of $\varepsilon_{ij}(\omega, \boldsymbol{k})$ (Sect. 2.3.1). Owing to the aforesaid, the region of applicability of the inequalities (2.1.35 – 37) is unknown for complex $\boldsymbol{k}$. It is nevertheless possible to give some hints on this point. In the first place, if spatial dispersion is weak we may expect that the inequalities (2.1.35 – 37) also remain valid in a certain region of relatively small complex values of $\boldsymbol{k}$ considered in this case; this is simply deduced from the consideration that these inequalities apply as well to the case where $\mathrm{Im}\{\boldsymbol{k}\} = 0$ and spatial dispersion is ignored. Secondly, as will be explained in Sect. 2.2.3, the equations

$$\begin{aligned} &u = \varepsilon_{ij}(\omega, \boldsymbol{k}, \boldsymbol{B}_{\mathrm{ext}}) s_i s_j = 0\,, \\ &|\varepsilon_{ij}(\omega, \boldsymbol{k}, \boldsymbol{B}_{\mathrm{ext}})| = 0\,, \\ &|\varepsilon_{ij}^{-1}(\omega, \boldsymbol{k}, \boldsymbol{B}_{\mathrm{ext}})| = 0 \end{aligned} \tag{2.1.38}$$

are necessary conditions for the existence of solutions of the Coulomb problem (meaning Coulomb excitons corresponding to longitudinal waves, "fictitious" longitudinal waves and "polarization waves"; see Sect. 2.2.2). In a medium in equilibrium the Coulomb excitons are always damped although damping may be weak; this conclusion results quite naturally from physical considerations. It means that for Coloumb excitons the imaginary part of the frequency is negative ($\mathrm{Im}\{\omega\} = \omega'' < 0$), and in the region $\omega'' > 0$, (2.1.38) has no solution that corresponds to a Coulomb exciton. In this respect it seems natural to assume that (2.1.35 – 37) are valid for all complex values of the wave vector $\boldsymbol{k}$ which are of physical interest. If this is true, then the tensors $\varepsilon_{ij}(\omega, \boldsymbol{k})$ and $\varepsilon_{ij}^{-1}(\omega, \boldsymbol{k})$ can find equivalent application.

The preceding argument can be reinforced, evidently, if it is based from the very beginning on the properties of the tensor ε_{ij}^{-1} whose lack of singularities for all $\boldsymbol{k}$ and $\omega'' > 0$ for an equilibrium system is due to the principle of causality. Thus, for instance, (2.1.36) follows, from this point of view, immediately from (2.1.9) in the case of analyticity of the tensor ε_{ij}. To the best of our knowledge, a proper analysis of this problem has not been carried out. But, by virtue of the aforesaid, it seems to us to be no less reliable to base our reasoning on the analytic properties of the tensor $\varepsilon_{ij}(\omega, \boldsymbol{k}, \boldsymbol{B}_{\mathrm{ext}})$, at least in the long-wavelength region (optics and, in general, under the condition that $ka \ll 1$).

At the end of this subsection we wish to emphasize the possibility of transferring a number of the results for the electrodynamics of continuous media and optics to the mechanics of continuous media, to acoustics, and, generally, to the theory of linear systems. This conclusion is evident as the direct physical meaning of the quantities $\boldsymbol{D}$, $\boldsymbol{E}$, etc., mentioned above is in most cases inessential; use is made only of the linearity of the relation (2.1.6), of some symmetry properties and the like. If, however, we do not look at this problem from the viewpoint of formal analogy, we can readily find points of contact with electrodynamics in the mechanics of continuous media as well. This is also true, for instance, for the problem of spatial dispersion. Nevertheless we do not consider it expedient to dwell here on the effects of spatial dispersion in the propagation of sound in fluids and solids [2.13, 14], on dispersion relations in acoustics [2.15], etc.

2.1.3 The Approximation of Classical Crystal Optics. The Tensor $\varepsilon_{ij}(\omega,\boldsymbol{k})$ in an Isotropic Medium

Neglecting spatial dispersion, in a nonmagnetic medium (i.e., if $\mu_{ij} = \delta_{ij}$) the tensor $\varepsilon_{ij}(\omega,\boldsymbol{k})$ goes over to the tensor $\varepsilon_{ij}(\omega)$ used in classical crystal optics. Since in the optical range considered the medium is always nonmagnetic (disregarding ferromagnetic substances), we have in practice no limitations (see below). Moreover, we should stress the fact that the existence of the limiting transition

$$\varepsilon_{ij}(\omega,\boldsymbol{k}\to 0) = \varepsilon_{ij}(\omega) \tag{2.1.7}$$

is not at all trivial though quite natural. As a matter of fact there exist quantities which are still functions of $\boldsymbol{s} = \boldsymbol{k}/k$ even when $\boldsymbol{k}\to 0$ and thus are not analytic functions of k if $k\to 0$. In the case of $\varepsilon_{ij}(\omega,\boldsymbol{k})$, however, regularity when $\boldsymbol{k}\to 0$ results from physical considerations, since there are no reasons of either a theoretical or an experimental nature for the induction $\boldsymbol{D}(\omega,\boldsymbol{k})$ as a function of the field $\boldsymbol{E}(\omega,\boldsymbol{k})$ to depend on the magnitude or direction of $\boldsymbol{k}$ in the limiting case of sufficiently long wavelengths, i.e., in passing over to a homogeneous field.

In the approximation of classical crystal optics we have, according to (2.1.18),

$$\varepsilon_{ij}(-\omega,\boldsymbol{B}_{\text{ext}}) = \varepsilon_{ij}^{*}(\omega^{*},\boldsymbol{B}_{\text{ext}}) , \tag{2.1.39}$$

$$\varepsilon_{ij}(\omega,\boldsymbol{B}_{\text{ext}}) = \varepsilon_{ji}(\omega, -\boldsymbol{B}_{\text{ext}}) . \tag{2.1.40}$$

If, in addition, there is no external magnetic field ($\boldsymbol{B}_{\text{ext}} = 0$), then

$$\varepsilon_{ij}(\omega) = \varepsilon_{ji}(\omega) = \varepsilon_{ij}'(\omega) + \mathrm{i}\varepsilon_{ij}''(\omega) . \tag{2.1.41}$$

The Hermitian tensors ε_{ij}' and ε_{ij}'' are obviously real and symmetrical

$$\begin{aligned} \varepsilon_{ij}'(\omega) &= \mathrm{Re}\{\varepsilon_{ij}(\omega)\} , \\ \varepsilon_{ij}''(\omega) &= \mathrm{Im}\{\varepsilon_{ij}(\omega)\} . \end{aligned} \tag{2.1.42}$$

According to (2.1.39), on the real axis

$$\begin{aligned} &\varepsilon_{ij}(-\omega)=\varepsilon_{ij}^{*}(\omega)\,; \qquad \varepsilon_{ij}'(-\omega)=\varepsilon_{ij}'(\omega)\,, \\ &\varepsilon_{ij}''(-\omega)=-\varepsilon_{ij}''(\omega)\,, \qquad \omega''=\operatorname{Im}\{\omega\}=0\,. \end{aligned} \tag{2.1.43}$$

By virtue of (2.1.42, 43) the dispersion relations (2.1.27) of classical crystal optics (if $\boldsymbol{B}_{\text{ext}}=0$) assume the form

$$\begin{aligned} &\varepsilon_{ij}'(\omega)-\delta_{ij}=\frac{2}{\pi}\,⨍_{0}^{\infty}\frac{x\varepsilon_{ij}''(x)\,dx}{x^{2}-\omega^{2}}\,, \\ &\varepsilon_{ij}''(\omega)=-\frac{2\omega}{\pi}\,⨍_{0}^{\infty}\frac{\varepsilon_{ij}'(x)-\delta_{ij}}{x^{2}-\omega^{2}}\,dx\,. \end{aligned} \tag{2.1.44}$$

A few remarks on the notation are in order. In the electrodynamics of continuous media the field equations are most frequently written in the form

$$\begin{aligned} &\operatorname{rot}\boldsymbol{H}=\frac{1}{c}\frac{\partial\boldsymbol{D}}{\partial t}+\frac{4\pi}{c}\boldsymbol{j}_{\text{ext}}\,, \qquad \operatorname{div}\boldsymbol{D}=4\pi\rho_{\text{ext}}\,, \\ &\operatorname{rot}\boldsymbol{E}=-\frac{1}{c}\frac{\partial\boldsymbol{B}}{\partial t}\,, \qquad \operatorname{div}\boldsymbol{B}=0\,. \end{aligned} \tag{2.1.45}$$

For a nonmagnetic medium the magnetic field strength $\boldsymbol{H}$ and magnetic induction $\boldsymbol{B}$ coincide (this is how a nonmagnetic medium is defined). The systems of equations (2.1.1, 45) are then identical. Moreover, the vector $\boldsymbol{D}$ in (2.1.1) is so defined that the current induced by the fields $\boldsymbol{E}$ and $\boldsymbol{B}$ is related entirely to $\boldsymbol{D}$; in (2.1.45) only the current induced by the field $\boldsymbol{E}$ is related to $\boldsymbol{D}$. Another notation is also frequently used: for instance, in (2.1.45), $(1/c)\,\partial\boldsymbol{D}/\partial t$ is replaced by $(4\pi/c)\boldsymbol{j}_{\text{cond}}+(1/c)\,\partial\boldsymbol{D}/\partial t$, where $\boldsymbol{j}_{\text{cond}}$ is the conduction current density and $\boldsymbol{D}$ is the part of the induction which is in phase with the field $\boldsymbol{E}$ [in (2.1.41) it is then assumed that $D_i(\omega)=\varepsilon_{ij}'E_j(\omega)$]. If the medium is magnetic, the correspondence between the systems of equations (2.1.1, 45) is as follows: $\partial\boldsymbol{D}/\partial t$ from (2.1.1) corresponds to the vector $\partial\boldsymbol{D}/\partial t+4\pi c\operatorname{rot}\boldsymbol{M}$ from (2.1.45), and $4\pi\boldsymbol{M}=\boldsymbol{B}-\boldsymbol{H}$. The magnetic properties of the medium, arising from the nonidentical nature of $\boldsymbol{B}$ and $\boldsymbol{H}$, are, in the case of the system (2.1.1), completely included in $\boldsymbol{D}$, as was indicated in Sect. 2.1.1.

If (2.1.45) is applied in linear electrodynamics we set

$$\begin{aligned} &B_i(\omega)=\mu_{ij}(\omega)H_j(\omega)\,, \\ &\mu_{ij}(\omega)=\delta_{ij}+4\pi\chi_{ij}^{(\mathrm{M})}(\omega)\,, \end{aligned} \tag{2.1.46}$$

where μ_{ij} and $\chi_{ij}^{(\mathrm{M})}$ are the magnetic permeability and the magnetic susceptibility, respectively.

If spatial dispersion is neglected, the system (2.1.1) must be equivalent to (2.1.45) for a magnetic medium as well. From this it follows that for a magnetic medium, the relation (2.1.7) loses its validity and the tensors $\varepsilon_{ij}(\omega)$ and $\mu_{ij}(\omega)$ are both related to $\varepsilon_{ij}(\omega,\boldsymbol{k}\to 0)$. The corresponding limiting transition is traced in [Ref. 2.1, § 3]; it is not necessary, for our purposes, to discuss this problem in much detail, since the introduction of the function μ_{ij} makes sense only for a relatively narrow frequency band near the point $\omega/k=0$, where $1/k$ is the characteristic dimension of the field inhomogeneity in the medium. This is why in optics (ignoring ferromagnetic materials [2.8]), we should take $\mu_{ij}(\omega)=\delta_{ij}$ [Ref. 2.7, § 60].

We shall, however, present the relation between $\varepsilon_{ij}(\omega,\boldsymbol{k}\to 0)$ and $\varepsilon_{ij}(\omega)$, $\mu_{ij}(\omega)$ for an isotropic medium where we can put

$$\varepsilon_{ij}=\varepsilon(\omega)\,\delta_{ij} \quad \text{and} \quad \mu_{ij}=\mu(\omega)\,\delta_{ij}\,, \quad \text{while}$$

$$\varepsilon_{ij}(\omega,\boldsymbol{k})=\varepsilon_{\mathrm{tr}}(\omega,\boldsymbol{k})(\delta_{ij}-s_is_j)+\varepsilon_{\mathrm{lo}}(\omega,\boldsymbol{k})s_is_j$$

[see (2.1.48) below, where $s_i=k_i/k$ and $\varepsilon_{\mathrm{tr}}$ and $\varepsilon_{\mathrm{lo}}$ actually depend only on k^2]. In this case, as $k\to 0$,

$$1/\mu(\omega)=1-(\omega^2/c^2k^2)[\varepsilon_{\mathrm{tr}}(\omega,k)-\varepsilon_{\mathrm{lo}}(\omega,k)]\,,$$

$$\varepsilon(\omega)=\varepsilon_{\mathrm{lo}}(\omega,k\to 0)\,.$$

Furthermore, a limiting transition to $k\to 0$ is implied in the expression for μ, or, physically, a transition to low values of k. From the aforegoing it is already clear that at low frequencies and long wavelengths (formally, as $\omega\to 0$ and $k\to 0$), it is much more convenient to introduce ε and μ, rather than the quantities $\varepsilon_{\mathrm{tr}}$ and $\varepsilon_{\mathrm{lo}}$. The situation, however, is quite different in optics, as we have already stressed. It proves expedient, in general, to use only the tensor $\varepsilon_{ij}(\omega,\boldsymbol{k})$, without introducing the magnetic permeability tensor. The fact that refusing to introduce μ for a static field is inconvenient could be perceived from the very beginning: the use of $\varepsilon_{ij}(\omega,\boldsymbol{k})$ alone is based on the feasibility of taking into consideration the effect of the field $\boldsymbol{B}$ on $\boldsymbol{D}$, expressing $\boldsymbol{B}$ in terms of $\boldsymbol{E}$, by means of the equation

$$\boldsymbol{B}(\omega,\boldsymbol{k})=(c/\omega)\boldsymbol{k}\times\boldsymbol{E}(\omega,\boldsymbol{k})\,.$$

Evidently, as $\omega\to 0$, this equation can be applied only when simultaneously performing the limiting transition $k\to 0$ [Ref. 2.1, §§ 2 and 3].

We emphasize that apart from the field equations in the form (2.1.1 or 45), we can evidently, by redefining the fields $\boldsymbol{E}$, $\boldsymbol{D}$, $\boldsymbol{B}$ and $\boldsymbol{H}$, write any number of other equivalent systems of equations, with corresponding relations between the vectors $\boldsymbol{E}$, $\boldsymbol{D}$, $\boldsymbol{B}$ and $\boldsymbol{H}$. In particular, the relations used for a magnetic medium in [2.16] are

$$D_i = \varepsilon_{ij}(\omega, \boldsymbol{k})E_j + \alpha_{ij}(\omega, \boldsymbol{k})H_j,$$
$$B_i = \beta_{ij}(\omega, \boldsymbol{k})E_j + \mu_{ij}(\omega, \boldsymbol{k})H_j,$$

in which, of course, the tensor ε_{ij} does not coincide with the tensor ε_{ij} in (2.1.6).

It seems to us, however, simplest to apply the system (2.1.1) directly in optics, at least for nonmagnetic media on which we shall concentrate our attention, employing the tensor $\varepsilon_{ij}(\omega, \boldsymbol{k})$ as defined by (2.1.6) and taking into account the influence of the magnetic field as well. It proves convenient to apply the system (2.1.45) in the range of low frequencies and long wavelengths, in which case the properties of a linear medium can be fully described by introducing the tensors $\varepsilon_{ij}(\omega)$ and $\mu_{ij}(\omega)$.

Let us now consider the case of an isotropic medium (gas, liquid, and, under certain conditions, polycrystals as well). Such a medium, by definition, does not possess any preferential directions. If spatial dispersion is neglected the tensor ε_{ij} cannot depend on $\boldsymbol{k}$ nor on $\boldsymbol{s} = \boldsymbol{k}/k$, see (2.1.7). We are therefore obliged to use only one unit tensor of rank two, δ_{ij}, and, consequently,

$$\varepsilon_{ij}(\omega) = \varepsilon(\omega)\delta_{ij}. \tag{2.1.47}$$

If, however, spatial dispersion is taken into account, we have for an isotropic but nongyrotropic medium

$$\varepsilon_{ij}(\omega, \boldsymbol{k}) = \varepsilon_{\mathrm{tr}}(\omega, k^2)(\delta_{ij} - s_i s_j) + \varepsilon_{\mathrm{lo}}(\omega, k^2)s_i s_j, \tag{2.1.48}$$

where $k = k' + \mathrm{i}k''$, $\boldsymbol{k} = k\boldsymbol{s}$ (homogeneous waves), and $\varepsilon_{\mathrm{tr}}$ and $\varepsilon_{\mathrm{lo}}$ depend, apart from ω, only on $k^2 = (k' + \mathrm{i}k'')^2$ as there is no gyrotropy, see (2.1.23); in (2.1.48) we have also made use of the fact that applying the vector $\boldsymbol{s}$ makes it possible to form another tensor of rank two, namely, $s_i s_j$. The notation of (2.1.48) becomes clear when we consider the longitudinal and transverse electric fields $\boldsymbol{E}_\parallel$ and $\boldsymbol{E}_\perp$:

$$\boldsymbol{E}_\parallel = E_\parallel \boldsymbol{s}, \qquad \boldsymbol{s} \cdot \boldsymbol{E}_\perp = 0. \tag{2.1.49}$$

We thus have from (2.1.48)

$$\begin{aligned} \varepsilon_{ij}E_{\parallel,j} &= \varepsilon_{\mathrm{lo}}E_{\parallel,i}, \\ \varepsilon_{ij}E_{\perp,j} &= \varepsilon_{\mathrm{tr}}E_{\perp,i}. \end{aligned} \tag{2.1.50}$$

An isotropic nongyrotropic medium is thus characterized by the two functions $\varepsilon_{\mathrm{tr}}(\omega, k^2)$ and $\varepsilon_{\mathrm{lo}}(\omega, k^2)$, the transverse and longitudinal dielectric constants.

As a result of (2.1.7), we have

$$\varepsilon_{\mathrm{tr}}(\omega, 0) = \varepsilon_{\mathrm{lo}}(\omega, 0) = \varepsilon(\omega). \tag{2.1.51}$$

In the other case, the tensor $\varepsilon_{ij}(\omega, 0)$ would depend on $\boldsymbol{s}$. For weak spatial dispersion we have as a first approximation

$$\begin{aligned} \varepsilon_{\mathrm{tr}}(\omega, k^2) &= \varepsilon(\omega) + \alpha_{\mathrm{tr}}(\omega) k^2 , \\ \varepsilon_{\mathrm{lo}}(\omega) &= \varepsilon(\omega) + \alpha_{\mathrm{lo}}(\omega) k^2 , \end{aligned} \tag{2.1.52}$$

where α_{tr} and α_{lo} are of the order of a^2, a being the characteristic dimension. In a condensed nonmetallic medium a is, for example, the interatomic distance. In an isotropic plasma a is the Debye radius [Ref. 2.2, § 8]; in a gas and in metals a can be regarded as the mean free path of the molecules or electrons, etc.

The tensor $\varepsilon_{ij}^{-1}(\omega, \boldsymbol{k})$ of an isotropic medium can also be written in a form analogous to (2.1.48, 52). Note that (2.1.52) and the analogous relations for $\varepsilon_{ij}^{-1}(\omega, \boldsymbol{k})$ correspond to the material equations

$$\begin{aligned} \boldsymbol{D}(\omega) &= \varepsilon(\omega)\boldsymbol{E}(\omega) + a(\omega) \operatorname{rot} \operatorname{rot} \boldsymbol{E}(\omega) + b(\omega) \operatorname{grad} \operatorname{div} \boldsymbol{E}(\omega) , \\ \boldsymbol{E}(\omega) &= \boldsymbol{D}(\omega)/\varepsilon(\omega) + c(\omega) \operatorname{rot} \operatorname{rot} \boldsymbol{D}(\omega) + d(\omega) \operatorname{grad} \operatorname{div} \boldsymbol{D}(\omega) , \end{aligned} \tag{2.1.53}$$

where $\boldsymbol{D}(\omega)$ and $\boldsymbol{E}(\omega)$ depend, of course, on $\boldsymbol{r}$ as well.

The set (2.1.53) represents the most general relations between two vectors in an isotropic and nongyrotropic medium where derivatives of at most the second order are taken into account. The medium is assumed to be homogeneous because, generally speaking, the tensor $\varepsilon_{ij}(\omega, \boldsymbol{k})$ can be introduced only under this condition. In (2.1.53) the assumption that the medium is homogeneous is manifested by the fact that the factors a, b, c and d are considered to be independent of the coordinates $\boldsymbol{r}$. If the dependence of $\varepsilon(\omega)$ on $\boldsymbol{r}$ would not change the form of (2.1.53), then the dependence of the other factors on $\boldsymbol{r}$ would require the introduction of other terms (such terms appear if rot rot $(a\boldsymbol{E})$ is written in place of a rot rot $\boldsymbol{E}$, etc.).

An isotropic medium can be gyrotropic since its elements (e.g., certain molecules in a gas or in a solution) do not have to have a center of symmetry. In this case isotropy means that all directions are equivalent in the sense that normal waves of a given type have the same values of $\tilde{n}(\omega)$ and the same polarization for all values of $\boldsymbol{s}$.

The presence of gyrotropy in an isotropic medium corresponds formally to the appearance of some pseudoscalar g, i.e., of a quantity that changes its sign in the case of mirror reflection. Such a pseudoscalar is equivalent to a completely antisymmetric tensor of rank three, $\gamma_{ijl} = g e_{ijl}$, where e_{ijl} is a known completely antisymmetric pseudotensor of rank three ($e_{123} = 1$, $e_{132} = -1$, $e_{112} = 0$, etc.; e_{ijl} remains unchanged in the case of mirror reflection). If we make use of the tensor γ_{ijl} we can add to (2.1.48) the term

$$\mathrm{i}\gamma_{ijl}(\omega, k^2) k_l , \qquad \gamma_{ijl} = g(\omega, k^2) e_{ijl} . \tag{2.1.54}$$

This expression (by virtue of the fact that $\gamma_{ijl} = -\gamma_{jil}$) satisfies the condition (2.1.21); the factor i is introduced for convenience (Sect. 3.2.1).

For an isotropic homogeneous, but gyrotropic, medium keeping only the first derivatives with respect to the coordinates

$$\begin{aligned} \boldsymbol{D}(\boldsymbol{r},\omega) &= \varepsilon(\omega)\boldsymbol{E}(\boldsymbol{r},\omega) + \gamma(\omega)\,\mathrm{rot}\,\boldsymbol{E}(\boldsymbol{r},\omega)\,, \\ \boldsymbol{E}(\boldsymbol{r},\omega) &= \frac{\boldsymbol{D}(\boldsymbol{r},\omega)}{\varepsilon(\omega)} + \delta(\omega)\,\mathrm{rot}\,\boldsymbol{D}(\boldsymbol{r},\omega)\,, \end{aligned} \tag{2.1.55}$$

where γ and δ are pseudoscalars; in a nongyrotropic medium such pseudoscalars do not exist. As a result, terms containing first derivatives are absent in (2.1.53). The case of an isotropic gyrotropic medium will be treated in more detail in Sects. 3.2.1, 4.1 and 4.5.3. Properties of anisotropic gyrotropic media are mentioned, too.

2.2 Normal Electromagnetic Waves in a Medium

2.2.1 Wave Equation and Dispersion Equation

The name normal electromagnetic waves will be given to all solutions of the homogeneous field equations (2.1.1) which propagate as $\exp[\mathrm{i}(\boldsymbol{k}\cdot\boldsymbol{r} - \omega t)]$. In other words, they are solutions of (2.1.1) with $\boldsymbol{j}_{\mathrm{ext}} = 0$ and $\rho_{\mathrm{ext}} = 0$, so that

$$\begin{aligned} \mathrm{rot}\,\boldsymbol{B} &= \frac{1}{c}\frac{\partial \boldsymbol{D}}{\partial t}\,, \qquad \mathrm{div}\,\boldsymbol{D} = 0\,, \\ \mathrm{rot}\,\boldsymbol{E} &= -\frac{1}{c}\frac{\partial \boldsymbol{B}}{\partial t}\,, \qquad \mathrm{div}\,\boldsymbol{B} = 0\,. \end{aligned} \tag{2.2.1}$$

We shall seek a solution to this system in the form

$$\begin{aligned} \boldsymbol{E} &= \boldsymbol{E}_0 \mathrm{e}^{\mathrm{i}(\boldsymbol{k}\cdot\boldsymbol{r} - \omega t)}\,, \qquad \boldsymbol{D} = \boldsymbol{D}_0 \mathrm{e}^{\mathrm{i}(\boldsymbol{k}\cdot\boldsymbol{r} - \omega t)}\,, \qquad \boldsymbol{B} = \boldsymbol{B}_0 \mathrm{e}^{\mathrm{i}(\boldsymbol{k}\cdot\boldsymbol{r} - \omega t)}\,, \\ \boldsymbol{E}_0 &= \mathrm{const}\,, \qquad \boldsymbol{D}_0 = \mathrm{const}\,, \qquad \boldsymbol{B}_0 = \mathrm{const}\,, \end{aligned} \tag{2.2.2}$$

which is equivalent to working with Fourier transforms. Sometimes these solutions do not constitute a complete system of normal waves; namely, solutions of the type (2.2.2) are insufficient if the dispersion equation (2.2.22) (see below) has multiple essential roots; this problem will be discussed in Sect. 2.2.3. Now, however, we substitute (2.2.2) into (2.2.1) and obtain

$$\boldsymbol{D} = -\frac{c}{\omega}(\boldsymbol{k} \times \boldsymbol{B})\,, \tag{2.2.3}$$

$$\boldsymbol{k} \cdot \boldsymbol{D} = 0 \,, \tag{2.2.4}$$

$$\boldsymbol{B} = \frac{c}{\omega}(\boldsymbol{k} \times \boldsymbol{E}) \,, \tag{2.2.5}$$

$$\boldsymbol{k} \cdot \boldsymbol{B} = 0 \,. \tag{2.2.6}$$

Here and in the following we could just as well write $\boldsymbol{E}_0$, $\boldsymbol{D}_0$ and $\boldsymbol{B}_0$ or $\boldsymbol{E}(\omega,\boldsymbol{k}), \boldsymbol{D}(\omega,\boldsymbol{k})$ and $\boldsymbol{B}(\omega,\boldsymbol{k})$ instead of $\boldsymbol{E}$, $\boldsymbol{D}$ and $\boldsymbol{B}$. Obviously, when $\omega \neq 0$ (2.2.4) and (2.2.6) follow from (2.2.3) and (2.2.5), respectively.

Eliminating the field $\boldsymbol{B}$ from (2.2.3 and 5) we obtain the *wave equation*

$$\boldsymbol{D} = -\frac{c^2}{\omega^2}[\boldsymbol{k} \times (\boldsymbol{k} \times \boldsymbol{E})] = \frac{c^2}{\omega^2}[k^2\boldsymbol{E} - \boldsymbol{k} \cdot (\boldsymbol{k} \cdot \boldsymbol{E})] \,. \tag{2.2.7}$$

The more general equation obtained directly from (2.2.1) is also called the wave equation. Thus

$$\operatorname{rot}\operatorname{rot}\boldsymbol{E} + \frac{1}{c^2}\frac{\partial^2 \boldsymbol{D}}{\partial t^2} = 0 \,. \tag{2.2.8}$$

When we substitute (2.2.2) into (2.2.8), the latter, of course, assumes the form of (2.2.7). Even though it is obvious, to stress the fact that waves of the type (2.2.2) with single values of ω and $\boldsymbol{k}$ can be solutions of (2.2.1) only in the case when the vector $\boldsymbol{D}(\omega,\boldsymbol{k})$ is related to $\boldsymbol{E}(\omega,\boldsymbol{k})$ for only the same values of ω and $\boldsymbol{k}$. This is precisely what occurs if (2.1.6) is valid, i.e., the tensor $\varepsilon_{ij}(\omega,\boldsymbol{k})$ can be introduced as a well-defined quantity.

Substituting (2.1.6) into (2.2.7) we have

$$\left[\frac{\omega^2}{c^2}\varepsilon_{ij}(\omega,\boldsymbol{k}) - k^2\delta_{ij} + k_i k_j\right] E_j(\omega,\boldsymbol{k}) = 0 \,. \tag{2.2.9}$$

This homogeneous system of algebraic equations has a nontrivial solution, $\boldsymbol{E}(\omega,\boldsymbol{k}) \neq 0$, only if the determinant vanishes:

$$\left|\frac{\omega^2}{c^2}\varepsilon_{ij}(\omega,\boldsymbol{k}) - k^2\delta_{ij} + k_i k_j\right| = 0 \,. \tag{2.2.10}$$

If we use the tensor ε_{ij}^{-1} instead of ε_{ij} we obtain

$$\left|\frac{\omega^2}{c^2}\delta_{ij} - k^2\varepsilon_{ij}^{-1}(\omega,\boldsymbol{k}) + k_i k_l \varepsilon_{lj}^{-1}(\omega,\boldsymbol{k})\right| = 0 \,. \tag{2.2.11}$$

In all cases for which $|\varepsilon_{ij}| \neq 0$, (2.2.10 and 11) are equivalent. The condition $|\varepsilon_{ij}| = 0$, however, is associated with a very interesting class of solutions, namely, longitudinal waves for which $\boldsymbol{D} = 0$ even when $\boldsymbol{E} \neq 0$ (Sect. 2.2.2). In order to analyze general problems it is thus advantageous to use (2.2.9, 10). On the other hand, when investigating nonlongitudinal waves it is convenient to apply (2.2.11) and the system (2.2.7) where the vector $\boldsymbol{E}$ is expressed in terms of $\boldsymbol{D}$. By virtue of (2.2.4) which follows directly from (2.2.7), in a system of coordinates oriented with respect to $\boldsymbol{s} = \boldsymbol{k}/k$ (we consider homogeneous waves) the vector $\boldsymbol{D}$ has, in fact, only two components. This results in a considerable simplification of (2.2.7, 11), see Sect. 2.2.3, and (2.2.45, 46).

The relation between ω and $\boldsymbol{k}$ is given by the dispersion equation (2.2.10) whose solution has the form

$$\omega_l = \omega_l(\boldsymbol{k}), \quad l = 1, 2, \ldots, \tag{2.2.12}$$

where the subscript l corresponds to the various normal waves.

If solutions of equations like (2.2.2) are substituted into the initial system of equations (2.1.1) rather than into (2.2.1) we obtain, instead of (2.2.3, 9)

$$\begin{aligned} \boldsymbol{D} &= -\frac{c}{\omega}(\boldsymbol{k} \times \boldsymbol{B}) - \mathrm{i}\frac{4\pi}{\omega}\boldsymbol{j}_{\text{ext}}, & \boldsymbol{k}\cdot\boldsymbol{D} &= 4\pi\rho_{\text{ext}}, \\ \boldsymbol{B} &= \frac{c}{\omega}(\boldsymbol{k} \times \boldsymbol{E}), & \boldsymbol{k}\cdot\boldsymbol{B} &= 0; \end{aligned} \tag{2.2.13}$$

$$\left[\frac{\omega^2}{c^2}\varepsilon_{ij}(\omega, \boldsymbol{k}) - k^2\delta_{ij} + k_i k_j\right] E_j = -\mathrm{i}\frac{4\pi\omega}{c^2} j_{\text{ext}, i}, \tag{2.2.14}$$

where $\boldsymbol{E}$, $\boldsymbol{j}_{\text{ext}}$, ρ_{ext}, etc., are the Fourier transforms $\boldsymbol{E}(\omega, \boldsymbol{k})$, $\boldsymbol{j}_{\text{ext}}(\omega, \boldsymbol{k})$, $\rho_{\text{ext}}(\omega, \boldsymbol{k})$, etc.

If there are external sources $\boldsymbol{j}_{\text{ext}}$ the system (2.2.14) is inhomogeneous and has solutions of the type of (2.2.2) with ω and $\boldsymbol{k}$ that do not satisfy the dispersion equation (2.2.10). As mentioned in Sect. 2.1.1, it is this fact that makes it possible to regard ω and $\boldsymbol{k}$ in $\varepsilon_{ij}(\omega, \boldsymbol{k})$ as independent variables. If the density $\boldsymbol{j}_{\text{ext}}(\omega, \boldsymbol{k})$ is nonzero for ω and $\boldsymbol{k}$ that satisfy the dispersion equation (2.2.10), then the system has no meaning. This is clear since this case corresponds to a sharp resonance between external influence and normal oscillations (waves) in the medium.

A problem that may arise in this connection is how to set up the field $\boldsymbol{E}(\omega, \boldsymbol{k})$ with ω and $\boldsymbol{k}$ linked by the dispersion equation. Under such conditions we may, for instance, put $\boldsymbol{j}_{\text{ext}}(\omega, \boldsymbol{k}) = 0$ and $\rho_{\text{ext}}(\omega, \boldsymbol{k}) = 0$ and assume that the field is set up by sources outside the medium (strictly speaking, this corresponds to the transition to a spatially limited problem). Another method makes use of the fact that $\varepsilon_{ij}(\omega, \boldsymbol{k})$ is an analytic function of its arguments. It is then sufficient to know ε_{ij} for the values of ω and $\boldsymbol{k}$ that do not satisfy the dispersion equation in order to determine ε_{ij} uniquely, and particularly for values of $\omega_l = \omega_l(\boldsymbol{k})$ which satisfy this equation. The behavior of a medium in the presence of resonances may be elucidated by investigating a problem for which initial or boundary conditions are given [Ref. 2.1, § 5].

In crystal optics we deal, as a rule, only with solutions of the homogeneous equations (2.2.1) so that we concentrate our attention on normal waves. If the total system of normal waves has been found, it can be used to expand any other solutions of (2.2.1), particularly the wave packets and wave pulses (trains) that are of physical interest.

For inhomogeneous plane waves of the type (2.2.2)

$$\begin{aligned} \boldsymbol{k} &= \boldsymbol{k}' + \mathrm{i}\boldsymbol{k}'' , \\ \boldsymbol{k}' &= k'\boldsymbol{s}' , \qquad s' = 1 , \\ \boldsymbol{k}'' &= k''\boldsymbol{s}'' , \qquad s'' = 1 , \end{aligned} \tag{2.2.15}$$

where the vectors $\boldsymbol{s}'$ and $\boldsymbol{s}''$ and the quantities $\boldsymbol{k}'$ and $\boldsymbol{k}''$ are real, and the vectors $\boldsymbol{s}'$ and $\boldsymbol{s}''$ have different directions.

For homogeneous plane waves, however,

$$\boldsymbol{s}' = \boldsymbol{s}'' = \boldsymbol{s} , \qquad \boldsymbol{k} = (k' + \mathrm{i}k'')\boldsymbol{s} , \qquad s = 1 . \tag{2.2.16}$$

For inhomogeneous waves the field has the form

$$\boldsymbol{E} = \boldsymbol{E}_0 \mathrm{e}^{-\boldsymbol{k}''\cdot\boldsymbol{r}} \mathrm{e}^{\mathrm{i}(\boldsymbol{k}'\cdot\boldsymbol{r} - \omega t)} . \tag{2.2.17}$$

It is obvious that in this case the plane of equal amplitudes $\boldsymbol{k}''\cdot\boldsymbol{r} = \text{const}$, and the plane of equal phases, $\boldsymbol{k}'\cdot\boldsymbol{r} = \text{const}$, do not coincide. Inhomogeneous waves appear in a natural way when solving the boundary value problem. For example, in the case of inclined incidence of a plane wave on a plate (or on a half-space) of absorbing material, the wave within the plate (the refracted wave) will be inhomogeneous: the direction of $\boldsymbol{k}'$ is determined by the law of refraction, and the vector $\boldsymbol{k}''$ is directed along the normal to the interface.

In an isotropic medium without spatial dispersion $\varepsilon_{ij} = \varepsilon(\omega)\,\delta_{ij}$ and the dispersion equation (2.2.10) for longitudinal waves assumes the form

$$\begin{aligned} &\varepsilon(\omega) = 0 , \qquad \text{and} \\ &k_i k_j E_j = k^2 E_i . \end{aligned} \tag{2.2.18}$$

For transverse waves we have

$$\begin{aligned} &\varepsilon(\omega) = \varepsilon'(\omega) + \mathrm{i}\varepsilon''(\omega) = \frac{c^2k^2}{\omega^2} \equiv \frac{c^2}{\omega^2}(\boldsymbol{k}' + \mathrm{i}\boldsymbol{k}'')^2 , \\ &\boldsymbol{k}\cdot\boldsymbol{E} \equiv (\boldsymbol{k}' + \mathrm{i}\boldsymbol{k}'')\cdot\boldsymbol{E} = 0 , \qquad \boldsymbol{E} = \boldsymbol{E}' + \mathrm{i}\boldsymbol{E}'' \end{aligned} \tag{2.2.19}$$

or

$$\frac{\omega^2}{c^2}\varepsilon'(\omega) = k'^2 - k''^2, \quad \frac{\omega^2}{c^2}\varepsilon''(\omega) = 2\boldsymbol{k}'\boldsymbol{k}'',$$
$$\boldsymbol{k}'\cdot\boldsymbol{E}' - \boldsymbol{k}''\cdot\boldsymbol{E}'' = 0, \quad \boldsymbol{k}'\cdot\boldsymbol{E}'' + \boldsymbol{k}''\cdot\boldsymbol{E}' = 0. \tag{2.2.20}$$

If the directions of $\boldsymbol{s}'$ and $\boldsymbol{s}''$ – see (2.2.16) – are fixed, k' and k'' can be expressed in terms of $\varepsilon'(\omega)$ and $\varepsilon''(\omega)$; in this case the vector $\boldsymbol{E}' + \mathrm{i}\boldsymbol{E}''$ is subject to two conditions so that we may speak of transverse waves only in a very conditional sense. Only when absorption is neglected so that we can put $\varepsilon'' = 0$, are the vectors $\boldsymbol{s}'$ and $\boldsymbol{s}''$ orthogonal and solutions $\boldsymbol{E}' \neq 0$, $\boldsymbol{E}'' = 0$ and $\boldsymbol{E}' = 0$, $\boldsymbol{E}'' \neq 0$ exist for which the vector $\boldsymbol{E}$ is perpendicular to both vectors $\boldsymbol{s}'$ and $\boldsymbol{s}''$.

It is evident, even from the simplest example of an isotropic medium without spatial dispersion, that a consideration of inhomogeneous waves involves appreciable difficulties. Nevertheless, the problem can still be regarded as relatively elementary. In investigating an anisotropic medium, however, the dispersion equation (2.2.10) for inhomogeneous waves generally becomes unwieldy even when $\varepsilon_{ij}(\omega,\boldsymbol{k}) = \varepsilon_{ij}(\omega)$. If spatial dispersion is taken into account, the situation becomes substantially more complicated [6].

The use of inhomogeneous waves, although necessary when solving certain boundary value problems (Chap. 5) is fortunately of no interest when investigating the physical properties of the medium itself at points distant from its boundaries. In Chaps. 1–4, 6 of the present book, we therefore use only homogeneous waves of the type (2.2.16). For these waves, provided they are normal,

$$\boldsymbol{k} = \frac{\omega}{c}\tilde{n}(\omega,\boldsymbol{s})\boldsymbol{s},$$
$$\tilde{n}(\omega,\boldsymbol{s}) = n(\omega,\boldsymbol{s}) + \mathrm{i}\varkappa(\omega,\boldsymbol{s}) \tag{2.2.21}$$

and the dispersion equation (2.2.10) is conveniently written in the form

---

[6] Equation (2.1.48) is written for homogeneous waves. If it is given for real $\boldsymbol{k}$ values in the form

$$\varepsilon_{ij}(\omega,\boldsymbol{k}) = \varepsilon_{\mathrm{tr}}(\omega,k^2)\left(\delta_{ij} - \frac{k_i k_j}{k^2}\right) + \varepsilon_{\mathrm{lo}}(\omega,k^2)\frac{k_i k_j}{k^2}, \quad k^2 \equiv k_r k_r, \quad (r = 1,2,3)$$

then with an analytic continuation into the domain of complex values of k_i it maintains its form for inhomogeneous waves if the components k_i of the vector $\boldsymbol{k}$ are arbitrary complex numbers within wide limits (for complex $\boldsymbol{k}$ the quantity $k^2 = k_r k_r$ can vanish for $\boldsymbol{k} \neq 0$). If we do not proceed, however, from the principle of analyticity, other expressions of $\varepsilon_{ij}(\omega,\boldsymbol{k})$ could also be formulated which reduce to (2.1.48) for real $\boldsymbol{k}$ values or for homogeneous waves.

$$\left|\tilde{n}^2(\omega,s)(\delta_{ij}-s_is_j)-\varepsilon_{ij}\left(\omega,\frac{\omega}{c}\tilde{n}(\omega,s)s\right)\right|=\varepsilon_{ij}s_is_j\tilde{n}^4$$

$$-[(\varepsilon_{ij}s_is_j)\varepsilon_{ll}-\varepsilon_{il}\varepsilon_{lj}s_is_j]\tilde{n}^2+|\varepsilon_{ij}|=0\,. \tag{2.2.22}$$

This is the fundamental equation of crystal optics. In classical crystal optics $\varepsilon_{ij}=\varepsilon_{ij}(\omega)$ and (2.2.22) becomes quadratic with respect to $\tilde{n}^2$; this form of (2.2.22) is frequently called *Fresnel's equation.*

In the solutions $\tilde{n}_l(\omega,s)$ of the dispersion equation, $\boldsymbol{k}$ and ω are interrelated but this relationship is not, of course, unique. A more exact statement is that for normal waves, with given ω, s, and l, the vector $\boldsymbol{k}=(\omega/c)\tilde{n}_l(\omega,s)s$ is fixed. Conversely, if the vector $\boldsymbol{k}=ks=(\omega/c)\tilde{n}_l(\omega,s)s$ is given, this means that the quantity $\omega\tilde{n}_l(ks)$ is also given with a fixed s. Hence we can determine the frequencies $\omega_l(ks)$ of the normal waves, i.e., the solution of the dispersion equation is expressed in the form of (2.2.12).

In the case of homogeneous normal waves satisfying (2.2.21), Eqs. (2.2.3 – 7) have the form

$$\boldsymbol{D}=-\tilde{n}(s\times\boldsymbol{B})\,, \tag{2.2.23}$$

$$s\cdot\boldsymbol{D}=0\,, \tag{2.2.24}$$

$$\boldsymbol{B}=\tilde{n}(s\times\boldsymbol{E})\,,\qquad s\cdot\boldsymbol{B}=0\,, \tag{2.2.25}$$

$$\boldsymbol{D}=\tilde{n}^2(\omega,s)[\boldsymbol{E}-s\cdot(s\cdot\boldsymbol{E})]\,. \tag{2.2.26}$$

The equation $\omega\tilde{n}(s\cdot\boldsymbol{D})=0$ is obtained directly from (2.2.4). The factor $\omega\tilde{n}$ is omitted when passing over to (2.2.24) and likewise for the equation $s\cdot\boldsymbol{B}=0$ since we assumed that $\omega\neq0$, and at $\tilde{n}=0$ the inductions $\boldsymbol{D}$ and $\boldsymbol{B}$ vanish by virtue of (2.2.3, 5). Moreover, as we have already indicated, when $\omega\neq0$ the initial equations (2.2.4, 6) are consequences of two other field equations, namely, (2.2.3, 5); it is, however, convenient to write (2.2.4, 6), as well as the equations $s\cdot\boldsymbol{D}=0$ and $s\cdot\boldsymbol{B}=0$ separately.

2.2.2 Transverse and Longitudinal Waves, "Fictitious" Longitudinal Waves and "Polarization Waves". Real, Coulomb and Mechanical Excitons

Normal waves in an anisotropic medium for which the refractive index $\tilde{n}$ is a solution to the dispersion equation (2.2.22) are, generally speaking, neither transverse nor longitudinal, where the terms "transverse" and "longitudinal" characterize the direction of the electric field vector $\boldsymbol{E}$ with respect to the direction s of propagation. The induction vector $\boldsymbol{D}$ is always transverse in normal waves because $s\cdot\boldsymbol{D}=0$, see the field equation (2.2.24). Nevertheless transverse and longitudinal waves exist for certain directions and they play an important role.

For transverse waves we have according to (2.2.24, 26),

$$\begin{aligned} &\boldsymbol{D} = \tilde{n}^2(\omega, \boldsymbol{s})\boldsymbol{E}\,, \qquad \boldsymbol{s}\cdot\boldsymbol{E} = 0\,, \\ &\boldsymbol{s}\cdot\boldsymbol{D} = 0\,, \qquad D_i = \varepsilon_{ij}(\omega, \boldsymbol{k})E_j\,, \end{aligned} \tag{2.2.27}$$

where the condition $\boldsymbol{s}\cdot\boldsymbol{E} = 0$ defines the transverse waves [if $\boldsymbol{s}\cdot\boldsymbol{E} = 0$ and $\tilde{n}^2$ is finite then the condition $\boldsymbol{s}\cdot\boldsymbol{D} = 0$ also follows from the first equation of (2.2.27)].

Equations (2.2.27) impose four conditions on the three quantities E_i and thus have no solutions in the general case. This can be readily confirmed if the direction $\boldsymbol{s}$ of propagation is taken to be the z axis. Then, according to (2.2.27), the following relations apply to transverse waves:

$$\begin{aligned} &E_3 = 0\,, \\ &D_3 = \varepsilon_{31}E_1 + \varepsilon_{32}E_2 = 0\,, \\ &D_1 = \varepsilon_{11}E_1 + \varepsilon_{12}E_2 = \tilde{n}^2E_1\,, \qquad \text{and} \\ &D_2 = \varepsilon_{21}E_1 + \varepsilon_{22}E_2 = \tilde{n}^2E_2\,. \end{aligned}$$

In other words, the vectors $\boldsymbol{D}$ and $\boldsymbol{E}$ are parallel only in exceptional cases in an arbitrary anisotropic medium whereas this is precisely what must be true for transverse waves, see (2.2.27a).

Searching for transverse waves is thus associated with finding the eigenvectors of the matrix (tensor) ε_{ij}. Those eigenvectors $\boldsymbol{e}_i$ which are orthogonal to $\boldsymbol{s}$ satisfy the system (2.2.27). In the absence of spatial dispersion, ε_{ij} is independent of $\boldsymbol{s}$ and this simply results in reducing the tensor $\varepsilon_{ij}(\omega)$ to its diagonal form. Let us consider, for example, a rhombic crystal. In this case the principal axes of the tensor $\varepsilon_{ij}(\omega)$ coincide for any ω with the twofold symmetry crystal axes x, y, z. This concerns the crystal classes D_2 and D_{2h} (using the Schönflies designation which is more customary for physicists [2.7, 17]; for conversion of this notation to international symbols see Appendix A and [2.18]). In class C_{2v} the z axis lies along a twofold symmetry axis and the axes x, y are perpendicular to the symmetry planes. In this coordinate system (x, y, z) the diagonal elements ε_x, ε_y and ε_z of ε_{ij} differ from one another (if there is no random degeneracy, as we have presupposed). All waves whose vector $\boldsymbol{E}$ is directed along any of the axes x, y, z are transverse; the vector $\boldsymbol{s}$ then lies in one of the coordinate planes. For these transverse waves

$$\tilde{n}^2_{1,2,3} = \varepsilon_{x,y,z}(\omega)\,. \tag{2.2.28}$$

In the case of tetragonal crystals, if we take the z axis in the direction of a fourfold-symmetry crystal axis (without this interpretation it is impossible to distinguish symmetry axes from rotation-reflection axes), $\varepsilon_x(\omega) = \varepsilon_y(\omega) = \varepsilon_\perp(\omega)$ and the vector $\boldsymbol{E}$ for transverse waves with $s_x = s_y = 0$, $s_z = 1$ can be arbitrarily oriented in the xy plane (degeneracy). In cubic crystals $\varepsilon_x = \varepsilon_y = \varepsilon_z = \varepsilon(\omega)$ and transverse waves with arbitrary polarization can be propagated in any direction. The propagation of transverse waves in the presence of spatial dispersion will be discussed in Sects. 4.1 – 3.

For longitudinal waves, by definition,

$$\boldsymbol{E}(\omega,\boldsymbol{k}) = E(\omega,\boldsymbol{k})\boldsymbol{s} \,. \tag{2.2.29}$$

Therefore according to (2.2.23, 25)

$$\boldsymbol{D} = 0 \,, \qquad \boldsymbol{B} = 0 \,. \tag{2.2.30}$$

If we write the equation $\boldsymbol{D} = 0$ in expanded form as

$$D_i(\omega,\boldsymbol{k}) = \varepsilon_{ij}(\omega,\boldsymbol{k})E_j(\omega,\boldsymbol{k}) = 0 \,, \tag{2.2.31}$$

it becomes evident that longitudinal waves with $\boldsymbol{E} \neq 0$ can exist only if

$$|\varepsilon_{ij}(\omega,\boldsymbol{k})| = 1/|\varepsilon_{ij}^{-1}(\omega,\boldsymbol{k})| = 0 \,. \tag{2.2.32}$$

This represents a necessary condition for longitudinal waves to appear. It may, however, also be satisfied by waves which cannot be called longitudinal. In fact, as is clear from the dispersion equation (2.2.22), a root $\tilde{n}^2 = 0$ appears if $|\varepsilon_{ij}| = 0$ [this is also obvious, e.g., from the equation $\boldsymbol{D} = \tilde{n}^2\boldsymbol{E}$ for transverse waves, see (2.2.27)]. Of course, if $\tilde{n} = 0$ the vector $\boldsymbol{k} = (\omega/c)\tilde{n}\boldsymbol{s} = 0$ and it is impossible to speak of polarizing the wave directly (the direction of $\boldsymbol{E}$ with respect to $\boldsymbol{s}$). However, it suffices to consider a weakly nonhomogeneous medium for the separation of the waves into longitudinal and transverse ones to be meaningful at $\tilde{n} = 0$ as well (the direction of the vector $\boldsymbol{E}$ at the point $\tilde{n} = 0$ is determined by its direction near this point; for details see [2.2]). In other words, polarization can be introduced also for $\tilde{n} = 0$ as the result of a certain limiting transition or by a more accurate physical statement of the problem (introduction of weak nonhomogeneity, taking boundaries into consideration and polarization of the waves incident upon such a boundary, etc.). Another fact is also important. Since the tensor $\varepsilon_{ij}(\omega,\boldsymbol{k})$ is not generally Hermitian, the solutions of the system of equations $\varepsilon_{ij}E_j = 0$ need not be real. This means that we should consider the nonhomogeneous longitudinal waves

$$\boldsymbol{E}(\omega,\boldsymbol{k}) = A(\omega,\boldsymbol{k})\boldsymbol{k} \,, \qquad \boldsymbol{k} = \boldsymbol{k}' + \mathrm{i}\boldsymbol{k}'' \,.$$

By virtue of (2.2.3, 5) $\boldsymbol{D} = 0$ and $\boldsymbol{B} = 0$ for such waves, see (2.2.30), but they can be called longitudinal only in a conditional sense. In the following, unless an explicit statement to the contrary is made, only homogeneous waves of the form (2.2.29) will be called longitudinal. As is obvious from the above, (2.2.32) is a necessary but not sufficient condition for the appearance of such waves.

In an arbitrary anisotropic medium, longitudinal, as well as transverse, waves can be propagated only in certain exceptional directions. This is clear even within the scope of classical crystal optics. For example, a rhombic crystal in the system of principal axes x, y, z is characterized by

$$D_x = \varepsilon_x(\omega)E_x, \quad D_y = \varepsilon_y(\omega)E_y, \quad D_z = \varepsilon_z(\omega)E_z. \tag{2.2.33}$$

Longitudinal waves ($\boldsymbol{D} = 0$) can evidently exist in this case only if one of the values of $\varepsilon_i(\omega)$ vanishes because, without random degeneracy, all these quantities differ for the given frequency [this is equivalent to the condition (2.2.32) since $|\varepsilon_{ij}| = \varepsilon_x\varepsilon_y\varepsilon_z$]. The vector $\boldsymbol{E}$ in the case of longitudinal waves is thus automatically directed along one of the principal axes for which $\varepsilon_i(\omega) = 0$. In the presence of degeneracy new possibilities arise. For example, for cubic crystals $\varepsilon_x = \varepsilon_y = \varepsilon_z$ and, obviously, longitudinal waves may be propagated in any direction, provided that

$$\varepsilon(\omega_{\parallel}) = 0.$$

This equation determines the frequency $\omega_{\parallel}$ of the longitudinal waves; the value of $k = \omega\tilde{n}/c$ remains arbitrary if spatial dispersion is neglected. Since under these conditions, the frequency $\omega_{\parallel}$ is independent of $\boldsymbol{k}$, the group velocity is $d\omega/d\boldsymbol{k} = 0$ and we can speak of oscillations rather than of traveling waves.

In crystals of the triclinic or monoclinic systems the principal axes of a tensor of rank two are not fixed with respect to the lattice. This follows from symmetry conditions. In the case of a monoclinic crystal this is true only for two axes; the third axis is fixed. Under such conditions it should be borne in mind that the tensor $\varepsilon_{ij}(\omega)$ is equivalent to two symmetric real tensors ε'_{ij} and ε''_{ij}, see (2.1.42). Each of these tensors has real orthogonal eigenvectors (principal axes) but in general their directions do not coincide and they are also frequency dependent (dispersion of axes). If no single principal axis is common to ε'_{ij} and ε''_{ij}, longitudinal waves cannot exist at all. If we do assume the existence of a longitudinal wave and take the z axis along the direction of the field vector $\boldsymbol{E}$ of this wave, then in such a system, $D_z = \varepsilon_{xz}E_z = 0$, $D_y = \varepsilon_{yz}E_z = 0$ and $D_z = \varepsilon_{zz}E_z = 0$, and the direction of z may be taken as the principal axis of ε_{ij} for which the eigenvalue vanishes.

We recall that with $\boldsymbol{B}_{\text{ext}} \neq 0$ (magnetoactive medium) the tensors $\varepsilon'_{ij}(\omega, \boldsymbol{B}_{\text{ext}})$ and $\varepsilon''_{ij}(\omega, \boldsymbol{B}_{\text{ext}})$ are Hermitian but not real. Consequently, the eigenvectors of these tensors are, in general, *complex* and therefore the tensors cannot be diagonalized in any of the coordinate system (x, y, z). The appearance of longitudinal waves under these conditions is an exception and not the rule (in a magnetoactive plasma where the field $\boldsymbol{B}_{\text{ext}}$ is a symmetry axis, a longitudinal wave can exist only if it is propagated along the field). These examples are sufficient to show that in the general case [when use is made of $\varepsilon_{ij}(\omega, \boldsymbol{k})$] the existence of longitudinal waves is by no means a fact; they arise mainly in exclusive directions. For many media displaying spatial dispersion Sects. 4.1 – 3 offer convincing evidence of this. Of special interest is the question of the propagation of waves in directions which are very close to those for which longitudinal waves may exist. The problem has been analyzed for a magnetoactive plasma [2.2], and we shall deal with it to some extent in Sect. 4.5.3.

It is clear that within the scope of classical crystal optics only two normal waves can, in general, be propagated in an anisotropic medium in a given arbitrary direction. This conclusion follows, in the general case, directly from the dispersion equation (2.2.22) which, if $\varepsilon_{ij} = \varepsilon_{ij}(\omega)$, is quadratic with respect to $\tilde{n}^2$. The roots of this equation, $\tilde{n}_1^2$ and $\tilde{n}_2^2$, correspond to the two waves mentioned above, which, in many cases, are called the ordinary and the extraordinary waves (the solutions with $\tilde{n}_{1,2}$ and $-\tilde{n}_{1,2}$ refer to waves propagated in opposite directions so that, if $\boldsymbol{s}$ is given, we can speak of only two normal waves). If there is no degeneracy, a third (longitudinal) wave may appear, but only in certain directions.

a) Coulomb Excitions: Longitudinal and "Fictitious" Longitudinal Waves

So far we have dealt with transverse and longitudinal normal waves, i.e., waves satisfying the field equations (2.2.1) or, more specifically, (2.2.23 – 26). These and all other normal waves which are neither longitudinal (or transverse) can exist. In the optical range, the normal waves, except the purely longitudinal ones, are simply light waves in a medium. In the quantized state when absorption is neglected, they are equivalent to "photons in a medium" having frequency $\hbar\omega$ and momentum $\hbar\boldsymbol{k} = (\hbar\omega/c)n\boldsymbol{s}$. The quanta of longitudinal (plasma) waves are usually called plasmons. According to another terminology, which we shall adhere to (Chap. 1), all normal waves (thus including the longitudinal ones) are to be called "real excitons" or simply "excitons". Real excitons may also be called "polaritons".

Also of interest for various reasons are waves which are the solutions, not of the complete system of field equations (2.2.1 or 23 – 26), but only of the homogeneous equations of the Coulomb problem:

$$\mathrm{rot}\,\boldsymbol{E} = 0\,, \qquad \mathrm{div}\,\boldsymbol{D} = 0\,.$$

Applied to a homogeneous plane wave s, (2.2.16), this means (if $k \neq 0$) that

$$\boldsymbol{E} = E\boldsymbol{s}\,, \qquad \boldsymbol{s}\cdot\boldsymbol{D} = 0\,. \tag{2.2.34}$$

Solutions which satisfy these conditions will be called "Coulomb excitons".

Coulomb excitons may be divided into three types. The first type corresponds to the longitudinal waves, (2.2.29, 30 and 32), which, at the same time, are also solutions of the complete system of field equations and which satisfy the Coulomb condition (2.2.34). The reason for this coincidence of real and Coulomb excitons is trivial: for longitudinal waves $\boldsymbol{D}=0$ and $\boldsymbol{B}=0$, see (2.2.30). For Coulomb excitons of the second and third types

$$\boldsymbol{D} \neq 0 . \tag{2.2.35}$$

From the field equation[7] $\boldsymbol{D} = -(c/\omega)(\boldsymbol{k}\times\boldsymbol{B}) = -\tilde{n}(\boldsymbol{s}\times\boldsymbol{B})$ – see (2.2.3 and 23) – it is obvious that if $\boldsymbol{D}\neq 0$ and $\boldsymbol{k}$ (or $\tilde{n}$) is finite, the magnetic induction $\boldsymbol{B}\neq 0$ and therefore, by virtue of the field equation $\boldsymbol{B}=(c/\omega)(\boldsymbol{k}\times\boldsymbol{E}) = \tilde{n}(\boldsymbol{s}\times\boldsymbol{E})$, the electric field $\boldsymbol{E}$ is not longitudinal. Thus, Coulomb excitions of the type (2.2.34, 35) do not satisfy all field equations, i.e., they are not identical with real excitons except in the limiting case $k\to\infty$. If $k/\omega\to\infty$ and if there are no external sources, the magnetic induction $\boldsymbol{B}$ and the transverse electric field $\boldsymbol{E}_\perp$ tend to zero; otherwise the induction $\boldsymbol{D}$ would grow without limit, see (2.2.3, 5). This is no longer true if corresponding sources $\boldsymbol{j}_{\mathrm{ext}}$ and ρ_{ext} exist, in which case any arbitrarily short-wavelength fields $\boldsymbol{B}$ and $\boldsymbol{E}_\perp$ can be produced.

Coulomb excitons of the second type (we shall call them "fictitious" longitudinal waves) satisfy the conditions (2.2.34 and 35); moreover, by definition, their electric field $\boldsymbol{E}\neq 0$:

$$\boldsymbol{E} = E\boldsymbol{s} \neq 0 , \quad \boldsymbol{s}\cdot\boldsymbol{D}=0 , \quad \boldsymbol{D}\neq 0 . \tag{2.2.36}$$

The tensor $\varepsilon_{ij}(\omega,\boldsymbol{k})$ relates $\boldsymbol{D}$ and $\boldsymbol{E}$ for any fields which can be produced by external sources in a homogeneous medium. This refers, in particular, to the Coulomb field $\boldsymbol{E}(\omega,\boldsymbol{k})$ even if it does not satisfy a homogeneous field equation. For the case of crystals the situation is somewhat more complicated since the concept of the tensor $\varepsilon_{ij}(\omega,\boldsymbol{k})$ itself must now be defined more accurately (Sect. 3.1).

As mentioned before in bounded media the quantity $\boldsymbol{k}$ may be complex. For example, in an isotropic medium we obtain $k^2\varepsilon(\omega)=0$ from the conditions $\boldsymbol{E}=E\boldsymbol{s}$, $\boldsymbol{k}\cdot\boldsymbol{D}=0$, $\boldsymbol{D}=\varepsilon(\omega)\boldsymbol{E}$ in the Coulomb limit. This relation gives us $\varepsilon(\omega)=0$ (bulk longitudinal waves) and $k_1^2+k_2^2+k_3^2=0$ (surface waves). The existence of this type of surface Coulomb wave follows also from the equation $k^2c^2/\omega^2=\varepsilon(\omega)$ as $c\to\infty$ and with finite ω and $\varepsilon(\omega)$. These surface waves should be taken into account in using Coulomb excitons when solving boundary problems [2.20, 21].

---

[7] The refractive index $\tilde{n}$ is introduced in accordance with (2.2.21) so that $\boldsymbol{k}=k\boldsymbol{s}=(\omega/c)\tilde{n}\boldsymbol{s}$ for normal waves, i.e., waves satisfying the dispersion equation (2.2.22). Here and elsewhere, when this cannot give rise to misunderstandings, n will also be used to denote the quantity $\tilde{n}=ck/\omega$ for any homogeneous waves where $\boldsymbol{k}=k\boldsymbol{s}$.

For "fictitious" longitudinal bulk waves ($k^2 \neq 0$), (2.2.36), we can write

$$s_i D_i = s_i \varepsilon_{ij} E_j = \varepsilon_{ij} s_i s_j E = 0$$

which, if $E \neq 0$, leads to

$$\varepsilon_{ij}(\omega, \boldsymbol{k}) s_i s_j = 0\,, \qquad \boldsymbol{k} = k\boldsymbol{s}\,. \tag{2.2.37}$$

Since this condition relates ω and $\boldsymbol{k}$, we have good reason to call it the dispersion equation of "fictitious" longitudinal waves. The solution of this equation will be denoted by $\omega_{\|}'(\boldsymbol{k})$. In the limit of classical crystal optics (2.2.37) has the form

$$\varepsilon_{ij}(\omega) s_i s_j = 0\,. \tag{2.2.38}$$

The solution $\omega_{\|}'(\boldsymbol{s})$ of this equation, in general, depends on the direction of $\boldsymbol{s}$. This means that the frequency $\omega_{\|}'(\boldsymbol{k})$ satisfying (2.2.37) is not an analytic function of $\boldsymbol{k}$ for $\boldsymbol{k} \to 0$.

From the physical viewpoint this result is quite reasonable. In an anisotropic medium the induction $\boldsymbol{D}$ and polarization $\boldsymbol{P} = (\boldsymbol{D} - \boldsymbol{E})/4\pi$ are, for most directions, not parallel to the vector $\boldsymbol{E}$, this being true also for arbitrarily long wavelengths. Moreover, for a Coulomb field without sources we have

$$\boldsymbol{D} = \boldsymbol{E}_{\|} + 4\pi \boldsymbol{P}\,, \qquad \boldsymbol{E}_{\|} = \boldsymbol{E} = E\boldsymbol{s}\,, \qquad \boldsymbol{s} \cdot \boldsymbol{D} = E + 4\pi(\boldsymbol{s} \cdot \boldsymbol{P}) = 0\,, \qquad \text{i.e.,}$$

$$\boldsymbol{E}_{\|} = -4\pi \boldsymbol{s} \cdot (\boldsymbol{s} \cdot \boldsymbol{P})\,. \tag{2.2.39}$$

Thus, if $\boldsymbol{k} \to 0$ the field $\boldsymbol{E}_{\|}$ is also not parallel to $\boldsymbol{P}$ and, consequently, it depends on $\boldsymbol{s}$. However, the field $\boldsymbol{E}_{\|}$ is associated with a definite amount of energy, and therefore the frequency $\omega_{\|}'$ of the wave's oscillations depends on $\boldsymbol{E}_{\|}$ and, consequently, on $\boldsymbol{s}$ as well.

We emphasize that condition (2.2.37) is necessary for the wave to be a "fictitious" longitudinal wave with $\boldsymbol{E} \neq 0$ and in which $\boldsymbol{D} \neq 0$. Obviously, for longitudinal waves with $\boldsymbol{D} = 0$ the condition (2.2.37) is always satisfied, too. For example, let us consider a rhombic crystal in classical crystal optics in the system of principal axes of the tensor $\varepsilon_{ij}(\omega)$. Instead of (2.2.38), we have

$$\varepsilon_1(\omega) s_1^2 + \varepsilon_2(\omega) s_2^2 + \varepsilon_3(\omega) s_3^2 = 0\,. \tag{2.2.38a}$$

In this case (with $\varepsilon_1 \neq \varepsilon_2 \neq \varepsilon_3$), longitudinal waves can be propagated only along the principal axes; in that case (2.2.38a) is also satisfied.

The proof of this statement in obvious in the general case as well. If a longitudinal wave exists, then the condition $\varepsilon_{ij} s_j = 0$ is satisfied (recall that in a homogeneous longitudinal wave $\boldsymbol{E} = E\boldsymbol{s}$ and $D_i = \varepsilon_{ij} E_j = 0$), i.e., the weaker condition (2.2.37) is also fulfilled. In general, it could be contended that longitudinal waves (with $\boldsymbol{D} = 0$) represent a special case of "fictitious" longitudinal waves.

From (2.2.37, 38), and in a particularly obvious way from (2.2.38a), it is clear that the direction $\boldsymbol{s}$ of "fictitious" longitudinal waves may, in principle, be arbitrary. But for this to occur, certain stricter conditions must be fulfilled. For example, it is evident from (2.2.38a) that the sign of at least one of the values of ε_i must be negative. We should mention here that besides the homogeneous type we can consider nonhomogeneous "fictitious" longitudinal waves as well, but we shall not dwell on these.

b) Coulomb Excitions: "Polarization Waves"

Coulomb excitons of the third (and last) type are characterized by

$$\boldsymbol{D} \neq 0\,, \quad \boldsymbol{s} \cdot \boldsymbol{D} = 0\,, \quad \boldsymbol{E}(\omega,\boldsymbol{k}) = 0\,. \tag{2.2.40}$$

We shall call these Coulomb excitons "polarization waves" because their polarization $\boldsymbol{P} = \boldsymbol{D}/4\pi \neq 0$. Since $E_i(\omega,\boldsymbol{k}) = \varepsilon_{ij}^{-1}(\omega,\boldsymbol{k})D_j(\omega,\boldsymbol{k})$, the relation

$$|\varepsilon_{ij}^{-1}(\omega,\boldsymbol{k})| = 1/|\varepsilon_{ij}(\omega,\boldsymbol{k})| = 0 \tag{2.2.41}$$

represents a necessary condition for the existence of "polarization waves". As already mentioned, the "polarization waves" satisfy the complete system of homogeneous field equations only in the limit $k \to \infty$. The fact that these waves may exist for the Coulomb problem becomes particularly clear from (2.2.40): at $sP = 0$, the field $E_{\|} = 0$, but the polarization $\boldsymbol{P}$ may differ from zero. Physically, this is associated with resonances, or poles of permittivity, see (2.2.43) below. 'Polarization waves' also have a simple meaning from the microscopic point of view, say, for the oscillator model (Sect. 5.2).

Equation (2.2.41) represents a necessary and sufficient condition for the existence of a solution of the type

$$\boldsymbol{D}(\omega,\boldsymbol{k}) \neq 0\,, \quad \boldsymbol{E}(\omega,\boldsymbol{k}) = 0\,. \tag{2.2.42}$$

Polarization waves are a particular case of these solutions for which, in addition, $\boldsymbol{s} \cdot \boldsymbol{D} = 4\pi(\boldsymbol{s} \cdot \boldsymbol{P}) = 0$. The condition (2.2.41) is thus by no means sufficient for the solutions to be polarization waves.

We shall call waves satisfying the condition (2.2.42) *mechanical excitons.* This term will be applied to all solutions (either exact or approximate) for the medium considered or for the same medium at the temperature $T = 0$ that are obtained in the absence of macroscopic (long-wavelength) fields $\boldsymbol{E}(\omega,\boldsymbol{k})$ and $\boldsymbol{B}(\omega,\boldsymbol{k})$ or if their effect is neglected. Mechanical excitons will be discussed in greater detail in Chap. 6. The physical significance of this concept consists precisely in the fact that waves (solutions) are considered which have no connection with the macroscopic fields $\boldsymbol{E}(\omega,\boldsymbol{k})$ and $\boldsymbol{B}(\omega,\boldsymbol{k})$. Consequently, mechanical excitons may play the role of states of the unperturbed problem when calculating the current induced in the system by the action of a macroscopic field.

It may seem at first sight that the concept of the mechanical exciton has no exact meaning since no system of equations has been indicated whose solutions correspond to mechanical excitons. This conclusion is, however, erroneous. Special types of mechanical excitons (polarization waves) follow from the conditions (2.2.40, 41). In the case of mechanical excitons of the type (2.2.42) the homogeneous Coulomb problem has to be considered. In fact we have to find the nontrivial solution $\boldsymbol{D} \neq 0$ of the system of equations $E_i(\omega,\boldsymbol{k}) = \varepsilon_{ij}^{-1}(\omega,\boldsymbol{k})D_j(\omega,\boldsymbol{k}) = 0$ that can always be taken as a solution to the Coulomb equations $\mathrm{i}\boldsymbol{k}\cdot\boldsymbol{D} = 4\pi\rho_{\mathrm{ext}}(\omega,\boldsymbol{k})$ since the density ρ_{ext} of the external sources can be an arbitrarily given quantity. In the general case mechanical excitons (their frequencies ω_{m} and their polarization) are also obtained if the tensor $\varepsilon_{ij}(\omega,\boldsymbol{k})$ is known for the medium or for some approximate model of this medium, both for the given temperature and for $T=0$.

In the case of a rhombic crystal, if $\boldsymbol{k}\to 0$, the condition (2.2.41) assumes the form, recalling (2.2.33),

$$|\varepsilon_{ij}^{-1}(\omega,\boldsymbol{k})| = 1/\varepsilon_x(\omega)\,\varepsilon_y(\omega)\,\varepsilon_z(\omega) = 0\,. \tag{2.2.43}$$

The frequencies ω_p of the 'polarization waves' thus correspond in this case to the poles of the principal values of $\varepsilon_i(\omega)$; the vector $\boldsymbol{D} = 4\pi\boldsymbol{P}$ in 'polarization waves' is directed – see (2.2.33) – along the one of the principal axes for which $\varepsilon_i(\omega_p) = \infty$. By virtue of the condition $\boldsymbol{s}\cdot\boldsymbol{D} = 0$ it follows from this that the only existing 'polarization waves' are those whose propagation vector $\boldsymbol{s}$ lies in one of the coordinate planes xy, xz or yz. The frequencies and polarization of mechanical excitons of the type (2.2.42) are the same in our example as those of the 'polarization waves', but the vector $\boldsymbol{s}$ can be arbitrarily directed. Within the scope of the Coulomb problem, of course, these mechanical excitons can exist (be exact solutions) only in the presence of external charges having density

$$\rho_{\mathrm{ext}}(\omega,\boldsymbol{k}) = \frac{\mathrm{i}}{4\pi}\boldsymbol{k}\cdot\boldsymbol{D}(\omega,\boldsymbol{k}) = \frac{\mathrm{i}k}{4\pi}\boldsymbol{s}\cdot\boldsymbol{D}\,.$$

The frequencies ω_m and the eigenfunctions $\Psi_{k,m}$ of the mechanical excitons are used in order to calculate $\varepsilon_{ij}(\omega,\boldsymbol{k})$. In doing so we find that the poles of the expression for $\varepsilon_{ij}(\omega,\boldsymbol{k})$ obtained as a first approximation are at frequencies ω equal to ω_m or to the sum of two frequencies ω_m, etc., (Sect. 6.1).

The significance of the Coulomb excitons consists in the fact that their frequencies $\omega_{\|}(\boldsymbol{k})$, $\omega_{\|}'(\boldsymbol{k})$ and $\omega_p(\boldsymbol{k})$ are closely related to certain characteristic frequencies of real excitons, $\omega_l(\boldsymbol{k})$, i.e., to the shape of the corresponding dispersion curves $\tilde{n}_l(\omega,\boldsymbol{s})$.

Table 2.1. Waves of the type $\boldsymbol{D} = \boldsymbol{D}_0 \exp[\mathrm{i}(\boldsymbol{k}\cdot\boldsymbol{r} - \omega t)]$, $\boldsymbol{E} = \boldsymbol{E}_0 \exp[\mathrm{i}(\boldsymbol{k}\cdot\boldsymbol{r} - \omega t)]$, $\boldsymbol{k} = k\boldsymbol{s} = (\omega/c) \times \tilde{n}(\omega,\boldsymbol{s})\boldsymbol{s}$ and $s = 1$ (monochromatic homogeneous waves) in an anisotropic medium

General case of normal waves $\boldsymbol{D} = \tilde{n}^2[\boldsymbol{E} - \boldsymbol{s}(\boldsymbol{s}\cdot\boldsymbol{E})]\,, \quad \boldsymbol{s}\cdot\boldsymbol{D} = 0\,,$ $\lvert \tilde{n}^2(\delta_{ij} - s_i s_j) - \varepsilon_{ij}(\omega,\boldsymbol{k}) \rvert = 0\,.$	
Transverse normal waves $\boldsymbol{D} = \tilde{n}^2\boldsymbol{E}\,, \quad \boldsymbol{s}\cdot\boldsymbol{E} = 0\,, \quad \boldsymbol{s}\cdot\boldsymbol{D} = 0\,.$	Normal electromagnetic waves (real excitons), i.e., solutions of the homogeneous field equations
Longitudinal normal waves $\boldsymbol{D} = 0\,, \quad \boldsymbol{E} = E\boldsymbol{s} \neq 0\,,$ $\lvert \varepsilon_{ij}(\omega,\boldsymbol{k}) \rvert = 0\,.$	
"Fictitious" longitudinal waves $\boldsymbol{D} \neq 0\,, \quad \boldsymbol{s}\cdot\boldsymbol{D} = 0\,, \quad \boldsymbol{E} = E\boldsymbol{s} \neq 0\,,$ $\varepsilon_{ij}(\omega,\boldsymbol{k}) s_i s_j = 0\,.$	Coulomb excitons, i.e., solutions of the homogeneous Coulomb problem $\boldsymbol{s}\cdot\boldsymbol{D} = 0$, $\boldsymbol{E} = E\boldsymbol{s}$
"Polarization waves" $\boldsymbol{D} \neq 0\,, \quad \boldsymbol{s}\cdot\boldsymbol{D} = 0\,, \quad \boldsymbol{E} = 0\,,$ $\lvert \varepsilon_{ij}^{-1}(\omega,\boldsymbol{k}) \rvert = 0\,.$	
Mechanical excitons of the general form $\boldsymbol{D} \neq 0\,, \quad \boldsymbol{E} = 0\,,$ $\lvert \varepsilon_{ij}^{-1}(\omega,\boldsymbol{k}) \rvert = 0\,.$	Mechanical excitons, i.e., solutions to the Coulomb problem without the macroscopic (long-wavelength) field $\boldsymbol{E}(\omega,\boldsymbol{k})$, or if its effects are ignored

For longitudinal waves which are simultaneously Coulomb and real excitons, this is evident [for longitudinal waves the refractive index $\tilde{n}_{\|}(\omega_{\|},\boldsymbol{s}) = ck/\omega_{\|}$], i.e., it is obtained when the equation

$$\omega_{\|} = \omega_{\|}(k\boldsymbol{s}) = \omega_{\|}[(\omega/c)\tilde{n}(\omega,\boldsymbol{s})\boldsymbol{s}]$$

is solved with respect to $k = (\omega_{\|}/c)\tilde{n}(\omega_{\|},\boldsymbol{s})$. Equation (2.2.37), determining the frequencies $\omega_{\|}$ of the 'fictitious' longitudinal waves, is also a condition for $\tilde{n}(\omega,\boldsymbol{s})$ becoming infinite for finite components of the tensor $\varepsilon_{ij}(\omega,\boldsymbol{k})$. This follows from the dispersion equation (2.2.22), since $\varepsilon_{ij}s_i s_j$ is the coefficient of $\tilde{n}^4$. The same result can be obtained, however, directly from the initial wave equation (2.2.26): if $\boldsymbol{D} \neq 0$ and the polarization of the field $\boldsymbol{E}$ approaches the longitudinal field $\boldsymbol{E} = E\boldsymbol{s}$, then $\tilde{n}^2 \to \infty$. The 'polarization waves' also correspond to a pole of $\tilde{n}^2$, in this case when $\lvert \varepsilon_{ij}(\omega,\boldsymbol{k}) \rvert \to \infty$, see (2.2.26, 40, 41). It should be emphasized that the zone of sharp growth (resonance) of $n = \mathrm{Re}\{\tilde{n}\}$, in general, corresponds to an absorption band, i.e., a zone of growth of $\varkappa = \mathrm{Im}\{\tilde{n}\}$. Finding the poles of $\tilde{n}^2$ is therefore of direct physical significance. If we know the frequencies of the Coulomb excitons we thus have essential information, not only about the longitudinal waves, but also on the zeros and poles of the dispersion curves for nonlongitudinal real excitons (normal waves).

Homogeneous waves of various types and the names assigned to them are listed in Table 2.1.

2.2.3 Multiple Roots of the Dispersion Equation

Up to now we have considered only monochromatic solutions of the homogeneous field equations (2.2.1) which depend on $\boldsymbol{r}$ and t according to the law $\mathrm{const} \cdot \exp[\mathrm{i}(\boldsymbol{k} \cdot \boldsymbol{r} - \omega t)]$, see (2.2.1). No other solutions of this type exist for a vacuum, and it is usually assumed that the situation remains unchanged if we consider the propagation of waves in a medium. In any case, we have not found a hint in any lecture notes on electrodynamics as to the necessity of considering other solutions as well. As a matter of fact, even in classical crystal optics solutions of the type (2.2.2) with constant amplitude do not always constitute a complete system of normal waves. Solutions of the type (2.2.2) are, for example, insufficient in the case of propagation of waves along the so-called singular optical axes [2.7, 22–24].

In classical crystal optics, disregarding the possibility of the appearance of longitudinal waves, two normal waves with the refractive indices $\tilde{n}_1$ and $\tilde{n}_2$ may be propagated in a given direction $\boldsymbol{s}$. Here $\tilde{n}_1^2$ and $\tilde{n}_2^2$ are roots of the quadratic (with respect to $\tilde{n}^2$) Eqs. (2.2.22), and the polarization of the waves is determined by the set of equations, see (2.2.9),

$$[\varepsilon_{ij}(\omega) - \tilde{n}^2(\omega,\boldsymbol{s})(\delta_{ij} - s_i s_j)]E_j = 0\,, \quad j = 1,2,3]\,, \tag{2.2.44}$$

where E_j denotes components of the field $\boldsymbol{E} = \boldsymbol{E}_0 \exp[\mathrm{i}(\omega \tilde{n} \boldsymbol{s} \cdot \boldsymbol{r}/c - \omega t)]$ itself or the components of its amplitude $\boldsymbol{E}_0$.

Substituting the values of $\tilde{n}_1$ and $\tilde{n}_2$ in (2.2.44) we obtain two linearly independent solutions $\boldsymbol{E}_1$ and $\boldsymbol{E}_2$. It is clear from both formal and physical considerations that two linearly independent solutions $\boldsymbol{E}_1$ and $\boldsymbol{E}_2$ of some kind must exist for any $\tilde{n}_1$ and $\tilde{n}_2$. At the same time, two solutions of type (2.2.2) of the system (2.2.44) may not exist for the multiple root $\tilde{n}_1 = \tilde{n}_2$.

In order to discuss this problem, as well as many other cases, it is more convenient to use the equations for $\boldsymbol{D}$ rather than those for $\boldsymbol{E}$. This is due to the fact that in a coordinate system whose z axis is directed along $\boldsymbol{s}$ the vector $\boldsymbol{D}$ has only two components, D_x and D_y, since $\boldsymbol{s} \cdot \boldsymbol{D} = 0$; we consider here solutions of the homogeneous field equations (2.2.23–26).

In the coordinate system used the wave equation and the dispersion equation have the following form, see (2.2.26) for $s_1 = s_2 = 0$, $s_3 = 1$,

$$D_\alpha = \tilde{n}^2 \varepsilon_{\alpha\beta}^{-1} D_\beta \equiv 1/\tilde{m}^2 \nu_{\alpha\beta} D_\beta \,, \tag{2.2.45}$$

$$|\tilde{m}^2 \delta_{\alpha\beta} - \nu_{\alpha\beta}| = \tilde{m}^4 - (\nu_{11} + \nu_{22}) \tilde{m}^2 + \nu_{11} \nu_{22} - \nu_{12} \nu_{21} = 0 \,, \tag{2.2.46}$$

in which we introduced the notation

$$\varepsilon_{\alpha\beta}^{-1} = \nu_{\alpha\beta} \,, \qquad 1/\tilde{n}^2 = \tilde{m}^2 \,, \qquad \alpha, \beta = x, y = 1, 2 \tag{2.2.47}$$

(we recall that the z axis, or axis 3, of our coordinate system is directed along $\boldsymbol{s}$).

The dispersion equation (2.2.46) has a multiple root $\tilde{m}_1^2 = \tilde{m}_2^2$ if

$$(\nu_{11} - \nu_{22})^2 + 4 \nu_{12}^2 = 0 \,, \tag{2.2.48}$$

where the symmetry of the tensor $\nu_{\alpha\beta}$ is taken into account ($\nu_{\alpha\beta} = \nu_{\beta\alpha}$). There are two possible cases. If the matrix $\nu_{\alpha\beta} = \varepsilon_{\alpha\beta}^{-1}$ is diagonalized, we then have, under the condition (2.2.48) for the corresponding axes x_0 and y_0

$$\nu_{12,0} = \nu_{21,0} = 0 \,, \qquad \nu_{11,0} = \nu_{22,0} = \nu_0 \,, \tag{2.2.49}$$

i.e., degeneracy. Then, evidently, $\tilde{m}^2 = \nu_0$, and we can set

$$E_{1,x_0} = 1 \,, \quad E_{1,y_0} = 0 \,, \quad E_{2,x_0} = 0 \,, \quad E_{2,y_0} = 1 \,, \quad \tilde{m}_{1,2}^2 = \nu_0 \,. \tag{2.2.50}$$

The physical interpretation in this case is that the medium is optically isotropic in a given direction, and that two linearly independent solutions $\boldsymbol{E}_1$ and $\boldsymbol{E}_2$ exist at the same time; these vectors can be considered as being not only plane polarized, as in (2.2.50), but elliptically polarized as well.

Let us assume now that the matrix $\nu_{\alpha\beta}$ is not reduced to diagonal form. We then have for the multiple root (which we call an essential multiple root)

$$\tilde{m}_{1,2}^2 = 1/\tilde{n}_0^2 = (\nu_{11} + \nu_{22})/2 \,, \qquad \nu_{11} - \nu_{22} = \pm 2\mathrm{i}\, \nu_{12} \tag{2.2.51}$$

and the system (2.2.45) has only one solution; this solution corresponds to circular polarization, since

$$\frac{D_2}{D_1} = \frac{1/\tilde{n}_0^2 - \nu_{11}}{\nu_{12}} = \mp \mathrm{i} \,, \qquad \frac{E_2}{E_1} = \mp \mathrm{i} \,,$$

where the sign is determined according to which one of the solutions $\nu_{11} - \nu_{22} = \pm 2\mathrm{i}\, \nu_{12}$ of (2.2.48) we take into consideration. In the present case (classical crystal optics)

$$\nu_{\alpha\beta}(\omega)=\nu_{\beta\alpha}(\omega)=\nu'_{\alpha\beta}(\omega)+\mathrm{i}\,\nu''_{\alpha\beta}\,,\qquad \nu'_{\alpha\beta}=\mathrm{Re}\{\nu_{\alpha\beta}\}\,,\qquad \nu''_{\alpha\beta}=\mathrm{Im}\{\nu_{\alpha\beta}\}\,. \tag{2.2.52}$$

In the absence of absorption (at $\nu''_{\alpha\beta}=0$) the real symmetric matrix $\nu_{\alpha\beta}=\nu'_{\alpha\beta}$ is therefore always diagonalized and the case (2.2.51) cannot occur. This can also be proved if the matrices $\nu'_{\alpha\beta}$ and $\nu''_{\alpha\beta}$ have common principal axes. This, precisely, is the situation for crystals of all systems except for the lowest: rhombic, monoclinic and triclinic. In fact, in crystals of higher symmetries (tetragonal, hexagonal and trigonal), the three-dimensional second-rank tensors have two equal principal values (uniaxial crystals) which correspond to an ellipsoid of revolution. Under similar conditions, as is readily seen, the two-dimensional tensors $\nu'_{\alpha\beta}$ and $\nu''_{\alpha\beta}$ have common axes. For crystals of lower systems, however, the principal axes of these two-dimensional tensors do not, in general, coincide. Consequently, if absorption in these crystals is taken into account, (2.2.48) can be satisfied for certain directions and there will exist only one circularly polarized solution of the type $\boldsymbol{E}=\boldsymbol{E}_0\exp\{\mathrm{i}[\omega\tilde{n}\boldsymbol{s}\cdot\boldsymbol{r}/c-\omega t]\}$. Such a direction, for which an essential multiple root (2.2.51) exists, is called a singular, or circular, optical axis.

We have assumed here and in the following that the frequency ω is real, so that the condition $\nu''_{\alpha\beta}=0$ means that no heat is released, see (2.1.29). Singular optical axes may also occur in a magnetoactive medium for which $\varepsilon_{ij}(\omega,\boldsymbol{B}_{\mathrm{ext}})=\varepsilon'_{ij}+i\varepsilon''_{ij}=\varepsilon_{ij}(\omega,\boldsymbol{B}_{\mathrm{ext}})$ and, at a given $\boldsymbol{B}_{\mathrm{ext}}$, the Hermitian tensors ε'_{ij} and ε''_{ij} are not real. The same applies to the tensors $\nu'_{\alpha\beta}$ and $\nu''_{\alpha\beta}$. Here, even without absorption (in which case the tensor $\nu''_{\alpha\beta}$ vanishes) it is impossible to diagonalize the tensor $\nu'_{\alpha\beta}$ in an arbitrary coordinate system x_0, y_0. Nevertheless, if $\nu''_{\alpha\beta}=0$, no essential multiple root can exist since for a multiple root

$$(\nu_{11}-\nu_{22})^2+\nu_{12}\nu_{21}=(\nu'_{11}-\nu'_{22})^2+4\,|\nu'_{12}|^2=0$$

[in the transition to (2.2.48) we assumed that $\nu_{12}=\nu_{21}$; now, however, $\nu'_{21}=(\nu'_{12})^*$]. This condition is evidently satisfied only in the case of degeneracy, see (2.2.49). Even in an absorbing magnetoactive medium essential multiple roots may appear; moreover, for a magnetoactive plasma this possibility is, in fact, realized for certain values of the parameters [2.2, 11].

We are not concerned here with an investigation of the propagation of waves along the singular axes in crystals [4.22–24] or in a plasma [2.2, 11–25]. Here we wish only to emphasize the important fact that even without spatial dispersion essential multiple roots of the dispersion equation can appear to which only one solution of the type (2.2.2) corresponds. The second linearly independent solution is easily found by solving the initial differential equations (2.2.1).

This point becomes quite clear when we use the example of the natural oscillations of an oscillator, described by $\ddot{x}+2\gamma\dot{x}+\omega_i^2x=0$. Substituting the solution $x = A\exp(\mathrm{i}\omega t)$ we obtain $\omega_{1,2} = \mathrm{i}\gamma \pm \sqrt{\omega_i^2 - \gamma^2}$. For the multiple root, if $\omega_{1,2} = \mathrm{i}\gamma$ and $\omega_i = \pm\gamma$, apart from the solution $x_1 = A\exp(-\gamma t)$, there also exists a solution of the form $x_2 = Bt\exp(-\gamma t)$ which is also obtained from the solutions $x_1 = A_1\exp(\mathrm{i}\omega_i t)$ and $x_2 = E_1\exp(\mathrm{i}\omega_2 t)$ by means of the limiting transition

$$\lim_{\omega_1\to\omega_2} C\frac{\mathrm{e}^{\mathrm{i}\omega_1 t} - \mathrm{e}^{\mathrm{i}\omega_2 t}}{\omega_1 - \omega_2} = \frac{d}{d\omega} C\mathrm{e}^{\mathrm{i}\omega t} = \mathrm{i}Ct\mathrm{e}^{\mathrm{i}\omega t}.$$

This procedure proves to be most convenient for determining the second linearly independent solution for the essential multiple root of the dispersion equation. Near the multiple root (i.e., for corresponding values of the parameters) we consider, along with the solutions $\boldsymbol{E} = \boldsymbol{E}_{0,1,2}\exp\{\mathrm{i}[\omega\tilde{n}_{1,2}\boldsymbol{s}\cdot\boldsymbol{r}/c - \omega t]\}$, the solution depending linearly on them

$$\boldsymbol{E} = \frac{\mathrm{e}^{-\mathrm{i}\omega t}}{\tilde{n}_2 - \tilde{n}_1}(\boldsymbol{E}_{0,2}\mathrm{e}^{\mathrm{i}\omega\tilde{n}_2\boldsymbol{s}\cdot\boldsymbol{r}/c} - \boldsymbol{E}_{0,1}\mathrm{e}^{\mathrm{i}\omega\tilde{n}_2\boldsymbol{s}\cdot\boldsymbol{r}/c}). \tag{2.2.53}$$

Near the singular optical axis we can set

$$\tilde{n}_{1,2} = \tilde{n} + \delta\tilde{n}_{1,2}, \quad \boldsymbol{E}_{0,1,2} = \boldsymbol{E}_0 + \delta\boldsymbol{E}_{1,2},$$

where $\delta\tilde{n}_{1,2}$ and $\delta E_{1,2}$ are small quantities which vanish at the axis itself. Then, in the limit (approaching the singular axis), (2.2.53) proves to be the sum of the solutions of the old type $\boldsymbol{E}_1 = \boldsymbol{E}_{0,1}\exp\{\mathrm{i}[\omega\tilde{n}\boldsymbol{s}\cdot\boldsymbol{r}/c - \omega t]\}$ and the second (new) solution of the type

$$\boldsymbol{E}_2 = \boldsymbol{E}_{0,2}\cdot(\boldsymbol{s}\cdot\boldsymbol{r})\,\mathrm{e}^{\mathrm{i}(\omega\tilde{n}\boldsymbol{s}\cdot\boldsymbol{r}/c - \omega t)}. \tag{2.2.54}$$

Both for crystals of lower systems and magnetoactive media an essential multiple root appears only in the presence of absorption (see above)[8]. In the absence of spatial dispersion this result can be understood physically from the nature of the solution (2.2.54). For the sake of clarity, let us consider the

---

[8] We do encounter one case of the appearance of multiple roots at a given polarization even in dealing with the simplest problem of the propagation of transverse waves in a nonabsorbing isotropic medium. Here the wave equation is of the form

$$(d^2E_{x,y})/(dz^2) + (\omega^2/c^2)\,\varepsilon(\omega)E_{x,y} = 0$$

and at $n^2 = \varepsilon = 0$, the field is

$$E = E^{(1)}_{0,x,y} + E^{(2)}_{0,x,y}z\,.$$

However, the roots merge for waves propagated (at $n \neq 0$) in different directions. Besides, if absorption is taken into account, $\varepsilon = \varepsilon' + \mathrm{i}\varepsilon'' = \tilde{n}^2 \neq 0$.

concrete problem of a wave coming from the region $z<0$ and normally incident on a medium occupying the half-space $z \geqslant 0$. Then the solution (2.2.54) has the form $\boldsymbol{E}_2 = \boldsymbol{E}_{0,2} z \exp\{i[\omega \tilde{n} z/c - \omega t]\}$ and if $\tilde{n}$ is real it increases with the depth of penetration. This is, however, impossible since in the region $z \geqslant 0$ there are, by assumption, no sources of field. At the same time the assumption that the $\boldsymbol{E}_2$ wave is not at all excited on the boundary is contradictory to the boundary conditions. In the case of absorption, however, $\tilde{n} = n + i\varkappa$, and without spatial dispersion $\varkappa > 0$ (if $n > 0$, see Sect. 2.3.3). This is why (2.2.54) contains the factor $\exp[-(\omega/c)\varkappa z]$ which, for sufficiently large z, leads always to a reduction of the amplitudes of solutions of the type $\boldsymbol{E}_2$ despite the presence of z as a factor.

When spatial dispersion is taken into account new roots can exist besides the two roots (solutions) of the dispersion equation already considered. This provides the possibility for new essential multiple roots to appear, even in the absence of absorption. This problem will be dealt with in Sects. 4.1, 2. Here, it should be noted that under all conditions known to us we need to deal only with double roots; this fact is simply due to the weakness of spatial dispersion in crystal optics. When considering waves which are proportional to the factor $\exp[i(\omega \tilde{n} \boldsymbol{s} \cdot \boldsymbol{r}/c - \omega t)]$ it is sufficient for double roots to consider new solutions of the type (2.2.54).

The fact that in crystal optics we sometimes have to use solutions of the type (2.2.54) is quite interesting and relatively unknown although such a possibility was mentioned a long time ago [2.26]. We have therefore discussed this problem in detail. Moreover, we should stress that singular optical axes are encountered only in exceptional cases since they appear only for special values of the parameters, including the frequency. We particularly have in mind real values of the frequency which are the only ones of interest in optical problems. In the case of complex frequencies the appearance of multiple roots is more probable under certain conditions and this is important when obtaining the dispersion relations for $\tilde{n}$ (Sect. 2.2.5). Note that solutions of the type (2.2.54) are met with not only in the case of singular optical axes but also when considering inhomogeneous waves [2.27].

In conclusion, we make one remark of a general nature. When we say that a solution of the type $\boldsymbol{E}_0 \exp[i(\boldsymbol{k} \cdot \boldsymbol{r} - \omega t)]$ forms a complete set of solutions (excluding the case of the essential multiple root), then, of course, we have in mind the set of homogeneous equations (2.2.1) and a medium homogeneous with respect to both space and time. Though it is known that this approach is very fruitful and justified, it is obviously limited. If, however, initial or boundary conditions are to be taken into account the solution of the problem does not necessarily lead to monochromatic waves. Near the boundaries or immediately after the beginning of the process other solutions also exist; only asymptotically, when the field "forgets" the initial or boundary conditions, can we restrict ourselves to considering monochromatic waves. We shall not take the influence of initial conditions into account [Ref. 2.1, § 5], the problem of the role of the boundaries will, however, be discussed in Sect. 4.5.

2.2.4 Separating the Transverse Field $E_\perp$ and the Tensor $\varepsilon_{\perp,ij}$

Normal waves in an anisotropic medium are, in general, neither longitudinal nor transverse. Hence, any division of the field into a transverse and a longitudinal component is irrelevant in investigating most problems of the electrodynamics and optics of continuous anisotropic media. For long wavelengths, moreover, the physical difference between transverse and longitudinal fields is of a secondary importance, since if $k \to 0$ (infinitely long wavelength) the concept of the polarization of waves completely loses its meaning. There is therefore no necessity whatsoever for such a division. On the other hand, a division of the field into a longitudinal and a transverse component has a distinct mathematical and physical meaning and it is sometimes convenient for an anisotropic medium as well. As this division appears, in one way or another, rather frequently in the literature, we shall discuss it.

The electric field $\boldsymbol{E}$ is divided into a longitudinal ($\boldsymbol{E}_\parallel$) and a transverse ($\boldsymbol{E}_\perp$) part as follows:

$$\boldsymbol{E} = \boldsymbol{E}_\parallel + \boldsymbol{E}_\perp , \qquad \mathrm{rot}\,\boldsymbol{E}_\parallel = 0 , \qquad \mathrm{div}\,\boldsymbol{E}_\perp = 0 \tag{2.2.55}$$

or, in the case of monochromatic homogeneous plane waves,

$$\begin{aligned} &\boldsymbol{E}(\omega,\boldsymbol{k}) = \boldsymbol{E}_\parallel(\omega,\boldsymbol{k}) + \boldsymbol{E}_\perp(\omega,\boldsymbol{k}) , \\ &\boldsymbol{E}_\parallel = E_\parallel \boldsymbol{s} = (\boldsymbol{E}\cdot\boldsymbol{s})\cdot\boldsymbol{s} = (\boldsymbol{E}_\parallel\cdot\boldsymbol{s})\cdot\boldsymbol{s} , \qquad \boldsymbol{s}\cdot\boldsymbol{E}_\perp = 0 . \end{aligned} \tag{2.2.56}$$

The magnetic induction $\boldsymbol{B}$ is always transverse ($\mathrm{div}\,\boldsymbol{B} = 0$). The electrical induction $\boldsymbol{D}$ is transverse only if there are no external sources, so that $\mathrm{div}\,\boldsymbol{D} = 0$, i.e., $\boldsymbol{s}\cdot\boldsymbol{D} = 0$. Particularly in the latter case it is sometimes tempting to eliminate the field $\boldsymbol{E}_\parallel$. In the case of homogeneous waves in a medium we have

$$\boldsymbol{s}\cdot\boldsymbol{D} = \varepsilon_{ij}s_iE_j = \varepsilon_{ij}s_iE_{\parallel,j} + \varepsilon_{ij}s_iE_{\perp,j} = 0 . \tag{2.2.57}$$

By virtue of (2.2.56, 57),

$$\varepsilon_{ij}s_iE_{\parallel,j} = \varepsilon_{ij}s_is_j(s_lE_l) = \varepsilon_{ij}s_is_jE_\parallel$$

and

$$E_{\parallel,i} = -\frac{\varepsilon_{lj}s_lE_{\perp,j}}{\varepsilon_{rt}s_rs_t}s_i , \qquad E_i = E_{\perp,i} - \frac{\varepsilon_{lj}s_lE_{\perp,j}}{\varepsilon_{rt}s_rs_t}s_i , \tag{2.2.57a}$$

where $\varepsilon_{ij} = \varepsilon_{ij}(\omega,\boldsymbol{k})$, and $\boldsymbol{E}$, $\boldsymbol{E}_\parallel$ and $\boldsymbol{E}_\perp$ denote the Fourier transforms $\boldsymbol{E}(\omega,\boldsymbol{k})$, etc.

With the aid of (2.2.57), the wave equation can be written for only two transverse components of the field $\boldsymbol{E}$, and this is convenient, for example, when investigating the propagation of waves in an inhomogeneous medium whose tensor ε_{ij} is given, for example, in a magnetoactive plasma [2.2].

In the general case, however, the elimination of the field $\boldsymbol{E}_{\parallel}$ leads only to unjustified complications. This can be illustrated by using the introduction and application of the tensor $\varepsilon_{\perp,ij}(\omega,\boldsymbol{k})$ as an example. Its determination involves the possibility of establishing the following relation for fields of the type (2.2.57):

$$D_i = \varepsilon_{ij}E_{\parallel,j} + \varepsilon_{ij}E_{\perp,j} = \left(\varepsilon_{ij} - \frac{\varepsilon_{il}\varepsilon_{mj}s_l s_m}{\varepsilon_{rt}s_r s_t}\right)E_{\perp,j}$$

$$\equiv \varepsilon_{\perp,ij}E_{\perp,j} = E_i + 4\pi\chi_{\perp,ij}E_{\perp,j}, \tag{2.2.58}$$

$$\varepsilon_{\perp,ij} = \varepsilon_{ij} - \frac{\varepsilon_{il}\varepsilon_{mj}s_l s_m}{\varepsilon_{rt}s_r s_t},$$

where we introduce another quantity, $\chi_{\perp,ij}$:

$$4\pi\chi_{\perp,ij} = \varepsilon_{ij} - \delta_{ij} - \frac{\varepsilon_{il}\varepsilon_{mj}s_l s_m - \varepsilon_{lj}s_l s_i}{\varepsilon_{rt}s_r s_t}$$

$$= \varepsilon_{\perp,ij} - \delta_{ij} + \frac{\varepsilon_{lj}s_l s_i}{\varepsilon_{rt}s_r s_t}, \tag{2.2.59}$$

$$P_i = \frac{D_i - E_i}{4\pi} = \chi_{\perp,ij}E_{\perp,j}.$$

Evidently,

$$\varepsilon_{\perp,ij}s_j = 0, \qquad \varepsilon_{\perp,il}\eta_{lj} = \varepsilon_{\perp,ij},$$
$$E_{\perp,i} = \eta_{ij}E_j, \qquad \chi_{\perp,ij}s_j = 0, \tag{2.2.60}$$

where the projection tensor is

$$\eta_{ij} = \delta_{ij} - s_i s_j. \tag{2.2.61}$$

By virtue of (2.2.58 – 61) we can write

$$D_i = \varepsilon_{\perp,ij}E_{\perp,j} = \varepsilon_{\perp,ij}E_j = (\delta_{ij} + 4\pi\chi_{\perp,il}\eta_{lj})E_j = (\delta_{ij} + 4\pi\chi_{\perp,ij})E_j. \tag{2.2.62}$$

We should remember, of course, that these relations apply only to fields of the type (2.2.57) for which $\varepsilon_{lj}s_l E_j = 0$. In such fields the role of ε_{ij} is thus played by the tensor

$$\tilde{\varepsilon}_{ij} = \delta_{ij} + 4\pi\chi_{\perp,il}\eta_{lj}, \qquad D_i = \tilde{\varepsilon}_{ij}E_j, \tag{2.2.63}$$

where the tilde over ε_{ij} indicates the limitation of the class of fields.

Instead of (2.2.63) used in [2.28, 30], the relations

$$\tilde{\tilde{\varepsilon}}_{ij} = \delta_{ij} + 4\pi\chi_{\perp,il}\left(\delta_{lj} + \frac{4\pi\chi_{\perp,mj}s_l s_m}{1-4\pi\chi_{\perp,rt}s_r s_l}\right), \qquad P_i = \chi_{\perp,ij}E_{\perp,j} \tag{2.2.64}$$

were given in [2.29]. They are obtained if (2.2.27) is written in the form, (2.2.56, 59),

$$s_i D_i = s_i E_i + 4\pi s_i P_i = s_i E_{\parallel,i} + 4\pi\chi_{\perp,ij}s_i E_{\perp,j} = 0 \quad \text{or} \tag{2.2.65}$$

$$E_{\parallel,i} = E_i - E_{\perp,i} = -4\pi\chi_{\perp,lj}s_l s_i E_{\perp,j}, \qquad E_i = (\delta_{ij} - 4\pi\chi_{\perp,lj}s_l s_i)E_{\perp,j}. \tag{2.2.66}$$

Further,

$$\begin{aligned}\chi_{\perp,lj}s_l s_i E_{\perp,j} &= \chi_{\perp,lj}s_l s_i E_j - s_i(\chi_{\perp,lm}s_l s_m)(s_j E_j)\\ &= [\chi_{\perp,lj}s_l s_i - \delta_{ij}(\chi_{\perp,lm}s_l s_m)]E_j + (\chi_{\perp,lm}s_l s_m)E_{\perp,i}.\end{aligned}$$

Combining this with (2.2.66) we obtain

$$E_{\perp,i} = \eta_{ij}E_j = \left(\delta_{ij} + \frac{4\pi\chi_{\perp,lj}s_l s_i}{1-4\pi\chi_{\perp,rt}s_r s_t}\right)E_j \tag{2.2.67}$$

and, after substituting and reindexing, we have

$$D_i = \delta_{ij}E_j + 4\pi\chi_{\perp,ij}E_{\perp,j} = \tilde{\tilde{\varepsilon}}_{ij}E_j, \tag{2.2.68}$$

where $\tilde{\tilde{\varepsilon}}_{ij}$ is determined by (2.2.64). At the same time it is clear from (2.2.67) that under the adopted conditions the tensor

$$\delta_{ij} + \frac{4\pi\chi_{\perp,lj}s_l s_i}{1-4\pi\chi_{\perp,rt}s_r s_t}$$

is equivalent to the projection tensor $\eta_{ij} = \delta_{ij} - s_i s_j$.

This can be readily proved, directly as well, if we apply both tensors to the field

$$E_i = (\delta_{ij} - 4\pi\chi_{\perp,lj}s_l s_i)E_{\perp,j},$$

which satisfies the initial conditions (2.2.57, 65).

Hence, $\tilde{\tilde{\varepsilon}}_{ij} = \tilde{\varepsilon}_{ij}$ if applied to fields of the type (2.2.57, 65), see (2.2.63, 64). Moreover, if $\chi_{\perp,ij}$ in (2.2.64) is defined as in (2.2.59), then $\chi_{\perp,ij}s_j = 0$ and $\chi_{\perp,ij} = \chi_{\perp,il}\eta_{lj}$, see (2.2.60). In this case, evidently,

$$\tilde{\tilde{\varepsilon}}_{ij} = \delta_{ij} + 4\pi\chi_{\perp,ij} = \delta_{ij} + 4\pi\chi_{\perp,il}\eta_{lj} = \tilde{\varepsilon}_{ij}\,.$$

Therefore, the introduction of the tensor $\tilde{\tilde{\varepsilon}}_{ij}$ has some meaning only if $\chi_{\perp,ij}$ is defined by an expression different from (2.2.59). Such a possibility exists because the tensor

$$\chi'_{\perp,ij} = \chi_{\perp,il} + a(\omega,\boldsymbol{k})s_i s_j\,, \qquad \chi_{\perp,ij}s_i = 0\,, \tag{2.2.64a}$$

when applied to $\boldsymbol{E}_\perp$ yields the same result as the tensor $\chi_{\perp,ij}$. The tensor $\tilde{\varepsilon}_{ij}$, of course, is the same for both $\chi_{\perp,ij}$ and $\chi'_{\perp,ij}$. As we shall see in Sect. 5.1, it is actually possible in the microscopic theory to obtain $\chi'_{\perp,ij}$ different from $\chi_{\perp,ij}$. We know, however, of no basis for choosing one or another value of $\chi'_{\perp,ij}$ in (2.2.64a), i.e., for a unique choice of the scalar function $a(\omega,\boldsymbol{k})$. As a matter of fact the tensor $\tilde{\varepsilon}_{ij}$ is independent of the form of function a for fields of the class (2.2.57, 65). For general fields, with $\boldsymbol{s}\cdot\boldsymbol{D} \neq 0$, no reasons are evident for preferring one or another expression for $a(\omega,\boldsymbol{k})$. In other words, the tensor $\tilde{\varepsilon}_{ij}$ acts in some subspace of the vectors $\boldsymbol{E}$ and $\boldsymbol{D}$, and its generalization for the entire vector space is ambiguous or, in any case, requires additional considerations. On the contrary, if ε_{ij} is known, we can obtain a unique expression for $\varepsilon_{\perp,ij}$ and $\tilde{\varepsilon}_{ij}$, see (2.2.58, 63). Hence, there is no basis, to our knowledge, for applying the tensor $\tilde{\varepsilon}_{ij}$ or $\tilde{\tilde{\varepsilon}}_{ij}$ in a meaningful way to fields of a more general nature than a field with $\boldsymbol{s}\cdot\boldsymbol{D} = 0$. To avoid confusion we shall not employ the notation $\tilde{\varepsilon}_{ij}$ and $\tilde{\varepsilon}_{ij}$ and $\tilde{\tilde{\varepsilon}}_{ij}$ at all, and confine ourselves to the introduction of the tensors $\varepsilon_{\perp,ij}$ and $\chi_{\perp,ij}$.

The tensors $\varepsilon_{\perp,ij}(\omega,\boldsymbol{k})$ and $\chi_{\perp,ij}(\omega,\boldsymbol{k})$ enable $\boldsymbol{D}$ and $\boldsymbol{P}$ to be expressed in terms of the transverse field $\boldsymbol{E}_\perp$, but only for fields of the type (2.2.57), i.e., fields whose induction $\boldsymbol{D}$ is transverse. It is thereby evident that the tensors $\varepsilon_{\perp,ij}$ and $\chi_{\perp,ij}$ are of a less general nature than the tensor $\varepsilon_{ij} = \delta_{ij} + 4\pi\chi_{ij}$ which relates $\boldsymbol{D}$ and $\boldsymbol{E}$ for any fields.

Another factor is even more important: the comparatively complex structure of the tensors $\varepsilon_{\perp,ij}$ and $\chi_{\perp,ij}$, and the possibility that they may possess singularities. Indeed, assume the tensor $\varepsilon_{ij}(\omega,\boldsymbol{k})$ to be a regular function of its variables. The tensors $\varepsilon_{\perp,ij}$ and $\chi_{\perp,ij}$ will nevertheless possess singularities (poles) at values of ω and $\boldsymbol{k}$ that satisfy the condition $\varepsilon_{rt}s_r s_t = 0$, i.e., the condition (2.2.37) for the appearance of "fictitious" longitudinal waves. This is to be expected because "fictitious" longitudinal waves have the properties $\boldsymbol{E} = \boldsymbol{E}_\parallel$ and $\boldsymbol{D} \neq 0$ which is possible for a finite field $\boldsymbol{E}_\perp$ only if $\varepsilon_{\perp,ij} \to \infty$. Condition (2.2.37) may be satisfied even in the case of no spatial dispersion and no absorption. Finally, $\varepsilon_{\perp,ij}$ and $\chi_{\perp,ij}$ depend in general on $\boldsymbol{s}$ if $\boldsymbol{k}\to 0$, i.e., they are nonanalytic functions for $\boldsymbol{k}$ if $\boldsymbol{k}\to 0$. This makes the application of the tensors $\varepsilon_{\perp,ij}$ and $\chi_{\perp,ij}$ in place of ε_{ij} and χ_{ij} unreasonable, at any rate in the case of general investigations.

2.2.5 The Dispersion Relations for a Complex Refractive Index. Inequalities for the Region of Transparency

In Sect. 2.1.2 we considered the dispersion relations for the components of the tensor $\varepsilon_{ij}(\omega,\boldsymbol{k})$. In optics, however, it is not ε_{ij} but the complex index of refraction $\tilde{n}_l(\omega,\boldsymbol{s}) = n_l(\omega,\boldsymbol{s}) + \mathrm{i}\varkappa_l(\omega,\boldsymbol{s})$ that is determined directly for normal waves of various types (subscript l). This poses the problem [2.11, 32] of obtaining dispersion relations for $\tilde{n}$.

In the case of an isotropic medium without spatial dispersion

$$\varepsilon_{ij}(\omega) = \varepsilon(\omega)\delta_{ij}\,, \qquad \tilde{n}^2 = \varepsilon(\omega)\,, \qquad \tilde{n}^2 - \varkappa^2 = \varepsilon'(\omega)\,, \qquad 2n\varkappa = \varepsilon''(\omega)\,, \tag{2.2.69}$$

and the dispersion relations (2.1.44) assume the form

$$\begin{aligned} n^2(\omega) - \varkappa^2(\omega) - 1 &= \frac{4}{\pi}\,⨍_0^\infty \frac{x\,n(x)\varkappa(x)\,dx}{x^2-\omega^2}\,, \\ n(\omega)\varkappa(\omega) &= -\frac{\omega}{\pi}\,⨍_0^\infty \frac{n^2(\pi) - \varkappa^2(x) - 1}{x^2-\omega^2}\,dx\,. \end{aligned} \tag{2.2.70}$$

In this case the transition from ε_{ij} to $\tilde{n}^2$ occurs, so to speak, automatically. However, in an anisotropic medium this is not the case and one cannot conclude in advance that the relations (2.2.70) could be written for n_l and $\varkappa_l$. As a matter of fact, as we shall see, this cannot always be done, not even in the absence of spatial dispersion. Let us consider this problem in more detail [2.11].

The dispersion relations (2.2.70) can be written for a certain function $f(\omega) = u + \mathrm{i}v$, provided this function does not possess any singularities in the upper half-plane and on the real axis of the complex variable ω. If $f = \varepsilon_{ij}$, the absence of singularities results from very general considerations dealt with in Sect. 2.1.2. If we assume, however, that $\varepsilon(\omega)$ has no singularities, then this will also be true for the function $\varepsilon^2(\omega)$; however, the dispersion relations for ε^2 and for analogous functions of ε do not yield any new aspects compared with those for ε. Even in the case of the function $\sqrt{\varepsilon(\omega)} = \tilde{n}(\omega)$, it is not certain that the dispersion relations can be written since a branch point may appear. If the medium is in thermodynamic equilibrium, $\varepsilon(\omega)$ has no zeros in the upper half-plane and none on the real axis ([2.7] and Sect. 2.1.2). Hence, $\sqrt{\varepsilon(\omega)}$ has no branch point in this region and we can write the dispersion relations for $\tilde{n} = n + \mathrm{i}\varkappa$ as for $\tilde{n}^2$. They have the form of (2.2.70), $n^2 - \varkappa^2$ being replaced by n, and $2n\varkappa$ by $\varkappa$.

Consequently, one sees that the dispersion relations (2.2.70) are valid for $\tilde{n}_l^2(\omega,\boldsymbol{s})$ if this function has no singularities for $\omega = \omega' + \mathrm{i}\omega''$ and $\omega'' \geqslant 0$.

The function $\tilde{n}_l(\omega,\boldsymbol{s})$ is a root of the dispersion equation (2.2.22). In the presence of spatial dispersion in (2.2.22), $\varepsilon_{ij} = \varepsilon_{ij}[\omega, (\omega/c)\tilde{n}(\omega,\boldsymbol{s})\boldsymbol{s}]$ and the roots of this equation may have, in principle, various singularities. We cannot obtain any result unless we assume the type of dependence of ε_{ij} on $\tilde{n}$. It is thereby clear that when spatial dispersion is taken into account we cannot use dispersion relations of the form of (2.2.70) without resorting to a special analysis.

Within the scope of classical crystal optics the problem is much simpler since ε_{ij} in (2.2.22) is independent of $\tilde{n}$, and this equation is then a simple quadratic equation of $\tilde{n}^2$. Hence we can find out whether the roots of these equations possess singularities such as poles or branch points.

If ε_{ij} has no singularities, then the coefficients of (2.2.22) must be analytic. The root $\tilde{n}^2$ may thus become infinite only if the coefficient of $\tilde{n}^4$ becomes zero, i.e., under the well-known condition $\varepsilon_{ij}s_is_j = 0$, see (2.2.37, 38). According to (2.1.35) this condition does not hold for a medium in equilibrium without spatial dispersion (if $\omega'' \geqslant 0$). Therefore $\tilde{n}^2$ may possess only branch-point singularities. This is what happens when an essential multiple root $\tilde{n}^2$ exists (Sect. 2.2.4). With the exception of the lowest systems (rhombic, monoclinic and triclinic), essential multiple roots cannot occur when considering crystals. Thus, for all crystals not belonging to the lowest systems, i.e., for optically isotropic and uniaxial crystals, the dispersion relations (2.2.70) apply to either of the normal waves 1 and 2 (ordinary and extraordinary waves). In the case of crystals of the lowest systems (optically biaxial crystals) the validity of these dispersion relations is not guaranteed, even in the general case, since singular optical axes may appear in accordance with the essential multiple roots. It should be emphasized here that this holds for the case of possible multiple roots appearing in the half-plane $\omega'' \geqslant 0$, and not only on the real axis $\omega'' = 0$. It is therefore practically impossible to exclude the presence of a multiple root from the analysis of experimental data[9].

An essential multiple root certainly appears for $\omega'' \geqslant 0$ in the case of a magnetoactive plasma for a large range of values of the parameters (electron concentration and the like) and plays an appreciable role in investigating waves in a nonhomogeneous medium, which are propagated at a small angle with respect to the direction of the magnetic field ([2.11] and [Ref. 2.2, § 28]).

Dispersion relations can be written even when a multiple root is present, but only with integration along a contour that goes around the branch point [2.11]. However, these relations are not of any real interest.

When we take spatial dispersion into account, the probability of $\tilde{n}^2$ possessing singularities increases to a considerable degree (this follows, in particular, from remarks made in Sects. 2.1.2, 4). At the same time, one should

---

[9] It is possible, in principle, to determine the function $\tilde{n}(\omega, s)$ experimentally for complex frequencies ω as well, for which we use a field of the type $E = E_0 = E_0 \exp(\omega'' t) \exp(-\mathrm{i}\omega' t)$ or, more correctly, a pulse whose shape is close to the indicated one for a sufficiently long time interval t. However, such measurements cannot feasibly be performed over a sufficiently large range of values of the variable ω.

not draw the conclusion that the relations (2.2.70) do not apply or are inaccurate in the case of spatial dispersion. On the contrary, these relationships can be retained for some normal waves. Thus, in the case of weak spatial dispersion, which is our sole interest here, the relations (2.2.70) with $n = n_{1,2}$ and $\varkappa = \varkappa_{1,2}$ may turn out to be fully applicable to the two normal waves 1 and 2, which exist without spatial dispersion as well.

Within this framework the experimental proof of relations (2.2.70) for some normal waves of type l verifies the absence of (or, more precisely, the small contribution made by) the singularities of $\tilde{n}_l^2(\omega, s)$ if $\omega'' \geqslant 0$. Conversely, a violation of (2.2.70) indicates the presence of such singularities. It does not seem, however, that any great promise is offered by verifying (2.2.70) since these are integral relations and, because the accuracy of measurements is inevitably limited, these relations are usually not very sensitive. This problem, however, will be touched on again in Sect. 4.7.

Let us turn now to another problem and discuss some of the inequalities resulting from the dispersion relations for $\varepsilon_{ij}(\omega, \boldsymbol{k})$ in the region of transparency.

The inequalities considered in the following for an anisotropic nongyrotropic medium [2.33] are generalizations of the results which are known for isotropic media without spatial dispersion [Ref. 2.7, § 64]. In the latter case

$$\frac{\partial \varepsilon(\omega)}{\partial \omega} > 0\,, \qquad \frac{\partial \varepsilon(\omega)}{\partial \omega} > \frac{2[1-\varepsilon(\omega)]}{\omega}\,, \qquad \varepsilon(\omega) = \varepsilon'(\omega)\,, \qquad \varepsilon''(\omega) = 0\,. \tag{2.2.71}$$

From (2.2.71) we obtain the following inequality for the group velocity

$$v_{\mathrm{gr}} = \frac{d\omega}{dk} = \frac{c}{d(\omega n)/d\omega} < c\,. \tag{2.2.72}$$

These inequalities are proved and discussed for a more general case in what immediately follows. We have used the inequality sign in (2.2.71, 72) and the following, although in the limiting case of empty space the equality signs would apply, i.e., we could use the $\geqslant$ and $\leqslant$ signs instead.

In the region of transparency, absorption may be neglected by definition, and the vector $\boldsymbol{k} = (\omega/c)\tilde{n}(\omega, s)\boldsymbol{s}$ is real for the waves being considered. Under similar conditions $\mathrm{Im}\{\varepsilon_{ij}(x, \boldsymbol{k})\} = 0$ at least for the components of ε_{ij} that we are interested in. As is evident from the dispersion equation (2.2.22), the root $\tilde{n}$ of this equation can be real only if the $\varepsilon_{ij}[\omega(\omega/c)\tilde{n}\boldsymbol{s}]$ determining the value of $\tilde{n}$ are real.

Let us now consider the first of the dispersion relations (2.1.27)

$$\mathrm{Re}\{\varepsilon_{ij}(\omega, \boldsymbol{k})\} - \delta_{ij} = \frac{1}{\pi} \fint_{-\infty}^{+\infty} \frac{\mathrm{Im}\{\varepsilon_{ij}(x, \boldsymbol{k})\}}{x - \omega}\, dx\,. \tag{2.2.73}$$

Since $\mathrm{Im}\{\varepsilon_{ij}(\omega,\boldsymbol{k})\}=0$ for the frequencies ω corresponding to the region of transparency, the fact that the denominator in the integrand of (2.2.73) vanishes cannot be significant, i.e., in the region of transparency this integral cannot be improper. Strictly speaking, this argument is inaccurate since there is always some absorption (though it may be very small) and this fact was used in Sect. 2.1.2 in order to obtain a series of relations. On the other hand, the contribution from the region of transparency to the integral of (2.2.73) must in any case be very small. The purpose of these considerations was only to prove that the integral of (2.2.73) can be differentiated with respect to the parameter ω, the angular frequency, in the region of transparency. Proceeding in this manner, which is quite legitimate [as follows from the above and from the nature of the results, one of which is the inequality (2.2.72)], we have

$$\frac{\partial}{\partial\omega}\mathrm{Re}\{\varepsilon_{ij}(\omega,\boldsymbol{k})\}=\frac{1}{\pi}\fint_{-\infty}^{+\infty}\frac{\mathrm{Im}\{\varepsilon_{ij}(x,\boldsymbol{k})\}}{(x-\omega)^2}dx\,. \tag{2.2.74}$$

Analogously, in the case of a nongyrotropic transparent medium, we differentiate the first of the relations (2.1.27a) and obtain

$$\frac{\partial}{\partial\omega}\varepsilon'_{ij}(\omega,\boldsymbol{k})=\frac{4\omega}{\pi}\fint_{0}^{\infty}\frac{x\varepsilon''_{ij}(x,\boldsymbol{k})}{(x^2-\omega^2)^2}dx\,. \tag{2.2.75}$$

If a medium in equilibrium is subject to the action of a field, heat must necessarily be evolved and not absorbed. In an arbitrary monochromatic field [ω and $\boldsymbol{k}$ being real, see (2.1.29)]

$$q=\frac{\omega}{8\pi}\varepsilon''_{ij}(\omega,\boldsymbol{k})E_{0j}E^*_{0i}>0\,. \tag{2.2.76}$$

In the case under consideration the tensor ε''_{ij} is both real and symmetric, see (2.1.22a). However, we need pay no attention here to this fact since we can show that for a tensor ε''_{ij} the quadratic form, (2.2.76) can be reduced to the form $q=\lambda_1\,|E_1|^2+\lambda_2\,|E_2|^2+\lambda_3\,|E_3|^2>0$. Hence it follows that all the coefficients $\lambda_i>0$ (the field $\boldsymbol{E}$ being arbitrary). Introducing the determinant $D(\lambda)=|\varepsilon''_{ij}-\lambda\delta_{ij}|$ (for simplicity, we assume $\omega>0$) we can formulate the condition (2.2.76) as the requirement for all roots of the equation $D(\lambda)=0$ to be positive. Hence we obtain the following necessary and sufficient conditions that lead to (2.2.76):

$$D(\lambda=0)=|\varepsilon''_{ij}|>0\,,\qquad M_{ii}>0\,,\qquad \varepsilon''_{ii}>0\,, \tag{2.2.77}$$

where M_{ij} is a minor of the element ε''_{ij} in the determinant $|\varepsilon''_{ij}|$. Instead of the inequalities (2.2.77) we could also have used other equivalent conditions, e.g., the inequalities

$$\varepsilon''_{11} > 0 , \quad \begin{vmatrix} \varepsilon''_{11} & \varepsilon''_{12} \\ \varepsilon''_{21} & \varepsilon''_{22} \end{vmatrix} > 0 , \quad |\varepsilon''_{ij}| > 0 ,$$

which are frequently applied.

Using the last inequality from (2.2.77) together with (2.2.75) we obtain

$$\frac{\partial}{\partial \omega} \varepsilon'_{ii}(\omega, \boldsymbol{k}) > 0 , \tag{2.2.78}$$

since after adding the diagonal elements the integrand of (2.2.75) is always positive.

Similarly, we obtain still another inequality if the first of the relations (2.1.27a) is multiplied by ω^2 and then differentiated with respect to ω. As a result we obtain

$$\frac{\partial}{\partial \omega} [\omega^2(\varepsilon'_{ii} - \delta_{ii})] > 0 \qquad \text{or}$$

$$\frac{\partial \varepsilon'_{ii}(\omega, \boldsymbol{k})}{\partial \omega} > \frac{2[3 - \varepsilon'_{ii}(\omega, \boldsymbol{k})]}{\omega} . \tag{2.2.79}$$

If we use the first two of the inequalities (2.2.77) together with (2.2.75), we arrive, in the general case, at cumbersome relationships. Let us restrict ourselves to one remark. Under particular conditions (in the absence of spatial dispersion this applies to all kinds of crystals except triclinic and monoclinic ones) the principal axes of the tensors ε'_{ij} and ε''_{ij} coincide for all frequencies; these tensors are real and thus, in a certain coordinate system, they are diagonalizable. Under these conditions, from (2.2.75), from an analogous expression for $(\partial/\partial\omega)[\omega^2(\varepsilon'_{ij} - \delta_{ij})]$, and the inequalities (2.2.77), the inequality (2.2.71) follows for each of the principal values of ε_1, ε_2 and ε_3 of the tensor $\varepsilon_{ij} = \varepsilon'_{ij}$.

For an isotropic medium without spatial dispersion and without absorption $\varepsilon_{ij} = \varepsilon(\omega)\delta_{ij} = \varepsilon'(\omega)\delta_{ij}$ and the inequalities (2.2.78, 79) go over to the inequality (2.2.71). In this case the refractive index of transverse waves is $n^2(\omega) = \varepsilon(\omega)$ and therefore $2n(dn/d\omega) = d\varepsilon/d\omega$. Hence, instead of (2.2.71) we can write

$$\frac{dn}{d\omega} > 0 , \qquad \omega \frac{dn}{d\omega} > \left(\frac{1}{n} - n\right) . \tag{2.2.71a}$$

Here we have already taken into account the fact that the refractive index n can be taken as positive, in so far as we consider waves propagated in the positive direction (if $\omega > 0$ and the waves are propagated along the z axis, the field is proportional to $\exp[i(\omega nz/c - \omega t)]$). If $n > 1$, by virtue of the first inequality of (2.2.71a) $d\omega n/d\omega = n + \omega(dn/d\omega) > 1$; if however $0 < n < 1$, then from the second inequality of (2.2.71a) it again follows that $d\omega n/d\omega > 1$. Hence inequality (2.2.72) is obtained directly.

The conclusion that the group velocity $v_{gr} = d\omega/dk = c/(d\omega n/d\omega)$ of the waves does not exceed the velocity c of light in empty space indicates only the possibility of consistently applying the expression $d\omega/dk$ as the velocity of signal propagation in a real transparent medium. The following question may arise: the dispersion relations are obtained by taking the principle of causality into account, but they are by no means related to the requirement of relativistic invariance. Why then is a limitation on the speed of the signal obtained? In this case we have also made use of the dispersion equation $n^2 = \varepsilon$ which is equivalent to taking the field equations into consideration. The field equations are relativistically invariant. Together with the demand for satisfying the principle of causality this is exactly what leads to the inequality $v_{gr} \leqslant c$.

The opinion that the limitation of the speed of any signal by c follows only from the requirement of relativistic invariance is a rather widespread error. It is easy to construct relativistically invariant equations according to which the signal could have any speed. The equations of the electrodynamics of continuous media can be taken as an example of this; they are, as is well known, relativistically invariant or, better, covariant. We can consider them for an isotropic nondispersive medium with $0 < \varepsilon < 1$. Here $n = \sqrt{\varepsilon} < 1$ and $v_{gr} = v_{ph} = c/n > c$, v_{ph} being the phase velocity of the waves.

Such a medium, however, cannot exist precisely because a signal could be propagated in it at a speed higher than that of light in empty space, which, by virtue of the relativistic law of addition of velocities, leads to a violation of the principle of causality (the reference system could be then be chosen in such a way that the effect is prior to its cause ([2.34] and [Ref. 2.35, § 6]). So, both relativistic invariance and the principle of causality are required to show that $v_{gr} \leqslant c$.

In an anisotropic medium, even without spatial dispersion, it is not so simple to go from the inequalities (2.2.78, 79) and similar ones to the inequality $v_{gr} \leqslant c$. In the case of uniaxial crystals inequality (2.2.72) is proved in [2.33]. It is undoubtedly possible to obtain a more general result as well, but this has not yet been done and is not really of any particular interest. The problem of the group-velocity vector and certain of its properties will be discussed in Sect. 2.3.2.

2.3 Energy Relations and Other Equations for Waves in an Anisotropic Medium

2.3.1 The Law of Energy Conservation in the Electrodynamics of Media Displaying Spatial Dispersion

When discussing energy propagation in electrodynamics attention is focussed on finding Poynting's theorem. In order to obtain it we multiply the first of the equations (2.1.1) by $\boldsymbol{E}$ and the third by $\boldsymbol{B}$, and then subtract one from the other. Applying the identity $\operatorname{div}(\boldsymbol{E}\times\boldsymbol{B}) = \boldsymbol{B}\operatorname{rot}\boldsymbol{E} - \boldsymbol{E}\operatorname{rot}\boldsymbol{B}$ we arrive at

$$\frac{1}{4\pi}\boldsymbol{E}\frac{\partial \boldsymbol{D}}{\partial t} + \frac{1}{4\pi}\boldsymbol{B}\frac{\partial \boldsymbol{B}}{\partial t} = -\frac{c}{4\pi}\operatorname{div}(\boldsymbol{E}\times\boldsymbol{B}) - \boldsymbol{j}_{\text{ext}}\cdot\boldsymbol{E}\,. \tag{2.3.1}$$

After integrating over the volume V which is bounded by the surface S we obtain

$$\frac{1}{4\pi}\int_V\left(\boldsymbol{E}\frac{\partial \boldsymbol{D}}{\partial t} + \boldsymbol{B}\frac{\partial \boldsymbol{B}}{\partial t}\right)dV = -\frac{c}{4\pi}\oint_S(\boldsymbol{E}\times\boldsymbol{B})_n dS - \int \boldsymbol{j}_{\text{ext}}\cdot\boldsymbol{E}\,dV\,. \tag{2.3.2}$$

Let us now consider a field which tends to zero at infinity so that the surface integral in (2.3.2) can be neglected as the volume increases ($V\to\infty$). Then

$$\frac{1}{4\pi}\int\left(\boldsymbol{E}\frac{\partial \boldsymbol{D}}{\partial t} + \boldsymbol{B}\frac{\partial \boldsymbol{B}}{\partial t}\right)dV$$

represents the sum of field energy variation and evolved heat per unit time[10] which is exactly equal to the power $-\int \boldsymbol{j}_{\text{ext}}\cdot\boldsymbol{E}\,dV$ of the sources.

The surface integral in (2.3.2) can actually also be neglected if $V\to\infty$, in the case of a monochromatic field:

$$\begin{aligned}\boldsymbol{E} &= \tfrac{1}{2}(\boldsymbol{E}_0\mathrm{e}^{\mathrm{i}(\boldsymbol{k}\cdot\boldsymbol{r}-\omega t)} + \boldsymbol{E}_0^*\,\mathrm{e}^{-\mathrm{i}(\boldsymbol{k}\cdot\boldsymbol{r}-\omega t)})\,,\\ \boldsymbol{B} &= \tfrac{1}{2}(\boldsymbol{B}_0\mathrm{e}^{\mathrm{i}(\boldsymbol{k}\cdot\boldsymbol{r}-\omega t)} + \boldsymbol{B}_0^*\,\mathrm{e}^{-\mathrm{i}(\boldsymbol{k}\cdot\boldsymbol{r}-\omega t)})\,,\end{aligned} \tag{2.3.3}$$

where ω and $\boldsymbol{k}$ are real. For the field (2.3.3) the surface integral does not actually vanish; however it does not grow with V, while the volume integrals do increase. Further, for the field, (2.3.3), after averaging with respect to the "high-frequency" ω (i.e., for time intervals much larger than $2\pi/\omega$) every-

[10] In an alternating field in the presence of absorption (dissipation) the energetic meaning of various terms of Poynting's theorem is, in general, ambiguous (see below). In the given case, however, we take as the only basis the fact that the work done by the external sources is performed to change the ordered motion (energy) and the random motion (heat) in the medium.

thing is effectively static, and the energy of the field does not change. This makes it clear that in the field (2.3.3) the average heat evolved per unit time and unit volume of the medium amounts to

$$q = \frac{1}{4\pi V}\overline{\int_V E\frac{\partial D}{\partial t}dV}$$

$$= \frac{\mathrm{i}\omega}{16\pi}[\varepsilon_{ij}^*(\omega,k)E_{0j}^*E_{0i} - \varepsilon_{ij}(\omega,k)E_{0j}E_{0i}^*]$$

$$= \frac{\mathrm{i}\omega}{16\pi}[\varepsilon_{ij}^*(\omega,k) - \varepsilon_{ji}(\omega,k)]E_{0j}^*E_{0i} = \frac{\omega}{8\pi}\varepsilon_{ij}''(\omega,k)E_{0j}E_{0i}^*, \tag{2.3.4}$$

where we have made use of the symmetry property, (2.1.18), and the Hermiticity of the tensors ε_{ij}' and ε_{ij}'' (Sect. 2.1.2; the reindexing from ij to ji, etc., is commonly used and will be done in the following without additional explanations). Averaging over time, indicated here by a bar, is equivalent to dropping all terms containing the factor $\exp(\pm 2\mathrm{i}\omega t)$; moreover, by virtue of the constancy of E_0 and B_0, $\partial E_0/\partial t = 0$ and $\partial B_0/\partial t = 0$. In the following, if this does not induce confusion, the average values will no longer be indicated by bars.

If ω and k are real, (2.3.4) obtained for the heat would be just the same in the absence of spatial dispersion. In the latter case ε_{ij}'' is independent of k, and therefore this vector can be taken to be complex as well.

Naturally, this poses the problem of obtaining an expression for q in the region of complex k in the presence of spatial dispersion. Moreover, the above considerations are rather special (as regards integration over the volume, etc.) and we began with them only in order to arrive at (2.3.4) in the simplest way [2.1].

Let us turn now to a much more general problem, namely, that of considering relation (2.3.1) in the case of a quasi-monochromatic field [2.36]. Strictly speaking, these are precisely the fields that we encounter in practice even if we speak of a monochromatic field. Introducing the latter is justified only if the respective results are valid for a sufficiently "long" pulse of waves as well, i.e., for a quasi-monochromatic field of the form

$$E(r,t) = \tfrac{1}{2}[E_{00}(r,t)\mathrm{e}^{\mathrm{i}(k\cdot r - \omega t)} + E_{00}^*(r,t)\mathrm{e}^{-\mathrm{i}(k^*\cdot r - \omega t)}]$$

$$= \tfrac{1}{2}[E_0(r,t)\mathrm{e}^{\mathrm{i}(k'\cdot r - \omega t)} + E_0^*(r,t)\mathrm{e}^{-\mathrm{i}(k'\cdot r - \omega t)}]$$

$$= \int [g(\tilde{\omega},\tilde{k})\mathrm{e}^{\mathrm{i}(\tilde{k}\cdot r - \tilde{\omega}t)} + g^*(\tilde{\omega},\tilde{k})\mathrm{e}^{-\mathrm{i}(\tilde{k}^*\cdot r - \tilde{\omega}t)}]\,d\tilde{\omega}\,d\tilde{k}\,. \tag{2.3.5}$$

Here $E_0 = E_{00}\exp(-k''\cdot r)$, the function $E_{00}(r,t)$ varies slowly with time for times of the order of $2\pi/\omega$ and over a path of the order of $2\pi/k'$ and $2\pi/k''$, where $k = k' + \mathrm{i}k''$. Here the frequency ω is assumed real (this corresponds to the usual statement of a problem in optics; but it would also be possible, of course, to discard this restriction[11]). The slowness of variation of $\boldsymbol{E}_{00}(\boldsymbol{r}, t)$ indicates that the function $g(\tilde{\omega}, \tilde{\boldsymbol{k}})$ has a sharp maximum near the values $\tilde{\omega} = \omega$ and $\tilde{\boldsymbol{k}} = \boldsymbol{k}$. Due to this fact it is usually possible, for example, to place the following under the integral sign:

$$\tilde{\omega}\varepsilon_{ij}(\tilde{\omega}, \tilde{\boldsymbol{k}}) = \omega(\varepsilon_{ij})_0 + \left(\frac{\partial \omega \varepsilon_{ij}}{\partial \omega}\right)_0 \Omega + \omega \left(\frac{\partial \varepsilon_{ij}}{\partial k_l}\right)_0 \cdot q_l \equiv a_{ij}, \qquad (2.3.6)$$

where $\tilde{\omega} = \omega + \Omega$, $\tilde{\boldsymbol{k}} = \boldsymbol{k} + \boldsymbol{q}$ and the subscript 0 indicates that the quantity refers to $\tilde{\omega} = \omega$ and $\tilde{\boldsymbol{k}} = \boldsymbol{k}$.

Integration with respect to $\tilde{\omega}$ in (2.3.5) is performed along the real axis from 0 to ∞, and the integral with respect to $\tilde{\boldsymbol{k}}$ has to be taken over the contour or region containing point $\boldsymbol{k}$. It is inessential for the following to specify this contour. But it must be borne in mind that the function $\varepsilon_{ij}(\tilde{\omega}, \tilde{\boldsymbol{k}})$ is assumed to exist and to be expandable in a series near the values of ω and $\boldsymbol{k}$; moreover, all the integrals being considered are assumed to exist. When calculating them it is assumed that it suffices when the terms written down in (2.3.6) are taken into account.

Obviously, taking into account the properties of (2.1.18), we have

$$\begin{aligned} \frac{\partial D_i}{\partial t} = \int [& -\mathrm{i}\tilde{\omega}\varepsilon_{ij}(\tilde{\omega}, \tilde{\boldsymbol{k}}) g_i(\tilde{\omega}, \tilde{\boldsymbol{k}}) \mathrm{e}^{\mathrm{i}(\tilde{\boldsymbol{k}} \cdot \boldsymbol{r} - \omega t)} \\ & + \mathrm{i}\tilde{\omega}\varepsilon^*_{ij}(\tilde{\omega}, \tilde{\boldsymbol{k}}) g^*_i(\tilde{\omega}, \tilde{\boldsymbol{k}}) \mathrm{e}^{-\mathrm{i}(\tilde{\boldsymbol{k}}^* \cdot \boldsymbol{r} - \tilde{\omega} t)}] \, d\tilde{\omega} \, d\tilde{\boldsymbol{k}} . \end{aligned} \qquad (2.3.7)$$

We then substitute (2.3.5 and 7) into the expression $\boldsymbol{E}(\partial \boldsymbol{D}/\partial t)$ and average over the time, the interval being large compared with $2\pi/\omega$, but small compared to the characteristic time of variation of the amplitude $\boldsymbol{E}_{00}(\boldsymbol{r}, t)$. Making use of (2.3.6) as well, we arrive at

$$\begin{aligned} \overline{\boldsymbol{E}\frac{\partial \boldsymbol{D}}{\partial t}} = \int \mathrm{e}^{-2k''r} (& -\mathrm{i} a_{ij} g_j(\Omega, \boldsymbol{q}) g^*_i(\Omega', \boldsymbol{q}') \exp\{\mathrm{i}[(\boldsymbol{q} - \boldsymbol{q}'^*) \cdot \boldsymbol{r} - (\Omega - \Omega')t]\} \\ & + \mathrm{i} a^*_{ij} g^*_j(\Omega', \boldsymbol{q}) g_i(\Omega', \boldsymbol{q}') \exp\{\mathrm{i}[(\boldsymbol{q}^* - \boldsymbol{q}') \cdot \boldsymbol{r} - (\Omega - \Omega')t]\} \\ & \times d\Omega \, d\Omega' \, d\boldsymbol{q} \, d\boldsymbol{q}' . \end{aligned} \qquad (2.3.8)$$

Next, we readily see that [2.36, 37]

$$\begin{aligned} E_{0i}E^*_{0j} = 4 \int & g_i(\Omega, q) g^*_j(\Omega', \boldsymbol{q}') \mathrm{e}^{-2k''r} \exp\{\mathrm{i}[(\boldsymbol{q} - \boldsymbol{q}'^*) \cdot \boldsymbol{r} - (\Omega - \Omega')t]\} \\ & \times d\Omega \, d\Omega' \, d\boldsymbol{q} \, d\boldsymbol{q}' , \end{aligned}$$

---

[11] Apart from this we should keep in mind that (2.3.5) does not mean that the field varies according to the law $\exp(\pm \mathrm{i}\omega t)$; on the contrary, this expression is also quite useful when, e.g., $\boldsymbol{E}_0 = \boldsymbol{E}_{00} \exp(\gamma t)$, provided that $|\gamma| \ll \omega$.

$$E_{0i}\frac{\partial E^*_{0j}}{\partial t} = 4\mathrm{i}\int \Omega' g_i(\Omega,\boldsymbol{q})g^*_j(\Omega',\boldsymbol{q}')\,\mathrm{e}^{-2\boldsymbol{k}''\cdot\boldsymbol{r}}\exp\{\mathrm{i}[(\boldsymbol{q}-\boldsymbol{q}'^*)\cdot\boldsymbol{r}-(\Omega-\Omega')t]\}$$
$$\times\, d\Omega\, d\Omega'\, d\boldsymbol{q}\, d\boldsymbol{q}'\,. \tag{2.3.9}$$

Here we shall not derive analogous expressions for $(\partial/\partial t)(E_{0i}E^*_{0j})$ and $(\partial/\partial x_i)\times(E_{0i}E^*_{0j})$ as they are readily obtained by differentiating the integral representation of the product $E_{0i}E^*_{0j}$ given in (2.3.9).

Taking (2.3.8, 9) into account, we can bring (2.3.1), after high-frequency averaging, to the form

$$\frac{1}{16\pi}\frac{\partial}{\partial t}\left[\left(\frac{\partial(\omega\varepsilon'_{ij})}{\partial\omega}\right)_0 E_{0j}E^*_{0i}+B_{0i}B^*_{0i}\right]+\frac{1}{8\pi}(\omega\varepsilon''_{ij})_0E_{0j}E^*_{0i}$$
$$+\frac{\mathrm{i}}{16\pi}\left(\frac{\partial\omega\varepsilon''_{ij}}{\partial\omega}\right)_0\left(\frac{\partial E_{0i}}{\partial t}E^*_{0j}-\frac{\partial E^*_{0i}}{\partial t}E_{0j}\right)$$
$$-\frac{\mathrm{i}\omega}{16\pi}\left(\frac{\partial\varepsilon''_{ij}}{\partial k_l}\right)_0\left(\frac{\partial E_{0i}}{\partial x_l}E^*_{0j}-\frac{\partial E^*_{0i}}{\partial x_l}E_{0j}\right)=-\frac{c}{16\pi}\operatorname{div}(\boldsymbol{E}^*_0\times\boldsymbol{B}_0+\boldsymbol{E}_0\times\boldsymbol{B}^*_0)$$
$$+\frac{\omega}{16\pi}\left(\frac{\partial\varepsilon'_{ij}}{\partial k_l}\right)_0\left[\frac{\partial}{\partial x_l}(E_{0j}E^*_{0i})+2k''_l E_{0j}E^*_{0i}\right]-A\,, \tag{2.3.10}$$

where $A=\boldsymbol{j}_{\mathrm{ext}}\cdot\boldsymbol{E}$ and the notation

$$\left(\frac{\partial\varepsilon_{ij}}{\partial k_l}\right)_0=\left(\frac{\partial\varepsilon'_{ij}}{\partial k_l}\right)_0+\mathrm{i}\left(\frac{\partial\varepsilon''_{ij}}{\partial k_l}\right)_0$$

is introduced, both tensors, $(\partial\varepsilon'_{ij}/\partial k_l)_0$ and $(\partial\varepsilon''_{ij}/\partial k_l)_0$ being Hermitian. Obviously, only if $\boldsymbol{k}$ is real can we take, e.g., $(\partial\varepsilon'_{ij}/\partial k_l)_0$ to be the result of differentiating the tensor $\varepsilon'_{ij}=\varepsilon_{ij}-\mathrm{i}\varepsilon''_{ij}$. For the sake of clarity we have not introduced new notation; this should not lead to any misunderstanding [we can also express the result directly by $\partial\varepsilon_{ij}/\partial k_l$, see (2.3.11)]. Note that the next to last term in (2.3.10) can be reduced to the form [recall that $E_{0,i}(\boldsymbol{r},t)=E_{00,i}(\boldsymbol{r},t)\exp(-\boldsymbol{k}''\cdot\boldsymbol{r})$]:

$$-\frac{\partial}{\partial x_l}S^{(1)}_l,\quad S^{(1)}_l=-\frac{\omega}{16\pi}\left(\frac{\partial\varepsilon'_{ij}}{\partial k_l}\right)_0(E_{00,j}E^*_{00,i})$$
$$=-\frac{\omega}{16\pi}\operatorname{Re}\left\{\left(\frac{\partial\varepsilon_{ij}}{\partial k_l}\right)_0(E_{00,j}E^*_{00,i})\right\}. \tag{2.3.11}$$

The vector $\boldsymbol{S}^{(1)}$ appears in (2.3.10) on the same footing with the high-frequency average of Poynting's vector

$$\boldsymbol{S}^{(0)} = \frac{c}{16\pi}(\boldsymbol{E}_0^* \times \boldsymbol{B}_0 + \boldsymbol{E}_0 \times \boldsymbol{B}_0^*) = \frac{c}{8\pi}\mathrm{Re}\{\boldsymbol{E}_0 \times \boldsymbol{B}_0^*\}. \tag{2.3.12}$$

For a gyrotropic or magnetoactive medium, high-frequency averaging is essential since the vector $\boldsymbol{S}^{(0)} = (c/4\pi)(\boldsymbol{E} \times \boldsymbol{B})$ is spatially rotated in this process and its instantaneous direction has no special physical meaning [Ref. 2.2, § 24].

In the case of absorption, (2.3.10) contains many terms and it is by no means simple to explain their physical meaning even if spatial dispersion is disregarded [2.37]. In this connection it is particularly important to consider first the case of a nonabsorbing transparent medium in which a wave with real ω and $\boldsymbol{k}$ is propagated.

As is clear from (2.3.4),

$$\varepsilon''_{ij}(\omega, \boldsymbol{k}) = 0\,, \tag{2.3.13}$$

applies to a nonabsorbing transparent medium in so far as the amplitudes E_{0j} in (2.3.4) are arbitrary (no assumption was made in passing over to (2.3.4) that we are concerned with normal waves). A nonabsorbing medium is not necessarily transparent, of course; an isotropic medium with $\varepsilon(\omega) = \varepsilon'(\omega) < 0$, $\varepsilon''(\omega) = 0$ may serve as an example; in this case $\tilde{n} = \mathrm{i}\varkappa = \mathrm{i}\sqrt{|\varepsilon'|}$.

We see from the inequalities (2.1.30, 33) that if the frequency $\omega \neq 0$ is real the tensor $\varepsilon''_{ij}(\omega, \boldsymbol{k})$ cannot vanish; in other words, there is always some absorption, however weak; these, precisely, were the considerations used to derive the inequalities. Absorption can, however, be so weak that under given concrete physical conditions (e.g., if the thickness of the specimen is given) the medium may be regarded as being practically transparent. Precisely this is implied when considering the application of (2.3.13).

If (2.3.13) is satisfied, (2.3.10) assumes the form

$$\frac{1}{16\pi}\frac{\partial}{\partial t}\left[\left(\frac{\partial \omega \varepsilon'_{ij}}{\partial \omega}\right)_0 E_{0i}E_{0j}^* + \boldsymbol{B}_0 \cdot \boldsymbol{B}_0^*\right] = -\operatorname{div}(\boldsymbol{S}^{(0)} + \boldsymbol{S}^{(1)})\,, \tag{2.3.14}$$

where

$$S_l^{(1)} = -\frac{\omega}{16\pi}\left(\frac{\partial \varepsilon'_{ij}}{\partial k_l}\right)_0 E_{0j}E_{0i}^* \tag{2.3.15}$$

since at $\boldsymbol{k} = \boldsymbol{k}'$, the amplitudes $\boldsymbol{E}_{00}$ and $\boldsymbol{E}_0$ coincide.

Equation (2.3.14) represents the energy conservation law, the high-frequency average of the energy density being equal to

$$W = W' = \frac{1}{16\pi}\left(\frac{\partial[\omega\varepsilon'_{ij}(\omega,\boldsymbol{k})]}{\partial\omega} E_{0j}E^*_{0i} + B_{0i}B^*_{0i}\right). \tag{2.3.16}$$

If the field considered represents a normal-wave packet, i.e., if there are no sources, the derivatives of $E_{0j}E^*_{0i}$ and other quantities with respect to t and x_i are related. In fact, if $\boldsymbol{k}'' = 0$, it can be readily shown by means of (2.3.9) that

$$\frac{\partial}{\partial t}(E_{0i}E^*_{0j}) = -\left(\frac{\partial\omega}{\partial k_i}\right)_0 \frac{\partial}{\partial x_i}(E_{0i}E^*_{0j}).$$

Since $\boldsymbol{v}_{gr} = d\omega/d\boldsymbol{k}$ is the vector of group velocity, the relation obtained has a clear interpretation. In the approximation (2.3.6) the normal-wave packet moves without distortion at the velocity $\boldsymbol{v}_{gr}$. The product $E_{0i}E^*_{0j}$ is therefore a function of t and x_i according to the law $f(\boldsymbol{r} - \boldsymbol{v}_{gr}t)$. Differentiating this leads us directly to the above relation. If we consider a packet for which $\tilde{\boldsymbol{k}} = \tilde{\boldsymbol{k}}(\tilde{\omega})$, $\boldsymbol{k} = \boldsymbol{k}'$, then, analogously, we have

$$\frac{\partial}{\partial x_i}(E_{0j}E^*_{0i}) = -\left(\frac{dk_i}{d\omega}\right)_0 \frac{\partial}{\partial t}(E_{0j}E^*_{0i}).$$

Under these conditions summation of the terms of (2.3.14) containing $(\partial/\partial t)(E_{0j}E^*_{0i})$ and $\operatorname{div}\boldsymbol{S}^{(1)}$ yields

$$\frac{1}{16\pi}\frac{\partial}{\partial t}\left[\left(\frac{\partial\omega\varepsilon'_{ij}}{\partial\omega}\right)_0 + \omega\left(\frac{\partial\varepsilon'_{ij}}{\partial k_l}\right)\left(\frac{\partial k_l}{\partial\omega}\right)_0\right] E_{0i}E^*_{0j}.$$

The same expression is obtained [2.38] if the energy density of the electric field is taken to be equal to

$$\frac{1}{16\pi}\left\{\frac{d}{d\omega}[\omega\varepsilon'_{ij}(\omega,\boldsymbol{k})]\right\}_0 E_{0i}E^*_{0j}.$$

It is clear, however, from the aforesaid that (2.3.16) is to be considered as the energy density for a normal-wave packet, and not some other expression.

It should be noted that quantities averaged over time were indicated above with a bar. There is no necessity for this in the case of quantities such as W, $\boldsymbol{S}^{(1)}$ and $\boldsymbol{S}^{(0)}$, which are already defined as time averages (or, more exactly, as high-frequency averages).

At the same time, as is clear from (2.3.14), $\boldsymbol{S} = \boldsymbol{S}^{(0)} + \boldsymbol{S}^{(1)}$ – see (2.3.12 and 15) – is the average energy flux. Here Poynting's vector $\boldsymbol{S}^{(0)}$ is the average energy flux of the electromagnetic field and the vector $\boldsymbol{S}^{(1)}$, appearing when spatial dispersion is taken into account [2.33, 36, 39], is the average energy flux transferred by the particles of the medium. The flux $\boldsymbol{S}^{(1)}$ vanishes if spatial dispersion is neglected, since this is equivalent to omitting the gradient terms in the equations of motion of the particles.

Let us explain this by using the example of a plasma where the role of the equation of motion is played by the kinetic equation of the distribution function $f(t,\boldsymbol{r},\boldsymbol{v})$

$$\frac{\partial f}{\partial t}+\boldsymbol{v}\nabla_r f+\frac{e}{m}[\boldsymbol{E}+(\boldsymbol{v}\times\boldsymbol{B})/c]\,\nabla_v f+I=0\,, \tag{A}$$

where I is the collision integral, and normalization is given by $\int f d\boldsymbol{v}=N$. Ignoring spatial dispersion in calculating the current density $\boldsymbol{j}=e\int \boldsymbol{v} f d\boldsymbol{v}$ means neglecting the term $\boldsymbol{v}\nabla_r f$ in (A). This is clear at once when we take into consideration the fact that in the case of monochromatic waves $\boldsymbol{v}\nabla_r f=\mathrm{i}\boldsymbol{k}\cdot\boldsymbol{v}f$ and that there are no further terms containing $\boldsymbol{k}$.

The law of energy conservation is obtained when (A) is multiplied by $mv^2/2$ and integrated with respect to the velocity. Thus

$$\frac{\partial}{\partial t}\left(\frac{E^2+B^2}{8\pi}+\int\frac{mv^2}{2}f d\boldsymbol{v}\right)=-\operatorname{div}\left[\frac{c}{4\pi}(\boldsymbol{E}\times\boldsymbol{B})+\int\frac{mv^2}{2}\boldsymbol{v}f d\boldsymbol{v}\right]-\int\frac{mv^2}{2}I d\boldsymbol{v}\,. \tag{B}$$

This equation is given without derivation [2.33, 37] as its nature (law of energy conservation) is evident. It is important to stress that the particle energy flux $\boldsymbol{S}^{(1)}(f)=\int(mv^2/2)\boldsymbol{v}f d\boldsymbol{v}$ enters (B) only if the term $\boldsymbol{v}\nabla_r f$ is taken into account in (A), as mentioned above. Analogous conclusions can be drawn when considering a continuous medium, e.g., with the aid of the equations of hydrodynamics or of the theory of elasticity. To avoid misunderstandings let us stress the following: the vector $\boldsymbol{S}^{(1)}$ defined by (2.3.15) is interpreted through (2.3.14) as an energy flux. The latter refers to the case of a transparent medium ($\varepsilon''_{ij}=0$; $\boldsymbol{k}$ is real). The vector $\boldsymbol{S}^{(1)}(f)$ obtained from (B) must therefore coincide with $\boldsymbol{S}^{(1)}$ – see (2.3.15) – only under the above conditions.

In Sect. 2.3.2 a relation linking the energy flux $\boldsymbol{S}$ with the group velocity $\boldsymbol{v}_{\mathrm{gr}}=d\omega/d\boldsymbol{k}$ in a transparent medium will be derived. We can also obtain expressions for the energy density W and the energy flux $\boldsymbol{S}=\boldsymbol{S}^{(0)}+\boldsymbol{S}^{(1)}$ for a transparent medium (or, more exactly, for the region of transparency). These expressions are valid for any dependence of the field on the coordinates and on time (the point is that W and $\boldsymbol{S}$ are represented in the form of integrals with respect to frequency, which contain functions $\varepsilon_{ij}(\omega,\ldots)$ and Fourier components of the field [2.38]). Since in taking spatial dispersion into consideration, the dielectric constant $\varepsilon'_{ij}(\omega,\boldsymbol{k})$ takes into account the influence, not only of the electric field, but of the magnetic field as well (Sect. 2.1.1), the expression

$$W'_E=\frac{1}{16\pi}\,\frac{\partial[\omega\varepsilon'_{ij}(\omega,\boldsymbol{k})]}{\partial\omega}E_{0j}E^*_{0i}$$

cannot be regarded as the average energy density of only the electric field. At the same time, as mentioned above, expression (2.3.16), i.e., $W=W'=W'_E+(1/16\pi)B_{0i}B_{0j}$, for a transparent medium, represents the total energy density of a quasi-monochromatic electromagnetic field.

Returning to the general case of an absorbing medium, we can write (2.3.10) for a monochromatic field, ω and $\boldsymbol{k}$ being real and $\boldsymbol{E}_0 = \boldsymbol{E}_{00} = \text{const}$ in (2.3.5). Such a field can also be produced in an absorbing medium if sources are present. We then have

$$q \equiv \frac{1}{8\pi}(\omega\varepsilon_{ij}'')_0 E_{0j}E_{0i}^* = -A \equiv -\overline{\boldsymbol{j}_{\text{ext}}\cdot\boldsymbol{E}}\,. \tag{2.3.17}$$

Thus we arrive at a result equivalent to (2.3.4). If, however, the field is not monochromatic, the conditions are complicated even if spatial dispersion is ignored and (2.3.10) assumes the form, see (2.3.12, 16):

$$\frac{\partial W'}{\partial t} + \frac{1}{8\pi}(\omega\varepsilon_{ij}'')_0 E_{0j}E_{0i}^* + \frac{\mathrm{i}}{16\pi}\left(\frac{\partial\omega\varepsilon_{ij}''}{\partial\omega}\right)_0\left(\frac{\partial E_{0i}}{\partial t}E_{0j}^* - \frac{\partial E_{0j}^*}{\partial t}E_{0i}\right) = -\operatorname{div}\boldsymbol{S}^{(0)} - A\,. \tag{2.3.18}$$

It is impossible here to interpret the various terms of the left-hand side of the equation as energy variation and dissipation. It is sufficient to assert that the quantity W' can no longer be assumed equal to the energy density W and, in the presence of absorption, may be negative. The expression $q = (1/8\pi)(\omega\varepsilon_{ij}'') \times E_{0j}E_{0i}^*$ is, in general, not equal to the work done by the friction forces, i.e., it cannot be identified with the evolved heat. This can be readily shown for the special case of a plasma or some other model (for instance, an array of oscillators [2.40, 41]).

Let us consider, for example, an isotropic plasma within the scope of elementary theory where the ordered electron velocity is described by the equation $m\partial\boldsymbol{u}/\partial t = e\boldsymbol{E} - m\nu\cdot\boldsymbol{u}$. Then

$$\varepsilon = 1 - \frac{4\pi e^2 N}{m\omega(\omega + \mathrm{i}\nu)}\,,$$

$$W_E' = \left(\frac{\partial\omega\varepsilon'}{\partial\omega}\right)_0 \frac{\boldsymbol{E}_0\cdot\boldsymbol{E}_0^*}{16\pi} = \left(1 + \frac{4\pi e^2 N(\omega_0^2 - \nu_0^2)}{m(\omega^2+\nu^2)^2}\right)\frac{\boldsymbol{E}_0\cdot\boldsymbol{E}_0^*}{16\pi}$$

and both W_E' and $W' = W_E' + B_0 B_0^*/16\pi$ can be negative. The work done by frictional forces is then

$$mN\nu\overline{u^2} = q - \frac{8\pi e^2 N\nu^2}{m(\omega^2+\nu^2)^2}\frac{\partial}{\partial t}\left(\frac{\boldsymbol{E}_0\cdot\boldsymbol{E}_0^*}{16\pi}\right) + \mathrm{i}\frac{e^2 N\omega\nu}{2m(\omega^2+\nu^2)^2}\left(\frac{\partial\boldsymbol{E}_0}{\partial t}\boldsymbol{E}_0^* - \frac{\partial\boldsymbol{E}_0^*}{\partial t}\boldsymbol{E}_0\right),$$

$$q = \frac{e^2 N\nu}{2m(\omega^2+\nu^2)}\boldsymbol{E}_0\cdot\boldsymbol{E}_0^* = \frac{1}{8\pi}(\omega\varepsilon'')_0\boldsymbol{E}_0\cdot\boldsymbol{E}_0^*\,.$$

For such a plasma, the time-average (or, more exactly, the high-frequency average) of the energy density is equal to the sum of the energy density of the electric field and the kinetic energy of the particles associated with this field. Thus

$$W_E = \left(1 + \frac{4\pi e^2 N}{m(\omega^2 + \nu^2)}\right) \frac{\boldsymbol{E}_0 \cdot \boldsymbol{E}_0^*}{16\pi}.$$

This expression is always positive, of course. The model of a plasma as well as a more general oscillator model are dealt with in more detail in [2.40, 41]. These references also cover certain other problems associated with the derivation of expressions for the energy density and evolved heat in the electrodynamics of dispersive and absorbing media.

It is sufficient here to make one general remark. The phenomenological equation (2.3.10) or (2.3.18) contains as the characteristic parameters of the medium only the dielectric constant $\varepsilon_{ij}(\omega, \boldsymbol{k})$ and its derivatives with respect to ω and $\boldsymbol{k}$. But ε_{ij} determines the 'response' of the system to an electromagnetic field, and this 'response' may also be the same under conditions in which the available energy in the system differs. In fact, even with such a simple example of a linear system as an electric circuit with capacitance C, inductance L and resistance R we can readily see that the impedance of the circuit, $Z(\omega) = R + \mathrm{i}X$, if $R \neq 0$, can be the same for several combinations of values of L and C, i.e., in the case of various amounts of available energy in the circuit (see Fig. 2.2, where $Z = R$ when $L = \varkappa R$, and $C = \varkappa/R$ for any value of $\varkappa$).

There is no such ambiguity if absorption is absent; in the case of (2.3.14) this follows simply from the fact that only two terms of a different nature remain. This is clear from a physical viewpoint as well, because in a conservative system the energy relations follow from the dynamical ones. The same situation is encountered in the case of a monochromatic field with real ω and $\boldsymbol{k}$, where only two terms from (2.3.17) remain. This picture is changed, however, even in the case of a monochromatic field with real ω but complex $\boldsymbol{k}$. Then $\boldsymbol{E}_0 = \boldsymbol{E}_{00} \exp(-\boldsymbol{k}'' \cdot \boldsymbol{r})$, $\boldsymbol{E}_{00} = \text{const}$, and (2.3.10) assumes the form

$$q \equiv \frac{1}{8\pi}(\omega \varepsilon_{ij}'')_0 E_{0j} E_{0i}^* = -\operatorname{div} \boldsymbol{S}^{(0)} - A\,. \tag{2.3.19}$$

There is no longer a basis for interpreting the expression for q as heat. Even in the absence of absorption, but when spatial dispersion is taken into account

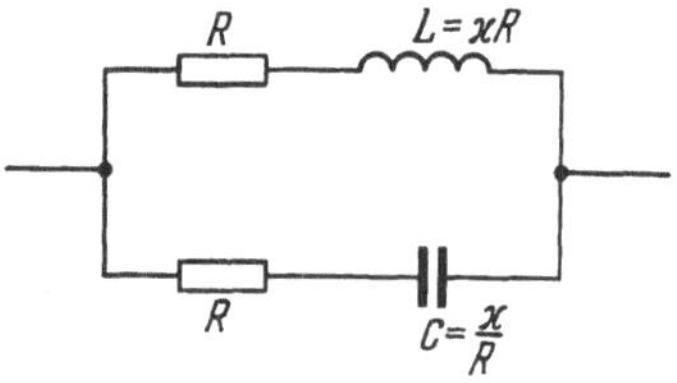

Fig. 2.2. Electrical circuit with the impedance $Z = R$ and arbitrary $\varkappa$

and $\boldsymbol{k}$ is complex, the tensor ε_{ij} is no longer Hermitian and $\varepsilon_{ij}'' \neq 0$. Let us assume, for instance, $\varepsilon_{ij}(\omega,\boldsymbol{k}) = \varepsilon_{ij}^{(0)}(\omega) + \alpha_{ijlm}(\omega)k_l k_m$ with real $\varepsilon_{ij}^{(0)}$ and α_{ijlm} (at a real frequency ω). Then, if $\boldsymbol{k}$ is real, ε_{ij}'' must vanish and if the reality of $\boldsymbol{k}$ is compatible with the field equations, the medium is transparent. But if $\boldsymbol{k}$ is complex, $\varepsilon_{ij}'' \neq 0$ even though there is no dissipation mechanism and the field with complex $\boldsymbol{k}$ is produced by external sources.

Thus, in the case of spatial dispersion, (2.3.4 and 17) can be interpreted as describing the heat evolved per unit volume only if ω and $\boldsymbol{k}$ are real. Moreover, only in the case of transparent medium is W the energy density – see (2.3.16) – and $\boldsymbol{S}$ the energy flux, see (2.3.12, 15). In the absence of absorption but if $\boldsymbol{k}$ is complex, as is the case with total internal reflection, the problem of the energy balance is also quite simple (Sects. 2.3.3, 4.5.5, 6), but with absorption and complex $\boldsymbol{k}$ (normal waves in an absorbing medium) the energy relations become more complicated. We have only the law of energy conservation, (2.3.10), but we cannot give any energy interpretation for the various terms of this equation. The causes of this situation have already been explained. In addition to this, it can be mentioned that in a dissipative system thermodynamic concepts such as internal or free energy are inapplicable. We could therefore have expected it to be impossible to attribute a simple energetic meaning to the various terms in the phenomenological law of energy conservation. It then becomes clear why in Sect. 2.1.2 the principle of growth of entropy was taken into account when using the tensor $\varepsilon_{ij}(\omega,\boldsymbol{k})$ only when ω and $\boldsymbol{k}$ were real. The same can be done here, and the energy relations for a dissipative system (or for a nonabsorbing medium with complex $\boldsymbol{k}$) can only be investigated in detail as the result of a more comprehensive analysis of the system's properties, requiring more than just a knowledge of the dielectric constant $\varepsilon_{ij}(\omega,\boldsymbol{k})$.

It is important here to emphasize that the lack of knowledge of energy density, energy flux and amount of heat, all referring to the interior of the medium, in no way prevents analysis of a wide class of experiments. In optics we usually deal with the reflection and transmission coefficients or the change of phase of waves passing through plates, etc. Since these measurements are made outside the medium, we do not need to know anything about local energy relations inside.

2.3.2 Quadratic Functions of the Normal Wave Amplitudes

A great many relations can be established, which link the amplitudes of normal waves and which are quadratic functions of these amplitudes. For this purpose, we begin with the field equations (2.2.3, 5). For the sake of convenience these equations are written once more, together with their complex conjugate expressions (ω is assumed to be real) from (2.2.3 and 5), namely

$$\boldsymbol{D}_0 = -\frac{c}{\omega}(\boldsymbol{k}\times\boldsymbol{B}_0)\,, \qquad (2.2.3)$$

$$\boldsymbol{B}_0 = \frac{c}{\omega}(\boldsymbol{k} \times \boldsymbol{E}_0), \tag{2.2.5}$$

$$\boldsymbol{D}_0^* = -\frac{c}{\omega}(\boldsymbol{k}^* \times \boldsymbol{B}_0^*), \tag{2.2.3 a}$$

$$\boldsymbol{B}_0^* = \frac{c}{\omega}(\boldsymbol{k}^* \times \boldsymbol{E}_0^*). \tag{2.2.5 a}$$

These relations are valid for fields of the form

$$\boldsymbol{E} = \boldsymbol{E}_0 \mathrm{e}^{\mathrm{i}(\boldsymbol{k}\cdot\boldsymbol{r} - \omega t)}, \quad \boldsymbol{D} = \boldsymbol{D}_0 \mathrm{e}^{\mathrm{i}(\boldsymbol{k}\cdot\boldsymbol{r} - \omega t)}, \quad \boldsymbol{B} = \boldsymbol{B}_0 \mathrm{e}^{\mathrm{i}(\boldsymbol{k}\cdot\boldsymbol{r} - \omega t)}, \tag{2.3.20 a}$$

for the amplitudes of these fields $\boldsymbol{E}_0$, $\boldsymbol{D}_0$ and $\boldsymbol{B}_0$, and also for the amplitudes $\boldsymbol{E}_0$, $\boldsymbol{D}_0$ and $\boldsymbol{B}_0$ of real fields of the form

$$\boldsymbol{E}(\boldsymbol{r}, t) = \tfrac{1}{2}(\boldsymbol{E}_0 \mathrm{e}^{\mathrm{i}(\boldsymbol{k}\cdot\boldsymbol{r} - \omega t)} + \boldsymbol{E}_0^* \mathrm{e}^{-\mathrm{i}(\boldsymbol{k}^*\cdot\boldsymbol{r} - \omega t)}). \tag{2.3.20 b}$$

As previously indicated, an investigation of quadratic relations (in particular, energy relations) becomes more convenient if we use real fields to begin with. We shall do this now. The difference between the amplitudes $\boldsymbol{E}_0$ and $\boldsymbol{E}_{00}$ was discussed in Sect. 2.3.1, but in the following, for the sake of simplicity, we shall use only expressions of the type (2.3.20 b) where the amplitudes are independent of r and t.

Multiplying (2.2.3) by $\boldsymbol{E}_0$ and (2.2.5) by $\boldsymbol{B}_0$ we obtain (for $\omega \neq 0$)

$$\boldsymbol{D}_0 \cdot \boldsymbol{E}_0 = \boldsymbol{B}_0 \cdot \boldsymbol{B}_0 = \frac{c}{\omega}\boldsymbol{k} \cdot (\boldsymbol{E}_0 \times \boldsymbol{B}_0). \tag{2.3.21}$$

In a transparent medium, i.e., if $\boldsymbol{k}$ is real we have, moreover[12]

$$\boldsymbol{D}_0 \cdot \boldsymbol{E}_0^* = \boldsymbol{B}_0 \cdot \boldsymbol{B}_0^* \quad (\boldsymbol{k} = \boldsymbol{k}^* = \boldsymbol{k}'). \tag{2.3.22}$$

This follows from (2.2.3) and (2.2.5 a). From (2.2.3) and (2.2.5), after multiplying by $\boldsymbol{E}_0^*$ or $\boldsymbol{B}_0^*$ and subsequent addition or subtraction, we have

$$\omega(\boldsymbol{D}_0 \cdot \boldsymbol{E}_0^* \pm \boldsymbol{B}_0 \cdot \boldsymbol{B}_0^*) = c\boldsymbol{k} \cdot (\boldsymbol{E}_0^* \times \boldsymbol{B}_0 \pm \boldsymbol{E}_0 \times \boldsymbol{B}_0^*). \tag{2.3.23}$$

It follows from this and from (2.3.22) that for a transparent medium

$$\boldsymbol{E}_0 \times \boldsymbol{B}_0^* = \boldsymbol{E}_0^* \times \boldsymbol{B}_0 \quad (\boldsymbol{k} = \boldsymbol{k}^* = \boldsymbol{k}'). \tag{2.3.24}$$

---

[12] Equations applying solely to transparent media will be frequently supplemented by the equation $\boldsymbol{k} = \boldsymbol{k}^* = \boldsymbol{k}'$; this indicates that the $\boldsymbol{k}$ vector is real (in the general case, $\boldsymbol{k} = \boldsymbol{k}' + \mathrm{i}\boldsymbol{k}''$, $\boldsymbol{k}'$ and $\boldsymbol{k}''$ being real vectors).

We can likewise find from (2.2.3, 5) and (2.2.3a, 5a) that

$$\omega(\boldsymbol{D}_0 \times \boldsymbol{B}_0^* \pm \boldsymbol{D}_0^* \times \boldsymbol{B}_0) = -\,c[(\boldsymbol{k} \times \boldsymbol{B}_0) \times \boldsymbol{B}_0^* \pm (\boldsymbol{k} \times \boldsymbol{E}_0) \times \boldsymbol{D}_0^*]\,, \tag{2.3.25}$$

$$\omega(D_{0i}B_{0j}^* \pm B_{0i}D_{0j}^*) = [(\boldsymbol{k} \times \boldsymbol{B}_0)_i B_{0j}^* \pm (\boldsymbol{k} \times \boldsymbol{E}_0)_i D_{0j}^*]\,. \tag{2.3.26}$$

For the sake of completeness, we also present some other obvious relations following from (2.2.3 and 5):

$$\boldsymbol{k} \cdot \boldsymbol{D}_0 = 0\,, \quad \boldsymbol{k} \cdot \boldsymbol{B}_0 = 0\,, \quad \boldsymbol{B}_0 \cdot \boldsymbol{D}_0 = 0\,, \quad \boldsymbol{B}_0 \cdot \boldsymbol{E}_0 = 0\,. \tag{2.3.27}$$

Equations of the types (2.3.21, 23), etc., can also be written, of course, for the complex conjugate expressions.

If there is no monochromatic field of the type (2.3.20), but a wave packet instead, then relations (2.3.21 – 27) apply to each of the Fourier components of the packet. The amplitudes $\boldsymbol{E}_0$, $\boldsymbol{D}_0$ and $\boldsymbol{B}_0$ depend on $\omega(\boldsymbol{k})$ and $\boldsymbol{k}$. Hence, the relations obtained from (2.3.21 – 27) as a result of differentiating them with respect to $\boldsymbol{k}$ (i.e., to k_l, $l = 1, 2, 3$) can be referred to an arbitrary quasi-monochromatic packet only if the derivatives of the amplitudes with respect to $\boldsymbol{k}$ can be cancelled out. A series of such relations does, in fact, exist [2.33, 39, 42].

a) The Conservation of Energy for a Field

We differentiate (2.3.23) with respect to k_l and, since

$$D_{0i}(\omega, \boldsymbol{k}) = \varepsilon_{ij}(\omega, \boldsymbol{k}) E_{0j}(\omega, \boldsymbol{k})\,, \qquad \omega = \omega(\boldsymbol{k})\,, \tag{2.3.28}$$

we obtain

$$\begin{aligned} &\frac{\partial \omega}{\partial k_l}\left(\frac{\partial[\omega \varepsilon_{ij}(\omega, \boldsymbol{k})]}{\partial \omega} E_{0j}E_{0i}^* \pm B_{0i}B_{0i}^*\right) \\ &\quad = c[(\boldsymbol{E}_0^* \times \boldsymbol{B}_0)_l \pm (\boldsymbol{E}_0 \times \boldsymbol{B}_0^*)_l] - \omega \frac{\partial \varepsilon_{ij}}{\partial k_l} E_{0j}E_{0i}^* \\ &\quad\quad + \left\{[c(\boldsymbol{k} \times \boldsymbol{E}_0^*) \mp \omega \boldsymbol{B}_0^*]\frac{d\boldsymbol{B}_0}{dk_l} \pm [c(\boldsymbol{k} \times \boldsymbol{E}_0) - \omega \boldsymbol{B}_0]\frac{d\boldsymbol{B}_0^*}{dk_l}\right\} \\ &\quad\quad - \left\{[c(\boldsymbol{k} \times \boldsymbol{B}_0)_i + \omega \varepsilon_{ij}E_{0j}]\frac{dE_{0i}^*}{dk_l} + [\pm c(\boldsymbol{k} \times \boldsymbol{B}_0)_j + \omega \varepsilon_{ij}E_{0i}^*]\frac{dE_{0j}}{dk_l}\right\}. \end{aligned} \tag{2.3.29}$$

In the case of a transparent medium $\boldsymbol{k} = \boldsymbol{k}^*$ and $\varepsilon_{ij} = \varepsilon'_{ij} = \varepsilon_{ji}^*$. Then, if we take the field equations (2.2.3, 5, 3a, 5a) into account, (2.3.29), taken with the upper sign, assumes the form

$$W\boldsymbol{v}_{\rm gr} = \boldsymbol{S}\,, \tag{2.3.30}$$

where

$$W = \frac{1}{16\pi}\left\{\frac{\partial}{\partial\omega}[\omega\varepsilon_{ij}(\omega,\boldsymbol{k})]E_{0j}E_{0i}^{*} + B_{0i}B_{0i}^{*}\right\},$$

$$\boldsymbol{v}_{\rm gr} = \frac{d\omega}{d\boldsymbol{k}}, \quad \boldsymbol{S} = \boldsymbol{S}^{(0)} + \boldsymbol{S}^{(1)}, \quad \boldsymbol{k} = \boldsymbol{k}^{*} = \boldsymbol{k}'\,, \quad \varepsilon_{ij} = \varepsilon_{ij}'\,, \tag{2.3.31}$$

$$\boldsymbol{S}^{(0)} = \frac{c}{16\pi}[(\boldsymbol{E}_0\times\boldsymbol{B}_0^{*}) + (\boldsymbol{E}_0^{*}\times\boldsymbol{B}_0)]\,, \quad S_l^{(1)} = -\frac{\omega}{16\pi}\frac{\partial\varepsilon_{ij}(\omega,\boldsymbol{k})}{\partial k_l}E_{0i}^{*}E_{0j}\,.$$

The meaning of the quantities W, $\boldsymbol{S}^{(0)}$ and $\boldsymbol{S}^{(1)}$ was discussed in Sect. 2.3.1; these quantities are the high-frequency averages of the energy density and energy flux of a quasi-monochromatic packet. From this point of view, (2.3.30) is quite natural: as a first approximation (i.e., if only the first derivatives with respect to ω and $\boldsymbol{k}$ are taken into account) the quasi-monochromatic packet of normal waves travels as a whole, without diffusion, at the group velocity $\boldsymbol{v}_{\rm gr}$. The total energy flux $\boldsymbol{S}$, made up of the electromagnetic flux $\boldsymbol{S}^{(0)}$ and the flux $\boldsymbol{S}^{(1)}$, must evidently be equal to the product of the energy density by the group velocity $\boldsymbol{v}_{\rm gr}$. It is, perhaps, necessary on the basis of independent considerations to show that the group velocity

$$\boldsymbol{v}_{\rm gr} = d\omega/d\boldsymbol{k} \tag{2.3.32}$$

(meaning that $v_{{\rm gr},i} = \partial\omega(\boldsymbol{k})/\partial k_i$).

b) The Vector of Group Velocity

The derivation of this expression is well known [Ref. 2.2, § 34] but it is advisable to consider it and, besides, to derive some additional valuable relations.

Let us represent the electric field of a quasi-monochromatic normal-wave packet in the form (here we need not consider a real field)

$$\begin{aligned}\boldsymbol{E}(\boldsymbol{r},t) &= \int g(\tilde{\boldsymbol{k}})\,{\rm e}^{{\rm i}[\tilde{\boldsymbol{k}}\cdot\boldsymbol{r} - \tilde{\omega}(\tilde{\boldsymbol{k}})t]}d\tilde{\boldsymbol{k}} \\ &= {\rm e}^{{\rm i}(\boldsymbol{k}\cdot\boldsymbol{r}-\omega t)}\int g(\boldsymbol{q})\,{\rm e}^{{\rm i}(\boldsymbol{r}-[d\omega/d\boldsymbol{k}]t)\boldsymbol{q}}d\boldsymbol{q} = \boldsymbol{A}\left(\boldsymbol{r} - \frac{d\omega}{d\boldsymbol{k}}t\right){\rm e}^{{\rm i}(\boldsymbol{k}\cdot\boldsymbol{r}-\omega t)}\,.\end{aligned} \tag{2.3.33}$$

Here $\boldsymbol{k}$ and $\omega = \omega(\boldsymbol{k})$ are the "carrier" wave vector and the angular frequency of the pulse

$$\tilde{\boldsymbol{k}} = \boldsymbol{k} + \boldsymbol{q}\,, \quad \tilde{\omega} = \omega + \frac{d\omega}{d\boldsymbol{k}}\boldsymbol{q}\,, \quad q \ll k\,, \quad \left(\frac{d\omega}{d\boldsymbol{k}}\right)_i \equiv \frac{\partial\omega}{\partial k_i} \equiv \left(\frac{\partial\omega}{\partial\tilde{k}_i}\right)_{\tilde{\boldsymbol{k}}=\boldsymbol{k}}\,.$$

In (2.3.33) only the first derivatives of ω with respect to k_i are taken into account. This indicates that we are considering a quasi-monochromatic pulse (of the duration $T \gg 2\pi/\omega$), whose spreading (change of shape) is neglected. The amplitude A in (2.3.33) is obviously constant if $r = (d\omega/dk)t = \text{const}$, i.e., the pulse moves with the group velocity (2.3.32).

In order to make sure that the spreading of the pulse can be neglected, we should first investigate the nature of this spreading. Such an analysis is itself of interest. Here, however, we confine ourselves to the following remark (for details see [Ref. 2.2, § 21] and [2.43]). It is clear, from general considerations, that the longer the path L the pulse travels, the greater the spreading. In the case of a homogeneous medium the time the pulse takes to travel the distance L is evidently equal to[13]

$$\Delta t_{gr} = \frac{L}{v_{gr}} = \frac{L}{d\omega/dk} = L\frac{dk}{d\omega} = \frac{dkL}{d\omega} = \frac{d\varphi}{d\omega}, \tag{2.3.34}$$

where, for simplicity, we consider a unidimensional pulse (packet) traveling with group velocity $d\omega/dk$; $\varphi = kL$ is evidently the phase shift the wave is subject to along the path L. If the time $\Delta t_{gr} = d\varphi/d\omega$ is independent of the carrier frequency ω the pulse will not spread. But, in general, $\Delta t_{gr} = \Delta t_{gr}(\omega) = d\varphi(\omega)/d\omega$ and the spreading of the pulse is characterized by the derivative $d^2\varphi/d\omega^2$. As follows from a more detailed analysis, the front and rear edges of a long pulse (wave train) whose duration is $T \gg 2\pi/\omega$ are broadened by an amount of the order of

$$\tau = \sqrt{\frac{d^2\varphi}{d\omega^2}} = \sqrt{L\frac{d^2k}{d\omega^2}}, \qquad k = \frac{\omega}{c}n(\omega). \tag{2.3.35}$$

---

[13] We arrive immediately at (2.3.34) when considering a linear pulse whose field at $z = 0$ is equal to

$$E(z=0,t) = \int q(\tilde{\omega})\exp(-\mathrm{i}\tilde{\omega}t)\,d\tilde{\omega}$$

which, after having traveled along the path L equals

$$E(z=L,t) = \int R(\tilde{\omega})g(\tilde{\omega})\exp\{-\mathrm{i}[\tilde{\omega}t - \varphi(\tilde{\omega})]\}d\tilde{\omega}.$$

In the case of no absorption $R(\tilde{\omega}) = 1$ and for a quasi-monochromatic pulse

$$E(z=0,t) = A(t)\exp(-\mathrm{i}\omega t), \qquad \varphi(\tilde{\omega}) = \varphi(\omega) + (d\varphi/d\omega)\Omega.$$

Then, as is readily seen in the approximation being considered

$$E(z=L,t) = A(t - d\varphi/d\omega)\exp\{-\mathrm{i}[\omega t - \varphi(\omega)]\}$$

and the pulse takes the time $\Delta t_{gr} = d\varphi/d\omega$ to travel the distance L.

The spreading width of a long pulse is independent of T or $d\varphi/d\omega$; the period $2\pi/\omega$ of the oscillations is also of no consequence, though this is less evident. Consequently, τ depends only on $d^2\varphi/d\omega^2$. This being the case, we can obtain (2.3.35) simply from considerations of dimensionality. The shape which the pulse acquires upon spreading depends on its initial shape. Calculations have been carried out for a rectangular pulse ("cut-off" sine curve) [Ref. 2.2, § 21] (see also [2.43]; special features, appearing when additional waves are taken into account, were discussed in [2.44]). As long as $\tau \ll T$ the signal can be regarded as not being spread in first approximation. By means of (2.3.35) it is easy to ascertain whether the spreading of the pulse is sufficiently weak; it is then possible to use the concept of the group velocity without making additional stipulations.

A complication arises in the case of absorption or of a nonabsorbing but nontransparent medium, i.e., when k is complex or imaginary. In this case (2.3.30) not only loses its validity, but it is also impossible to give a simple interpretation of the energy relations and the concept of group velocity has, in general, no meaning. We note that the vector $\boldsymbol{v}_{\mathrm{gr}} = d\omega/d\boldsymbol{k}$ is complex if $\boldsymbol{k}$ is complex and ω real, whereas the group velocity, according to its definition, is a real vector. By analogy with the case of a transparent medium we can try to determine the group velocity of a nontransparent medium as follows:

$$\boldsymbol{v}'_{\mathrm{gr}} = \frac{d\omega}{d\boldsymbol{k}'}, \quad \boldsymbol{k} = \boldsymbol{k}' + \mathrm{i}\boldsymbol{k}''. \tag{2.3.36a}$$

Then, for an isotropic medium and a packet of homogeneous waves

$$\boldsymbol{v}'_{\mathrm{gr}} = \frac{d\omega}{dk'}\boldsymbol{s} = \frac{c}{n + \omega\dfrac{dn}{d\omega}}\boldsymbol{s}, \quad \boldsymbol{k} = \frac{\omega}{c}[n(\omega) + \mathrm{i}\varkappa(\omega)]\boldsymbol{s},$$
$$k' = \frac{\omega}{c}n(\omega), \quad s^2 = 1. \tag{2.3.37a}$$

Equations (2.3.36a and 37a) are transformed, of course, into the respective expressions for the case of no absorption. This requirement is satisfied by the expression

$$\boldsymbol{v}''_{\mathrm{gr}} = \mathrm{Re}\left\{\frac{d\omega}{d\boldsymbol{k}}\right\}, \tag{2.3.36b}$$

or, for an isotropic medium, by

$$\boldsymbol{v}''_{\mathrm{gr}} = \mathrm{Re}\left\{\frac{c}{d(\omega\tilde{n})/d\omega}\boldsymbol{s}\right\} = \frac{c\dfrac{d(\omega n)}{d\omega}}{(d\omega n/d\omega)^2 + (d\omega\varkappa/d\omega)^2}\boldsymbol{s}. \tag{2.3.37b}$$

Such determinations of the group velocity are not only unfounded but also clearly restricted. It is sufficient to point out that the quantity v'_{gr} may exceed the velocity of light in empty space, c. In fact $v'_{gr} > c$ if $n + \omega(dn/d\omega) = d\omega n/d\omega < 1$, a case encountered in the region of anomalous dispersion. But since this region is characterized by considerable absorption and strong distortion of the pulse, the concept of group velocity as the velocity of the signal as a whole cannot be applied.[14] This indicates only that in an absorbing medium (and, in general, if $\boldsymbol{k}$ is complex) the problem of the motion of a pulse requires a more detailed analysis even in an approximation such as (2.3.33), not to speak of the more general problem which is solved by investigating the Fourier integral. In the case of an absorbing isotropic medium such an investigation has been made for a particular shape of pulse (a "cut-off" sine curve) in [2.43, 45, 46].

Calculation of the energy flow velocity v_E based on the classical damping oscillator microscopic model of a dielectric [see (1.2.3)] was carried out in [2.41]. This velocity was determined in [2.41] to be the Poynting vector divided by the stored energy density $\tilde{W}$ in the monochromatic wave:

$$\tilde{W} = \frac{M}{2V}(\dot{r}^2 + \omega_{0s}^2 r^2) + \frac{1}{8\pi}(\varepsilon_0 E^2 + H^2) .$$

It is evident that v_E defined in this way must be equal to v_{gr} for a nonabsorbing dielectric, but the two velocities differ in the presence of absorption. It was shown in [2.41] that $v_E = c/(n + 2\omega\varkappa/\gamma)$, $(n + i\varkappa)^2 = \varepsilon(\omega)$. For zero damping $(\gamma \to 0)$ $n + 2\omega\varkappa/\gamma \to d(\omega n)/d\omega$ and $v_E \to v_{gr}$. The current time-of-flight measurements for polaritons in semiconductors necessitate further study of this problem.

In an isotropic medium there is no problem of the direction of a symmetrical pulse or ray, because this direction coincides with $\boldsymbol{s}$; this is already evident from considerations of symmetry [we consider homogeneous waves, the carrier wave being characterized by the wave vector $\boldsymbol{k} = (\omega/c)(n + i\varkappa)\boldsymbol{s}$]. If the medium is anisotropic we must determine not only the character of the motion of the pulse, propagated in a certain direction, but the direction as well. The problem of reflection of a finite-duration pulse perpendicularly incident on a semi-infinite medium, with spatial dispersion, was considered by *Agrawal* et al. [2.47] (see also [2.48]).

---

[14] More exactly, the concept of group velocity is usually not applicable to absorbing media. If, however, the absorption coefficient is independent of the frequency within the limits of the spectral width of the pulse, i.e., if $\omega(\boldsymbol{k}) = \omega(\boldsymbol{k}')$, the pulse, though attenuated, shows the same spreading as without absorption. As we can see from (2.3.33), the velocity of the pulse in this case is equal to $\boldsymbol{v}'_{gr} = d\omega/d\boldsymbol{k}'$. In the absence of absorption and if $\boldsymbol{k}$ is real we always have $v_{gr} = d\omega/dk \leqslant c$ (as shown in Sect. 2.2.5).

In the case of sufficiently long pulses (long compared with the wavelength) to which the "sharp" function $g(\tilde{\boldsymbol{k}})$ corresponds, we can use (2.3.33) as a first approximation, where the function

$$A\left(\boldsymbol{r} - \frac{d\omega}{d\boldsymbol{k}}t\right) = \int g(\boldsymbol{q}) \exp\left[\mathrm{i}\left(\boldsymbol{r} - \frac{d\omega}{d\boldsymbol{k}}t\right)\boldsymbol{q}\right] d\boldsymbol{q}$$

is assumed to have a definite meaning (or simply to be analytic) in the region of complex values of $\boldsymbol{r} - (d\omega/d\boldsymbol{k})t$. Owing to the complexity of the vector $d\omega/d\boldsymbol{k}$, the argument $\boldsymbol{r} - (d\omega/d\boldsymbol{k})t$ does not vanish, in general, since $\boldsymbol{r}$ and t are real by definition. This fact specifies the absorbing medium in the given case, the pulse is "displaced as a whole" in the complex direction, i.e., with real $\boldsymbol{r}$ its shape is altered even in the approximation (2.3.33). The character of motion of the pulse and, particularly, the direction of the beam in the absorbing medium (see [2.43, 46] and the remarks following (2.3.40)) can be determined, in principle, without difficulties in the approximation (2.3.33) or in the similar approximation (2.3.40) which is used in the following to determine the beam direction in a transparent medium.

In many cases, however, if the absorption is relatively weak, the group velocity and the direction of a ray can be simply determined to a good approximation when absorption is neglected, or more conveniently, using (2.3.36, 37).

In the case of a transparent medium (real $\boldsymbol{k}$) we have

$$\begin{aligned} &\omega = \omega(\boldsymbol{k}) = \omega(k_x, k_y, k_z) = \omega(k, \alpha, \beta)\,, \qquad \alpha \equiv s_x = k_x/k\,, \\ &\beta \equiv s_y = k_y/k\,, \qquad \gamma \equiv s_z = k_z/k\,, \qquad \alpha^2 + \beta^2 + \gamma^2 = 1\,. \end{aligned} \tag{2.3.38}$$

Hence

$$\begin{aligned} \frac{\partial\omega}{\partial k_x} &= \frac{\partial\omega}{\partial k}\alpha + \frac{\partial\omega}{\partial\alpha}\frac{1-\alpha^2}{k} - \frac{\partial\omega}{\partial\beta}\frac{\alpha\beta}{k}\,, \\ \frac{\partial\omega}{\partial k_y} &= \frac{\partial\omega}{\partial k}\beta + \frac{\partial\omega}{\partial\beta}\frac{1-\beta^2}{k} - \frac{\partial\omega}{\partial\alpha}\frac{\alpha\beta}{k}\,, \\ \frac{\partial\omega}{\partial k_z} &= \frac{\partial\omega}{\partial k}\gamma - \frac{\partial\omega}{\partial\alpha}\frac{\alpha\gamma}{k} - \frac{\partial\omega}{\partial\beta}\frac{\beta\gamma}{k}\,. \end{aligned} \tag{2.3.39}$$

It is, however, more convenient to express $\partial\omega/\partial k_i$ in terms of n, taking into account that $\boldsymbol{k} = (\omega/c)n(\omega, \alpha, \beta)\boldsymbol{s}$. Thus

$$\frac{\partial k}{\partial k_x} = \alpha \equiv s_x \equiv \frac{k_x}{k} = \frac{1}{c}\frac{\partial\omega}{\partial k_x}n + \frac{\omega}{c}\frac{\partial n}{\partial\omega}\frac{\partial\omega}{\partial k_x} + \frac{\omega}{c}\frac{\partial n}{\partial\alpha}\frac{\partial\alpha}{\partial k_x}$$

$$+\frac{\omega}{c}\frac{\partial n}{\partial \beta}\frac{\partial \beta}{\partial k_x}=\frac{1}{c}\frac{\partial(\omega n)}{\partial \omega}\frac{\partial \omega}{\partial k_x}+\frac{\omega}{c}\frac{\partial n}{\partial \alpha}\frac{1-\alpha^2}{k}-\frac{\omega}{c}\frac{\partial n}{\partial \beta}\frac{\alpha\beta}{k} \tag{2.3.39a}$$

and we can proceed similarly with $\partial k/\partial k_y$, and $\partial k/\partial k_z$. Since $ck/\omega = n$ we obtain

$$v_{\mathrm{gr},x}=\frac{\partial \omega}{\partial k_x}=\frac{\alpha+\frac{1}{n}\left[-\frac{\partial n}{\partial \alpha}(1-\alpha^2)+\frac{\partial n}{\partial \beta}\alpha\beta\right]}{\frac{1}{c}\frac{\partial(\omega n)}{\partial \omega}},$$

$$v_{\mathrm{gr},y}=\frac{\partial \omega}{\partial k_y}=\frac{\beta+\frac{1}{n}\left[\frac{\partial n}{\partial \alpha}\alpha\beta-\frac{\partial n}{\partial \beta}(1-\beta^2)\right]}{\frac{1}{c}\frac{\partial(\omega n)}{\partial \omega}}, \tag{2.3.39b}$$

$$v_{\mathrm{gr},z}=\frac{\partial \omega}{\partial k_z}=\frac{\gamma+\frac{1}{n}\left[\frac{\partial n}{\partial \alpha}\alpha\gamma+\frac{\partial n}{\partial \beta}\beta\gamma\right]}{\frac{1}{c}\frac{\partial(\omega n)}{\partial \omega}}.$$

Further, as can be readily seen,

$$v_{\mathrm{gr}}=\frac{c}{\frac{\partial(\omega n)}{\partial \omega}}\times\sqrt{1+\frac{1}{n^2}\left[\left(\frac{\partial n}{\partial \alpha}\right)^2(1-\alpha^2)+\left(\frac{\partial n}{\partial \beta}\right)^2(1-\beta^2)-2\alpha\beta\frac{\partial n}{\partial \alpha}\frac{\partial n}{\partial \beta}\right]}. \tag{2.3.39c}$$

It is clear from (2.3.39b and c) that the direction of the group velocity, i.e., the ratio $v_{\mathrm{gr},i}/v_{\mathrm{gr}}$, does not depend on $\partial(\omega n)/\partial\omega$ and is determined only by the values of α, β, n, $\partial n/\partial\alpha$ and $\partial n/\partial\beta$.

The form of the expression for $v_{\mathrm{gr},i}$ becomes particularly simple in a system of coordinates where $\alpha=\beta=0$ and $\gamma=1$, i.e., where the z axis coincides with the wave normal. In this case

$$v_{\mathrm{gr},x}=-\frac{\partial n/\partial\alpha}{\frac{n}{c}\frac{\partial(\omega n)}{\partial\omega}},\quad v_{\mathrm{gr},y}=-\frac{\partial n/\partial\beta}{\frac{n}{c}\frac{\partial(\omega n)}{\partial\omega}},\quad v_{\mathrm{gr},z}=\frac{c}{\partial(\omega n)/\partial\omega},$$

$$v_{gr} = \frac{c}{\partial(\omega n)/\partial\omega}\sqrt{1+\frac{1}{n^2}\left[\left(\frac{\partial n}{\partial\alpha}\right)^2+\left(\frac{\partial n}{\partial\beta}\right)^2\right]}, \tag{2.3.39d}$$

$$\frac{v_{gr,x}}{v_{gr,z}} = -\frac{\partial n/\partial\alpha}{n}, \quad \frac{v_{gr,y}}{v_{gr,z}} = -\frac{\partial n/\partial\beta}{n}.$$

Since the direction of $\boldsymbol{v}_{gr}$ does not depend on $\partial n/\partial\omega$ it is clear that this direction can also be found if the statement of the problem differs from that given above. Let us consider a sufficiently broad (compared with the wavelength) beam (ray) of light, having the fixed frequency ω so that the field $\boldsymbol{E}$ can be written in the form

$$\boldsymbol{E}(\boldsymbol{r},t) = e^{-i\omega t}\int g(\alpha,\beta)\exp\left[i\frac{\omega}{c}n(\omega,\alpha,\beta)(\alpha x+\beta y+\gamma z)\right]d\alpha\, d\beta.$$

If we now take the z axis in the direction of the "carrier"-wave vector of the beam we can assume that the field varies slowly in the directions of x and y (broad beam) and therefore we can write

$$\begin{aligned}\boldsymbol{E}(\boldsymbol{r},t) &= \exp\left[i\left(\frac{\omega}{c}nz-\omega t\right)\right]\int g(\Delta\alpha,\Delta\beta)\\ &\quad\times\exp\left\{i\frac{\omega}{c}\left[\left(nx+\frac{\partial n}{\partial\alpha}z\right)\Delta\alpha+\left(ny+\frac{\partial n}{\partial\beta}z\right)\Delta\beta\right]\right\}d\Delta\alpha\, d\Delta\beta\\ &= \exp\left[i\left(\frac{\omega}{c}nz-\omega t\right)\right]A\left(nx+\frac{\partial n}{\partial\alpha}z,\ ny+\frac{\partial n}{\partial\beta}z\right),\end{aligned} \tag{2.3.40}$$

where, n, $\partial n/\partial\alpha$ and $\partial n/\partial\beta$ are taken for $\alpha=\beta=0$ and $\gamma=1$. Hence it becomes clear that the direction of the beam along whith the field differs from zero is determined by the ratios $x/z = -(\partial n/\partial\alpha)/n$ and $y/z = -(\partial n/\partial\beta)/n$. This result agrees with (2.3.39d).

In the case of an absorbing medium and, in general, with complex $\tilde{n} = n + i\varkappa$, (2.3.40) is also valid [2.43, 45] in the corresponding approximation. Here, evidently

$$|E_i| = e^{-(\omega/c)\varkappa z}\left|A_i\left(\tilde{n}x+\frac{\partial\tilde{n}}{\partial\alpha}z,\ \tilde{n}y+\frac{\partial\tilde{n}}{\partial\beta}z\right)\right|.$$

If $\tilde{n}$ is complex, the function $|A_i|$ does not remain constant along any direction even in this approximation, and the statement of the problem of the beam direction has to be specified more accurately. As a concrete example we shall consider a crystal plate arranged in the region between the planes $z = 0$ and $z = l$. A light beam is assumed to be normally incident to this plate, the field being $E_i = \exp(-\mathrm{i}\omega t)A(\tilde{n}x, \tilde{n}y)$ at the point of entry (at $z = +0$) and the amplitude being

$$|E_i| = \mathrm{e}^{-(\omega/c)\varkappa l}\left|A_i\left(\tilde{n}x + \frac{\partial \tilde{n}}{\partial \alpha}l,\ \tilde{n}y + \frac{\partial \tilde{n}}{\partial \beta}l\right)\right|$$

at the point of exit (at $z = l$, but still inside the plate as we do not consider reflection from the surface). The light beam is described by the function $|A_i(\tilde{n}x, \tilde{n}y)|$ which is significant only in a certain region of variation of x and y and which, for example, might have the form $|A_i(x,y,0)| = \exp(-x^2/a^2 - y^2/b^2)$. At the exit the function $|A_i(x,y,l)|$ changes and its maximum is characterized by the condition $\partial|A_i|/\partial x = \partial|A_i|/\partial y = 0$. The latter expressions may be used to determine the point x, y at which the beam emerges from the plate. A similar method can be applied with the necessary generalizations to a beam containing Fourier components corresponding to nonhomogeneous waves.

In reflection and refraction of a light beam of finite diameter at an interface, and especially under the condition of attenuated total reflection, many other effects can exist. Among these, the so-called Good-Hänchen shift [2.64, 65] has been most extensively investigated. Due to this effect, the point where the line of the reflected beam intersects with the reflecting interfacial plane is displaced from the intersection point, in the same plane, of the incident beam (the shifts in the plane of incidence and perpendicular to it are called the longitudinal and transverse shifts). The shifts are of the order of 10^{-4} cm for the optical range of the spectrum, but increase in the region of resonances [2.66 – 68] of the dielectric constants for the contacting media. In the "Goos-Hänchen" geometry [2.64, 65] the crystal is cut so as to permit normal incidence at the lower left surface. The laser beam in this geometry is incident from vacuum at the lower left surface and propagates to the upper reflecting interface where it is nearly totally internally reflected, since $\theta \approx \theta_c$.

Lateral Good-Hänchen shifts, as in [2.69], can be increased by factors of 10 – 100 and can even be "negative" for a laser frequency near an excition resonance in semiconductors, where spatial dispersion is taken into account (Fig. 2.3).

Let us now return to the transparent medium. Whereas the direction of the vector $\boldsymbol{v}_{\mathrm{gr}}$ does not depend on $\partial n/\partial\omega$, the projection of $\boldsymbol{v}_{\mathrm{gr}}$ on $\boldsymbol{k} = k\boldsymbol{s}$ is independent of $\partial n/\partial\alpha$ and $\partial n/\partial\beta$ and is completely determined by the values of $\partial(\omega n)/\partial\omega$. This is immediately clear from (2.3.39d) where $v_{\mathrm{gr},z} = \boldsymbol{v}_{\mathrm{gr}} \cdot \boldsymbol{s}$. In the general form the same result is obtained from (2.3.39b) after multiplying $\boldsymbol{v}_{\mathrm{gr}}$ by $\boldsymbol{s} = \boldsymbol{k}/k = \{\alpha, \beta, \gamma\}$:

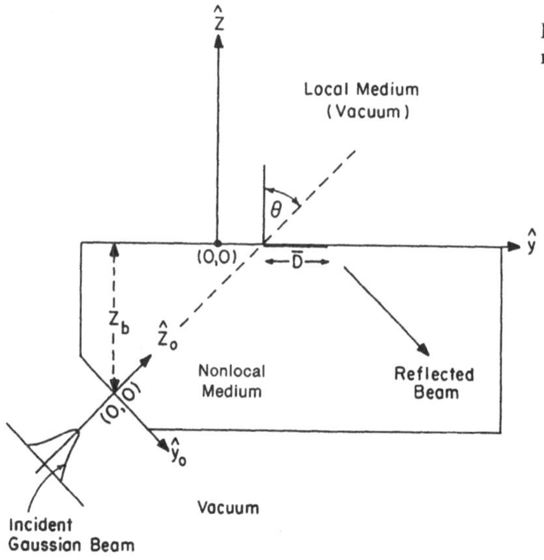

Fig. 2.3. The "Goos-Hänchen" geometry (from [2.69])

$$v_{\mathrm{gr},k} = \frac{\boldsymbol{k}}{k}\,\frac{d\omega}{d\boldsymbol{k}} = v_{\mathrm{gr}}\cos\psi = \frac{\partial\omega}{\partial k} = \frac{c}{\partial(\omega n)/\partial\omega}\,, \tag{2.3.41}$$

where ψ is the angle made by $\boldsymbol{v}_{\mathrm{gr}}$ and $\boldsymbol{k} = (\omega/c)n(\boldsymbol{k}/k)$ and, evidently $n = n(\omega, \boldsymbol{s})$ is the refractive index of the wave considered; recall also that we adopted the notation

$$\frac{d\omega}{d\boldsymbol{k}} = \mathrm{grad}_k\,\omega = \left\{\frac{\partial\omega}{\partial k_x}, \frac{\partial\omega}{\partial k_y}, \frac{\partial\omega}{\partial k_z}\right\}.$$

Equation (2.3.41) results directly from the definition of the derivative of $\omega(k\boldsymbol{s})$ with respect to the direction of the unit vector $\boldsymbol{s}$.

According to its definition – this is formally clear from (2.3.30) – the group velocity $\boldsymbol{v}_{\mathrm{gr}}$ is the velocity of energy propagation (signal velocity) and therefore $v_{\mathrm{gr}} \leqslant c$, i.e.,

$$\frac{c}{v_{\mathrm{gr},k}} = \frac{\partial(\omega n)}{\partial\omega} > 1\,. \tag{2.3.42}$$

This inequality was discussed from another point of view in Sect. 2.2.5 (the signs $\geqslant$ or $\leqslant$ are replaced in relations of the type (2.3.42) by $>$ or $<$ since the equality sign applies solely to empty space). It will be shown in Sect. 2.3.3 that in the absence of spatial dispersion, the angle ψ between $\boldsymbol{v}_{\mathrm{gr}}$ and $\boldsymbol{k}$ is acute. Then, as is clear from (2.3.41, 42)

$$\frac{\partial(\omega n)}{\partial\omega} > 1 , \quad \varepsilon_{ij} = \varepsilon_{ij}(\omega) . \tag{2.3.43}$$

The relation (2.3.30) was obtained from (2.3.29) and, due to the absence of derivatives with respect to k_l, it is universal in the sense that it applies to any quasi-monochromatic normal-wave packet[15]. If $\boldsymbol{k}$ is complex, or even if $\boldsymbol{k}$ is real but the lower sign was taken in (2.3.29), the derivative with respect to k_l cannot be canceled and the relations obtained are evidently useless.

c) The Conservation of Momentum for a Field

Apart from (2.3.30), representing the law of energy conservation, we know of another relation [2.33, 39, 42] which is related to the conservation of momentum. To obtain it we differentiate (2.3.25) and its complex conjugate (with the upper sign) with respect to k_l. If $\boldsymbol{k}$ is real we arrive at the result

$$T_{ij} = -g_i v_{\mathrm{gr},j} , \tag{2.3.44}$$

where

$$\begin{aligned} g_i &= \frac{1}{16\pi c}[(\boldsymbol{D}_0\times\boldsymbol{B}_0^*)_i + (\boldsymbol{D}_0^*\times\boldsymbol{B}_0)_i] + \frac{k_i}{16\pi}\frac{\partial\varepsilon_{jl}(\omega,\boldsymbol{k})}{\partial\omega}E_{0j}^*E_{0l} , \\ T_{ij} &= T_{ij}^{(0)} + T_{ij}^{(1)} , \quad T_{ij}^{(0)} = \frac{B_{0i}B_{0j}^* + B_{0i}^*B_{0j} + E_{0i}D_{0j}^* + E_{0i}^*D_{0j}}{16\pi} \\ &\quad - \frac{\delta_{ij}}{32\pi}(2\boldsymbol{B}_0\cdot\boldsymbol{B}_0^* + \boldsymbol{E}_0\cdot\boldsymbol{D}_0^* + \boldsymbol{E}_0^*\cdot\boldsymbol{D}_0) , \\ T_{ij}^{(1)} &= -\frac{k_i}{16\pi}\frac{\partial\varepsilon_{ml}(\omega,\boldsymbol{k})}{\partial k_j}E_{0m}^*E_{0l} . \end{aligned} \tag{2.3.45}$$

The vector g_i and the tensor T_{ij}, introduced here are frequently considered to be the time-averaged density of the electromagnetic pulse and the stress tensor, respectively. If we disregard the part proportional to $\partial\varepsilon_{jl}/\partial\omega$ and due to spatial dispersion, $\boldsymbol{g}$ has a form corresponding to the choice of the energy-momentum tensor of Minkowski form. It is clear today, however, that such a choice for the four-dimensional energy-momentum tensor of the electromagnetic field is incorrect (see [Ref. 2.7, §§ 16 and 56] as well as [2.49]). However, under some conditions (in particular, when a wave train is propagated through a medium), the vector $\boldsymbol{g}$ – see (2.3.45) – can be interpreted as the average density of the total momentum of the electromagnetic field and the medium.

---

[15] The fact that the packet is quasi-monochromatic is used since we have taken only first derivatives with respect to ω and k_l into account; the fact that we are considering a normal-wave packet is taken into account by (2.2.3, 5), where $j_{\mathrm{ext}} = 0$ and $\rho_{\mathrm{ext}} = 0$ (no sources), and also by (2.3.29), where $\omega = \omega(\boldsymbol{k})$.

Thus, (2.3.44) undoubtedly represents the law of conservation of momentum of the field [2.49]. The problem of the momentum of the field and the forces is not commonly encountered in crystal optics. Hence, we shall not dwell on this problem in more detail; various of its aspects are clear from [2.31, 49, 50] and the references contained therein (special mention may be made of the problem of Van der Waals forces between bodies under conditions when some of them are crystalline [2.51]).

We note that the relations (2.3.30, 44) were obtained in [2.39, 42] where the four-dimensional form of the notation of field equations is used, and applied directly to a transparent medium. As is clear from the aforesaid, this is justified. One cannot object to the use of four-dimensional, i.e., explicitly relativistically invariant expressions in macroscopic electrodynamics. The transition to four-dimensional notation and its treatment in the electrodynamics of continuous media seems to us, however, more complex than the derivation [2.33] of the formulas we are interested in. Hence we shall not make any use of four-dimensional notation in this book.

d) Some Applications of the Poynting Theorem

Let us now consider four special cases which illustrate the application of the Poynting theorem. In Sect. 2.1.3 we established the following expressions characterizing an isotropic nongyrotropic medium with weak spatial dispersion:

$$\begin{aligned} &\boldsymbol{D} = \varepsilon(\omega)\boldsymbol{E} + a(\omega)\,\mathrm{rot}\,\mathrm{rot}\,\boldsymbol{E} + b(\omega)\,\mathrm{grad}\,\mathrm{div}\,\boldsymbol{E}\,, \\ &\boldsymbol{D} = \boldsymbol{D}(\omega, \boldsymbol{r})\,, \qquad \boldsymbol{E} = \boldsymbol{E}(\omega, \boldsymbol{r})\,, \\ &\varepsilon_{ij}(\omega, \boldsymbol{k}) = (\varepsilon + ak^2)\,\delta_{ij} - (a+b)\,k_i k_j\,. \end{aligned} \tag{2.3.46}$$

We assume, for simplicity, that ε, a and b are independent of ω or, more exactly, that this dependence is negligibly weak. These quantities, ε, a and b, are then real (otherwise the reality of $\boldsymbol{E}$ would not guarantee the reality of $\boldsymbol{D}$). Then Poynting's theorem (2.3.1) assumes the form

$$\begin{aligned} &\frac{\partial W}{\partial t} = -\,\mathrm{div}\,\boldsymbol{S} - \boldsymbol{j}_{\mathrm{ext}}\boldsymbol{E}\,, \\ &8\pi W = \varepsilon E^2 + B^2 + a(\mathrm{rot}\,\boldsymbol{E})^2 - b(\mathrm{div}\,\boldsymbol{E})^2\,, \\ &\boldsymbol{S} = \boldsymbol{S}^{(0)} - \frac{a}{4\pi}\left(\boldsymbol{E}\times\mathrm{rot}\frac{\partial \boldsymbol{E}}{\partial t}\right) + \frac{b}{4\pi}\boldsymbol{E}\,\mathrm{div}\frac{\partial \boldsymbol{E}}{\partial t}\,, \\ &\boldsymbol{S}^{(0)} = \frac{c}{4\pi}(\boldsymbol{E}\times\boldsymbol{B})\,. \end{aligned} \tag{2.3.47}$$

In the case of a monochromatic field, (2.3.3), with real ω and $\boldsymbol{k}$, we obtain after high-frequency averaging (omitting the bars indicating averaging over W and $\boldsymbol{S}$)

$$\begin{aligned} 16\pi W &= \varepsilon \boldsymbol{E}_0 \cdot \boldsymbol{E}_0^* + \boldsymbol{B}_0 \cdot \boldsymbol{B}_0^* + a(\boldsymbol{k}\times\boldsymbol{E}_0)\cdot(\boldsymbol{k}\times\boldsymbol{E}_0^*) - b(\boldsymbol{k}\cdot\boldsymbol{E}_0)\cdot(\boldsymbol{k}\cdot\boldsymbol{E}_0^*)\,, \\ 16\pi \boldsymbol{S} &= c(\boldsymbol{E}_0^*\times\boldsymbol{B}_0 + \boldsymbol{E}_0\times\boldsymbol{B}_0^*) \\ &\quad - 2a\omega\boldsymbol{k}\cdot(\boldsymbol{E}_0\cdot\boldsymbol{E}_0^*) + (a+b)\,\omega[\boldsymbol{E}_0\cdot(\boldsymbol{k}\cdot\boldsymbol{E}_0^*) + \boldsymbol{E}_0^*\cdot(\boldsymbol{k}\cdot\boldsymbol{E}_0)]\,. \end{aligned} \tag{2.3.48}$$

On the other hand, according to (2.3.15 and 46)

$$\begin{aligned} 16\pi \boldsymbol{S}^{(1)} &= -\,\omega\left(\frac{\partial\varepsilon'_{ij}}{\partial\boldsymbol{k}}\right)E_{0j}E_{0i}^* \\ &= -\,2a\omega\boldsymbol{k}\cdot(\boldsymbol{E}_0\cdot\boldsymbol{E}_0^*) + (a+b)\,\omega[\boldsymbol{E}_0\cdot(\boldsymbol{k}\cdot\boldsymbol{E}_0^*) + \boldsymbol{E}_0^*\cdot(\boldsymbol{k}\cdot\boldsymbol{E}_0)]\,. \end{aligned} \tag{2.3.49}$$

Equations (2.3.48, 49) agree with each other, as expected.

With the upper sign, (2.3.23) can be written in the form, see (2.3.31),

$$W_0 = \frac{\boldsymbol{D}_0\cdot\boldsymbol{E}_0^* + \boldsymbol{B}_0\cdot\boldsymbol{B}_0^*}{16\pi} = \frac{\boldsymbol{k}\cdot\boldsymbol{S}^{(0)}}{\omega}\,. \tag{2.3.50}$$

If there is no frequency dispersion

$$W_0 = \frac{\boldsymbol{D}_0\cdot\boldsymbol{E}_0^* + \boldsymbol{B}_0\cdot\boldsymbol{B}_0^*}{16\pi} = W = \frac{1}{16\pi}\left\{\frac{\partial}{\partial\omega}[\omega\varepsilon_{ij}(\omega,\boldsymbol{k})]E_{0j}E_{0i}^* + \boldsymbol{B}_0\cdot\boldsymbol{B}_0^*\right\}$$

and therefore, according to (2.3.50), $W = W_0 = \boldsymbol{k}\cdot\boldsymbol{S}^{(0)}/\omega$. On the other hand, when both spatial and frequency dispersion are absent we obtain (2.3.30): $W_0\boldsymbol{v}_{gr} = \boldsymbol{S}^{(0)}$. In this case we have

$$\boldsymbol{k}\cdot\boldsymbol{v}_{gr}/\omega = 1\,, \qquad \varepsilon_{ij}(\omega,\boldsymbol{k}) = \varepsilon_{ij}\,. \tag{2.3.51}$$

Since $\omega/k = v_{ph}$ is the phase velocity and $\boldsymbol{k}\cdot\boldsymbol{v}_{gr}/k = v_{gr,k}$, (2.3.51) means that $v_{gr,k} = v_{ph}$.

e) The Boundary Conditions for a Gyrotropic Medium

As the next example we shall consider a gyrotropic medium with a relation of the type of (2.1.55). For the sake of simplicity we shall assume that there is no frequency dispersion but that the medium can be nonhomogeneous. Specifically, we shall employ the relation

$$\boldsymbol{D} = \varepsilon\boldsymbol{E} + \gamma_{\mathrm{I}}\operatorname{rot}\boldsymbol{E} + \operatorname{rot}(\gamma_{\mathrm{II}}\boldsymbol{E}) = \varepsilon\boldsymbol{E} + \gamma\operatorname{rot}\boldsymbol{E} + [\nabla\gamma_{\mathrm{II}}\boldsymbol{E}]\,, \tag{2.3.51 a}$$

where the factors ε, γ_{I}, γ_{II} and $\gamma = \gamma_{\mathrm{I}} + \gamma_{\mathrm{II}}$, not to mention the fields $\boldsymbol{D}$ and $\boldsymbol{E}$, may depend upon the coordinates.

Making use of (2.3.51a) enables us, incidentally, to illustrate the appearance of the densities σ and $\boldsymbol{i}$ in the boundary condition, (2.1.2). Integrating the equation $\operatorname{div} \boldsymbol{D} = 0$ along a direction perpendicular to the broadened interface, and letting the width of the broadening tend to zero, we obtain the boundary condition (where $\boldsymbol{n}$ is the normal to the interface):

$$D_{2n} - D_{1n} = \gamma_{\mathrm{II}}^{(2)} \operatorname{rot}_n \boldsymbol{E}^{(2)} - \gamma_{\mathrm{II}}^{(1)} \operatorname{rot}_n \boldsymbol{E}^{(1)} \equiv 4\pi\sigma . \tag{2.3.51 b}$$

The emergence of the density σ is evidently associated with the fact that in employing (2.3.51a) the integrals

$$\int_0^l \frac{\partial D_x}{\partial x} dz \quad \text{and} \quad \int_0^l \frac{\partial D_y}{\partial x} dz$$

cannot be neglected even when the thickness l of the transition layer between the media tends to zero (here the normal to the interface is made coincident with the z axis; an analogous application of the first equation of (2.1.1) leads to an expression for the current density $\boldsymbol{i}$).

Substituting (2.3.51a) into (2.3.1) we obtain

$$\frac{\partial}{\partial t}\left(\frac{\varepsilon E^2 + B^2 + \gamma(\boldsymbol{E} \cdot \operatorname{rot} \boldsymbol{E})}{8\pi}\right)$$

$$= -\frac{c}{4\pi} \operatorname{div}\left[(\boldsymbol{E} \times \boldsymbol{B}) - \frac{\gamma}{2c}\left(\boldsymbol{E} \times \frac{\partial \boldsymbol{E}}{\partial t}\right)\right] + \nabla(2\gamma_{\mathrm{II}} - \gamma)$$

$$\times \left(\frac{\boldsymbol{E}}{8\pi} \times \frac{\partial \boldsymbol{E}}{\partial t}\right) - \boldsymbol{j}_{\mathrm{ext}} \cdot \boldsymbol{E} .$$

It can be seen from this expression that when $\gamma \neq 0$ (i.e., when spatial dispersion is taken into account) an additional term $\gamma(\boldsymbol{E} \operatorname{rot} \boldsymbol{E})/8\pi$ appears in the expression for the energy density and the term $-(\gamma/2c)[\boldsymbol{E} \times (\partial \boldsymbol{E}/\partial t)]$ in the expression for the energy flux density. Another term which appears is

$$\boldsymbol{A} = \frac{1}{8\pi} \nabla(2\gamma_{\mathrm{II}} - \gamma)\left(\boldsymbol{E} \times \frac{\partial \boldsymbol{E}}{\partial t}\right),$$

which is proportional to $\nabla(2\gamma_{\mathrm{II}} - \gamma) \equiv \nabla(\gamma_{\mathrm{II}} - \gamma_{\mathrm{I}})$ and, consequently, is localized near the interface between two media (or, in principle, at other places where the medium is inhomogeneous). If $\boldsymbol{A} \neq 0$, as could occur when

$\gamma_{\rm I} \mp \gamma_{\rm II}$, absorption (or evolution) of energy would be observed at the interface. Such a result is unusual, of course, but in principle may be possible, for instance, in exciting surface waves. Nevertheless, the emergence of the term A seems to be suspicious and poses the question: will it disappear in the general form as a consequence of requiring the condition

$$\gamma_{\rm I} = \gamma_{\rm II} = \gamma/2 .$$

The observance of such an equality is precisely what was asserted in [2.52] on the basis of the requirement that the Poynting theorem should be of the form of the conservation law $\partial W/\partial t + \operatorname{div} S = 0$. Such a line of reasoning seems to us to be insufficiently substantiated [2.53] in the presence of dispersion and absorption, as well as the possibility that surface waves may appear (see above). Nevertheless, the equality $\gamma_{\rm I} = \gamma_{\rm II}$ turns out to be actually valid in the general form, because it follows from the principle of symmetry of the kinetic coefficients [2.54–56]. One should also taken into consideration the relationships which define the ellipticity of linearly polarized light reflected from a gyrotropic medium (discussed in [2.57]).

With respect to the resonance region, we could have used an analogous expansion for $\boldsymbol{E}$ instead of (2.3.51 a) and obtained corrections to the "old" boundary conditions. It is more important in this case, however, to obtain additional boundary conditions, which are required in connection with the appearance of new solutions (Sect. 4.5.). This problem has not yet been sufficiently investigated for gyrotropic media (nevertheless, see Sects. 4.5.6, 9).

In the case of an anisotropic, gyrotropic and inhomogeneous medium, and far from resonance, the material equation relating the vectors $\boldsymbol{D}$ and $\boldsymbol{E}$ is of the form [2.52, 54]

$$D_i(\boldsymbol{r},t) = \varepsilon_{ij}(\boldsymbol{r}) E_j(\boldsymbol{r},t) + \gamma_{ijl}(\boldsymbol{r}) \nabla_l E_j(\boldsymbol{r}) + \tfrac{1}{2} E_j(\boldsymbol{r}) \nabla_l \gamma_{ijl}(\boldsymbol{r}) . \tag{2.3.51 c}$$

At the plane boundary of two media ($z = 0$) and under conditions when the tensor γ_{ijl} depends only on z and is subject to a discontinuity at $z = 0$ ($\gamma = \gamma^{(1)}$ at $z < 0$, and $\gamma = \gamma^{(2)}$ at $z > 0$), the value of the surface charge σ, appearing in the boundary condition for normal components of the induction, is determined from the relationship

$$4\pi\sigma \equiv \frac{1}{2}\left\{\gamma^{(2)}_{1j3}\frac{\partial E^{(2)}_j}{\partial x} + \gamma^{(2)}_{2j3}\frac{\partial E^{(2)}_j}{\partial y} - \gamma^{(1)}_{1j3}\frac{\partial E^{(1)}_j}{\partial x} - \gamma^{(1)}_{2j3}\frac{\partial E^{(1)}_j}{\partial y}\right\} . \tag{2.3.51 d}$$

f) The Boundary Conditions for the Optically Nonlinear Medium in the Center of Inversion

We point out that with spatial dispersion taken into account surface currents and charges may occur in the boundary conditions for the fields in nonlinear optics as well. To illustrate this we consider, for the sake of simplicity, an isotropic medium with a center of inversion. In such media the nonlinear part of the polarization, leading to the generation of the second harmonics, is of the form

$$\boldsymbol{P}^{\mathrm{NL}} = \frac{1}{4\pi}[\alpha \boldsymbol{E} \cdot \operatorname{div} \boldsymbol{E} + \beta(\boldsymbol{E} \cdot \nabla) \cdot \boldsymbol{E} + \gamma \boldsymbol{E} \times \operatorname{rot} \boldsymbol{E}] , \qquad (2.3.52)$$

so that

$$\boldsymbol{D} = \varepsilon \boldsymbol{E} + 4\pi \boldsymbol{P}^{\mathrm{NL}} .$$

If this relation is substituted into (2.3.1) with $j_{\mathrm{ext}} = 0$, and it is assumed that the medium is infinite and that there is no frequency dispersion, then (2.3.1) takes the form of the energy conservation law

$$\frac{\partial W_0}{\partial t} + \operatorname{div} \boldsymbol{S}_0 = 0 , \qquad (2.3.53)$$

where

$$W_0 = \frac{1}{8\pi}(\varepsilon E^2 + B^2 + 2\alpha E^2 \operatorname{div} \boldsymbol{E}) \quad \text{and} \quad \boldsymbol{S}_0 = \frac{c}{4\pi}(\boldsymbol{E} \times \boldsymbol{B}) - \frac{\alpha \boldsymbol{E}}{8\pi} \frac{\partial E^2}{\partial t}$$

only when $\alpha = -\beta = -\gamma$ [2.58]. If, however, the medium is inhomogeneous or, specifically, confined, then the factors α, β and γ depend on the coordinates. Then, as in the case of gyrotropy considered above, the new terms of the expression for the induction that are to be taken into account are those quadratic ones with respect to the field, but proportional to the gradients of certain material constants [2.59].

Thus

$$\boldsymbol{D} = \varepsilon \boldsymbol{E} + \alpha[\boldsymbol{E} \cdot \operatorname{div} \boldsymbol{E} - (\boldsymbol{E} \cdot \nabla) \cdot \boldsymbol{E} - \boldsymbol{E} \times \operatorname{rot} \boldsymbol{E}] + \boldsymbol{E} \cdot (\boldsymbol{E} \cdot \nabla \rho) + E^2 \nabla \mu , \qquad (2.3.54)$$

where ρ and μ are certain new parameters, for instance, characterizing nonlinear properties of the interface between the media. If we substitute into Poynting's theorem (2.3.1) the expression we have written for $\boldsymbol{D}$, then this relation takes the form of the conservation law $\partial W/\partial t + \operatorname{div} \boldsymbol{S} = 0$ only when $\rho = \alpha + 2\mu$. Here $W = W_0 + W_1$, $W_1 = E^2(\boldsymbol{E} \cdot \nabla \cdot \rho)/4\pi$ and $\boldsymbol{S} = \boldsymbol{S}_0$. An interesting feature of the expression for W is the occurrence of a surface energy density [at the abrupt boundary, the corresponding terms are proportional to the delta function $\delta(z)$]. We also note that if the terms proportional to $\nabla \rho$ and $\nabla \mu$ are omitted in the expression used for $\boldsymbol{D}$, nonphysical sources of energy appear on the interface between media.

In particular, the application of a material relation for $\boldsymbol{D}$ enables the boundary conditions to be determined for fields of the second harmonic.

These boundary conditions are required to calculate the intensity of second-harmonic generation that occurs when light is reflected from the surfaces of media having a center of inversion (for a more detailed discussion, see [2.59]).

g) Spatial Dispersion and Orthogonality of Normal Waves

In the foregoing we have considered expressions which were quadratic with respect to the normal wave amplitudes, both amplitudes referring to one and the same wave. Apart from these there also exist important relations linking the amplitudes and the fields themselves for various waves.

Consider two normal waves which will be distinguished by the indices 1 and 2; both are assumed to propagate in the direction $\boldsymbol{s}$

$$\begin{aligned} \boldsymbol{D}_{1,2} &= \boldsymbol{D}_{0,1,2}\, e^{i([\omega_{1,2}/c]\tilde{n}_{1,2}\boldsymbol{s}\cdot\boldsymbol{r} - \omega_{1,2}t)}, \\ \boldsymbol{E}_{1,2} &= \boldsymbol{E}_{0,1,2}\, e^{i([\omega_{1,2}/c]\tilde{n}_{1,2}\boldsymbol{s}\cdot\boldsymbol{r} - \omega_{1,2}t)}, \\ \boldsymbol{D}_{1,2} &= \tilde{n}_{1,2}^2[\boldsymbol{E}_{1,2} - \boldsymbol{s}\cdot(\boldsymbol{s}\cdot\boldsymbol{E}_{1,2})], \quad \boldsymbol{s}\cdot\boldsymbol{D}_{1,2} = 0. \end{aligned} \tag{2.3.55}$$

It follows directly that

$$\boldsymbol{D}_1\cdot\boldsymbol{D}_2 = \tilde{n}_1^2\boldsymbol{E}_1\cdot\boldsymbol{D}_2 = \tilde{n}_2^2\boldsymbol{E}_2\cdot\boldsymbol{D}_1. \tag{2.3.56}$$

If $\tilde{n}_1^2 \neq \tilde{n}_2^2$ and we make use of the symmetry relation (2.1.21) with $\boldsymbol{k} = (\omega/c)\tilde{n}\boldsymbol{s}$ and $\boldsymbol{B}_{\text{ext}} = 0$, i.e., of the fact that $\varepsilon_{ij}[\omega, (\omega/c)\tilde{n}\boldsymbol{s}] = \varepsilon_{ji}[\omega, -(\omega/c)\tilde{n}\boldsymbol{s}]$, and of the analogous expression for ε_{ij}^{-1} we obtain from (2.3.56):

$$\begin{aligned} \boldsymbol{D}_1\cdot\boldsymbol{D}_2\left(\frac{1}{\tilde{n}_1^2} - \frac{1}{\tilde{n}_2^2}\right) &= \left[\varepsilon_{ij}\left(\omega_2, \frac{\omega_2}{c}\tilde{n}_2\boldsymbol{s}\right) - \varepsilon_{ij}\left(\omega_1, -\frac{\omega_1}{c}\tilde{n}_1\boldsymbol{s}\right)\right]E_{1i}E_{2j} \\ &= \left[\varepsilon_{ij}^{-1}\left(\omega_1, \frac{\omega_1}{c}\tilde{n}_1\boldsymbol{s}\right) - \varepsilon_{ij}^{-1}\left(\omega_2, -\frac{\omega_2}{c}\tilde{n}_2\boldsymbol{s}\right)\right]D_{1j}D_{2i}. \end{aligned} \tag{2.3.57}$$

In the absence of spatial dispersion and if $\omega_1 = \omega_2$ it follows from (2.3.57) that

$$\boldsymbol{D}_1\cdot\boldsymbol{D}_2 = 0 \quad \text{at} \quad \omega_1 = \omega_2, \quad \varepsilon_{ij} = \varepsilon_{ij}(\omega). \tag{2.3.58}$$

We should not forget that the vector $\boldsymbol{D}$ is complex, in general, i.e., $\boldsymbol{D} = \boldsymbol{D}' + i\boldsymbol{D}''$.

By analogous considerations that led to (2.3.54) we obtain the relation ($\boldsymbol{D} = \boldsymbol{D}' + i\boldsymbol{D}''$, $\boldsymbol{D}^* = \boldsymbol{D}' - i\boldsymbol{D}''$)

$$\boldsymbol{D}_1 \cdot \boldsymbol{D}_2^* \left(\frac{1}{\tilde{n}_1^2} - \frac{1}{(\tilde{n}_2^2)^*} \right) = \left[\varepsilon_{ij}^* \left(\omega_2, \frac{\omega_2}{c} \tilde{n}_2 s \right) - \varepsilon_{ij} \left(\omega_1, -\frac{\omega_1}{c} \tilde{n}_1 s \right) \right] E_{1i} E_{2j}^*$$

$$= \left\{ \varepsilon_{ij}^{-1} \left(\omega_1, \frac{\omega_1}{c} \tilde{n}_1 s \right) - \left[\varepsilon_{ji}^{-1} \left(\omega_2, -\frac{\omega_2}{c} \tilde{n}_2 s \right) \right]^* \right\} D_{1j} D_{2i}^* . \qquad (2.3.59)$$

If $\omega_1 = \omega_2$, if there is no spatial dispersion and absorption is neglected, the tensor ε_{ij} is Hermitian and independent of $\tilde{n}$.

Under such conditions

$$\boldsymbol{D}_1 \cdot \boldsymbol{D}_2^* = 0 \quad \text{at} \quad \omega_1 = \omega_2 , \quad \varepsilon_{ij} = \varepsilon'_{ij}(\omega) . \qquad (2.3.60)$$

If spatial dispersion is taken into account then, even in the case of a transparent medium (the index $\tilde{n} = n$ is real and the tensor $\varepsilon_{ij} = \varepsilon'_{ij}$ is Hermitian), the condition $\boldsymbol{D}_1 \cdot \boldsymbol{D}_2^* = 0$ is violated – in (2.3.59) we assumed that $\tilde{n}_1^2 \neq \tilde{n}_2^2$.

The violation of the conditions $\boldsymbol{D}_1 \cdot \boldsymbol{D}_2 = 0$ and $\boldsymbol{D}_1 \cdot \boldsymbol{D}_2^* = 0$ is one of the characteristic features of crystal optics with spatial dispersion (Sect. 4.1).

h) The Reciprocity Theorem

In concluding this subsection, we shall discuss the theorem of reciprocity which relates the fields $\boldsymbol{E}^{(1)}(\boldsymbol{r}, \omega)$ and $\boldsymbol{E}^{(2)}(\boldsymbol{r}, \omega)$ produced by external sources which correspond to the current densities $\boldsymbol{j}_{\text{ext}}^{(1)}(\boldsymbol{r}, \omega)$ and $\boldsymbol{j}_{\text{ext}}^{(2)}(\boldsymbol{r}, \omega)$.

The fields produced by the sources $\boldsymbol{j}_{\text{ext}}^{(1)}$ and $\boldsymbol{j}_{\text{ext}}^{(2)}$ satisfy the equations

$$\begin{aligned} \operatorname{rot} \boldsymbol{B}^{(1)} &= \mathrm{i} \frac{\omega}{c} \boldsymbol{D}^{(1)} + \frac{4\pi}{c} \boldsymbol{j}_{\text{ext}}^{(1)} , \\ \operatorname{rot} \boldsymbol{E}^{(1)} &= -\frac{\mathrm{i}\omega}{c} \boldsymbol{B}^{(1)} , \end{aligned} \qquad (2.3.61\,\text{a})$$

$$\begin{aligned} \operatorname{rot} \boldsymbol{B}^{(2)} &= \mathrm{i} \frac{\omega}{c} \boldsymbol{D}^{(2)} + \frac{4\pi}{c} \boldsymbol{j}_{\text{ext}}^{(2)} , \\ \operatorname{rot} \boldsymbol{E}^{(2)} &= -\frac{\mathrm{i}\omega}{c} \boldsymbol{B}^{(2)} . \end{aligned} \qquad (2.3.61\,\text{b})$$

Here the Fourier components with respect to ω are evidently used, but the dependence of all quantities on $\boldsymbol{r}$ is preserved.

Multiplying (2.3.61 a) by $\boldsymbol{E}^{(2)}$ and $\boldsymbol{B}^{(2)}$, respectively, and (2.3.61 b) by $-\boldsymbol{E}^{(1)}$ and $-\boldsymbol{B}^{(1)}$, respectively, adding them and applying the identity $\operatorname{div}(\boldsymbol{A} \times \boldsymbol{B}) = \boldsymbol{B} \times \operatorname{rot} \boldsymbol{A} - \boldsymbol{A} \times \operatorname{rot} \boldsymbol{B}$, we obtain

$$\operatorname{div}(\boldsymbol{E}^{(1)}\times\boldsymbol{B}^{(2)}-\boldsymbol{E}^{(2)}\times\boldsymbol{B}^{(1)})$$
$$=\frac{4\pi}{c}(\boldsymbol{j}^{(1)}_{\text{ext}}\cdot\boldsymbol{E}^{(2)}-\boldsymbol{j}^{(2)}_{\text{ext}}\cdot\boldsymbol{E}^{(1)})+\frac{\mathrm{i}\omega}{c}(\boldsymbol{D}^{(1)}\cdot\boldsymbol{E}^{(2)}-\boldsymbol{E}^{(1)}\cdot\boldsymbol{D}^{(2)})\,. \qquad (2.3.62)$$

If this expression is integrated over the volume, the first term becomes a surface integral and vanishes (the field is assumed to decrease sufficiently fast at infinity; it is readily seen that the discontinuity surface also makes no contribution to the integral). As a result we obtain the reciprocity theorem

$$\int\boldsymbol{j}^{(1)}_{\text{ext}}(\boldsymbol{r},\omega)\cdot\boldsymbol{E}^{(2)}(\boldsymbol{r},\omega)\,d\boldsymbol{r}=\int\boldsymbol{j}^{(2)}_{\text{ext}}(\boldsymbol{r},\omega)\cdot\boldsymbol{E}^{(1)}(\boldsymbol{r},\omega)\,d\boldsymbol{r}\,, \qquad (2.3.63)$$

provided

$$\int(\boldsymbol{D}^{(1)}\cdot\boldsymbol{E}^{(2)}-\boldsymbol{E}^{(1)}\cdot\boldsymbol{D}^{(2)})\,d\boldsymbol{r}=0\,. \qquad (2.3.64)$$

In the absence of spatial dispersion and an external field ($\boldsymbol{B}_{\text{ext}}=0$), the tensor $\varepsilon_{ij}(\omega)=\varepsilon_{ji}(\omega)$ and (2.3.64) is obviously satisfied since

$$\boldsymbol{E}^{(1)}\cdot\boldsymbol{D}^{(2)}=\varepsilon_{ij}E^{(2)}_jE^{(1)}_i=\varepsilon_{ji}E^{(2)}_jE^{(1)}_i=\boldsymbol{D}^{(1)}\cdot\boldsymbol{E}^{(2)}\,;$$

the medium is assumed to be nonmagnetic and without spatial dispersion, but there is no difficulty in generalizing to the case of a magnetic medium [Ref. 2.2, § 29]. If $\boldsymbol{B}_{\text{ext}}=0$, but spatial dispersion is taken into account, then (2.3.64) is also valid because this equation expresses the principle of symmetry of the kinetic coefficients in macroscopic electrodynamics [Ref. 2.7, § 83]. For the sake of completeness we prove (2.3.64) directly from (2.1.21) with $\boldsymbol{B}_{\text{ext}}=0$, i.e., with the condition $\varepsilon_{ij}(\omega,\boldsymbol{k})=\varepsilon_{ji}(\omega,-\boldsymbol{k})$. According to (2.1.4 and 6)

$$D_i(\boldsymbol{r},\omega)=\int\hat{\varepsilon}_{ij}(\omega,\boldsymbol{r}-\boldsymbol{r}')E_j(\boldsymbol{r}',\omega)\,d\boldsymbol{r}'\,,$$

$$\hat{\varepsilon}_{ij}(\omega,\boldsymbol{r}-\boldsymbol{r}')=\int_0^\infty\hat{\varepsilon}_{ij}(\tau,\boldsymbol{r}-\boldsymbol{r}')\,\mathrm{e}^{\mathrm{i}\omega\tau}d\tau\,,$$

$$\varepsilon_{ij}(\omega,\boldsymbol{k})=\int\hat{\varepsilon}_{ij}(\omega,\boldsymbol{R})\,\mathrm{e}^{-\mathrm{i}\boldsymbol{k}\cdot\boldsymbol{r}}d\boldsymbol{R}\,.$$

We see from this that the condition $\varepsilon_{ij}(\omega,\boldsymbol{k})=\varepsilon_{ji}(\omega,-\boldsymbol{k})$ implies that

$$\hat{\varepsilon}_{ij}(\omega,\boldsymbol{R})=\hat{\varepsilon}_{ji}(\omega,-\boldsymbol{R})\,,\qquad \boldsymbol{R}=\boldsymbol{r}-\boldsymbol{r}'\,. \qquad (2.3.65)$$

For an inhomogeneous medium we have to use (2.1.3), and the symmetry condition

$$\hat{\varepsilon}_{ij}(t,t',\boldsymbol{r},\boldsymbol{r}')=\hat{\varepsilon}_{ji}(t,t',\boldsymbol{r}',\boldsymbol{r}) \qquad (2.3.66)$$

is obtained from the principle of symmetry of the kinetic coefficients, or, if the medium is homogeneous in time, we find

$$\hat{\varepsilon}_{ij}(\omega,\boldsymbol{r},\boldsymbol{r}') = \hat{\varepsilon}_{ji}(\omega,\boldsymbol{r}',\boldsymbol{r})\,. \tag{2.3.67}$$

If the medium is spatially homogeneous, (2.3.67) becomes (2.3.65). From (2.3.67) and

$$D_i(\boldsymbol{r},\omega) = \int \hat{\varepsilon}_{ij}(\omega,\boldsymbol{r},\boldsymbol{r}')E_j(\boldsymbol{r}',\omega)\,d\boldsymbol{r}'$$

we can readily see that (2.3.64) is satisfied. But, as we have already mentioned, we can start in fact from (2.3.64), and obtain the condition (2.3.67).

In the presence of a magnetic field $\boldsymbol{B}_{\text{ext}}$, (2.3.64) no longer applies and this is true also for the reciprocity theorem (2.3.63). If $\boldsymbol{B}_{\text{ext}} \neq 0$ the symmetry conditions, see (2.1.21), are of the form

$$\begin{aligned} &\varepsilon_{ij}(\omega,\boldsymbol{k},\boldsymbol{B}_{\text{ext}}) = \varepsilon_{ji}(\omega,-\boldsymbol{k},-\boldsymbol{B}_{\text{ext}})\,,\\ &\hat{\varepsilon}_{ij}(\omega,\boldsymbol{r},\boldsymbol{r}',\boldsymbol{B}_{\text{ext}}) = \hat{\varepsilon}_{ji}(\omega,\boldsymbol{r}',\boldsymbol{r},-\boldsymbol{B}_{\text{ext}})\,, \end{aligned} \tag{2.3.68}$$

and therefore the relations [Ref. 2.2, § 29]

$$\begin{aligned} \int [&\boldsymbol{D}^{(1)}(\boldsymbol{r},\omega,\boldsymbol{B}_{\text{ext}})\boldsymbol{E}^{(2)}\cdot(\boldsymbol{r},\omega,\boldsymbol{B}_{\text{ext}})\\ &-\boldsymbol{E}^{(1)}(\boldsymbol{r},\omega,-\boldsymbol{B}_{\text{ext}})\cdot\boldsymbol{D}^{(2)}(\boldsymbol{r},\omega,-\boldsymbol{B}_{\text{ext}})]\,d\boldsymbol{r} = 0\,, \end{aligned} \tag{2.3.69}$$

$$\int \boldsymbol{j}^{(1)}_{\text{ext}}(\boldsymbol{r},\omega)\cdot\boldsymbol{E}^{(2)}(\boldsymbol{r},\omega,\boldsymbol{B}_{\text{ext}})\,d\boldsymbol{r} = \int \boldsymbol{j}^{(2)}_{\text{ext}}(\boldsymbol{r},\omega)\cdot\boldsymbol{E}^{(1)}(\boldsymbol{r},\omega,-\boldsymbol{B}_{\text{ext}})\,d\boldsymbol{r} \tag{2.3.70}$$

are valid.

The generalized reciprocity theorem (2.3.70) is weaker than the common one (2.3.63) in the sense that in order to satisfy (2.3.70) we have to consider the field $\boldsymbol{E}^{(1)}$ in a medium in which the sign of $\boldsymbol{B}_{\text{ext}}$ is reversed, compared with the case where the field $\boldsymbol{E}^{(2)}$ due to the sources $\boldsymbol{j}^{(2)}_{\text{ext}}$ is calculated. (In certain special cases, of course, theorem (2.3.70) coincides with (2.3.63) even if $\boldsymbol{B}_{\text{ext}} \neq 0$ [Ref. 2.2, § 29].) The fact that the reciprocity theorem (2.3.63) is violated makes it possible to produce an optical valve in a magnetoactive medium, permitting waves to pass through only in one direction.

As we have already indicated in Sect. 2.1, for a medium traveling at the velocity $\boldsymbol{u}$ the symmetry condition has the form

$$\varepsilon_{ij}(\omega,\boldsymbol{k},\boldsymbol{B}_{\text{ext}},\boldsymbol{u}) = \varepsilon_{ji}(\omega,-\boldsymbol{k},-\boldsymbol{B}_{\text{ext}},-\boldsymbol{u})\,.$$

Therefore, evidently, the reciprocity theorem (2.3.70) is generalized as follows:

$$\int j_{\text{ext}}^{(1)}(r,\omega)\cdot E^{(2)}(r,\omega,B_{\text{ext}},u)\,dr$$
$$=\int j_{\text{ext}}^{(2)}(r,\omega)\cdot E^{(1)}(r,\omega,-B_{\text{ext}},-u)\,dr\,. \qquad (2.3.70\text{a})$$

Naturally, this theorem is also valid when $B_{\text{ext}}=0$, i.e., in the absence of an external magnetic field. In this case, but when $u \neq 0$ (traveling medium), the familiar reciprocity theorem (2.3.63) is, in general, also inapplicable.

The reciprocity theorem is usually applied to dipoles. Thus, in the case of electric point dipoles $j_{\text{ext}}^{(1,2)}=\mathrm{i}\omega p^{(1,2)}\delta(r-r_{1,2})$ and from (2.3.70) we have

$$p^{(1)}\cdot E^{(2)}(r_1,B_{\text{ext}})=p^{(2)}\cdot E^{(1)}(r_2,-B_{\text{ext}})\,, \qquad (2.3.70\text{b})$$

where, for example, $E^{(2)}(r_1,B_{\text{ext}})$ is the field E produced at point r_1 by dipole 2 located at point r_2 and having the dipole moment $p^{(2)}$; the external magnetic field is then B_{ext}. In many cases the use of the reciprocity theorem greatly facilitates the solution of electrodynamic problems [2.2, 60–62].

2.3.3 Certain Theorems for Ray and Wave Propagation in a Medium

Since in crystal optics the influence of spatial dispersion is characterized by a small parameter it is, as has been repeatedly indicated, important to focus attention upon qualitatively new aspects. In this connection we shall mention three theorems which are valid in cases where spatial dispersion is absent, and which can be violated by its presence [2.33].

The first theorem is the following: in the absence of spatial dispersion the angle ψ between v_{gr} and the wave vector k (direction of phase velocity) is acute, i.e., $|\psi|<\pi/2$. We consider, of course, a transparent medium where the vector v_{gr} corresponds to the signal velocity.

Without spatial dispersion, the energy flux $S=S^{(0)}$, and (2.3.30) has the form, see (2.3.34):

$$Wv_{\text{gr}}=S^{(0)}\,,\quad S^{(0)}=\frac{c}{16\pi}(E_0\times B_0^*+E_0^*\times B_0)=\frac{c}{8\pi}(E_0\times B_0^*)\,. \qquad (2.3.71)$$

Using the field equations (2.2.5 and 5a) for the normal waves under consideration, we can write

$$S^{(0)}=\frac{c}{16\pi}(E_0\times B_0^*+E_0^*\times B_0)$$
$$=\frac{c^2}{16\pi\omega}[(k^*+k)(E\cdot E_0^*)-E_0(k\cdot E_0^*)-E_0^*(k^*\cdot E_0)]\,. \qquad (2.3.72)$$

Here, of course, the vector $k=k'+\mathrm{i}k''$ is not considered to be real. Projecting (2.3.72) onto k' we obtain

$$\boldsymbol{k}'\cdot \boldsymbol{S}^{(0)} = \frac{c^2}{16\pi\omega}[2(k')^2(\boldsymbol{E}_0\cdot \boldsymbol{E}_0^*) - 2(\boldsymbol{k}'\cdot \boldsymbol{E}_0)(\boldsymbol{k}'\cdot \boldsymbol{E}_0^*) \\ - \mathrm{i}(\boldsymbol{k}'\cdot \boldsymbol{E}_0)(\boldsymbol{k}''\cdot \boldsymbol{E}_0^*) + \mathrm{i}(\boldsymbol{k}'\cdot \boldsymbol{E}_0^*)(\boldsymbol{k}''\cdot \boldsymbol{E}_0)] . \tag{2.3.73}$$

If the vectors $\boldsymbol{k}'$ and $\boldsymbol{k}''$ are parallel (collinear), i.e., if the waves are homogeneous and plane, then $\boldsymbol{k} = k\boldsymbol{s} = (k' + \mathrm{i}k'')\boldsymbol{s} = (\omega/c)\tilde{n}\boldsymbol{s} = (\omega/c)(n + \mathrm{i}\varkappa)\boldsymbol{s}$ and from (2.3.73) it is clear that with $\omega n \neq 0$

$$\boldsymbol{s}\cdot \boldsymbol{S}^{(0)} > 0 . \tag{2.3.74}$$

The equality $\boldsymbol{s}\cdot \boldsymbol{S}^{(0)} = 0$ would correspond to the longitudinal field $\boldsymbol{E}_0 = E_0\boldsymbol{s}$ when $\boldsymbol{S}^{(0)} = 0$.

Thus the time-averaged Poynting vector $\boldsymbol{S}^{(0)}$ always makes an acute angle with the direction of the wave vector $\boldsymbol{k} = k\boldsymbol{s}$ of homogeneous plane waves. In the region of transparency, without spatial dispersion, the vectors $\boldsymbol{v}_{\mathrm{gr}}$ and $\boldsymbol{S}^{(0)}$ are parallel, see (2.3.72). This constitutes a proof of the above statement about the acuteness of the angle ψ between $\boldsymbol{v}_{\mathrm{gr}}$ and $\boldsymbol{k} = k\boldsymbol{s}$.

The second theorem reads: in a medium in equilibrium, without spatial dispersion, the product $n\varkappa > 0$, i.e., $n = \mathrm{Re}\{\tilde{n}\}$ and $\varkappa = \mathrm{Im}\{\tilde{n}\}$ have the same sign (the frequency $\omega = \omega' > 0$ is assumed to be real). This indicates that for a wave propagating along the arbitrary direction z and having the form

$$\boldsymbol{E} = \boldsymbol{E}_0 \exp\left[\mathrm{i}\left(\frac{\omega}{c}nz - \omega t\right)\right]\exp\left(-\frac{\omega}{c}\varkappa z\right),$$

the amplitude decreases in the same direction, the wave is propagating. In an isotropic medium, $(n + \mathrm{i}\varkappa)^2 = \varepsilon' + \mathrm{i}\varepsilon''$ and, if this medium is in thermodynamic equilibrium, $\varepsilon'' > 0$. Hence we find $2n\varkappa = \varepsilon'' > 0$, and the proof is obvious. In the case of an anisotropic medium we consider the quadratic form

$$[\varepsilon_{ij}(\omega, \boldsymbol{B}_{\mathrm{ext}}) - \tilde{n}^2(\delta_{ij} - s_i s_j)]\, a_i a_j^* , \tag{2.3.75}$$

where a_i is an arbitrary complex vector. The dispersion equations (2.2.10, 22) are obtained when the determinant of the expression is set equal to zero. The solution of the dispersion equation can therefore be represented in the form

$$\tilde{n}^2 = \frac{\varepsilon_{ij} a_i a_j^*}{(\delta_{ij} - s_i s_j)\, a_i a_j^*} .$$

Moreover, as was shown in Sect. 2.1, for a medium in equilibrium $\mathrm{Im}\{u\} = \mathrm{Im}\{\varepsilon_{ij} a_i a_j^*\} > 0$ when $\omega = \omega' > 0$, see (2.1.32). It is also evident that $(\delta_{ij} - s_i s_j)\, a_i a_j^*$ is real. Thus [2.11, 33]

$$\mathrm{Im}\{\tilde{n}^2\} = 2n\varkappa > 0 , \quad \omega = \omega' > 0 . \tag{2.3.76}$$

This result is inapplicable to the case of spatial dispersion since then, for a complex $\boldsymbol{k}$ (when there is absorption, the vector $\boldsymbol{k}$ is always complex for normal waves in a medium in equilibrium if the frequency ω is real) it is impossible to guarantee the validity of $\mathrm{Im}\{u\} > 0$ (Sect. 2.1.2). For a medium not in equilibrium, the inequality $\mathrm{Im}\{u\} > 0$ may also be violated since, if waves are propagating in the medium, heat may be absorbed rather than liberated (this means that $\varepsilon'' < 0$ for an isotropic medium, for example, in the case of masers and lasers).

The third theorem refers to the propagation of normal waves in a medium without spatial dispersion and without absorption. In this case the quantity

$$\tilde{n}^2 = (n + \mathrm{i}\varkappa)^2 \tag{2.3.77}$$

is always real. This means that in a transparent medium, $\tilde{n}^2 = n^2$ and $\varkappa = 0$, and in a nonabsorbing nontransparent medium, $\tilde{n}^2 = -\varkappa^2$ and $n = 0$ (in the latter case, the field is proportional to the factor $\exp[-(\omega/c)/\varkappa z]$, i.e., it is damped without oscillations).

The reality of $\tilde{n}^2$ can be proved in various ways, but the proof is always based on the fact that under the specified conditions the tensor $\varepsilon_{ij} = \varepsilon_{ij}(\omega, \boldsymbol{B}_{\text{ext}}) = \varepsilon'_{ij}(\omega, \boldsymbol{B}_{\text{ext}})$ is Hermitian. Therefore, as shown in the discussion of (2.3.75), the quantity

$$\tilde{n}^2 = \frac{\varepsilon_{ij}(\omega, \boldsymbol{B}_{\text{ext}}) a_i a_j^*}{(\delta_{ij} - s_i s_j) a_i a_j^*}$$

is real. We arrive at the same result when we consider the dispersion equation in the form (2.2.46) because, since the tensor ε_{ij} is Hermitian, the inverse tensor ε_{ij}^{-1} must also be Hermitian. If the tensor $\gamma_{\alpha\beta} \equiv \varepsilon_{\alpha\beta}^{-1}$ is Hermitian, we readily deduce from the quadratic equation (2.2.46) that $\tilde{n}^2$ is in fact always real.

These three theorems are basically interrelated and any possible violations as a result of taking spatial dispersion into account are of the same, clearly physical, nature.

The electromagnetic energy flux given by Poynting's vector $\boldsymbol{S}^{(0)}$ always makes an acute angle with the wave vector $\boldsymbol{k} = k\boldsymbol{s}$, see (2.3.74). At the same time, in the absence of spatial dispersion, the entire energy flux amounts to $\boldsymbol{S}^{(0)}$. It is therefore natural that the wave is damped precisely in the direction assumed by the vector $\boldsymbol{k}$, see (2.3.76). In the case of a transparent medium with substantial spatial dispersion, when the group-velocity vector $\boldsymbol{v}_{\text{gr}}$ is introduced, it may be arbitrarily directed; its direction will coincide with the direction of the total energy flux $\boldsymbol{S} = \boldsymbol{S}^{(0)} + \boldsymbol{S}^{(1)}$. In the case of absorption it is, in general, impossible to use the concept of group velocity, but in a medium with spatial dispersion the energy flux, as before, cannot, of course, consist of only $\boldsymbol{S}^{(0)}$. The inequality (2.3.76) may therefore be violated.

In this connection it is very illustrative to consider the case of an isotropic medium where, if spatial dispersion is taken into account, the group velocity $v_{gr} = d\omega/dk$ may have either sign (if $v_{gr} < 0$ the group velocity vector is antiparallel to $\boldsymbol{k} = \omega n \boldsymbol{s}/c$). If absorption is sufficiently weak, the group velocity does not lose its meaning and, at the same time, if $v_{gr} < 0$ the energy flux is directed opposite to $\boldsymbol{k}$, i.e., the wave is damped in a direction opposite to the direction of the phase velocity. In this case $n\varkappa < 0$.

Without absorption, the average energy flux of a monochromatic normal wave must be spatially constant; this means that if spatial dispersion is ignored

$$S^{(0)} = \frac{c}{16\pi}(\boldsymbol{E}_0^* \times \boldsymbol{B}_0 + \boldsymbol{E}_0 \times \boldsymbol{B}_0^*) = \text{const}\,, \qquad \operatorname{div} \boldsymbol{S}^{(0)} = 0\,, \tag{2.3.78}$$

where $\boldsymbol{E}_0 = \boldsymbol{E}_{00} \exp(-\boldsymbol{k}'' \cdot \boldsymbol{r})$, $\boldsymbol{B}_0 = \boldsymbol{B}_{00} \exp(-\boldsymbol{k}'' \cdot \boldsymbol{r})$, $\boldsymbol{E}_{00} = \text{const}$ and $\boldsymbol{B}_{00} = \text{const}$ [we consider normal waves depending on $\boldsymbol{r}$ and t according to the law

$$\exp[\mathrm{i}(\boldsymbol{k} \cdot \boldsymbol{r} - \omega t)] = \exp(-\boldsymbol{k}'' \cdot \boldsymbol{r}) \exp[\mathrm{i}(\boldsymbol{k}' \cdot \boldsymbol{r} - \omega t)]\,,$$

see (2.3.5)].

Equations (2.3.78) are derived from the conservation law, (2.3.10), if both spatial dispersion and absorption are neglected and if the work done by external sources is assumed to be zero. To satisfy (2.3.78) it is necessary that either the medium be transparent (then $\boldsymbol{k}'' = 0$ and $\tilde{n}^2 = n^2 > 0$) or that $\boldsymbol{S}^{(0)} = 0$. As is readily evident from (2.3.72), the latter is the case if $\boldsymbol{k} = \mathrm{i}\boldsymbol{k}''$ and if the vector $\boldsymbol{E}_0$ is linearly polarized [i.e., $\boldsymbol{E}_0 = \boldsymbol{a} \exp(\mathrm{i}\varphi)$, $\boldsymbol{a}$ being a real vector]. The fact that the vector $\boldsymbol{k}$ is imaginary indicates that $\tilde{n} = \mathrm{i}\varkappa$, and $\tilde{n}^2 = -\varkappa^2$ is real.

In the presence of spatial dispersion the energy flux does not consist only of $\boldsymbol{S}^{(0)}$ and the condition that the energy flux must be constant does not imply that $\boldsymbol{S}^{(0)}$ is constant. In this case, even without absorption, the index $\tilde{n}^2$ may be complex. Such cases are known for magnetoactive plasmas [2.2] and for crystals [2.30, 63]; they will be dealt with in Sects. 4.1, 2. The proof that $\tilde{n}^2$ is real, given above, loses validity if spatial dispersion is taken into account, since then $\varepsilon_{ij} = \varepsilon_{ij}(\omega, \omega\tilde{n}\boldsymbol{s}/c)$, which gives a dispersion equation whose roots $\tilde{n}^2$ are not necessarily real.

In the problem of total internal reflection of waves perpendicularly incident on a medium occupying the half-space $z \geqslant 0$, the field must vanish as $z \to \infty$; this is also true for the energy flux. The high-frequency average of the total energy flux is therefore zero in such a problem. If, however, $\tilde{n}^2$ is complex, then $\boldsymbol{S}^{(0)} \neq 0$ and the flux of nonelectromagnetic origin, due to the fact that spatial dispersion is taken into account, must exactly compensate the flux $\boldsymbol{S}^{(0)}$. In a plasma, as shown in Sect. 2.3.1, the nonelectromagnetic energy flux is simply the kinetic-energy flux of the particles. In a transparent medium (or, more accurately, in the case of normal waves propagating in the given medium) the additional energy flux $\boldsymbol{S}^{(1)}$ is determined by (2.3.15, 31); the complexity of $\boldsymbol{k}$ complicates the conditions and therefore we do not use the notation $\boldsymbol{S}^{(1)}$. The problem of total internal reflection with spatial dispersion will be discussed again in Sects. 4.5.5, 6.

3. The Tensor $\varepsilon_{ij}(\omega, \boldsymbol{k})$ in Crystals

The dielectric-constant tensor $\varepsilon_{ij}(\omega, \boldsymbol{k})$, with spatial dispersion taken into consideration, is applied to crystals. It is shown that the formulation of the tensor $\varepsilon_{ij}(\omega, \boldsymbol{k})$ and its use within the framework of macroscopic electrodynamics are possible only under the condition $ka \ll 1$, a being the lattice constant. Hence, spatial dispersion is weak in crystal optics, and the function $\varepsilon_{ij}(\omega, \boldsymbol{k})$ or $\varepsilon_{ij}^{-1}(\omega, \boldsymbol{k})$ can be replaced by a series expansion in powers of ka. The features of such an expansion are discussed, as well as the form of the tensor $\varepsilon_{ij}(\omega, \boldsymbol{k})$ for gyrotropic and nongyrotropic crystals of various classes.

3.1 The Tensor $\varepsilon_{ij}(\omega, \boldsymbol{k})$ for Crystals

3.1.1 Introduction of the Tensor $\varepsilon_{ij}(\omega, \boldsymbol{k})$

Crystals are spatially inhomogeneous bodies since the lattice nodes are not equivalent to the other points. The utility of the tensor $\varepsilon_{ij}(\omega, \boldsymbol{k})$, which was introduced under the assumption of a homogeneous medium, is necessarily limited in crystals.

Before we turn to crystals we rewrite (2.1.3), which refers to an arbitrary linear medium. We assume that the medium is homogeneous with respect to time, i.e., the kernel $\hat{\varepsilon}_{ij}(t, t', \boldsymbol{r}, \boldsymbol{r}')$ of (2.1.3) is taken to be a function of only the interval $t - t'$. We adopt the Fourier transformation

$$\begin{aligned} \boldsymbol{E}(\boldsymbol{r}, t) &= \int \boldsymbol{E}(\boldsymbol{r}, \omega)\, \mathrm{e}^{-\mathrm{i}\omega t} d\omega\,, \\ \boldsymbol{E}(\boldsymbol{r}, t) &= \int \boldsymbol{E}(\omega, \tilde{\boldsymbol{k}})\, \mathrm{e}^{\mathrm{i}(\tilde{\boldsymbol{k}} \cdot \boldsymbol{r} - \omega t)} d\omega\, d\tilde{\boldsymbol{k}} \end{aligned}$$

and likewise for $\boldsymbol{D}(\boldsymbol{r}, t)$; in this case (2.1.3) becomes

$$\begin{aligned}
D_i(\boldsymbol{r}, \omega) &= \int \hat{\varepsilon}_{ij}(\omega, \boldsymbol{r}, \boldsymbol{r}') E_j(\boldsymbol{r}', \omega)\, d\boldsymbol{r}' , \\
\hat{\varepsilon}_{ij}(\omega, \boldsymbol{r}, \boldsymbol{r}') &= \int_0^\infty \hat{\varepsilon}_{ij}(\tau, \boldsymbol{r}, \boldsymbol{r}')\, e^{i\omega\tau} d\tau , \\
D_i(\omega, \boldsymbol{k}) &= \frac{1}{(2\pi)^3} \int D_i(\boldsymbol{r}, \omega)\, e^{-i\boldsymbol{k}\cdot\boldsymbol{r}} d\boldsymbol{r} = \int \varepsilon_{ij}(\omega, \boldsymbol{k}, \tilde{\boldsymbol{k}}) E_j(\omega, \tilde{\boldsymbol{k}})\, d\tilde{\boldsymbol{k}} , \\
\varepsilon_{ij}(\omega, \boldsymbol{k}, \tilde{\boldsymbol{k}}) &= \frac{1}{(2\pi)^3} \int \hat{\varepsilon}_{ij}(\omega, \boldsymbol{r}, \boldsymbol{r}') \exp[i(-\boldsymbol{k}\cdot\boldsymbol{r} + \tilde{\boldsymbol{k}}\cdot\boldsymbol{r}')]\, d\boldsymbol{r}\, d\boldsymbol{r}' .
\end{aligned} \tag{3.1.1}$$

For an unbounded homogeneous medium, $\hat{\varepsilon}_{ij}(\tau, \boldsymbol{r}, \boldsymbol{r}') = \hat{\varepsilon}_{ij}(\tau, \boldsymbol{r} - \boldsymbol{r}')$ and since $1/(2\pi)^3 \int \exp[i(\boldsymbol{k} - \tilde{\boldsymbol{k}})\boldsymbol{r}']\, d\boldsymbol{r}' = \delta(\boldsymbol{k} - \tilde{\boldsymbol{k}})$, we see that $\varepsilon_{ij}(\omega, \boldsymbol{k}, \tilde{\boldsymbol{k}}) = \varepsilon_{ij}(\omega, \boldsymbol{k}) \times \delta(\boldsymbol{k} - \tilde{\boldsymbol{k}})$. In other words, we arrive at (2.1.6), as expected.

A crystal represents an inhomogeneous medium whose properties remain unchanged in the case of displacement by an arbitrary lattice vector $\boldsymbol{a}$ (translational symmetry)[1]. In a crystal we therefore have

$$\varepsilon_{ij}(\omega, \boldsymbol{r}, \boldsymbol{r}') = \varepsilon_{ij}(\omega, \boldsymbol{r} + \boldsymbol{a}, \boldsymbol{r}' + \boldsymbol{a}) . \tag{3.1.2}$$

A function possessing the properties (3.1.2) can be written in the form

$$\varepsilon_{ij}(\omega, \boldsymbol{r}, \boldsymbol{r}') = \sum_{\boldsymbol{b}} g_b(\omega, \boldsymbol{r} - \boldsymbol{r}') \exp(-2\pi i \boldsymbol{b}\cdot\boldsymbol{r}') ,$$

where $\boldsymbol{b} = n_1\boldsymbol{b}_1 + n_2\boldsymbol{b}_2 + n_3\boldsymbol{b}_3$ is an arbitrary vector of the reciprocal lattice, the n_j are integers and the $\boldsymbol{b}_j$ are the three fundamental vectors of the reciprocal lattice, i.e., $\exp(2\pi i \boldsymbol{a}\cdot\boldsymbol{b}) = 1$. Substituting such an expression for $\varepsilon_{ij}(\omega, \boldsymbol{r}, \boldsymbol{r}')$ in (3.1.1) we obtain

$$\varepsilon_{ij}(\omega, \boldsymbol{k}, \tilde{\boldsymbol{k}}) = \sum_{\boldsymbol{b}} \varepsilon_{ij}^{\boldsymbol{b}}(\omega, \boldsymbol{k})\, \delta(\tilde{\boldsymbol{k}} - \boldsymbol{k} - 2\pi\boldsymbol{b}) , \tag{3.1.3}$$

where

$$\varepsilon_{ij}^{\boldsymbol{b}}(\omega, \boldsymbol{k}) = \int g_b(\omega, \boldsymbol{R}) \exp(-i\boldsymbol{k}\cdot\boldsymbol{r})\, d\boldsymbol{R}$$

and $\boldsymbol{R} = \boldsymbol{r} - \boldsymbol{r}'$. Obviously, if we account only for the term with $\boldsymbol{b} = 0$, then $\varepsilon_{ij}(\omega, \boldsymbol{k}, \tilde{\boldsymbol{k}}) = \varepsilon_{ij}^{\boldsymbol{b}=0}(\omega, \boldsymbol{k})\, \delta(\tilde{\boldsymbol{k}} - \boldsymbol{k})$ and $D_i(\omega, \boldsymbol{k}) = \varepsilon_{ij}^{\boldsymbol{b}=0}(\omega, \boldsymbol{k}) E_j(\omega, \boldsymbol{k})$.

It is clear that the dielectric-constant tensor $\varepsilon_{ij}(\omega, \boldsymbol{k})$ can be used in the electrodynamics of crystals if it is sufficient to retain only the one term with $\boldsymbol{b} = 0$ on the right-hand side of (3.1.3), or if in the corresponding relations all other terms can be expressed in terms of this first term. As we shall see now this is the situation encountered in crystal optics.

---

[1] It becomes necessary, of course, to deal with inhomogenous media of other types. A plasma, for instance, whose parameters depend on the coordinates is inhomogeneous [3.1].

According to (3.1.1, 3) the electric field $\boldsymbol{E}$ and the induction $\boldsymbol{D}$ are related as follows:

$$\begin{aligned} D_i(\omega,\boldsymbol{k}') &= \sum_{\boldsymbol{b}} \varepsilon_{ij}^{\boldsymbol{b}}(\omega,\boldsymbol{k}')E_j(\omega,\boldsymbol{k}'+2\pi\boldsymbol{b}) \\ &= \varepsilon_{ij}^{0}(\omega,\boldsymbol{k}')E_j(\omega,\boldsymbol{k}') + \sum_{\boldsymbol{b}'\neq 0} \varepsilon_{ij}^{\boldsymbol{b}'}(\omega,\boldsymbol{k}')E_j(\omega,\boldsymbol{k}'+2\pi\boldsymbol{b}')\,, \end{aligned} \tag{3.1.4}$$

where the wave vector is $\boldsymbol{k}'$ (not to be confused with $\boldsymbol{k}' = \mathrm{Re}\{\boldsymbol{k}\}$), since it is desirable to retain the symbol $\boldsymbol{k}$ only when applied to long wavelengths. These waves (whose wavelength λ is much larger than the lattice constant a) are considered in crystal optics. This means that in crystal optics the wave vectors $\boldsymbol{k}$ should satisfy the condition

$$k \ll 2\pi b \gtrsim 2\pi/a\,. \tag{3.1.5}$$

Even at long wavelengths, i.e., if in (3.1.4) $\boldsymbol{k}' = \boldsymbol{k}$, $D_i(\omega,\boldsymbol{k})$ depends on both $E_j(\omega,\boldsymbol{k})$ and $E_j(\omega,\boldsymbol{k}+2\pi\boldsymbol{b}')$. In considering crystals it is necessary to find the conditions under which it may be presumed that

$$D_i(\omega,\boldsymbol{k}) = \varepsilon_{ij}(\omega,\boldsymbol{k})E_j(\omega,\boldsymbol{k})\,. \tag{3.1.6}$$

This problem has already been discussed a long time ago ([3.2] and [Ref. 3.3, § 44]) but like other problems related to spatial dispersion it has reappeared in recent years [3.4, 5][2]. The transition from (3.1.4 to 6) could be made more or less automatically if the inequality $\varepsilon_{ij}^{\boldsymbol{b}'\neq 0}(\omega,\boldsymbol{k}+2\pi\boldsymbol{b}') \ll \varepsilon_{ij}^{0}(\omega,\boldsymbol{k}) \equiv \varepsilon_{ij}^{\boldsymbol{b}=0}(\omega,\boldsymbol{k})$ is satisfied. Unfortunately this assumption is not true for crystals. Let us assume, for example, that the field E_j has only one long-wavelength component, $E_j(\omega,\boldsymbol{k})$. Then, according to (3.1.4),

$$\begin{aligned} D_i(\omega,\boldsymbol{k}) &= \varepsilon_{ij}^{0}(\omega,\boldsymbol{k})E_j(\omega,\boldsymbol{k})\,, \\ D_i(\omega,\boldsymbol{k}') &= \varepsilon_{ij}^{\boldsymbol{b}}(\omega,\boldsymbol{k}')E_j(\omega,\boldsymbol{k})\,, \\ \boldsymbol{k} &= \boldsymbol{k}'+2\pi\boldsymbol{b}\,. \end{aligned} \tag{3.1.7}$$

If $b \sim 2\pi/a$ (but not if $b \gg 2\pi/a$), the Fourier components of $\boldsymbol{D}(\omega,\boldsymbol{k}')$ need not, in general, be small. This is evident from simple physical considerations. Imagine a capacitor filled with small dielectric spheres of radius a. If a voltage is applied to this capacitor, polarization and induction in it will have a constant component (analogous to the long-wavelength component of the

[2] Unfortunately, the books [3.2, 3] were not referred to in [3.4, 5]. Moreover, in [3.4] only the transverse component of the short-wavelength field was considered and the footnote [Ref. 3.4, p. 669] contains an error.

field) and intense short-wavelength components with spatial period l of the order of a. The components having period $l \ll a$ will have a small specific weight[3] and the components with period $l \sim a$ may also be small (compared with the constant component) if the dielectric constant ε of the spheres is not too different from that of the surrounding medium (or if the spheres with $\varepsilon \approx 1$ are in a vacuum). A crystal resembles this model, but the constituent "spheres", i.e., its atoms, have, in general, a comparatively high polarizability. We thus see that the quantity $\varepsilon_{ij}^{b \neq 0}(\omega, \boldsymbol{k}')$ is not at all small compared with $\varepsilon_{ij}^{b=0}(\omega, \boldsymbol{k}') \equiv \varepsilon_{ij}^{0}(\omega, \boldsymbol{k}')$ if we disregard the cases where $b \gg 2\pi/a$ (it will be shown in Sect. 5.1.1 that the components of ε_{ij}^{b} do, in fact, decrease rapidly, as b increases).

In the general case, short-wavelength components of the field $\boldsymbol{D}(\omega, \boldsymbol{k}')$ also exist in the crystal, in addition to the short-wavelength components $\boldsymbol{E}(\omega, \boldsymbol{k}')$. A field existing with only one Fourier component for either $\boldsymbol{D}$ or $\boldsymbol{E}$ can be obtained only if the external currents and charges [i.e., $\boldsymbol{j}_{\text{ext}}$ and ρ_{ext} in (2.1.1)] are suitably chosen.

The theory for the dielectric constant of crystals can be advanced only by stating the problem more specifically in some manner, making definite assumptions as to the external sources of the field. This is exactly what we intend to do. In full agreement with the statement of the problem both in crystal optics and for the electrodynamics of continuous media in general, we shall assume that the medium possesses neither external charges nor currents with short-wavelength Fourier components. We thus assume

$$\boldsymbol{j}_{\text{ext}}(\omega, \boldsymbol{k}') = 0\,, \quad \rho_{\text{ext}}(\omega, \boldsymbol{k}') = 0\,, \quad k' \gtrsim 2\pi b \gg k\,. \tag{3.1.8}$$

In crystal optics, i.e., in considering normal waves, we also have

$$\boldsymbol{j}_{\text{ext}}(\omega, \boldsymbol{k}) = 0\,, \quad \rho_{\text{ext}}(\omega, \boldsymbol{k}) = 0\,, \quad k \ll 2\pi b\,, \quad b \gtrsim 1/a\,. \tag{3.1.9}$$

It is, however, unnecessary to make use of (3.1.9) in order to include the tensor $\varepsilon_{ij}(\omega, \boldsymbol{k})$ in our considerations of crystals or of homogeneous media.

By virtue of (3.1.8) we have for the short-wavelength Fourier components [see (2.2.4, 7); $\boldsymbol{k}' = k'\boldsymbol{s}'$, $s' = 1$]:

$$\frac{\omega^2}{c^2} D_i(\omega, \boldsymbol{k}') - (k')^2 E_i(\omega, \boldsymbol{k}') + k_i' k_j' E_j(\omega, \boldsymbol{k}') = 0\,, \tag{3.1.10}$$

$$s_i' D_i(\omega, \boldsymbol{k}') = 0\,, \quad k' \approx 2\pi b \gg k\,, \quad \frac{k}{2\pi b} \lesssim \frac{ka}{2\pi} \ll 1\,. \tag{3.1.11}$$

---

[3] This is not at all evident if the spheres have sharp boundaries. But we have used spheres only for the sake of visualization. A representation much closer to a real crystal is a periodic medium in which the polarizability is modulated, but is not subject to abrupt changes. If at distances $l \ll a$ the variation in polarizability is negligible, then no appreciable field components appear with characteristic periods $l \ll a$.

Let us now show [3.3–5] that the transverse component $\boldsymbol{E}_{\perp}(\omega, \boldsymbol{k}')$ of the field $\boldsymbol{E}(\omega, \boldsymbol{k}')$ is small compared with $\boldsymbol{E}(\omega, \boldsymbol{k})$. In the case of homogeneous waves, see (2.2.60, 61),

$$(k')^2 E_i - k_i' k_j' E_j = (k')^2(\delta_{ij} - s_i' s_j') E_j \equiv (k')^2 \eta_{ij} E_j = (k')^2 E_{\perp,i}$$

and, as follows from (3.1.4, 10)

$$E_{\perp,i}(\omega, \boldsymbol{k}') = \frac{\omega^2}{c^2 (k')^2} D_i(\omega, \boldsymbol{k}') \sim \frac{\omega^2}{4\pi^2 c^2 b^2} D_i(\omega, \boldsymbol{k}')$$
$$\sim \frac{\omega^2}{4\pi^2 c^2 b^2} \varepsilon_{ij}^{-b}(\omega, \boldsymbol{k}') E_j(\omega, \boldsymbol{k}), \quad (3.1.12)$$
$$\boldsymbol{k}' = \boldsymbol{k} + 2\pi \boldsymbol{b} \approx 2\pi \boldsymbol{b}, \quad k \ll 2\pi b.$$

Here we have taken into account that the short-wavelength field $\boldsymbol{E}_{\parallel}(\omega, \boldsymbol{k}')$ created by the long-wavelength field $\boldsymbol{E}(\omega, \boldsymbol{k})$ and contributing to $\boldsymbol{D}(\omega, \boldsymbol{k}')$ is no larger in order of magnitude than the field $\boldsymbol{E}(\omega, \boldsymbol{k})$.

In crystal optics

$$\frac{\omega^2}{4\pi^2 c^2 b^2} |\varepsilon_{ij}^{-b}(\omega, \boldsymbol{k}')| \lesssim \frac{\omega^2 a^2}{4\pi^2 c^2} |\varepsilon_{ij}^{-b}| = \frac{a^2}{\lambda_0^2} |\varepsilon_{ij}^{-b}| \ll 1, \quad (3.1.13)$$

since the ratio of the lattice parameter $a \sim 3 \times 10^{-8}$ cm to the wavelength $\lambda_0 = 2\pi c/\omega \sim 5 \times 10^{-5}$ cm in vacuo is very small ($a^2/\lambda_0^2 \lesssim 10^{-6}$), and the quantity $|\varepsilon_{ij}^{-b}(\omega, \boldsymbol{k}')|$ is probably of order unity (in any case, there is no reason to assume it to be large). By virtue of (3.1.12, 13)

$$E_{\perp}(\omega, \boldsymbol{k}') \ll E(\omega, \boldsymbol{k}), \quad k \ll k' \simeq 2\pi b. \quad (3.1.14)$$

Thus, neglecting small terms of the order of $(a/\lambda_0)^2 \lesssim 10^{-6}$, the transverse short-wavelength field $\boldsymbol{E}_{\perp}(\omega, \boldsymbol{k}')$ may be disregarded in crystal optics. Neglecting the field $\boldsymbol{E}_{\perp}(\omega, \boldsymbol{k}')$ we can write (3.1.11 and 4) in the form

$$s_i' D_i(\omega, \boldsymbol{k}') = s_i' \varepsilon_{ij}^{-b}(\omega, \boldsymbol{k}') E_j(\omega, \boldsymbol{k})$$
$$+ \sum_{b' \neq -b} s_i' \varepsilon_{ij}^{b'}(\omega, \boldsymbol{k}') E_{\parallel,j}(\omega, \boldsymbol{k}' + 2\pi \boldsymbol{b}') = 0, \quad (3.1.15)$$

$$D_i(\omega, \boldsymbol{k}) = \varepsilon_{ij}^{0}(\omega, \boldsymbol{k}) E_j(\omega, \boldsymbol{k}) + \sum_{b \neq 0} \varepsilon_{ij}^{b}(\omega, \boldsymbol{k}) E_{\parallel,j}(\omega, \boldsymbol{k} + 2\pi \boldsymbol{b}), \quad (3.1.16)$$

where $\boldsymbol{k}' = \boldsymbol{k} + 2\pi \boldsymbol{b}$, $k \ll 2\pi b$ and $\boldsymbol{k}' = k' \boldsymbol{s}'$. The system (3.1.15) is formally infinite (recall that $\boldsymbol{b} = n_1 \boldsymbol{b}_1 + n_2 \boldsymbol{b}_2 + n_3 \boldsymbol{b}_3$; $n_1, n_2, n_3 = 0, \pm 1, \pm 2, \pm 3, \dots$)

but it can be truncated since with sufficiently large values of the vector $\boldsymbol{b}$ of the reciprocal lattice the coefficients ε_{ij}^{b} decrease rapidly, as $\boldsymbol{b}$ increases. The linear system (3.1.15) allows us, in general, to express the components $E_{\|,j}(\boldsymbol{k}' + 2\pi\boldsymbol{b}') = E_{\|,j}[\boldsymbol{k} + 2\pi(\boldsymbol{b}+\boldsymbol{b}')]$, $\boldsymbol{b} + \boldsymbol{b}' \neq 0$ in terms of $E_j(\omega,\boldsymbol{k})$.

Thus, all vectors $E_{\|,j}(\omega,\boldsymbol{k}+2\pi\boldsymbol{b})$, $\boldsymbol{b} \neq 0$ occurring in (3.1.16) are linearly related to $E_j(\omega,\boldsymbol{k})$, i.e., (3.1.16) can be transformed into (3.1.6). In this case a single condition has to be satisfied, namely the absence of nontrivial solutions of the homogeneous system of equations (3.1.15), for which the inhomogeneous term is proportional to $E_j(\omega,\boldsymbol{k})$. If such solutions existed, i.e., if the determinant of the homogeneous system (3.1.15) were equal to zero, this would reveal the possibility of peculiar short-wavelength excitations arising in the crystal. Such excitations might either be unrelated to $E_j(\omega,\boldsymbol{k})$, i.e., not caused by the long-wavelength field, or they might be built up by this field, within the scope of the system (3.1.14); the second case corresponds to infinite solutions $E_{\|,j}(\omega,\boldsymbol{k}' + 2\pi\boldsymbol{b}')$. The first possibility, regardless of whether it corresponds to reality or not is of no interest from our point of view; we are interested only in the induction caused by the field $E_j(\omega,\boldsymbol{k})$. The second possibility would imply some instability and obviously this cannot occur, at least not in a medium which is in equilibrium. We thus have every reason to assume that the short-wavelength field is expressed in crystal optics in terms of the long-range one, and that (3.1.6) can be applied, analogous to that used in a homogeneous medium.

The fact that only the longitudinal short-wavelength field $\boldsymbol{E}_{\|}$ appears in (3.1.15) is of vital importance. If the transverse field could not be neglected or taken approximately into account (see below), then, at $E_j(\omega,\boldsymbol{k}) = 0$, a system of the type (3.1.15) would relate the short-wavelength fields $E_{\|,j}$ and $E_{\perp,j}$ (see Sect. 2.2.4), but with $E_j(\omega,\boldsymbol{k}) \neq 0$ it would not allow the entire short-wavelength field E_j to be expressed in terms of $E_j(\omega,\boldsymbol{k})$. In other words, if the field $E_{\|,j}$ in system (3.1.15) were replaced by $E_j(\omega,\boldsymbol{k}'+2\pi\boldsymbol{b}')$, the system would contain three times as many unknowns ($E_{\|,j}$ and $E_{\perp,j}$) with the same number of equations, so that it could not be used for uniquely determining $E_j(\omega,\boldsymbol{k}+2\pi\boldsymbol{b})$ in terms of $E_j(\omega,\boldsymbol{k})$.

The situation can be summarized as follows. When there are no sources for short-wavelength induction and transverse fields [see conditions (3.1.8)], the short-wavelength field in the crystal is, to a good approximation, purely longitudinal (a Coulomb field). Additionally, the divergence of the short-wavelength part of the induction equals zero [condition (3.1.11)]. Hence, it can be assumed that the short-wavelength part of the induction as such is absent, by virtue of which the short-wavelength part of the electric field is expressed [see system (3.1.15)] in terms of its long-wavelength part $\boldsymbol{E}(\omega,\boldsymbol{k})$.

In (3.1.15, 16), the short-wavelength field $\boldsymbol{E}_{\perp}(\omega,\boldsymbol{k}')$ was completely suppressed. This is not necessary if $\boldsymbol{E}_{\perp}(\omega,\boldsymbol{k}')$ can somehow be expressed in terms of $\boldsymbol{E}(\omega,\boldsymbol{k})$ and $\boldsymbol{E}_{\|}(\omega,\boldsymbol{k}')$. This possibility arises under the conditions (3.1.13, 14) as a result of the application of the method of successive approximations – the field $\boldsymbol{E}_{\perp}(\omega,\boldsymbol{k}')$ is then determined by (3.1.10), see also (3.1.11, 12).

In order to make the picture clearer let us point out the following. The behavior of a crystal in an electromagnetic field is analogous in many respects to the dielectric-spheres model mentioned above. Under the action of a long-wavelength electric field $\boldsymbol{E}$ (the wavelength $\lambda \gg a$, a being the radius of the sphere) not only is a long-wavelength induction $\boldsymbol{D}$ produced in such a "medium", but also a short-wavelength polarization setting up a short-wavelength electric field. In an inhomogeneous medium the short-wavelength field $\boldsymbol{E}$, however, may, in turn, produce long-wavelength polarization [precisely as is indicated in (3.1.4, 16)]. As an illustration, consider a medium having $\varepsilon = \varepsilon_0 + \varepsilon_1 \cos bz$ which is subject to the action of the field $E = E_0 \exp(\mathrm{i}k'z)$. Then

$$D = \varepsilon E = \varepsilon_0 E_0 \exp(\mathrm{i}k'z) + (\varepsilon_1/2)\{\exp[\mathrm{i}(k'+b)z] + \exp[\mathrm{i}(k'-b)z]\}E_0 \,;$$

if $k' - b = k \ll b$ then the induction D has a long-wavelength Fourier component.

Hence, in general, the field equation cannot be decomposed into equations for the separate Fourier components. If, on the average (with respect to the assembly or over a long period of time) the constituent "spheres" (atoms) exist in a uniformly distributed manner in space, then the short-wavelength components vanish when we consider averaged quantities. This case corresponds to a spatially uniform medium, uniformity being viewed as a statistical feature. At a given instant of time such a statistically uniform medium is, of course, always spatially inhomogeneous, but the respective deviations from the average can be regarded as fluctuations. When these fluctuations are taken into account the long-wavelength field produces a short-wavelength field even in a medium which is, on the average, homogeneous.

Crystals also display fluctuations as a result of thermal motion or structural imperfections, but in contrast to homogeneous media the inhomogeneities persist even after averaging. At high temperatures this inhomogeneity may play a comparatively small role, and a solid crystal will differ only slightly from a liquid one (an anisotropic, but, on the average, spatially homogeneous medium). These facts are well known, but we mention them in order to emphasize that there are no insurmountable differences between crystals and such homogeneous media as liquids.

Returning to crystals or to the model of the orderly arranged "spheres" it can be easily verified that the longitudinal (Coulomb) short-wavelength field, created by the long-wavelength field and not by some kind of sources [i.e., if condition (3.1.8) is satisfied], is proportional to this long-wavelength field within the framework of the linear theory. Hence the long-wavelength induction created by this short-wavelength field will also be proportional to the long-wavelength field strength – a result which we have already obtained in a more formal and rigorous manner.

In the absence of external sources of a short-wavelength field we thus have the linear relation (3.1.6) $D_i(\omega, \boldsymbol{k}) = \varepsilon_{ij}(\omega, \boldsymbol{k}) E_j(\omega, \boldsymbol{k})$ linking the long-wavelength induction and the field in the crystal, where $\varepsilon_{ij}(\omega,\boldsymbol{k})$ plays the part of the dielectric-constant tensor of the crystal. This relation can be applied in the presence of long-wavelength field sources $\boldsymbol{j}_{\text{ext}}(\omega,\boldsymbol{k})$ and $\rho_{\text{ext}}(\omega,\boldsymbol{k})$ as well as in their absence, if we consider the propagation of the assembly of normal waves.

Owing to the inhomogeneity of the medium the long-wavelength field can be viewed as being modulated by the short-wavelength field; in a condensed medium the depth of modulation is, in general, large. The main difference between crystals and fluids is that in fluids the modulation is irregular (fluctuating) whereas in crystals the fluctuating component is accompanied by a regular modulation displaying spatial periods equal to or less than the lattice constant (or constants).

It should also be stressed that the lattice constant a plays a double role. On the one hand, the parameter a characterizes the spatial nonuniformity of the medium which, when passing over to $\varepsilon_{ij}(\omega,\boldsymbol{k})$, is unimportant in the above sense. On the other hand, the parameter a determines the 'radius of molecular action' R_0 – the interval of values R for which the kernel $\hat{\varepsilon}_{ij}(\omega,\boldsymbol{r},\boldsymbol{r}+\boldsymbol{R})$ in (3.1.4) differs appreciably from zero. When the radius R_0 tends to zero this means that spatial dispersion is neglected, as indicated above. If two formally independent parameters, the lattice constant a and the 'radius of molecular action' R_0, are introduced at the beginning, the picture may be clearer. In fact, for crystals usually $a \sim R_0$, or if we consider complex crystals with anomalously large lattice parameters, $a \gg R_0$. Evidently, the inequality $a \ll R_0$ (or, formally, a tending to zero if $R_0 \neq 0$) does not apply to real crystals.

Another important problem arises; namely, to what extent the tensor $\varepsilon_{ij}(\omega,\boldsymbol{k})$ of a crystal possesses the general properties established for the tensor $\varepsilon_{ij}(\omega,\boldsymbol{k})$ of the homogeneous medium (see, particularly, Sect. 2.1.2)?

If we make use of the principle of causality, the symmetry of the kinetic coefficients or the reality of $\boldsymbol{D}$ when $\boldsymbol{E}$ is real, it is obvious that no difference arises when we consider crystals. If, however, we are concerned with the energy relations, the situation appears more complex at first sight. The short-wavelength field created by the long-wavelength field being considered, possesses a certain energy and makes a contribution to the liberation of heat, etc.

In this connection the question arises as to whether we have accounted for the role of the short-wavelength field when analyzing the energy relations in Sect. 2.3. The answer is yes. Poynting's theorem (2.3.1), which has been taken to be fundamental in deriving the expressions of energy, amount of heat, and energy flux, in fact follows directly from the field equations (2.1.1) without any assumptions as to the contribution of the short-wavelength field components. This contribution is automatically taken into account if we use the corresponding macroscopic expressions. The latter is confirmed directly on the basis of the microscopic theory [Ref. 3.6, § 31][4]. Moreover, as is clear

[4] In [3.6] only a simple model for the medium is considered. It is therefore not superfluous to give a more detailed microtheoretical analysis of the problem of energy relations in the electrodynamics of media displaying spatial dispersion.

from the discussion, the difference between crystals and such uniform media as liquids is not so great as regards the problem of the short-wavelength field (i.e., the microfield). Therefore, if the energy relations discussed in Sect. 2.3 had not taken the short-wavelength field into account, they would, in general, almost always be inapplicable. These relations do indeed apply to fluids and to crystals; this applicability to crystals is limited to cases in which the condition (3.1.6) is satisfied. This condition, i.e., the possibility of introducing the tensor $\varepsilon_{ij}(\omega,\boldsymbol{k})$, is exactly what is assumed to be fulfilled in discussing all the problems dealt with in this book. It is thus justified to apply the tensor $\varepsilon_{ij}(\omega,\boldsymbol{k})$, linking the fields $\boldsymbol{D}(\omega,\boldsymbol{k})$ and $\boldsymbol{E}(\omega,\boldsymbol{k})$, to crystals – see (3.1.6) – in the same way as for a homogeneous medium.

As we have seen, (3.1.15, 16), when the tensor $\varepsilon_{ij}(\omega,\boldsymbol{k})$ is applied to crystals it must be expressed in terms of the tensors $\varepsilon_{ij}^{0}(\omega,\boldsymbol{k})$ and $\varepsilon_{ij}^{b\neq 0}(\omega,\boldsymbol{k})$, in a rather complicated way. One might therefore think that it is much more complicated to calculate $\varepsilon_{ij}(\omega,\boldsymbol{k})$ for crystals than for a homogeneous medium. This is not the case, however, since in order to find $\varepsilon_{ij}(\omega,\boldsymbol{k})$ it is not necessary to determine the tensor $\varepsilon_{ij}^{b}(\omega,\boldsymbol{k})$ in advance. Moreover, when applied to crystals, the usual methods of calculation immediately yield the tensor $\varepsilon_{ij}(\omega,\boldsymbol{k})$ and not the tensors $\varepsilon_{ij}^{b}(\omega,\boldsymbol{k})$ or $\varepsilon_{ij}^{b=0}(\omega,\boldsymbol{k}) \equiv \varepsilon_{ij}^{0}(\omega,\boldsymbol{k})$. This becomes still clearer from the calculations themselves (Sect. 6.1.1), but we find it useful to clarify the essential points now.

Both classical and quantum-mechanical determinations of $\varepsilon_{ij}(\omega,\boldsymbol{k})$ consist of calculating the current density induced (i.e., the induction $\boldsymbol{D}$) by the long-wavelength field $\boldsymbol{E}(\omega,\boldsymbol{k})$. The role of the unperturbed states (natural oscillations) of the system is played by the "mechanical excitons", the waves with $\boldsymbol{D} \neq 0$ for which $\boldsymbol{E}(\omega,\boldsymbol{k}) = 0$. The short-wavelength Coulomb field, however, is already accounted for in the nonperturbed problem; due to the influence of the perturbing action of the field $\boldsymbol{E}(\omega,\boldsymbol{k})$ the induction $\boldsymbol{D}(\omega,\boldsymbol{k})$ arises which represents and accounts for the action of the short-wavelength field. In other words, the value of $\boldsymbol{D}(\omega,\boldsymbol{k})$ is determined under conditions where not only $\boldsymbol{E}(\omega,\boldsymbol{k}) \neq 0$ but $\boldsymbol{E}(\omega,\boldsymbol{k}+2\pi\boldsymbol{b}) \neq 0$ as well. If, however, the quantity $\varepsilon_{ij}^{0}(\omega,\boldsymbol{k})$ is to be calculated, we would have to find $\boldsymbol{D}(\omega,\boldsymbol{k})$ with $\boldsymbol{E}(\omega,\boldsymbol{k}) \neq 0$, but $\boldsymbol{E}(\omega,\boldsymbol{k}+2\pi\boldsymbol{b}) = 0$, see (3.1.4). Only a special choice of the short-wavelength Fourier components of the external charge density $\rho_{\text{ext}}(\boldsymbol{k}+2\pi\boldsymbol{b})$ makes it possible to cancel the Coulomb field $\boldsymbol{E}(\omega,\boldsymbol{k}+2\pi\boldsymbol{b})$ with $\boldsymbol{b} \neq 0$ when $\boldsymbol{E}(\omega,\boldsymbol{k}) \neq 0$. Moreover, $\boldsymbol{D}(\omega,\boldsymbol{k})$ is usually calculated under the condition that $\rho_{\text{ext}}(\boldsymbol{k}+2\pi\boldsymbol{b}) = 0$, condition (3.1.8). The same situation is encountered when calculating the tensor $\varepsilon_{\perp,ij}$ with the aid of the wave functions of Coulomb excitons (Sect. 6.1.2). In this case the transverse field $\boldsymbol{E}_{\perp}$ also induces polarization caused by the longitudinal field $\boldsymbol{E}_{\parallel}$, thereby satisfying the condition $\boldsymbol{s} \cdot \boldsymbol{D} = 0$ (Sect. 2.2.4). In the microtheoretical calculation of $\varepsilon_{ij}(\omega,\boldsymbol{k})$ then, the crystal case is analogous to that of a homogeneous medium.

3.1.2 Weak Spatial Dispersion

The tensor $\varepsilon_{ij}(\omega,\boldsymbol{k})$ can be introduced and applied in the wave equation (2.2.9) for crystals only if, see (3.1.13),

$$(a/\lambda_0)^2 \ll 1 \,. \tag{3.1.17}$$

In optics, (3.1.17) can always be taken to be fulfilled, but this is not a sufficient condition for indicating the smallness of spatial dispersion. As has already been pointed out several times, spatial dispersion is, in fact, characterized by the parameter $a/\lambda = an/\lambda_0$. Obviously, in the case of large refractive indices n, the parameter an/λ_0 and thus spatial dispersion can be large despite the validity of inequality (3.1.17). Under such conditions, $\varepsilon_{ij}(\omega,\boldsymbol{k})$ may be a very complex function of $\boldsymbol{k}$ and if $\varepsilon_{ij}(\omega,\boldsymbol{k})$ is expanded in a series with respect to $\boldsymbol{k}$, the series will consist of many terms (the expansion parameter is an/λ_0). In this case the wave equation (2.2.9) may have many solutions, i.e., the dispersion equation (2.2.10) will have many roots $\omega_l(\boldsymbol{k})$. If spatial dispersion is ignored we have only two roots of the dispersion equation, which correspond to the ordinary and extraordinary waves, and, under certain conditions, a root $\omega_{\parallel}$ = const will also exist; it corresponds to the longitudinal wave.

In connection with the above, it is important to note that in crystal optics spatial dispersion is a weak effect so that the inequality

$$ka \sim \frac{a}{\lambda} = \frac{an}{\lambda_0} \ll 1 \tag{3.1.18}$$

is also satisfied. Under the condition (3.1.18) it is convenient, or necessary in most cases, to expand the tensor $\varepsilon_{ij}(\omega,\boldsymbol{k})$ or $\varepsilon_{ij}^{-1}(\omega,\boldsymbol{k})$ as a series with respect to $\boldsymbol{k}$, retaining only a few terms [3.7]. Let us thus take the expansion

$$\varepsilon_{ij}(\omega,\boldsymbol{k}) = \varepsilon_{ij}(\omega) + \mathrm{i}\gamma_{ijl}(\omega)k_l + a_{ijlm}(\omega)k_l k_m \tag{3.1.19}$$

or, if we have homogeneous plane waves with $\boldsymbol{k} = \omega\tilde{n}\boldsymbol{s}/c$,

$$\varepsilon_{ij}(\omega,\boldsymbol{k}) = \varepsilon_{ij}(\omega) + \mathrm{i}\gamma_{ijl}(\omega)\frac{\omega}{c}\tilde{n}s_l + a_{ijlm}(\omega)\left(\frac{\omega}{c}\right)^2 \tilde{n}^2 s_l s_m \,. \tag{3.1.20}$$

Analogously we have for the inverse tensor

$$\varepsilon_{ij}^{-1}(\omega,\boldsymbol{k}) = \varepsilon_{ij}^{-1}(\omega) + \mathrm{i}\delta_{ijl}(\omega)k_l + \beta_{ijlm}(\omega)k_l k_m \,, \tag{3.1.21}$$

or

$$\varepsilon_{ij}^{-1}(\omega,\boldsymbol{k}) = \varepsilon_{ij}^{-1}(\omega) + \mathrm{i}\,\delta_{ijl}(\omega)\frac{\omega}{c}\tilde{n}s_l + \beta_{ijlm}(\omega)\left(\frac{\omega}{c}\right)^2 \tilde{n}^2 s_l s_m\,. \qquad (3.1.22)$$

The use of the tensors ε_{ij} and ε_{ij}^{-1} in their general form or in the form (3.1.19–22) is equivalent within wide limits and the choice may be made according to convenience, except for the case where certain components of the tensors $\varepsilon_{ij}(\omega)$ and $\varepsilon_{ij}^{-1}(\omega)$ tend to infinity (increase rapidly). For instance, if any of the components of $\varepsilon_{ij}(\omega)$ tends to infinity, the expansions (3.1.19, 20) for the corresponding component of $\varepsilon_{ij}(\omega,\boldsymbol{k})$ would be inadequate since all terms would then be vanishingly small compared with $\varepsilon_{ij}(\omega)$. At the same time $\tilde{n}$ usually grows with ε_{ij}, i.e., the role played by spatial dispersion becomes more important. In such a case the expansion (3.1.21, 22) should be used; it is particularly effective in the case of decreasing $\varepsilon_{ij}^{-1}(\omega)$.

In an analogous way, in the region where $\varepsilon_{ij}^{-1}(\omega)$ increases rapidly, the expansion (3.1.19, 20) should be used instead of (3.1.21, 22).

In the following, use will be made of the expansions (3.1.19, 22). It is thus necessary to discuss the conditions of their applicability aside from the initial condition (3.1.18).

The expressions (3.1.19–22) are meaningful only if the functions $\varepsilon_{ij}(\omega,\boldsymbol{k})$ and $\varepsilon_{ij}^{-1}(\omega,\boldsymbol{k})$ permit a series expansion near the point $\boldsymbol{k} = 0$. This problem has already been dealt with in Sect. 2.1. Let us recall that as $\boldsymbol{k} \to 0$, the functions $\varepsilon_{ij}(\omega,\boldsymbol{k})$ and $\varepsilon_{ij}^{-1}(\omega,\boldsymbol{k})$ tend to $\varepsilon_{ij}(\omega)$ and $\varepsilon_{ij}^{-1}(\omega)$, respectively, and that they depend neither on the magnitude nor on the direction of $\boldsymbol{k}$. Furthermore, $\varepsilon_{ij}(\omega,\boldsymbol{k})$ and $\varepsilon_{ij}^{-1}(\omega,\boldsymbol{k})$ are integral quantities since they are obtained by summation over the spectrum (Sect. 6.1). The components of ε_{ij} will therefore have no singularities, not even if the integrand has certain weak singularities (e.g., at the natural frequencies of the unperturbed problem). As we have already mentioned in Chap. 1 and as will be discussed in Sect. 6.2, in the cases we know, the natural frequencies of a properly chosen unperturbed problem without degeneracy do not display any singularities as $\boldsymbol{k} \to 0$. In the presence of degeneracy the natural frequencies may be nonanalytic functions of $\boldsymbol{k}$ with $\boldsymbol{k} \to 0$ (see, e.g., Sect. 4.5.3, where we consider an isotropic gyrotropic medium). But even in this case ε_{ij} remains an analytic function of $\boldsymbol{k}$ (with $\boldsymbol{k} \to 0$).

We therefore have no reason to doubt the analyticity of the functions ε_{ij} and ε_{ij}^{-1} if $\boldsymbol{k} \to 0$.

Expansions of the type (3.1.19–22) in which only a few terms are retained, may, however, prove insufficient in the following peculiar situation. Let us assume, for instance, that $\varepsilon_{ij}(\omega,\boldsymbol{k}) = \varepsilon(\omega,\boldsymbol{k})\,\delta_{ij}$ and

$$\varepsilon(\omega,\boldsymbol{k}) = \varepsilon(\omega) + \frac{\nu k^2}{(\omega-\omega_l)/\omega_l - \mu k^2} \qquad (3.1.23)$$

[under certain conditions this expression approximately describes the behavior of $\varepsilon(\omega,\boldsymbol{k}]$ near a quadrupolar absorption line). So long as the term μk^2 of the expansion is unimportant we deal here with an expansion of the type (3.1.19). Otherwise

$$[\varepsilon(\omega,\boldsymbol{k}) - \varepsilon(\omega)]^{-1} = \frac{(\omega-\omega_l)/\omega_l - \mu k^2}{\nu k^2},$$

which corresponds neither to (3.1.19) nor to (3.1.21). An expression can readily be written which generalizes (3.1.23) for any arbitrary crystal along the lines of the phenomenological expansion (3.1.19)

$$\begin{aligned}\varepsilon_{ij}(\omega,\boldsymbol{k}) &= \varepsilon_{ij}(\omega) + \mathrm{i}\,\gamma_{ijl}(\omega)k_l + \alpha_{ijlm}(\omega,\boldsymbol{k})k_l k_m\,,\\ \alpha^{-1}_{ijlm}(\omega,\boldsymbol{k}) &= \xi_{ijlm}(\omega) + \mathrm{i}\,\eta_{ijlmn}(\omega)k_n + \zeta_{ijlmnp}(\omega)k_n k_p\,.\end{aligned} \tag{3.1.24}$$

In (3.1.19) we similarly may replace $\gamma_{ijl}(\omega)$ by $\gamma_{ijl}(\omega,\boldsymbol{k})$, where

$$\gamma^{-1}_{ijl}(\omega,\boldsymbol{k}) = \gamma^{-1}_{ijl}(\omega) + \mathrm{i}\,\mu_{ijlm}(\omega)k_m + \nu_{ijlmn}(\omega)k_m k_n\,. \tag{3.1.25}$$

Equation (3.1.24) is equivalent to (3.1.23) in the case of a nongyrotropic cubic crystal, where

$$\alpha_{ijlm}k_l k_m = \delta_{ij}\frac{\nu(\omega)k^2}{(\omega-\omega_l)/\omega_l - \mu k^2}\,.$$

In the more complex case where the combined expansion (3.1.24) is used, we obtain a cumbersome expression which is of very little practical interest. The fact that formulas of the type (3.1.23) or (3.1.24 and 25) should be used only in exceptional cases is more important. If (3.1.23) is generalized to some extent and written in the form

$$\varepsilon_{ij}(\omega,\boldsymbol{k}) = \varepsilon_{ij}(\omega) + \frac{\alpha_{ijlm}k_l k_m}{(\omega-\omega_l)/\omega_l + \mathrm{i}\,\nu' - \mu_{lm}k_i k_m}, \tag{3.1.23a}$$

then, with $\nu' = 0$, and $\boldsymbol{k}\to 0$ the tensor $\varepsilon_{ij}(\omega,\boldsymbol{k})$ at $\omega\to\omega_l$ may depend on $\boldsymbol{s} = \boldsymbol{k}/k$. In the case of a real frequency ω, however, there is always some absorption, i.e., in (3.1.23a) the parameter $\nu' \neq 0$. Therefore $\varepsilon_{ij}(\omega,\boldsymbol{k}\to 0) = \varepsilon_{ij}(\omega)$ and no singularities arise; this also follows from microtheoretical calculations ([3.8], and Sect. 6.3).

The expansions (3.1.19–22) have a/λ as their parameter, i.e., the coefficients γ and δ are of the order of a and the coefficients α, β, ν and μ are of the order of a^2. Therefore, as is evident from (3.1.23), the term μk^2 should be taken into account in the denominator only if $|\omega-\omega_l|/|\omega_l| \sim \mu k^2 \sim (2\pi a/\lambda)^2$, and if, in addition, $\nu k^2/[(\omega-\omega_l)/\omega_l] \sim \mu k^2 \sim \varepsilon(\omega) \sim 1$. The latter means that a quadrupolar line makes a contribution to ε which is comparable with the contribution of the dipole lines. The respective value of the frequency ratio is $|\omega-\omega_l|/\omega_l \sim 10^{-1}$ to 10^{-6}; this corresponds to an approach of $\Delta\lambda \approx 10^{-2}$ Å toward the center of the line in the case of weak absorption. Even in the case of similar conditions the term μk^2 in (3.1.23) or the dependence of α_{ijlm} on $\boldsymbol{k}$ in (3.1.24) is required only in a very narrow region near the line center. Outside this region (3.1.19–22) are applicable and the entire 'quadrupolar effect' is accounted for by the term $\beta_{ijlm}k_lk_m$ or $\alpha_{ijlm}k_lk_m$ which thus plays a dominant role.

Equation (3.1.25) corresponds in a sense to the consideration of higher-order effects when passing over to expressions of the type (3.1.23 and 24). The corresponding ranges of applicability of (3.1.19–22) are therefore limited in a quite natural manner. Perhaps such a higher-order effect has been encountered in experiments (Sect. 4.6.2).

Summing up, we may say that the use of the formulas (3.1.19–22) in crystal optics with spatial dispersion is consistent[5], although sometimes it is necessary to use the generalized forms (3.1.23, 24). In the literature we may find expressions for $\varepsilon_{ij}(\omega, \boldsymbol{k})$ of crystals, which seem to be even more general than (3.1.19–22). Thus, an expression of the type

$$\varepsilon_{ij}(\omega, \boldsymbol{k}) = \varepsilon_{ij}^{(0)} + \frac{g_{ij}}{\omega - \omega_l - F(\boldsymbol{k})} \tag{3.1.26}$$

is used in [3.10], $\varepsilon_{ij}^{(0)}$, and g_{ij} being constants. Actually, however, we should set $F(\boldsymbol{k}) = f + g_l k_l + h_{lm} k_l k_m$ in (3.1.26) as it goes beyond the limits of the basic approximation to make allowance for higher terms in k. As a result, (3.1.26) reduces to (3.1.21), and applies only near the line center. The formulas given in [3.11] and in Sect. 4.5.2 are convenient over a wide frequency range but, outside the limits of validity of the expansions (3.1.19–22), they are merely of extrapolational character. We believe that in some cases it is better not to make use of extrapolation formulas for $\varepsilon_{ij}(\omega, \boldsymbol{k})$, for which the weakness of spatial dispersion is explicitly taken into account. On the contrary, it proves convenient to apply extrapolation formulas for $\varepsilon_{ij}(\omega, \boldsymbol{k})$ near the dipole absorption line, e.g., by assuming

$$\varepsilon_{ij}(\omega, \boldsymbol{k}) = \varepsilon_{0ij} - \frac{\omega_{0ij}^2}{\omega^2 - \omega_l^2 + \mathrm{i}\nu\omega - \mu_{lm}k_lk_m}\,. \tag{3.1.27}$$

---

[5] If series of the type (3.1.19–22) are substituted in the dispersion equation (2.2.10, 22), as is done in [3.7], algebraic equations are obtained for the refractive index $\tilde{n}$, whose order increases with the number of terms of the series taken into consideration. The new roots $\tilde{n}$, however, lie at higher and higher values of k; if the condition $ka \ll 1$ is violated, see inequality (3.1.18), the macroscopic method is unsuitable. Hence, in determining the first few roots of $\tilde{n}$, which can be considered only in connection with the condition (3.1.18) and by virtue of the absorption (Sects. 4.1, 2), the restriction to the first terms of the series proves again to be justified. The remarks to the contrary in [3.9] therefore seem to be incorrect.

The tensors in (3.1.19 – 22) satisfy a series of relations which are consequences of the general symmetry properties of the tensor $\varepsilon_{ij}(\omega,\boldsymbol{k})$ discussed in Sect. 2.1.2. By virtue of (2.1.21) we thus have

$$\begin{aligned} &\varepsilon_{ij}(\omega) = \varepsilon_{ji}(\omega)\,, \qquad \varepsilon_{ij}^{-1}(\omega) = \varepsilon_{ji}^{-1}(\omega)\,, \qquad \alpha_{ijlm}(\omega) = \alpha_{jilm}(\omega)\,, \\ &\gamma_{ijl}(\omega) = -\gamma_{jil}(\omega)\,, \qquad \delta_{ijl}(\omega) = -\delta_{jil}(\omega)\,, \qquad \beta_{ijlm}(\omega) = \beta_{jilm}(\omega)\,. \end{aligned} \tag{3.1.28}$$

Moreover, it is always possible to choose the tensors α_{ijlm} and β_{ijlm} in such a way that $\alpha_{ijlm} = \alpha_{ijml}$ and $\beta_{ijlm} = \beta_{ijml}$ (in the following we assume that such a choice was made). We also recall that the magnetic induction $\boldsymbol{B}_{\text{ext}}$ of the external field is always assumed to be zero unless otherwise indicated.

If there is a center of symmetry and, in general, for a nongyrotropic medium it follows from (2.1.22) that

$$\gamma_{ijl} = 0\,, \qquad \delta_{ijl} = 0\,. \tag{3.1.29}$$

If there is no absorption and $\boldsymbol{k}$ is real, the tensor $\varepsilon_{ij}''(\omega,\boldsymbol{k}) = 0$ (Sect. 2.3.1) and, consequently, the tensor $\varepsilon_{ij}(\omega,\boldsymbol{k}) = \varepsilon_{ij}'(\omega,\boldsymbol{k})$ is Hermitian. By virtue of (3.1.28) the tensors $\varepsilon_{ij}(\omega)$, $\varepsilon_{ij}^{-1}(\omega)$, γ_{ijl}, α_{ijlm}, δ_{ijl} and β_{ijlm} are real in this case (when $\boldsymbol{B}_{\text{ext}} \neq 0$, this statement is, in general, no longer true; see the end of Sect. 3.2.1).

In connection with the use of the expressions of the type (3.1.19 – 27) we have already mentioned the dipole and quadrupole absorption lines. In fact, in the neighborhood of dipole lines it is usually sufficient to use the expansions (3.1.19, 20) or (3.1.21, 22, 27), whereas (3.1.23 – 25) are encountered primarily in the case of quadrupole lines. It should, however, be stressed that these series expansions in $\boldsymbol{k}$ are not multipole expansions. Moreover, the use of (3.1.19 – 22) is by no means restricted to the range of any lines. It should also be taken into account that the appearance of absorption lines (or of a pole $\tilde{n}^2 = n^2$ if absorption is neglected) in an arbitrary optically anisotropic medium for any direction of light propagation is not associated with a resonant growth of the components of $\varepsilon_{ij}(\omega,\boldsymbol{k})$; this is evident even from the formulas of classical crystal optics. A pole of $\tilde{n}^2$ does appear under the conditions (2.2.37, 38), but the aforesaid can be made even more clear by means of the well-known expression for $\tilde{n}^2$ of an extraordinary wave in a uniaxial crystal:

$$1/\tilde{n}^2 = (s_x^2 + s_y^2)/\varepsilon_z(\omega) + s_z^2/\varepsilon_\perp(\omega)\,.$$

In Chap. 1 we have already indicated the role played by the phenomenological approach to crystal optics; one of the features of this approach is an extended use of expansions of the type (3.1.19 – 28) and others, which are based on the weakness of spatial dispersion, condition (3.1.18). The problem of combining phenomenological (macroscopic) and microscopic procedures in physics is nowadays clear, and further comments are hardly required. We therefore think it is unnecessary to prove the irrelevance of statements encountered in the literature [3.12] on the "nonsubstantiality of the phenomenological introduction of spatial dispersion of the dielectric permeability" and the "incompatible contradiction" said to exist between the results of the phenomenological theory and the experimental facts.

3.2 The Tensor $\varepsilon_{ij}(\omega, \boldsymbol{k})$ for Crystal of Various Classes

3.2.1 Gyrotropic Crystals

The tensors $\varepsilon_{ij}(\omega, \boldsymbol{k})$ and $\varepsilon_{ij}^{-1}(\omega, \boldsymbol{k})$ and, of course, also the tensors $\varepsilon_{ij}(\omega)$, $\varepsilon_{ij}^{-1}(\omega)$, $\gamma_{ijl}(\omega)$, $\delta_{ijl}(\omega)$, $\alpha_{ijlm}(\omega)$ and $\beta_{ijlm}(\omega)$ from the expansions (3.1.19 – 22) are appreciably simplified by the presence of symmetry elements. Thus, if a symmetry center exists, $\gamma_{ijl} = \delta_{ijl} = 0$, see (3.1.29); in an isotropic nongyrotropic medium the tensor α_{ijlm} has only two independent components since, see (2.1.48, 52):

$$\varepsilon_{ij}(\omega, \boldsymbol{k}) = \varepsilon(\omega) + \alpha_{\mathrm{tr}}(\omega)(\delta_{ij} - s_i s_j) k^2 + \alpha_{\mathrm{lo}}(\omega) s_i s_j k^2 . \tag{3.2.1}$$

Assuming here $k^2 = (\omega/c)^2 \tilde{n}^2(\omega)$ and introducing the convention $(\omega/c)^2 \times \alpha_{\mathrm{tr}}(\omega) = \alpha_{\perp}$ and $(\omega/c)^2 \alpha_{\mathrm{lo}}(\omega) = \alpha_{\parallel}$, (3.2.1) and the wave equation (2.2.26) for transverse and longitudinal normal waves, see also (2.1.50), yield:

$$\tilde{n}_{\perp}^2 = \varepsilon(\omega)/[1 + \alpha_{\perp}(\omega)]^{-1} , \quad \tilde{n}_{\parallel}^2 = \varepsilon(\omega)/\alpha_{\parallel}(\omega) . \tag{3.2.2}$$

Near the pole of $\varepsilon(\omega)$ in the case of an isotropic nongyrotropic medium with weak spatial dispersion we should use an expansion of the type (3.1.21, 22), i.e.,

$$\varepsilon_{ij}^{-1} = \varepsilon^{-1}(\omega)\delta_{ij} + \beta_{\perp}(\omega)\tilde{n}^2(\delta_{ij} - s_i s_j) + \beta_{\parallel}(\omega)\tilde{n}^2 s_i s_j . \tag{3.2.3}$$

By virtue of the condition $\boldsymbol{s} \cdot \boldsymbol{D} = 0$ we evidently have $\boldsymbol{E} = [1/\varepsilon + \beta_{\perp}(\omega)\tilde{n}_{\perp}^2]\boldsymbol{D}$, and from the wave equation (2.2.26) we find

$$\beta_{\perp}\tilde{n}_{\perp}^4 + \frac{\tilde{n}_{\perp}^2}{\varepsilon} - 1 = 0 , \quad n_{\perp}^2 = -\frac{1}{2\beta_{\perp}\varepsilon} \pm \sqrt{\left(\frac{1}{2\beta_{\perp}\varepsilon}\right)^2 + \frac{1}{\beta_{\perp}}} . \tag{3.2.4}$$

This solution will be dealt with further in Sect. 4.2.2. The condition $\varepsilon(\omega_{\parallel}) = 0$ for the longitudinal wave is obtained from (3.2.3). In the case of the longitudinal wave, spatial dispersion is not taken into account when the expansion (3.2.3) is used.

As was noted in Sect. 2.1.3, (3.2.1, 2) correspond to the most general relation between the vectors $\boldsymbol{D}$ and $\boldsymbol{E}$ in an isotropic nongyrotropic medium if only derivatives up to the second order are taken into account.

Consequences following from the consideration of crystal symmetry are well known for the tensors ε_{ij} and γ_{ijl} [3.13 – 17]. For reasons of convenience, let us nevertheless recall these results (the symmetry properties for the inverse tensors are obviously the same as for the original tensors. Therefore the components of the inverse tensors are not given in the following).

A symmetric tensor of rank two, particularly $\varepsilon_{ij}(\omega)$ has no more than 6 independent components. For a characteristic second-order surface $\varepsilon_{ij}x_i x_j = 1$ this corresponds to the length of the three axes and to the three parameters (angles) determining the orientation of these axes. The symmetry of the tensor ε_{ij} is the same for all classes of a given crystal system (syngony).

This can easily be proved by determining ε_{ij} for the least symmetrical class of each system. In doing so, it is useful to keep in mind the following fact, which is clear from the properties of a second-order surface: in a plane perpendicular to threefold or higher symmetry axes, the surface cross section degenerates to a circle. The characteristic surface, for example, therefore degenerates to a sphere, i.e., $\varepsilon_{ij} = \varepsilon\delta_{ij}$, even for the least symmetrical class T of the cubic system (the crystals of class T have four threefold symmetry axes corresponding to the body diagonals of the cube).

In tetragonal, trigonal (rhombohedral) and hexagonal systems where the z axis is oriented along the axes of four-, three- or sixfold symmetry, respectively, the tensor ε_{ij} is reduced to the form

$$\begin{pmatrix} \varepsilon_1 & 0 & 0 \\ 0 & \varepsilon_1 & 0 \\ 0 & 0 & \varepsilon_3 \end{pmatrix} = \begin{pmatrix} \varepsilon_\perp & 0 & 0 \\ 0 & \varepsilon_\perp & 0 \\ 0 & 0 & \varepsilon_\parallel \end{pmatrix}$$

etc. (Table 3.1)[6]. The axes of ε_{ij} that are fixed by the crystal's symmetry are given in the column 'principal axes'. Of course, in the case of degeneracy the axes of the second-rank tensor ε_{ij} may be directed differently. For instance, in cubic crystals $\varepsilon_{ij}(\omega) = \varepsilon(\omega)\delta_{ij}$ for any system of axes.

By virtue of (3.1.28) the tensor γ_{ijl} (and δ_{ijl}) possesses the following properties:

$$\gamma_{xx,l} = \gamma_{yy,l} = \gamma_{zz,l} = 0\,,$$

$$\gamma_{xy,l} = -\gamma_{yx,l},\ \gamma_{yz,l} = -\gamma_{zy,l},\ -\gamma_{zx,l} = -\gamma_{xz,l}$$

$$(l = 1, 2, 3 \equiv x, y, z)\,.$$

[6] The simplification of tensors made possible by symmetry is described in great detail in [3.16].

Table 3.1. Properties of the tensor $\varepsilon_{ij}(\omega)$ and of the pseudotensor $f'_{ij}(\omega)$ ($\varepsilon_{ij} = \varepsilon_{ji}, f'_{ij} = f'_{ji}, \delta_{ijl} = e_{ijm} f'_{ml}$)

System	Principal axes	Tensor ε_{ij}	Pseudotensor f'_{ij}	Class[a]
Triclinic	Not fixed	$\begin{pmatrix} \varepsilon_{11} & \varepsilon_{12} & \varepsilon_{13} \\ \varepsilon_{12} & \varepsilon_{22} & \varepsilon_{23} \\ \varepsilon_{13} & \varepsilon_{23} & \varepsilon_{33} \end{pmatrix}$	$f'_{ij} = 0$	C_i
			$f'_{ij} = \begin{pmatrix} f'_{11} & f'_{12} & f'_{13} \\ f'_{12} & f'_{22} & f'_{23} \\ f'_{13} & f'_{23} & f'_{33} \end{pmatrix}$	C_1
Monoclinic	The y axis is directed along a twofold symmetry axis or perpendicular to a symmetry plane		$f'_{ij} = 0$	C_{2h}
			$f'_{ij} = \begin{pmatrix} f'_{11} & 0 & f'_{13} \\ 0 & f'_{22} & 0 \\ f'_{13} & 0 & f'_{33} \end{pmatrix}$	C_2
		$\begin{pmatrix} \varepsilon_{11} & 0 & \varepsilon_{13} \\ 0 & \varepsilon_{22} & 0 \\ \varepsilon_{13} & 0 & \varepsilon_{33} \end{pmatrix}$	$f'_{ij} = \begin{pmatrix} 0 & f'_{12} & 0 \\ f'_{12} & 0 & f'_{23} \\ 0 & f'_{23} & 0 \end{pmatrix}$	C_s
Rhombic	Axes x, y, z are directed along twofold symmetry axes; in class C_{2v} the x and y axes are perpendicular to the symmetry plane	$\begin{pmatrix} \varepsilon_{11} & 0 & 0 \\ 0 & \varepsilon_{22} & 0 \\ 0 & 0 & \varepsilon_{33} \end{pmatrix}$	$f'_{ij} = 0$	D_{2h}
			$f'_{ij} = \begin{pmatrix} f'_{11} & 0 & 0 \\ 0 & f'_{22} & 0 \\ 0 & 0 & f'_{33} \end{pmatrix}$	D_2
			$f'_{ij} = \begin{pmatrix} 0 & f'_{12} & 0 \\ f'_{12} & 0 & 0 \\ 0 & 0 & 0 \end{pmatrix}$	C_{2v}

Table 3.1 (continued)

System	Principal axes	Tensor ε_{ij}	Pseudotensor f'_{ij}	Class[a]
Tetragonal	In classes C_4, S_4 and C_{4h} only the z axis is fixed (fourfold symmetry axis). In classes D_4, C_{4v}, D_{2d} and D_{4h} all axes are fixed	$\begin{pmatrix} \varepsilon_\perp & 0 & 0 \\ 0 & \varepsilon_\perp & 0 \\ 0 & 0 & \varepsilon_\parallel \end{pmatrix}$	$f'_{ij} = 0$	C_{4h}, C_{4v} and D_{4h}
			$f'_{ij} = \begin{pmatrix} f'_\perp & 0 & 0 \\ 0 & f'_\perp & 0 \\ 0 & 0 & f'_\parallel \end{pmatrix}$	C_4 and D_4
			$f'_{ij} = \begin{pmatrix} f'_{11} & f'_{12} & 0 \\ f'_{12} & -f'_{11} & 0 \\ 0 & 0 & 0 \end{pmatrix}$	S_4
			$f'_{ij} = \begin{pmatrix} 0 & f'_{12} & 0 \\ f'_{12} & 0 & 0 \\ 0 & 0 & 0 \end{pmatrix}$	D_{2d}
Trigonal	In classes C_3 and C_{3i} only the z axis (threefold symmetry axis) is fixed. In classes D_3, C_{3v} and D_{3d} all axes are fixed	$\begin{pmatrix} \varepsilon_\perp & 0 & 0 \\ 0 & \varepsilon_\perp & 0 \\ 0 & 0 & \varepsilon_\parallel \end{pmatrix}$	$f'_{ij} = 0$	C_{3i}, C_{3v} and D_{3d}
			$f'_{ij} = \begin{pmatrix} f'_\perp & 0 & 0 \\ 0 & f'_\perp & 0 \\ 0 & 0 & f'_\parallel \end{pmatrix}$	C_3 and D_3
Hexagonal	In classes C_6, C_{3h} and C_{6h} only the z axis (sixfold symmetry axis) is fixed	$\begin{pmatrix} \varepsilon_\perp & 0 & 0 \\ 0 & \varepsilon_\perp & 0 \\ 0 & 0 & \varepsilon_\parallel \end{pmatrix}$	$f'_{ij} = 0$	C_{3h}, C_{6h}, C_{6v}, D_{3h}, D_{6h}
			$f'_{ij} = \begin{pmatrix} f'_\perp & 0 & 0 \\ 0 & f'_\perp & 0 \\ 0 & 0 & f'_\parallel \end{pmatrix}$	C_6 and D_6

Table 3.1 (continued)

System	Principal axes	Tensor ε_{ij}	Pseudotensor f'_{ij}	Class[a]
Cubic	All axes are fixed (axes x, y, z are twofold symmetry axes in classes T and T_h, and fourfold symmetry axes in classes O, T_d and O_h)	$\begin{pmatrix} \varepsilon & 0 & 0 \\ 0 & \varepsilon & 0 \\ 0 & 0 & \varepsilon \end{pmatrix}$	$f'_{ij} = 0$	T_h, T_d and O_h
			$f'_{ij} = \begin{pmatrix} f' & 0 & 0 \\ 0 & f' & 0 \\ 0 & 0 & f' \end{pmatrix}$	T and O
Isotropic medium	Arbitrary choice of axes	$\begin{pmatrix} \varepsilon & 0 & 0 \\ 0 & \varepsilon & 0 \\ 0 & 0 & \varepsilon \end{pmatrix}$	$f'_{ij} = 0$	With a center of symmetry
			$f'_{ij} = \begin{pmatrix} f' & 0 & 0 \\ 0 & f' & 0 \\ 0 & 0 & f' \end{pmatrix}$	With no center of symmetry

[a] The crystallographic classes are designated by Schönflies symbols [3.17]. (Appendix 1 includes a table listing the corresponding international notation).

Hence, in the general case, the tensors γ_{ijl} and δ_{ijl} have nine independent components which can be written in the form

$$\gamma_{ijl} = e_{ijm} g'_{ml}, \qquad \delta_{ijl} = e_{ijm} f'_{ml}, \tag{3.2.5}$$

where e_{ijm} is a unit pseudotensor of rank three ($e_{123} = 1$, $e_{213} = -1$, $e_{112} = 0$, etc.; e_{ijm} is not affected by mirror reflection) and f'_{ml} and g'_{ml} are pseudotensors of rank two. We may write

$$\gamma_{ijl} k_l = e_{ijm} g'_{ml} k_l = e_{ijm} g'_m, \qquad \delta_{ijl} k_l = e_{ijm} f'_{ml} k_l = e_{ijm} f'_m, \tag{3.2.6}$$

where we have introduced the pseudovectors (i.e., axial vectors) $\boldsymbol{g}'$ and $\boldsymbol{f}'$ as gyration vectors.

Neglecting terms which are quadratic in $\boldsymbol{k}$ in (3.1.19–22) we obtain

$$\begin{aligned} D_i &= \varepsilon_{ij}(\omega, \boldsymbol{k}) E_j = \varepsilon_{ij}(\omega) E_j - i(\boldsymbol{g}' \times \boldsymbol{E})_i, \\ E_i &= \varepsilon_{ij}^{-1}(\omega, \boldsymbol{k}) D_j = \varepsilon_{ij}^{-1}(\omega) D_j - i(\boldsymbol{f}' \times \boldsymbol{D})_i. \end{aligned} \tag{3.2.7}$$

After substituting into the wave equation (2.2.7 or 26), it becomes clear (Sect. 4.1) that in this equation only the scalar products $f'_{ij} s_i s_j$ or $f'_i s_i$ play a role. The refractive indices and the ratios of the components of the vector $\boldsymbol{D}$ are therefore independent of the antisymmetric part of f'_{ij}, i.e., in determining $\tilde{n}^2$ this tensor may be chosen in a symmetric form (see nevertheless below). For an isotropic medium and cubic crystals (without a center of symmetry, of course, as otherwise $g'_{ij} = f'_{ij} = 0$; one should also be aware of the fact that the tensors g'_{ij} and f'_{ij} may vanish even when there is no center of symmetry as is the case, e.g., with cubic crystals of the class T_d):

$$\begin{aligned} &g'_{ij} = \tilde{g}\delta_{ij}, \quad f'_{ij} = \tilde{f}\delta_{ij}, \quad \boldsymbol{g}' = g'\boldsymbol{s}, \quad \boldsymbol{f}' = f'\boldsymbol{s}, \quad g' = k\tilde{g}, \quad f' = k\tilde{f}, \\ &\boldsymbol{D} = \varepsilon\boldsymbol{E} - \mathrm{i}g'(\boldsymbol{s} \times \boldsymbol{E}), \quad \boldsymbol{E} = \frac{\boldsymbol{D}}{\varepsilon} - \mathrm{i}f'(\boldsymbol{s} \times \boldsymbol{D}). \end{aligned} \tag{3.2.8}$$

In this case not only $\boldsymbol{s} \cdot \boldsymbol{D} = 0$ but also $\boldsymbol{s} \cdot \boldsymbol{E} = 0$ if $\boldsymbol{D} \neq 0$. If, however, $\boldsymbol{D} = 0$ and $\boldsymbol{E} = E\boldsymbol{s}$ (longitudinal waves), the gyrotropy proves to be unimportant. The form of the symmetric tensor f'_{ij} (or, more exactly, the symmetric part of f'_{ij}) for various crystal classes is given in Table 3.1.

It does not follow from the aforesaid that the antisymmetric parts of the tensors g'_{ij} and f'_{ij} are of no interest. These parts, denoted by $g'_{ij,a}$ and $f'_{ij,a}$, can be reduced according to (3.2.6) to the pseudovectors $\boldsymbol{g}'_a$ and $\boldsymbol{f}'_a$ which have the form $(\boldsymbol{h} \times \boldsymbol{k})$, $\boldsymbol{h}$ being a vector connected with the crystal. Such a vector may evidently exist only in the case of crystals which possess a direction that is invariant (with respect to mirror reflection as well) under all symmetry transformations. In other words, the vector $\boldsymbol{h}$ exists only for pyroelectric crystals (classes C_1, C_s, C_2, C_{2v}, C_4, C_{4v}, C_3, C_{3v}, C_6, C_{6v} [Ref. 3.13, § 13]). Hence and from Table 3.1 it can be seen that the full tensor f'_{ij} is antisymmetric for crystals of the classes C_{3v}, C_{4v} and C_{6v} which do not usually belong to the gyrotropic crystals (in Table 3.1 only the symmetric part of the tensor f'_{ij} is given). It follows from (3.2.7) that the vector $\boldsymbol{f}_a$ gives rise only to a longitudinal component of $\boldsymbol{E}$ (we assume that $\boldsymbol{k} = \omega \boldsymbol{s}/c$). In the solution of the wave equation for a gyrotropic crystal (Sect. 4.1), the tensor $f_{ij,a}$ does not contribute either to the refractive index or to the ratio between the components of the vector $\boldsymbol{D}$, as has been mentioned before. The vector $\boldsymbol{E}$ is usually elliptically polarized in normal waves, also when $f_{ij} = f_{ij,a}$. In this case, however, the degree of ellipticity is proportional to k, i.e., we are concerned with an effect of the order of a/λ. Thus, if we restrict ourselves only to circular or almost circular polarization, crystals of the classes C_{3v}, C_{4v} and C_{6v} are nongyrotropic, and for all other crystals the tensor f_{ij} can be regarded as symmetric.

This will be assumed in what follows (and in Table 3.1 as well). Moreover, if we are concerned with effects of second order with respect to k we shall assume in the following that there are no terms linear with respect to k in (3.1.19 – 22) even though these terms may also give rise to second-order effects.

'Weak gyrotropy' (degree of ellipticity $\sim a/\lambda$), which in its form ought to be observed in crystals of classes C_{3v}, C_{4v} and C_{6v} (as was indicated in [3.18]) seems, however, worthy of attention. Of special interest is an analysis of the region near the resonances, where additional (new) waves may appear.

As previously emphasized, (3.1.28) and those relations stemming from them are valid, in general, only at $\boldsymbol{B}_{\text{ext}} = 0$. If, however, $\boldsymbol{B}_{\text{ext}} \neq 0$, in particular in antiferromagnetic materials, we obtain from (2.1.21) and (3.1.19)

$$\gamma_{ijl}(\omega, \boldsymbol{B}_{\text{ext}}) = -\gamma_{jil}(\omega, -\boldsymbol{B}_{\text{ext}}), \tag{3.1.28a}$$

and (3.1.28) cannot be applied for a given $\boldsymbol{B}_{\text{ext}}$. Upon real ω and $\boldsymbol{k}$ (transparent medium), $\varepsilon''(\omega, \boldsymbol{k}, \boldsymbol{B}_{\text{ext}}) = 0$ and, consequently, $\varepsilon_{ij}(\omega, \boldsymbol{k}, \boldsymbol{B}_{\text{ext}}) = \varepsilon^*_{ji}(\omega, \boldsymbol{k}, \boldsymbol{B}_{\text{ext}})$, i.e., ε_{ij} is Hermitian, see (2.3.4, 13). Hence

$$\gamma_{ijl}(\omega, \boldsymbol{B}_{\text{ext}}) = -\gamma^*_{jil}(\omega, \boldsymbol{B}_{\text{ext}}), \tag{3.1.28b}$$

i.e., (with fixed ω and $\boldsymbol{B}_{\text{ext}}$)

$$\gamma'_{ijl} = -\gamma'_{jil}, \quad \gamma''_{ijl} = \gamma''_{jil}, \quad \gamma_{ijl} = \gamma'_{ijl} + \mathrm{i}\,\gamma''_{ijl},$$
$$\gamma'_{ijl} = \mathrm{Re}\{\gamma_{ijl}\}, \quad \gamma''_{ijl} = \mathrm{Im}\{\gamma_{ijl}\}.$$

If $\boldsymbol{B}_{\text{ext}} = 0$, it is clear from (3.1.28a and b) that $\gamma''_{ijl} = 0$. Thus, we have returned to an ordinary isotropic medium in which the factors γ_{ijl} are real in the transparent region.

The peculiar effect of gyromagnetic birefringence [3.19] occurs in a transparent magnetic medium ($\boldsymbol{B}_{ext} \neq 0$) for which $\gamma''_{ijl} \neq 0$. The maximum number of independent components of γ''_{ijl} is 18, but is reduced if some symmetry elements are present (for example, the tensor γ''_{ijl} has only two independent components for the Cr_2O_3 crystal [3.19, 20]). Gyrotropic birefringence (discussed in [3.21] for the region of resonance) is manifested in its pure form, so to speak, in crystals for which $\gamma''_{ijl} \neq 0$, but $\gamma'_{ijl} = 0$, i.e., ordinary gyrotropy is absent. In this case, in particular, the refractive index changes when $\boldsymbol{k}$ is replaced by $-\boldsymbol{k}$ (i.e., in reversing the direction of wave propagation). The optical properties of magnetic crystals having optical activity were also discussed in [3.22, 23].

It is necessary to emphasize, of course, that in a magnetic medium under consideration the quantity $\gamma'_{ijl}(\boldsymbol{B}_{ext})$ is generally nonzero. For small values $\boldsymbol{B}_{ext}$ (e.g., near magnetic phase transition points, where vector $\boldsymbol{B}_{ext}$ is an order parameter or may be expressed in terms of it)

$$\gamma'_{ijl}(\boldsymbol{B}_{ext}, \boldsymbol{H}^0) = \gamma'_{ijl}(0,0) + \tilde{\gamma}^{(1)}_{ijlmn} B_{ext,m} B_{ext,n} + \tilde{\gamma}^{(2)}_{ijlmn} B_{ext,,} H^0_n + \dots \,.$$

Also taken into account here, for the sake of generality, is the weak external magnetic field $\boldsymbol{H}^{(0)}$. Tensors $\tilde{\gamma}^{(1)}$ and $\tilde{\gamma}^{(2)}$, like the ordinary gyrotropy tensor $\gamma'_{ijl}(0,0)$, are antisymmetric with respect to the first pair of indices and can be reduced to pseudotensors $\tilde{g}^{(1,2)}_{ijlm}$ of rank four [$\tilde{\gamma}^{(1)}_{ijlmn} = e_{ijk}\tilde{g}^{(2)}_{klmn}$, $\tilde{\gamma}^{(2)}_{ijlmn} = e_{ijk}\tilde{g}^{(2)}_{klmn}$; nonzero components for the symmetrical part of $\tilde{g}^{(1)}$ and $\tilde{g}^{(2)}$ are the same as for pseudotensor f'_{iklm} (Table 4.1)].

3.2.2 Nongyrotropic Crystals

If spatial dispersion is to be taken into account in nongyrotropic crystals we should consider terms of order k^2, i.e., the terms $\alpha_{ijlm}(\omega) k_l k_m$ or $\beta_{ijlm}(\omega) \times k_l k_m$ in (3.1.19 – 22).

We shall first consider the tensors α_{ijlm} for crystals with various symmetries (the symmetry of the tensor β_{ijlm} is the same as that of the tensor α_{ijlm}). Owing to the separate symmetry of these tensors with respect to the subscripts ij and lm, see (3.1.28), this tensor has in the general case 36 components (instead of the 81 components of an arbitrary fourth-rank tensor). If we consider a concrete case of crystal symmetry the number of independent components can be further reduced. The general principle is the requirement that the tensor components (physical quantities) be invariant with respect to transformations permitted by crystal symmetry (for details see [3.16, 24]).

Triclinic crystals either have no symmetry elements (class C_1) or have a symmetry center (class C_i). The presence or absence of a symmetry center, however, does not impose any conditions on a tensor of rank four. Consequently, in a triclinic system 36 independent coefficients α_{ijlm} remain.

Three coefficients can, in fact, be fixed by choosing the axes which in the given case are arbitrary. It seems more rational, however, to count the number of independent components without taking into account the possibility of a free choice of axes. It is sufficient to cite the example of the tensor $\varepsilon_{ij}(\omega)$ (or another second-rank tensor). The number of independent components of ε_{ij} in triclinic, monoclinic or rhombic systems is 6, 4 and 3, respectively (Table 3.1). At the same time in all these cases, ε_{ij} has three independent components in the system of principal axes. There is, however, a great difference between crystals of these classes, since in a rhombic crystal the principal axes are fixed, whereas in a triclinic crystal all principal axes have to be determined, this being equivalent to three more parameters. In the following we have indicated a possibility of choosing the principal axes (i.e., we mention cases where this choice is not fixed by symmetry considerations), but the number of independent components is counted without allowing for this fact.[7]

In monoclinic crystals of classes C_2 and C_{2h} there exists a twofold symmetry axis which is usually taken as the y axis. Rotation about this axis involves the coordinate transformation $x \to -x$, $z \to -z$. The components of tensors with an odd sum of the number of subindices x and of the number of subindices z change sign during this transformation; from a physical point of view, however, the two coordinate systems are completely equivalent; herein lies the significance of contending that we are dealing with a twofold symmetry axis. Consequently, the corresponding 16 components of the tensor α_{ijlm} vanish; in other words,

$$\left.\begin{array}{l} \alpha_{xyyy},\ \alpha_{zyyy},\ \alpha_{yyyx},\ \alpha_{yyyz},\ \alpha_{xxxy},\ \alpha_{yxxx}, \\ \alpha_{zzzy},\ \alpha_{yzzz},\ \alpha_{xzxy},\ \alpha_{xyxz},\ \alpha_{xxyz},\ \alpha_{yzxx}, \\ \alpha_{zxzy},\ \alpha_{zyzx},\ \alpha_{zzxy},\ \alpha_{xyzz} \end{array}\right\} = 0\,. \tag{3.2.9}$$

Class C_s has only a symmetry plane which is perpendicular to the y axis and, owing to symmetry, the components of α_{ijlm} that have an odd number of y subindices must vanish. This requirement leads to the same result as in the case of classes C_2 and C_{2n}. In monoclinic crystals the tensor α_{ijlm} therefore has 20 independent components. In monoclinic crystals the symmetry properties fix only one principal axis (the y axis) and by the choice of the other axes it is possible to reduce the number of components of the tensor α_{ijlm} by one.

The classes D_2 and D_{2h} of the rhombic system have three twofold symmetry axes. It is evident that in this case the following 12 components, each having even numbers of subindices x, y, and z (x, y, z being twofold axes), are nonvanishing:

---

[7] In [3.24] the independent components of the Young's modulus tensor were counted allowing for the choice of axes. This is why the numbers of independent components given in [3.24] and in [3.16] differ.

$$\begin{aligned}&\alpha_{xxxx},\ \alpha_{xxyy},\ \alpha_{xxzz},\ \alpha_{yyxx},\ \alpha_{zzxx},\ \alpha_{xyxy},\\&\alpha_{xzxz},\ \alpha_{yyyy},\ \alpha_{yyzz},\ \alpha_{zzyy},\ \alpha_{yzyz},\ \alpha_{zzzz}\end{aligned} \tag{3.2.10}$$

(moreover, of course, $\alpha_{xyxy} = \alpha_{yxxy} = \alpha_{xyyx}$, etc.).

The third class of rhombic systems, the class C_{2v}, has one twofold symmetry axis (the z axis) and two mutually perpendicular symmetry planes which contain this axis. Reflection in these planes, i.e., the transformations $x \to -x$ or $y \to -y$ leave unaffected only the tensor components that have an even number of subindices x and, at the same time, an even number of subindices y. Hence, the tensors of rank four require an even number of z subindices, i.e., we obtain again the system (3.2.10). Rhombic crystals thus have 12 independent components α_{ijlm}.

For the classes D_4, C_{4v}, D_{2d} and D_{4h} of the tetragonal system we have, apart from the symmetry elements corresponding to one of the classes of the rhombic system, the z axis as a fourfold symmetry or rotation-reflection axis. Rotation through an angle of $\pi/2$ about this axis should not alter the tensor components (physical properties) but results in the transformation $x \to y$, $y \to -x$. It follows that certain of the coefficients in (3.2.10) are equal to each other and only seven of them are independent (the remaining components α_{ijlm}, except for those obtained from the condition $\alpha_{ijlm} = \alpha_{jilm} = \alpha_{ijml}$, vanish in the chosen coordinate system):

$$\begin{aligned}&\alpha_{xxxx} = \alpha_{yyyy},\ \alpha_{zzzz},\ \alpha_{xxyy} = \alpha_{yyxx},\ \alpha_{xxzz} = \alpha_{yyzz}\,,\\&\alpha_{zzxx} = \alpha_{zzyy},\ \alpha_{xzxz} = \alpha_{yzyz},\ \alpha_{xyxy}\,.\end{aligned} \tag{3.2.11}$$

In the classes C_4, S_4 and C_{4h} of the tetragonal system, having only a fourfold symmetry or rotation-reflection axis (and, in class C_{4h} a symmetry plane perpendicular to this axis as well) we cannot proceed from (3.2.10). The same considerations as above lead us to the conclusion that these classes, besides the coefficients (3.2.11), have the following nonvanishing coefficients:

$$\alpha_{xyxx} = -\alpha_{xyyy},\ \alpha_{xxxy} = -\alpha_{yyyx},\ \alpha_{xzyz} = -\alpha_{yzxz}\,. \tag{3.2.12}$$

Thus, the classes C_4, S_4 and C_{4h} have ten independent coefficients. Furthermore, in these classes of crystal symmetry only the z axis is fixed so that there is one degree of freedom with regard to the choice of the coordinate system.

The classes of the cubic system have no nonvanishing components apart from those of the rhombic system, see (3.2.10). This conclusion is connected with the fact that all classes of the cubic system have (among others) symmetry elements of at least one of the classes of the rhombic system [Ref. 3.16, Table 21]. It should also be borne in mind that the axes x, y, z in all classes of the cubic system are fixed (twofold symmetry axes in the classes T, T_h and T_d; fourfold symmetry axes in the classes O and O_h). The number of independent components of the cubic system, however, decreases drastically. Thus even the existence of four axes, possessed by all classes of the cubic system (body diagonals of the cube) leads to the equivalence of all tensor components in the case of the substitution $xyz \to yzx \to zxy$. Therefore only four independent components remain:

$$\begin{aligned} &\alpha_1 = \alpha_{xxxx} = \alpha_{yyyy} = \alpha_{zzzz}\,, \qquad \alpha_2 = \alpha_{xxzz} = \alpha_{yyxx} = \alpha_{zzyy}\,, \\ &\alpha_3 = \alpha_{xyxy} = \alpha_{yzyz} = \alpha_{zxzx}\,, \qquad \alpha_4 = \alpha_{zzxx} = \alpha_{xxyy} = \alpha_{yyzz}\,. \end{aligned} \tag{3.2.13}$$

No further simplifications are possible for the classes T and T_h. For the classes T_d, O and O_h we have, moreover

$$\alpha_2 = \alpha_4 \qquad (\text{classes } T_d, O, O_h)\,. \tag{3.2.14}$$

The equality (3.2.14) results directly from (3.2.13) if we take into account that the axes of the cube xyz are fourfold rotation-reflection axes (class T_d) or fourfold symmetry axes (classes O and O_h).

In the case of an isotropic medium we have only two independent components α_{ijlm}, as is seen from (3.2.1). For an arbitrary Cartesian coordinate system we have in tensorial representation

$$\begin{aligned} &\alpha_{xxxx} = \alpha_{yyyy} = \alpha_{zzzz} = \alpha_1 \equiv \alpha_{\mathrm{lo}}(\omega)\,, \\ &\alpha_{xxyy} = \alpha_{xxzz} = \alpha_{yyxx} = \alpha_{yyzz} = \alpha_{zzxx} = \alpha_{zzyy} = \alpha_2 = \alpha_4 \equiv \alpha_{\mathrm{tr}}(\omega)\,, \\ &\alpha_{xyxy} = \alpha_{xzxz} = \alpha_{yzyz} = \alpha_3 = \alpha_1 - \alpha_2 \equiv \alpha_{\mathrm{lo}} - \alpha_{\mathrm{tr}}\,. \end{aligned} \tag{3.2.15}$$

Compared with cubic crystals of the classes T_d, O and O_h, the only difference consists in the existence of the relation $\alpha_3 = \alpha_1 - \alpha_2$.

To take into account the symmetry properties of crystals of the trigonal or hexagonal systems, simple analytic transformations are required. They will not be carried out here (such transformations, applied to some other tensors, may be found, e.g., in [3.16, 24]). This is, moreover, unnecessary since the symmetry properties of the tensor α_{ijlm} (including those mentioned above) may in fact be assumed to be known from the literature. The symmetry of the tensor α_{ijlm} is actually the same as that of the well-known piezo-optical tensor components π_{ijlm} which link the variation in ε_{ij}^{-1} with the stress tensor σ_{lm} (i.e., $\delta\varepsilon_{ij}^{-1} = \pi_{ijlm}\sigma_{lm}$; [3.16]). The symmetry properties of the elasto-optical tensor p_{ijlm} are, of course, the same (here we have $\delta\varepsilon_{ij}^{-1} = p_{ijlm}u_{lm}$, where u_{lm} is the deformation tensor). For trigonal and hexagonal systems the symmetry properties of the tensor α_{ijlm} can therefore be taken from [Ref. 3.16, Table 15]. This has not been done for other systems because they are readily deduced and because we wished to point out certain facts. All values of the α_{ijlm} are listed in Table 3.2.

Table 3.2. Properties of the tensor α_{ijlm} ($\alpha_{ijlm} = \alpha_{jilm} = \alpha_{ijml}$; the properties of β_{ijlm} are analogous to those of α_{ijlm})

System	Components of α_{ijlm} [a].
Triclinic	All 36 components of α_{ijlm} are nonvanishing.
Monoclinic	20 independent components are nonvanishing; the following are equal to zero: α_{xyyy}, α_{zyyy}, α_{yyyx}, α_{yyyz}, α_{xxxy}, α_{yxxx}, α_{zzzy}, α_{yzzz}, α_{xzxy}, α_{xyxz}, α_{xxyz}, α_{yzxx}, α_{zxzy}, α_{zyzx}, α_{zzxy}, α_{xyzz}.
Rhombic	The following 12 independent components are nonvanishing: α_{xxxx}, α_{yyyy}, α_{zzzz}, α_{xxyy}, α_{xxzz}, α_{yyxx}, α_{zzxx}, α_{xyxy}, α_{xzxz}, α_{yyzz}, α_{zzyy}, α_{yzyz}.
Tetragonal	In classes C_4, C_{4v}, D_{2d} and D_{4h} the following 7 independent components are nonvanishing: $\alpha_{xxxx} = \alpha_{yyyy}$, α_{zzzz}, $\alpha_{xxyy} = \alpha_{yyxx}$, $\alpha_{xxzz} = \alpha_{yyzz}$, $\alpha_{zzxx} = \alpha_{zzyy}$, $\alpha_{xzxz} = \alpha_{yzyz}$, α_{xyxy}. In the classes C_4, C_{4h} and S_4 three more independent components are nonvanishing, namely: $\alpha_{xyzz} = -\alpha_{xyyy}$, $\alpha_{xxxy} = -\alpha_{yyyx}$, $\alpha_{xzyz} = -\alpha_{yzxz}$.
Trigonal	Classes C_3 and C_{3i} have 12 independent components: $\alpha_{xxxx} = \alpha_{yyyy}$, $\alpha_{xxyy} = \alpha_{yyxx}$, $\alpha_{xxzz} = \alpha_{yyzz}$, $\alpha_{xxyz} = \alpha_{xyxz} = -\alpha_{yyyz}$, $\alpha_{xxxz} = -\alpha_{yyxz} = -\alpha_{xyyz}$, $\alpha_{xxxy} = -\alpha_{yyxy} = \alpha_{xyyy} = -\alpha_{xyxx}$, $\alpha_{zzxx} = \alpha_{zzyy}$, α_{zzzz}, $\alpha_{yzxx} = -\alpha_{yzyy} = \alpha_{xzxy}$, $\alpha_{xzxx} = -\alpha_{xzyy} = -\alpha_{yzxy}$, $\alpha_{yzyz} = \alpha_{xzxz}$, $\alpha_{yzxz} = -\alpha_{xzyz}$; besides, $\alpha_{xyxy} = \alpha_{xxxx} - \alpha_{xxyy}$. In addition, for classes D_3, C_{3v} and D_{3d}: $\alpha_{xxxz} = \alpha_{xxxy} = \alpha_{yzxz} = \alpha_{xzxx} = 0$.
Hexagonal	Classes C_6, C_{3h} and C'_{6h} have 8 independent components (all others, which are not listed, equal zero): $\alpha_{xxxx} = \alpha_{yyyy}$, $\alpha_{xxyy} = \alpha_{yyxx}$, $\alpha_{xxzz} = \alpha_{yyzz}$, $\alpha_{zzxx} = \alpha_{zzyy}$, α_{zzzz}, $\alpha_{xyyy} = \alpha_{xxxy} = -\alpha_{yyxy} = -\alpha_{xyxx}$, $\alpha_{yzyz} = \alpha_{xzxz}$, $\alpha_{yzxz} = -\alpha_{xzyz}$; besides, $\alpha_{xyxy} = \alpha_{xxxx} - \alpha_{xxyy}$. In addition to this, in classes D_6, C_{6v}, D_{3h} and D_{6h}: $\alpha_{xxyy} = \alpha_{yzxz} = 0$ (in connection with the above-mentioned symmetry relationships, other components also vanish).
Cubic	In classes T and T_h there are 4 independent nonzero components, namely: $\alpha_1 = \alpha_{xxxx} = \alpha_{yyyy} = \alpha_{zzzz}$, $\alpha_2 = \alpha_{xxzz} = \alpha_{yyxx} = \alpha_{zzyy}$, $\alpha_3 = \alpha_{xyxy} = \alpha_{yzyz} = \alpha_{zxzx}$, $\alpha_4 = \alpha_{zzxx} = \alpha_{xxyy} = \alpha_{yyzz}$. In addition, for classes O, T_d and O_h, $\alpha_2 = \alpha_4$.
Isotropic medium	The nonvanishing components are the same as for the cubic system but, besides $\alpha_2 = \alpha_4$, $\alpha_3 = \alpha_1 - \alpha_2$ as well (altogether we thus have two independent components).

[a] Apart from the components mentioned, the components obtained from the condition $\alpha_{ijlm} = \alpha_{jilm} = \alpha_{ijml}$ are also nonvanishing. The possibility of fixing the axes was not taken into account in this table; see the text and Table 3.1.

In the following we shall not consider simultaneously the first- and second-order terms of $\boldsymbol{k}$ owing to the fact that spatial dispersion, as mentioned above, is weak. We are interested therefore not in the general expressions (3.1.19–22) but in (3.2.7) for the gyrotropic medium and in the following expressions for a nongyrotropic medium:

$$\varepsilon_{ij}(\omega,\boldsymbol{k}) = \varepsilon_{ij}(\omega) + (\omega/c)^2 \alpha_{ijlm}(\omega)\tilde{n}^2 s_l s_m\,, \tag{3.2.16}$$

$$\varepsilon_{ij}^{-1}(\omega,\boldsymbol{k}) = \varepsilon_{ij}^{-1}(\omega) + (\omega/c)^2 \beta_{ijlm}(\omega)\tilde{n}^2 s_l s_m\,. \tag{3.2.17}$$

The tensor $\varepsilon_{ij}(\omega,\boldsymbol{k})$ can, of course, always be diagonalized by choosing the appropriate (principal) axes. If the tensor $\varepsilon_{ij}(\omega,\boldsymbol{k})$ is non-Hermitian, ε'_{ij} and ε''_{ij} should be considered separately, and the principal axes of these tensors (i.e., the eigenvectors which, in general, are complex) may not coincide. Unless otherwise indicated, for reasons of brevity we shall deal only with the tensor ε'_{ij} which is assumed to be real. With arbitrary $\boldsymbol{s}$ the direction of the principal axes coincides neither with $\boldsymbol{s}$ nor with the axes of the tensor $\varepsilon_{ij}(\omega)$; in all cases where the axes of the tensor $\varepsilon_{ij}(\omega)$ are fixed (i.e., in the absence of degeneracy taking place in cubic and uniaxial crystals), the axes of $\varepsilon_{ij}(\omega,\boldsymbol{k})$ are nearly the same as the axes of $\varepsilon_{ij}(\omega)$ owing to the smallness of the terms depending on $\boldsymbol{s}$ in (3.2.16, 17).

If spatial dispersion is accounted for in crystal optics, the principal axes of $\varepsilon_{ij}(\omega,\boldsymbol{k})$ whose direction coincides with $\boldsymbol{s}$ are naturally of great interest. For rhombic crystals these are the x, y, z axes, see (3.2.10) and Table 3.2. If, for instance, the vector $\boldsymbol{s}$ is directed along the x axis, the principal values of the tensor $\varepsilon_{ij}(\omega,\boldsymbol{k})$ are

$$\varepsilon_1 \equiv \varepsilon_{xx}(\omega,\boldsymbol{k}) = \varepsilon_{xx}(\omega) + \left(\frac{\omega}{c}\tilde{n}\right)^2 \alpha_{xxxx}\,,$$

$$\varepsilon_2 \equiv \varepsilon_{yy}(\omega,\boldsymbol{k}) = \varepsilon_{yy}(\omega) + \left(\frac{\omega}{c}\tilde{n}\right)^2 \alpha_{yyxx}\,,$$

$$\varepsilon_3 \equiv \varepsilon_{zz}(\omega,\boldsymbol{k}) = \varepsilon_{zz}(\omega) + \left(\frac{\omega}{c}\tilde{n}\right)^2 \alpha_{zzxx}\,.$$

In tetragonal crystals of the classes D_4, C_{4v}, D_{2d} and D_{4h}, for an $\boldsymbol{s}$ vector being directed along the x and y axes the tensor $\varepsilon_{ij}(\omega,\boldsymbol{k})$ can be diagonalized with dissimilar principal values. If, however, the vector $\boldsymbol{s}$ is directed along the z axis (along a fourfold symmetry axis),

$$\varepsilon_1 = \varepsilon_2 = \varepsilon_\perp(\omega) + \left(\frac{\omega}{c}\tilde{n}\right)^2 \alpha_{xxzz}\,,$$

$$\varepsilon_3 = \varepsilon_\parallel + \left(\frac{\omega}{c}\tilde{n}\right)^2 \alpha_{zzzz}\,.$$

Without dealing here with crystals of the order systems, we next discuss the case of cubic crystals. In this case (Table 3.2):

$$
\begin{aligned}
\varepsilon_{xx} &= \varepsilon + \left(\frac{\omega}{c}\tilde{n}\right)^2 (\alpha_1 s_x^2 + \alpha_4 s_y^2 + \alpha_2 s_z^2)\,,\\
\varepsilon_{xy} &= 2\left(\frac{\omega}{c}\tilde{n}\right)^2 \alpha_3 s_x s_y\,,\\
\varepsilon_{yy} &= \varepsilon + \left(\frac{\omega}{c}\tilde{n}\right)^2 (\alpha_2 s_x^2 + \alpha_1 s_y^2 + \alpha_4 s_z^2)\,,\\
\varepsilon_{xz} &= 2\left(\frac{\omega}{c}\tilde{n}\right)^2 \alpha_3 s_x s_z\,,\\
\varepsilon_{zz} &= \varepsilon + \left(\frac{\omega}{c}\tilde{n}\right)^2 (\alpha_4 s_x^2 + \alpha_2 s_y^2 + \alpha_1 s_z^2)\,,\\
\varepsilon_{yz} &= 2\left(\frac{\omega}{c}\tilde{n}\right)^2 \alpha_3 s_y s_z
\end{aligned}
\tag{3.2.18}
$$

(in addition to this, $\alpha_2 = \alpha_4$ for the classes O, T_d and O_h; the factor 2 in the expression for ε_{xy}, ε_{xz} and ε_{yz} is due to the summation in (3.2.16) of terms proportional to $s_x s_y$ and $s_y s_x$). This shows that the x, y, z axes of the cube are the principal axes of the tensor, if the vector $\boldsymbol{s}$ is directed along any of these axes. The corresponding second-degree surface with $\alpha_2 = \alpha_4$ degenerates to a surface of revolution (ellipsoid or hyperboloid). If the vector $\boldsymbol{s}$ coincides with a body diagonal of the cube ($|s_x| = |s_y| = |s_z| = 1/\sqrt{3}$), then

$$
\begin{aligned}
\varepsilon_{xx} = \varepsilon_{yy} = \varepsilon_{zz} &= \varepsilon + \frac{1}{3}\left(\frac{\omega}{c}\tilde{n}\right)^2 (\alpha_1 + \alpha_2 + \alpha_4)\,,\\
|\varepsilon_{xy}| = |\varepsilon_{xz}| = |\varepsilon_{yz}| &= 2\left(\frac{\omega}{c}\tilde{n}\right)^2 \frac{\alpha_3}{3}\,.
\end{aligned}
\tag{3.2.19}
$$

4. Spatial Dispersion in Crystal Optics

The wide range of optical phenomena discussed in this chapter result from taking spatial dispersion into account. The dispersion of normal waves in gyrotropic and nongyrotropic crystals is considered for the region of exciton resonances, in particular, under conditions when additional light waves are formed and when it is necessary to account for dissipation. A theory of optical anisotropy of cubic crystals is developed for the region of dipole and quadrupole transitions. Discussed in detail is the effect of the external constant electric and magnetic fields, as well as the mechanical stresses, on the optical properties of crystals. It is shown how an application of the symmetry principle of kinetic coefficients enables the terms to be quite simply separated in the expression for the tensor $\varepsilon_{ij}(\omega, \boldsymbol{k})$. These terms lead not only to gyrotropy, but to electrogyrations, the electric analogue of the Faraday effect, gyrotropy induced by external stresses, etc. The problem of additional boundary conditions (ABC), posed by the formation of additional waves, is discussed in detail. The use of the model of an isolated resonance enables dispersion of gyrotropy to be examined, and the most general form of ABC to be determined. A number of examples shows how the application of ABC permits the reflection and refraction of light waves at the interfaces of media to be investigated in the presence of additional waves. In conclusion, the chapter reviews experimental investigations of the effects of spatial dispersion in gyrotropic and nongyrotropic crystals.

4.1 Gyrotropic Crystals

The problem of spatial dispersion in crystal optics consists, as indicated before, in investigating the propagation, reflection and refraction of various kinds of normal waves in crystals, based on the use of the tensor $\varepsilon_{ij}(\omega, \boldsymbol{k})$.

Formally speaking, this field of investigation is wider than in the case of classical crystal optics since a series of new problems and questions arises (including, e.g., the necessity of investigating the optical anisotropy of cubic crystals). In fact, however, the situation is simpler, primarily due to the smallness of spatial dispersion. As a result, the only problems that are of interest are those in which spatial dispersion leads to qualitatively new effects, or, in any case, when it yields nonnegligible corrections to the formulas of classical crystal optics. Hence, the following discussion, dealing with a number of problems, is of a fragmentary nature and consists of a treatment of known effects in whose analysis it proves essential to take spatial dispersion into account.

In this connection the following fact should also be kept in mind. In the first place, the phenomenological study of natural optical activity (gyrotropy), though this phenomenon belongs to the spatial dispersion effects in crystal physics, began long ago and has been treated in detail in several monographs [4.1 – 3]. Hence, we confine our discussion of gyrotropic crystals to the problem of the dispersion of optical activity [4.4 – 7], to corrections in the boundary conditions [4.8 – 10], as well as to the problem of new waves. These waves have been analyzed in detail in their application to both gyrotropic and nongyrotropic crystals only, to the best of our knowledge, in [4.11, 12]. Secondly, even if spatial dispersion (including gyrotropy as well) is ignored, an analysis of the propagation of light in absorbing crystals, particularly in the case of low symmetry, proves to be very cumbersome [4.2, 3]. Moreover, special cases exist. They include the propagation of light along singular optical axes (Sect. 2.2.3 and [4.3, 13, 14]), where we cannot confine our considerations to plane waves of the type (2.2.2). The role of spatial dispersion has not yet been subjected to a detailed discussion in the case of absorbing crystals and complex $\boldsymbol{k}$. Various problems of this wide field will be dealt with in Sects. 4.5 and 5.3.

Finally, it is necessary to emphasize that spatial dispersion in a nongyrotropic medium [or, more exactly, second-order effects proportional to $(a/\lambda)^2$] in the case of crystal optics even for a transparent or almost transparent medium is a problem that has been dealt with in comparatively few papers. In this connection we may assume that far from all the aspects of spatial dispersion that are of interest in crystal optics have been noted and discussed. Thus, the relatively narrow scope of the discussion of spatial dispersion in crystal optics is due to the present state of research in this field and not only to the weakness of this dispersion.

4.1.1 Normal Waves in Gyrotropic Crystals

Spatial dispersion in gyrotropic crystals manifests itself in terms of first-order smallness with respect to a/λ. In the expansions of the tensors $\varepsilon_{ij}(\omega,\boldsymbol{k})$ and $\varepsilon_{ij}^{-1}(\omega,\boldsymbol{k})$ in power series of the wave vector, terms quadratic in a/λ can be neglected to a first approximation.

In the following we intend to investigate the propagation of waves in the vicinity of the absorption band and shall therefore use the expansion of the inverse dielectric tensor. The role played by terms quadratic with respect to a/λ and the features of normal wave propagation in the vicinity of frequencies where the expansion (4.1.1) is inapplicable (region of longitudinal waves, see Sect. 2.2.2), have to be considered separately. Accordingly we use the following expansion

$$\varepsilon_{ij}^{-1}(\omega,\boldsymbol{k}) = \varepsilon_{ij}^{-1}(\omega) + \mathrm{i}\tilde{n}\frac{\omega}{c}\delta_{ijl}s_l\,. \tag{4.1.1}$$

In the spectral range under consideration the induction vector is nonvanishing for normal waves ($\boldsymbol{D} \neq 0$), as follows from the analysis of Sect. 2.2.2. The dispersion of these waves is therefore conveniently investigated in a coordinate system where the z axis lies along the direction of $\boldsymbol{s}$ and the induction vector, by virtue of the conditions $\boldsymbol{s}\cdot\boldsymbol{D} = 0$, is characterized by the equality $D_3 \equiv D_z = 0$. The equations satisfied by the nonvanishing components of the induction vector – see (2.2.26, 45) and (4.1.1) – have the following form in this coordinate system

$$\begin{aligned}
&\left[\frac{1}{\tilde{n}^2} - \varepsilon_{xx}^{-1}(\omega)\right]D_x - \varepsilon_{xy}^{-1}(\omega)\,D_y = \mathrm{i}\tilde{n}\,\delta_{123}\frac{\omega}{c}D_y\,,\\
&-\varepsilon_{xy}^{-1}(\omega)\,D_x + \left[\frac{1}{\tilde{n}^2} - \varepsilon_{yy}^{-1}(\omega)\right]D_y = -\mathrm{i}\tilde{n}\,\delta_{123}\frac{\omega}{c}D_x\,.
\end{aligned} \tag{4.1.2}$$

(Here we may, of course, replace δ_{123} by $f'_{33} = f'_{33}s_{33}^2$ or, in the invariant formulation, by $f'_{ij}s_is_j$). The directions of the x and y axes are chosen along the principal axes of the two-dimensional tensor $\varepsilon_{\alpha\beta}^{-1}$ ($\alpha, \beta = x, y$), the principal values of this tensor being denoted by $1/n_{01}^2$ and $1/n_{02}^2$. (For simplicity we have omitted the tilde on $n_{01,2}$ although in the general case $n_{01,2}$ represents the complex quantity $\tilde{n}_{01,2}$.) This tensor is clearly assumed to be diagonalizable. Equations (4.1.2) are then simplified and assume the form

$$\begin{aligned}
&\left(\frac{1}{n_{01}^2} - \frac{1}{\tilde{n}^2}\right)D_x + \mathrm{i}\,\delta_{123}\tilde{n}\frac{\omega}{c}D_y = 0\,,\\
&-\mathrm{i}\,\delta_{123}\tilde{n}\frac{\omega}{c}D_x + \left(\frac{1}{n_{02}^2} - \frac{1}{\tilde{n}^2}\right)D_y = 0\,.
\end{aligned} \tag{4.1.3}$$

The condition that the determinant of this system has to vanish yields a third-order equation with respect to $\tilde{n}^2$:

$$\left(\frac{1}{\tilde{n}^2} - \frac{1}{n_{01}^2}\right)\left(\frac{1}{\tilde{n}^2} - \frac{1}{n_{02}^2}\right) = \delta_{123}^2\frac{\omega^2}{c^2}\tilde{n}^2\,. \tag{4.1.4}$$

If we set $\delta_{123}=0$ in (4.1.3) the solutions of this equation, $\tilde{n}^2_{1,2}=n^2_{01,2}$, will be solutions to the Fresnel equation (2.2.22) with $\varepsilon_{ij}=\varepsilon_{ij}(\omega)$, which is thus particularly simple in the coordinate system used. If, however, $\delta_{123}\neq 0$, then the roots of (4.1.4) for a given direction of $\boldsymbol{s}$ determine not two but three values of the refractive index, $\tilde{n}_1$, $\tilde{n}_2$, $\tilde{n}_3$ (from two values $\pm\sqrt{\tilde{n}^2}$ we always choose the one that corresponds to damping of the wave in the direction of energy propagation, whereas the choice of the $\tilde{n}$ value with the other sign simply corresponds to a change in sign of $\boldsymbol{s}$).

In investigating (4.1.4) we shall distinguish between frequency ranges far away from and close to the resonance. In a frequency range sufficiently far away from a resonance the right-hand side of (4.1.4) is small. For one of the roots (e.g., for $\tilde{n}_3$) the quantity $\mathrm{Re}\{\tilde{n}_3\}=n_3$ will therefore be very large, $n_3^2\approx c^2/\omega^2\delta^2_{123}n^2_{01}n^2_{02}\gg 1$. In fact, since $\delta_{123}\sim a\sim 10^{-3}\ \lambda_0$ in the frequency range under consideration, where the values of n_{01} and n_{02} are of the order of 1, n_3 is of the order of 10^3 and the corresponding wavelength in the medium is $\lambda=\lambda_0/n_3\approx 10^{-7}$ to 10^{-8} cm. This indicates that in crystals, in the optical frequency range, waves with the refractive index $\tilde{n}_3$ cannot, in general, be investigated by means of the tensor $\varepsilon_{ij}(\omega,\boldsymbol{k})$. The corresponding solution cannot be regarded as meaningful.

The roots $\tilde{n}_1$ and $\tilde{n}_2$, however, can be determined with sufficient accuracy by replacing the quality $\tilde{n}^2$ in the right side of (4.1.4) by n^2_{01} if the considered root of (4.1.4) is close to n^2_{01} and by n^2_{02} if this root is close to n^2_{02}. Let us calculate, for example, the correction to the root n^2_{01}. If this correction, $\Delta\tilde{n}^2_1$ is assumed to be considerably smaller than the quantity $|n^2_{01}-n^2_{02}|\sim 1$, then we can seek it in the form

$$\tilde{n}^2_1=n^2_{01}+\Delta\tilde{n}^2_1\,. \tag{4.1.5}$$

Substituting this equation into (4.1.4) we obtain

$$\Delta\tilde{n}^2_1=\delta^2_{123}\frac{\omega^2}{c^2}\,\frac{n^8_{01}}{n^2_{01}-n^2_{02}}\,n^2_{02}\,. \tag{4.1.6a}$$

It can likewise be shown that the correction

$$\Delta\tilde{n}^2_2=\delta^2_{123}\frac{\omega^2}{c^2}\,\frac{n^8_{02}}{n^2_{02}-n^2_{01}}\,n^2_{01}\,. \tag{4.1.6b}$$

In the case under investigation the correction to the refractive indices caused by the presence of gyrotropy is thus of the order of $(a/\lambda)^2$. But changes of the same order of magnitude occur if, in the expansions of the tensors $\varepsilon_{ij}(\omega,\boldsymbol{k})$ or $\varepsilon^{-1}_{ij}(\omega,\boldsymbol{k})$, terms are taken into account which are quadratic with respect to the wave vector. Since second-order effects with respect to $(a/\lambda)^2$ are very weak it is, in general, difficult to conduct experimental investigations of the gyrotropy in the case under discussion in which $|n_{01}^2 - n_{02}^2| \sim 1$. If we are interested, however, not in the changes of $\tilde{n}^2$ but in the polarization of normal waves, the situation is quite different. With $\delta_{123} = 0$ the normal waves are, in fact, linearly polarized, whereas if $\delta_{123} \neq 0$ they are elliptically polarized, the ratio between minor and major axes being of the order of (a/λ). Let us consider, e.g., the propagation of waves $\tilde{n}^2 = \tilde{n}_1^2$. From (4.1.3) it follows that in such a wave

$$D_y^{(1)} = \mathrm{i}\rho_1 D_x^{(1)}, \tag{4.1.7a}$$

where

$$\rho_1 = \delta_{123}\frac{\omega}{c}n_{01}^3 n_{02}^2 \frac{1}{n_{01}^2 - n_{02}^2}. \tag{4.1.7b}$$

Analogously, in a wave such that $\tilde{n}^2 = \tilde{n}_2^2$,

$$D_y^{(2)} = -\mathrm{i}\frac{1}{\rho_2}D_x^{(2)}, \tag{4.1.8a}$$

where

$$\rho_2 = \delta_{123}\frac{\omega}{c}n_{02}^3 n_{01}^2 \frac{1}{n_{01}^2 - n_{02}^2}. \tag{4.1.8b}$$

Thus it follows from (4.1.7a and 8a) that if absorption is ignored normal waves are, in fact, elliptically polarized, the axes of the ellipses coinciding with the x and y axes. These ellipses are rotated through 90° relative to each other, with opposite directions of revolution. The quantities $\rho_{1,2} \sim a/\lambda_0$ determine the ratio of the semiaxes of the ellipses. From (4.1.7b, 8b) we find that

$$\rho_1/\rho_2 = n_{01}/n_{02} \neq 1, \tag{4.1.9}$$

i.e., the ellipses have somewhat different ratios of semiaxes. In the presence of absorption the quantities n_{01}, n_{02} and δ_{123} are complex. Therefore, when absorption is taken into account, the parameters $\rho_{1,2}$ are complex too, a fact that causes the principal axes of the ellipses to be turned with respect to the x and y directions.

Making use of (4.1.7a, 8a) and ignoring absorption, we find that

$$\boldsymbol{D}^{(1)}\cdot\boldsymbol{D}^{*(2)} = D_x^{(1)}D_x^{*(2)} + D_y^{(1)}D_y^{*(2)} = (1-\rho_1/\rho_2^*)D_x^{(1)}D_x^{*(2)} \neq 0. \tag{4.1.10}$$

Thus in the normal waves under investigation the induction vectors $\boldsymbol{D}^{(1)}$ and $\boldsymbol{D}^{(2)}$ generally cease to be orthogonal when gyrotropy is taken into account. This fact is a consequence of the more general conclusion drawn in Sect. 2.3.2, which contends that the induction vectors in normal electromagnetic waves even with different $\tilde{n}$ in a medium are no longer orthogonal to each other when spatial dispersion is taken into account.

An exception in this sense is the case in which the vector $\boldsymbol{s}$ is directed along the optical axis and when $n_{01} = n_{02} \equiv n_0$. Then

$$\tilde{n}_{1,2}^2 = n_0^2 \pm \frac{\omega}{c}\delta_{123} n_0^5 , \tag{4.1.11 a}$$

$$D_y^{(1)} = \mathrm{i} D_x^{(1)} , \tag{4.1.11 b}$$

$$D_y^{(2)} = -\mathrm{i} D_x^{(2)} , \tag{4.1.11 c}$$

so that the ellipses degenerate into circles and the induction vectors $\boldsymbol{D}^{(1)}$ and $\boldsymbol{D}^{(2)}$ remain orthogonal:

$$D_x^{(1)} D_x^{*(2)} + D_y^{(1)} D_y^{*(2)} = 0 .$$

Up to this point we have considered only scalar products of the form $\boldsymbol{D}^{(1)} \cdot \boldsymbol{D}^{*(2)}$. It was shown in Sect. 2.3.2 that when spatial dispersion is ignored and if $n_{01} \neq n_{02}$ a product of the form $\boldsymbol{D}^{(1)} \cdot \boldsymbol{D}^{(2)}$ also vanishes. In a gyrotropic medium, independent of the direction of propagation of light,

$$\boldsymbol{D}^{(1)} \cdot \boldsymbol{D}^{(2)} = D_x^{(1)} D_x^{(2)} + D_y^{(1)} D_y^{(2)} = (1 + \varrho_1/\varrho_2) D_x^{(1)} D_x^{(2)} \neq 0 \tag{4.1.10a}$$

is readily obtained from (4.1.7a, 8a, 11b, c) [if the vector $\boldsymbol{s}$ is directed along the optical axis, then $\varrho_1 = \varrho_2 = 1$; see (4.1.11b, c)]. Thus, in the indicated sense, the orthogonality of $\boldsymbol{D}^{(1)}$ and $\boldsymbol{D}^{(2)}$ is violated if spatial dispersion is taken into account. All these conclusions can also be drawn, of course, on the basis of the relations (2.3.54, 56). In fact, by means of (4.1.1) we find that

$$\varepsilon_{ij}^{-1}\left(\omega, \frac{\omega}{c}\tilde{n}_1 \boldsymbol{s}\right) - \varepsilon_{ij}^{-1}\left(\omega, -\frac{\omega}{c}\tilde{n}_2 \boldsymbol{s}\right) = \mathrm{i}\,\delta_{ijl} s_l \frac{\omega}{c}(n_{01} + n_{02}) ,$$

$$\varepsilon_{ij}^{-1}\left(\omega, \frac{\omega}{c}\tilde{n}_1 \boldsymbol{s}\right) - \left[\varepsilon_{ij}^{-1}\left(\omega, \frac{\omega}{c}\tilde{n}_2 \boldsymbol{s}\right)\right]^* = \mathrm{i}\,\delta_{ijl} s_l \frac{\omega}{c}(n_{01} - n_{02}) ,$$

these relations being accurate to terms of the order of a/λ. Substituting these expressions into (2.3.54, 56) we find that $\boldsymbol{D}^{(1)} \cdot \boldsymbol{D}^{(2)} \neq 0$ irrespective of the direction of $\boldsymbol{s}$. The scalar product $\boldsymbol{D}^{(1)} \cdot \boldsymbol{D}^{*(2)}$ vanishes in agreement with the aforesaid only if $n_{01} = n_{02}$.

According to (4.1.11a), if light is propagated along the optical axis the refractive indices are subject to changes of the order of a/λ_0, the angle of rotation φl of the polarization plane, in traveling along path l being determined by the expression

$$\varphi l = \frac{\pi}{\lambda_0}(\tilde{n}_1 - \tilde{n}_2)\, l = \frac{\omega^2}{2c^2} n_0^4 \delta_{123} l \,. \qquad (4.1.12)$$

We should recall that this expression was obtained under the assumption that $(\omega/c)\,|\delta_{123}|\, n_0^3 \ll 1$.

Results based on the case in which only the two roots $\tilde{n}_1$ and $\tilde{n}_2$ of (4.1.6) need be considered, with the exception, perhaps, of the problem of the nonorthogonality of the solutions $\boldsymbol{D}^{(1)}$ and $\boldsymbol{D}^{(2)}$ – see (4.1.9, 10) – are well known and they are discussed here only in order to emphasize the limited applicability of these results and to correlate them with the more complex situation encountered near resonances.

Note that in general a gyrotropy, caused by terms of the oder of $(a/\lambda)^3$, may also be of interest. The point is the use of the expansion [4.15]

$$\varepsilon_{ij}(\omega, \boldsymbol{k}) = \varepsilon_{ij}(\omega) + \mathrm{i}\gamma_{ijl}(\omega) k_l + \alpha_{ijlm}(\omega) k_l k_m + \mathrm{i}\gamma_{ijlmn}(\omega) k_l k_m k_n + \dots ,$$

$$\varepsilon_{ij}^{-1}(\omega, \boldsymbol{k}) = \varepsilon_{ij}^{-1}(\omega) + \mathrm{i}\delta_{ijl}(\omega) k_l + \beta_{ijlm}(\omega) k_l k_m + \mathrm{i}\delta_{ijlmn} k_l k_m k_n$$

taking the last term into account, cf. (3.1.19). As mentioned previously, $\gamma_{ijl}(\omega) = 0$ for certain crystals, notwithstanding the absence of a center of symmetry (specifically, this concerns crystals of the classes C_{3v}, C_{4v}, C_{3h} and D_{3h}, see below). In such cases, all rotation of the polarization plane is determined by the term $\gamma_{ijlmn}(\omega) k_l k_m k_n$, i.e., a term of the order of $(a/\lambda)^3$. Moreover, the coefficients $\gamma_{ijl}(\omega)$ are not of the resonance type near the quadrupole lines, whereas the coefficients $\gamma_{ijlmn}(\omega)$ may be resonance functions of ω. In this case, evidently, the role of terms of the order of $(a/\lambda)^3$ becomes more important.

The tensors γ_{ijlmn} and δ_{ijlmn} can be expressed in terms of the pseudotensors of rank four, g'_{slmn} and f'_{slmn}:

$$\gamma_{ijlmn} = e_{ijs} g'_{slmn} , \qquad \delta_{ijlmn} = e_{ijs} f'_{slmn} \,.$$

Thus, taking gyrotropy of the second order into account, the components of the vectors g' and f' assume the following form

$$g'_i = g'_{il} k_l + f'_{ilmn} k_l k_m k_n ,$$

$$f'_i = f'_{il} k_l + f'_{ilmn} k_l k_m k_n \,.$$

The rotatory power of crystals is determined, as mentioned above, by the product $\boldsymbol{k} \cdot \boldsymbol{f}'$, i.e., by the symmetrical parts of the pseudotensors f'_{il} and f'_{ilmn}. There is no gyrotropy in the crystal classes C_i, C_{2h}, D_{2h}, C_{3i}, D_{3d}, C_{4h}, D_{4h}, C_{6h}, D_{6h}, T_h and O_h, having a center of symmetry, because for them all components $f'_{il} = 0$, in the same way as $f'_{ilmn} = 0$. The nonzero components of the symmetrical part of the pseudotensor f'_{ilmn} and the relations between these

Table 4.1. The pseudotensor f'_{ilmn}

System	Components of the symmetrical part of pseudotensor f'_{ilmn} (only nonzero components are listed)
Triclinic	$f'_{xxxx}, f'_{yyyy}, f'_{zzzz}, f'_{xxxy}, f'_{xyyy}, f'_{xxxz}, f'_{yyyz}, f'_{xzzz}, f'_{yzzz}, f'_{xxyy}, f'_{xxzz}, f'_{yyzz}, f'_{xyyz}, f'_{xxyz}, f'_{xyzz}$.
Monoclinic	In class C_s: $f'_{xxxy}, f'_{yyyz}, f'_{zzzy}, f'_{yyyx}, f'_{xxzy}, f'_{zzxy}$. In class C_2: $f'_{xxxx}, f'_{yyyy}, f'_{zzzz}, f'_{xzzz}, f'_{zxxx}, f'_{xxzz}, f'_{yyzz}, f'_{xxyy}, f'_{xzyy}$.
Rhombic	In class D_2: $f'_{xxxx}, f'_{yyyy}, f'_{zzzz}, f'_{xxyy}, f'_{xxzz}, f'_{yyzz}$. In class C_{2v}: $f'_{xxxy}, f'_{xyyy}, f'_{xyzz}$.
Tetragonal	In class C_4: $f'_{xxxx} = f'_{yyyy}, f'_{zzzz}, f'_{xxxy} = -f'_{xyyy}, f'_{xxyy}, f'_{xxzz} = f'_{yyzz}$. In class D_4: $f'_{xxxx} = f'_{yyyy}, f'_{zzzz}, f'_{xxyy}, f'_{xxzz} = f'_{yyzz}$. In class S_4: $f'_{xxxx} = -f'_{yyyy}, f'_{xxxy} = f'_{xyyy}, f'_{xyzz}, f'_{xxzz} = -f'_{yyzz}$. In class D_{2d}: $f'_{xxxx} = -f'_{yyyy}, f'_{xxzz} = -f'_{yyzz}$.
Trigonal	In class C_3: $f'_{xxxx} = f'_{yyyy} = 3f'_{xxyy}, f'_{zzzz}, f'_{xxzz} = f'_{yyzz}, f'_{yyyz} = -f'_{xxyz}, f'_{xxxz} = -f'_{xyyz}$. In class D_3 (x axis parallel to an axis of twofold symmetry): $f'_{xxxx} = f'_{yyyy} = 3f'_{xxyy}, f'_{zzzz}, f'_{xxzz} = f'_{yyzz}, f'_{yyyz} = -f'_{xxyz}$. In class C_{3v} (x axis perpendicular to a symmetry plane): $f'_{xxxz} = -f'_{xyyz}$.
Hexagonal	In class D_6: $f'_{xxxx} = f'_{yyyy} = 3f'_{xxyy}, f'_{zzzz}, f'_{xxzz} = f'_{yyzz}$. In class C_6: $f'_{xxxx} = f'_{yyyy} = 3f'_{xxyy}, f'_{zzzz}, f'_{xxzz} = f'_{yyzz}$. In class C_{6v} all components of $f'_{ilmn} = 0$. In class C_{3h}: $f'_{yyyz} = -f'_{xxyz}, f'_{xxxz} = -f'_{xyyz}$. In class D_{3h} (x axis parallel to an axis of twofold symmetry): $f'_{yyyz} = -f'_{xxyz}$.
Cubic	In class T: $f'_{xxxx} = f'_{yyyy} = f'_{zzzz}, f'_{xxyy} = f'_{xxzz} = f'_{yyzz}$. In class O: $f'_{xxxx} = f'_{yyyy} = f'_{zzzz}, f'_{xxyy} = f'_{xxzz} = f'_{yyzz}$. In class T_d all components of $f_{ilmn} = 0$.
Isotropic medium	With a center of symmetry all components of $f'_{ilmn} = 0$. With no center of symmetry: $f'_{xxxx} = f'_{yyyy} = f'_{zzzz}$.

components for noncentrosymmetrical crystals were found in [4.16] and are listed in Table 4.1.

In classes C_{3v}, C_{4v}, C_{3h} and D_{3h}, in which all the components of the pseudotensor $f'_{il}=0$, notwithstanding the nonzero components of the pseudotensor f'_{ijlmn}, there is no rotation of the polarization plane in the direction of the principal optical axes. Hence, in these classes, rotatory power is always accompanied by birefringence. Note that in the classes C_{6v} and T_d, in which $f'_{ilmn}=0$, the rotary power is manifested only in the form of a pseudotensor of a higher rank, f'_{ilmnsp}, which has one independent component in each of thes classes [4.16].

In conclusion, we shall consider the dispersion law for polaritons in gyrotropic media in the region of their spectrum where the retarded interaction is negligible. Under these circumstances, the dispersion law coincides with the one for long-wavelength ($k \ll 1/a$) Coulomb excitons and can be derived on the basis of (4.1.4) by passing to the limit $c \to \infty$ at finite values of ω and $\boldsymbol{k}$. As a matter of fact, we can rewrite (4.1.4) in the following form

$$\left(\frac{\omega^2}{c^2k^2}-\frac{1}{n_{01}^2}\right)\left(\frac{\omega^2}{c^2k^2}-\frac{1}{n_{02}^2}\right)=\delta_{123}^2k_3^2\,.$$

As a result of passing to the limit we obtain

$$\frac{1}{n_{01}^2(\omega)\,n_{02}^2(\omega)}=\delta_{123}^2k_3^2\,, \tag{4.1.4a}$$

which is precisely what determines the required dispersion law $\omega=\omega(k)$. In particular, in isotropic media, in cubic crystals or in the propagation of light along the optical axis ($n_{01}=n_{02}$) for the resonance region [$\omega \approx \omega_l$, see (4.1.13)], where $n_0^2 \approx -A\,\omega_l/(\omega-\omega_l)$, it follows from (4.1.4a) that

$$\omega=\omega_l \pm \alpha k\,, \quad \text{where} \quad \alpha=A\,\omega_l\delta_{123}\,.$$

For this case Fig. 4.1 offers a general idea of the polariton spectrum.

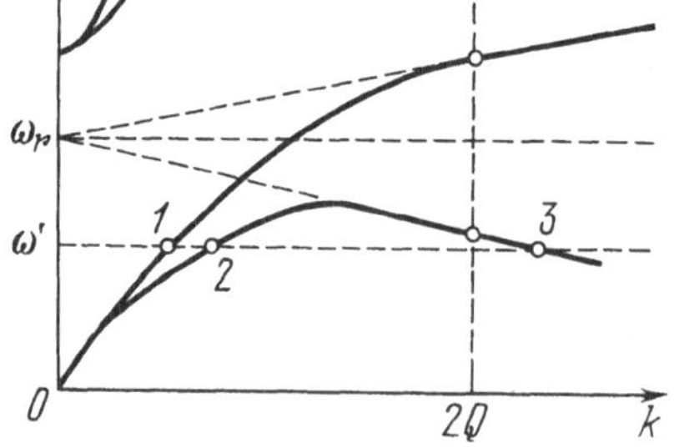

Fig. 4.1. Dependence of the polariton frequency on the wave vector in gyrotropic cubic crystals in the resonance region $\omega_r=\omega_l$. 1, 2 and 3 are points on the curve of the function $\omega(k)$, which determine the values of k_i, where $i=1, 2, 3$, for three polaritons having the same frequency ω'. Also indicated (points) that determine the polariton frequencies at $k=2Q$

The resonances of the refractive indices n_{01} and n_{02} do not, in general, coincide in anisotropic crystals. Therefore, for example in the resonance region of index n_{01}, where the dependence of n_{02} on ω can be neglected (the case of strong anisotropy), a relation of the form

$$\omega = \omega_l + \beta k_3^2, \quad \text{where} \quad \beta = \text{const}$$

follows from (4.1.4a), i.e., a quadratic rather than a linear dependence of ω on k. If the resonances of the quantities n_{01} and n_{02} are close together (the case of weak anisotropy), then the dispersion law $\omega = \omega(k)$ can also be derived from (4.1.4a), but the expression for $\omega = \omega(k)$ is more cumbersome in this case, and we do not give it here. It should be emphasized that the results obtained for $\omega = \omega(k)$ when $c \to \infty$ are independent of the model and can be utilized for the frequency region where dipole transitions of any nature (electronic or vibratory excitons, etc.)[1] occur.

4.1.2 A New Wave Near the Absorption Lines in a Gyrotropic Medium

Near resonances of a definite frequency range the values n_{01}^2 and n_{02}^2 (or one of them) may become comparatively large so that the right side of (4.1.4) should be calculated more accurately when determining the roots of this equation. It may be found in particular that all three roots of (4.1.4) correspond to comparatively large wavelengths so that all three solutions (as mentioned in [4.11]) can be considered within the framework of the macroscopic approach. We shall discuss this case in greater detail, assuming at first that there is no absorption so that n_{01}^2, n_{02}^2 and δ_{123} are real. Then (4.1.4) becomes a third-order equation for $\tilde{n}^2$ with real coefficients and possesses either three real, or one real and two complex-conjugate roots, according to the frequency. For definiteness we assume $\delta_{123} > 0$ and the wave to be propagated along the optical axis, i.e., $n_{01} = n_{02} = n_0$. Let us confine our problem to the vicinity of some particular resonance $\omega \approx \omega_l$ and assume that in this frequency range, i.e., at $|\omega - \omega_l| \ll \omega_l$,

$$n_0^2(\omega) = \varepsilon_0(\omega) = \varepsilon_{00} - \frac{2A\omega_l^2}{\omega^2 - \omega_l^2} \approx \varepsilon_{00} - \frac{A}{\xi}, \tag{4.1.13}$$

where $\xi = (\omega - \omega_l)/\omega_l$, $A = 2\pi e^2 N_{\text{eff}}/m\omega_l^2$; e is the charge and m the mass of a free electron, N_{eff}/N is the oscillator strength, N being the total number of electrons per unit volume and N_{eff} that fraction of the electrons which "effectively" determines the optical properties of the medium in the spectral region being considered. For this case Fig. 4.2 shows the shape of the curves $n^2(\xi) =$

[1] An allied problem is the gyrotropy of sonic waves [4.17, 18]; for its investigation by the Bragg reflection method, see [4.19].

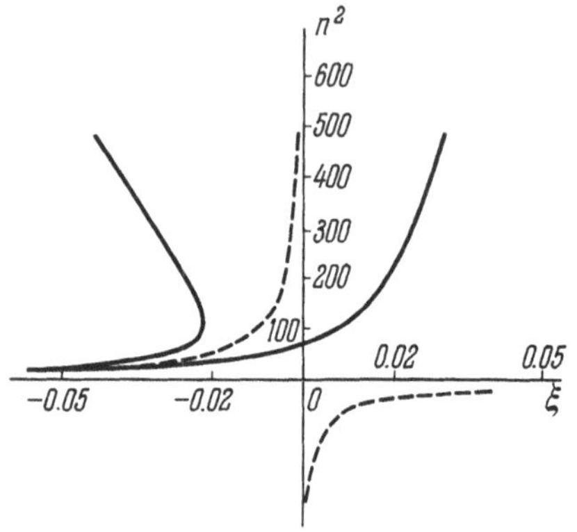

Fig. 4.2

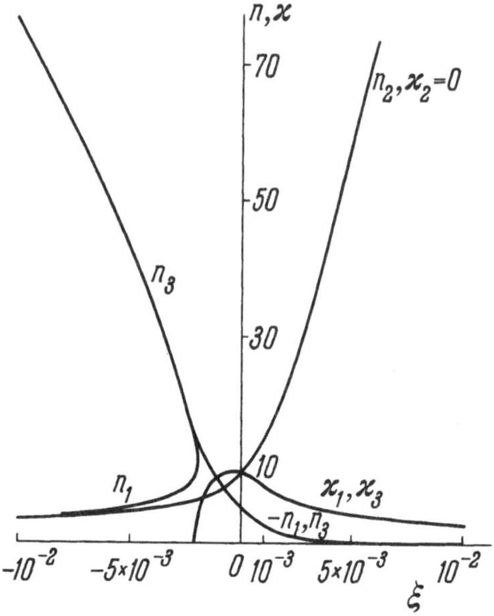

Fig. 4.3

Fig. 4.2. Dependence of n^2 on $\xi = (\omega - \omega_l)/\omega_l$ in the vicinity of the frequency $\omega = \omega_l$ in a gyrotropic medium whose absorption is ignored ($\delta = 0$; see (4.1.13a)). The values taken for the parameters are: $A = 1$, $\varepsilon_{00} = n_{00}^2 = 1$ and $\omega_l \delta_{123}/c = 10^{-3}$

Fig. 4.3. Dependence of n and $\varkappa$ ($\tilde{n} = n + i\varkappa$) on $\xi = (\omega - \omega_l)/\omega_l$ in the vicinity of the frequency $\omega = \omega_l$ in a gyrotropic medium whose absorption is ignored ($\delta = 0$). The values taken for the parameters are: $A = 0.1$, $\varepsilon_{00} = n_{00}^2 = 3$ and $\omega \delta_{123}/c = 10^{-3}$

$n^2[(\omega - \omega_l)/\omega_l]$, obtained as the result of solving (4.1.4) (see also Fig. 4.3 where n and $\varkappa$ are shown as functions of ξ).

If damping is ignored and (4.1.13) is applied we then have either three real solutions $\tilde{n}_{1,2,3}^2$ or one real solution $\tilde{n}_2^2 = n_2^2$ and two complex-conjugate solutions $\tilde{n}_1^2 = n_3^{*2}$, according to the value of ω. Equation (4.1.4) determines the quantity $\tilde{n}$ except for the sign. This sign should be chosen in such a way that the imaginary part of $\tilde{n}$ is positive and corresponds to the damping of the wave in the direction of s. For the complex-conjugate solutions we thus have $n_1 = -n_3$, $\varkappa_1 = \varkappa_3$. The shape of the curve $n_0^2(\omega)$ is shown by the dashed lines in Fig. 4.2. It follows from this illustration that the values of n^2 on the upper branch satisfy the inequality $n^2 > n_0^2$. Hence, as follows from (4.1.3, 4), the upper branch satisfies the relation

$$\frac{D_x}{D_y} = -\frac{\mathrm{i}\delta_{123}(\omega/c)n}{(1/n_0^2 - 1/n^2)} = -\mathrm{i}\,,$$

i.e., this branch of solutions n^2 corresponds to clockwise circular polarization and, conversely, the lower branch of solutions n^2 to counterclockwise circular polarization. Figure 4.2 remains unchanged if we assume $\delta_{123}<0$ but the polarization of the solutions corresponding to the upper and lower branches is reversed. It is an interesting feature of the dispersion curves shown in Fig. 4.2 that on the right-hand side of the turning point there exists only one real solution of (4.1.4) and on the left-hand side of this point there are three real solutions. Furthermore, the presence of a turning point and, consequently, of a multiple root of (4.1.4) is of particular interest in connection with what we have said in Sect. 2.2.3. In this connection we point out that Fig. 4.2 was plotted assuming that there is no absorption; this, however, can be particularly intense in this frequency range. This changes the shape of the dispersion curves, especially in the case where the turning point corresponding to the multiple root lies in a region of appreciable absorption. Note that a frequency ω_m satisfying the relation $n_0^2(\omega_m)=2^{2/3}/3(\delta_{123}\omega_m/c)^{2/3}$ corresponds to the multiple root (i.e., to the turning point). In this case the multiple root[2] is $n_m^2=[2/(\omega/c)\delta_{123}]^{2/3}$, the smaller root being $n^2=n_m^2/4$. Assuming that $n_0^2(\omega)\approx 0.2\,\omega_1/|\omega-\omega_l|$ and $(\omega/c)\delta_{123}\approx 10^{-3}$ we obtain from this $|\omega_m-\omega_l|/\omega_l\approx 4\times 10^{-3}$ which, if $\omega_l\approx 3\times 10^4$ cm^{-1}, corresponds to $|\omega_m-\omega_l|\approx 100$ cm^{-1}. An estimate of the quantity $(\omega/c)\delta_{123}$ directly from experimental data is discussed in Sect. 4.6. Since the exciton absorption linewidths for many, e.g., molecular, crystals [4.20] measure several tens of cm^{-1} at the temperature of liquid hydrogen or helium, we may expect the "three-wave effect" to become observable in such crystals at low temperatures. In the frequency range to the right of the turning point (Fig. 4.2), the character of the dispersion is considerably altered if absorption is taken into account.

In order to illustrate the aforesaid, we substitute into (4.1.4), instead of (4.1.13), the more general relation

$$n_0^2(\omega)=\varepsilon_0(\omega)=\varepsilon_{00}-\frac{2A\omega_l^2}{\omega^2-\omega_l^2+\mathrm{i}\nu\omega}\approx n_{00}^2-\frac{A}{\xi+\mathrm{i}\delta}\,,\qquad n_{00}^2\equiv\varepsilon_{00}\,, \tag{4.1.13a}$$

where ν is the 'effective' collision frequency causing damping of the waves in the medium, $\delta=\nu/2\omega_l$. The results of calculations in which (4.1.13a) was used with $\delta\neq 0$ are illustrated in Fig. 4.4 and are discussed in more detail in Sect. 4.6.1.

If only terms of the order of k are taken into account, new waves may also appear [4.21] in magnetic (antiferromagnetic) crystals for which $\gamma'_{ijl}=0$, but $\gamma''_{ijl}\neq 0$ (see the end of Sect. 3.2.1).

Note that we have made no mention whatsoever of liquid crystals. Of course, many of the general expressions apply to liquid-crystal media as well.

---

2 Recall that the condition $p^3=-q^2$, with $3p=b-a^2/3$ and $2q=2a^3/27-ab/3+c$ corresponds to the multiple root of the third-order equation $x^3+ax^2+bx+c=0$.

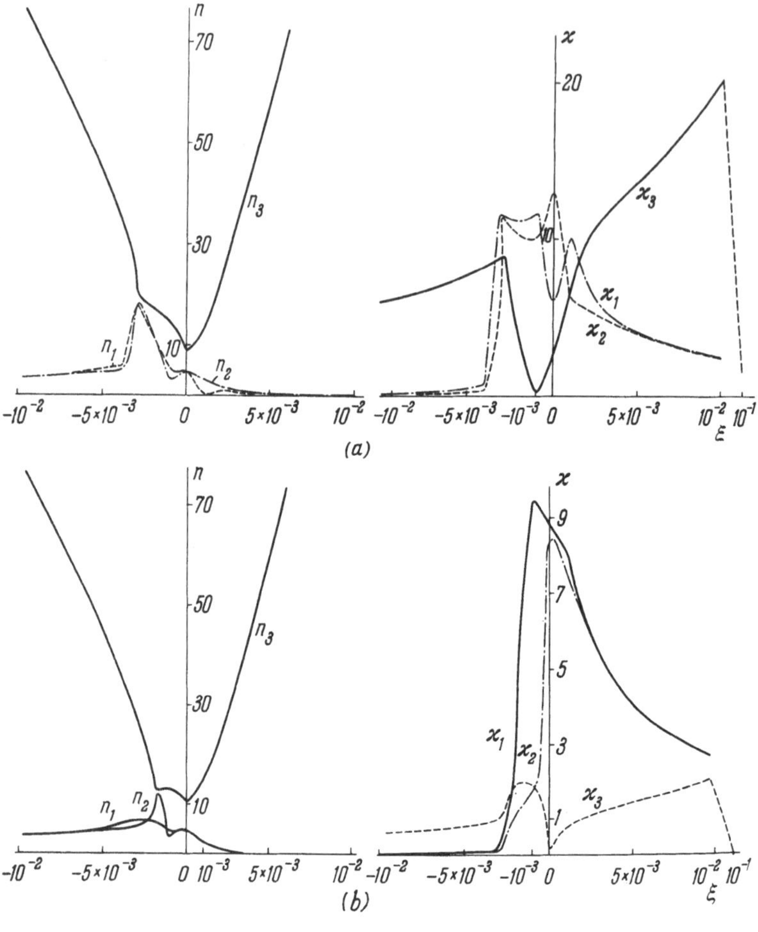

Fig. 4.4a, b. Dependence of n and $\varkappa$ on ξ in the vicinity of the frequency $\omega = \omega_l$ in a gyrotropic medium whose absorption is taken into account. The values taken for the parameters are: $A = 0.1$, $\varepsilon_{00} = n_{00}^2 = 3$ and $\omega_l \delta_{123}/c = 10^{-3}$, with $\delta = 10^{-3}$ for **(a)** and $\delta = 10^{-4}$ for **(b)**

But, as a whole, the gyrotropy of liquid crystals requires special investigation and is undoubtedly of great interest (we mention only that the optical rotatory power found in cholesteric liquid crystals may exceed that of ordinary crystals by 4 or 5 orders of magnitude). The optics of liquid crystals are dealt with in [4.22, 23] which also lists other references. Another interesting reference [4.24] deals with a theoretical analysis of special features of light propagation when both the helical structure of cholesteric liquid crystals and the intrinsic gyrotropy of the molecules in the region of molecular resonances are taken into account.

4.2 Nongyrotropic Crystals

4.2.1 Normal Waves

In nongyrotropic crystals the effects of spatial dispersion are determined by the tensor β_{ijlm} since in these crystals the tensor $\delta_{ijl} = 0$ and, consequently – see (3.1.21) – the inverse dielectric constant tensor will be given by

$$\varepsilon_{ij}^{-1}(\omega,\boldsymbol{k}) = \varepsilon_{ij}^{-1}(\omega) + \beta_{ijlm}(\omega)k_l k_m . \tag{4.2.1}$$

As in the case of gyrotropic crystals (Sect. 4.1), we put the z axis along the direction of the wave vector $\boldsymbol{k}$. In the new coordinate system, where $k_1 = k_2 = 0$ and $k_3 = k$, we then have

$$\varepsilon_{ij}^{-1}(\omega,\boldsymbol{k}) = \varepsilon_{ij}^{-1}(\omega) + \beta_{ij33}(\omega)\frac{\omega^2}{c^2}\tilde{n}^2 \tag{4.2.2}$$

for the inverse tensor and, consequently, (2.2.45), describing the transverse components of the induction vector which, in the case of a gyrotropic medium may be given in the form (4.1.2), can now be written in the following form:

$$\begin{aligned}
&\left[\frac{1}{\tilde{n}^2} - \varepsilon_{11}^{-1}(\omega) - \frac{\omega^2}{c^2}\tilde{n}^2\beta_{1133}(\omega)\right]D_1 \\
&\qquad - \left[\varepsilon_{12}^{-1}(\omega) + \frac{\omega^2}{c^2}\tilde{n}^2\beta_{1233}(\omega)\right]D_2 = 0 , \\
&- \left[\varepsilon_{21}^{-1}(\omega) + \frac{\omega^2}{c^2}\tilde{n}^2\beta_{2133}(\omega)\right]D_1 \\
&\qquad + \left[\frac{1}{\tilde{n}^2} - \varepsilon_{22}^{-1}(\omega) - \frac{\omega^2}{c^2}\tilde{n}^2\beta_{2233}(\omega)\right]D_2 = 0 .
\end{aligned} \tag{4.2.3}$$

Let us assume as in Sect. 4.1 that the two-dimensional tensor $\varepsilon_{\alpha\beta}^{-1}$ $(\alpha,\beta = 1,2)$ can be diagonalized. Then we can simplify further by orientating the x and y axes along the principal axes of this tensor. In this case, the quantities $\varepsilon_{21}^{-1}(\omega) = \varepsilon_{12}^{-1}(\omega)$ vanish and $\varepsilon_{11}^{-1}(\omega)$ and $\varepsilon_{22}^{-1}(\omega)$ are simply related to the solutions of Fresnel's dispersion equation (2.2.22) when spatial dispersion is neglected. In fact, if $\beta_{ijlm} = 0$ the roots n_{01}^2 and n_{02}^2 of the dispersion equation corresponding to system (4.2.3) have the form (if $\varepsilon_{12}^{-1}(\omega) = 0$):

$$\varepsilon_{11}^{-1}(\omega) = 1/n_{01}^2 , \quad \varepsilon_{22}^{-1}(\omega) = 1/n_{02}^2 .$$

As a result, the dispersion equation required to determine the quantities $\tilde{n}^2$, corresponding to normal electromagnetic waves with $\boldsymbol{D} \neq 0$, is obtained under the condition that the determinant of the system (4.2.3) vanishes and is of the form

$$\left(\frac{1}{\tilde{n}^2} - \frac{1}{n_{01}^2}\right)\left(\frac{1}{\tilde{n}^2} - \frac{1}{n_{02}^2}\right) - \frac{\omega^2}{c^2}\tilde{n}^2 \beta_{1133}\left(\frac{1}{\tilde{n}^2} - \frac{1}{n_{02}^2}\right)$$
$$- \frac{\omega^2}{c^2}\tilde{n}^2 \beta_{2233}\left(\frac{1}{\tilde{n}^2} - \frac{1}{n_{01}^2}\right) + \frac{\omega^4}{c^4}\tilde{n}^4(\beta_{1133}\beta_{2233} - \beta_{1233}\beta_{2133}) = 0 \,. \tag{4.2.4}$$

This is a fourth-degree equation for the quantity $\tilde{n}^2$. It follows that in the frequency range under consideration four waves can be propagated with the same frequency, but with different wavelengths and different polarizations. It is significant that these four normal waves are not mutually orthogonal (Sect. 2.3.2) and it is, in general, impossible to separate them into two groups of mutually orthogonal waves of the same polarization. A more detailed investigation of the polarization of normal waves can be conducted more conveniently by specifying the symmetry of the crystal.

If the two-dimensional tensor $\varepsilon_{\alpha\beta}^{-1}$ cannot be diagonalized, as may be the case with biaxial crystals (triclinic, monoclinic, and rhombic systems), the equation used to determine the quantity $\tilde{n}^2$ can, of course, be obtained in a way similar to that used to derive (4.2.4). It is, however, more cumbersome and therefore will not be given here. We only point out that it might be of particular interest to analyze this equation taking account of spatial dispersion only in the case where multiple roots of the dispersion equation are obtained without taking spatial dispersion into consideration (Sect. 2.2.3).

4.2.2 Isotropic Medium and New Waves Near Dipole Absorption Lines

First we shall consider the simplest case of an isotropic medium where $\beta_{1233} = \beta_{2133} = 0$, $\beta_{1133} = \beta_{2233} = \beta$ and, regardless of the direction of $\boldsymbol{s}$, $n_{01} = n_{02} = n_0$ (Sect. 3.2.2). Under these conditions (4.2.4) can be decomposed into two identical equations, each having the form

$$\frac{1}{\tilde{n}^2} - \frac{1}{n_0^2} - \frac{\omega^2}{c^2}\beta\tilde{n}^2 = 0 \,. \tag{4.2.5}$$

From this it readily follows that

$$\tilde{n}_{1,2}^2 = (n + \mathrm{i}\varkappa)_{1,2}^2 = -\frac{1}{2\varepsilon_0\beta'} \pm \sqrt{\left(\frac{1}{2\varepsilon_0\beta'}\right) + \frac{1}{\beta'}} \,, \tag{4.2.6}$$

where

$$\varepsilon_0(\omega) = n_0^2, \quad \beta' = \frac{\omega^2}{c^2}\beta.$$

Considering a range of frequencies in the vicinity of some single absorption line, we use (4.1.13, 13a) to find n_0^2. It is then clear from (4.2.6) that if

$$|\varepsilon_0^2\beta'| \ll 1 \tag{4.2.7}$$

we may set

$$\tilde{n}_1^2 = \varepsilon_0(1 - \varepsilon_0^2\beta' + \dots), \quad \tilde{n}_2^2 = -\frac{1}{\varepsilon_0\beta'} - \varepsilon_0 + \dots. \tag{4.2.8}$$

For $\beta' \sim 4 \cdot 10^{-6}$ and in the absence of absorption, condition (4.2.7) assumes the form $\varepsilon_0 = n_0^2 \ll 10^3$, which, if $\varepsilon_{00} \approx 1$ and $A \approx 0.1$, see (4.1.13), leads to the inequality $|\xi| = |\omega - \omega_l|/\omega_l \gg 10^{-4}$. In the visible part of the spectrum, where $\omega_l \approx 10^{15}$ Hz $\approx 2 \times 10^4$ cm^{-1}, this indicates that the expressions (4.2.8) are suitable even at distances $\Delta\omega = |\omega - \omega_l| \gg 2 \cdot 10^{-4}\, \omega_l \approx 4$ cm^{-1} from the center of the absorption line. If, however, we put $A \approx 1$ (which, evidently, does not correspond to reality) we obtain the inequality $\Delta\omega \gg 40$ cm^{-1}.

In this frequency range $\tilde{n}_1^2 \approx n_0^2$ as follows from (4.2.8). The root $\tilde{n}_2^2$, however, is very large and $\varepsilon_0 \approx 1$, as is true far from the line, $|\tilde{n}_2^2| \approx 1/|\beta'| \approx 2 \cdot 10^5$. Then $\lambda = \lambda_0/n_2 \approx 10^{-7}$ cm and the expansion (4.2.1), in which we limited ourselves to terms of order k^2, no longer applies. The new root $\tilde{n}_2$ of the dispersion equation therefore is meaningful only close to the line, in the range where $\lambda = \lambda_0/|\tilde{n}_2| \gg a \approx 3 \times 10^{-8}$ cm, i.e., as long as $|\tilde{n}_2| \ll \lambda_0/a$.

In certain cases when spatial dispersion is taken into consideration near absorption lines the shape of the curves $\tilde{n}^2(\omega)$ is even altered qualitatively. This becomes particularly marked when absorption is negligibly small, and we may set $\delta = 0$. Such a case is illustrated in Fig. 4.5a and b [4.11, 12]. In both cases the dashed line represents the limiting curve of (4.1.13) with $A = 1$ (since usually $\varepsilon_{00} \sim 1$ and we are interested in the region $|\varepsilon_0| \gg 1$, in Fig. 4.5a and b we have put $\varepsilon_{00} = 0$ to simplify matters). The shape of the curves in Figs. 4.5a and b reveals that the character of dispersion without absorption depends on the sign of β'. If $\beta' \geqslant 0$ the quantity $\tilde{n}^2$ is real everywhere so that the medium is transparent if $\tilde{n}^2 > 0$, and if $n^2 = -\varkappa^2 < 0$ the wave is completely reflected from the medium. If, however, $\beta' < 0$, the values of $\tilde{n}^2 = (n + i\varkappa)^2$ are complex in the region $|\varepsilon_0| > 1/2\sqrt{|\beta'|}$ even without absorption. Thus, if $\beta' > 0$, spatial dispersion without absorption leads to the appearance of a new wave propagated in a region where, if $\beta' = 0$, total internal reflection occurs (Sect. 4.5). If, however, $\beta' < 0$, then to the left of the turning point two waves may be propagated having the same frequency and polarization, but dissimilar refractive indices and, therefore, different wavelengths [4.11, 12].

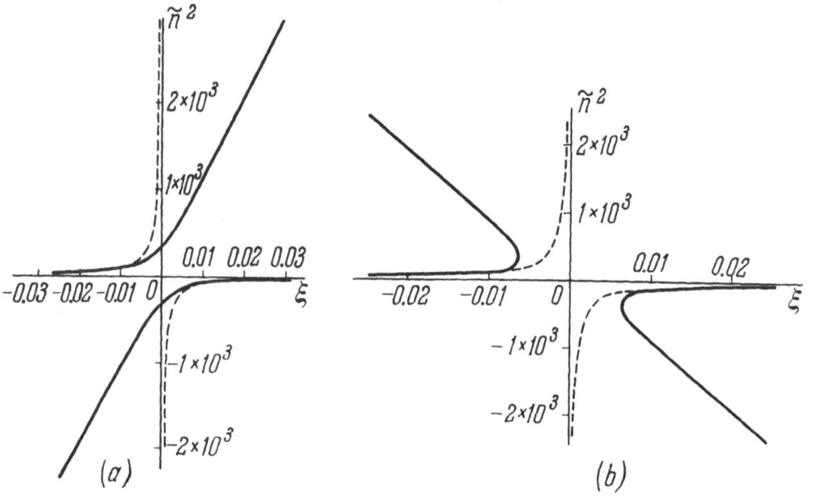

Fig. 4.5 a, b. Dependence of $\tilde{n}^2$ on ξ in the vicinity of the frequency $\omega = \omega_l$ in a nongyrotropic medium with absorption neglected ($\delta = 0$). The values taken for the parameters are: $A = 1$ and $\varepsilon_{00} = 0$, with $\beta' = 10^{-5}$ for **(a)** $\beta' = -10^{-5}$ for **(b)**

Note that in plotting Fig. 4.5 a and b the values taken for the parameters A and $|\beta'|$ are, in general, much too large. In the following we therefore consider more realistic examples.

In cases involving exciton absorption lines, the sign of β' is determined by the sign of the 'effective mass' of the exciton. This becomes evident if, instead of (4.1.13, 13 a), we consider a more general expression for ε that takes spatial dispersion as well as time dispersion into account:

$$\varepsilon(\omega, \boldsymbol{k}) = \varepsilon_{00} + \frac{4\pi e^2 N_{\text{eff}}/m}{\omega_l^2(k) - \omega^2 - \mathrm{i}\,\omega\nu}. \tag{4.2.9}$$

Expression (4.2.9) describes the function $\varepsilon(\omega, k)$ in the vicinity of an isolated resonance. Equation (4.2.9) can be used if the longitudinal-transverse splitting [i.e., the difference $\omega_\parallel - \omega_\perp$, $\omega_\perp \equiv \omega_l(0)$, $\varepsilon(\omega_\parallel, 0) = 0$] and the half-width ν of the resonance are small compared to the quantity $|\omega - \omega_l'|$, where ω_l' is the frequency of the nearest resonance. If $\omega - \omega_l \gg \omega_\parallel - \omega_\perp$ and $\omega - \omega_l \gg \nu$, the dielectric constant $\varepsilon(\omega, k) \to \varepsilon_{00} \equiv \varepsilon_\infty$. If, however, $\omega \ll \omega_l$, then $\varepsilon(\omega, k) \to \varepsilon_{00} + 4\pi e^2 N_{\text{eff}}/m\,\omega_l^2(k) \equiv \varepsilon_0$. This limiting value of $\varepsilon(\omega, k)$, like the value ε_∞, depends only slightly on k, and this dependence is usually neglected. In notation that is indicated and frequently used in the literature, (4.2.9) can be rewritten in the form

$$\varepsilon(\omega, k) = \varepsilon_\infty + \frac{(\varepsilon_0 - \varepsilon_\infty)\,\omega_l^2(k)}{\omega_l^2(k) - \omega^2 - \mathrm{i}\,\omega\nu}. \tag{4.2.9a}$$

Expressions of the form of (4.2.9) are obtained from simple assumptions on the basis of the microscopic theory (Sects. 6.1 – 3). Expanding the energy $\hbar\omega_l(k)$ of a 'mechanical' exciton in a power series of the wave vector, we obtain for an isotropic nongyrotropic medium

$$\hbar\omega_l(k) = \hbar\omega_l(0) + \frac{\hbar^2 k^2}{2m_{\text{exc}}} + \dots,$$

where m_{exc} is the effective mass of the mechanical exciton. Hence, for $|\omega_l(0) - \omega| \ll \omega_0$ and $\omega v \ll \omega_0^2$, where $\omega_0^2 = 4\pi e^2 N_{\text{eff}}/m$,

$$\frac{1}{\varepsilon(\omega,k)} \approx \frac{1}{\varepsilon_0(\omega)} + \frac{\hbar\omega_l(0)}{4\pi e^2 N_{\text{eff}}/m} \frac{k^2}{m_{\text{exc}}}$$

and, in accordance with (4.2.2), see also Sect. 3.2.2, and (4.1.13) we obtain

$$\beta = \frac{c^2}{\omega^2}\beta' = \frac{\hbar\omega_l(0)}{4\pi e^2 N_{\text{eff}}} \frac{m}{m_{\text{exc}}} = \frac{1}{2} \frac{\hbar}{A m_{\text{exc}}\omega_l(0)}. \tag{4.2.9b}$$

Thus, in the approximation (4.2.9) the sign of β [or of β', see (4.2.6)] does in fact coincide with the sign of the effective mass of the mechanical exciton.

Note also that when the quantity A, (4.1.13), is directly proportional to the oscillator strength, the quantity β may in certain cases be practically independent of the oscillator strength. The fact is that sometimes the effective mass of a mechanical exciton proves to be inversely proportional to the oscillator strength so that $N_{\text{eff}} m_{\text{exc}}/m \sim N$, where N is the concentration of valence electrons. Such a situation (as was shown in [4.25]) is encountered in spectral regions where the main contribution to the interaction between individual molecules (or cells) of the crystal is made by the electrostatic dipole-dipole interaction, whose intensity determines the bandwidth of the mechanical exciton and, consequently, its effective mass. Under these conditions

$$\beta' = \beta\frac{\omega^2}{c^2} = \frac{\hbar\omega_l(0)\,\omega^2}{\omega_0^2 m c^2},$$

where $\omega_0 = \sqrt{4\pi e^2 N/m}$ is the 'plasma' frequency. For example, in molecular crystals $N \approx 3\times10^{22}$ to 3×10^{23}, $\omega_0^2 \approx 8\times10^{31}$ to 8×10^{32} (Hz)^2, and with $\omega_l(0) \approx 3\times10^{15}$ Hz we obtain $\beta' \approx 10^{-6}$ to 10^{-7}. Hence, it is also obvious that the parameter β' could be larger than the values obtained if, in the case being considered, the dipole interactions were weak and the main and by no means small contribution to the exciton-band width was not made by dipole-dipole but by other interactions, for example, exchange interactions. In this case the quantity A, being proportional to the square of the matrix element of the dipole moment, could be small whereas m_{exc} could be of the order of m (such a situation arises for semiconductors, where $m = m_e + m_h$, m_e, m_h – electron and hole effective masses). Then the parameter β' [Eq. (4.2.9b) with $m_{exc} = m$ and $\omega_l(0) \approx 3\times10^{15}$ Hz] would be equal to $\beta' \approx 2\times10^{-6}/A$ and with $A \approx 0.1$ we would have $\beta' \approx 10^{-5}$, or with $A \approx 10^{-2}$, even $\beta' \approx 10^{-4}$. But the characteristic parameter determining the effect of spatial dispersion in the presence of damping is the product $A\sqrt{\beta'}$, (4.2.10), rather than β'. This product is not usually small. We shall assume for the following estimates that the parameter $|\beta'|$ lies within the limits $10^{-5} \lesssim |\beta'| \lesssim 10^{-7}$. Moreover $A \ll 1$ in all the cases we know; hence the values taken for the parameters A and $|\beta'|$ in plotting the curves in Figs. 4.5a and b are, generally speaking, much too large. With more reasonable values of A and $|\beta'|$ the possibility of the appearance of a new wave near a resonance depends to a still greater extent on the amount of damping.

Without absorption, spatial dispersion exerts a great influence even at $4\varepsilon_0^2|\beta'| \approx 1$, i.e., if $|\xi| \sim |\xi_k| \equiv 2A\sqrt{|\beta'|}$. Therefore, if

$$\delta = \frac{\nu}{2\omega_l} \ll |\xi_k| \equiv 2A\sqrt{|\beta'|}, \tag{4.2.10}$$

absorption is unimportant for frequencies with $|\xi| = |\omega - \omega_l|/\omega_l \gtrsim |\xi_k|$ as it exerts a weak effect on $\mathrm{Re}\{\varepsilon_0\}$ and, at the same time, $|\mathrm{Im}\{\varepsilon_0\}| \ll |\mathrm{Re}\{\varepsilon_0\}|$. With $A \lesssim 0.1$ to 0.01 and $|\beta'| \approx 10^{-5}$ to 10^{-7}, the value of $|\xi_k| \approx 10^{-3}$ to 10^{-6}. At the same time, near a resonance, the value of δ usually lies in the interval $10^{-5} \lesssim \delta \lesssim 10^{-2}$ as follows, e.g., from experimental investigations of molecular and semiconductor crystals at low temperatures. As a rule we therefore have instead of (4.2.10) the inverse inequality $\delta \gg |\xi_k| = 2A\sqrt{|\beta'|}$.

It follows from this that when spatial dispersion is taken into account, absorption has to be considered in an analysis of the propagation of light near resonances of $\varepsilon_0(\omega)$ since absorption, as we have seen, substantially reduces the effect of spatial dispersion.

Using (4.2.6) and assuming that $\varepsilon_0(\omega)$ is complex and given by (4.1.13a), we obtain $\tilde{n} = n + \mathrm{i}\varkappa$, where

$$n_\pm^2 = \frac{I_1^\pm}{2} + \frac{1}{2}\sqrt{(I_1^\pm)^2 + (I_2^\pm)^2},$$

$$\varkappa_\pm^2 = -\frac{I_1^\pm}{2} + \frac{1}{2}\sqrt{(I_1^\pm)^2 + (I_2^\pm)^2},$$

$$I_1^\pm = \frac{A\xi - \varepsilon_{00}(\xi^2 + \delta^2) \pm M}{2\beta'[(\varepsilon_{00}\xi - A)^2 + \varepsilon_{00}^2\delta^2]}, \quad I_2^\pm = \frac{A\delta \pm N}{2\beta'[(\varepsilon_{00}\xi - A)^2 + \varepsilon_{00}^2\delta^2]}, \tag{4.2.11}$$

$$M + \mathrm{i}N = \sqrt{\{[(\xi^2 + \delta^2)\varepsilon_{00} - A\xi] - \mathrm{i}A\delta\}^2 + 4\beta'[(\varepsilon_{00}\xi - A)^2 + \varepsilon_{00}^2\delta^2]^2}.$$

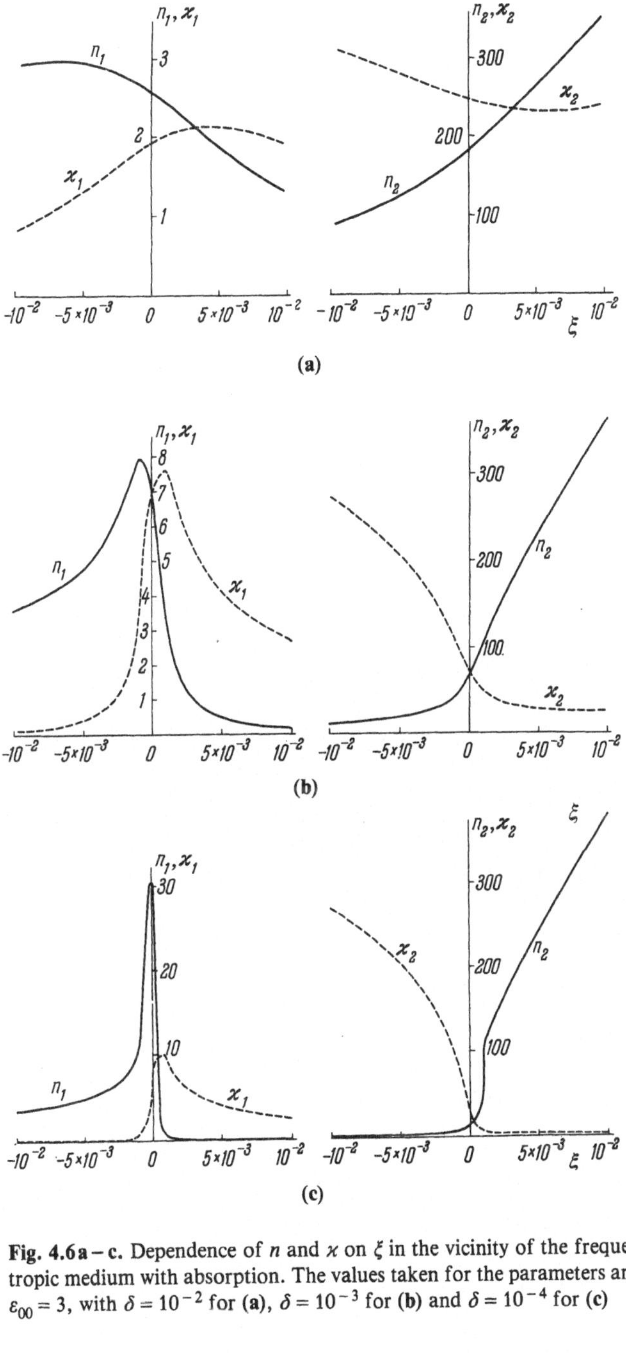

Fig. 4.6a – c. Dependence of n and $\varkappa$ on ξ in the vicinity of the frequency $\omega = \omega_l$ in a nongyrotropic medium with absorption. The values taken for the parameters are: $A = 0.1$, $\beta' = 10^{-6}$ and $\varepsilon_{00} = 3$, with $\delta = 10^{-2}$ for **(a)**, $\delta = 10^{-3}$ for **(b)** and $\delta = 10^{-4}$ for **(c)**

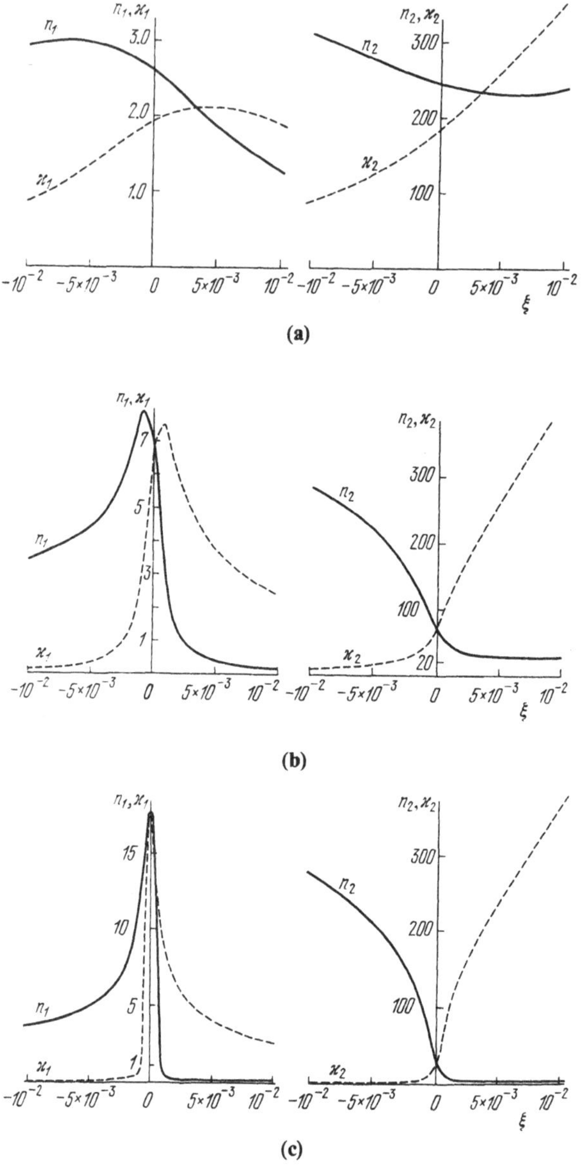

Fig. 4.7a–c. As in Fig. 4.6, but with $\beta' = -10^{-6}$

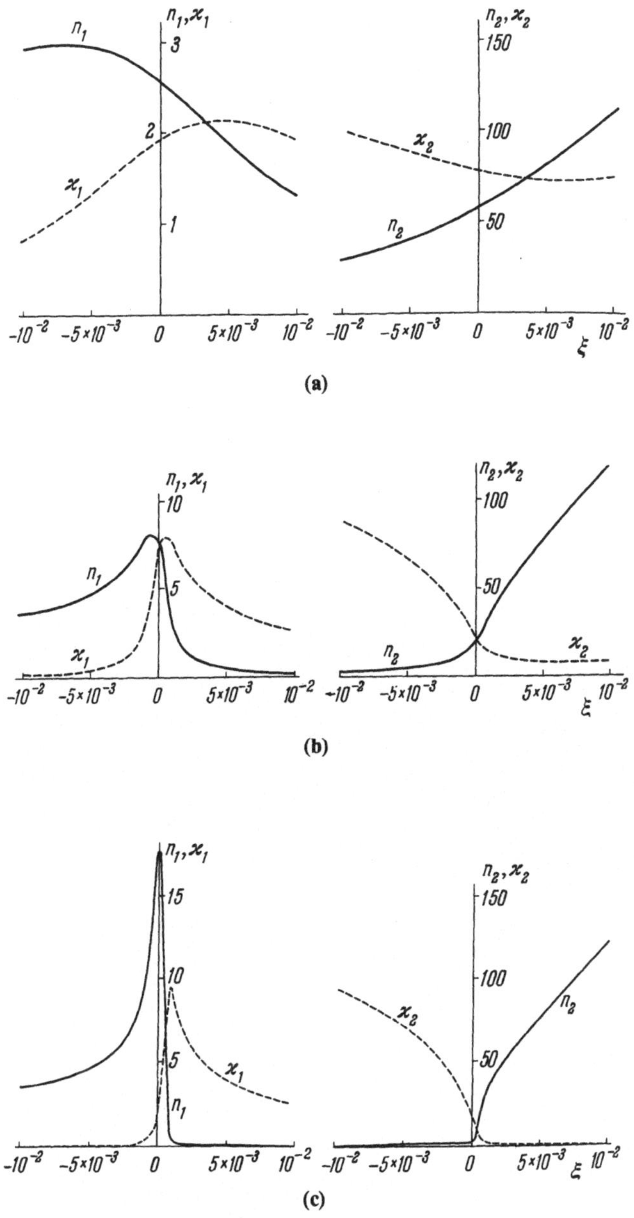

Fig. 4.8 a – c. As in Fig. 4.6, but with $\beta' = 10^{-5}$

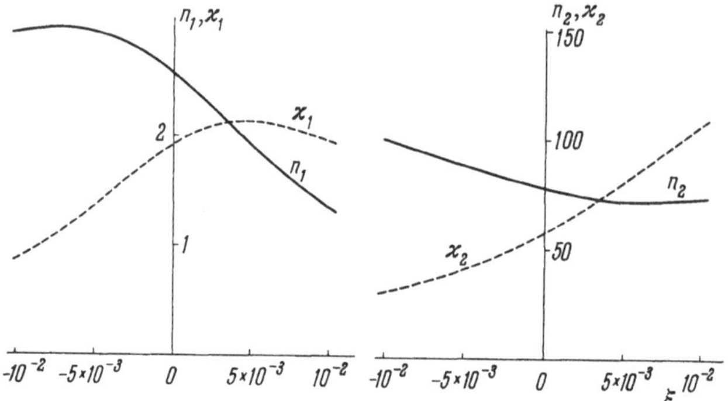

(a)

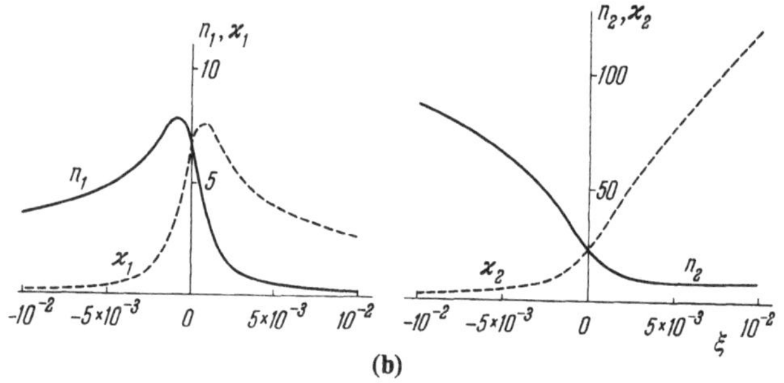

(b)

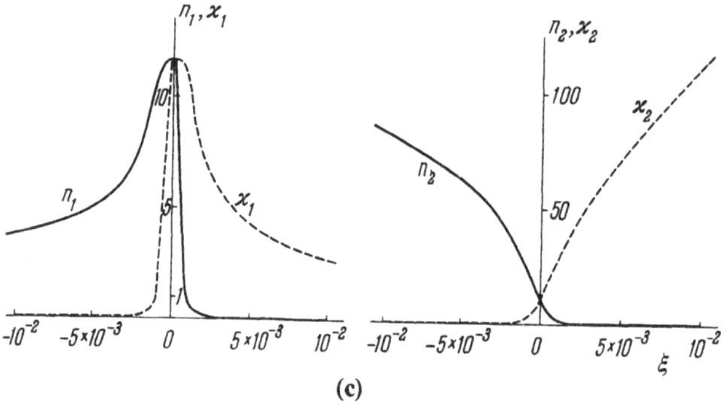

(c)

Fig. 4.9a–c. As in Fig. 4.6, but with $\beta' = -10^{-5}$

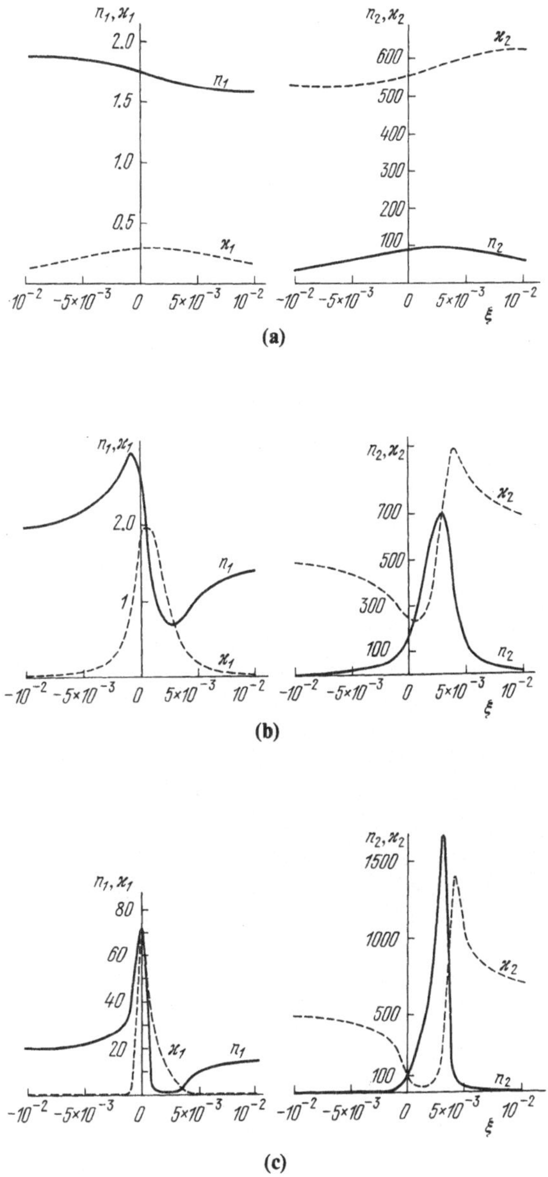

Fig. 4.10a – c. Dependence of n and $\varkappa$ on ξ for the following parameters: $A = 0.01$, $\varepsilon_{00} = 3$ and $\beta' = 10^{-6}$, with $\delta = 10^{-2}$ for **(a)**, $\delta = 10^{-3}$ for **(b)** and $\delta = 10^{-4}$ for **(c)**

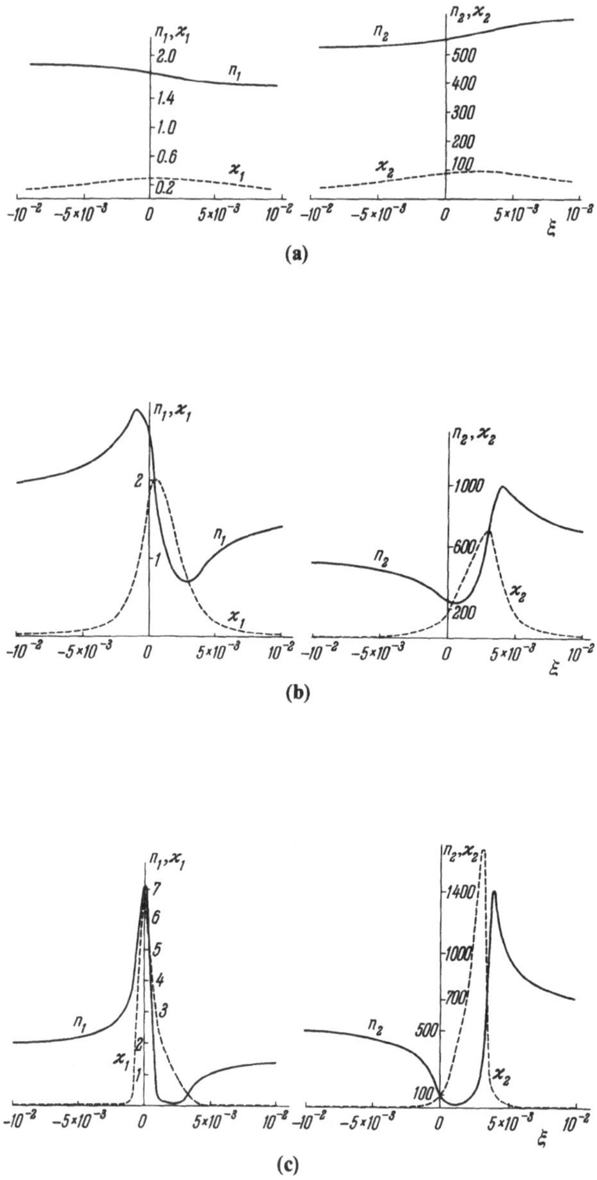

Fig. 4.11a–c. As in Fig. 4.10, but $\beta' = -10^{-6}$

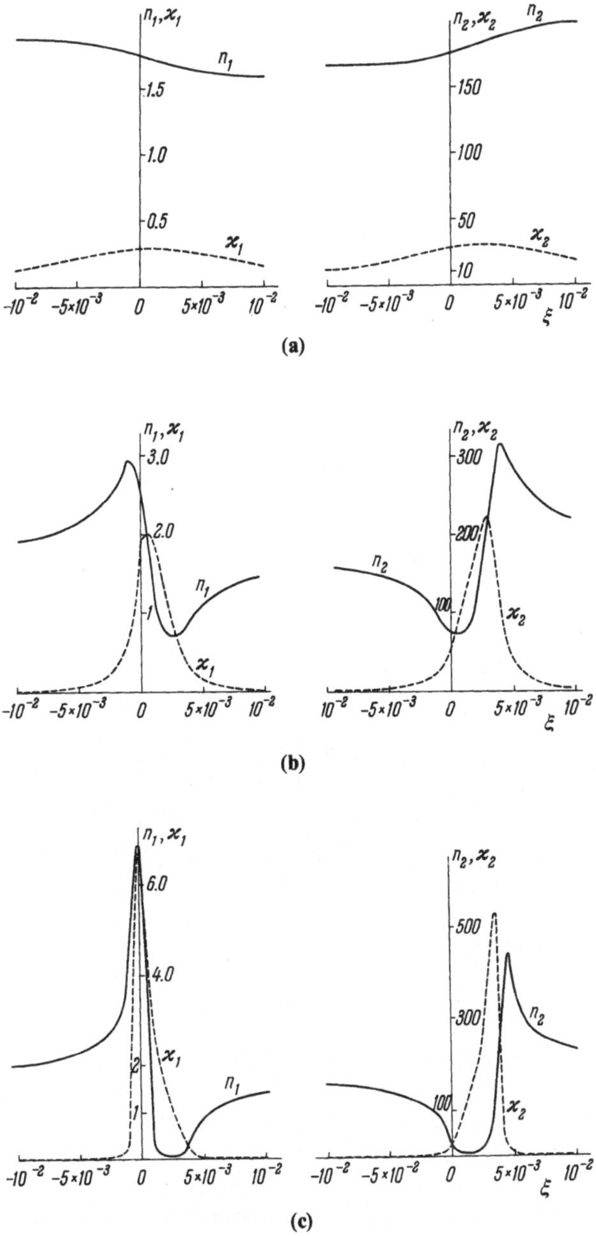

Fig. 4.12a–c. As in Fig. 4.10, but with $\beta' = -10^{-5}$

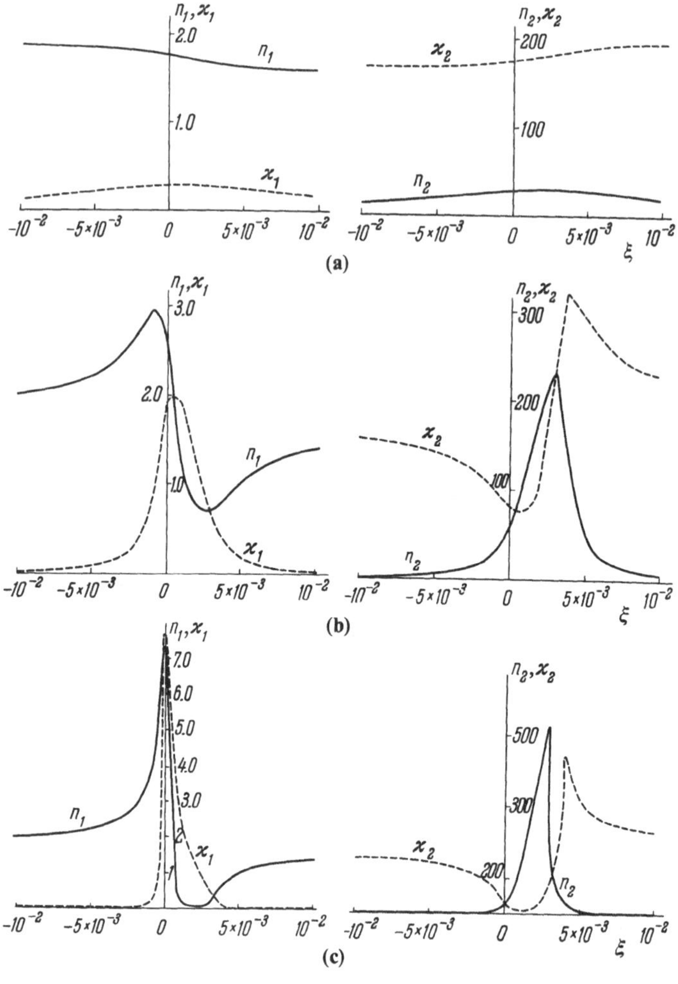

Fig. 4.13a – c. As in Fig. 4.10, but with $\beta' = 10^{-5}$

In (4.2.11) we should take the positive value of the root. The results of calculating $n(\omega)$ and $\varkappa(\omega)$ for various values of β', A and δ are illustrated in Figs. 4.6 – 13. From the curves given in these figures we can see that with real values of δ, the new wave is more strongly damped than the corresponding ordinary wave [new waves are understood to be waves that correspond to those branches of the $n^2(\omega)$ curves for which $n^2(\omega)$ does not tend to ε_{00} as ω departs from the resonance region]. On the other hand, it follows from these figures that in the case of small δ and large $|\beta'|$, even for ordinary waves when spatial dispersion is taken into account, the frequency dependence of the absorption coefficient, like the function $n^2(\omega)$, no longer satisfies well-known formulas [4.26], which are obtained from (4.1.13a) for $n(\omega)$ and $\varkappa(\omega)$. For small $|\beta'|$ and sufficiently large δ, however, the shape of the curves $n(\omega)$ and $\varkappa(\omega)$ is reestablished for the ordinary wave. As regards the new wave, either $n(\omega)$ or $\varkappa(\omega)$ become so large that the consideration of this wave is no longer justified.

Since, in general, the damping parameter δ increases with the temperature, it can be said that as the temperature is raised, we go over to the dispersion curves of classical crystal optics. The lower the oscillator strength and the greater the effective mass of the exciton, the sooner this occurs, of course (for a further discussion of the relevant problems, see Sect. 4.5.8).

It thus follows from the above that spatial dispersion effects can be observed near dipole absorption lines [i.e., lines for which $A \neq 0$ in (4.1.13)] in nongyrotropic media only under very special and favorable conditions. The effects of spatial dispersion can also be observed near quadrupole lines because, at sufficiently low temperatures, the width of these lines is sometimes less than the width of the dipole absorption lines.

Moreover, in comparing the curves of Figs. 4.6 – 13 with experimental curves we should note that in crystals the 'effective frequency' ν of collisions and, thereby, the quantity δ are in fact functions of the frequency (see Sect. 5.3.1 and also [4.27] where $\nu(\omega)$ has been calculated for the classical model of crystals, as well as [4.28] where $\nu(\omega)$ has been calculated for molecular crystals). For example, in those cases for which light is damped because of electron-phonon interaction, the character of the function $\delta(\omega)$ is closely related to the shape of the energy band of the mechanical exciton and the phonon spectrum. The function $\delta(\omega)$ then always decreases drastically as $|\omega - \omega_l(0)|$ increases. The form of the function $\delta(\omega)$ becomes particularly significant at low temperatures. Thus, for example, for a mechanical exciton of positive effective mass the value of $\delta(\omega)$ in the vicinity of an exciton band with $\omega \leqslant \omega_l(0)$ is much smaller than $\delta(\omega)$ in the case of $\omega \geqslant \omega_l(0)$ (Sect. 6.3.2). Hence it is clear that the experimental curves $n(\omega)$ and, $\varkappa(\omega)$ of crystals may deviate considerably from those shown in Figs. 4.6 – 13, where δ has been assumed independent of ω. To illustrate the influence of the ω dependence of δ on the shape of the dispersion curves, curves are given in Fig. 4.14 which represent the discontinuous function $\delta(\xi)$. The curves obtained possess a characteristic asymmetry (with respect to large n if $\xi < 0$ and small n if $\xi > 0$) and thereby resemble the experimental curves of a number of crystals (see, e.g., the experimental investigation [4.29] and others). A further analysis of experimental data will be given in Sect. 4.6 in connection with the problem of the experimental detection of new waves. Here we wish to mention still another effect caused by the ω dependence of δ. It can be observed even without accounting for spatial dispersion. Applying (4.1.13a) to $\varepsilon_0(\omega) = (n + \mathrm{i}\varkappa)^2$ we obtain

$$2n\varkappa = A\,\delta/(\xi^2 + \delta^2)\,, \qquad \xi = (\omega - \omega_l)/\omega_l\,.$$

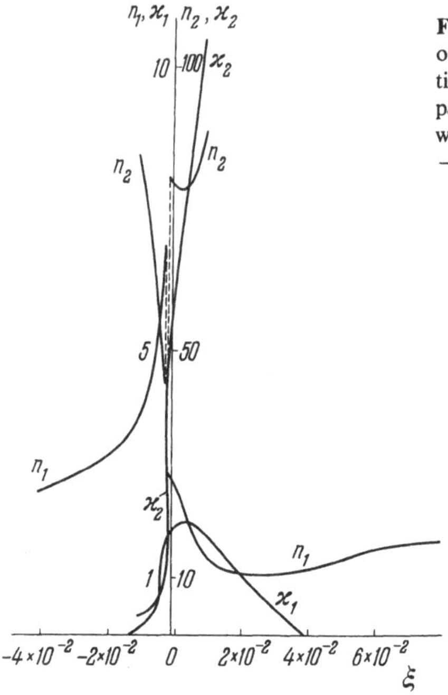

Fig. 4.14. Dependence of n and $\varkappa$ on ξ in the vicinity of $\omega = \omega_l$ in a nongyrotropic medium with absorption taken into account. The values taken for the parameters are: $A = 0.1$, $\varepsilon_{00} = 4$ and $\beta' = -10^{-5}$, with $\delta(\xi) = 10^{-3}$ at $\xi \leqslant -2\times10^{-3}$, $\delta(\xi) = 10^{-2}$ at $-2\times10^{-3} < \xi < 10^{-2}$ and $\delta(\xi) = 10^{-3}$ at $\xi \geqslant 10^{-2}$

Let us assume first that δ is independent of ω and integrate with respect to ω from $-\infty$ to $+\infty$:

$$\int_{-\infty}^{+\infty} 2n(\omega)\varkappa(\omega)d\omega = A\,\omega_l \int_{-\infty}^{+\infty} \frac{\delta d\xi}{\xi^2+\delta^2} = A\,\pi\,\omega_l\,. \tag{4.2.12}$$

This relation indicates that the area below the curve $2n(\omega)\cdot\varkappa(\omega)$ for an isolated line is determined mainly by the oscillator strength, and is independent of δ. This conclusion applies fairly well to gases but in the region of exciton absorption it can be appreciably violated. The fact is that in the vicinity of an isolated exciton transition, at low temperatures, the function $\delta(\omega)$ is, as a rule, strongly dependent on the frequency and assumes comparatively large values only of the immediate proximity of the frequency $\omega \approx \omega_l$. The left side of (4.2.12) must therefore depend not only on A but also on the character of $\delta(\omega)$. Let us illustrate this by an example. Assume for simplicity that

$$\delta(\omega) = 0 \quad \text{if} \quad |\xi| > \xi_0 \quad \text{and} \quad \delta(\omega) = \delta_0 \quad \text{if} \quad |\xi| < \xi_0\,,$$

where $2\xi_0\omega_l(0)$ is the absorption linewidth. Then, evidently,

$$\int_{-\infty}^{+\infty} 2n(\omega)\varkappa(\omega)d\omega = 2A\,\omega_l \arctan(\xi_0/\delta_0)\,.$$

At low temperatures the quantity ξ_0 may be of the order of δ_0 so that the value of $\arctan(\xi_0/\delta_0)$ may deviate substantially from $\pi/2$. As the temperature rises and the ratio ξ_0/δ_0 increases, the value of $\arctan(\xi_0/\delta_0)$ may, however, approach $\pi/2$ and, as a result, the area under the curve $2n(\omega)\varkappa(\omega)$ practically ceases to vary with temperature and, as for gases, becomes independent of δ.

From this it follows, first of all, that the function $\delta(\omega)$ leads to an asymmetry of the dispersion curves and, secondly, may lead to a change in the area below the curve $2n(\omega)\varkappa(\omega)$ when the temperature is changed. When analyzing experimental data it should be borne in mind that both of these effects are completely independent of spatial dispersion.

As an illustration, we mention the results [4.30] of investigations into light absorption in a GaSe crystal in the vicinity of an exciton line with a transition energy $E \approx 2.1$ eV [4.30] [3]. As the temperature was lowered (when $T < T_0$, where $T_0 \approx 100$ to 150 K) a substantial reduction (by a factor of 1.5 to 2) was observed in the amount of integral exciton absorption S. Spatial dispersion, however, can hardly have appreciable effects in this case because the exciton line in GaSe corresponds to a very low oscillator strength ($f \lesssim 10^{-2}$), whereas its half-width is quite large, being of the order of 10^{-2} eV [$\delta = \nu/2\omega_l \approx 5 \times 10^{-3}$, see (4.2.10)]. At the same time, it becomes important to consider an additional wave in the region of exciton transitions with sufficiently large oscillator strengths and with small exciton effective masses, and this can entirely alter the nature of the dependence of integral absorption on the temperature. This problem is discussed in Sect. 4.5.7.

Group-Velocity Vector

To conclude this subsection we shall discuss the problem of the direction of the group velocity of normal waves in a transparent medium with spatial dispersion. As has already been shown in Sect. 2.3.3, since the direction of the group velocity coincides with the direction of the total energy flux $\boldsymbol{S} = \boldsymbol{S}^{(0)} + \boldsymbol{S}^{(1)}$, the vector $\boldsymbol{v}_{\mathrm{gr}}$ can, in principle, make any angle with $\boldsymbol{k}$ if spatial dispersion is taken into account.

In the case of an isotropic medium (this results from symmetry considerations) this means that the group velocity can be oriented either parallel or antiparallel to the direction of the wave vector, $\boldsymbol{s} = \boldsymbol{k}/k$. We arrive at the same conclusion when using the expression for the vector $\boldsymbol{v}_{\mathrm{gr}}$ in an isotropic medium, see (2.3.37a and b),

$$\boldsymbol{v}_{\mathrm{gr}} \equiv \frac{d\omega}{d\boldsymbol{k}} = \frac{d\omega}{dk}\boldsymbol{s} = \frac{c}{n + \omega\dfrac{dn}{d\omega}}\boldsymbol{s}\,.$$

[3] This work [4.30] contains references to earlier investigations in which, at sufficiently low temperatures, the dependence of the integral exciton absorption on the temperature was observed (see also Sect. 4.6.2).

The vector $\boldsymbol{v}_{\mathrm{gr}}$ can, of course, be directed antiparallel to $\boldsymbol{s}$ only if $dn/d\omega < 0$. As is evident, e.g., from Figs. 4.2–13, this applies to the new waves. However, we still have to ascertain the sign of the sum $(n+\omega dn/d\omega)$. For this purpose we perform a calculation which is based on (4.2.6).

In a frequency range where $|\varepsilon_0^2\beta'| \ll 1$ the refractive indices of the ordinary and the new wave, are given by $\tilde{n}_1^2(\omega) \approx \varepsilon_0(\omega)$ and $\tilde{n}_2^2(\omega) \approx -1/\varepsilon_0(\omega)\beta'$, respectively, see (4.2.7, 8). Regardless of the sign of $\beta', n_1+\omega dn_1/d\omega > 0$ holds for the ordinary nonabsorbed wave, i.e., the vector $\boldsymbol{v}_{\mathrm{gr}}$ is directed along $\boldsymbol{s}$. However, for the new wave, if $\beta' < 0$ and $\tilde{n}_2 = n_2$, the inequality $\boldsymbol{v}_{\mathrm{gr}}\boldsymbol{s} < 0$ may hold. In fact, assuming $n_2^2(\omega) = -1/\varepsilon_0\beta'$, we have

$$n_2 + \omega\frac{dn_2}{d\omega} = \frac{1}{n_2}\left(n_2^2 + \frac{1}{2}\omega\frac{dn_2^2}{d\omega}\right) = \frac{1}{n_2|\beta'|\varepsilon_0^2}\left[\varepsilon_0(\omega) - \frac{\omega}{2}\,\frac{d\varepsilon_0(\omega)}{d\omega}\right].$$

Therefore, when $\varepsilon_0(\omega)$ is described by (4.1.13),

$$\varepsilon_0(\omega) = \varepsilon_{00} - A/\xi \approx -A/\xi\,, \qquad \xi = (\omega-\omega_l)/\omega_l\,,$$

and if $\xi < 0$ we find

$$\varepsilon_0(\omega) - \frac{\omega}{2}\,\frac{d\varepsilon_0(\omega)}{d\omega} = \frac{A}{|\xi|} - \frac{1}{2}\,\frac{\omega}{\omega_l}\,\frac{A}{\xi^2} \approx -\frac{1}{2}\,\frac{A}{\xi^2} < 0\,,$$

since in the frequency range being considered $|\xi| \ll 1/2$ and $\omega/\omega_l \approx 1$.

In an isotropic medium for which spatial dispersion is taken into account, the angle ψ between $\boldsymbol{v}_{\mathrm{gr}}$ and the wave vector $\boldsymbol{k}$ of the new wave is thus equal, in general, to π.

For an anisotropic medium $\boldsymbol{v}_{\mathrm{gr}}$ may be calculated from (2.3.39) but we do not wish to discuss this problem here. We only note that in the case of spatial dispersion in an anisotropic medium the angle ψ may, in general, assume all values between $\psi = 0$ and $\psi = \pi$ for different directions of wave propagation.

In addition to what was said in Sects. 2.3.2 and 3 we wish to make one further remark concerning the relation between the direction of $\boldsymbol{v}_{\mathrm{gr}}$ and the sign of the absorption coefficient $\varkappa$ (we consider, of course, a weakly absorbing medium where the concept of the group velocity is still applicable). For reasons of simplicity we shall confine ourselves to an isotropic medium.

In the absence of spatial dispersion in a medium in equilibrium, the product $n\varkappa > 0$ and therefore the amplitude of a propagating wave decreases in the same direction in which the wave travels. This is closely connected with the fact that in an isotropic medium without spatial dispersion the group-velocity vector $\boldsymbol{v}_{\mathrm{gr}}$ is necessarily directed along $\boldsymbol{k}$; at the same time, the energy of the wave in a medium in equilibrium also decreases in the direction of pulse propagation. For waves whose group velocity is antiparallel to $\boldsymbol{k}$, as may be the case when spatial dispersion is taken into account, the situation is quite different. The wave amplitude should then decrease in the direction opposite to $\boldsymbol{k}$, so that in the wave

$$\boldsymbol{E} = \boldsymbol{E}_0 \exp[\mathrm{i}(\omega n z/c - \omega t) - \omega \varkappa z/c] ,$$

$\varkappa$ must be negative if $n > 0$ and positive if $n < 0$. This fact proves to be essential when we consider refraction and reflection of waves at the boundary of a crystal, and spatial dispersion is taken into account. Since, however, the equation used to determine $\tilde{n}_l$ enables us to find the complex quantity $\tilde{n}_l = n_l + \mathrm{i}\varkappa_l$, the sign of the ratio $\varkappa_l/n_l$ cannot be chosen arbitrarily so that in this connection there is no necessity to compute $\boldsymbol{v}_{\mathrm{gr}}$ for normal waves (the required procedure in the absence of absorption is described in Sect. 4.5.9).

4.3 Cubic Nongyrotropic Crystals

4.3.1 Optical Anisotropy: Dipole Transitions

It has already been pointed out in Chap. 1 that the optical anisotropy of cubic crystals which was theoretically dealt with in [4.11, 32, 33, 43, 44] has been observed experimentally [4.35][4] by studying Cu_2O crystals at low temperatures in the range of the quadrupole transition at $\lambda = 6125$ Å. We do not consider here the so-called latent anisotropy of cubic crystals due to the presence of anisotropic centers in the crystals. These centers result from local lattice defects (e.g., various kinds of F centers, etc.). If spatial dispersion is neglected and there are no oriented external influences, cubic crystals with anisotropic centers (impurities) behave like optically isotropic bodies (for more detail, see [4.31]).

Optical anisotropy in cubic crystals may become apparent not only near quadrupole transitions but near dipole transitions as well, and, in general, far away from all transitions. A dipole transition is, as usual, understood to be one at the frequency of which the dielectric constant [in the case of a cubic crystal this is the scalar function $\varepsilon_0(\omega)$] becomes infinite, if absorption and spatial dispersion are ignored. It follows from this definition that all dipole transitions correspond to nonvanishing oscillator strengths, see (4.1.13).

The dielectric constant tensor becomes infinite at the frequencies of quadrupole transitions only if spatial dispersion is taken into account. We have already emphasized in Sect. 3.1.2 that series expansions in k_i of the tensors $\varepsilon_{ij}(\omega, \boldsymbol{k})$ and $\varepsilon_{ij}^{-1}(\omega, \boldsymbol{k})$ are not generally expansions with respect to multipoles. When investigating these expansions, the nature of the transition affects primarily the frequency dependence of the coefficients. Far away from

[4] Absorption spectra were investigated in [4.35]. Experimental results for the optical anisotropy of cubic crystals in the transparent region are discussed in Sect. 4.6.2.

the transitions the frequency dependence is weak and we may use ε_{ij} or ε_{ij}^{-1} equally well. Near the dipole transition frequencies in cubic crystals, where $\varepsilon_0(\omega)$ may assume large values, it is necessary to use the expansions of the inverse tensor $\varepsilon_{ij}^{-1}(\omega,\boldsymbol{k})$, i.e., an expansion of the type (3.1.21), in order to account for spatial dispersion. Since in nongyrotropic crystals the tensor $\delta_{ijl}(\omega)=0$, we have

$$\varepsilon_{ij}^{-1}(\omega,\boldsymbol{k})=\delta_{ij}\varepsilon_0^{-1}(\omega)+\beta'_{ijlm}(\omega)\tilde{n}^2 s_l s_m \tag{4.3.1}$$

[the tensor $\beta_{ijlm}=(c/\omega)^2\beta'_{ijlm}$ appears in (3.1.21); the tensor $\varepsilon_{ij}^{-1}(\omega)$, as formulated in (4.3.1), already takes cubic symmetry into account]. As we have shown in Sect. 3.2.2, the tensor β'_{ijlm} can be considerably simplified by making the axes x, y and z coincides with the fourfold symmetry axes. In this case, for crystals of classes T and T_h, the tensor β'_{ijlm} is determined by the four scalars β'_1, β'_2, β'_3 and β'_4 where

$$\begin{aligned}&\beta'_1=\beta'_{xxxx}=\beta'_{yyyy}=\beta'_{zzzz}\,; &&\beta'_2=\beta'_{yyxx}=\beta'_{xxzz}=\beta'_{zzyy}\,;\\ &\beta'_3=\beta'_{xyxy}=\beta'_{yzyz}=\beta'_{zxzx}\,; &&\beta'_4=\beta'_{zzxx}=\beta'_{xxyy}=\beta'_{yyzz}\,.\end{aligned} \tag{4.3.2}$$

In addition, in the classes T_d, O and O_h we have $\beta'_2=\beta'_4$.

For the sake of simplicity we shall consider in the following the character of the polarization and the dispersion of electromagnetic waves in cubic crystals, which pertain to the most symmetrical classes T_d and O_h[5]. The results obtained may be used not only near dipole transitions but far away from all transitions (range of transparency) as well. In the latter case, however, the optical anisotropy is less than that near resonances. As follows from (2.2.26), and also from (4.3.1, 2) the equations for the components of the vector $\boldsymbol{D}$ are of the form

$$\frac{1}{\tilde{n}^2}D_1=\left(\frac{1}{\varepsilon_0}+\beta'_2\tilde{n}^2\right)D_1-s_1s_iD_i\left(\frac{1}{\varepsilon_0}+\beta'_2\tilde{n}^2\right)+\tilde{\beta}\tilde{n}^2s_1^2D_1-\tilde{\beta}\tilde{n}^2s_1D_is_i^3\,,$$

$$\frac{1}{\tilde{n}^2}D_2=\left(\frac{1}{\varepsilon_0}+\beta'_2\tilde{n}^2\right)D_2-s_2s_iD_i\left(\frac{1}{\varepsilon_0}+\beta'_2\tilde{n}^2\right)+\tilde{\beta}\tilde{n}^2s_2^2D_2-\tilde{\beta}\tilde{n}^2s_2D_is_i^3\,,$$

$$\frac{1}{\tilde{n}^2}D_3=\left(\frac{1}{\varepsilon_0}+\beta'_2\tilde{n}^2\right)D_3-s_3s_iD_i\left(\frac{1}{\varepsilon_0}+\beta'_2\tilde{n}^2\right)+\tilde{\beta}\tilde{n}^2s_3^2D_3-\tilde{\beta}\tilde{n}^2s_3D_is_i^3\,, \tag{4.3.3}$$

where

$$\tilde{\beta}=\beta'_1-\beta'_2-2\beta'_3\,. \tag{4.3.4}$$

---

[5] Crystals belonging to class O are gyrotropic. Consequently, the tensor $\delta_{ijl}\neq 0$ in these crystals, see (4.1.1). Anisotropy investigations should, in general, be carried out by taking the tensors δ_{ijl} and β_{ijlm} into account simultaneously.

It is a natural consequence of the set of equations (4.3.3) and also of the initial equation (2.2.26) that solutions of this set satisfy the transversality condition of the vector $\boldsymbol{D}$, i.e., $\boldsymbol{s} \cdot \boldsymbol{D} = 0$. This fact can be used to cancel components in (4.3.3) which are proportional to $s_i D_i$. After this, however, the simplified set of equations is to be considered with the additional condition $s_i D_i = 0$. This is of no advantage compared with the application of system (4.3.3).

Several special cases are of interest.

(a) Assume that the vector $\boldsymbol{s}$ is directed along one of the cube's edges, e.g., along the z axis. The equations for the components of the vector $\boldsymbol{D}$ are then simplified to

$$\frac{1}{\tilde{n}^2} D_1 = \left(\frac{1}{\varepsilon_0} + \beta_2' \tilde{n}^2\right) D_1 , \quad \frac{1}{\tilde{n}^2} D_2 = \left(\frac{1}{\varepsilon_0} + \beta_2' \tilde{n}^2\right) D_2 , \quad D_3 = 0 .$$

It follows that in the case being considered the quantity $\tilde{n}^2$ is independent of polarization, i.e., of the direction of $\boldsymbol{D}$, and is determined by the equation

$$1/\tilde{n}^2 = 1/\varepsilon_0 + \beta_2' \tilde{n}^2 . \tag{4.3.5}$$

This type of equation has already been considered earlier in analyzing the dispersion of electromagnetic waves in an isotropic medium with spatial dispersion taken into account (Sect. 4.2.2).

(b) Let us now assume $\boldsymbol{s}$ to be directed along one of the cube's body diagonals: $|s_1| = |s_2| = |s_3| = 1/\sqrt{3}$. If $s_i D_i = 0$ we obtain from (4.3.3)

$$\frac{1}{\tilde{n}^2} D_i = \left(\frac{1}{\varepsilon_0} + \beta_2' \tilde{n}^2 + \frac{1}{3} \tilde{\beta} \tilde{n}^2\right) D_i , \quad i = 1, 2, 3 .$$

Consequently, in this case as well, the quantity $\tilde{n}^2$ is independent of the polarization and, as in case (a), two values of $\tilde{n}^2$, obtained from the equation

$$1/\tilde{n}^2 = 1/\varepsilon_0 + (\beta_2' + \tilde{\beta}/3) \tilde{n}^2 , \tag{4.3.6}$$

correspond to each of the two independent polarizations. From what has been said under items (a) and (b) it follows that seven of the directions being considered (3 fourfold symmetry axes and 4 body diagonals of the cube) are optical axes of the crystal.

(c) Let $\boldsymbol{s}$ be directed along a diagonal of a face and, e.g., $|s_1| = |s_2| = 1/\sqrt{2}$ and $s_3 = 0$. Then, as follows from (4.3.3), the equations for D_i, $i = 1, 2, 3$ have the following form

$$\frac{1}{\tilde{n}^2} D_1 = \left(\frac{1}{\varepsilon_0} + \beta_2' \tilde{n}^2 + \frac{1}{2} \tilde{\beta} \tilde{n}^2\right) D_1 , \quad D_2 = -D_1 ,$$

$$\frac{1}{\tilde{n}^2} D_3 = \left(\frac{1}{\varepsilon_0} + \beta_2' \tilde{n}^2 \right) D_3 .$$

In the direction being considered the values of $\tilde{n}^2$ are thus strongly dependent on the polarization. If $D_3 \neq 0$ and $\tilde{\beta} \neq 0$ then $D_1 = D_2 = 0$ and, therefore,

$$1/\tilde{n}^2 = 1/\varepsilon_0 + \beta_2' \tilde{n}^2 . \tag{4.3.7}$$

In this connection we wish to mention that this equation for the determination of $\tilde{n}^2$ remains valid in the more general case when $|s_1| \neq |s_2|$, but, as before, $s_3 = 0$; again we have $D_1 = D_2 = 0$. If, on the contrary, $D_3 = 0$ and $|s_1| = |s_2| = 1/\sqrt{2}$, the equation for $\tilde{n}^2$ is

$$1/\tilde{n}^2 = 1/\varepsilon_0 + (\beta_2' + \tfrac{1}{2}\tilde{\beta}) \tilde{n}^2 . \tag{4.3.8}$$

It is easy to set up the equation for $\tilde{n}^2$ for a wave with $D_3 = 0$ and in the case when, at $s_3 = 0$, $|s_1| \neq |s_2|$. It follows, as a matter of fact, from the transversality condition for $\boldsymbol{D}$ that

$$D_2 = -s_1 D_1 / s_2 .$$

Substituting this expression for D_2 in the first equation of the set (4.3.3), we find that the equation for the determination of $\tilde{n}^2$ can be given in the form

$$1/\tilde{n}^2 = 1/\varepsilon_0 + (\beta_2' + 2\tilde{\beta} s_1^2 s_2^2) \tilde{n}^2 \tag{4.3.9}$$

which is transformed to (4.3.8) if $s_1^2 = s_2^2 = 1/2$.

Thus, we generally have two values of the refractive index for each polarization. An exception to this rule may occur only if, e.g., at certain values of s_1^2, s_2^2 and $s_3 = 0$, the expression $\beta_2' + 2\tilde{\beta} s_1^2 s_2^2$ vanishes. Here it is a necessary condition that the quantities β_2' and $\tilde{\beta}$ are of opposite sign and that $|\beta_2'/\tilde{\beta}| < 2$. Then, with $\beta_2' + 2\tilde{\beta} s_1^2 s_2^2 \to 0$, one of the solutions of (4.3.9), namely, that which corresponds to the new wave, tends to infinity ($\tilde{n}_2 \to \infty$). Note, besides, that for waves with $D_3 = 0$ the refractive index $\tilde{n}$ in case (c) coincides with that of case (a).

The formulas obtained for $\tilde{n}^2$ clearly prove that the influence of spatial dispersion depends considerably on the direction of $\boldsymbol{s}$ and on the polarization. Within the scope of the phenomenological approach [4.11] used here the values of β_2' and $\tilde{\beta}$ cannot be computed. Our consideration of the optical anisotropy in cubic crystals of classes T_d and O_h, far away from absorption lines and close to dipole transition lines, proves to be not only sufficiently complete but also considerably simpler that the corresponding calculations made with the exciton wave functions [4.36]. Such calculations, however, enable us to get explicit expressions for the quantities β. This, of course, also applies to crystals of other symmetry classes.

In the immediate proximity of a resonance (pole) the quantity $\varepsilon_0(\omega)$ should be regarded as being complex. Consequently, accounting for spatial dispersion leads to the observation that in this frequency range both the real and the imaginary parts of the refractive index depend on the direction of $\boldsymbol{s}$ and on the polarization. In cubic crystals, in the vicinity of dipole absorption lines, anisotropy can thus exist not only with respect to dispersion, but also with respect to absorption. This more or less evident fact has been emphasized only because it is a natural consequence of the aforesaid.

In cubic crystals near dipole lines accounting for spatial dispersion, as in an isotropy medium (Sect. 4.2.2), may in many cases lead to an appreciable change in the shape of the dispersion curves (due to the appearance of a new wave). It may also lead to distinct anisotropy of the optical properties. Consequently, these effects of spatial dispersion are by no means small.

4.3.2 Optical Anisotropy and Quadrupole Transitions

Let us now discuss the anisotropy of light dispersion and absorption near quadrupole absorption lines, where $\varepsilon_0(\omega)$ is a smoothly varying function, but where at least one of the components of the tensor $\alpha_{ijlm}(\omega)$ has a resonance (pole). For this purpose we make use of the following expansion of the tensor $\varepsilon_{ij}(\omega,\boldsymbol{k})$, see (3.1.20),

$$\varepsilon_{ij}(\omega,\boldsymbol{k}) = \delta_{ij}\varepsilon_0(\omega) + \alpha_{ijlm}(\omega)s_l s_m \tilde{n}^2 . \qquad (4.3.10)$$

This tensor $\alpha_{ijlm}(\omega)$ differs from the identically denoted tensor in (3.1.20) by the factor $(\omega/c)^2$. Also, the fact that in nongyrotropic crystals the tensor $\gamma_{ijl} = 0$ was taken into consideration. Assuming that the main light absorption in a crystal is in connection with quadrupole absorption lines, we may assume that $\varepsilon_0(\omega)$ is a real function and that the tensor $\alpha_{ijlm}(\omega)$ is complex:

$$\alpha_{ijlm}(\omega) = \alpha'_{ijlm}(\omega) + \mathrm{i}\,\alpha''_{ijlm}(\omega) .$$

Since in cubic crystals both tensors α'_{ijlm} and α''_{ijlm} are simplified at the same time, provided the coordinate axes coincide with the fourfold symmetry axes. The presence of absorption ($\alpha''_{ijlm} \neq 0$) does not complicate the consideration. The nonvanishing components of the tensor α_{ijlm} are given by (3.2.13). As in Sect. 4.3.1, we confine ourselves for simplicity to the anisotropy of absorption and dispersion in the vicinity of quadrupole absorption lines in cubic crystals belonging to the classes T_d and O_h. In these crystals the nonvanishing components of $\alpha_{ijlm}(\omega)$ are determined by the three scalars α_1, α_3, $\alpha_2 = \alpha_4$ [see Sect. 3.2.2 and, in particular, the notation in (3.2.13)].

In accordance with (2.2.9), we therefore obtain for the components of the vector $\boldsymbol{E}$ the following set of equations [see also (3.2.18) with $\alpha_2 = \alpha_4$]:

$$\tilde{n}^2 E_1 = (\varepsilon_0 + \alpha_2\tilde{n}^2)E_1 + \tilde{\alpha}\tilde{n}^2 s_1^2 E_1 + 2\alpha_3(\boldsymbol{E}\cdot\boldsymbol{s})\tilde{n}^2 s_1 + (\boldsymbol{E}\cdot\boldsymbol{s})\tilde{n}^2 s_1 , \qquad (4.3.11)$$

etc., where

$$\tilde{\alpha} = \alpha_1 - \alpha_2 - 2\alpha_3 . \tag{4.3.12}$$

Let us again consider some special cases.

a) Let the vector s be directed along one of the cube's edges, e.g., along the z axis ($s_3 = 1$, $s_1 = s_2 = 0$). In this case the set (4.3.11) assumes the form

$$\tilde{n}^2 E_1 = (\varepsilon_0 + \alpha_2 \tilde{n}^2) E_1 , \quad \tilde{n}^2 E_2 = (\varepsilon_0 + \alpha_2 \tilde{n}^2) E_2 , \quad (\varepsilon_0 + \alpha_1 \tilde{n}^2) E_3 = 0 .$$

We thus have a transverse wave ($E_3 = 0$) for which, irrespective of the polarization,

$$\tilde{n}^2 = (n + \mathrm{i}\varkappa)^2 = \frac{\varepsilon_0}{1 - \alpha_2' - \mathrm{i}\alpha_2''} . \tag{4.3.13}$$

A longitudinal wave may also exist ($E_1 = E_2 = 0$, $E_3 \neq 0$) for which

$$\tilde{n}^2 = -\frac{\varepsilon_0}{\alpha_1' + \mathrm{i}\alpha_1''} . \tag{4.3.14}$$

b) Let us now assume the vector s to be directed along one of the body diagonals of the cube: $|s_1| = |s_2| = |s_3| = 1/\sqrt{3}$. In this case, using (4.3.11), we obtain the following equation for the components of E:

$$\tilde{n}^2 E_i = (\varepsilon_0 + \alpha_2 \tilde{n}^2 + \tfrac{1}{3}\tilde{\alpha}\tilde{n}^2) E_i + (1 + 2\alpha_3)(E \cdot s)\tilde{n}^2 s_i , \quad i = 1, 2, 3 . \tag{4.3.15}$$

If $E = Es$ (longitudinal wave), $E_i = s_i(E \cdot s)$ and therefore

$$\tilde{n}^2 = -\frac{3\varepsilon_0}{\alpha_1 + 2\alpha_2 + 4\alpha_3} . \tag{4.3.16}$$

For transverse waves $E \cdot s = 0$. In this case, independent of the polarization,

$$\tilde{n}^2 = (n + \mathrm{i}\varkappa)^2 = \frac{\varepsilon_0}{1 - \tilde{\alpha}/3 - \alpha_2} , \quad \alpha_2 = \alpha_2' + \mathrm{i}\alpha_2'' , \quad \tilde{\alpha} = \tilde{\alpha}' + \mathrm{i}\tilde{\alpha}'' . \tag{4.3.17}$$

c) Assume now that the vector s is directed along a diagonal of a face of the cube, for example, $|s_1| = |s_2| = 1/\sqrt{2}$, $s_3 = 0$. We then obtain from (4.3.11)

$$\tilde{n}^2 E_i = (\varepsilon_0 + \alpha_2 \tilde{n}^2) E_i + \frac{\tilde{\alpha}}{2} \tilde{n}^2 E_i + (2\alpha_3 + 1)\tilde{n}^2 (E \cdot s) s_i \tag{4.3.18a}$$

and

$$\tilde{n}^2 E_3 = (\varepsilon_0 + \alpha_2 \tilde{n}^2) E_3 . \tag{4.3.18b}$$

In this case the values of $\tilde{n}^2$ depend on the polarization for transverse waves as well. In fact, as follows from (4.3.18), we have:

I) For waves with $E_3 = 0$ and $\boldsymbol{E} \cdot \boldsymbol{s} = 0$, i.e., transverse waves, polarized in the facial plane

$$\tilde{n}^2 = \frac{\varepsilon_0}{1 - \alpha_2 - \tilde{\alpha}/2} . \tag{4.3.19}$$

II) For waves with $E_3 \neq 0$ and $\boldsymbol{E} \cdot \boldsymbol{s} = 0$, i.e., transverse waves polarized perpendicularly to the facial plane,

$$\tilde{n}^2 = \frac{\varepsilon_2}{1 - \alpha_2} . \tag{4.3.20}$$

For waves with $E_3 \neq 0$ and $\boldsymbol{E} \cdot \boldsymbol{s} = 0$ this relation remains valid even when $|s_1| \neq |s_2|$ and $s_3 = 0$ since, in this case, the equation for E_3 again has the form of (4.3.18b).

III) In the case of a longitudinal wave where $E_3 = 0$ and $\boldsymbol{E} = E\boldsymbol{s}$

$$\tilde{n}^2 = -\frac{\varepsilon_0}{\tilde{\alpha}/2 + \alpha_2 + 2\alpha_3} . \tag{4.3.21}$$

This phenomenological consideration of the anisotropy of cubic crystals in the region of quadrupole absorption shows that the optical properties of crystals of the classes T_d and O_h are characterized not only by ε_0, but also by the values of the three quantities α_1, α_2, α_3 which are directly related when spatial dispersion is taken into account. Up to now we have not specified the nature of these excited states in the crystal whose existence gives rise to the resonant behavior of the functions $\varepsilon_0(\omega)$ and $\alpha_{ijlm}(\omega)$. Such a specification enables us to establish a number of rules. For example the selection rules establish some additional relations linking the quantities α_1, α_2 and α_3 for individual transitions.

Since we are primarily interested in the exciton states, we shall discuss the classification of the states that we shall deal with in analyzing quadrupolar transitions. Transitions whose probabilities are proportional to k^2 can, of course, be classified without having recourse to the concepts of excitons and wave functions. In fact, these transitions correspond to radiation by the scalar source [in this case only longitudinal waves are obtained; see (4.3.27) and the following], by a quadrupole or a magnetic dipole. In other words, as in the case of dipole emission, there is no particular necessity to have recourse to quantum notation. We shall, however, use it because it will be required when we discuss exciton lines.

4.3.3 Classification of the States of 'Mechanical' Excitons with $k = 0$ and Selection Rules for Quadrupole Transitions

In crystals steady states and, in particular, exciton-excited states can be classified according to the irreducible representations of the space groups [4.37 – 41]. Each space group contains a subgroup of parallel translations which comprise all possible parallel translations displacing the lattice in such a way that it coincides with itself. The complete space group is obtained from this subgroup by adding N elements ("rotary" elements) which contain rotations and reflections, N being equal to the number of elements of the group of the corresponding class of crystals. Each space group element may be regarded as the product of one of the elements of the translation subgroup and one of the 'rotary' elements. If the space group does not contain essential screw axes and glide planes all 'rotary' elements taken together form a point group, the group of the corresponding crystal class ([4.39] and Appendix B).

Since we are dealing here only with weak spatial dispersion so that the tensors $\varepsilon_{ij}(\omega, \boldsymbol{k})$ or $\varepsilon_{ij}^{-1}(\omega, \boldsymbol{k})$ may be expanded in a power series with respect to the wave vector, the tensors $\varepsilon_{ij}(\omega)$, $\alpha_{ijlm}(\omega)$, etc., occurring in an expansion of the type (3.1.19) are determined by the properties of states of the 'mechanical' excitons with $\boldsymbol{k} = 0$. The wave functions of the exciton states with $\boldsymbol{k} = 0$ are invariant with respect to the elements of the translation subgroup, i.e., the crystal can in this case be regarded as being homogeneous, see (1.1.12) and Appendix A. The corresponding exciton states can therefore be classified according to the irreducible point-group representations of the crystal class characterizing the directional symmetry in the crystal. This is due to the fact that the point group of a crystal is isomorphic to the factor group of the translation subgroup (see [4.38, 39, 41] and Appendix B, which gives the basic definitions). We shall make use of this kind of exciton state classification in the following.

Let us consider as an example Cu_2O-type crystals in greater detail; they belong to the most symmetrical class O_h of the cubic system. The characters of the irreducible representations of the O_h group are listed in Table 4.2 (in the notation of [4.42], see Appendix A). The second column of this table indicates how the corresponding wave functions are transformed by the symmetry operations of group O_h. Thus, for example, we may see from Table 4.2 that the three wave functions which correspond to a triply degenerate exciton term (at $\boldsymbol{k} = 0$) and which possess the symmetry of the irreducible representation F_2 are transformed like the symmetrized products of unlike components of two polar vectors (x_1, y_1, z_1) and (x_2, y_2, z_2). Table 4.2 only indicates how linear and quadratic combinations of polar vector components are transformed. The combinations of three or more polar vector components correspond to the rest of the irreducible representations. For example, the representation A_2' corresponds to a pseudoscalar, i.e., to an expression like $\boldsymbol{r}_1 \cdot (\boldsymbol{r}_2 \times \boldsymbol{r}_3)$, etc.

As in the case of the lowest electron terms of polyatomic molecules [Ref. 4.42, § 98], an empirical rule is usually applied to crystals. It is verified by

Table 4.2. Characters of the irreducible representations of group O_h

Irreducible represen-tations	Expressions to be transformed according to irreducible repre-sentations (x, y, z are the com-ponents of polar vector $\boldsymbol{r}$)	Symmetry operations									
		E	$3C_4^2$	$6C_4$	$6C_2$	$8C_3$	I	$3IC_4^2$	$6IC_4$	$6IC_2$	$8IC_3$
A_1	$x_1x_2=y_1y_2+z_1z_2$ (scalar)	1	1	1	1	1	1	1	1	1	1
A_2	–	1	1	−1	−1	1	1	1	1	−1	−1
E	$x_1x_2-z_1z_2$, $2y_1y_2-x_1x_2-z_1z_2$	2	2	0	0	−1	2	2	0	0	−1
F_1	$\boldsymbol{r}_1\times\boldsymbol{r}_2$ (pseudovector)	3	−1	1	−1	0	3	−1	1	−1	0
F_2	$x_1y_2+x_2y_1$, $x_1z_2+x_2z_1$, $y_1z_2+y_2z_1$	3	−1	−1	1	0	3	−1	−1	1	0
A_1'	–	1	1	1	1	1	−1	−1	−1	−1	−1
A_2'	–	1	1	−1	−1	1	−1	−1	1	1	−1
E'	–	2	2	0	0	−1	−2	−2	0	0	1
F_1'	x, y, z (vector)	3	−1	1	−1	0	−3	1	−1	1	0
F_2'	–	3	−1	−1	1	0	−3	1	1	−1	0

model calculations (see, e.g., [4.37] and Sect. 6.2), according to which the wave function of the ground state of the crystal possesses full symmetry with respect to symmetry transformations. In the case under consideration this indicates that the ground state possesses the symmetry of the irreducible representation A_1 as we shall assume hereafter. If we take into account that the dipole moment operator transforms like a polar vector, the matrix element of the dipole moment operator will thus be nonvanishing only for transitions from the ground state to exciton states whose wave functions are transformed at $\boldsymbol{k}=0$ according to the irreducible representation F_1' (Table 4.2). Nonvanishing transition oscillator strengths correspond to the nonvanishing matrix elements of the dipole-moment operator so that the dielectric constant $\varepsilon_0(\omega)$, ignoring absorption and spatial dispersion, becomes infinite at the transition frequency.

Note that the classification of 'mechanical' exciton states used here is simpler than and dissimilar to the classification of the Coulomb exciton states. In particular, in the classification of the latter, even if $\boldsymbol{k}=0$, the direction $\boldsymbol{s}=\boldsymbol{k}/k$ proves to be essential. The classification of states of these excitons with respect to their symmetry should be carried out separately, in general, for each direction $\boldsymbol{s}$. This is due to the fact that when the Coulomb interaction and thereby also the influence of the long-wavelength field are fully taken into account, e.g., in the case of the group O_h, the F_1' representation is subject to splitting because a representation corresponding to longitudinal excitons is singled out in this representation. The energies of the exciton states corresponding to the rest of the representations of the O_h group remain unchanged for $\boldsymbol{k}=0$, when the Coulomb interaction is fully accounted for (Sect. 6.2).

When using (4.3.1) for the tensor ε_{ij}^{-1}, the results obtained in Sect. 4.3.1 enable us to take into account the spatial dispersion that is associated with the contribution of the exciton bands whose wave functions, if $\boldsymbol{k}=0$, are transformed according to the irreducible representation F_1'. The exciton bands whose wave functions at $\boldsymbol{k}=0$ are transformed according to irreducible representations unlike the F_1' representation occur in the formulas for $\varepsilon_{ij}(\omega,\boldsymbol{k})$ only if spatial dispersion is taken into account. Thus, for example, in a similar manner as in the case of atoms and molecules [4.39], only the exciton bands whose wave functions are transformed at $\boldsymbol{k}=0$ like products of the components of two polar vectors contribute to quadrupole absorption and emission of light. The set of these products in the O_h group forms the reducible representation V^2, this being the sum of irreducible representations (Table 4.2 and [4.39]):

$$V^2=A_1+E+F_1+F_2\,. \tag{4.3.22}$$

Here A_1 is the representation according to which the scalar product $\mathrm{I}=x_1x_2+y_1y_2+z_1z_2$ of two polar vectors is transformed, E is the doubly degenerate representation according to which the two linearly independent combinations $\mathrm{II}=x_1x_2-z_1z_2$ and $\mathrm{III}=2y_1y_2-x_1x_2-z_1z_2$ are transformed, F_1 is the triply degenerate representation according to which the three components $\mathrm{IV} = y_1z_2 - y_2z_1$, $\mathrm{V} = z_1x_2 - z_2x_1$ and $\mathrm{VI} = x_1y_2 - x_2y_1$ of the vector product of two polar vectors are transformed, and F_2 is the triply degenerate representation according to which the three symmetrical combinations $\mathrm{VII} = y_1z_2 + y_2z_1$, $\mathrm{VIII} = z_1x_2 + z_2x_1$ and $\mathrm{IX} = x_1y_2 + x_2y_1$ are transformed. Multipoles of higher order correspond to the representations A_1', A_2, A_2', E' and F_2' (Table 4.2) which are, however, of no especial interest. Also, since we call all absorption modes whose probability is proportional to k^2 quadrupole absorption, this concept, in addition to true quadrupole transitions (representations E and F_2), also includes magnetic dipole (representation F_1) and scalar (representation A_1) transitions.

Each of the quadrupolar exciton states may, in general, contribute to quadrupole absorption of light by crystals. In a frequency range which corresponds to the vicinity of a discrete resonance, however, only a single band of exciton states can be taken into consideration. In such an approximation, as will be shown in Sect. 6.1, in applying the matrix elements of the transition from the ground state 0 to an excited state of the L-th band ($\boldsymbol{k} = 0$), we find

$$\alpha_{ijlm}(\omega) \sim \frac{1}{2}\sum_{\rho} (\langle 0,L_\rho|\hat{T}_{il}|0\rangle\langle 0|\hat{T}_{jm}|0,L_\rho\rangle + \langle 0,L_\rho|\hat{T}_{im}|0\rangle\langle 0,|\hat{T}_{jl}|0,L_\rho\rangle)\,. \tag{4.3.23}$$

The operator $\hat{T}_{ij}$ is given by

$$\hat{T}_{ij} = \sum_{\alpha=1}^{N} (p_i^\alpha r_j^\alpha + r_j^\alpha p_i^\alpha)\,, \tag{4.3.24}$$

where ρ is the number of the degenerate exciton state ($\boldsymbol{k} = 0$) in the L-th band, $\boldsymbol{r}^\alpha$ and $\boldsymbol{p}^\alpha$ are the coordinate and momentum operators of the α-th electron in the crystal. The quantities $\hat{T}_{ij}$ are transformed like products of the components of two polar vectors. Consequently the components of $\alpha_{ijlm}(\omega)$ are determined by the contribution only of the exciton states whose wave functions are transformed according to one of the irreducible representations A_1, E, F_1 or F_2.

It is obvious that in applying (4.3.23, 24) we go beyond the limits of purely phenomenological description. We do not, however, use the quantum-mechanical expression itself for $\varepsilon_{ij}(\omega,\boldsymbol{k})$ here, but consider only a factor of the form (4.3.23) whose appearance in this expression is to be understood from general considerations. In the case of a degenerate exciton term we shall choose the wave functions so that they are transformed, apart from a numerical factor, like the corresponding linear combinations of products of polar vector components, as shown in Table 4.2. For example, in the case of a doubly degenerate term, the wave functions Ψ_E^{II} and Ψ_E^{III} are chosen such that under symmetry transformations of group O_h they transform like $\sqrt{3}\,\mathrm{II} = \sqrt{3}(x_1x_2 - z_1z_2)$ and $\mathrm{III} = 2y_1y_2 - x_1x_2 - z_1z_2$ (the factor $\sqrt{3}$ results from normalization of the basic functions). It is clear that the scalar product

$\int \Psi_E^{\text{II}} \Psi_E^{\text{III}} d\boldsymbol{r}$ of the functions Ψ_E^{II} and Ψ_E^{III} is zero. (If $\boldsymbol{k}=0$ all functions may be regarded as being real; this was taken into account in the formulation of the scalar product; if the coordinate system is rotated through the angle $\pi/2$ about the y axis the function Ψ_E^{II} changes its sign whereas the function Ψ_E^{III} remains unchanged. The invariant quantity, i.e., the scalar product, would therefore have to change sign under this coordinate transformation and must thus be equal to zero). Analogously we can prove the mutual orthogonality of the functions $\Psi_{F_1}^{\text{IV}}$, $\Psi_{F_1}^{\text{V}}$, $\Psi_{F_1}^{\text{VI}}$, etc. In the following we shall make use of the following identities:

$$\begin{aligned}
&x_1x_2=\tfrac{1}{2}\text{II}-\tfrac{1}{6}\text{III}+\tfrac{1}{3}\text{I}\,, && y_1y_2=\tfrac{1}{3}\text{III}+\tfrac{1}{3}\text{I}\,,\\
&z_1z_2=-\tfrac{1}{2}\text{II}-\tfrac{1}{6}\text{III}+\tfrac{1}{3}\text{I}\,, && x_1y_2=\tfrac{1}{2}\text{IX}+\tfrac{1}{2}\text{VI}\,,\\
&y_1z_2=\tfrac{1}{2}\text{VII}+\tfrac{1}{2}\text{IV}\,, && y_1x_2=\tfrac{1}{2}\text{IX}-\tfrac{1}{2}\text{VI}\,,\\
&z_1y_2=\tfrac{1}{2}\text{VII}-\tfrac{1}{2}\text{IV}\,, && x_1z_2=\tfrac{1}{2}\text{VIII}-\tfrac{1}{2}\text{V}\,.
\end{aligned} \tag{4.3.25}$$

These obvious relationships express any of the products of coordinates of two polar vectors in terms of the linear combinations of these products which, according to our choice of the exciton wave functions at $\boldsymbol{k}=0$, are transformed in the same way as these wave functions (we are, of course, referring to wave functions which correspond to the representations A_1, E, F_1 and F_2).

Next, we shall calculate the components of the tensor $\alpha_{ijlm}(\omega)$, determined by the contributions of the various exciton states.

I) Let us assume, for example, that the exciton state L at $\boldsymbol{k}=0$ is transformed according to the nondegenerate representation A_1, according to which the ground state of the system is also transformed. Then, applying (4.3.24, 25) we obtain

$$\left\langle 0,L\left|\sum_{\alpha=1}^{N} r_i^\alpha p_j^\alpha\right|0\right\rangle=\frac{\delta_{ij}}{3}\left\langle 0,L\left|\sum_{\alpha=1}^{N} \boldsymbol{r}^\alpha\cdot\boldsymbol{p}^\alpha\right|0\right\rangle . \tag{4.3.26}$$

Hence, see also (3.2.13, 14),

$$\begin{aligned}
&\alpha_1\equiv\alpha^{A_1}\neq 0\,,\quad \alpha_2=0\,,\quad \alpha_3=\alpha^{A_1}/2\,,\\
&\tilde{\alpha}\equiv\alpha_1-\alpha_2-2\alpha_3=0\,,
\end{aligned} \tag{4.3.27}$$

where, according to (4.3.11),

$$(\tilde{n}^2-\varepsilon_0)\boldsymbol{E}=(1+\alpha^{A_1})\tilde{n}^2(\boldsymbol{E}\cdot\boldsymbol{s})\boldsymbol{s}\,.$$

It is clear that the excited state being considered is important only when longitudinal waves are investigated, for which $\tilde{n}^2=-\varepsilon_0/\alpha^{A_1}$ independent of the direction of $\boldsymbol{s}$, although for transverse waves $\boldsymbol{E}\cdot\boldsymbol{s}=0$ and $\tilde{n}^2=\varepsilon_0$.

II) Let us next consider the case of the doubly degenerate term E. For wave functions which correspond to the representation E we have

$$\int \Psi_E^{\mathrm{II}}(2\hat{T}_{22}-\hat{T}_{11}-\hat{T}_{33})\,\Psi_0 dr = \int \Psi_E^{\mathrm{III}}(\hat{T}_{11}-\hat{T}_{33})\,\Psi_0 dr = 0\,.$$

In accordance with the identities (4.3.25) we therefore have

$$\langle \Psi_E|\hat{T}_{ij}|\Psi_0\rangle = \delta_{ij}M_{jj}\,, \tag{4.3.28}$$

where M_{jj} is a diagonal element and not the trace, and where

$$\begin{aligned}
M_{11}(\mathrm{II}) &\equiv \langle \Psi_E^{\mathrm{II}}|\hat{T}_{11}|\Psi_0\rangle = \tfrac{1}{2}\langle \Psi_E^{\mathrm{II}}|\hat{T}_{11}-\hat{T}_{33}|\Psi_0\rangle \equiv \tfrac{1}{2}M_1\,,\\
M_{22}(\mathrm{II}) &= 0\,,\quad M_{33}(\mathrm{II}) = -\tfrac{1}{2}M_1\,,\\
M_{11}(\mathrm{III}) &\equiv \langle \Psi_E^{\mathrm{III}}|\hat{T}_{11}|\Psi_0\rangle\\
&= -\tfrac{1}{6}\langle \Psi_E^{\mathrm{III}}|2\hat{T}_{22}-\hat{T}_{11}-\hat{T}_{33}|\Psi_0\rangle \equiv -\tfrac{1}{6}M_2\,,\\
M_{22}(\mathrm{III}) &= \tfrac{1}{3}M_2\,,\quad M_{33}(\mathrm{III}) = -\tfrac{1}{6}M_2\,.
\end{aligned} \tag{4.3.29}$$

To derive (4.3.29) we made use of the relation

$$\langle \Psi_E^{\mathrm{II,III}}|\hat{T}_{11}+\hat{T}_{22}+\hat{T}_{33}|\Psi_0\rangle = 0\,,$$

which is satisfied by virtue of the fact that the operator $(\hat{T}_{11}+\hat{T}_{22}+\hat{T}_{33})$ and the functions $\Psi_E^{\mathrm{II,III}}$ are transformed according to dissimilar irreducible representations of the group O_h.

It will be readily seen that

$$M_2 = \sqrt{3}M_1\,. \tag{4.3.30}$$

As a matter of fact, after performing a g operation (i.e., rotation about the z axis through the angle $\pi/2$) under the integral sign of the expression:

$$M_1 = \langle \Psi_E^{\mathrm{II}}|\hat{T}_{11}-\hat{T}_{33}|\Psi_0\rangle\,,$$

taking into account the invariance of the integral and using the relations:

$$\begin{aligned}
&g\Psi_0 = \Psi_0\,,\\
&g(\hat{T}_{11}-\hat{T}_{33}) = (\hat{T}_{22}-\hat{T}_{33}) \equiv \tfrac{1}{2}(\hat{T}_{11}-\hat{T}_{33}) + \tfrac{1}{2}(2\hat{T}_{22}-\hat{T}_{11}-\hat{T}_{33})\,,
\end{aligned}$$

as well as

$$g\Psi_E^{\mathrm{II}} = \frac{1}{2}\Psi_E^{\mathrm{II}} + \frac{\sqrt{3}}{2}\Psi_E^{\mathrm{III}}\,,$$

we obtain

$$M_1 = \tfrac{1}{4}(M_1 + \sqrt{3}M_2)\,, \quad \text{i.e.,} \quad M_2 = \sqrt{3}M_1\,.$$

Next we apply (4.3.23), as well as (4.3.29, 30), and find that

$$\begin{aligned} \alpha_1 &\equiv \alpha^E \neq 0\,, \quad \alpha_2 = 0\,, \quad \alpha_3 = -\tfrac{1}{4}\alpha^E\,, \\ \tilde{\alpha} &\equiv \alpha_1 - \alpha_2 - 2\alpha_3 = \tfrac{3}{2}\alpha^E\,. \end{aligned} \tag{4.3.31}$$

In this case the system of equations (4.3.11) is still cumbersome:

$$\frac{\tilde{n}^2 - \varepsilon_0}{\tilde{n}^2}E_1 = \frac{3\alpha^E}{2}E_1 s_1^2 + \left(1 - \frac{\alpha^E}{2}\right)(\boldsymbol{E}\cdot\boldsymbol{s})s_1\,, \text{ etc.} \tag{4.3.32}$$

For arbitrary $\boldsymbol{s}$ the expression for $\tilde{n}^2$ will therefore be derived by means of perturbation theory. For the moment we wish to mention only a consequence of the relations (4.3.13, 17, 19 and 20). The presence of a quadrupolar exciton state of this type in the crystal has no influence on either dispersion or absorption if the light is propagated along the edge of the cube, and, conversely, it manifests itself with any polarization provided the light is propagated along the body diagonals of the cube. If the vector $\boldsymbol{s}$ is directed along the diagonal of a cube face, an exciton state of type E becomes apparent only if the electrical field vector lies in the plane of the face. When applied to absorption, these conclusions agree with those drawn in [4.43]. The problem of the anisotropy of dispersion, i.e., of the dependence of $\tilde{n}$ on $\boldsymbol{s}$ was not dealt with in [4.43].

III) Let us now turn to the case of the triply degenerate term F_1 whose wave function is transformed as a pseudovector. In this case, in agreement with our choice of the basis and also by virtue of (4.3.25),

$$\langle \Psi_{F_1} | \hat{T}_{ij} | \Psi_0 \rangle = (1 - \delta_{ij})M_{ij}\,, \tag{4.3.33}$$

where, of course, we do not sum over i and j, and

$$\begin{aligned} M_{23}(\mathrm{IV}) &\equiv \langle \Psi_{F_1}^{\mathrm{IV}} | \hat{T}_{23} | \Psi_0 \rangle \\ &= \tfrac{1}{2}\langle \Psi_{F_1}^{\mathrm{IV}} | \hat{T}_{23} - \hat{T}_{32} | \Psi_0 \rangle = -M_{32}(\mathrm{IV})\,, \\ M_{12}(\mathrm{IV}) &= M_{21}(\mathrm{IV}) = M_{13}(\mathrm{IV}) = M_{31}(\mathrm{IV}) = 0\,, \\ M_{13}(\mathrm{V}) &\equiv \langle \Psi_{F_1}^{\mathrm{V}} | \hat{T}_{13} | \Psi_0 \rangle \\ &= -\tfrac{1}{2}\langle \Psi_{F_1}^{\mathrm{V}} | \hat{T}_{31} - \hat{T}_{13} | \Psi_0 \rangle = -M_{31}(\mathrm{V})\,, \\ M_{12}(\mathrm{V}) &= M_{21}(\mathrm{V}) = M_{23}(\mathrm{V}) = M_{32}(\mathrm{V}) = 0\,, \\ M_{12}(\mathrm{VI}) &\equiv \langle \Psi_{F_1}^{\mathrm{VI}} | \hat{T}_{12} | \Psi_0 \rangle \\ &= \tfrac{1}{2}\langle \Psi_{F_1}^{\mathrm{VI}} | \hat{T}_{12} - \hat{T}_{21} | \Psi_0 \rangle = -M_{21}(\mathrm{VI})\,, \\ M_{23}(\mathrm{VI}) &= M_{32}(\mathrm{VI}) = M_{13}(\mathrm{VI}) = M_{31}(\mathrm{VI}) = 0\,. \end{aligned} \tag{4.3.34}$$

Moreover, it can readily be seen that

$$M_{23}(\mathrm{IV}) = M_{31}(\mathrm{V}) = M_{12}(\mathrm{VI}) \equiv M(F_1)\,. \tag{4.3.35}$$

Let us now prove, for example, the validity of the first of these equations. For this purpose under the integral sign

$$M_{23}(\mathrm{IV}) = \tfrac{1}{2}\langle \Psi_{F_1}^{\mathrm{IV}} | \hat{T}_{23} - \hat{T}_{32} | \Psi_0 \rangle$$

we perform a g operation (as in case II, rotation about the z axis through the angle $\pi/2$). Then again taking the invariance of the integral into account and using the relations $g\Psi_0 = \Psi_0$, $g(\hat{T}_{23} - \hat{T}_{32}) = (\hat{T}_{31} - \hat{T}_{13})$ and $g\Psi_{F_1}^{\mathrm{IV}} = \Psi_{F_1}^{\mathrm{V}}$, we directly obtain the desired relationship. Then applying (4.3.23, 33 – 35), we find that

$$\begin{aligned} &\alpha_1 = 0\,, \quad \alpha_2 = \alpha^{F_1} \neq 0\,, \quad \alpha_3 = -\alpha^{F_1}/2\,, \\ &\tilde{\alpha} = \alpha_1 - \alpha_2 - 2\alpha_3 = 0\,. \end{aligned} \tag{4.3.36}$$

Since here, according to (4.3.11),

$$(\tilde{n}^2 - \varepsilon_0)E_i = \alpha^{F_1}\tilde{n}^2 E_i + (1 - \alpha^{F_1})\tilde{n}^2(\boldsymbol{E}\cdot\boldsymbol{s})s_i\,, \qquad i = 1, 2, 3\,,$$

we arrive at the conclusion that both absorption and dispersion are completely isotropic. For transverse waves $\boldsymbol{E}\cdot\boldsymbol{s} = 0$ and

$$\tilde{n}^2 = \frac{\varepsilon_0}{1 - \alpha'^{F_1} - \mathrm{i}\,\alpha''^{F_1}}\,. \tag{4.3.37}$$

For the longitudinal wave, however, there are no solutions in the vicinity of F_1 transitions in the frequency range under investigation, where $\varepsilon_0 \neq 0$.

IV) As a final example, we shall consider the situation arising in the case where the exciton term at $\boldsymbol{k} = 0$ is triply degenerate and corresponds to the representation F_2 (Table 4.2).

By analogy with (4.3.33) we have in this case

$$\langle \Psi_{F_2} | \hat{T}_{ij} | \Psi_0 \rangle = (1 - \delta_{ij})M_{ij}\,, \tag{4.3.38}$$

where

$$\begin{aligned} M_{23}(\mathrm{VII}) &\equiv \langle \Psi_{F_2}^{\mathrm{VII}} | \hat{T}_{23} | \Psi_0 \rangle \\ &= \tfrac{1}{2}\langle \Psi_{F_2}^{\mathrm{VII}} | \hat{T}_{23} + \hat{T}_{32} | \Psi_0 \rangle = M_{32}(\mathrm{VII})\,, \\ M_{12}(\mathrm{VII}) &= M_{21}(\mathrm{VII}) = M_{13}(\mathrm{VII}) = M_{31}(\mathrm{VII}) = 0\,, \end{aligned}$$

$$\begin{aligned} M_{31}(\mathrm{VIII}) &\equiv \langle \Psi_{F_2}^{\mathrm{VIII}} | \hat{T}_{31} | \Psi_0 \rangle \\ &= \tfrac{1}{2} \langle \Psi_{F_2}^{\mathrm{VIII}} | \hat{T}_{31} + \hat{T}_{13} | \Psi_0 \rangle = M_{13}(\mathrm{VIII}) \,, \\ M_{12}(\mathrm{VIII}) &= M_{21}(\mathrm{VIII}) = M_{23}(\mathrm{VIII}) = M_{32}(\mathrm{VIII}) = 0 \,, \\ M_{12}(\mathrm{IX}) &\equiv \langle \Psi_{F_2}^{\mathrm{IX}} | \hat{T}_{12} | \Psi_0 \rangle \\ &= \tfrac{1}{2} \langle \Psi_{F_2}^{\mathrm{IX}} | \hat{T}_{12} + \hat{T}_{21} | \Psi_0 \rangle = M_{21}(\mathrm{IX}) \,, \\ M_{23}(\mathrm{IX}) &= M_{32}(\mathrm{IX}) = M_{13}(\mathrm{IX}) = M_{13}(\mathrm{IX}) = 0 \,. \end{aligned} \tag{4.3.39}$$

Moreover, as with (4.3.35) the equation

$$M_{23}(\mathrm{VII}) = M_{31}(\mathrm{VIII}) = M_{12}(\mathrm{IX}) \equiv M(F_2) \tag{4.3.40}$$

is satisfied. From this, and on the basis of (4.3.23, 24), we find that

$$\begin{aligned} &\alpha_1 = 0 \,, \quad \alpha_2 = \alpha^{F_2} \neq 0 \,, \quad \alpha_3 = \alpha^{F_2}/2 \,, \\ &\tilde{\alpha} = \alpha_1 - \alpha_2 - 2\alpha_3 = -2\alpha^{F_2} \,. \end{aligned} \tag{4.3.41}$$

Consequently, as follows from (4.3.13, 17, 19 and 20) the quadrupole transition being considered leads to an appreciable anisotropy in light absorption and dispersion. If the vector s coincides with an edge of the cube, then

$$\tilde{n}^2 = \frac{\varepsilon_0}{1 - \alpha'^{F_2} - \mathrm{i}\alpha''^{F_2}} \,,$$

independent of the polarization. If the vector s coincides with a body diagonal of the cube, then

$$\tilde{n}^2 = \frac{\varepsilon_0}{1 - (\alpha'^{F_2} + \mathrm{i}\alpha''^{F_2})/3} \,,$$

independent of the polarization. However, if s lies in a diagonal of cube face, then

$$\tilde{n}^2 = \varepsilon_0$$

if the wave is polarized in the plane of the face and, consequently, no quadrupole absorption occurs; for waves polarized perpendicularly to the plane of the face

$$\tilde{n}^2 = \frac{\varepsilon_0}{1 - \alpha'^{F_2} - \mathrm{i}\alpha''^{F_2}} \,.$$

In the general case of arbitrary s the system (4.3.11) is to be solved to determine $\tilde{n}^2$; if (4.3.41) holds, this system assumes the form

$$\frac{\tilde{n}^2-\varepsilon_0}{\tilde{n}^2}E_1 = \alpha^{F_2}E_1 - 2\alpha^{F_2}E_1 s_1^2 + (1+\alpha^{F_2})(E\cdot s)s_1 \,, \tag{4.3.42}$$

etc. It is quite cumbersome to calculate $\tilde{n}^2$ exactly from (4.3.42) taking the terms α^2, α^3, etc., into account. Generally speaking, this would correspond to an accuracy exceeding that of the initial expressions. In the approximation considered for nonlongitudinal waves it is in most cases reasonable to determine only corrections to $\tilde{n}^2$ which are linear in α. For this purpose it suffices to apply perturbation theory. Let us introduce the notation

$$L_{ij}^{(0)} = \alpha^{F_2}\delta_{ij} + (1+\alpha^{F_2})s_i s_j \,, \qquad L_{ij}^{(1)} = -2\alpha^{F_2}s_i^2\delta_{ij}$$

and rewrite the set (4.3.42) in the form

$$(L^{(0)}+L^{(1)})E = \rho E \,, \qquad \rho = (\tilde{n}^2-\varepsilon_0)/\tilde{n}^2 \,.$$

In zeroth approximation $(L^{(0)}-\rho)E = 0$, isotropy is complete and the waves may be either strictly transverse $(\rho_0^{\perp} = \alpha^{F_2})$ or strictly longitudinal $(\rho_0'' = 1+2\alpha^{F_2})$.

The correction to the zeroth approximation is

$$\rho_1 = \frac{1}{|E_0|^2}(E_0, L^{(1)}E_0) = -2\alpha^{F_2}(s_1^2e_1^2+s_2^2e_2^2+s_3^2e_3^2) \,,$$

where $e = E_0/|E_0|$ is the polarization vector of the wave. For transverse waves we thus have

$$\frac{\tilde{n}^2-\varepsilon_0}{\tilde{n}^2} = \alpha^{F_2} - 2\alpha^{F_2}s_i^2e_i^2 \qquad \text{or}$$

$$\tilde{n}_{\perp}^2 = \varepsilon_0 + \alpha^{F_2}\varepsilon_0(1-2s_i^2e_i^2) \,, \tag{4.3.43}$$

which are accurate to terms of the first order in α. For longitudinal waves $(e = \pm s)$

$$\frac{\tilde{n}^2-\varepsilon_0}{\tilde{n}^2} = 1+2\alpha^{F_2} - 2\alpha^{F_2}(s_1^4+s_2^4+s_3^4) \qquad \text{or}$$

$$\tilde{n}_{\parallel}^2 = -\frac{\varepsilon_0}{2\alpha^{F_2}(1-s_1^4-s_2^4-s_3^4)} \,.$$

The fact that $\tilde{n}^2$ becomes infinite when the vector $\boldsymbol{s}$ lies in one of the cube's edges corresponds to the absence of a longitudinal wave in the frequency range where $\varepsilon_0 \neq 0$.

It is sometimes convenient to consider the optical properties of a medium in a spherical coordinate system (Fig. 4.15) for polarizations in two mutually perpendicular directions, $\boldsymbol{e}^p$ and $\boldsymbol{e}^s$. Here, $\boldsymbol{e}^p$ corresponds to polarization in the meridian plane containing the z axis and the direction of light propagation. (The superscript s referring to the polarization $\boldsymbol{e}^s$ has nothing to do with the vector $\boldsymbol{s} = \boldsymbol{k}/k$.)

The components of the vectors $\boldsymbol{e}^p$ and $\boldsymbol{e}^s$ are determined by the relations

$$\begin{aligned} &e_1^s = \sin\varphi\,, \quad e_2^s = -\cos\varphi\,, \\ &e_3^s = 0\,, \quad e_1^p = -\cos\theta\cos\varphi\,, \\ &e_2^p = -\cos\theta\sin\varphi\,, \quad e_3^p = \sin\theta\,, \end{aligned} \tag{4.3.44}$$

whereas the components of the vector $\boldsymbol{s} = \boldsymbol{k}/k$ are evidently

$$s_1 = \sin\theta\cos\varphi\,, \quad s_2 = \sin\theta\sin\varphi\,, \quad s_3 = \cos\theta\,. \tag{4.3.45}$$

Substituting (4.3.44) into (4.3.43) we obtain for the s and p polarizations

$$\begin{aligned} &\tilde{n}_s^2 = \varepsilon_0 + \alpha^{F_2}\varepsilon_0(1 - \sin^2\theta\sin^2 2\varphi)\,, \\ &\tilde{n}_p^2 = \varepsilon_0 + \frac{\alpha^{F_2}\varepsilon_0}{4}(\sin^2 2\theta\sin^2 2\varphi + 4\cos^2 2\theta)\,. \end{aligned} \tag{4.3.46}$$

With $\tilde{n} = n + \mathrm{i}\varkappa$, we obtain

$$\begin{aligned} &n_s^2 = \varepsilon_0 + \alpha'^{F_2}\varepsilon_0(1 - \sin^2\theta\sin^2 2\varphi)\,, \\ &\varkappa_s = \tfrac{1}{2}\alpha''^{F_2}\sqrt{\varepsilon_0}(1 - \sin^2\theta\sin^2 2\varphi)\,, \end{aligned} \tag{4.3.47}$$

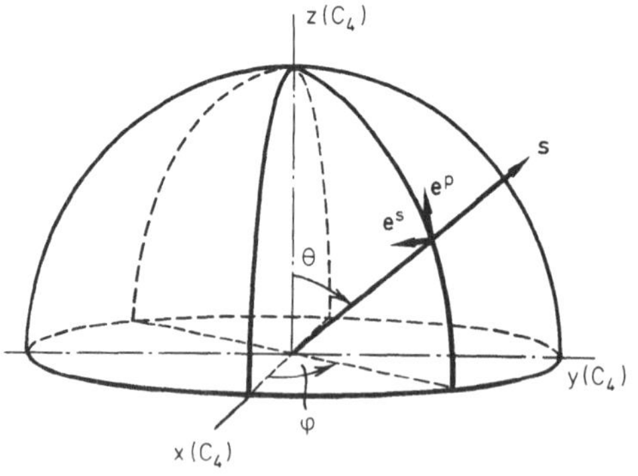

Fig. 4.15. Spherical coordinate system

$$n_p^2 = \varepsilon_0 + \frac{\alpha'^{F_2}\varepsilon_0}{4}(\sin^2 2\theta \sin^2 2\varphi + 4\cos^2 2\theta)\,,$$

$$\varkappa_p = \frac{\alpha''^{F_2}\sqrt{\varepsilon_0}}{8}(\sin^2 2\theta \sin^2 2\varphi + 4\cos^2 2\theta)\,,$$

which are accurate to terms of first order in α; these formulas refer to a level that is transformed according to the representation F_2.

Analogously, to linear order in α, we may obtain an expression for the complex refractive index $\tilde{n}$ in the case of a doubly degenerate exciton term (representation E) discussed above. For quasi-transverse waves (i.e., waves which are transverse if $\alpha = 0$) we obtain from (4.3.32):

$$\tilde{n}^2 = \varepsilon_0 + \frac{3\alpha^E\varepsilon_0}{2}(s_1^2e_1^2 + s_2^2e_2^2 + s_3^2e_3^2) \qquad \text{or} \tag{4.3.48}$$

$$\tilde{n}_s^2 = \varepsilon_0 + \frac{3\alpha^E\varepsilon_0}{4}\sin^2\theta \sin^2 2\varphi\,,$$
$$\tilde{n}_p^2 = \varepsilon_0 + \frac{3\alpha^E\varepsilon_0}{16}\sin^2 2\theta(3 + \cos^2 2\varphi)\,. \tag{4.3.49}$$

Hence

$$n_s^2 = \varepsilon_0 + \frac{3\alpha'^E\varepsilon_0}{4}\sin^2\theta \sin^2 2\varphi\,,$$

$$\varkappa_s = \frac{3\alpha''^E\sqrt{\varepsilon_0}}{8}\sin^2\theta \sin^2 2\varphi\,,$$

$$n_p^2 = \varepsilon_0 + \frac{3\alpha'^E\varepsilon_0}{16}\sin^2\theta(3 + \cos^2 2\varphi)\,, \tag{4.3.50}$$

$$\varkappa_p = \frac{3\alpha''^E\sqrt{\varepsilon_0}}{32}\sin^2 2\theta(3 + \cos^2 2\varphi)\,.$$

4.3.4 New Waves Near Quadrupole Absorption Lines. Longitudinal Waves

Up to this point, we have implicitly assumed in employing (4.3.10) that the components of the tensor α_{ijlm} depend only on the frequency, but not on the wave vector $\boldsymbol{k}$ of the light wave. As has already been shown in Sect. 3.1.2, the expansion

$$\varepsilon_{ij}(\omega, \boldsymbol{k}) = \delta_{ij}\varepsilon_0(\omega) + \alpha_{ijlm}(\omega, \boldsymbol{k})s_i s_j \tilde{n}^2 \tag{4.3.51}$$

is more general, see (3.1.24). It can, however, be shown (Sect. 6.1) that (4.3.23) remains valid in this more general case as well. This relation serves as the basis for the selection rules for quadrupole transitions (Sect. 4.3.3). This fact enables us to analyze the consequence of taking spatial dispersion into account by applying the expansion (4.3.51). As a matter of fact, since (4.3.23) is still valid, the nonzero components of $\alpha_{ijlm}(\omega, \boldsymbol{k})$ for various quadrupole transitions turn out to be the same as in the case considered in Sect. 4.3.3. Hence, all conclusions drawn as to the nature of the anisotropy near the various quadrupole transition lines remain unchanged, as do all the relations involving the quantity $\tilde{n}^2$, see equations of the form of (4.3.13, 14), etc. Since, however, in the case under consideration the relevant quantities α_1, α_2 and α_3 are themselves functions of $\tilde{n}^2$, see (3.1.24), the above-mentioned relations become equations for $\tilde{n}^2$.

In order to elucidate what we have said, let us consider a concrete example, e.g., the vicinity of a quadrupole transition of symmetry E, assuming at first that the vector $\boldsymbol{s}$ lies in one of the body diagonals of the cube. In this case $\tilde{n}^2$ will be independent of the polarization of light and satisfy (4.3.17):

$$\tilde{n}^2 = \varepsilon_0 + \tfrac{1}{2}\tilde{n}^2 \alpha^E(\omega, \boldsymbol{k}) \, . \tag{4.3.52}$$

Let us now put, in accordance with (3.1.23),

$$\alpha^E(\omega, \boldsymbol{k}) = \frac{2\nu}{(\omega - \omega_l)/\omega_l - \mu\tilde{n}^2} \equiv \frac{2\nu}{\xi - \mu\tilde{n}^2} \, . \tag{4.3.52a}$$

Then we see from (4.3.52) that in the case being considered two types of transverse waves may be propagated in the crystal; they have the same frequency and polarization but different refractive indices, $\tilde{n}_1$ and $\tilde{n}_2$:

$$\tilde{n}^2_{1,2} = \frac{1}{2}\left(\frac{\xi}{\mu} + \varepsilon_0 - \frac{\nu}{\mu}\right) \pm \frac{1}{2}\sqrt{\left(\frac{\xi}{\mu} + \varepsilon_0 - \frac{\nu}{\mu}\right)^2 - 4\varepsilon_0\frac{\xi}{\mu}} \, . \tag{4.3.53}$$

A relation of this type also applies to quadrupolar excitons belonging to the representations F_1 and F_2, except that ν and μ then assume other values.

Generally speaking, additional waves may also appear near quadrupole absorption lines [4.44]. The problem of their possible experimental detection and, in this connection, the role played by absorption, ignored in (4.3.53), will be discussed in Sect. 4.6. Here we mention only that additional solutions for $\tilde{n}^2$ may arise in the vicinity of quadrupole lines only when the waves are polarized in such a way that they are subject to quadrupole absorption. In the case of waves propagating along a face diagonal of the cube, for example, a new wave may thus arise near an exciton transition corresponding to the representation F_2 only if the vector $\boldsymbol{E}$ is perpendicular to the face (Sect. 4.3.3); for the vector $\boldsymbol{E}$ in the face plane, $\nu = 0$, see (4.3.52). Some results for the

Table 4.3. Anisotropy of absorption and dispersion in the vicinity of quadrupolar transitions in cubic crystals of class O_h

Direction of *s*	Symmetry of the quadrupolar exciton			
	A_1	E	F_1	F_2
Along an edge of the cube	This transition has no influence on the character of propagation of transverse waves	No absorption and no additional waves	Absorption and additional waves exist independently of polarization and the direction of propagation of the waves	Absorption and additional waves exist independently of the polarization
Along a body diagonal of the cube		Absorption and additional waves exist independently of polarization		Absorption and additional waves exist independently of the polarization
Along a face diagonal		Absorption and additional waves exist if the electric field **E** lies in a face plane		Absorption and additional waves exist if the electric field **E** is perpendicular to a force plane

anisotropy of the optical properties of crystals of the class O_h in the vicinity of quadrupolar absorption lines are listed in Table 4.3. This table also illustrateswhat has been said about the conditions under which additional waves may appear.

At the end of this section, we shall deal briefly with the problem of longitudinal waves which may propagate in the vicinity of certain quadrupole absorption lines. As follows from the system of equations (4.3.11), in the vicinity of the exciton states whose wave functions are transformed according to the irreducible representation A_1 (Table 4.2), longitudinal waves may be propagated independent of the direction of $\boldsymbol{s}$. In fact, for these exciton states, see (4.3.27), the quantity $\tilde{\alpha}=0$ and the system of equations (4.3.11) permit the solution $\boldsymbol{E}=E\boldsymbol{s}$, and then

$$\tilde{n}^2=-\varepsilon_0/(\alpha_2+2\alpha_3)=-\varepsilon_0/2\alpha_3=-\varepsilon_0/\alpha^{A_1}.$$

For exciton states of the symmetry F_1, $\tilde{\alpha}$ is also equal to zero, see (4.3.36), but in this case the relation $\alpha_2+2\alpha_3=0$ is satisfied at the same time. No longitudinal waves may therefore appear in this frequency range, irrespective of $\boldsymbol{s}$ [we assume $\varepsilon_0(\omega)\neq 0$]. In the vicinity of the frequency range of exciton states of symmetries E and F_2, longitudinal waves may propagate only in certain directions. For the exciton states of symmetry E, it follows from (4.3.32) that

$$\tilde{n}_\parallel^2=-\varepsilon_0/\alpha^E, \quad \text{if } \boldsymbol{s} \text{ is parallel to one of the cube's edges, and}$$

$$\tilde{n}_\parallel^2=-4\varepsilon_0/\alpha^E, \quad \text{if } \boldsymbol{s} \text{ is parallel to a face diagonal.}$$

For the exciton states of symmetry F_2, it follows from (4.3.42) that

$$\tilde{n}_\parallel^2=-\varepsilon_0/\alpha^{F_2}, \quad \text{if } \boldsymbol{s} \text{ is parallel to a face diagonal, and}$$

$$\tilde{n}_\parallel^2=-3\varepsilon_0/4\alpha^{F_2}, \quad \text{if } \boldsymbol{s} \text{ is parallel to a body diagonal of the cube.}$$

4.4 Influence of Mechanical Stresses and External Electric and Magnetic Fields

4.4.1 Anisotropy of the Optical Properties and Selection Rules in the Presence of External Influences

Many experiments [4.45 – 48] have been carried out to investigate the external influences on the features of the tensor $\varepsilon_{ij}(\omega,\boldsymbol{k})$ and on the exciton states of a crystal. In dealing with cubic crystals, one is faced with the artificial anisotropy of the optical properties of the crystal near exciton absorption lines and their dependence on the nature of deformation, directions of the magnetic or electric field, etc. A number of these problems have been studied with Cu_2O-type crystals [4.48 – 52] and with others [4.52a].

In the phenomenological crystal optics with spatial dispersion, the influences of external effects may be analyzed with the aid of the tensors in (3.1.19, 21), resulting from the study of external fields and mechanical stresses. There are no essential differences in the formulation of the problem, as compared with ordinary crystal optics accounting for external effects on the tensor $\varepsilon_{ij}(\omega)$ [4.34]. Of greatest interest to us, however, is the investigation of new features that appear when spatial dispersion is taken into account. Nevertheless, for illustration, we also deal with effects which appear even when spatial dispersion is ignored.

External static or quasi-static effects generally lower the symmetry of the crystal and, as a result, the restrictions imposed by the symmetry conditions on the components of the tensors $\varepsilon_{ij}(\omega,\boldsymbol{k},\boldsymbol{E}^{(0)},\boldsymbol{H}^{(0)},\sigma^{(0)}_{lm})$ or directly on $\varepsilon_{ij}(\omega,\boldsymbol{E}^{(0)},\boldsymbol{H}^{(0)},\sigma^{(0)}_{lm})$, $\gamma_{ijl}(\omega,\boldsymbol{E}^{(0)},\boldsymbol{H}^{(0)},\sigma^{(0)}_{lm})$, $\alpha_{ijlm}(\omega,\boldsymbol{E}^{(0)},\boldsymbol{H}^{(0)},\sigma^{(0)}_{lm})$, etc., differ from those in the case of $\boldsymbol{E}^{(0)}=\boldsymbol{H}^{(0)}=\sigma^{(0)}_{lm}=0$. [Here $\boldsymbol{E}^{0}$ and $\boldsymbol{H}^{(0)}$ are the strengths of the external electric and magnetic fields, respectively, $\sigma^{(0)}_{ij}$ is the stress tensor; the medium is assumed to be nonmagnetic so that we need not distinguish between the induction $\boldsymbol{B}^{(0)}$ and the field $\boldsymbol{H}^{(0)}$.]

Taking into account the dependence of ε_{ij} on $\boldsymbol{E}^{(0)}$, $\boldsymbol{H}^{(0)}$ and $\sigma^{(0)}_{lm}$ enables us, as in Sect. 4.3, to determine the polarization of normal electromagnetic waves and the anisotropy of dispersion and absorption in the range of exciton absorption lines in the presence of external effects. For crystals of the various classes this problem is simple to treat, but is nevertheless a cumbersome task. We shall therefore confine ourselves to a few relevant problems.

Let us consider, for example, the influence of a constant electric field on the exciton lines of a Cu_2O crystal (class O_h). We shall assume here, for simplicity, that the electric field lies in the direction of a fourfold symmetry axis. If the z axis coincides with the field direction, the operator of the field effect, accurate to first-order with respect to the field, is $\hat{H}'=-E^{(0)}\hat{P}_z$, where $\hat{\boldsymbol{P}}$ is the crystal's dipole moment operator. As we infer from the operator $\hat{H}'$, the presence of field lowers the symmetry of the crystal to the group C_{4v} whose characters of irreducible representations are given in Table 4.4. Comparing the characters of the group C_{4v} with the characters of the group O_h (Table 4.2), it is readily seen that instead of the dipole absorption lines a doublet should appear under the influence of the electric field (the representation F_1', irreducible in O_h, becomes reducible in C_{4v} so that $F_1'=A_1+E$).

According to Table 4.4, one of the components of the doublet is polarized along the field and the other perpendicular to the field, so that the crystal becomes anisotropic (uniaxial), even if spatial dispersion is ignored. Then the nonvanishing components of the tensor $\varepsilon_{ij}(\omega)$ are $\varepsilon_{11}=\varepsilon_{22}$, ε_{33}.

The quadrupole exciton states are also subject to interesting transformations[6]. Particularly, the nondegenerate exciton state whose symmetry corresponds to the representation A_1 of the group O_h assumes the symmetry

[6] A number of problems connected with the influence of a constant electric field on quadrupole exciton lines of Cu_2O may be found in [4.49].

Table 4.4. Characters of the irreducible representations of group C_{4v}

Irreducible representations	Expressions which are transformed according to the irreducible representations (x, y, z are the components of polar vector $\boldsymbol{r}$)	Symmetry operations				
		E	C_4^2	$2C_4$	$2IC_4^2$	$2IC_4$
A_1	$z, z_1z_2, x_1x_2+y_1y_2$	1	1	1	1	1
A_2	$x_1y_2-x_2y_1$	1	1	1	−1	−1
B_1	$x_1x_2-y_1y_2$	1	1	−1	1	−1
B_2	$x_1y_2+x_2y_1$	1	1	−1	−1	1
E	x, y	2	−2	0	0	0

of the representation A_1 of the group C_{4v}, in the presence of an electric field directed along the z axis, and its wave function is transformed like the z component of a polar vector. As a result, transitions to this state from the ground state (symmetry A_1) become allowed[7] in the dipole approximation (Table 4.4) and the component ε_{33} becomes complex. Independent of the amount of spatial dispersion the intensity of such lines increases with the electric field strength. It is evident that in the vicinity of these exciton lines, when the crystal behaves as a uniaxial one if spatial dispersion is ignored, we have ($\varepsilon_{33}=\varepsilon_{11}+\Delta\varepsilon_{33}$)

$$\varepsilon_{ij}(\omega,\boldsymbol{E}^{(0)})=\varepsilon_{11}(\omega,\boldsymbol{E}^{(0)})\,\delta_{ij}+\Delta\varepsilon_{33}(\omega,\boldsymbol{E}^{(0)})\,\delta_{i3}\,\delta_{j3}\,.$$

The equations for the electric field components in a light wave, in agreement with (2.2.9), therefore assume the form

$$\tilde{n}^2E_i=\varepsilon_{11}E_i+\Delta\varepsilon_{33}\delta_{i3}E_3+\tilde{n}^2(\boldsymbol{E}\cdot\boldsymbol{s})s_i\,.$$

From this it directly follows that two types of waves may be propagated in an arbitrary direction $\boldsymbol{s}$, as is usually observed with uniaxial crystals. For one of them the electric vector $\boldsymbol{E}$ is perpendicular to the plane through the z axis and the vector $\boldsymbol{s}$, where $\tilde{n}^2=n_s^2=\varepsilon_{11}$ and $\varkappa\equiv\varkappa_s=0$. For the other wave

$$\tilde{n}^2=\tilde{n}_p^2=\varepsilon_{11}\frac{1+\Delta\varepsilon_{33}/\varepsilon_{11}}{1+\Delta\varepsilon_{33}\cos^2\theta/\varepsilon_{11}}\,.$$

Introducing the notation $\Delta\varepsilon_{33}=\Delta\varepsilon_{33}'+\mathrm{i}\Delta\varepsilon_{33}''$ we obtain

$$n_p^2=\varepsilon_{11}+\Delta\varepsilon_{33}'\sin^2\theta\,,\qquad \varkappa_p=\frac{\Delta\varepsilon_{33}''}{2\sqrt{\varepsilon_{11}}}\sin^2\theta\,.$$

[7] The possibility of a "flare-up" of exciton lines in Cu_2O in the presence of an eletric field has been indicated in [4.52].

It has been taken into account that $\varkappa_p^2 \ll n_p^2$ if the external fields $E^{(0)}$ are weak. None of these formulas contains the factor k^2, and hence it is obvious that reference is made to dipole transitions.

An analogous situation ("flare-up") is encountered in the case of quadrupole lines, whose wave functions in the absence of a field have the symmetry of the representations E, F_1 and F_2 because these terms are split by an electric field (Appendix B)

$$E \to A_1 + B_1, \quad F_1 \to E + A_2, \quad F \to E + B_2.$$

At the same time the transitions to exciton states of the symmetry of representations A_1 and E from the crystal's ground state (representation A_1) are allowed in the dipole approximation, as can be seen from Table 4.4.

The transitions to exciton states having the symmetry of the irreducible representations B_1, B_2 and A_2 of the group C_{4v} in the presence of an electric field, remain quadrupole transitions and the method discussed in Sect. 4.3 is sufficient for an analysis of the corresponding anisotropy of the optical properties.

In the group C_{4v}, quantities of the type $x_1y_2 - x_2y_1$ are transformed according to the representation A_2, quantities of the type $x_1x_2 - y_1y_2$ according to B_1 and quantities of the type $x_1y_2 + x_2y_1$ according to B_2. Hence, using (4.3.23), it can readily be seen that in the vicinity of exciton lines of the type A_2 only the following components of the tensor α_{ijlm} will differ from zero:

$$\alpha_{1122} = \alpha_{2211} \equiv \alpha(A_2), \quad \alpha_{1212} = \alpha_{2121} = \alpha_{2112} = \alpha_{1221} = -\alpha(A_2)/2;$$

in the vicinity of exciton lines of type B_1 we have

$$\alpha_{1111} = \alpha_{2222} \equiv \alpha(B_1); \quad \alpha_{1212} = \alpha_{2121} = \alpha_{1221} = \alpha_{2112} = -\alpha(B_1)/2$$

and, in the vicinity of the B_2 lines, the components

$$\alpha_{1122} = \alpha_{2211} = \alpha(B_2), \quad \alpha_{1212} = \alpha_{2112} = \alpha_{2121} = \alpha_{1221} = \alpha(B_2)/2.$$

A knowledge of these nonzero components of the tensor α_{ijlm} enables us to carry out a complete analysis of the anisotropy of the optical properties of crystals in the vicinity of lines of the types A_2, B_1 and B_2. Since the corresponding calculations are analogous to those made in Sect. 4.3.3, we shall give only the results obtained using perturbation theory (Sect. 4.3.4).

For quasi-transverse waves (i.e., waves which are transverse if $\alpha_{ijlm} = 0$) we have the following formulas

$$\tilde{n}^2 = \varepsilon_0 + \varepsilon_0 \alpha(A_2) [\boldsymbol{e} \cdot \boldsymbol{s}]_z^2 \quad \text{for transitions of the type } A_2,$$

$$\tilde{n}^2 = \varepsilon_0 + \varepsilon_0 \alpha(B_1)(e_x s_x - e_y s_y)^2 \quad \text{for transitions of the type } B_1,$$

$$\tilde{n}^2 = \varepsilon_0 + \varepsilon_0 \alpha(B_2)(e_x s_y + e_y s_x)^2 \quad \text{for transitions of the type } B_2.$$

Hence, it follows that:

I) For transitions of type A_2:

$$n_s^2=\varepsilon_0+\varepsilon_0\alpha'(A_2)\sin^2\theta\,,\qquad \varkappa_s=\tfrac{1}{2}\sqrt{\varepsilon_0}\,\alpha''(A_2)\sin^2\theta\,,$$

$$n_p^2=\varepsilon_0\,,\qquad \varkappa_p=0\,.$$

II) For transitions of type B_1:

$$n_s^2=\varepsilon_0+\varepsilon_0\alpha'(B_1)\sin^2\theta\sin^2 2\varphi\,,$$

$$\varkappa_s=\tfrac{1}{2}\sqrt{\varepsilon_0}\,\alpha''(B_1)\sin^2\theta\sin^2 2\varphi\,,$$

$$n_p^2=\varepsilon_0+\tfrac{1}{4}\varepsilon_0\alpha'(B_1)\sin^2 2\theta\cos^2 2\varphi\,,$$

$$\varkappa_p=\tfrac{1}{8}\sqrt{\varepsilon_0}\,\alpha''(B_1)\sin^2 2\theta\cos^2 2\varphi\,.$$

III) For transitions of type B_2:

$$n_s^2=\varepsilon_0+\varepsilon_0\alpha'(B_2)\sin^2\theta\cos^2 2\varphi\,,$$

$$\varkappa_s=\tfrac{1}{2}\sqrt{\varepsilon_0}\,\alpha''(B_2)\sin^2\theta\cos^2 2\varphi\,,$$

$$n_p^2=\varepsilon_0+\tfrac{1}{4}\varepsilon_0\alpha'(B_2)\sin^2 2\theta\sin^2 2\varphi\,,$$

$$\varkappa_p=\tfrac{1}{8}\sqrt{\varepsilon_0}\,\alpha''(B_2)\sin^2 2\theta\sin^2 2\varphi\,.$$

In the above expressions $\alpha=\alpha'+\mathrm{i}\alpha''$ and the argument of α [e.g., $\alpha(A_2)$] indicates the state to which the value of α corresponds. The choice of the angles and directions can be seen from Fig. 4.15. The same method can be used to study effects due to the influence of a constant magnetic field or mechanical stresses. In these cases we make use of the following general principle in order to determine a new group of crystal symmetry: crystals which are subject to external influences possess only the symmetry elements that are common for the Hamiltonian of the crystal without external influences and the part of the Hamiltonian that depends on the influences (magnetic field, stresses, etc.).

4.4.2 The Explicit Dependence of the Tensor $\varepsilon_{ij}(\omega,k)$ on the Strength of Weak External Fields

So far we have not considered the problem of the explicit dependence of optical effects on the magnitudes of the applied external influences. If the external effect is sufficiently weak, the explicit dependence on the external influences can be expressed by expanding the tensors in (3.1.19 – 22) in a power series in the quantities $E^{(0)}$, $H^{(0)}$, $\sigma_{ij}^{(0)}$ in the same way as is done in ordinary crystal optics [4.34]. Since accounting for spatial dispersion leads to certain special features, we shall discuss some of these, confining the consideration to the effects caused by external electric and magnetic fields and to those due to both types of fields simultaneously.

In the presence of weak external electric and magnetic fields, (3.1.19) should be replaced by the expansion

$$\begin{aligned}\varepsilon_{ij}(\omega,\boldsymbol{k},\boldsymbol{E}^{(0)},\boldsymbol{H}^{(0)}) = {} & \varepsilon_{0ij}(\omega)+\mathrm{i}\gamma_{ijl}(\omega)k_l+\alpha_{ijlm}(\omega)k_lk_m \\ & +A_{ijl}(\omega)E_l^{(0)}+A_{ijl}^{(1)}(\omega)H_l^{(0)}+A_{ijlm}^{(1)}(\omega)H_l^{(0)}k_m \\ & +A_{ijlm}^{(2)}(\omega)H_l^{(0)}E_m^{(0)}+A_{ijlm}^{(3)}(\omega)E_l^{(0)}k_m+A_{ijlm}^{(4)}(\omega)H_l^{(0)}H_m^{(0)} \\ & +A_{ijlm}^{(5)}(\omega)E_l^{(0)}E_m^{(0)}+A_{ijlmn}(\omega)E_l^{(0)}H_m^{(0)}k_n+\dots,\end{aligned} \tag{4.4.1}$$

where $\boldsymbol{H}^{(0)}$ is an axial vector, and $\boldsymbol{k}$ and $\boldsymbol{E}^{(0)}$ are polar vectors, A_{ijl}, $A_{ijlm}^{(3)}$, $A_{ijlm}^{(4)}$ and $A_{ijlm}^{(5)}$ are ordinary tensors and $A_{ijl}^{(1)}$, $A_{ijlm}^{(1)}$, $A_{ijlm}^{(2)}$ and A_{ijlmn} are pseudotensors.

The principle of symmetry of the kinetic coefficients requires, see (2.1.21), that the tensor (4.4.1) satisfies the following relation (as already mentioned, we do not distinguish between $\boldsymbol{B}^{(0)}$ and $\boldsymbol{H}^{(0)}$ since the medium is assumed to be nonmagnetic; for gyrotropy tensor in magnetic medium see Sect. 3.2.1 and [4.53]):

$$\varepsilon_{ij}(\omega,\boldsymbol{k},\boldsymbol{E}^{(0)},\boldsymbol{H}^{(0)})=\varepsilon_{ji}(\omega,-\boldsymbol{k},\boldsymbol{E}^{(0)},-\boldsymbol{H}^{(0)}). \tag{4.4.2}$$

Hence, it follows that

$$\begin{aligned} &\gamma_{ijl}=-\gamma_{jil}, \quad A_{ijl}=A_{jil}, \quad A_{ijl}^{(1)}=-A_{jil}^{(1)}, \\ &A_{ijlm}^{(2)}=-A_{jilm}^{(2)}, \quad A_{ijlm}^{(3)}=-A_{jilm}^{(3)}, \\ &A_{ijlm}^{(1)}=A_{jilm}^{(1)}, \quad A_{ijlm}^{(4)}=A_{jilm}^{(4)}, \quad A_{ijlm}^{(5)}=A_{jilm}^{(5)}, \\ &A_{ijlmn}=A_{jilmn}. \end{aligned} \tag{4.4.2a}$$

The condition of no absorption necessitates that the dielectric-constant tensor be Hermitian: $\varepsilon_{ij}=\varepsilon_{ji}^*$. In this case the tensors in (4.4.1) which are antisymmetric with respect to ij are purely imaginary. This fact is taken in account for the term with γ_{ijl} only. Obviously all these terms, as does the term containing γ_{ijl}, lead to the rotation of the plarization plane.

Note also that the pseudotensor $A_{ijl}^{(1)}$ (and also $B_{ijl}^{(1)}$ in (4.4.4)) stimulates a Faraday effect in the crystal optics:

$$A_{ijl}^{(1)}=e_{ijm}A_{ml}, \quad B_{ijl}^{(1)}=e_{ijm}B_{ml}, \tag{4.4.3}$$

where A_{ml} and B_{ml} are tensors of rank two and, in general, nonsymmetric; as always e_{ijm} is a completely antisymmetric unit pseudotensor or rank three.

The symmetry of the crystals considerably reduces the number of independent components of the tensors occurring in (4.4.1). The components of these tensors should be invariant with repect to substitutions of the coordinate system corresponding to any of the crystal symmetry operations. If follows, in particular, that in crystals possessing an inversion center the tensors of rank three vanish: $\gamma_{ijl} = A_{ijl} = 0$. Analogously we arrive at the conclusion that for these crystals the pseudotensors $A^{(1)}_{ijlm}$ and $A^{(2)}_{ijlm}$ of rank four also vanish since they change sign in an inversion. A relation like (4.4.1) can also be written for the inverse dielectric-constant tensor $\varepsilon^{-1}_{ij}(\omega, \boldsymbol{k}, \boldsymbol{E}^{(0)}, \boldsymbol{H}^{(0)})$:

$$\begin{aligned}\varepsilon^{-1}_{ij}(\omega, \boldsymbol{k}, \boldsymbol{E}^{(0)}, \boldsymbol{H}^{(0)}) = {} & \varepsilon^{-1}_{0ij}(\omega) + \mathrm{i}\delta_{ijl}(\omega)k_l + \beta_{ijlm}(\omega)k_l k_m \\ & + B_{ijl}(\omega)E^{(0)}_l + B^{(1)}_{ijl}(\omega)H^{(0)}_l + B^{(1)}_{ijlm}(\omega)H^{(0)}_l k_m \\ & + B^{(2)}_{ijlm}(\omega)H^{(0)}_l E^{(0)}_m + B^{(3)}_{ijlm}(\omega)E^{(0)}_l k_m + B^{(4)}_{ijlm}(\omega)H^{(0)}_l H^{(0)}_m \\ & + B^{(5)}_{ijlm}(\omega)E^{(0)}_l E^{(0)}_m + B_{ijlmn}H^{(0)}_l k_m E^{(0)}_n \dots \end{aligned} \qquad (4.4.4)$$

The terms of the form $A^{(1)}_{ijlm}(\omega)H^{(0)}_l k_m$, $B^{(1)}_{ijlm}(\omega)H^{(0)}_l k_m$ in (4.4.1, 4) corresponds to so-called magnetic-field-induced spatial dispersion.

For conducting media, terms of the type $B^{(6)}_{ijl}I_l$ and $B^{(7)}_{ijlm}I_l k_m$ should be taken into account in (4.4.1, 4), where $\boldsymbol{I}$ is the current vector. It follows from the principle of symmetry of the kinetic coefficients that

$$B^{(6)}_{ijl} = -B^{(6)}_{jil}, \quad B^{(7)}_{ijlm} = B^{(7)}_{jilm}.$$

Thus, the tensor $B^{(6)}_{ijl}$ is nonzero only in gyrotropic media [4.54]; the calculation of $B^{(7)}_{ijlm}$ can be found in [4.55, 56]. The terms containing $B^{(6)}_{ijl}$ lead to the rotation of the polarization plane of light (the electrical analog of the Faraday effect). This effect was first observed in tellurium [4.57].

All that has been said above on the limitations connected with the principle of symmetry of the kinetic coefficients and the crystal remains valid. Moreover, the following should be noted. Expansions of the type (4.4.1 or 4) are valid only if the coefficients $A_{ij\dots}(\omega)$ and $B_{ij\dots}(\omega)$ are small when the tensors ε_{ij} and ε^{-1}_{ij} are analytic functions of the external electric and magnetic field strengths. It is otherwise impossible to limit ourselves to terms of lowest order in $\boldsymbol{E}^{(0)}$ or $\boldsymbol{H}^{(0)}$. This is precisely the situation that may be encountered in the vicinity of a degenerate term in the case where the degeneracy is removed by an external field. This may occur, for instance, in cubic crystals possessing a symmetry center in the vicinity of a degenerate level, for which a linear Stark effect may occur[8]. In the following it is assumed that the above expansions apply.

When the tensors

$$\varepsilon_{ij}(\omega, \boldsymbol{k}, \boldsymbol{E}^{(0)}, \dots) \quad \text{or} \quad \varepsilon^{-1}_{ij}(\omega, \boldsymbol{k}, \boldsymbol{E}^{(0)}, \dots)$$

cannot be simply expanded in a series in the appropriate variable, there are usually no essential singular points, and these tensors can be represented as ratios of polynomials [for an expansion with respect to $\boldsymbol{k}$ (see, e.g. (3.1.23)]. In the case of the given example we should evidently begin with the relation

$$\varepsilon(\omega,\boldsymbol{E}^{(0)})=\frac{[\omega^2-\omega_l^2(0)]^2-2a[\omega^2-\omega_l^2(0)-\mu^2|\boldsymbol{E}^{(0)}|^2]}{[\omega^2-\omega_l^2(0)]^2-\mu^2|\boldsymbol{E}^{(0)}|^2}.$$

The situation is more complicated in the case of a semiconductor in a strong external magnetic field, when cyclotron resonance can be observed. Here, for instance, we can assume for the initial cubic crystal

$$\varepsilon_{ij}(\omega,\boldsymbol{k},\boldsymbol{B}_{\text{ext}})=\varepsilon\delta_{ij}, \quad \text{but} \quad \varepsilon=\varepsilon(\omega,k,\boldsymbol{k}\cdot\boldsymbol{B}_{\text{ext}}/k),$$

the dependence of ε on $\boldsymbol{B}_{\text{ext}}$ being of square-root type near the resonance. The dependence of ε on $\boldsymbol{k}\cdot\boldsymbol{B}_{\text{ext}}/k$, i.e., on the angle between the directions of wave propagation and the external magnetic field, results in noncoincidence of the directions of the phase and group velocities [4.58].

Let us now consider the changes in or the appearance of gyrotropy resulting from the influence of the external electric field $\boldsymbol{E}^{(0)}$. To lowest order in $\boldsymbol{E}^{(0)}$, this effect is described by the term $A^{(3)}_{ijlm}(\omega)E^{(0)}_l k_m$ in the expansion (4.4.1). It is convenient to consider the effect of the electric field on the gyrotropy on the basis of the expressions, see (3.2.5)

$$\begin{aligned}
\varepsilon_{ij}(\omega,\boldsymbol{k},\boldsymbol{E}^{(0)})&=\varepsilon_{ij}(\omega,\boldsymbol{E}^{(0)})+\mathrm{i}\gamma_{ijl}(\omega,\boldsymbol{E}^{(0)})k_l,\\
\gamma_{ijl}(\omega,\boldsymbol{E}^{(0)})&=e_{ijm}g'_{ml}(\omega,\boldsymbol{E}^{(0)}),\\
g'_{ml}(\omega,\boldsymbol{E}^{(0)})&=g'_{ml}(\omega)+h^{(1)}_{mlr}(\omega)E^{(0)}_r+h^{(2)}_{mlrs}(\omega)E^{(0)}_r E^{(0)}_s,
\end{aligned} \tag{4.4.5}$$

where use is made, obviously, of the expansion of the pseudotensor $g'_{ml}(\omega,\boldsymbol{E}^{(0)})$ in a series with respect to the field, limiting the expansion to quadratic terms. Analogous expressions can be written for the tensors ε^{-1}_{ij}, δ_{ijl} and f'_{ml} (Sect. 3.2.1). The effect of the electric field on gyrotropy has already been investigated in [4.59] and references therein. Gyrotropy can also appear or be changed by the effect of stress $\sigma^{(0)}_{ij}$. In this case it proves expedient to employ the tensor

$$\gamma_{ijl}(\omega,\sigma^{(0)}_{rs})=e_{ijm}g'_{ml}(\sigma^{(0)}_{rs})$$

---

8 To elucidate the aforesaid, consider a medium for which, if $\boldsymbol{E}^{(0)}=0$, then $\varepsilon=1-2a\times[\omega^2-\omega_l^2(0)]^{-1}$ and, in the presence of a field, $\omega_l=\omega_l(0)\pm\mu|\boldsymbol{E}^{(0)}|/2\omega_l(0)$. Then in a first approximation

$$\varepsilon(\boldsymbol{E}^{(0)})=1-a[\omega^2-\omega_l^2(0)+\mu|\boldsymbol{E}^{(0)}|]^{-1}-a[\omega^2-\omega_l^2(0)-\mu|\boldsymbol{E}^{(0)}|]^{-1}. \qquad \text{Hence}$$

$$\varepsilon^{-1}(\omega,\boldsymbol{E}^{(0)})\approx 1+\frac{2a}{\omega^2-\omega_l^2(0)+2a}+\frac{2a\mu^2|\boldsymbol{E}^{(0)}|^2}{[\omega^2-\omega_l^2(0)-2a]^2[\omega^2-\omega_l^2(0)]}.$$

It is obvious that in this expansion the term containing $|\boldsymbol{E}^{(0)}|^2$ increases without limit, as ω tends to $\omega_l(0)$.

and its expansion into a series with respect to $\sigma_{rs}^{(0)}$. This effect (the appearance of gyrotropy as an effect of stresses) was investigated in [4.60], using the crystal CuCl as an example.

The Magnetic-Field Inversion Effect

As has already been shown in Sect. 2.2.2, a knowledge of the dielectric tensor makes it possible to determine the $\boldsymbol{k}$ dependence of the natural frequencies of the Coulomb problem which, disregarding absorption and spatial dispersion, correspond to poles of $n^2(\omega)$ and, therefore, determine the position of the lines in the absorption spectrum. If absorption and spatial dispersion are taken into account, the absorption lines becomes blurred and form more or less broad bands. If, however, the absorption linewidth is substantially narrower than the width of the exciton band (Sect. 6.2.1), then the line position in the spectrum is approximately determined by the value of the natural frequency of the Coulomb problem taken for a value of $\boldsymbol{k}$ which corresponds to the wave vector $\boldsymbol{k}_0 = 2\pi \boldsymbol{s}/\lambda_0$ of light in vacuo (for details, see Sect. 6.3). For the sake of simplicity, we do not consider effects of other resonances.

In the presence of the external influences, the dependence of the frequencies of 'fictitious' longitudinal waves and the frequencies of the 'polarization waves' on the wave vector, according to (2.2.37, 41) can be obtained from

$$\varepsilon_{ij}(\omega, \boldsymbol{k}, \boldsymbol{E}^{(0)}, \boldsymbol{H}^{(0)}) s_i s_j = 0 , \quad \boldsymbol{E} \neq 0 \tag{4.4.6}$$

and

$$|\omega_{ij}^{-1}(\omega, \boldsymbol{k}, \boldsymbol{E}^{(0)}, \boldsymbol{H}^{(0)}| = 0 , \quad \boldsymbol{k} \cdot \boldsymbol{D} = 0 , \quad \boldsymbol{D} \neq 0 , \tag{4.4.7}$$

respectively.

If only the external magnetic field is nonzero ($\boldsymbol{H}^{(0)} \neq 0, \boldsymbol{E}^{(0)} = 0$), then in the crystals possessing an inversion center, since $A_{ijlm}^{(1)} = B_{ijlm}^{(1)} = 0$ (recall that these pseudotensors change sign under inversion and are therefore equal to zero), spatial dispersion, as follows from (4.4.1, 4) is only a second-order effect with respect to $\boldsymbol{k}$, whereas for crystals without an inversion center the expansions (4.4.1, 4) also contain components which are linear in $\boldsymbol{k}$. The presence of components linear in $\boldsymbol{H}^{(0)}$ in (4.4.1, 4) leads, in general, to the effect of 'magnetic field inversion' since

$$\varepsilon_{ij}(\omega, \boldsymbol{k}, \boldsymbol{H}^{(0)}) \neq \varepsilon_{ij}(\omega, \boldsymbol{k}, -\boldsymbol{H}^{(0)}) .$$

This possibility, however, is also clear from (4.4.2). The effect of magnetic field inversion manifests itself in the fact that the frequencies of the Coulomb problem and, consequently, the position of the absorption lines are changed. Since, however, the results of investigating optical effects involving the terms $A_{ijl}^{(1)} H_l^{(0)}$ or $B_{ijl}^{(1)} H_l^{(0)}$ are well known [Ref. 4.53, § 82] we shall deal here more thoroughly with the role played by the components in (4.4.1, 4), which are affected by spatial dispersion.

For this purpose we represent the pseudotensors $A^{(1)}_{ijlm}$ and $B^{(1)}_{ijlm}$ as sums

$$A^{(1)}_{ijlm} = A^{(1)s}_{ijlm} + A^{(1)a}_{ijlm} \tag{4.4.8a}$$

$$B^{(1)}_{ijlm} = B^{(1)s}_{ijlm} + B^{(1)a}_{ijlm}, \tag{4.4.8b}$$

where

$$A^{(1)s}_{ijlm} = A^{(1)s}_{ijml}, \quad A^{(1)a}_{ijlm} = -A^{(1)a}_{ijml}, \quad \text{etc.}$$

The symmetry properties of pseudotensors $A^{(1)s}_{ijlm}$ and $B^{(1)s}_{ijlm}$, with respect to rotation and permutation of the subscripts, are completely analogous to the tensor α_{ijlm} (see Table 3.2 for the crystal classes without an inversion center).

The pseudotensors $A^{(1)a}_{ijlm}$ and $B^{(1)a}_{ijlm}$ can be represented in the following form

$$A^{(1)a}_{ijlm} = e_{lmn} C_{nij}, \tag{4.4.9a}$$

$$B^{(1)a}_{ijlm} = e_{lmn} D^{(1)}_{nij}. \tag{4.4.9b}$$

Since the completely antisymmetric unit matrix e_{lmn} is a pseudotensor, C_{nij} and $D^{(1)}_{nij}$ are ordinary tensors of rank three, $C_{nij} = C_{nji}$ and $D^{(1)}_{nij} = D^{(1)}_{nji}$. Consequently, the symmetry properties of the tensors C_{nij} and $D^{(1)}_{nij}$ are exactly the same as those of the tensor characterizing the magnitude of the piezoelectric effect [Ref. 4.34, Chap. 7]. In view of the fact that the nonvanishing components of a tensor of this type for various crystal classes have been indicated in [4.34], we shall not go into further details here.

Making use of (4.4.8, 9), we obtain

$$A^{(1)}_{ijlm} H^{(0)}_l k_m = C_{nij} (\boldsymbol{H}^{(0)} \times \boldsymbol{k})_n + A^{(1)s}_{ijlm} H^{(0)}_l k_m, \tag{4.4.10a}$$

$$B^{(1)}_{ijlm} H^{(0)}_l k_m = D^{(1)}_{nij} (\boldsymbol{H}^{(0)} \times \boldsymbol{k})_n + B^{(1)s}_{ijlm} H^{(0)}_l k_m. \tag{4.4.10b}$$

Next we consider the influence of an external magnetic field on the frequencies of 'fictitious' longitudinal waves. For this purpose we substitute (4.4.1) into (4.4.6) with $\boldsymbol{E}^{(0)} = 0$, and arrive at the conclusion that the frequencies sought will satisfy

$$\begin{aligned} &s_i \varepsilon_{0ij}(\omega) s_j + s_i s_j \alpha_{ijlm} k_l k_m + \boldsymbol{C} \cdot (\boldsymbol{H}^{(0)} \times \boldsymbol{k}) \\ &\quad + s_i s_j A^{(1)s}_{ijlm} H^{(0)}_l k_m + s_i s_j A^{(4)}_{ijlm} H^{(0)}_l H^{(0)}_m = 0, \end{aligned} \tag{4.4.11}$$

where the components of the vector $\boldsymbol{C}$ are

$$C_n = C_{nij} s_i s_j, \quad n = 1, 2, 3. \tag{4.4.12}$$

Let us confine ourselves to the vicinity of one of the resonances and set

$$\varepsilon_{0ij}(\omega) = \varepsilon^0_{0ij} + g_{ij}/[\omega^2 - \omega_L^2(0)] , \tag{4.4.13}$$

where $\hbar\omega_L(0)$ denotes the energy of a 'mechanical' exciton in the L-th band at $\boldsymbol{k} = 0$. Now substituting (4.4.13) into (4.4.11) and restricting ourselves to only the components linear in $\boldsymbol{H}^{(0)}$ we obtain for the frequency of the 'fictitious' longitudinal wave (i.e., for one of the Coulomb excitons)

$$\omega(\boldsymbol{k},\boldsymbol{H}^{(0)}) = \omega_L(0) - \frac{g(s)}{2\,\omega_L(0)\,\varepsilon_0^2(s)}[\varepsilon_0(s) - \boldsymbol{C}\cdot(\boldsymbol{H}^{(0)}\times\boldsymbol{k})]$$
$$\times A^{(1)s}_{ijlm} s_i s_j H^{(0)}_l k_m - \alpha_{ijlm} s_i s_j k_l k_m] , \tag{4.4.14}$$

where $g(s) = g_{ij}s_is_j$, $\varepsilon_0(s) = \varepsilon^0_{0ij}s_is_j$. It follows from (4.4.14) that $\omega(\boldsymbol{k},\boldsymbol{H}^{(0)}) \neq \omega(\boldsymbol{k}, -\boldsymbol{H}^{(0)})$ where this effect (the effect of "magnetic field inversion") is brought about by taking spatial dispersion into account. The point is that in obtaining (4.4.14), the components which are proportional to the tensor $A^{(1)}_{ijl}$, see (4.4.1), were dropped because $s_is_jA^{(1)}_{ijl}H^{(0)} = 0$ by virtue of (4.4.4).

4.4.3 Influence of Magnetic and Electric Fields on Cadmium Sulfide (CdS) Crystals

Let us now consider in greater detail the influence of a magnetic field inversion on the frequency of 'fictitious' longitudinal waves in CdS-type crystals (space group C^4_{6v}). The characters of the point group C_{6v} are given in Table 4.5 (using the notation of [4.42]). It follows from this table that the states of the 'mechanical' excitons at $\boldsymbol{k} = 0$, which can be excited in the dipole approximation from the ground state by light, possess the symmetry of the representations A_1 and E_1 of the group C_{6v}. Therefore, if at $\boldsymbol{k} = 0$ the wave functions of an exciton in the L-th band are transformed according to the representation A_1, then $g_{ij} = g_{A_1}\delta_{i3}\delta_{j3}$; if, however, they are transformed according to E_1, then $g_{ij} = g_{E_2}(\delta_{i1}\delta_{j1} + \delta_{i2}\delta_{j2})$. For crystals belonging to class C_{6v} the non-vanishing components of the tensor C_{nij} are the following [4.34]

$$C_{333}, \quad C_{223} = C_{232} = C_{113} = C_{131}, \quad C_{311} = C_{322}. \tag{4.4.15}$$

As follows from (4.4.12) the components of the vector $\boldsymbol{C}$ are

$$C_1 = 2C_{113}s_1s_3, \quad C_2 = 2C_{223}s_2s_3, \quad C_3 = C_{311}(s_1^2+s_2^2) + C_{333}s_3^2. \tag{4.4.16}$$

Thus the vector $\boldsymbol{C}$, like the gyration vector $f'_i = f'_{ij}s_j$ (Sect. 3.2.1), is therefore a function of the direction of the vector $\boldsymbol{s}$ and, in general, is not directed along an optical axis. Its direction is along an optical axis only when the vector $\boldsymbol{s} = \boldsymbol{k}/k$ is either perpendicular to the optical axis or parallel to it. Let us also

Table 4.5. Characters of irreducible representations of group C_{6v}

Irreducible representations	Symmetry operations					
	E	C_2	$2C_3$	$2C_6$	$3\sigma_v$	$3\sigma_{v'}$
A_1; z	1	1	1	1	1	1
A_2	1	1	1	1	−1	−1
B_2	1	−1	1	−1	1	−1
B_1	1	−1	1	−1	−1	1
E_2	2	2	−1	−1	0	0
E_1; x, y	2	−2	−1	1	0	0

note that for various exciton terms certain of the components of (4.4.15) may vanish. To clarify this problem we need an appropriate theory which would allow us to express the components of the tensor C_{nij} in terms of the wave functions of 'mechanical' excitons at $\boldsymbol{k}=0$. Since this theory is only under development, we confine ourselves to one remark. For the exciton terms for which $C_{113}=C_{223}=0$, the vector $\boldsymbol{C}$ lies in the optical axis, for any direction of $\boldsymbol{s}$. If, however, $C_{311}=C_{333}=0$, the vector $\boldsymbol{C}$ is perpendicular to the optical axis, irrespective of the direction of $\boldsymbol{s}$.

Let us now analyze the term of (4.4.15) that is associated with the pseudotensor $A^{(1)s}_{ijlm}$. For this purpose we determine the nonzero components of the pseudotensor $A^{(1)s}_{ijlm}$ for crystals of the class C_{6v}. In determining the additional relations imposed on the components of $A^{(1)s}_{ijlm}$ by the symmetry properties, we should recall that upon any rotation the pseudotensor $A^{(1)s}_{ijlm}$ behaves like a true tensor. Hence, we can make use of the limitations imposed on $A^{(1)s}_{ijlm}$ by the presence of a sixfold symmetry axis. Since $A^{(1)s}_{ijlm}$ is similar to the tensor α_{ijlm} with regard to the permutation of its subscripts, the nonzero components of $A^{(1)s}_{ijlm}$ can be found only among the components having the same subscripts as the nonzero components of α_{ijlm} for the crystal class C_6. The final choice of the nonzero components of $A^{(1)s}_{ijlm}$ is, however, based on the fact that in the class C_{6v}, these components correspond to the nonzero components of α_{ijlm} for the class C_6 which change sign upon reflection from the symmetry planes σ_v, and $\sigma_{v'}$ of the group C_{6v}. This is associated with the fact that, in this case, the components of the pseudotensor $A^{(1)s}_{ijlm}$ remain unchanged.

On the basis of the above considerations we assume that the x and y axes lie in the orthogonal symmetry planes σ_v and $\sigma_{v'}$. Using Table 3.2 we find that the pseudotensor $A^{(1)s}_{ijlm}$ possesses the following nonzero components:

$$A^{(1)s}_{1112}=A^{(1)s}_{1222}=-A^{(1)s}_{2212}=-A^{(1)s}_{1211}\,,\qquad A^{(1)s}_{2313}=-A^{(1)s}_{1323}\,. \tag{4.4.17}$$

Therefore

$$A^{(1)s}_{ijlm}s_is_jH^{(0)}_lk_m=-\left[A^{(1)s}_{1112}(s_3^2-1)+2A^{(1)s}_{2313}s_3^2\right](\boldsymbol{k}\times\boldsymbol{H}^{(0)})_3\,. \tag{4.4.18}$$

From (4.4.14, 18) it follows, in particular, that the terms linear with respect to $\boldsymbol{H}^{(0)}$ in the energy equation of the Coulomb exciton vanish if the vector $\boldsymbol{s}$ lies in the direction of the magnetic field. This effect was observed in [4.61] for CdS, in [4.61a] for $LiIO_3$ and recently in [4.61b]. Investigated in the latter was the transmission of light in CdS and CdSe in the region of polariton branch intersection, where this effect is dramatically enhanced.

In order to interpret the effect of magnetic field inversion, it has been proposed in [4.61] that the presence of an external magnetic field should lead to the appearance of the additional term $(\hbar/cm_{\text{exc}})[\boldsymbol{d}\cdot(\boldsymbol{H}^{(0)}\times\boldsymbol{k})]$ in the exciton energy expression, which describes the interaction between the exciton dipole moment $\boldsymbol{d}$ and the electric field $(\hbar/cm_{\text{exc}})[\boldsymbol{k}\times\boldsymbol{H}^{(0)}]$. *Gross* et al. [4.61] (see also [4.62, 63]) are of the opinion that such an electric field is set up in the presence of an external magnetic field in a coordinate system which is fixed to the travelling exciton; for more recent results consult [4.63a] (CdS and CdSe, magnetoluminescence and magnetoreflectance) and [4.63b] (theory).

This approach, however, is not sufficiently general, even if we ignore the fact that the exciton's group velocity is equal to $d\omega/d\boldsymbol{k}$ and that only in a special case is it equal to $\hbar\boldsymbol{k}/m_{\text{exc}}$. As a matter of fact, the sum of the terms in (4.4.14) which are linear with respect to $\boldsymbol{H}^{(0)}$, if we take (4.4.18) into account, can be presented in the following concise form

$$\boldsymbol{C}\cdot(\boldsymbol{H}^{(0)}\times\boldsymbol{k})+A^{(1)s}_{ijlm}s_i s_j H^{(0)}_l k_m=\tilde{\boldsymbol{C}}\cdot(\boldsymbol{H}^{(0)}\times\boldsymbol{k})\,,$$

where the components of the vector $\tilde{\boldsymbol{C}}$ are

$$\tilde{C}_1=C_1\,,\qquad \tilde{C}_2=C_2\,,\qquad \tilde{C}_3=C_3-(A^{(1)s}_{1112}(1-s_3^2)-2A^{(1)s}_{2313}s_3^2)\,.$$

Hence, as is clear from (4.4.14), a term of the type of $\tilde{\boldsymbol{C}}\cdot(\boldsymbol{H}^{(0)}\times\boldsymbol{k})$ actually does appear in the exciton-energy expression $\hbar\omega(\boldsymbol{k}\cdot\boldsymbol{H}^{(0)})$, but the vector $\tilde{\boldsymbol{C}}(\boldsymbol{s})$ need not be directed along an optical axis, even in a uniaxial crystal.

We recall that the frequencies of the "fictitious" longitudinal waves determine the positions of the absorption or luminescence lines for light propagated only in directions in which the light waves (real excitons) are not exactly transverse. For CdS-type crystals this is the wave vector $\boldsymbol{k}$ directed neither parallel nor perpendicular to the optical axis [only in this case does (4.4.14) apply]. Otherwise the positions of absorption lines are determined by the frequencies of the polarization waves. Therefore we should also consider the influence of an external magnetic field on the frequency dispersion of the polarization waves.

Let us restrict ourselves to one particular case of the CdS-type crystal, assuming that the vectors $\boldsymbol{H}^{(0)}$, $\boldsymbol{k}$ and the optical axis form a triad of orthogonal coordinate axes. According to this choice let, e.g., $k_2=k_3=0$ and $H_1^{(0)}=H_3^{(0)}=0$. Since we are interested in whether the expressions for the frequencies of the polarization waves contain terms of the order of $H^{(0)}$ and

$kH^{(0)}$, we shall employ the following expression for the tensor ε_{ij}^{-1} in the vicinity of dipole lines:

$$\varepsilon_{ij}^{-1}(\omega, \boldsymbol{k}, \boldsymbol{H}^{(0)}) = \varepsilon_{0ij}^{-1} + e_{ijm}[\boldsymbol{h} \times \boldsymbol{k}]_m + e_{ijm} B_{ml} H_l^{(0)} + B_{ijlm}^{(1)} H_l^{(0)} k_m \,. \quad (4.4.19)$$

This expression accounts for the second equation of (4.4.3) and the fact that, in crystals belonging to the class C_{6v}, the gyration vector $\boldsymbol{f}'$ (Sect. 3.2.1) is equal to $(\boldsymbol{h} \times \boldsymbol{k})$, $\boldsymbol{h}$ being a vector directed along the optical axis. With the help of this and the fact that in the crystals being considered the tensor B_{ml} is diagonal, (4.4.19) may be expressed as

$$\varepsilon_{ij}^{-1}(\omega, \boldsymbol{k}, \boldsymbol{H}^{(0)}) = \varepsilon_{0ij}^{-1} + e_{ij2} hk + e_{ij2} B_{22} H^{(0)} - D_{3ij}^{(1)} H^{(0)} k + B_{ij21}^{(1)} H^{(0)} k \,. \quad (4.4.20)$$

As has been shown in Sect. 2.2.2, the electric field strength $\boldsymbol{E} = 0$ in polarization waves, but the induction vector $\boldsymbol{D} \neq 0$ and $\boldsymbol{k} \cdot \boldsymbol{D} = 0$. In the case under consideration all these conditions indicate that for frequencies ω equal to the frequencies of the polarization waves, the set of homogeneous equations

$$\varepsilon_{ij}^{-1}(\omega, \boldsymbol{k}, \boldsymbol{H}^{(0)}) D_j = 0 \quad (4.4.21)$$

must have a nontrivial solution under the additional condition $D_1 = 0$. Since the properties of $D_{nij}^{(1)}$ are precisely analogous to those of the tensor C_{nij} and the properties of the pseudotensor $B_{ijlm}^{(1)s}$ to those of the pseudotensor $A_{ijlm}^{(1)s}$, see (4.4.10), we may use (4.4.15, 17) to arrive at the conclusion that in the crystals of class C_{6v} under consideration the components $D_{3ij}^{(1)} \neq 0$ only if $i = j = 1$ or $i = j = 2$ or $i = j = 3$, and the components $B_{ij12}^{(1)s} \neq 0$ only if $i = j = 1$ or $i = j = 2$. Hence (recalling also that $\varepsilon_{0ij}^{-1} = \delta_{ij} \varepsilon_{0ii}^{-1}$, with no summation, of course, with respect to the subscripts i) we can write the set of equations (4.4.21) under the additional condition $D_1 = 0$, in the form

$$\begin{gathered} -(hk + B_{22} H^{(0)}) D_3 = 0 \,, \\ [\varepsilon_{022}^{-1}(\omega) - D_{322}^{(1)} H^{(0)} k + B_{2212}^{(1)s} H^{(0)} k] D_2 = 0 \,, \\ [\varepsilon_{033}^{-1}(\omega) - D_{333}^{(1)} H^{(0)} k] D_3 = 0 \,. \end{gathered} \quad (4.4.22)$$

It follows from (4.4.22) that the set (4.4.21) always has a solution that corresponds to $D_3 = 0$, $D_2 \neq 0$. This indicates that among the solutions of the Coulomb problem there is always a solution, i.e., a polarization wave, which is propagated perpendicular to the optical axis and which is also polarized perpendicular to this axis ($D_1 = D_3 = 0$, $D_2 \neq 0$, $k_1 \neq 0$, $k_2 = k_3 = 0$). The position of the absorption line that corresponds to the solution of the Coulomb problem, as may be seen from (4.4.22), is determined by the equation

$$\varepsilon_{022}^{-1}(\omega) - D_{322}^{(1)} H^{(0)} k + B_{2212}^{(1)s} H^{(0)} k = 0 \,. \quad (4.4.23)$$

In the vicinity of a resonance (we mean a resonance corresponding to an exciton state that is transformed at $\boldsymbol{k}=0$ according to the representation E_1 of group C_{6v}; see Table 4.5), we have

$$\varepsilon_{022}^{-1}(\omega) \approx \frac{\omega^2-\omega_L^2(0)}{g_2} .$$

Hence, according to (4.4.23), this case yields

$$\omega_L(k, \boldsymbol{H}^{(0)}) = \omega_L(0) + \frac{g_2}{2\,\omega_L(0)}(D_{322}^{(1)} H^{(0)} k - B_{2212}^{(1)s} H^{(0)} k) ,$$

where, as for the 'fictitious' longitudinal waves, see (4.4.14), the "effect of a magnetic field inversion" occurs only when spatial dispersion is taken into consideration.

Let us now analyze the dependence of the frequencies of waves with $D_2=0$ on $\boldsymbol{k}$ and $\boldsymbol{H}^{(0)}$. If follows from (4.4.22) that with $D_1=D_2=0$, $D_3 \neq 0$ and $k_2=k_3=0$, $k_1 \neq 0$ polarization waves could exist if the quantities h and B_{22} vanish for some reason, so that the first equation of (4.4.22) is satisfied if $D_3 \neq 0$. The corresponding frequency of a polarization wave, for which the vector $\boldsymbol{D}=4\pi\boldsymbol{P}$ (Sect. 2.2.2) is directed along the optical axis, can be determined from

$$\varepsilon_{033}^{-1}(\omega) - D_{333} H^{(0)} k = 0 .$$

If, however, $h \neq 0$ and $B_{22} \neq 0$, then, as follows from (4.4.22), no polarization waves with $D_1=D_2=0$, $D_3 \neq 0$ exist in the case being considered. Hence, in the vicinity of the frequencies of mechanical excitons which are polarized at $\boldsymbol{k}=0$ along the optical axis (representation A_1, Table 4.5), taking spatial dispersion into account or a magnetic field, $\boldsymbol{k}$ being in any direction, we are not concerned with polarization waves, but with 'fictitious' longitudinal waves, whose frequencies are expressed by (4.4.6). Thus, in general, the effect of magnetic field inversion may take place near any of the resonances of the tensor $\varepsilon_{0ij}(\omega)$ (as has been indicated before, these resonances correspond to 'mechanical' exciton frequencies at $\boldsymbol{k}=0$, which are transformed according to the representations A_1 and E_1 of group C_{6v}).

Quadratic Magneto-Stark Effect

Let us now consider the case where external electric and magnetic fields exist simultaneously. As before, we consider a CdS-type crystal and assume the magnetic field to be in the y direction, the vector $\boldsymbol{k}$ in the x direction and the external electric field along the optical axis in this case the z axis. According to (4.4.4) [instead of (4.4.20)] we then have to linear order in $H^{(0)}$:

$$\varepsilon_{ij}^{-1}(\omega,k,E^{(0)},H^{(0)}) = \varepsilon_{0ij}^{-1} + e_{ij2}hk + e_{ij2}B_{22}H^{(0)} - (D_{3ij}^{(1)} - B_{ij21}^{(1)s})H^{(0)}k$$
$$+ B_{ij3}E^{(0)} + B_{ij23}^{(2)}H^{(0)}E^{(0)} + B_{ij31}^{(3)}E^{(0)}k$$
$$+ B_{ij33}^{(5)}(E^{(0)})^2 + B_{ij213}H^{(0)}kE^{(0)} + \dots \quad (4.4.24)$$

Note that the tensor B_{ijl} resembles the tensor C_{lij} as regards the symmetry properties, and the tensor $B_{ijlm}^{(5)}$ resembles the tensor α_{ijlm}. As in the case of $A_{ijlm}^{(1)}$, see (4.4.8), we represent the pseudotensor $B_{ijlm}^{(2)}$ in the following form:

$$B_{ijlm}^{(2)} = B_{ijlm}^{(2)s} + B_{ijlm}^{(2)a} . \quad (4.4.25)$$

Since, however, like $A_{ijlm}^{(2)}$, the pseudotensor $B_{ijlm}^{(2)}$ is antisymmetric with respect to the first pair of subscripts, see (4.4.2a),

$$B_{ijlm}^{(2)s} = e_{ijn}D_{nlm}^{(2)} , \qquad B_{ijlm}^{(2)a} = e_{ijn}e_{lmn'}B_{nn'}^{(2)} , \quad (4.4.26)$$

where the tensor $D_{nij}^{(2)} = D_{nji}^{(2)}$. It therefore possesses exactly the same symmetry properties as the tensor C_{nij}, and $B_{nn'}^{(2)}$ is a pseudotensor of rank two. Like the pseudotensor f'_{ij} (Sect. 3.2.1), for the crystals of class C_{6v} under consideration, the pseudotensor $B_{nn'}^{(2)}$ is completely antisymmetric

$$B_{nn'}^{(2)} = e_{nn'n''}B_{n''}^{(2)} ,$$

where the vector $\boldsymbol{B}^{(2)}$, like $\boldsymbol{h}$, see (4.4.19), is directed along the optical axis. Therefore

$$B_{nn'}^{(2)} = e_{nn'3}B^{(2)}$$

and, consequently,

$$B_{ijlm}^{(2)a} = e_{ijn}e_{lmn'}e_{nn'3}B^{(2)} \equiv (e_{ij1}e_{lm2} - e_{ij2}e_{lm1})B^{(2)} .$$

If follows directly from these relations that in the case under consideration the tensor $B_{ijlm}^{(2)a}$ has only the following nonzero components:

$$B_{2331}^{(2)a} = -B_{3123}^{(2)a} = -B_{3231}^{(2)a} = -B_{2313}^{(2)a} = B_{1323}^{(2)a} = B_{3132}^{(2)a} = B_{3213}^{(2)a} = -B_{1332}^{(1)a} . \quad (4.4.27)$$

Thus, taking into account previously made remarks and making use of (4.4.26, 27) and (4.4.15),

$$B_{ijlm}^{(2)}H_l^{(0)}E_m^{(0)} = e_{ij2}\tilde{B}H^{(0)}E^{(0)} , \quad (4.4.28)$$

where $\tilde{B} = D_{223}^{(2)} - B_{3123}^{(2)a}$.

Note that in the special case we considered (4.4.28) can be obtained in a much simpler way if we take into account that when an electric field is applied along the hexagonal axis the symmetry of CdS remains unchanged, so that the nonvanishing components of the pseudotensor $B^{(2)}_{ijl3}$ correspond to the nonvanishing components of the pseudotensor B'_{ijl} of rank three. This circumstance can also be used to find the nonvanishing components of the tensors B_{ij3}, $B^{(3)}_{ij3l}$, $B^{(5)}_{ij33}$, and of the pseudotensor B_{ijlm3}, since the tensors B_{ij3}, $B^{(5)}_{ij33}$ are similar to a symmetric tensor of rank two, the tensor $B^{(3)}_{ij3l}$ is similar to a tensor of rank three, to which the gyrotropy is due, and the pseudotensor B_{ijlm3} is completely analogous to the pseudotensor of rank four, $B^{(1)}_{ijlm}$.

These considerations make it easy to write equations for determining the frequencies of the polarization waves in the presence of external electric and magnetic fields. For the sake of simplicity, we consider here only the case of waves which are polarized perpendicular to the optical axis ($D_1 = D_3 = 0$, $D_2 \neq 0$). The equations then needed to determine the frequencies are of the form

$$\begin{aligned}\varepsilon^{-1}_{022}(\omega) - (D^{(1)}_{322} - B^{(1)s}_{2221}) H^{(0)} k + B_{223} E^{(0)} \\ + B^{(5)}_{2233} (E^{(0)})^2 + B_{22213} H^{(0)} E^{(0)}_2 k = 0 \,. \end{aligned} \tag{4.4.29}$$

Near the frequency of a mechanical exciton polarized perpendicular to the optical axis (representation E_1), $\varepsilon^{-1}_{022}(\omega) \approx [\omega^2 - \omega^2_L(0)]/g_{E_1}$. As follows from (4.4.26), the position of the absorption line is determined by

$$\begin{aligned}\omega(k, H^{(0)}, E^{(0)}) = \omega_L(0) - \frac{g_{E_1}}{2\,\omega_L(0)} [(D^{(1)}_{322} - B^{(1)s}_{2221}) H^{(0)} k \\ - B_{223} E^{(0)} - B^{(5)}_{2233} (E^{(0)})^2 - B_{22213} H^{(0)} k E^{(0)}] \,. \end{aligned} \tag{4.4.30}$$

It follows from (4.4.30) that if the values of $\boldsymbol{k}$, $\boldsymbol{H}^{(0)}$ and $\boldsymbol{E}^{(0)}/E^{(0)}$ are fixed, and $E^{(0)}$ is allowed to vary, the frequency of the absorption lines describes a parabola, whose position depends on the magnetic field strength and changes, in particular, if the field direction is reversed. This kind of effect was actually observed with CdS crystals [4.62, 63] and will certainly occur in many other cases (for GaSe see [4.63c]).

According to [4.62, 63] the effect of an absorption frequency shift is interpreted in a way already mentioned, based on the fact that in the presence of a magnetic field in a coordinate system linked to a moving exciton, an additional electric field arises whose strength in "effective" mass approximation is equal to $(\hbar/cm_{\mathrm{exc}})\, \boldsymbol{k} \times \boldsymbol{H}^{(0)}$. The external field which acts on the electron and the hole in an exciton is therefore equal to $\boldsymbol{E}_{\mathrm{eff}} = \boldsymbol{E}^{(0)} + (\hbar/cm_{\mathrm{exc}})\, \boldsymbol{k} \times \boldsymbol{H}^{(0)}$. In media displaying a quadratic Stark effect, it causes a frequency shift along a parabola, if $\boldsymbol{E}^{(0)}$ is varied. Determination of the position of the parabola enables the 'effective' mass of the Coulomb exciton to be determined for given $\boldsymbol{H}^{(0)}$ and $\boldsymbol{k}$. In fact, the experiment does not yield the 'effective' exciton mass but only the ratio between the coefficients of (4.4.30), in particular, the ratio B_{22213}/B_{2233} or, in the vicinity of transitions polarized along the optical axis (representation A_1), the ratio B_{33213}/B_{3333}. Though these ratios have the dimensionality of the quantity $\hbar/cm_{\text{exc}}$ it is, in general, impossible to reduce them only to this quantity. It is obvious that we must resort to microtheory to find the relations for the ratios of the parameters obtained in such experiments as [4.62, 63, 63c] and the characteristics of the 'mechanical' exciton such as its 'effective' mass. A modification of this theory can be found in [4.64] where it is shown that quantities of the form B_{22213}/B_{2233} are expressed not only in terms of the effective mass of the exciton, but in terms of components of the tensor ε_{0ij} as well.

4.5 Boundary Conditions in the Case of Spatial Dispersion Near a Separate Resonance (Absorption Line)

Up to this point the propagation of waves has been considered only for the case of an unbounded medium or the presence of boundaries has been ignored. In crystal optics, however, we are actually concerned with bounded media. It is true that in certain cases, e.g., when studying the problem of exciting real excitons by allowing light, x rays or charged particles to pass through the crystal (Sect. 6.4), the boundaries play an unimportant role. In typical problems of crystal optics (reflection of light from a crystal, light transmission through a crystal sheet, etc.), however, it is of prime importance to take the boundary into consideration.

If we have sharp interfaces we must supplement the field equations (2.1.1) by the boundary conditions (2.1.2) obtained from the former by passing to the limit:

$$\begin{aligned} &E_{1t}=E_{2t}, \quad [\boldsymbol{n},\boldsymbol{B}_2-\boldsymbol{B}_1]=\frac{4\pi}{c}(\boldsymbol{i}+\boldsymbol{i}_{\text{ext}}), \\ &B_{1n}=B_{2n}, \quad D_{2n}-D_{1n}=4\pi(\sigma+\sigma_{\text{ext}}). \end{aligned} \tag{2.1.2}$$

To solve arbitrary boundary value problems it is sufficient to know these conditions and the general relation (2.1.3) linking $\boldsymbol{D}(\boldsymbol{r},t)$ and $\boldsymbol{E}(\boldsymbol{r},t)$

$$D_i(\boldsymbol{r},t)=\int_{-\infty}^{t}dt'\int d\boldsymbol{r}'\,\varepsilon_{ij}(t,t',\boldsymbol{r},\boldsymbol{r}')E_j(\boldsymbol{r}',t'). \tag{2.1.3}$$

As has been stressed in Sect. 2.1.1, Eq. (2.1.3) is not, however, equivalent to introducing the tensor $\varepsilon_{ij}(\omega,\boldsymbol{k})$. In fact, when passing over from (2.1.3) to (2.1.4–6) we assumed that the kernel $\hat{\varepsilon}(t,t',\boldsymbol{r},\boldsymbol{r}')$ of the integral in (2.1.3) depends only the differences $t-t'$ and $\boldsymbol{r}-\boldsymbol{r}'$. The first of these assumptions is justified if the problem is homogeneous with respect to time and the presence of boundaries in this respect is inessential. On the other hand, strictly speaking, it is impossible in the case of a bounded medium to assume that the kernel $\mathscr{E}(t,t',\boldsymbol{r},\boldsymbol{r}')$ depends only on the difference $\boldsymbol{r}-\boldsymbol{r}'$ since this kernel must also depend on the distance from the boundary.

If the point $\boldsymbol{r}$ is fixed, the kernel $\mathscr{E}_{ij}(\tau,\boldsymbol{r},\boldsymbol{r}')$ is usually appreciably large only in a certain region with dimensions of the order of $a \ll \lambda$ in the vicinity of the point $\boldsymbol{r}$ (we consider nonmetallic condensed media, of course; see Chap. 1, and Sects. 3.1.1 and 4.2.1). It is therefore obvious that we can take $\mathscr{E}_{ij}$ in (2.1.3) to be a function of $\boldsymbol{r}-\boldsymbol{r}'$, except for a layer having thickness of the order of a adjacent to the interface (crystal surface). If, however, we are not interested in this subsurface layer and the field in it, we may proceed in the following way: we assume that the kernel $\mathscr{E}_{ij}$ in (2.1.3) depends only on $\boldsymbol{r}-\boldsymbol{r}'$ (or, in the case of crystals, we may use the procedure described in Sect. 3.1.1) thus making use of the tensor $\varepsilon_{ij}(\omega,\boldsymbol{k})$, and accounting for the boundary effects, if necessary, by supplementing the common boundary conditions (2.1.2) in some way. Simple considerations readily lead us to the result that when spatial dispersion is taken into account, additional boundary conditions (ABC's), if the general relation (2.1.3) is unknown, may actually become necessary.

In the first place, the use of (3.1.19 – 22) is equivalent to raising the order of the differential equations of the field, a fact which may occasion the appearance of new waves. Besides, it is obvious that equations of higher order require more boundary conditions. Secondly, we arrive at the very same results independently of the above conclusions, when we take into account that "new" waves, e.g., longitudinal (plasma) waves appear. As is well known, the boundary conditions (2.1.2) are only just sufficient to solve the problem of reflection and refraction of "ordinary" waves in the interface between media, say a crystal and empty space. If, however, even a single new wave can be additionally propagated in the crystal then the problem of reflection and refraction becomes insufficiently determined when we use only the common boundary conditions (2.1.2). If we add the appropriate ABC, however, any boundary value problem becomes fully determined. It is evident from the aforesaid that in this case we take only bulk waves of the type (2.2.2) or (2.2.54) into consideration. Solutions of the surface-wave type (particularly those which vary considerably within a distance of the order of a) and, in general, fields associated with the surface are not considered here. Surface-wave solutions of the "macroscopic type" (Chap. 5) can be obtained within the framework of the method described in the same way as bulk waves.

For the ordinary statement of a problem in crystal optics such an approach is quite natural, even though it involves certain well-known limitations. For example, in using ABC for a problem of a crystal-optical plate and ignoring pure surface effects, the plate thickness should be assumed sufficiently large compared to the parameter a, whose value may increase when approaching resonance (Sect. 4.5.3). The nature of ABC is determined by the physical properties of the medium and its surface and, therefore, is not general. A complete solution of the problem of ABC can therefore be found only if we make specific assumptions about the properties of the medium and its surface. Thus, for example, in the case of a semi-infinite plasma, from whose surface electrons are specularly reflected, we may use the kinetic equation [4.65] of a plasma to determine ABC's that enable us to solve the problem of reflection and refraction of waves, when the plasma is described by means of the tensor $\varepsilon_{ij}(\omega, \boldsymbol{k})$. In this case ABC's are of the form $E_n = 0$, E_n being the normal component of the field $\boldsymbol{E}$ at the inner interface (in the case of incident waves whose field $\boldsymbol{E}$ lies in the plane of incidence; for a review of the state in metal optics near the plasma frequency see [4.65a, b]).[9]

This example, which refers to a comparatively simple medium like a plasma, shows us that finding ABC is a problem of the same degree of complexity as calculating $\varepsilon_{ij}(\omega, \boldsymbol{k})$. Precisely this complexity necessitates, or, more exactly, indicates the expedience of dividing the boundary-value problem into two parts. First we should try to obtain certain general relations or to find the form of ABC by introducing unknown coefficients, and then these coefficients may be specifically defined in some way. In other words, we are concerned with the same usual problem as the problem of the phenomenological introduction of the tensor $\varepsilon_{ij}(\omega, \boldsymbol{k})$ and the specification of the above-mentioned coefficients is analogous to calculating or determining certain properties of the tensor $\varepsilon_{ij}(\omega, \boldsymbol{k})$ within the framework of microtheory.

In linear electrodynamics the general nature of ABC is immediately clear. These conditions must link $\boldsymbol{E}$ and $\boldsymbol{D}$ of the surface fields in a linear manner. Moreover, this connection must be homogeneous (we assume that the surface has no field sources) and the number of ABC's must be precisely such that the problem of reflection and refraction of waves is completely determined. This shows us that the number of ABC's depends on the number of new waves being considered.

Starting from these considerations and ignoring the derivatives we can write ABC in the form

$$D_i(0) + \Gamma'_{ij} E_j(0) = 0 \, , \tag{4.5.1}$$

where Γ'_{ij} is a certain tensor and the argument 0 indicates that the values of the fields refer to the surface. The conditions (4.5.1) represent three ABC's. If, however, the number of new waves is less than three, then we need to take only one or two of the components of the vector equation (4.5.1) (Sect. 4.5.1). If we consider a case, however, in which three ABC's are insufficient, we would have to use ABC's containing derivatives. This will be discussed in more detail below. For the moment it seems to us more important to stress another fact,

[9] In [4.65] it was shown that near the plasma surface the field equations admit additional solutions which cannot be reduced to (2.2.2 and 54) and which, of course, cannot be taken into account if the tensor $\varepsilon_{ij}(\omega, \boldsymbol{k})$ is used.

namely, the insufficient certainty of ABC of the type (4.5.1) which, in this or another form, are given and applied in [4.11, 12, 52a (Chaps. 2, 4), 66–68]. This, in fact, poses the questions of how Γ'_{ij} depends on ω and $\boldsymbol{k}$ and, in general, about the specific definition of this tensor. A full answer to these questions can be given only by microtheory (see [4.52a, 65, 69–71] and Sect. 4.5.3, and references therein). It is possible, however, to make an appreciable advance in a much simpler way if we consider the vicinity of a separate resonance (absorption line) in a phenomenological or semiphenomenological manner [4.4, 72]. This is what we shall do.

4.5.1 The Tensor $\varepsilon_{ij}(\omega,\boldsymbol{k})$ Near an Isolated Resonance

To solve the problem, we use the equation of motion for an anisotropic harmonic oscillator

$$\rho_{ij}\frac{d^2x_j}{dt^2}+r_{ij}\frac{dx_j}{dt}+k_{ij}x_j=e_{ij}E_j\,,$$

where, with the effective charge described by the tensor e_{ij}, we intend to take into consideration such effects as the difference between the field $\boldsymbol{E}_{\text{eff}}$ acting on the oscillator and the field $\boldsymbol{E}$. Furthermore, if we take an assembly of independent oscillators as a model of the solid under investigation, the electric polarization $P_i=eNx_i$, N being the concentration of oscillators. Hence the equation of polarization for such a model is

$$\rho_{ij}\frac{\partial^2P_j}{\partial t^2}+r_{ij}\frac{\partial P_j}{\partial t}+k_{ij}P_j=eNe_{ij}E_j\,. \tag{4.5.2}$$

A real crystal differs from the model of independent oscillators primarily by the fact that the elastic force and the force produced by the field $\boldsymbol{E}$ cannot be simply reduced to $k_{ij}P_j$ and $eNe_{ij}E_j$. The first two terms of (4.5.2) specifically define the time dependence of the polarization and, if we consider an isolated resonance, these terms remain unchanged. This, as a matter of fact, is exactly how we determine the region of an isolated resonance. Whether the corresponding time dependence or the equivalent frequency dependence is satisfied is a problem of the feasibility of confining our consideration to a single resonance. When studying the effects of gyrotropy in the region of an exciton B transition of a CdS crystal, it is necessary to account for four adjacent exciton bands, of which only three are dipolarly allowed [4.73a]. In this case, both quadrupole and magnetic dipole polarization should be taken into account.[10] As we know from experiments and microtheoretical considerations

[10] The polariton effect in degenerate valance-band semiconductors has been discussed in [4.73b].

(Sect. 6.1) it is quite permissible, in many cases, to speak of an isolated resonance.

By virtue of the aforesaid, the relationship between $\boldsymbol{P}(\boldsymbol{r},t)$ and $\boldsymbol{E}(\boldsymbol{r},t)$ in the region of an isolated resonance can naturally be written in the form

$$\rho_{ij}\frac{\partial^2 P_j(\boldsymbol{r},t)}{\partial t^2} + \int \mathcal{K}_{ij}(\boldsymbol{r},\boldsymbol{r}')P_j(\boldsymbol{r}',t)\,d\boldsymbol{r}' = \int \Lambda_{ij}(\boldsymbol{r},\boldsymbol{r}')E_j(\boldsymbol{r}',t)\,d\boldsymbol{r}' , \tag{4.5.3}$$

where, for simplicity, we omitted the dissipation term $r_{ij}(\partial P_j/\partial t)$. Of course, (4.5.3) is a special case of the general linear relation (2.1.3) between $\boldsymbol{D}=\boldsymbol{E}+4\pi\boldsymbol{P}$ and $\boldsymbol{E}$.

Since, in general, the functions $\mathcal{K}_{ij}(\boldsymbol{r},\boldsymbol{r}')$ and $\Lambda_{ij}(\boldsymbol{r},\boldsymbol{r}')$ cannot be reduced to $\delta(\boldsymbol{r}-\boldsymbol{r}')$, the "force" at point $\boldsymbol{r}$ is determined by the magnitudes of the electric and magnetic field vectors not only at the point $\boldsymbol{r}$ but in its neighborhood as well. Equation (4.5.3) therefore accounts for spatial dispersion. As we have seen in Sect. 3.1, spatial dispersion in crystals can be assumed to be weak, and far from the boundaries (i.e., for an unbounded medium), we can assume that $\mathcal{K}_{ij}(\boldsymbol{r},\boldsymbol{r}')=\mathcal{K}_{ij}(\boldsymbol{r}-\boldsymbol{r}')$ and similarly $\Lambda_{ij}(\boldsymbol{r},\boldsymbol{r}')=\Lambda_{ij}(\boldsymbol{r}-\boldsymbol{r}')$. Therefore, if we expand $P_j(\boldsymbol{r}',t)$ and $E_j(\boldsymbol{r}',t)$ in a series close to the point $\boldsymbol{r}'=\boldsymbol{r}$ and restrict ourselves to the main terms we find, instead of (4.5.3),

$$\begin{aligned}\rho_{ij}\frac{\partial^2 P_j(\boldsymbol{r},t)}{\partial t^2} &+ \beta_{ij}P_j(\boldsymbol{r},t) + \gamma_{ijl}\frac{\partial P_j(\boldsymbol{r},t)}{\partial x_l} + \alpha_{ijlm}\frac{\partial^2 P_j(\boldsymbol{r},t)}{\partial x_l \partial x_m}\\ &= \Lambda^{(0)}_{ij}E_j(\boldsymbol{r},t) + \Lambda^{(1)}_{ijl}\frac{\partial E_j(\boldsymbol{r},t)}{\partial x_l} + \Lambda^{(2)}_{ijlm}\frac{\partial^2 E_j(\boldsymbol{r},t)}{\partial x_l \partial x_m} .\end{aligned} \tag{4.5.4}$$

The tensors ρ_{ij}, β_{ij}, γ_{ijl}, etc., occurring in this equation are practically independent of ω in the frequency region under consideration; they are determined by the type of exciton transition and the crystal symmetry[11]. It is clear that for nongyrotropic crystals the tensors $\gamma_{ijl}=\Lambda^{(1)}_{ijl}=0$, whereas in the case of gyrotropic crystals, we can, as a first approximation, set $\alpha_{ijlm}=\Lambda^{(2)}_{ijlm}=0$.

The solution of the system (4.5.4) may be sought in the form of plane waves

$$\begin{aligned}\boldsymbol{P}(\boldsymbol{r},t) &= \boldsymbol{P}(\omega,\boldsymbol{k})\exp[\mathrm{i}(\boldsymbol{k}\cdot\boldsymbol{r}-\omega t)] ,\\ \boldsymbol{E}(\boldsymbol{r},t) &= \boldsymbol{E}(\omega,\boldsymbol{k})\exp[\mathrm{i}(\boldsymbol{k}\cdot\boldsymbol{r}-\omega t)] .\end{aligned} \tag{4.5.5}$$

Substituting (4.5.5) into (4.5.4) we find that

$$P_i(\omega,\boldsymbol{k}) = \Delta^{-1}_{il}(\omega,\boldsymbol{k})\Lambda_{lj}(\omega,\boldsymbol{k})E_j(\omega,\boldsymbol{k}) , \tag{4.5.6}$$

---

[11] Our approach is practically a generalization of the model [4.72, 74] for the case of an anisotropic medium when spatial dispersion is taken into account [4.4] and also [4.75].

where

$$\Delta_{ij}(\omega,\boldsymbol{k}) = -\rho_{ij}\omega^2+\beta_{ij}+\mathrm{i}\gamma_{ijl}k_l-\alpha_{ijlm}k_lk_m\,, \tag{4.5.7a}$$

$$\Lambda_{ij}(\omega,\boldsymbol{k}) = \Lambda_{ij}^{(0)}+\mathrm{i}\Lambda_{ijl}^{(1)}k_l-\Lambda_{ijlm}^{(2)}k_lk_m\,. \tag{4.5.7b}$$

Relation (4.5.6) determines the part of the polarization which is due to the contribution of the exciton transition under consideration. In the frequency region under investigation, the induction vector may therefore be represented in the following form

$$D_i(\omega,\boldsymbol{k}) = \varepsilon_{00,ij}E_j(\omega,\boldsymbol{k})+4\pi P_i(\omega,\boldsymbol{k})\,. \tag{4.5.8}$$

In this frequency region, the tensor $\varepsilon_{00\,ij}$ may be assumed to be independent of ω as are the tensors ρ_{ij}, β_{ij}, γ_{ijl}, etc. The tensor $\varepsilon_{00\,ij}-\delta_{ij}$ takes into account the contribution to the complete dielectric-constant tensor of the natural oscillations whose frequencies lie outside the frequency range under consideration. Substituting (4.5.6) into (4.5.8) we find that the complete tensor $\varepsilon_{ij}(\omega,\boldsymbol{k})$ is determined by

$$\varepsilon_{ij}(\omega,\boldsymbol{k}) = \varepsilon_{00\,ij}+4\pi\Delta_{il}^{-1}(\omega,\boldsymbol{k})\Lambda_{lj}(\omega,\boldsymbol{k})\,. \tag{4.5.9}$$

From this relation we can also determine the tensor $\varepsilon_{ij}^{-1}(\omega,\boldsymbol{k})$ which is equal to

$$\varepsilon_{ij}^{-1}(\omega,\boldsymbol{k}) = (\hat{\varepsilon}_{00})_{ij}^{-1}-4\pi(\hat{\Delta}\,\hat{\varepsilon}_{00}+4\pi\,\hat{\Delta})_{il}^{-1}\Lambda_{lm}(\hat{\varepsilon}_{00})_{mj}^{-1}\,, \tag{4.5.9a}$$

where $\hat{\varepsilon}_{00}$ and $\hat{\Delta}$ are the tensors $\varepsilon_{00\,ij}$ and Δ_{ij}.

It follows from these relations that the poles of the dielectric-constant tensor correspond to the roots of the equation

$$|\Delta_{il}(\omega,\boldsymbol{k})| = 0\,. \tag{4.5.10}$$

This result is quite natural. In fact if we substitute $\boldsymbol{E}(\boldsymbol{r},t) = 0$ into (4.5.4) the equations thus obtained describe, by definition (Sect. 2.2.2), the natural oscillations of the dipole moment of 'mechanical' excitons. Condition (4.5.10), however, determines the natural frequencies of these 'mechanical' excitons which thereby correspond precisely to poles of the tensor $\varepsilon_{ij}(\omega,\boldsymbol{k})$.

Generally speaking, (4.5.10) determines three branches of the frequencies $\omega_l(\boldsymbol{k})$, $l = 1,2,3$. Therefore, in this case, (4.5.9) contains the sum of three resonances. Such a description is justified if these resonances lie close to one another or if at $\boldsymbol{k} = 0$ the 'mechanical' exciton states are degenerate (cubic crystals, etc.). Otherwise, it is more convenient to consider the contribution to the tensor $\varepsilon_{ij}(\omega,\boldsymbol{k})$ from one nondegenerate band of a 'mechanical' exciton. Let us assume, for example, that at $\boldsymbol{k} = 0$ this band corresponds to a linearly polarized oscillation such that $\boldsymbol{P}(\boldsymbol{r},t) = \boldsymbol{e}P(\boldsymbol{r},t)$. Then (4.5.10) should be replaced by

$$\rho\frac{\partial^2 P}{\partial t^2}+\int\mathscr{K}(\boldsymbol{r},\boldsymbol{r}')P(\boldsymbol{r}',t)\,d\boldsymbol{r}'=\int\Lambda_j(\boldsymbol{r},\boldsymbol{r}')E_j(\boldsymbol{r}',t)\,d\boldsymbol{r}'\;, \tag{4.5.11}$$

where ρ is a scalar, $\mathscr{K}$ is a scalar function, and Λ is a vector function of $\boldsymbol{r}$ and $\boldsymbol{r}'$. Since in the approximation under consideration the vector $\boldsymbol{e}$ is the only one that is independent of $\boldsymbol{r}$ and $\boldsymbol{r}'$, we have $\Lambda(\boldsymbol{r},\boldsymbol{r}')=\boldsymbol{e}\Lambda(\boldsymbol{r},\boldsymbol{r}')$. Performing the same calculation as above we obtain instead of (4.5.6):

$$\boldsymbol{P}(\omega,\boldsymbol{k})=\frac{\boldsymbol{e}\Lambda(\omega,\boldsymbol{k})}{\Delta(\omega,\boldsymbol{k})}\boldsymbol{e}\cdot\boldsymbol{E}(\omega,\boldsymbol{k})\,,\qquad\text{where}$$

$$\Delta(\omega,\boldsymbol{k})=-\rho\omega^2+\beta+\mathrm{i}\gamma_m k_m-\alpha_{mn}k_m k_n\,,$$

$$\Lambda(\omega,\boldsymbol{k})=\Lambda^{(0)}+\mathrm{i}\Lambda_m^{(1)}k_m-\Lambda_{mn}^{(2)}k_m k_n\,.$$

In the neighborhood of one nondegenerate "mechanical" exciton band we therefore have

$$\varepsilon_{ij}(\omega,\boldsymbol{k})=\varepsilon_{00ij}+\frac{4\pi\Lambda(\omega,\boldsymbol{k})e_i e_j}{\Delta(\omega,\boldsymbol{k})}\,.$$

4.5.2 Examples

Dielectric Tensor for Gyrotropic Cubic Crystals

To illustrate the foregoing we first investigate the tensors $\varepsilon_{ij}(\omega,\boldsymbol{k})$ and $\varepsilon_{ij}^{-1}(\omega,\boldsymbol{k})$ in gyrotropic cubic crystals. In cubic crystals (Sect. 3.2.1)

$$\rho_{ij}=\rho\delta_{ij}\,,\quad \beta_{ij}=\beta\delta_{ij}\,,\quad \Lambda_{ij}=\Lambda^{(0)}\delta_{ij}\,,\quad \gamma_{ijl}=\gamma e_{ijl}\,,\quad \Lambda_{ijl}^{(1)}=\Lambda^{(1)}e_{ijl}\,, \tag{4.5.12}$$

where e_{ijl} is a completely antisymmetric unit pseudotensor of rank three. Assuming $\alpha_{ijlm}=\Lambda_{ijlm}^{(2)}=0$ we find that

$$\Delta_{ij}^{-1}=\frac{\beta-\rho\omega^2}{(\beta-\rho\omega^2)^2-\gamma^2k^2}\delta_{ij}-\mathrm{i}\gamma\frac{e_{ijl}k_l}{(\beta-\rho\omega^2)^2-\gamma^2k^2}-\frac{\gamma^2k_ik_j}{(\beta-\rho\omega^2)[(\beta-\rho\omega^2)^2-\gamma^2k^2]}\,. \tag{4.5.12a}$$

For the frequency region where

$$|\beta-\rho\omega^2|\gg|\gamma|k\,, \tag{4.5.13}$$

we obtain, instead of (4.5.12),

$$\Delta_{ij}^{-1}(\omega,\boldsymbol{k}) = \frac{\delta_{ij}}{\beta-\rho\omega^2} - \mathrm{i}\gamma\frac{e_{ijl}k_l}{(\omega-\rho\omega^2)^2}\,. \tag{4.5.13a}$$

Hence, in this case, the tensor $\varepsilon_{ij}(\omega,\boldsymbol{k})$, see (4.5.5, 7), has the form

$$\varepsilon_{ij}(\omega,\boldsymbol{k}) = \left(\varepsilon_{00} + \frac{4\pi\Lambda^{(0)}/\rho}{\omega_L^2-\omega^2}\right)\delta_{ij} + \mathrm{i}\left[\frac{4\pi\Lambda^{(1)}/\rho}{\omega_L^2-\omega^2} - \frac{4\pi\gamma\Lambda^{(0)}/\rho^2}{(\omega_L^2-\omega^2)^2}\right]e_{ijl}k_l\,, \tag{4.5.13b}$$

where $\hbar\omega_L \equiv \hbar\sqrt{\beta/\rho} \equiv \hbar\omega_L(0)$ is the energy of the 'mechanical' exciton in the L-th band at $\boldsymbol{k}=0$. Since we assumed $\Lambda^{(0)} \neq 0$, it is obvious that (at $\boldsymbol{k}=0$) this is a triply degenerate exciton term whose wave functions are transformed like the components of a polar vector (Table 4.2). In the immediate vicinity of the frequency $\omega=\omega_L(0)$ we should replace (4.5.13b) by an expansion of the tensor $\varepsilon_{ij}^{-1}(\omega,\boldsymbol{k})$. In this expansion, as before, we shall not take terms quadratic in $\boldsymbol{k}$ into account.

Making use of (4.5.12) we find that in this approximation

$$\left(\hat{\Delta} + \frac{4\pi}{\varepsilon_{00}}\hat{\Delta}\right)_{ij}^{-1} = \frac{\delta_{ij}}{\beta-\rho\omega^2+4\pi\Lambda^{(0)}/\varepsilon_{00}} - \mathrm{i}\frac{(\gamma+4\pi\Lambda^{(1)}/\varepsilon_{00})e_{ijl}k_l}{(\beta-\rho\omega^2+4\pi\Lambda^{(0)}/\varepsilon_{00})^2}\,. \tag{4.5.14}$$

or

$$\varepsilon_{ij}^{-1}(\omega,\boldsymbol{k}) = \frac{\delta_{ij}}{\varepsilon_0(\omega)} - \frac{\mathrm{i}}{\varepsilon_0^2(\omega)}\left(\frac{4\pi\Lambda^{(1)}/\rho}{\omega_L^2-\omega^2} - \frac{4\pi\gamma\Lambda^{(0)}/\rho^2}{(\omega_L^2-\omega^2)^2}\right)e_{ijl}k_l\,, \tag{4.5.15a}$$

where

$$\varepsilon_0(\omega) = \varepsilon_{00} + \frac{4\pi\Lambda^{(0)}/\rho}{\omega_L^2-\omega^2}\,. \tag{4.5.15b}$$

If $|\omega_L^2-\omega^2| \ll 4\pi\Lambda^{(0)}/\rho\varepsilon_{00}$, (4.5.15) can be simplified and assumes the form

$$\varepsilon_{ij}^{-1}(\omega,\boldsymbol{k}) = \frac{\omega_L^2-\omega^2}{4\pi\Lambda^{(0)}/\rho}\delta_{ij} + \frac{\gamma/\rho}{4\pi\Lambda^{(0)}/\rho}e_{ijl}k_l\,. \tag{4.5.16}$$

By virtue of (4.5.9a) we therefore have for the tensor $\varepsilon_{ij}^{-1}(\omega,\boldsymbol{k})$:

$$\varepsilon_{ij}^{-1}(\omega,\boldsymbol{k}) = \left(\varepsilon_{00} + \frac{4\pi\Lambda^{(0)}/\rho}{\omega_L^2-\omega^2}\right)^{-1}\delta_{ij}$$

$$-\frac{1}{(\varepsilon_{00})^2}\,\frac{\dfrac{4\pi\Lambda^{(1)}}{\rho}(\omega_L^2-\omega^2) - \dfrac{4\pi}{\rho}\,\dfrac{\Lambda^{(0)}}{\rho}\gamma}{(\omega_L^2-\omega^2+4\pi\Lambda^{(0)}/\rho\varepsilon_{00})^2}e_{ijl}k_l\,. \tag{4.5.15}$$

Comparing (4.5.15a) and (4.1.1), as well as (4.1.12), we see that the quantity in front of the tensor $e_{ijl}k_l$ in (4.5.15a) is proportional to the rotation of the polarization plane of light in a crystal and describes its dependence on ω. If $\Lambda^{(1)}=0$ and under the condition $|\omega^2-\omega_L^2| \gg 4\pi\Lambda^0/\rho\varepsilon_{00}$, the amount of rotation $\varphi(\omega)$ then has the same sign for both sides of the frequency ω_L. In this case the empirical formula given in [4.76] is valid (it is sometimes called Chandrasekhar's formula)

$$\varphi(\omega)=\frac{K\omega^2}{(\omega^2-\omega_L^2)^2}, \tag{4.5.17}$$

where K is a proportionality factor:

$$K=\frac{2\pi}{c^2}\,\frac{\Lambda^{(0)}\gamma}{\rho^2}. \tag{4.5.17a}$$

It can be assumed, at least in certain cases, that $\Lambda^{(1)}$ vanishes for crystals which lose their optical activity upon dissolution. For the case of molecular crystals this has been proven in [4.5, 77]. In a certain sense this conclusion also follows for the model considered here. For this purpose we assume that we enlarge the lattice constant of the crystal so that the interaction between the molecules of the crystal tends to zero. In this case the "mechanical" exciton band becomes narrower and, in the limit, it degenerates to a single level which no longer depends on the wave vector $\boldsymbol{k}$. This last fact indicates that the quantities γ_{ijl} and α_{ijlm} tend to zero. The amount of rotation is then independent of the interaction between the molecules and is completely determined by the coefficient $\Lambda_{ijl}^{(1)}$. Thus, this coefficient characterizes the individual rotatory power of the molecules that make up the crystal.

At the same time, of course, even for a medium consisting of nongyrotropic molecules, the quantity $\Lambda_{ijl}^{(1)}$ could be nonzero when the interaction between the cells of the crystal is strong enough.

It should be noted that for the frequency regions of semiconductors where the main contribution to dispersion is made by interband transitions of electrons rather than by excitons or phonons, the relations (4.5.3, 4) and consequently (4.5.17), are, in general, no longer valid. The approximation of an isolated resonance, employed above, becomes unsuitable in this case because for various photons having energy $\hbar\omega$, varying between quite wide limits, and for a very wide forbidden band Δ, "resonating" $v\to c$ transitions of electrons can always be found in the valence band $E_v(\boldsymbol{k})$ and in the conduction band $E_c(\boldsymbol{k})$ [for these transitions, $\hbar\omega=E_c(\boldsymbol{k})-E_v(\boldsymbol{k})$]. It is known [4.78] that when $\omega\to\Delta/\hbar$ the circumstance just mentioned leads to a total absence of resonances in absorption. The shape of the light-absorption edge in the region $\omega \geqslant \Delta/\hbar$ depends, in this case, on the type of bands and their structure. Thus, for instance, in the region of a so-called straight gap Δ, corresponding to an allowed dipole transition, when the effective masses of the electron and holes are isotropic, the absorption coefficient $\varkappa(\omega) \sim (\hbar\omega - \Delta)^{1/2}$. The application of the Kramers-Konig formula leads to the conclusion that when $\omega \lesssim \Delta/\hbar$, the dielectric constant $\varepsilon(\omega)$ may be represented in the form

$$\varepsilon(\omega) \approx \varepsilon_0 - a(\Delta - \hbar\omega)^{1/2},$$

where ε_0 and a are certain quantities independent of ω.

In the case of an isolated exciton or phonon resonance, the rotatory power, according to (4.5.17), is $\varphi(\omega) \sim d\varepsilon/d\omega$. If we use the same relation for semiconductors of the type mentioned above, we obtain

$$\varphi(\omega) \sim d\varepsilon/d\omega \sim (\Delta - \hbar\omega)^{-1/2}.$$

This relation was derived in [4.79] as the result of a systematic calculation of the dielectric-constant tensors of certain semiconductors, taking the structure of their bands and spatial dispersion into account [4.80]. Thus the rotatory power $\varphi(\omega)$ is proportional also in this case to $d\varepsilon/d\omega$ (for a consideration of the Coulomb interaction of the electron and hole see [4.81]).

Note that the factor K in (4.5.17) is related to the oscillator strength of the transition and the form of the energy band of the 'mechanical' exciton at small values of $\boldsymbol{k}$. This is evident if we consider, as follows from (4.5.13b) and (4.1.13), that

$$4\pi A^{(0)}/\rho = \omega_p^2 f, \tag{4.5.17b}$$

where f is the oscillator strength and ω_p^2 is the square of the "plasma" frequency (Sect. 4.2.2). The meaning of the quantity γ/ρ may be elucidated if we consider the $\boldsymbol{k}$ dependence of the frequencies of "mechanical" excitons, neglecting terms of higher than first order in $\boldsymbol{k}$. Substituting $\boldsymbol{E}(\boldsymbol{r},t) = 0$ in (4.5.4) and taking (4.5.12) into account we find that in cubic crystals mechanical excitons are governed by the relation

$$(\beta - \rho\omega^2)\boldsymbol{P} + \mathrm{i}\gamma(\boldsymbol{P}\times\boldsymbol{k}) = 0. \tag{4.5.18}$$

Hence it follows that for longitudinal waves ($\boldsymbol{P} = P\boldsymbol{k}/k$) the frequency of a mechanical exciton is given by

$$\omega_\parallel(\boldsymbol{k}) = \omega_L(0) = \sqrt{\beta/\rho}. \tag{4.5.19}$$

If we assume that the z direction coincides with $\boldsymbol{k}$ we find that for transverse waves $P_z = 0$ and, consequently, by virtue of (4.5.18)

$$(\beta - \rho\omega^2)P_x + \mathrm{i}\gamma P_y k = 0, \qquad -\mathrm{i}\gamma P_x k + (\beta - \rho\omega^2)P_y = 0.$$

We set the determinant of this system equal to zero and obtain the following expression for the frequencies of transverse 'mechanical' excitons

$$\omega_{\pm}(\boldsymbol{k}) = \omega_L(0) \pm \frac{\gamma}{2\rho\,\omega_L(0)}|\boldsymbol{k}| + \text{quadratic terms in } \boldsymbol{k}\,, \tag{4.5.20}$$

where the signs + and − correspond to right-handed and left-handed circularly polarized polarization waves (according to the sign of γ):

$$\frac{P_x}{P_y} = \mp \mathrm{i}\frac{\gamma}{|\gamma|}\,.$$

Relation (4.5.20) was previously obtained (Sect. 4.1.2) by passing to the limit $c \to \infty$.

Regardless of the fact that the tensor $\varepsilon_{ij}^{-1}(\omega,\boldsymbol{k})$ at $\omega \approx \omega_L$ is analytic if $\boldsymbol{k} \to 0$, see (4.5.15), the 'mechanical'-exciton frequencies, (4.5.20), are non-analytic functions of $\boldsymbol{k}$. This is due to the fact that the state of the 'mechanical' exciton under consideration becomes degenerate at $\boldsymbol{k} = 0$. At the same time, if there is no degeneracy, the mechanical-exciton frequencies are analytic functions of $\boldsymbol{k}$ if $\boldsymbol{k} \to 0$ (see, e.g., Sect. 6.2.1).

Let us also point out that in uniaxial crystals, when the vector $\boldsymbol{k}$ is directed along the optical axis, the equation for the frequencies $\omega_{\pm}(\boldsymbol{k})$ retains the form (4.5.20), the corresponding mechanical excitons being transverse circularly polarized waves, i.e., "polarization" waves. A solution also exists corresponding to a longitudinal wave. If, however, $\boldsymbol{k}$ deviates even slightly from the optical axis, the terms linear in $\boldsymbol{k}$ vanish in the expressions for the normal-wave frequencies. If the quadratic terms in $\boldsymbol{k}$ are omitted in (4.5.20), the resulting expression for the frequencies $\omega_{\pm}(\boldsymbol{k})$, accurate to terms linear with respect to $\boldsymbol{k}$, applies not only to isotropic media, but also to cubic crystals which, in this approximation, are completely isotropic.

Hence it follows that the factor K in (4.5.17), which determines the amount of rotation of the polarization plane in gyrotropic cubic crystals, can in fact be expressed in terms of the oscillator strength f of the transition, its frequency $\omega_L(0)$, the plasma frequency ω_p and the factor of the linear $\boldsymbol{k}$ term in the expression for the 'mechanical' exciton frequency. Besides the factor K in (4.5.17) can be determined from an experimental investigation of the dependence of the optical rotatory power of the crystal on the light frequency. The oscillator strength and the position of the resonance $\omega_L(0)$ can be determined independently from data on the frequency dependence of the refractive index. The application of (4.5.17 a, b) therefore makes it possible to determine the function $\omega_{\pm}(\boldsymbol{k})$ at small values of $\boldsymbol{k}$ (Sect. 5.2). Note also that the above considerations, in the case of gyrotropic cubic crystals, apply also to uniaxial gyrotropic crystals, provided the vector $\boldsymbol{k}$ lies along the optical axis; the quantity γ is then replaced by the component γ_{zxy}. Finally we should note that

(4.5.17), as already indicated, is valid only at a certain distance from the frequency $\omega_L(0)$. This is because on deducing this relation we actually assumed that the third wave originating in the immediate proximity of $\omega = \omega_L(0)$ is insignificant (Sect. 4.5.9). In accordance with what has been said in Sect. 4.1.3, this assumption is justified if $\omega < \omega_m$, where ω_m is the frequency which corresponds to the turning point.

Let us also mention that to an accuracy of terms linear in $\boldsymbol{k}$ the tensors defined by (4.5.13 b) and (4.5.15 or 15 a) are invertible so that their application in calculations of the optical activity yields identical results. The formulas for the tensor $\varepsilon_{ij}^{-1}(\omega, \boldsymbol{k})$ which we derived above, however, serve mainly to determine the values of the tensor $\delta_{ijl}(\omega)$ in (4.1.1). Comparing (4.1.1) and (4.5.16) we find that at $\omega \approx \omega_L$ this tensor is directly related only to γ and does not contain the factor $\Lambda^{(1)}$. Only in the case of gyrotropic crystals, which lose their gyrotropy upon dissolution, does a determination of the factor γ enable the amount of rotation $\varphi(\omega)$ of the polarization plane, due to the contribution of the resonance under consideration to be completely derived within the framework of our model.

Dielectric Tensor for Isotropic Nongyrotropic Media

Let us now consider the case of an isotropic nongyrotropic medium. Equation (4.5.12) applies to such a medium in which $\gamma = \Lambda^{(1)} = 0$. Hence, in (4.5.7) for the tensors $\hat{\Delta}$ and $\hat{\Lambda}$ we should account for terms quadratic with respect to $\boldsymbol{k}$. In an isotropic medium, see (3.2.1) and Table 3.2,

$$\begin{aligned} \alpha_{ijlm} k_l k_m &= \alpha_{\text{tr}} k^2 \delta_{ij} + (\alpha_{\text{lo}} - \alpha_{\text{tr}}) k_i k_j , \\ \Lambda_{ijlm}^{(2)} k_l k_m &= \Lambda_{\text{tr}}^{(2)} k^2 \delta_{ij} + (\Lambda_{\text{lo}}^{(2)} - \Lambda_{\text{tr}}^{(2)}) k_i k_j . \end{aligned}$$

Consequently, if we also take (4.5.12) into account, we obtain

$$\Delta_{ij}^{-1}(\omega, \boldsymbol{k}) = \frac{1}{\beta - \rho\omega^2 - \alpha_{\text{tr}} k^2} \delta_{ij} + \frac{(\alpha_{\text{lo}} - \alpha_{\text{tr}}) k_i k_j}{(\beta - \rho\omega^2 - \alpha_{\text{lo}} k^2)(\beta - \rho\omega^2 - \alpha_{\text{tr}} k^2)} . \tag{4.5.21}$$

The substitution of this expression into (4.5.9) yields the required relation

$$\begin{aligned} \varepsilon_{ij}(\omega, \boldsymbol{k}) &= \left(\varepsilon_{00} - \frac{4\pi(\Lambda^{(0)} - \Lambda_{\text{tr}}^{(2)} k^2)/\rho}{\omega_L^2 - \omega^2 - \alpha_{\text{tr}} k^2} \right) \delta_{ij} \\ &\quad + \left[\frac{\dfrac{4\pi\Lambda^{(0)}}{\rho^2} (\alpha'_{\text{lo}} - \alpha'_{\text{tr}})}{(\omega_L^2 - \omega^2 - \alpha'_{\text{tr}} k^2)(\omega_L^2 - \omega^2 - \alpha'_{\text{lo}} k^2)} - \frac{\dfrac{4\pi(\Lambda_{\text{lo}}^{(2)} - \Lambda_{\text{tr}}^{(2)})}{\rho}}{\omega_L^2 - \omega^2 - \alpha'_{\text{tr}} k^2} \right] k_i k_j \\ &\equiv \varepsilon^{\perp}(\omega) \delta_{ij} + [\varepsilon^{\parallel}(\omega, \boldsymbol{k}) - \varepsilon^{\perp}(\omega, \boldsymbol{k})] s_i s_j . \end{aligned} \tag{4.5.22}$$

The meaning of $\varepsilon^{\perp}$ and $\varepsilon^{\parallel}$is clear from (4.5.22). Here we also use the notation $\alpha'_{ijlm} = \alpha_{ijlm}/\rho$.

In a similar manner we can determine the nature of the dependence of the tensor $\varepsilon_{ij}(\omega, \boldsymbol{k})$ or $\varepsilon_{ij}^{-1}(\omega, \boldsymbol{k})$ on ω and $\boldsymbol{k}$ for a medium of any other symmetry.

To sum up we may say that the procedure proposed above is suitable in all the frequency regions where spatial and time dispersion of the dielectric-constant tensor are caused by a single 'mechanical'-exciton band. In this frequency region the tensors $\varepsilon_{ij}(\omega, \boldsymbol{k})$ or $\varepsilon_{ij}^{-1}(\omega, \boldsymbol{k})$ are known up to certain constant parameters which can be determined from experiments. Hence, the description of spatial and frequency dispersion applied here is more detailed than that which follows from expansions of the type (3.1.19 – 22) where certain frequency functions like $\gamma_{ijl}(\omega)$ or $\alpha_{ijlm}(\omega)$ remain unknown. This becomes particularly clear if we consider the above example of a gyrotropic medium. Of course a more detailed description is obtained here when some model is used or, more precisely, when additional assumptions are made as to the nature of the relation between $\boldsymbol{P}$ and $\boldsymbol{E}$ (Sect. 4.5.1).

We emphasize, however, that this assumption concerning the relation between fields $\boldsymbol{P}(r)$ and $\boldsymbol{E}(r)$, see (4.5.4), is sufficiently general and correctly describes the circumstances for the region of an isolated exciton or phonon resonance.

It is of interest to find the form of the kernel in the integral relation (2.1.3), which corresponds to the approximation (4.5.4) for an isotropic non-gyrotropic medium. In this case (4.5.4) can be rewritten (for $\hat{A}^{(2)} = 0$) in the form

$$\frac{\hbar\omega_L}{m_e^*}\Delta\boldsymbol{P} - \frac{\partial^2\boldsymbol{P}}{\partial t^2} - \omega_L^2\boldsymbol{P} - \nu\frac{\partial\boldsymbol{P}}{\partial t} = \frac{\omega_p^2 f}{4\pi}\boldsymbol{E}\,. \tag{4.5.4a}$$

Additionally introduced in (4.5.4a) is a term proportional to ν, accounting for dissipation. Also, the following notation is used:

$$m_e^* = -\hbar\omega_L\rho/\alpha\,, \qquad \Delta = \frac{\partial^2}{\partial x_i \partial x_i}\,.$$

The quantity m_e^* denotes the effective mass of a "mechanical" exciton since for normal waves of frequency Ω and with the wave vector $\boldsymbol{k}$ satisfying (4.5.4a) at $\boldsymbol{E} = 0$ (and $\nu = 0$), the frequency $\Omega \approx \omega_L + \hbar k^2/2m_e^*$. Hence, (4.5.4a) and (4.5.4) are sometimes said to be equivalent to an approximation of the effective mass of the exciton.

If follows from (4.5.4a) that in an unbounded medium for the monochromatic field $\boldsymbol{E}(\boldsymbol{r}, t) = \boldsymbol{E}(\boldsymbol{r})\exp(-\mathrm{i}\omega t)$, the polarization is

$$\boldsymbol{P}(\boldsymbol{r}, t) = \boldsymbol{P}(\boldsymbol{r})\,\mathrm{e}^{-\mathrm{i}\omega t}\,,$$

with

$$P(r) = \frac{\omega_p^2 f}{4\pi} \int G(r-r',\omega) E(r') dr' ,$$

where G is a Green's function for (4.5.4a), satisfying the equation

$$\hat{L} G(r,r';\omega) = -\delta(r-r') \frac{m_e^*}{\hbar\omega_L} ,$$

where the operator

$$\hat{L} = \Delta - \Gamma^2 , \quad \Gamma^2 = \frac{m_e^*}{\hbar\omega_L} (\omega_L^2 - \omega^2 - \mathrm{i}\nu\omega) .$$

It can readily be seen that a solution of the equation for G, which decreases as $|r-r'|$ increases, is of the form

$$G(r,r';\omega) = \frac{m_e^*}{\hbar\omega_L} \frac{1}{4\pi} \frac{\mathrm{e}^{-\Gamma|r-r'|}}{|r-r'|} ,$$

in which $\mathrm{Re}\{\Gamma\} > 0$. At $\nu = 0$ and when $\omega \to \omega_L$, the quantity $|\Gamma| \to 0$. This means that in the approximation $\nu = 0$, the range of action of the kernel ε_{ij} in (2.1.3) becomes unbounded. If, however, we take into account the fact that $\nu \neq 0$, then $|\Gamma| \approx |m_e^* \nu/\hbar|^{1/2}$ and, for example, at $|m_e^*| = m_0$ and $\nu = 10^{-4}\omega_L$, $|\Gamma| \approx 10^{-2}(m_0\omega_L/\hbar)^{1/2}$, where $\omega_L \approx 10^{15}\,\mathrm{s}^{-1}$ (m_0 being the mass of a free electron). Hence, in this case, the above-mentioned range of action is $a = 1/|\Gamma| \approx 300$ Å. The value of a decreases at higher values of ν so that in all real cases the quantity a remains small compared to the wavelength λ.

Within the framework of the model under investigation, the kernel of the integral equation (2.1.3) is of the form

$$\varepsilon_{ij}(\omega;r,r') = \varepsilon_{00ij}\delta(r-r') + \omega_p^2 f G(r-r';\omega)$$

and thus, as could be expected for an unbounded medium, depends only on the difference of the coordinates $r-r'$.

4.5.3 Boundary Conditions

General Discussion

Next, let us turn to an analysis of the additional boundary conditions (ABC's) on the basis of the expressions established above which are applicable in the vicinity of an isolated resonance. In contrast to (4.5.4), the basic relation (4.5.3 or 11) is also valid close to the surface of the medium. Assuming that

$P(r',t)=P(r')\exp(-i\omega t)$ and $E(r',t)=E(r')\exp(-i\omega t)$, and that the functions $P(r')$ and $E(r')$ depend sufficiently smoothly on r', we shall expand these functions in (4.5.3) in a power series about the point $r=r'$. As a result, instead of (4.5.4), at the surface of the crystal we obtain the following relation (the argument 0 corresponds to the crystal surface):

$$(-\omega^2\rho_{ij}+\beta'_{ij})P_j(0)+\gamma'_{ijl}\frac{\partial P_j(0)}{\partial x_l}=\Lambda'^{(0)}_{ij}E_j(0)+\Lambda'^{(1)}_{ijl}\frac{\partial E_j(0)}{\partial x_l}\,. \qquad (4.5.23)$$

It is significant here that $\beta'_{ij}\neq\beta_{ij}$, $\gamma'_{ijl}\neq\gamma_{ijl}$, $\Lambda'^{(0)}_{ij}\neq\Lambda^{(0)}_{ij}$ and $\Lambda'^{(1)}_{ijl}\neq\Lambda^{(1)}_{ijl}$. Thus, in media having an inversion center, where $\Lambda^{(1)}_{ijl}=\gamma_{ijl}=0$, the tensors γ'_{ijl} and $\Lambda'^{(1)}_{ijl}$ are generally nonzero. This last is due to the fact that in the presence of a boundary surface, even a nongyrotropic crystal is not invariant with respect to inversion. For reasons mentioned previously (long wavelengths), terms containing higher derivatives of $P(r)$ and $E(r)$ are not taken into account in (4.5.23).

It was actually assumed in deriving (4.5.23) that the depth l of the region of the subsurface layer (where the values of the tensors β_{ij}, ρ_{ij}, etc., appearing in (4.5.4), differ significantly from their bulk values) is much less than wavelength λ of any normal solution. In certain crystals, however, the inequality $l/\lambda \ll 1$ is not valid for the frequency regions of excitons having a large radius. In this case, a transition layer of thickness l on the surface of the crystal should be taken into account with better accuracy. The optical properties (polarizability) of this layer differ from the optical properties in the bulk of the crystal. If, for instance, the energy of an exciton in this layer differs substantially from that of an exciton in the bulk, then the contribution of the excitons to the polarizability of the transition layer decreases in the frequency region under consideration (model of a "dead" layer [4.82], Sect. 4.5.7). Other situations in the transition layer are also possible, of course; these correspond, for example, to a change in the polarization of the exciton, or its oscillator strength, etc. In all cases mentioned, the boundary condition (4.5.23) should be applicable at $z=l$ rather than at $z=0$.

Relation (4.5.23) can be rewritten in the form

$$P_i(0)+\frac{c}{\omega}T^{-1}_{ijl}\frac{\partial P_j(0)}{\partial x_l}+\Gamma_{ij}E_j(0)+\Gamma_{ijl}\frac{\partial E_j(0)}{\partial x_l}=0\,. \qquad (4.5.23\,\mathrm{a})$$

Although the classical oscillator model of the crystal was used in deriving the preceding relation, this relation is evidently more general. The tensors T^{-1}_{ijl}, Γ_{ij} and Γ_{ijl} appearing in this relation can be found within the framework of microtheory only when some model of the crystal is employed. If, however, we do not resort to microtheory, these tensors, as well as the dielectric-constant tensor, appear as new phenomenological quantities, being functions of the frequency ω in general.

We could, of course, have refrained from introducing ABC, (4.5.23 a), by making use of the more general relation (4.5.3). For fields varying with time according to a harmonic law, this relation can also be written in the form

$$P_i(\boldsymbol{r}) = \int \chi_{ij}(\omega, \boldsymbol{r}, \boldsymbol{r}') E_j(\boldsymbol{r}') d\boldsymbol{r}' , \tag{4.5.23 b}$$

where the tensor integral operator $\hat{\chi} \equiv (-\omega^2 \hat{\rho} + \hat{\mathscr{K}})^{-1} \hat{\Lambda}$ contains complete information on the conditions at the boundary. Far away from the boundaries and for plane waves of the type (4.5.5), the relation (4.5.23b) is equivalent to (4.5.6). We emphasize that both (4.5.23 a and b) are determined by the properties of the excited states of the crystal and depend strongly on the conditions at the boundary. Generally, these conditions can vary within quite wide limits. Hence, the values of the tensors appearing in (4.5.23 a), as well as the explicit form of (4.5.23 b) at the boundary of the crystal, are not at all universal.

The tensors $(c/\omega) T_{ijl}^{-1}$ and Γ_{ijl} have the dimensionality of length. If the conditions at the surface of the crystal are such that the typical length in the region of resonance under consideration is of the order of the lattice constant a, then in (4.5.23 a) the terms proportional to T_{ijl}^{-1} and Γ_{ijl} make only a small contribution, of the order of $a/\lambda \approx 10^{-3}$ (λ being the wavelength corresponding to the additional solution), and they can be omitted in (4.5.23 a). In this case, (4.5.23 a) is simplified and assumes the form

$$P_i(0) + \Gamma_{ij} E_j(0) = 0 .$$

It is clear, see (4.5.22, 23), that the quantity $\Lambda_{ij}^{\prime(0)}$, like $\Lambda_{ij}^{(0)}$, is nonzero only for dipole oscillations. For the same reason, Γ_{ij} is nonzero only in the vicinity of dipole exciton-absorption lines. In the opposite limiting case, when it is precisely T_{ijl}^{-1} and Γ_{ijl} which are relatively large, (4.5.23 a) can be written in the form

$$\frac{\partial P_i(0)}{\partial x_l} + t_{ijlm} \frac{\partial E_l(0)}{\partial x_m} = 0 ,$$

where $t_{ijlm} = (\omega/c) T_{ijk} \Gamma_{klm}$.

Further, in the discussion of certain specific models for excitons, it will be shown how ABC of the type (4.5.23 a) can be obtained by making use of microtheory. Here we note only that (4.5.23 a) yields three additional conditions at once. This is exactly the number of additional conditions required in considering boundary-value problems when we take into account the three bands of the "mechanical" exciton whose polarization forms a triad of noncoplanar coordinate axes. If, however, at $\boldsymbol{k} = 0$ the frequencies of the "mechanical" excitons are sufficiently distant from one another, then, in deriving ABC, it proves more convenient to consider only a single band and to employ (4.5.11). In this case, instead of three relations (4.5.23), we obtain a single relation, which, in the case of bands of dipole excitons, is of the form [recall that under the present conditions $\boldsymbol{P} = P(\boldsymbol{r})\boldsymbol{e}$]

$$P(0) + \frac{c}{\omega T} \frac{dP}{dz}\bigg|_0 + \Gamma(\boldsymbol{E} \cdot \boldsymbol{e}) = 0\,, \qquad \text{where} \tag{4.5.23c}$$

$$\frac{1}{T} = \frac{\omega}{c} \frac{\gamma'}{\beta' - \omega^2 \rho} \quad \text{and} \quad \Gamma = \frac{\Lambda'^{(0)}}{\beta' - \omega^2 \rho}\,.$$

In the vicinity of a resonance $\varepsilon_{0ij}(\omega)$, the relation $\omega^2 \approx \beta/\rho$ is valid. Therefore, assuming that $\Lambda'^{(0)} \approx \Lambda^{(0)}$ and making use of the relation $4\pi\Lambda^{(0)}/\rho = \omega_p^2 f$ (Sect. 4.5.2), where f is the oscillator strength and ω_p is the "plasma" frequency, we obtain

$$\Gamma \approx \frac{\omega_p^2 f}{4\pi} \frac{\rho}{\beta' - \beta}\,.$$

If $|\beta' - \beta| \sim \beta$, then

$$\Gamma \approx \frac{1}{4\pi}\left(\frac{\omega_p}{\omega_L}\right)^2 f\,,$$

where $\hbar\omega_L = \hbar\sqrt{\beta/\rho}$ is the energy of the "mechanical" exciton at $\boldsymbol{k} = 0$. If, however, $|\beta' - \beta| < \beta$, which is more likely to be true, then Γ increases and any further refinement of the estimates can be accomplished only if we investigate a specific model of the exciton. For $|\beta' - \beta| \to 0$, the quantity $\Gamma \to \infty$, so that ABC, (4.5.23c), assumes the form

$$\boldsymbol{E} \cdot \boldsymbol{e} = 0\,.$$

If, however, $\Gamma_{ij} \to \infty$, then instead of (4.5.23a) we obtain ABC

$$\boldsymbol{E} = 0\,.$$

For quadrupolar "mechanical" excitons, whose contribution to the polarization P is linear with respect to the wave vector, the factor $\Lambda_{ij}'^{(0)} = 0$. Hence, in the additional boundary conditions it is necessary to take into account the following terms of the expansion with respect to $\boldsymbol{k}$, bringing these boundary conditions to the form

$$P_i(0) + \frac{cT_{ijl}^{-1}}{\omega} \frac{\partial P_j}{\partial x_l}\bigg|_0 + \Gamma_{ijl} \frac{\partial E_j(0)}{\partial x_l} = 0\,.$$

In this relation, as in (4.5.23a), T_{ijl} and Γ_{ijl} are tensors of rank three, independent of ω in the frequency region under consideration.

For the degenerate exciton state region and especially in cases when there are dipolar, quadrupolar or magnetic dipolar states among states close in energy, the problem of finding the ABC becomes considerably more complicated. It is hardly reasonable under such conditions to search for general phenomenological ABS's. The number of unknown constants substantially increases and it is therefore of greater advantage to resort at once to microtheory [4.81a, 168].

As mentioned above, the tensors T_{ijl}^{-1}, Γ_{ij} and Γ_{ijl} can be found for specific crystals only on the basis of microtheory. The symmetry properties of these tensors, however, are determined by the symmetry of a finite, rather than infinite, crystal, and can be applied. Thus, for instance, in a uniaxial crystal bounded by a plane perpendicular to a threefold or higher axis of symmetry (z axis), the tensor Γ_{ij} has the following nonzero components: $\Gamma_{xx}=\Gamma_{yy}\equiv\Gamma_t$ and $\Gamma_{zz}\equiv\Gamma_n$. This is due to the fact that the above-mentioned symmetry axis is not affected by the presence of a limiting surface. The same components of Γ_{ij} are nonzero if the plane touches an isotropic medium or is perpendicular to one of the seven optical axes of a cubic crystal. The only nonzero components of Γ_{ijl} in the case under investigation are $\Gamma_{xxz}=\Gamma_{yyz}\equiv\Gamma_{tz}$ and $\Gamma_{zzz}\equiv\Gamma_{nz}$.

To illustrate the "action" of the boundary conditions we shall consider in the following a number of specific examples. We find it appropriate here to point out again that the additional boundary conditions (4.5.23) were actually derived under the assumption that the typical length close to the surface of the crystal, over which quantities of the form

$$\varphi(\boldsymbol{r})=\int\mathscr{K}_{ij}(\boldsymbol{r},\boldsymbol{r}')(x-x_1)^n(y-y_1)^m\dots d\boldsymbol{r}'$$

(where $n,m,\dots=1,2,\dots$) vary appreciably (these being the quantities in terms of which the tensors of (4.5.23) are expressed at $\boldsymbol{r}=0$), is small compared to the wavelength in the medium. Otherwise the thickness of the surface layer cannot be neglected and we are faced with the problem of the propagation of electromagnetic waves in a nonhomogeneous medium. The exact solution of this problem is extremely complicated. It proves convenient therefore to divide the subsurface layer along its depth into several parallel layers with constant optical properties in each layer [4.82]. We shall return to this question again in Sect. 4.5.7.

Next we shall consider, in a more general statement, the problem of taking the boundaries into account in media having spatial dispersion. We shall discuss methods for solving this problem, which are based both on the introduction of ABC of some kind and by the use of various approximations for the kernel $\varepsilon_{ij}(t-t';\boldsymbol{r},\boldsymbol{r}')$ of the integral relation appearing in (2.1.3). It is clear that in order to find this kernel within the framework of microtheory it is necessary to calculate the wave functions and energies of the ground and excited states of the crystal, taking the boundary into account. Then, using the properties of these states, we determine $\boldsymbol{P}(\boldsymbol{r})$ for the field $\boldsymbol{E}(\boldsymbol{r},t)$ set up by the propagating electromagnetic wave, where $\boldsymbol{P}(\boldsymbol{r})$ is the polarization produced in the crystal in the presence of excitation. In this case,

$$P_i(\boldsymbol{r},t) = \int \chi_{ij}(t-t';\boldsymbol{r},\boldsymbol{r}')E_j(\boldsymbol{r}',t')d\boldsymbol{r}'dt' ,$$

so that

$$\varepsilon_{ij}(t-t';\boldsymbol{r},\boldsymbol{r}') = \delta(t-t')\delta(\boldsymbol{r}-\boldsymbol{r}') + 4\pi\chi_{ij}(t-t';\boldsymbol{r},\boldsymbol{r}') .$$

It follows from the expression for P_i that in calculating the states of the crystal mentioned above, the macroscopic part of the electric field should be disregarded in the equations of motion (or in the Hamiltonian). This means that in calculating the polarizability χ_{ij} in a bounded medium, "mechanical" excitons should be used as in calculating the dielectric constant of unbounded media. If, however, use is made of the solution of the complete Coulomb problem, then only the transverse part of the field manifests itself as a perturbation. Hence,

$$P_i(\boldsymbol{r},t) = \int \chi_{ij}^{\perp}(t-t';\boldsymbol{r},\boldsymbol{r}')E_j^{\perp}(\boldsymbol{r}',t')d\boldsymbol{r}'dt' .$$

This method for calculating the polarization is simpler because corrections to the effective local transverse field are of order $(a/\lambda)^2$ and in the region of exciton resonance can be neglected in most cases.

It was emphasized above that the kernel $\tilde{\varepsilon}_{ij}(t-t';\boldsymbol{r},\boldsymbol{r}')$ contains, in particular, all the information about the properties of the subsurface region of the crystal that is required for solving problems in crystal optics of bounded media when their spatial dispersion is taken into account. The form of this kernel is not universal, however, and may be substantially modified, even for the same crystal, upon changes in the properties of the transition layer, which always exists at the boundary between two media. If, for instance, we consider a boundary toward vacuum, the properties of the transition layer may depend essentially on the chemical processes at the surface in which the remaining gases participate, and on the manufacturing process used to prepare the specimens. These properties may change when the surface is treated, e.g., by application of a thin film. Consequently, there seems to be no prospect for finding an expression for the kernel $\varepsilon_{ij}(t-t';\boldsymbol{r},\boldsymbol{r}')$, on the basis of a consistent application of microtheory, which would be suitable for all possible structures of the transition layer. A semiphenomenological way of searching for the kernel $\varepsilon_{ij}(t-t';\boldsymbol{r},\boldsymbol{r}')$ may turn out to be more fruitful. This method is analogous to one that has been used for a long time in investigating optical phenomena in metals in the region of an anomalous skin effect [4.83, 84]. It also proves necessary to use (2.1.3). The form of the kernel $\tilde{\varepsilon}_{ij}$, when we have a sharp boundary, can be expressed in terms of a phenomenological parameter, namely the amplitude of electron reflection from the boundary of the metal.

The Transition ("Dead") Layer Problem

For the exciton region of the spectrum, such a phenomenological parameter could be the amplitude of the exciton wave reflected from the surface of the crystal [4.69, 85, 86]. This factor, however, characterizes only the asymptotic form of the exciton wave function (i.e., its form at some distance from the crystal boundary). In the boundary region the exciton wave function can, in general, contain various kinds of additional rapidly diminishing terms, considered by introducing the transition layer mentioned above. The existence of these terms can be due to the presence of surface exciton levels or to the lack of sharpness of the boundary or to the fact that the structure of the exciton is revealed in the interaction of an exciton even with a sharp boundary (i.e., the fact that the exciton is, for example, a bound state of an electron and a hole, and the nature of its interaction with the boundary may depend upon the binding energy of the electron and the hole in the exciton [4.87]). It is clear from the aforesaid that the depth of the transition layer depends upon the type of exciton even in cases when the surface exciton levels are absent. Thus, in the case of Frenkel excitons, the corresponding depth of the layer for quadrupolar excitons is a quantity relating to several lattice constants. The fact that dipole Frenkel excitons are 'mechanical' dipole excitons, with the interaction being due to only the nearby surroundings, should be taken into account. Hence, for these excitons as well, the corresponding depth of the layer is evidently also small, notwithstanding the fact that for Coulomb excitons (for which the wave vector $k \sim 2\pi/\lambda_0$, λ_0 being the wavelength of light in vacuum) the effective region of interaction, according to calculations in [4.88], has a size of the order of 20 Å. The situation can be different for Wannier-Mott excitons in semiconductors. Here, even when the long-wavelength part of the field is ignored, the thickness of the transition layer is evidently of the order of the Bohr radius of the exciton, i.e., $l \approx 100$ Å (Sect. 4.6.2).

In connection with this situation, in [4.82] (whose results, as well as those of other experimental investigations along this line, are dealt with in more detail in Sect. 4.6.2) it is proposed that the exciton part of polarization be entirely ignored in a certain transition layer (called the "dead" layer) of thickness l (this phenomenological parameter is selected by comparing the calculations of the reflection spectrum with experimental data). Since the work function for removing an electron from a semiconductor is usually much greater than the binding energy of the electron and the hole in an exciton, values of the wave functions of the electron and the hole in an exciton in vacuum, i.e., also at the surface of the crystal, are small. In this connection, it was assumed in [4.89] that these values vanish at the boundary of the crystal. From this it immediately follows that the relation $\psi(z) = 0$ for the exciton wave function also holds. Calculations carried out in [4.89] support the assumption that $\psi(z) \approx 0$ at the boundary toward vacuum and to a depth of the order of the Bohr radius of the exciton $a_0 = \hbar^2 \varepsilon(0)/\mu e^2$, where $\varepsilon(0)$ is the static dielectric constant, $\mu = (m_e^{-1} + m_h^{-1})^{-1}$, and m_e and m_h are the masses of the electron and hole. The result obtained for the asymptotic form of $\psi(z)$ at $z > a_0$ is not trivial. At $z \gg a_0$, as shown in [4.89], the function $\psi(z) \approx \cos Qz$, Q being the wave vector of the exciton. This form of the wave function indicates that in the region where $Qz \ll 1$ and $z \approx l \gtrsim a_0$, the exciton part of the polarization satisfies the condition $dP/dz|_l = 0$ [recall that the function $\psi(z) \sim \sin Qz$ was used in [4.12], leading to ABC: $P(0) = 0$]. ABC of the form $P(0) = 0$ was used in [4.82] at the boundary of the "dead" layer l in calculations of the reflection spectra. In more recent calculations [4.90, 91] consideration has been given to a more extensive range of feasible ABC at the boundary of the "dead" layer, within the framework of the relation $\alpha P + \beta dP/dz = 0$. In [4.90] the reflection and transmission factors of light were calculated for the exciton range ($E_{exc} \approx 2.8$ eV) of the spectrum for films of the crystal ZnSe dependent on the film thickness, the frequency and the thickness of the "dead" layer. The results obtained in [4.90] depend essentially upon the choice of values for the parameters α, β and l.

Attempts have been made in recent years to advance a theory of the "dead" layer [4.92–94]. Since, in its precise formulation, the solution of such a problem encounters considerable difficulties, various kinds of approximations are usually resorted to. In order to elucidate their essence and the meaning of the results they lead to, we shall consider an equation that should be satisfied by the wave function of a crystal in an external electromagnetic field of frequency ω.

We shall, as previously, proceed from the Schrödinger time equation for the wave function Ψ of a crystal:

$$i\hbar \frac{\partial \Psi}{\partial t} = (\hat{H}_0 + \hat{W})\Psi ,$$

where $\hat{H}_0$ is the Hamiltonian of the crystal, $\hat{W}$ is the operator of interaction with the electromagnetic field, i.e.,

$$\hat{W} = -\frac{e}{c}\sum_i \dot{r}_i A(r_i, t) ,$$

r_i is the coordinate of the i-th electron, A is the vector potential of the field, and $A \propto \exp(i\omega t)$. The state of the crystal perturbed by the field can be found in the form

$$\Psi = \Psi_0 + Y(r_e, r_h)e^{i\omega t} ,$$

where Ψ_0 is the wave function of the ground state, and Y is a function of the electron and hole coordinates in the Wannier-Mott exciton. Retaining in the Schrödinger equation only terms of the first order of smallness with respect to the perturbance $\hat{W}$, and using the effective-mass approximation, we can rewrite this equation as follows:

$$\left[E_g-\hbar\omega-\frac{\hbar^2}{2m}\Delta_R-\frac{\hbar^2}{2\mu}\Delta_r-V(r)\right]Y(R,r)=M(r)E(R)\,,$$

where E_g is the gap energy, $V(r)$ is the interaction potential between the electron and hole, $\boldsymbol{r}=\boldsymbol{r}_e-\boldsymbol{r}_h$, $\boldsymbol{R}=(m_e\boldsymbol{r}_e+m_h\boldsymbol{r}_h)/m$, $m=m_e+m_h$, $\mu^{-1}=m_e^{-1}+m_h^{-1}$, E is the electric field of a light wave, and $M(r)$ is the transition dipole density (for details in the derivation of the equation for $Y(\boldsymbol{r}_e,\boldsymbol{r}_h)$ see [4.92]). In [4.92] the problem under investigation is simplified by going over to a one-dimensional exciton model with point-like interaction V. It is precisely in place of the above equation for Y that the following equation is considered:

$$\left[E_g-\hbar\omega-\frac{\hbar^2}{2m}\frac{\partial^2}{\partial z^2}-\frac{\hbar^2}{2\mu}\frac{\partial^2}{\partial x^2}-\hbar U_0\delta(x)\right]Y(z,x)=M_0\delta(x)E(z)\,,$$

where $z=(m_e z_e+m_h z_h)/m$ and $x=z_e-z_h$.

Since the electromagnetic field $\boldsymbol{E}$ should comply with Maxwell's equations, the equation

$$\Delta_R\boldsymbol{E}+\frac{\omega^2}{c^2}\boldsymbol{D}=0$$

should be taken into account along with the above equation for Y. In the case under study $\boldsymbol{D}=\varepsilon\boldsymbol{E}+4\pi\boldsymbol{P}$, where ε is the background part of the dielectric polarization, whereas $\boldsymbol{P}$ is the exciton part of the polarization. Thus, $P(R)=\int M(r)\,Y(R,r)\,dr$, so that at $M(r)=M_0\delta(r)$ we have $P(R)=M_0Y(R,0)$. Within the framework of the one-dimensional model, the equation for E, which should be considered together with the equation for $Y(z,x)$, is evidently of the form

$$\frac{\partial^2E}{\partial z^2}+\varepsilon\frac{\omega^2}{c^2}E=-\frac{4\pi\omega^2}{c^2}M_0Y(z,0)\,.$$

Used in [4.92] as the boundary condition for $Y(z,x)$ is the following: $Y=0$ at $z_e=0$ or $z_h=0$. This is equivalent to the vanishing of function Y at $x=mz/m_e$ and $x=-mz/m_h$. The polarizability $\chi(z,\omega)$ of the crystal, depending on z and ω, was calculated under the assumption (the adiabatic approximation [4.93, 93a]) that $m_e \ll m_h$, and for a value ω within the region of exciton resonance. It was shown, as a result of these calculations, that the polarizability of the crystal actually does depend on the distance to the boundary. Furthermore, this relationship extends over macroscopic distances of the order of one or two Bohr radii of the exciton, and it corresponds in its nature to the suppression of exciton resonance at the boundary, thereby qualitatively confirming the initial idea of *Thomas* and *Hopfield* [4.82]. Of interest is the fact that the calculations under discussion also indicate the dependence of the "dead" layer depth on the frequency, and at $\omega > \omega_T$, the dependence $\chi(z)$ is even nonmonotonic. To what extent these results are not associated with features of the model, but are general, can be shown only by subsequent, more precise investigations [4.94].

The quantities α, β and l (the last being the thickness of the "dead" layer) are proposed in all the works mentioned above as phenomenological parameters, and determined by comparing calculations with experimental data, as has been previously indicated. These parameters can be found (calculated) for a particular type of exciton only within the framework of microtheory. A few examples are available for crystals with Frenkel excitons [4.70, 95, 105] and with Wannier-Mott excitons [4.89, 96]. There is no doubt that in the future many more investigations will deal with the microtheory of ABC.

In connection with the aforesaid, we should point out that some time ago a number of papers were published that postulated the identity of the kernel in integral relation (2.1.3) with that in the analogous integral relation for an unbounded medium [4.97, 98]. Here the relation (2.1.3) was found to be valid; this, naturally, allowed these authors to manage without introducing ABC. In view of the aforesaid, this circumstance is not surprising. The assumption mentioned above is actually equivalent to a particular ABC (not containing additional constants). Such ABC cannot be obtained from realistic models. As has been shown in [4.98], the law of conservation of energy is violated; when applied the surface appears to be a source of energy. As has been mentioned, the approximation used in [4.97, 98] for a semi-infinite medium $(z > 0)$ agrees with the relation

$$P_i(\boldsymbol{r}) = \int_{z>0} \chi_{ij}^0(\boldsymbol{r}-\boldsymbol{r}')E_j(\boldsymbol{r}')d\boldsymbol{r}' ,$$

where $\chi_{ij}^0(\boldsymbol{r}-\boldsymbol{r}')$ is the kernel appearing in the integral relation (2.1.3) for an unbounded medium, see also (2.1.5). At the same time the boundary condition $\boldsymbol{P}|_{z=0} = 0$ satisfies the following expression for χ_{ij} (see [4.100, 101] where the relation between the type of ABC and certain simple expressions for χ_{ij} are also discussed)

$$\begin{aligned}\chi_{ij}(\boldsymbol{r},\boldsymbol{r}') = {} & \chi_{ij}^0(|x-x'|, |y-y'|, |z-z'|) \\ & - \chi_{ij}^0(|x-x'|, |y-y'|, |z+z'|) .\end{aligned}$$

If, however, $dP/dz|_0 = 0$, then

$$\begin{aligned}\chi_{ij}(\boldsymbol{r},\boldsymbol{r}') = {} & \chi_{ij}^0(|x-x'|, |y-y'|, |z-z'|) \\ & + \chi_{ij}^0(|x-x'|, |y-y'|, |z+z'|) .\end{aligned}$$

Theoretically, no additional constant appears in either case. It is clear, of course, that such a situation can occur only in special limiting cases. In general, however, the kernel $\chi_{ij}(\boldsymbol{r},\boldsymbol{r}')$ cannot be written for bounded media only in terms of the kernel χ^0_{ij}, which characterizes the bulk properties of the medium.

ABC for Molecular Crystals and the Surface Current Appearance

In order to illustrate how the problem of ABC can be solved for one or another medium on the basis of a consistent application of the microscopic theory, and also to examine the role played by surface excitons and to elucidate the microscopic nature of phenomenological constants appearing in ABC, we shall discuss the simplest model of a molecular cubic crystal having a single (isotropic) molecule in each unit cell [4.95].

Consider the crystal to be semi-infinite and bounded by a vacuum along the plane $z=0$ [in this case the coordinates of the molecules are $\boldsymbol{n}=(n_1a, n_2a, ma)$, where $-\infty \leqslant n_{1,2} \leqslant \infty$ and $m=1,2,\dots$]. For this crystal we shall examine polarization vibrations that are set up upon normal incidence of a light wave of frequency ω on the crystal. If the frequency ω is close to the frequency Ω of the allowed molecular dipole transition, then the vibrations that are set up are Frenkel excitons and can be considered in classical terms (for substantiation see Sect. 6.2). It is clear that in the case under consideration the vibrations of the dipole moments of the molecules are polarized in the same way as the electric field of the light wave; hence the vector subindices are not indicated in the following.

If u_n is the charge displacement in the n-th molecule, then taking intermolecular interaction into account, we have the following system of equations for the quantities u_n:

$$-M\omega^2 u_n = -M\Omega^2 u_n - \sum_{n'}{}' A_{nn'} u_{n'} + eE_n^{\perp}\,, \tag{4.5.24}$$

where e and M are the effective charge and mass of the molecular oscillator, $E_n^{\perp}$ is the amplitude of the transverse part of the electric macrofield at the point $\boldsymbol{n}$, and $A_{nn'}$ is the matrix element of thermomolecular interaction. $A_{nn'}$ determines the energy of the dipole-dipole resonance interaction between molecules $\boldsymbol{n}$ and $\boldsymbol{n}'$ (this energy is found when retarded interaction is neglected). The summation with respect to $\boldsymbol{n}'$, appearing in the system (4.5.24), divided by e is equal to the whole longitudinal electric field set up by the dipole moments $p_{n'}=eu_{n'}$ (where $\boldsymbol{n}' \neq \boldsymbol{n}$) at point $\boldsymbol{n}$.

In an unbounded medium the Ewald procedure [4.74] enables the macroscopic part $\boldsymbol{E}_n^{\parallel}$ to be separated out of this whole field so that the system (4.5.24) relates the polarization $\boldsymbol{P}_n = \boldsymbol{p}_n/a^3$ of the medium to the whole macrofield $\boldsymbol{E}_n = E_n^{\perp} + E_n^{\parallel}$ in the form $\boldsymbol{P}_n = \chi(\omega,\boldsymbol{k})\boldsymbol{E}_n$. This permits the dielectric constant $\varepsilon(\omega,\boldsymbol{k}) = 1+4\pi\chi(\omega,\boldsymbol{k})$ to be determined (see Sect. 6.1.2, where the more general case of an anisotropic molecular crystal is discussed). A generalization of the Ewald procedure has not yet been carried out for bounded media. Consequently, we shall restrict ourselves below to a discussion of the case of normal incidence of light, when the longitudinal part of the field $E^{\parallel}=0$, so that $E=E^{\perp}+E^{\parallel}=E^{\perp}$.

For normal incidence $E_n \equiv E_{n_1,n_2,m}=E_m$ and, since the displacement u_n actually depends only on the subindex m, $u_n=u_m$. Hence

$$\sum_{n'}{}' A_{nn'}u_{n'}=\sum_{m'}\tilde{A}_{mm'}u_{m'}\,,$$

where

$$\tilde{A}_{mm'}=\sum_{n_1',n_2'}A_{n_1n_2m;n_1'n_2'm'}\,,$$

a quantity determining the interaction energy of the dipole $p=e\cdot 1$ cm (where e is the charge of the electron) located in the plane $m=$ const, with a two-dimensional lattice of dipoles of the same length and orientation, located at the sites of a lattice with $m'=$ const (at $m=m'$, the action of the dipole on itself should be neglected, as indicated by the primed summation sign). Thus (4.5.24) can be rewritten as follows:

$$M(\omega^2-\tilde{\Omega}^2)u_m=\sum_{m'\neq m}\tilde{A}_{mm'}u_{m'}-eE_m\,, \tag{4.5.24a}$$

where $\tilde{\Omega}^2=\Omega^2+\tilde{A}_{mm}/M$. As shown in [4.102], the quantity $\tilde{A}_{mm'}=\tilde{A}_{m'm}$ decreases exponentially with an increase of $|m-m'|$. Consequently, in solving the system of equations (4.5.24a) only terms with $m'=m\pm1$ need be taken into account in the summation with respect to m' (recalling that $A_{-1,0}=0$). Solving the system of equations for the quantities u_m, we can obtain a relation between the polarization of unit volume $P_m=eu_m/a^3$ and the electric field strength:

$$P_m=\sum_{m'}\chi_{mm'}E_{m'}\,,$$

which is analogous to (2.1.3). The application of such a relation, accounting for spatial dispersion, enables the amplitude of all the fields to be determined. We can also proceed in a different way. We can combine Maxwell's "ideal" equations with the system (4.5.24a) and seek the solution of the complete system of equations thus obtained in the form of a superposition of plane waves. This is the approach which we shall discuss here because, within its framework, ABC appears in the most elementary way. First, however, we shall generalize, to some extent, the model under investigation. Namely, we shall assume that the molecules in the boundary layer $m=0$ are "damaged", i.e., they differ from the molecules in the bulk both in the value of their natural frequency $(\tilde{\Omega}_0 \neq \tilde{\Omega})$ and in the values of their effective mass and charge $(M_0 \neq M$ and $e_0 \neq e)$.

At a distance from the boundary $(m > 0)$, in an approximation to the nearest two-dimensional lattices (see above), (4.5.24a) is of the form

$$M(\omega^2 - \tilde{\Omega}^2)u_m = \tilde{A}_{m,m+1}(u_{m+1} + u_{m-1}) - eE_m,$$

whereas, for a plane wave $E_m = E\exp(\mathrm{i}kma)$, the polarization of unit volume is

$$P_m = -\frac{(e^2/a^3M)E_m}{\omega^2 - \tilde{\Omega}^2 - 2\tilde{A}_{m,m+1}\cos ka} \equiv \chi(\omega,k)E_m.$$

Hence, the dielectric constant for long wavelengths (i.e., when $ka \ll 1$) is

$$\varepsilon(\omega,k) = \varepsilon_{00} + \frac{\omega_p^2}{\Delta + \beta k^2}, \qquad \text{where} \tag{4.5.24b}$$

$$\omega_p^2 = 4\pi e^2/Ma^3,\ \Delta = \omega_\perp^2 - \omega^2,\ \omega_\perp^2(0) = \tilde{\Omega}^2 + 2\tilde{A}_{m,m+1},$$
$$\beta = -\tilde{A}_{m,m+1}a^2 = \hbar\omega_\perp(0)/m^*_{\mathrm{exc}},$$

$\omega_\perp(0)$ being the resonance frequency. It follows from Maxwell's equations that the possible values of k are determined by the condition

$$k^2c^2/\omega^2 = \varepsilon(\omega,k),$$

which leads to two values of the wave vector: k_1 and k_2.

Hence, in the medium

$$E_m = E_1\mathrm{e}^{\mathrm{i}k_1ma} + E_2\mathrm{e}^{\mathrm{i}k_2ma},$$

so that for u_m we also obtain

$$u_m \equiv \frac{a^3P_m}{e} = \frac{a^3}{e}[\chi(\omega,k_1)E_1\mathrm{e}^{\mathrm{i}k_1ma} + \chi(\omega,k_2)E_2\mathrm{e}^{\mathrm{i}k_2ma}]. \tag{4.5.25}$$

By virtue of the aforesaid, the equation relating u_0 and u_1 is of the following form

$$M_0(\omega^2 - \tilde{\Omega}_0^2)u_0 = \tilde{A}_{0,1}u_1 - e_0E_0. \tag{4.5.25a}$$

Here, for the sake of simplicity, we ignore the possible difference between the value of $\tilde{A}_{0,1}$ and its bulk value. Such an assumption does not contradict the inequality $e_0 \neq e$ only for those cases when the dipole-dipole interaction does not make any appreciable contribution to the width of the exciton band (as, for example, in a naphthalene crystal).

Assuming now that (4.5.25) is valid at $m = 0$ as well, we substitute the expressions for u_0 and u_1 into (4.5.25 a). This substitution leads to the required relation between the amplitudes E_1 and E_2, i.e., to ABC.

In the case when the boundary molecules ($m = 0$) do not differ from the "bulk" ones ($M_0 = M$, $\Omega_0 = \Omega$ and $e_0 = e$), we obtain

$$u_{-1} = 0 ,$$

where u_{-1} is the value of the function u_m, formally found at $m = -1$. Hence, in this case, ABC is of the form

$$P_1 \mathrm{e}^{-ik_1 a} + P_2 \mathrm{e}^{-ik_2 a} = 0 ,$$

where P_1 and P_2 are the amplitudes of exciton polarization in the waves 1 and 2. Since, by assumption, $k_1 a \ll 1$ and $k_2 a \ll 1$, ABC can also be written in the form

$$P(0) - a \left. \frac{dP}{dz} \right|_0 = 0 ,$$

where $P(z) = P_1 \exp(\mathrm{i} k_1 z) + P_2 \exp(\mathrm{i} k_2 z)$ is the total exciton polarization. In this case, obviously, the obtained ABC does not differ in practice from ABC of the form $P(0) = 0$. If we take into consideration the fact that the effective charge, frequency and mass of a boundary molecule differ from those of molecules located at some distance from the boundary, ABC assumes a more general form, see also (4.5.23 c),

$$\alpha P(0) + \beta \left. \frac{dP}{dz} \right|_0 = \gamma E(0) , \qquad (4.5.26)$$

where

$$\alpha = (M_0 - M)\,\omega^2 + M \tilde{\Omega}^2 - M_0 \tilde{\Omega}_0^2 ,$$
$$\beta = -A_{01} a , \quad \gamma = (e_0 - e)\, e/a^3 . \qquad (4.5.26\text{a})$$

Relation (4.5.26) can also be written as follows:

$$P(0) + \frac{c}{\omega T} \left. \frac{dP}{dz} \right|_0 + \Gamma E(0) = 0 , \qquad (4.5.26\text{b})$$

where, for the model under investigation, the quantities T and Γ are determined by

$$T = \frac{\alpha}{\beta} \frac{c}{\omega}, \quad \Gamma = \frac{\gamma}{\alpha} \equiv -\frac{\omega_p^2}{4\pi} \frac{\eta}{T}, \quad \eta = \frac{c}{\omega} \frac{\gamma}{\beta} \frac{Ma^3}{e^2}.$$

From these relations we can readily find the condition under which the ABC equation is reduced to a more simple one. Thus, at the frequency ω, where $\alpha = 0$, the ABC equation has the form $\beta dP/dz|_0 = \gamma E(0)$. If, on the contrary, the quantity $|\beta| \ll |\alpha/k_i|$, for $i = 1, 2, \ldots$, i.e., it is relatively small, ABC assumes the form $\alpha P(0) = \gamma E(0)$, etc.

We point out that at $\gamma \neq 0$ (i.e., at $e_0 \neq e$), surface currents appear at the boundary of the crystal. As a matter of fact, according to (4.5.25), the polarization per unit volume at a distance from the boundary (i.e., at $m \neq 0$) is determined by the relation $P(m) = eu_m/a^3$. But at $m = 0$ the polarization per unit volume is equal to $e_0 u_0/a^3$. Hence, in the general case,

$$P(m) = \frac{eu_m}{a^3} + \frac{(e_0 - e)u_0}{a^3} \delta_{m0}$$

or, if we go over to the continuous variable $z = ma$ and take (4.5.25) into account,

$$P(z) = \chi(\omega, k_1) E_1 \mathrm{e}^{\mathrm{i}k_1 z} + \chi(\omega, k_2) E_2 \mathrm{e}^{\mathrm{i}k_2 z} + \frac{e_0 - e}{a^2} u_0 \delta(z).$$

Thus, the density of the surface current $j(\omega)$ is determined by the relation

$$j(\omega) = \mathrm{i}\omega \frac{e_0 - e}{a^2} u_0 = \mathrm{i}\omega \frac{e_0 - e}{e} a P^V(0),$$

where $P^V(0) = \chi(\omega, k_1) E_1 + \chi(\omega, k_2) E_2$ is the limit to which the "bulk" value of $P(z)$ tends when $z \to 0$. The presence of a surface current leads to a discontinuity in the tangential components of the magnetic field intensity. In this case, see (2.1.2),

$$H_t(+0) - H_t(-0) = 4\pi N P_t(0), \tag{4.5.26c}$$

where $N = \mathrm{i}\eta\mu$, and $\mu = (\omega/c)^2 \tilde{A}_{10} a^2/M$. This, as is pointed out in [4.103] should be kept in mind when seeking the amplitude of the waves and in examining the energy relations at the boundary (Sects. 2.1.3, 4.5.5).

In conclusion, we mention [4.104] which, in connection with the problem of ABC, discusses a model of a dielectric being reduced to a one-dimensional model. A more general examination of the case of oblique incidence is dealt with in [4.105]. These investigations did not take into account the difference in the properties of the boundary molecules from those of the molecules in the bulk of the crystal. This is the reason why the ABC equations found in [4.104, 105] are not sufficiently general.

Of interest, in connection with the aforesaid, is also the so-called exponential model of a one-dimensional dielectric discussed in [4.106]. This model takes into consideration the effect of the boundary on the intermolecular interaction of the oscillators. It is assumed, within the framework of this model, that

$$A_{mm'} = g\,[\exp(-\gamma|m-m'|) + r\exp(-\gamma|m+m'|)]\,,$$

where g, γ and r are certain phenomenological parameters. This microscopic model serves as the basis for establishing a macroscopic theory of an excitonic dielectric. An integral relation of the type (4.5.11) was obtained, which describes the coupled vibrations of a semilimited array of oscillators, and a nonhomogeneous boundary condition of the type of (4.5.23a) was derived. It is also interesting that surface currents and charges do not appear within the framework of the model discussed in [4.106] so that Maxwell's boundary conditions (2.1.2) can be applied, assuming $i = \sigma = 0$.

4.5.4 Reflected and Refracted Waves Near Dipole and Quadrupole Transition Frequencies in a Nongyrotropic Crystal

Let us consider the simplest case of a monochromatic, linearly polarized wave, normally incident from a vacuum on a plane surface of an isotropic nongyrotropic medium. We choose the z axis along the inner normal and write the complete set of boundary conditions. First the boundary conditions (2.1.2) are applied, assuming that the medium is nonmagnetic ($\boldsymbol{B} = \boldsymbol{H}$). In this approximation the conditions (2.1.2) read as follows [here (4.5.26c) is also taken into account]:

$$\begin{aligned}&\boldsymbol{E}_{1\mathrm{t}} = \boldsymbol{E}_{2\mathrm{t}}\,, \quad H_{1\mathrm{n}} = H_{2\mathrm{n}}\,, \quad D_{1\mathrm{n}} = D_{2\mathrm{n}}\,,\\ &\boldsymbol{H}_{1\mathrm{t}} - \boldsymbol{H}_{2\mathrm{t}} = 4\pi \boldsymbol{N}\boldsymbol{P}_{2\mathrm{t}}\,, \quad \boldsymbol{N}^* = -\boldsymbol{N}\,.\end{aligned} \tag{4.5.27}$$

The additional boundary conditions, by virtue of (4.5.26), see also (4.5.26b), may be written in the following form (using the notation: $\Gamma_{11} = \Gamma_{22} \equiv \Gamma_{\mathrm{t}}$, $\Gamma_{33} = \Gamma_{\mathrm{n}}$):

$$\begin{aligned}&\boldsymbol{P}_{2\mathrm{t}} + \frac{c}{\omega T_{\mathrm{t}}}\left.\frac{\partial \boldsymbol{P}_{2\mathrm{t}}}{\partial z}\right|_0 + \Gamma_{\mathrm{t}}\boldsymbol{E}_{2\mathrm{t}} = 0\,,\\ &P_{2\mathrm{n}} + \frac{c}{\omega T_{\mathrm{n}}}\left.\frac{\partial P_{2\mathrm{n}}}{\partial z}\right|_0 + \Gamma_{\mathrm{n}}E_{2\mathrm{n}} = 0\,.\end{aligned} \tag{4.5.28}$$

For definiteness we shall assume that the subscript 1 refers to vacuum and 2 to the dispersive medium, and that the electric vector of the incident wave lies in the x direction. In this case, when the dispersive medium occupies a half-space, the electric field in it, including the new waves, may be represented as follows:

$$E_{2x} = a_1 \mathrm{e}^{\mathrm{i}\omega\tilde{n}_1 z/c} + a_2 \mathrm{e}^{\mathrm{i}\omega\tilde{n}_2 z/c}, \quad E_{2y} = E_{2z} = 0 . \tag{4.5.29}$$

The second equation of (4.5.29) follows directly from (4.5.27). In the expression for E_{2x}, the quantities $\tilde{n}_1(\omega)$ and $\tilde{n}_2(\omega)$ are two refractive indices corresponding to two waves of the same polarization, which may propagate in a nongyrotropic medium in the vicinity of the resonance $\varepsilon_0(\omega)$ when spatial dispersion is taken into account (Sect. 4.2.2).

The electric field in the vacuum consists of the fields of the incident and reflected waves:

$$E_{1x} = \mathrm{e}^{\mathrm{i}\omega z/c} + b\mathrm{e}^{-\mathrm{i}\omega z/c}, \quad E_{1y} = E_{1z} = 0 , \tag{4.5.30}$$

where b is the amplitude of the reflected wave, whereas the amplitude of the incident wave is assumed equal to unity. Using the field equation (2.2.5) we obtain

$$H_{1x}(z) = H_{1z}(z) = 0 , \quad H_{1y}(z) = E_{1x}(z) ,$$

$$H_{2x}(z) = H_{2z}(z) = 0 , \quad H_{2y}(z) = \sum_{k=1}^{2} a_k \tilde{n}_k \mathrm{e}^{\mathrm{i}\omega\tilde{n}_k z/c} .$$

Therefore, the equations for determining the amplitudes b, a_1 and a_2, by virtue of (4.5.27, 28), have the form

$$1 + b = a_1 + a_2 , \tag{4.5.31a}$$

$$1 - b = [\tilde{n}_1 + N(\tilde{n}_1^2 - \varepsilon_{00})]\, a_1 + [\tilde{n}_2 + N(\tilde{n}_2^2 - \varepsilon_{00})]\, a_2 , \tag{4.5.31b}$$

$$\left[(\tilde{n}_1^2 - \varepsilon_{00})\left(1 + \mathrm{i}\frac{\tilde{n}_1}{T_\mathrm{t}}\right) + 4\pi\Gamma_\mathrm{t}\right] a_1 + \left[(\tilde{n}_2^2 - \varepsilon_{00})\left(1 + \mathrm{i}\frac{\tilde{n}_2}{T_\mathrm{t}}\right) + 4\pi\Gamma_\mathrm{t}\right] a_2 = 0 . \tag{4.5.31c}$$

Equation (4.5.31c) was obtained by applying the equation $D_i = \tilde{n}^2 \eta_{ij} E_j = \varepsilon_{00} E_i + 4\pi P_i$ to each of the normal waves, where $\boldsymbol{P}$ is the polarization caused by the exciton transition under consideration, and ε_{00} is determined by the contribution of the rest of the exciton transitions to the dielectric constant. Note that the second of the ABC equations, (4.5.28) in our case (normal incidence), is an identity, since $E_{2\mathrm{n}} = P_{2\mathrm{n}} = 0$.

Solving the system (4.5.31 a – c) we obtain

$$a_1 = -\frac{\Delta_2}{\Delta_1}a_2 = \frac{2\Delta_2}{\Delta_2(\tilde{\Delta}_1+1)-\Delta_1(\tilde{\Delta}_2+1)}, \tag{4.5.32a}$$

$$b = \frac{\Delta_1(\tilde{\Delta}_2-1)-\Delta_2(\tilde{\Delta}_1-1)}{\Delta_2(\tilde{\Delta}_1+1)-\Delta_1(\tilde{\Delta}_2+1)}; \tag{4.5.32b}$$

$$\begin{aligned} \Delta_k &= (n_k^2-\varepsilon_{00})\left(1+\mathrm{i}\frac{\tilde{n}_k}{T_t}\right)+4\pi\Gamma, \\ \tilde{\Delta}_k &= \tilde{n}_k+N(\tilde{n}_k^2-\varepsilon_{00}), \qquad k=1,2. \end{aligned} \tag{4.5.33}$$

The reflection coefficient, equal to $|b|^2$, is determined by

$$R = |b|^2 = \left|\frac{\Delta_1(\tilde{\Delta}_2-1)-\Delta_2(\tilde{\Delta}_1-1)}{\Delta_2(\tilde{\Delta}_1+1)-\Delta_1(\tilde{\Delta}_2+1)}\right|^2. \tag{4.5.34}$$

If the frequency ω recedes from the resonance, the refractive index of the new wave increases drastically. Then $a_2 \sim 1/\tilde{n}_2^3 \to 0$, and the quantity R assumes its usual form

$$R = \left|\frac{\tilde{\Delta}_1-1}{\tilde{\Delta}_1+1}\right|^2 = \frac{(n_1-1)^2+\varkappa_1^2}{(n_1+1)^2+\varkappa_1^2}$$

only at $N=0$, when $\tilde{\Delta}_1 = \tilde{n}_1$[12]. If, however, the frequency ω approaches a resonance frequency, the reflection factor depends upon the values of coefficients β and δ, which determine the shape of the dispersion curves in the vicinity of the resonance (Figs. 4.5 – 13).

Then $\delta = 0$ and $\beta < 0$ (Fig. 4.5 b), and the frequency ω lies in the region where no real values of $\tilde{n}^2$ exist, $R = 1$. This conclusion follows from the fact that in this region of frequencies $\tilde{n}_2 = -\tilde{n}_1^*$, $\Delta_2 = \Delta_1^*$, $\tilde{\Delta}_2^* = -\tilde{\Delta}_1$ and $N^* = -N$. Hence, by virtue of (4.5.34),

$$R = \left|\frac{\Delta_1^*-\Delta_1-(\Delta_1^*\tilde{\Delta}_1+\Delta_1\tilde{\Delta}_1^*)}{\Delta_1^*-\Delta_1+(\Delta_1^*\tilde{\Delta}_1+\Delta_1\tilde{\Delta}_1^*)}\right| = 1.$$

Such a result was to be expected, of course, since in the absence of absorption ($\delta = 0$) but with complex $\tilde{n}$, the wave should be completely reflected by the medium (total internal reflection upon normal incidence; this case is encountered, also, for instance, in a plasma, see [4.108, 109]).

---

[12] If surface currents are taken into consideration, only small corrections, of the order of a/λ, are required here [4.102, 107].

When $\delta = 0$ and $\beta > 0$ (Fig. 4.5a), the quantity $\tilde{n}_1 = n_1$ has a real value in the vicinity of a resonance, whereas $\tilde{n}_2 \equiv \mathrm{i}\varkappa_2$ is purely imaginary. As long as $n_1 \sim \varkappa_2$, the quantity R is appreciably less than unity. If ω increases, however, the quantity n_1 becomes substantially greater than $|\tilde{n}_2|$ and $|\Delta_1| \gg |\Delta_2|$, so that the reflection factor assumes the value

$$R = \left| \frac{\tilde{\Delta}_2 - 1}{\tilde{\Delta}_2 + 1} \right|,$$

and $R = 1$ again if $\tilde{n}_2 = \mathrm{i}\varkappa_2$ when $\tilde{\Delta}_2 = \mathrm{i}\,|\tilde{\Delta}_2|$.

If we account for absorption ($\delta \neq 0$), then the quantity R in the vicinity of a resonance is always less than unity for any value of β. It should be noted that the nature of the function $R(\omega)$ depends strongly on the function $\delta(\omega)$. This may lead to drastic changes in the quantity $R(\omega)$ in the vicinity of a resonance when the temperature is low enough (in this connection, see also Sect. 4.6).

The case of oblique incidence can be investigated in a similar manner. If $E_{1n} \neq 0$, a longitudinal wave is excited in the medium, in addition to the two transverse waves of the same polarization. Here, in determining the amplitudes of all waves, the second of the ABC equations (4.5.28) also becomes essential, as do the surface currents and charges. Since, in the general case, explicit expressions for the relations between the amplitudes are quite cumbersome, we shall not give them here. Their derivation is entirely analogous to that described in textbooks [Ref. 4.53, § 66].

The Law of Energy Conservation

Making use of the values obtained above for the amplitudes of fields in vacuum and in a crystal, we shall consider whether the law of conservation of energy imposes any restrictions on the values of the phenomenological constants in the ABC equation (4.5.28). As mentioned previously, these quantities are determined by the nature of the interaction between excitons (of any kind) and the surface of the crystal. Hence, as a first approximation, the quantities can be assumed independent of possible dissipative processes both in the bulk of the crystal and on its surface. Assuming, therefore, that there is no dissipation of energy in the medium we shall use (4.5.24b) for the dielectric constant of the medium. This expression can be conveniently rewritten in the form

$$\varepsilon(\omega, k) = \varepsilon_{00} + \frac{\omega_p^2}{\Delta + \beta' n^2},$$

where $\beta' = (\omega^2/c^2)[\hbar\omega_0(0)/m^*_{\mathrm{exc}}]$. We also introduce the notation $\Phi_i = \Delta + \beta' n_i^2$, where $i = 1, 2$, and n_i is one of the solutions of the equation $\varepsilon(\omega, k) = n^2$ and will soon be necessary in our presentation. It can readily be shown that $\Phi_1 \Phi_2 = -\beta' \omega_p^2$ [4.103]. Making use next of the expression for the energy flux S in a spatially dispersive medium ($S = S^{(0)} + S^{(1)}$), see (2.3.12, 15), we find, for a flux $S_0^{(0)} = c/8\pi$ of energy incident from vacuum, that the reflected flux is $S_R^{(0)} = (c/8\pi)|b|^2$ and the energy flux in the medium carried away by the i-th wave at a real n_i equals

$$S_i = S_i^{(0)} + S_i^{(1)} = (c/8\pi) n_i |a_i|^2 (1 + \beta' \omega_p^2/\Phi_i^2) .$$

Thus, in the absence of energy dissipation [13], the law of conservation of energy is satisfied if the equation

$$1 = |b|^2 + n_1 |a_1|^2 \left(1 + \beta' \frac{\omega_p^2}{\Phi_1^2}\right) + n_2 |a_2|^2 \left(1 + \beta' \frac{\omega_p^2}{\Phi_2^2}\right) \equiv F^{\mathrm{I}} \tag{4.5.34a}$$

is fulfilled.

The preceding equation does not consider any losses whatsoever due to the excitation of surface waves. If the surface is "ideally" smooth, i.e., free of irregularities (roughness) or any other flaws, then, within the framework of linear electrodynamics, no surface waves can be excited by a bulk wave. In real crystals, however, this assumption may not be justified. In general, for light waves with frequencies in the region of longitudinal-transverse splitting, exactly where the frequencies of surface waves are located (Sect. 5.1), there may be energy losses due to the excitation of such waves. Lattice vibrations, initiating various kinds of Raman scattering processes with the participation of bulk and surface waves, may lead to an effect analogous to the energy losses just mentioned. As has already been emphasized, however, all of these processes, as well as the dissipation processes in the bulk of the crystal, can be neglected in determining the possible values of the phenomenological parameters appearing (4.5.23a), at any rate as a first approximation.

It was actually assumed in deriving (4.5.34a) that we were concerned with the frequency region in which both refractive indices are real. If we substitute the values of the amplitudes a_1, a_2 and b determined above into the right-hand side of this equation, we can show [4.103] that with arbitrary and real values of the parameters T and η, appearing in the ABC equation (4.5.26b), as well as in the boundary condition (4.5.26c), see also (4.5.27), the relation (4.5.34a) is transformed into an identity. The situation is similar in the frequency region in which one of the refractive indices (e.g., $\tilde{n}_2$) is imaginary ($\tilde{n}_2 = \mathrm{i}\,|\tilde{n}_2|$). In this case, wave 2 carries no energy and, instead of the condition (4.5.34a), the condition for the conservation of energy is of the form

$$1 = |b|^2 + n_1 |a_1|^2 \left(1 + \beta' \frac{\omega_p^2}{\Phi_1^2}\right) . \tag{4.5.34b}$$

---

[13] Energy fluxes at the boundary of a medium with dissipation, for ABC with $\Gamma = 0$, have been studied in [4.110].

Fulfillment of the conditions (4.5.34a and b) is an indication that the above-applied system of boundary conditions is internally self-consistent when dissipation processes are neglected as well, for example, as losses of energy due to the excitation of surface waves.

In a more general statement, an analysis of the kind of boundary conditions to be applied requires the use of the principle of symmetry of the kinetic coefficients or the application of microtheory.

4.5.5 Reflected and Refracted Waves Near a Dipole Transition Frequency in a Gyrotropic Crystal

As in Sect. 4.5.4 we shall consider the simplest case of normal incidence of a monochromatic linearly polarized wave from vacuum on a plane surface of an isotropic but gyrotropic medium.

It was shown in Sect. 4.5.4 that the ABC equation (4.5.24) is sufficient when we consider the propagation of waves in a nongyrotropic medium where spatial dispersion is taken into account to an accuracy of terms of the order of k^2.

In the case of gyrotropic media we confine ourselves to terms of the order of $\boldsymbol{k}$. The total number of normal waves is thereby diminished compared with the case where we also took terms of the order of $\boldsymbol{k}^2$ into account. Hence, the use of (4.5.28) in this approximation would result in superfluous ABC. This difficulty, however, is not of a fundamental nature. It suffices to account for terms quadratic with respect to $\boldsymbol{k}$ in a gyrotropic medium as well, along with the linear $\boldsymbol{k}$ terms, and neither of the conditions (4.5.28) is excessive. Among the normal waves, however, there will also be a wave whose refractive index has a much larger absolute value than those of the remaining waves. This fact is connected with the relative smallness of the coefficient of terms quadratic with respect to $\boldsymbol{k}$ and may be accounted for in the final results by a limiting transition to the approximation where terms linear with respect to $\boldsymbol{k}$ are used.

Let us illustrate this by way of a concrete example. We choose the coordinate axes in the same way as in Sect. 4.5.4. For $z>0$ the electric field in the medium can then be given in the form

$$\boldsymbol{E}_2(z)=\boldsymbol{x}E_{2x}(z)+\boldsymbol{y}E_{2y}(z)\,, \tag{4.5.35a}$$

where

$$\begin{aligned} E_{2x}(z) &= \sum_{k=1}^{4} a_k \mathrm{e}^{\mathrm{i}\omega\tilde{n}_k z/c}\,, \\ E_{2y}(z) &= \sum_{k=1}^{4} \mathrm{i}(-1)^k a_k \mathrm{e}^{\mathrm{i}\omega\tilde{n}_k z/c}\,. \end{aligned} \tag{4.5.35b}$$

In relation (4.5.35a) $\boldsymbol{x}$ and $\boldsymbol{y}$ are unit vectors in the x and y directions, respectively; $\tilde{n}_k$ $(k=1,\,2,\,3,\,4)$ are the refractive indices of the four normal waves which may be propagated in the medium under consideration when spatial dispersion is taken into account to an accuracy of terms quadratic with respect to $\boldsymbol{k}$, the subscripts $k=1,3$ correspond to waves of left-hand circular polarization and $k=2,4$ to waves of right-hand circular polarization (such a situation exists for excitons with negative effective mass and $\beta<0$, see Sect. 4.2.2; when $\beta>0$, three waves have the same circular polarization and this case can be examined in the same way).

In nongyrotropic media the refractive indices of these waves are equal pairwise (e.g., $\tilde{n}_1=\tilde{n}_2$, $\tilde{n}_3=\tilde{n}_4$). If, however, gyrotropy is taken into account, $\tilde{n}_1 \neq \tilde{n}_2$ and $\tilde{n}_3 \neq \tilde{n}_4$. Equations (4.5.35) also account for the fact that the amplitudes of the right-hand and left-hand polarized waves satisfy the relation $E_y/E_x = \pm \mathrm{i}$, respectively.

We shall assume that the vector $\boldsymbol{E}$ of the incident wave coincides with the x-axis and that its amplitude is unity. The field in vacuum ($z<0$) may therefore be represented by

$$E_{1x}=\mathrm{e}^{\mathrm{i}\omega z/c}+b\,\mathrm{e}^{-\mathrm{i}\omega z/c}, \quad E_{1y}=d\,\mathrm{e}^{-\mathrm{i}\omega z/c}. \tag{4.5.36}$$

From continuity of the tangential components of the electric field strength we find that

$$1+b=\sum_{k=1}^{4} a_k, \quad d=\mathrm{i}\sum_{k=1}^{4}(-1)^k a_k. \tag{4.5.37}$$

Similarly, the continuity of the tangential components of the vector $\boldsymbol{H}$ yields

$$1-b=\sum_{k=1}^{4} \tilde{n}_k a_k, \quad d=\mathrm{i}\sum_{k=1}^{4} a_k(-1)^k \tilde{n}_k. \tag{4.5.38}$$

Here and in what follows the surface currents and charges are neglected (Sects. 4.5.3 and 4). Hence, in (4.5.33), we should set $\Gamma_t = N = 0$.

From the ABC equation (4.5.28) we obtain two more relations:

$$\sum_{k=1}^{4} a_k \Delta_k = 0, \quad \sum_{k=1}^{4} a_k(-1)^k \Delta_k = 0, \tag{4.5.39}$$

where the values of Δ_k are determined by (4.5.33). If we solve our system of six equations we obtain

$$a_1=\frac{\Delta_3}{(\tilde{n}_1+1)\Delta_3-(\tilde{n}_3-1)\Delta_1}=-\frac{\Delta_3}{\Delta_1}a_3,$$

$$a_2=\frac{\Delta_4}{(\tilde{n}_2+1)\Delta_4-(\tilde{n}_4+1)\Delta_2}=-\frac{\Delta_4}{\Delta_2}a_4,$$

$$b=\frac{1}{2}\left[\frac{\Delta_3(1-\tilde{n}_1)-\Delta_1(1-\tilde{n}_3)}{\Delta_3(1+\tilde{n}_1)-\Delta_1(1+\tilde{n}_3)}+\frac{\Delta_4(1-\tilde{n}_2)-\Delta_2(1-\tilde{n}_4)}{\Delta_4(1+\tilde{n}_2)-\Delta_2(1+\tilde{n}_4)}\right], \tag{4.5.40}$$

$$d=\mathrm{i}\left[\frac{\Delta_3\tilde{n}_1-\Delta_1\tilde{n}_3}{\Delta_3(1+\tilde{n}_1)-\Delta_1(1+\tilde{n}_3)}-\frac{\Delta_4\tilde{n}_2-\Delta_2\tilde{n}_4}{\Delta_4(1+\tilde{n}_2)-\Delta_2(1+\tilde{n}_4)}\right].$$

The reflection coefficient $R(\omega)$ is equal to

$$R(\omega)=|b|^2+|d|^2.$$

The solution (4.5.40) is more general than that obtained in Sect. 4.5.5 and is transformed into it if we neglect the effect of gyrotropy and set $\tilde{n}_1=\tilde{n}_2$ and $\tilde{n}_3=\tilde{n}_4$. The quantity d then vanishes. If, however, gyrotropy is taken into account, then $d\neq 0$ which gives rise to elliptic polarization of the reflected wave, the ratio of the semiaxes being $\rho=|d|/|b|$.

When the gyrotropy is intense enough and spatial dispersion is sufficiently accurately described by terms linear with respect to k, then a limiting transition with respect to one of the refractive indices (say, $\tilde{n}_4$) should be performed: $\tilde{n}_4\to\infty$. In this case, as follows from (4.5.40), $\Delta_4\to\infty$ and, instead of (4.5.40), we obtain

$$a_2=\frac{1}{\tilde{n}_2+1},\qquad a_4=0,$$

$$a_1=\frac{\Delta_3}{(\tilde{n}_1+1)\Delta_3-(\tilde{n}_3+1)\Delta_1}=-\frac{\Delta_3}{\Delta_1}a_3,$$

$$b=\frac{1}{2}\left[\frac{\Delta_3(1-\tilde{n}_1)-\Delta_1(1-\tilde{n}_3)}{\Delta_3(1+\tilde{n}_1)-\Delta_1(1+\tilde{n}_3)}+\frac{1-\tilde{n}_2}{1+\tilde{n}_2}\right], \tag{4.5.41}$$

$$d=\mathrm{i}\left[\frac{\Delta_3\tilde{n}_1-\Delta_1\tilde{n}_3}{\Delta_3(1+\tilde{n}_1)-\Delta_1(1+\tilde{n}_3)}-\frac{\tilde{n}_2}{1+\tilde{n}_2}\right].$$

We arrive at the same result if, right at the beginning, only three solutions ($k=1,2,3$) are allowed for in (4.5.35), but the first boundary condition of (4.5.28) is written in the form

$$\left.P_{2\mathrm{t}}^{(-)}+\frac{c}{\omega T_\mathrm{t}}\frac{\partial P_{2\mathrm{t}}^{(-)}}{\partial z}\right|_0=0, \tag{4.5.42}$$

where the sign $(-)$ indicates that the corresponding vector $\boldsymbol{P}$ or $\boldsymbol{E}$ is the sum of the left-hand polarized solutions. Conversely, if in the medium there are two "right" and one "left" solutions we should replace (4.5.42) by the boundary condition

$$P_{2t}^{(+)}+\frac{c}{\omega T_t}\left.\frac{\partial P_{2t}^{(+)}}{\partial z}\right|_0=0\,. \tag{4.5.42a}$$

The boundary condition (4.5.42 or 42a) (with $T_t\to\infty$) can be obtained as shown in [4.111], on the basis of an analysis of (4.5.4) if only the first spatial derivatives are taken into account. To obtain ABC of a more general form, one must consider, as in Sect. 4.5.3 for a nongyrotropic medium, the effect of the crystal boundaries on the oscillator strengths of the transitions, and not only on the intermolecular interaction. Up to now we have supposed that a linearly polarized wave is incident on a plane-parallel plate. Also of some interest is the case of a circularly polarized wave. Let us assume for definiteness that when spatial dispersion is accounted for two "left" and one "right" waves may be propagated in the crystal. If the incident wave is a "right" one, then $a_1=a_3=0$ and the reflection coefficient is

$$R=\left|\frac{\tilde{n}_2-1}{\tilde{n}_2+1}\right|^2.$$

If the incident wave is a "left" wave, $a_2=0$, $a_1\neq 0$, $a_3\neq 0$ and

$$R=\left|\frac{(1-\tilde{n}_1)\Delta_3-(1-\tilde{n}_3)\Delta_1}{(1+\tilde{n}_1)\Delta_3-(1+\tilde{n}_3)\Delta_1}\right|^2.$$

It is of interest that if absorption is ignored and the frequency ω of the light lies to the right of the turning point (Fig. 4.2), $\tilde{n}_3=-\tilde{n}_1^*$ and $\Delta_3=\Delta_1^*$. In this frequency region the reflection coefficient of a "left" wave is therefore given by

$$R=\left|\frac{(1-\tilde{n}_1)\Delta_3-(1-\tilde{n}_3)\Delta_1}{(1+\tilde{n}_1)\Delta_3-(1+\tilde{n}_3)\Delta_1}\right|^2=\left|\frac{(\tilde{n}_1\Delta_1^*+\tilde{n}_1^*\Delta_1)-(\Delta_1^*-\Delta_1)}{(\tilde{n}_1\Delta_1^*+\tilde{n}_1^*\Delta_1)+(\Delta_1^*-\Delta_1)}\right|^2=1\,.$$

It should be mentioned that [4.112] contends that the application of ABC of the form of (4.5.28) to a gyrotropic medium makes the determination of the amplitudes of all waves an indeterminate problem when the crystal surfaces are taken into account. As we have shown above, this is a baseless assumption. Also discussed in [4.112] is a specific mechanism of surface losses which is due to the excitation of short-wavelength "mechanical" excitons. The surface losses may be accounted for in a formal manner if in the ABC equation (4.5.23a) the tensors Γ_{ij}, T_{ijl} and Γ_{ijl} are assumed complex. On the whole, however, the problem of surface losses still requires special investigations within the framework of both microtheory and macrotheory.

4.5.6 The Influence of a Nonhomogeneous Subsurface Layer on Light Reflection Near Exciton-Absorption Bands

The nature of the subsurface inhomogeneities depends on whether the potential energy of a 'mechanical' exciton increases or decreases when the center of the exciton approaches the surface of the crystal (here we use a model in which the exciton is viewed as a wave packet). A considerable increase in potential energy of the exciton in the subsurface layer leads to reduced polarizability of this layer. We can assume to some approximation in this case that in a subsurface layer (of thickness l) the polarization due to the "mechanical" exciton band under consideration is zero. If, however, the energy of the exciton is reduced in the subsurface layer, this may lead to the appearance of a surface exciton level (see Sect. 6.1.1 for more details concerning surface excitons).

Since the natural frequency of a surface exciton differs from the frequency $\omega_l(0)$ of a bulk exciton, it is evidently possible, in this case as well, to assume that the polarization vector $\boldsymbol{P}(\boldsymbol{r})$ in the subsurface layer is approximately equal to zero for the frequency region $\omega \approx \omega_l(0)$. Consequently, in both cases we can assume to some rough approximation that the surface of the crystal has a layer of thickness l in which the dielectric-constant tensor $\varepsilon_{ij} = \varepsilon_{00}\delta_{ij}$, see (4.1.13), for the frequency region under investigation. Here and in the following we consider, for the sake of simplicity, the vicinity of a dipole exciton band in an isotropic nongyrotropic medium. It is supposed that both time and spatial dispersion is manifested in thicker layers of the medium with all ensuing consequences.

Let us consider the reflection of light from such a double-layer medium.

We select, as previously, a coordinate system in which the z axis is directed along the inner normal to the surface of the crystal, and assume that a monochromatic electromagnetic wave, whose electric vector lies along the x axis, is normally incident from vacuum ($z<0$) upon the surface of the crystal. The electric field in vacuum (region I) is then:

$$E_x^{\mathrm{I}} = \mathrm{e}^{\mathrm{i}\omega z/c} + b\mathrm{e}^{-\mathrm{i}\omega z/c}, \quad E_y^{\mathrm{I}} = E_z^{\mathrm{I}} = 0 . \tag{4.5.43a}$$

In the subsurface layer ($l>z>0$, region II):

$$E_x^{\mathrm{II}} = c_1\mathrm{e}^{\mathrm{i}\omega n_0 z/c} + c_2\mathrm{e}^{-\mathrm{i}\omega n_0 z/c}, \quad E_y^{\mathrm{II}} = E_z^{\mathrm{II}} = 0 , \tag{4.5.43b}$$

where $n_0 = \sqrt{\varepsilon_{00}}$.

In the remaining part of the crystal ($z>l$, region III):

$$E_x^{\mathrm{III}} = a_1\mathrm{e}^{\mathrm{i}\omega \tilde{n}_1 z/c} + a_2\mathrm{e}^{\mathrm{i}\omega \tilde{n}_2 z/c}, \quad E_y^{\mathrm{III}} = E_z^{\mathrm{III}} = 0 . \tag{4.5.43c}$$

Making use of the continuity of the tangential components of $\boldsymbol{E}$ and $\boldsymbol{H}$, see (4.5.27), as well as the ABC equation (4.5.28), we obtain the following relations between the amplitudes:

$$\begin{aligned}
&1+b=c_1+c_2\,,\\
&1-b=n_0(c_1-c_2)\,,\\
&c_1\,\mathrm{e}^{\mathrm{i}\omega n_0 l/c}+c_2\,\mathrm{e}^{-\mathrm{i}\omega n_0 l/c}=a_1\,\mathrm{e}^{\mathrm{i}\omega\tilde{n}_1 l/c}+a_2\,\mathrm{e}^{\mathrm{i}\omega\tilde{n}_2 l/c}\,,\\
&n_0[c_1\,\mathrm{e}^{\mathrm{i}\omega n_0 l/c}-c_2\,\mathrm{e}^{-\mathrm{i}\omega n_0 l/c}]=a_1\tilde{n}_1\,\mathrm{e}^{\mathrm{i}\omega\tilde{n}_1 l/c}+a_2\tilde{n}_2\,\mathrm{e}^{\mathrm{i}\omega\tilde{n}_2 l/c}\,,\\
&\Delta_1 a_1\,\mathrm{e}^{\mathrm{i}\omega\tilde{n}_1 l/c}+\Delta_2 a_2\,\mathrm{e}^{\mathrm{i}\omega\tilde{n}_2 l/c}=0\,,
\end{aligned}\tag{4.5.44}$$

where Δ_1 and Δ_2 are determined by (4.5.33) at $k=1$ and $k=2$. It follows from these relations that the reflection coefficient of the double-layer medium under investigation is determined by the expression

$$R=|b|^2=\left|\frac{r_{12}\,\mathrm{e}^{-2\mathrm{i}\omega n_0 l/c}+r_{23}}{\mathrm{e}^{-2\mathrm{i}\omega n_0 l/c}+r_{12}r_{23}}\right|^2,\tag{4.5.45}$$

where

$$r_{12}=\frac{n_0-1}{n_0+1}\,,\tag{4.5.46a}$$

and

$$r_{23}=\frac{(\tilde{n}_1-n_0)\Delta_2-(\tilde{n}_2-n_0)\Delta_1}{(\tilde{n}_1+n_0)\Delta_2-(\tilde{n}_2+n_0)\Delta_1}\,.\tag{4.5.46b}$$

The quantities r_{12} and r_{23} have a simple physical meaning. The square of the modulus of r_{12} determines the reflection coefficient at the boundary between the vacuum and the semi-infinite medium II, and the square of the modulus or r_{23} determines the reflection coefficient at the boundary between the semi-infinite media II and III, allowing for the fact that two waves of the same frequency and polarization, but different refractive indices, may be propagated in medium III. Note that when the effects of spatial dispersion are ignored, the form of relation (4.5.45) remains unchanged and only the coefficient r_{23} has a different value [Ref. 4.53, problem No. 4 in § 66]. Naturally, for $l=0$, Eq. (4.5.45) is transformed into (4.5.34). However, if $l\approx 10^{-6}$ cm, the double phase shift, quantity $(2\omega/c)n_0 l$, is equal at $n_0\approx 2$ and $\lambda_0\approx 5\times 10^{-5}$ cm to $(2\omega/c)n_0 l=(4\pi/\lambda_0)\,l n_0\approx 1/2$ and consequently has to be taken into account. It should be pointed out that if the thickness l of the subsurface layer is not small ($l\approx 100$ Å), the inhomogeneity of this layer, considered in this section, can, in general, lead to distinctive features in the reflection of light even in the case when the effects of spatial dispersion are small and the new wave in medium III can be neglected. This circumstance should not be overlooked in attempts to reveal the effects of spatial dispersion in crystals by investigating light reflection ([4.82] and Sect. 4.6.2).

4.5.7 Transmission of Light Through a Plane-Parallel Plate (Nongyrotropic Medium)

In solving the problem of a plate we shall restrict ourselves to the simplest case in which a monochromatic linearly polarized electromagnetic wave is normally incident from vacuum on the front surface of a plane-parallel plate ($z=0$) of thickness l. Here we ascribe subscript 1 to the vacuum ($z<0$), subscript 2 to the dispersive medium, which we assume to be isotropic, and subscript 3 to the vacuum at $z>l$. For definiteness we assume that the electric vector in the incident wave lies in the x axis. Then, in all three regions, only the x component of the electric field strength vector is nonzero and

$$\begin{aligned} E_{1x}(z) &= \mathrm{e}^{\mathrm{i}\omega z/c} + b\,\mathrm{e}^{-\mathrm{i}\omega z/c}, \\ E_{2x}(z) &= \sum_{k=1}^{2} (a_k \mathrm{e}^{\mathrm{i}\omega\tilde{n}_k z/c} + b_k \mathrm{e}^{-\mathrm{i}\omega\tilde{n}_k z/c}), \\ E_{3x}(z) &= s\,\mathrm{e}^{\mathrm{i}\omega z/c}. \end{aligned} \tag{4.5.47}$$

These equations contain six unknown coefficients which can be determined by means of boundary conditions. Making use of the continuity of the tangential components of $\boldsymbol{E}$ and $\boldsymbol{H}$ we find, see (4.5.27),

$$\begin{aligned} 1+b &= \sum_{k=1}^{2} (a_k + b_k), \\ s\,\mathrm{e}^{\mathrm{i}\omega l/c} &= \sum_{k=1}^{2} (a_k \mathrm{e}^{\mathrm{i}\omega\tilde{n}_k l/c} + b_k \mathrm{e}^{-\mathrm{i}\omega\tilde{n}_k l/c}), \\ 1-b &= \sum_{k=1}^{2} \tilde{n}_k (a_k - b_k), \\ s\,\mathrm{e}^{\mathrm{i}\omega l/c} &= \sum_{k=1}^{2} \tilde{n}_k (a_k \mathrm{e}^{\mathrm{i}\omega\tilde{n}_k l/c} - b_k \mathrm{e}^{-\mathrm{i}\omega\tilde{n}_k l/c}). \end{aligned} \tag{4.5.48a}$$

Two more conditions are obtained in applying the ABC equation (4.5.28):

$$\begin{aligned} &\sum_k \Delta_k (a_k + b_k) = 0, \\ &\sum_k \Delta_k (a_k \mathrm{e}^{\mathrm{i}\omega\tilde{n}_k l/c} + b_k \mathrm{e}^{-\mathrm{i}\omega\tilde{n}_k l/c}) = 0. \end{aligned} \tag{4.5.48b}$$

When this system of equations is used to find the expressions for the amplitudes b and s, the latter enable us to determine the reflection coefficient $R=|b|^2$, as well as the transmission coefficient $T=|s|^2$; owing to energy absorption in the plate, $R+T<1$. The general formulas for R and T are cumbersome and will not be given here.

Numerical results for thin films ($l = 0.1$ µm and $l = 0.2$ µm) of the crystal ZnSe, according to the kind of ABC applied, were given in [4.90]; critical comparison between the theory and experiment for CdS was made in [4.112a].

The effect of boundaries (and thereby the form of the ABC) should be especially strong for extremely thin, but macroscopic films (of a thickness $L \approx 100-300$ Å, $L \gg a_B$, a_B is the Bohr radius of the exciton). Therefore the corresponding calculations were carried out recently in [4.112b], where the surface impedance of a thin, spatially dispersive, dielectric film deposited on a metal substrate was calculated for two commonly used forms of the additional boundary conditions (ABC) at the film-vacuum and film-substrate interfaces ($\boldsymbol{P} = 0$ or $d\boldsymbol{P}/dz = 0$). The reflectivity was calculated from the surface impedance for frequencies in the vicinity of the resonance frequency of the film, and numerical results were presented for the case of a ZnSe film on an Al substrate. Even without taking the dead-layer effect into account, the results for very thin films showed not only quantitative but also qualitative differences between the predictions of the two ABS's that should enable a choice to be made between them on the basis of experimental reflectivity data. So far such experimental data are not available for very thin films. It is clear from the aforesaid that they would be of great interest.

We point out that in the case of sufficiently thin plates (i.e., plates in which the absorption of normal waves can be neglected), the intensity of the light reflected by the plate and that passing through the plate are to be determined, depending on the frequency of the light, by the interference of both normal and "new" (additional) waves. In regions of the spectrum having only a single normal wave, the reflection minima correspond to the frequency values for which the interference conditions

$$2n(\omega)l = \lambda N, \qquad N = 1, 2, \ldots$$

are fulfilled, where $n(\omega)$ is the refractive index at the frequency ω, $\lambda = 2\pi c/\omega$, N is the order of interference and l is the thickness of the plate.

It follows, in particular, from this interference condition [4.113] that the range $d\lambda$ of wavelengths corresponding to changing the orders of interference N to $N+1$ at $N \gg 1$ is determined by the relation

$$\frac{2l}{\lambda^2}d\lambda = \frac{v_{gr}}{c}$$

where $v_{gr} = c(n - \lambda dn/d\lambda)^{-1}$ is the group velocity of light. Consequently, the observation of an interference pattern can be used to analyze the dispersion law $n = n(\omega)$ (see, e.g., [4.113, 114] in which the resonance region in CdS was investigated). The presence of an additional ("new") wave substantially complicates the interference pattern, leading to additional maxima and minima. This circumstance was effectively utilized for studying spatial dispersion in a number of experimental investigations (dealt with in Sect. 4.6). The formulas given below for R and T concern only the case of sufficiently thick plates, when the amplitudes of the waves reflected from the other surface of the plate can be neglected in the equations of system (4.5.48) which correspond to the boundary conditions at $z = 0$ (the general case, for example, was dealt with in [4.115]).

It is clear that the expression for R remains the same, in the approximation under consideration, as for the case of reflection from a semi-infinite medium, (4.5.34). The expression for T has the following form

$$T = |s|^2 = 16\left|\frac{(\Delta_2\tilde{n}_1 e^{\psi_1} - \Delta_1\tilde{n}_2 e^{\psi_2})(\Delta_2 - \Delta_1)}{[\Delta_2(1+\tilde{n}_1) - \Delta_1(1+\tilde{n}_2)]^2}\right|^2, \quad \text{where} \tag{4.5.49}$$

$\psi_i = \mathrm{i}(\omega/c)\tilde{n}_i l$, $\tilde{n}_j = n_j + \mathrm{i}\varkappa_j$ and $i = 1, 2$.

At some distance from the absorption band, where spatial dispersion does not have to be taken into account and where $\tilde{n}_2 \to \infty$ and, consequently, $\Delta_2 \to \infty$, Eq. (4.5.49) can be simplified. The transmission coefficient is then

$$T = 16\left|\frac{\tilde{n}_1^2}{(1+\tilde{n}_1)^2}\right|^2 e^{-2\omega\varkappa_1 l/c} \tag{4.5.50}$$

and decreases monotonically as the thickness increases.

Depending upon the amount of damping of the new wave ($\tilde{n} = \tilde{n}_2$), two cases are possible in the vicinity of an absorption band. If, for instance, $\varkappa_2 \equiv \mathrm{Im}\{\tilde{n}_2\} \gg \varkappa_1 \equiv \mathrm{Im}\{\tilde{n}_1\}$, terms in (4.5.49) which are proportional to the factor $\exp(\psi_2)$ can be neglected. In this case the transmission coefficient

$$T = 16\left|\frac{\Delta_2\tilde{n}_1(\Delta_2 - \Delta_1)}{[\Delta_2(1+\tilde{n}_1) - \Delta_1(1+\tilde{n}_2)]^2}\right|^2 e^{-2\omega\varkappa_1 l/c} \tag{4.5.51}$$

decreases monotonically as the plate thickness increases (as in the case when spatial dispersion is not taken into account). Here the difference between (4.5.51 and 50) consists of the different dependence of T on the refractive indices $\tilde{n}_1$ and $\tilde{n}_2$. If, however, the factors $\exp(\psi_1)$ and $\exp(\psi_2)$ are of the same order of magnitude, then according to (4.5.49), the quantity T is subject to nonmonotonic, oscillating decrease. Supposing, for instance, that the quantities $\varkappa_1$ and $\varkappa_2$ are small compared to n_1 and n_2, respectively. Then the presence of damping can be taken into account only approximately, relating $\varkappa_1$ and $\varkappa_2$ in the expressions for the phases ψ_i, see (4.5.49). In this approximation

$$T = \frac{16(\Delta_2-\Delta_1)^2}{[\Delta_2(1+n_1)-\Delta_1(1+n_2)]^4}\left[\Delta_2^2 n_1^2+\Delta_1^2 n_2^2 \exp\left[-\frac{2\omega}{c}(\varkappa_2-\varkappa_1)l\right]\right.$$
$$\left.-2\Delta_1\Delta_2 n_1 n_2 \exp\left[-\frac{\omega}{c}(\varkappa_2-\varkappa_1)l\right]\cos\frac{\omega}{c}(n_2-n_1)l\right]\mathrm{e}^{-2\omega\varkappa_1 l/c}. \tag{4.5.52}$$

The Integral Absorption

The results of calculations of the quantities $\varkappa_1$ and $\varkappa_2$, shown in Figs. 4.6–13, indicate that in the resonance region, both inequalities $\varkappa_2 > \varkappa_1$ and $\varkappa_1 > \varkappa_2$ can hold. Hence, in investigating light absorption in sufficiently thick plates, when $(2\omega/c)\varkappa_{1,2}l \gg 1$ and $\varkappa_1 \neq \varkappa_2$, the quantity T depends on l exponentially, so that the experimentally determined effective absorption factor

$$k(\omega) \equiv \frac{c}{2\omega}\ln\frac{T(0)}{T(l)}$$

is equal to $\varkappa_1(\omega)$, regardless of the kind of ABC applied, if $\varkappa_1(\omega) < \varkappa_2(\omega)$, and is equal to $\varkappa_2(\omega)$ if $\varkappa_1(\omega) > \varkappa_2(\omega)$. Since, in general, the curves of relations $\varkappa_1(\omega)$ and $\varkappa_2(\omega)$ may intersect, the effective absorption factor $k(\omega)$, at the frequency where $\varkappa_1(\omega) = \varkappa_2(\omega)$, may have a sharp bend. In [4.116], the integral absorption, i.e., the quantity

$$S = \int k(\omega)\, d\omega$$

for this case of sufficiently thick crystals, assuming that the dielectric constant of the crystal is of the form

$$\varepsilon(\omega, k) = \varepsilon_{00} + \frac{\omega_p^2}{\Omega_\perp^2 - \omega^2 - \mathrm{i}\omega\nu + \beta k^2},$$

where ν is the "effective" collision frequency, cf. (4.1.13a) and (4.5.24b). It was shown in [4.116] that if $\nu > \nu_0$, where $\nu_0 = (2\omega_p/c)\sqrt{\beta}$, the value of integral absorption

$$S = \frac{\pi\omega_p^2}{4\Omega_\perp^2\sqrt{\varepsilon_{00}}}$$

is independent of ν and, in fact, complies with (4.2.12), in which spatial dispersion is not taken into account.

If, however, $\nu < \nu_0$, then

$$S = \frac{\omega_p c}{2\Omega_\perp\sqrt{\varepsilon_{00}\beta}}\nu$$

so that at small values of ν the quantity S depends linearly on ν and, in this way tends to zero when $\nu \to 0$. It follows from the above relation that at small values of ν, the temperature dependence of S is determined by the temperature dependence of ν, as has been observed experimentally in [4.117].

4.5.8 Transmission of Light Through a Plane-Parallel Plate (Gyrotropic Medium)

In contrast to the preceding subsection, we now assume that the dispersive medium 2 is gyrotropic. However, we shall not take birefringence into account. This is justified if the medium is isotropic or when the plane $z=0$ is perpendicular to the optical axis. Then, in the case of normal incidence of monochromatic linearly polarized light on the plane $z=0$, the electric field strength in the plate, similar to (4.5.35 a), is equal to

$$E_2(z) = \boldsymbol{x}E_{2x}(z) + \boldsymbol{y}E_{2y}(z)\,, \tag{4.5.53}$$

where

$$\begin{aligned} E_{2x}(z) &= \sum_{k=1}^{4} a_k \mathrm{e}^{\mathrm{i}\omega\tilde{n}_k z/c} + \sum_{k=1}^{4} b_k \mathrm{e}^{-\mathrm{i}\omega\tilde{n}_k z/c}\,, \\ E_{2y}(z) &= \sum_{k=1}^{4} \mathrm{i}(-1)^k a_k \mathrm{e}^{\mathrm{i}\omega\tilde{n}_k z/c} - \sum_{k=1}^{4} \mathrm{i}(-1)^k b_k \mathrm{e}^{-\mathrm{i}\omega\tilde{n}_k z/c}\,. \end{aligned} \tag{4.5.53 a}$$

Equations (4.5.53 a) take into account the possibility of four waves being propagated in medium 2 (Sect. 4.5.5). In vacuum, when $z<0$, the electric field is determined by (4.5.36), whereas, when $z>l$, by

$$E_{3x}(z) = s_1 \mathrm{e}^{\mathrm{i}\omega z/c}\,, \qquad E_{3y}(z) = s_2 \mathrm{e}^{\mathrm{i}\omega z/c}\,. \tag{4.5.54}$$

Equations (4.5.36, 53 and 54) contain twelve unknown factors which can be found by applying boundary conditions. From (4.5.27, 28) at $z=0$ and $z=l$ we find that these boundary conditions are of the form

$$\begin{aligned} 1+b &= \sum_{k=1}^{4} (a_k+b_k)\,, \qquad d = \sum_{k=1}^{4} \mathrm{i}(-1)^k (a_k-b_k)\,, \\ 1-b &= \sum_{k=1}^{4} \tilde{n}_k (a_k-b_k)\,, \\ -d &= \sum_{k=1}^{4} \mathrm{i}(-1)^k (a_k+b_k)\tilde{n}_k\,, \\ \tilde{s}_1 &= \sum_{k=1}^{4} (\tilde{a}_k+\tilde{b}_k)\,, \qquad \tilde{s}_2 = \sum_{k=1}^{4} \mathrm{i}(-1)^k (\tilde{a}_k-\tilde{b}_k)\,, \end{aligned} \tag{4.5.55}$$

$$\tilde{s}_1 = \sum_{k=1}^{4} \tilde{n}_k(\tilde{a}_k - \tilde{b}_k)\,,$$

$$\tilde{s}_2 = \sum_{k=1}^{4} \mathrm{i}(-1)^k \tilde{n}_k(\tilde{a}_k + \tilde{b}_k)\,,$$

$$\sum_{k=1}^{4} \Delta_k(a_k + b_k) = 0\,, \qquad \sum_{k=1}^{4} \Delta_k(-1)^k(a_k - b_k) = 0\,,$$

$$\sum_{k=1}^{4} \Delta_k(\tilde{a}_k + \tilde{b}_k) = 0\,, \qquad \sum_{k=1}^{4} \Delta_k(-1)^k(\tilde{a}_k - \tilde{b}_k) = 0\,,$$

in which the notation of (4.5.33) is used, as well as

$$\tilde{a}_k = a_k \mathrm{e}^{\mathrm{i}\omega\tilde{n}_k l/c}\,, \qquad \tilde{b}_k = b_k \mathrm{e}^{-\mathrm{i}\omega\tilde{n}_k l/c}\,, \qquad \tilde{s}_k = s_k \mathrm{e}^{\mathrm{i}\omega l/c}\,.$$

Up to now it has been assumed that spatial dispersion in the plate is taken into consideration to an accuracy up to terms of the order of k^2. Hence the contributions of all four possible solutions of the dispersion equation have been retained in (4.5.53, 55). Since, however, a gyrotropic medium is being considered, we can pass over to a simpler case, in which only the contributions from three solutions are retained, as in Sect. 4.5.5. Note in this connection that both $|\mathrm{Re}\{\tilde{n}_4\}|$ and $|\mathrm{Im}\{\tilde{n}_4\}|$ of the complex index of refraction $\tilde{n}_4$ are anomalously large. This circumstance enables us to set $b_4 = \tilde{a}_4 = 0$ in (4.5.55). If we then eliminate $\tilde{b}_4$ and a_4 from the last four equations of the system (4.5.55), we obtain

$$\Delta_1 a_1 + \Delta_3 a_3 + \Delta_2 b_2 = 0\,, \qquad \Delta_1 \tilde{b}_1 + \Delta_2 \tilde{a}_2 + \Delta_3 \tilde{b}_3 = 0\,.$$

It follows from the same equations that the amplitudes a_4 and $\tilde{b}_4$ are inversely proportional to Δ_4 and, consequently, they are also small. Since $\Delta_4 \sim \tilde{n}_4^2$, and the first eight equations of (4.5.55) contain, along with terms of the order of a_4 and $\tilde{b}_4$, only the products $\tilde{n}_4 a_4$ or $\tilde{n}_4 \tilde{b}_4$, it is clear that in these equations, at anomalously large values of $\tilde{n}_4$, we can also put $a_4 \to 0$, $\tilde{n}_4 a_4 \to 0$, $b_4 \to 0$ and $\tilde{n}_4 \tilde{b}_4 \to 0$. Hence, in their final form the simplified system of equations for determining the unknown amplitudes can be written as

$$1 + b = \sum_{k=1}^{3} (a_k + b_k)\,, \qquad d = \sum_{k=1}^{3} \mathrm{i}(-1)^k(a_k - b_k)\,,$$

$$1 - b = \sum_{k=1}^{3} \tilde{n}_k(a_k - b_k)\,, \qquad -d = \sum_{k=1}^{3} \mathrm{i}(-1)^k(a_k + b_k)\tilde{n}_k\,,$$

$$\tilde{s}_1 = \sum_{k=1}^{3} (\tilde{a}_k + \tilde{b}_k)\,, \qquad \tilde{s}_2 = \sum_{k=1}^{3} \mathrm{i}(-1)^k(\tilde{a}_k - \tilde{b}_k)\,, \tag{4.5.55a}$$

$$\tilde{s}_1=\sum_{k=1}^{3}(\tilde{a}_k-\tilde{b}_k)\tilde{n}_k\,,\qquad \tilde{s}_2=\sum_{k=1}^{3}\mathrm{i}(-1)^k(\tilde{a}_k-\tilde{b}_k)\tilde{n}_k\,,$$

$$\Delta_1 a_1+\Delta_3 a_3=-\Delta_2 b_2\,,\qquad \Delta_1\tilde{b}_1+\Delta_3\tilde{b}_3=-\Delta_2\tilde{a}_2\,.$$

In this case, we should also set $a_4=b_4=0$ in (4.5.53a).

Thus, after eliminating the fourth solution, we must solve a system of equations with ten unknowns to find all the amplitudes. This is still a cumbersome task. Therefore, as in Sect. 4.5.7, we shall consider here only the case with sufficiently thick plates, where in the boundary conditions at $z=0$ we can neglect the amplitudes of waves reflected from the surface of the plate with $z=l$. In this approximation the reflection coefficient remains the same as for reflection from a semi-infinite medium (Sect. 4.5.4), and the factors a_k are determined by (4.5.41). For the remaining amplitudes we find

$$\begin{aligned}
\tilde{b}_1&=-\frac{\Delta_3(1-\tilde{n}_2)-\Delta_2(1+\tilde{n}_3)}{\Delta_3(1+\tilde{n}_1)-\Delta_1(1+\tilde{n}_3)}\tilde{a}_2\,,\\
\tilde{b}_2&=-\frac{1}{1+\tilde{n}_2}[\tilde{a}_1(1-\tilde{n}_1)+\tilde{a}_3(1-\tilde{n}_3)]\,,\\
\tilde{b}_3&=-\frac{\Delta_2(1+\tilde{n}_1)-\Delta_1(1-\tilde{n}_2)}{\Delta_3(1+\tilde{n}_1)-\Delta_1(1+\tilde{n}_3)}\tilde{a}_2
\end{aligned}\tag{4.5.56}$$

and

$$\begin{aligned}
\tilde{s}_1&=\frac{\tilde{n}_1+\tilde{n}_2}{1+\tilde{n}_2}\tilde{a}_1+\frac{\tilde{n}_2+\tilde{n}_3}{1+\tilde{n}_2}\tilde{a}_3\\
&\quad+\frac{\Delta_3(\tilde{n}_1+\tilde{n}_2)+\Delta_2(\tilde{n}_3-\tilde{n}_1)-\Delta_1(\tilde{n}_2+\tilde{n}_3)}{\Delta_3(1+\tilde{n}_1)-\Delta_1(1+\tilde{n}_3)}\tilde{a}_2\,,\\
\tilde{s}_2&=-\mathrm{i}\left(\frac{\tilde{n}_1+\tilde{n}_2}{1+\tilde{n}_2}\tilde{a}_1+\frac{\tilde{n}_2+\tilde{n}_3}{1+\tilde{n}_2}\tilde{a}_3\right.\\
&\quad\left.-\frac{\Delta_3(\tilde{n}_1+\tilde{n}_2)+\Delta_2(\tilde{n}_3-\tilde{n}_1)-\Delta_1(\tilde{n}_2+\tilde{n}_3)}{\Delta_3(1+\tilde{n}_1)-\Delta_1(1+\tilde{n}_3)}\tilde{a}_2\right).
\end{aligned}\tag{4.5.57}$$

At some distance from the absorption band, $\Delta_3\to\infty$ and $a_3\to 0$, see (4.5.41, 54). Hence

$$\frac{E_{3y}(z)}{E_{3x}(z)}=\frac{s_2}{s_1}=\frac{\tilde{s}_2}{\tilde{s}_1}=-\mathrm{i}\frac{\mathrm{e}^{\mathrm{i}\omega\tilde{n}_1 l/c}-\mathrm{e}^{\mathrm{i}\omega\tilde{n}_2 l/c}}{\mathrm{e}^{\mathrm{i}\omega\tilde{n}_1 l/c}+\mathrm{e}^{\mathrm{i}\omega\tilde{n}_2 l/c}}=\tan\varphi\,,$$

where $\varphi=(\omega/2c)(\tilde{n}_2-\tilde{n}_1)l$. If the refractive indices $\tilde{n}_1$ and $\tilde{n}_2$ can be considered real, then the electromagnetic field, in the region where $z>l$, turns out to be linearly polarized and the ratio φ/l determines the rotatory power of the plate per unit thickness.

The presence of three waves makes the ratio $\tilde{s}_2/\tilde{s}_1$ complex even when absorption can be neglected and all the quantities $\tilde{n}_k$, where $k = 1, 2, 3$, can be considered real. Consequently, the transmitted light is elliptically polarized in this case. If only two waves are taken into account, elliptical polarization appears only in the presence of absorption, which should be different for the "right" and "left" waves (circular dichroism). If absorption is neglected, the elliptic polarization is due to the fact that three waves, circularly polarized in different directions, emerge from the plate (two "left" and one "right" wave, or vice versa). As a matter of fact, by using (4.5.57) we find that the strength of the electric field when $z > l$ is

$$E_3(z) = A_1^- \, e^{i\omega z/c} + A_3^- \, e^{i\omega z/c} + A_2^+ \, e^{i\omega z/c}, \tag{4.5.58}$$

where $A = xA_x + yA_y$ and

$$\begin{aligned} A_{1x}^- &= iA_{1y}^- = \frac{\tilde{n}_1 + \tilde{n}_2}{1 + \tilde{n}_2} \, \frac{\Delta_3 \exp[i\omega(\tilde{n}_1 - 1)l/c]}{(\tilde{n}_1 + 1)\Delta_3 - (\tilde{n}_3 + 1)\Delta_1}, \\ A_{3x}^- &= iA_{3y}^- = -\frac{\tilde{n}_2 + \tilde{n}_3}{1 + \tilde{n}_2} \, \frac{\Delta_1 \exp[i\omega(\tilde{n}_3 - 1)l/c]}{(\tilde{n}_1 + 1)\Delta_3 - (\tilde{n}_3 + 1)\Delta_1}, \\ A_{2x}^+ &= -iA_{2y}^+ \\ &= \frac{\exp[i\omega(\tilde{n}_2 - 1)l/c]}{1 + \tilde{n}_2} \, \frac{\Delta_3(\tilde{n}_1 + \tilde{n}_2) + \Delta_2(\tilde{n}_3 - \tilde{n}_1) - \Delta_1(\tilde{n}_2 + \tilde{n}_3)}{(1 + \tilde{n}_1)\Delta_3 - (\tilde{n}_3 + 1)\Delta_1}. \end{aligned} \tag{4.5.58a}$$

For the region where $|\tilde{n}_1| \approx |\tilde{n}_2|$, $|\tilde{n}_3| \gg |\tilde{n}_1|$ and $|\tilde{n}_3| \gg |\tilde{n}_2|$ (Figs. 4.5–13), (4.5.58a) can be simplified. With an accuracy up to small terms of the order of $1/|\tilde{n}_3|$ we can write

$$\begin{aligned} A_{1x}^- &= iA_{1y}^- = \frac{\tilde{n}_1 + \tilde{n}_2}{1 + \tilde{n}_2} \, \frac{\exp[i\omega(\tilde{n}_1 - 1)l/c]}{1 + \tilde{n}_1} \left(1 + \frac{\Delta_1}{1 + \tilde{n}_1} \, \frac{1}{\tilde{n}_3}\right), \\ A_{2x}^+ &= -iA_{2y}^+ = \frac{\tilde{n}_1 + \tilde{n}_2}{1 + \tilde{n}_2} \, \frac{\exp[i\omega(\tilde{n}_2 - 1)l/c]}{1 + \tilde{n}_1} \left[1 + \frac{\tilde{n}_2 - \tilde{n}_1}{\tilde{n}_3} + \frac{\Delta_1}{(1 + \tilde{n}_1)\tilde{n}_3}\right], \\ A_{3x}^- &= iA_{3y}^- = -\frac{\Delta_1 \exp[i\omega(\tilde{n}_3 - 1)l/c]}{(1 + \tilde{n}_1)(1 + \tilde{n}_2)\tilde{n}_3}. \end{aligned}$$

In the transparent region the sum of waves with the amplitudes A_1^- and A_2^+ represents a linearly polarized wave, turned with respect to the x axis through the angle $\varphi = (\omega/2c)(n_2 - n_1)l$, with which a left-hand circularly polarized wave of amplitude A_3^- interferes. This circumstance leads to apparent circular dichroism of a magnitude that oscillates as the plate thickness changes. An investigation of this effect may yield valuable information about the dependence of the refractive index of the third wave on ω (in this connection, see also Sect. 4.6.1).

In concluding our discussion of the problem of the transmission of normal electromagnetic waves through a crystalline medium with spatial dispersion, we feel it necessary to make one remark of the selection of the roots of the dispersion equation for the refractive indices $\tilde{n}_l$ of normal waves. In accounting for spatial dispersion, the equation for determining the quantities $\tilde{n}_l$ remains, in fact, the equation for $\tilde{n}_2$ [see (4.1.4) for a gyrotropic medium and (4.2.4) for a nongyrotropic medium]. Consequently, (4.1.4) and (4.2.4) determine $\tilde{n}_l$ only up to a sign. It is clear that this sign should be chosen so that waves appearing at the surface of the crystal are damped below the surface. If, however, damping is neglected and, for instance, all $\tilde{n}_l$ have real values, then the choice of the sign for $\tilde{n}_l = n_l$ can be based on considerations discussed in Sects. 2.3.2, 3 and 4.2.2, and concerning the effect of spatial dispersion on the direction of the group velocity of normal waves. In the case of normal waves for which $\boldsymbol{v}_{gr} \cdot \boldsymbol{s} > 0$, the sign of n_l should be selected so that the direction of wave propagation, as well as the direction of $\boldsymbol{v}_{gr}$, make an acute angle with the inner normal to the surface of the crystal. Conversely, if $\boldsymbol{v}_{gr} \cdot \boldsymbol{s} < 0$, the direction $\boldsymbol{s}$ of wave propagation, in contrast to the direction of $\boldsymbol{v}_{gr}$, should make an obtuse angle with the inner normal to the interface. This is precisely the choice which meets the requirement that the energy in the medium should flow away from the interface. To illustrate the above, it proves expedient to consider the direction of a refracted wave in an isotropic transparent medium with oblique incidence of the wave from the vacuum (Fig. 4.16; in this connection see also [4.118]).

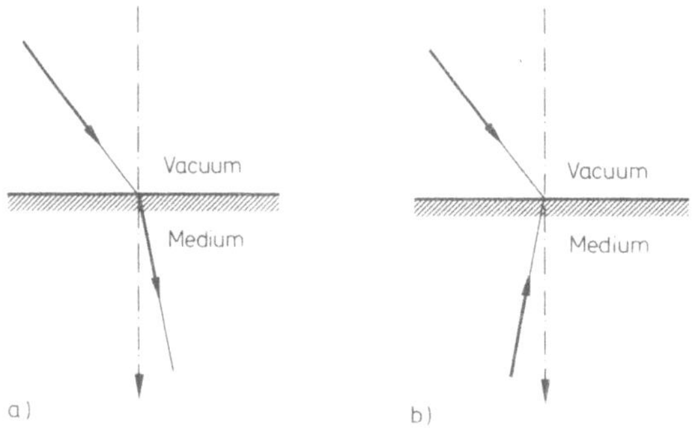

Fig. 4.16a, b. Direction of a refracted wave (vector $\boldsymbol{s}$) in an isotropic transparent medium with oblique incidence **(a)** $\boldsymbol{v}_{gr} \cdot \boldsymbol{s} > 0$ and **(b)** $\boldsymbol{v}_{gr} \cdot \boldsymbol{s} < 0$

4.6 Experimental Investigations of Spatial Dispersion in Crystals

4.6.1 Gyrotropic Crystals

The nature of the optical properties of gyrotropic crystals is a long-standing problem; it has been known that these properties depend on spatial dispersion of the order of a/λ ([4.119, 120] and Chap. 1). Nevertheless, experimental investigations of the optical properties of gyrotropic crystals in the vicinity of exciton absorption lines and, in general, in the frequency region of electronic transitions, were begun a relatively short time ago. In particular, there are still few data available on the frequency dependence of the angle of rotation of the polarization plane in crystals (especially at low temperatures) near various absorption bands. Among the first results obtained in investigating this problem are those in [4.76]. This research concerned several crystals belonging to the crystal class D_3 and possessing the symmetry properties of quartz (quartz, cinnabar and benzyl), as well as the cubic crystal of sodium chlorate belonging to class T. It was shown that the rotation outside the absorption bands, caused by the contribution of the excited states under investigation, is a function of the light frequency, and is governed by the relation (4.5.17), i.e., $\varphi(\omega) = K\omega^2/(\omega^2-\omega_L^2)^2$.

Subsequently, the gyrotropic dispersion in the vicinity of resonances of the dielectric constant was investigated for a considerably greater number of crystals. In particular, gyrotropy was studied in intrinsic semiconductors (tellurium, selenium, etc.) for the frequency regions in which it is due to interband transitions [4.121, 122]. In addition, gyrotropy was discovered in the region of vibrational transitions [4.123, 124], and extensive research was initiated on the rotary dispersion in liquid crystals. A detailed discussion of those results can be found in [4.125]. Here we only note that in investigating uniaxial crystals [4.76] the rotatory power was measured both for light propagating along the optical axis and light propagating perpendicular to this axis. When linearly polarized light is incident on the crystal at an angle to the optical axis and particularly when this is a right angle, elliptically polarized waves (Sect. 4.1), with different signs of rotation and various orientations of the axes of the ellipse, appear in the crystal instead of circularly polarized waves. In this case, as a wave passes through the crystal, a phase shift arises. In addition to the effect of gyration, ordinary birefringence contributes to this shift, too, thereby complicating the investigation of rotary dispersion to some extent. It has been shown in [4.76] that when the direction of light propagation is perpendicular to the optical axis in quartz, rotary dispersion is governed by a relation of the type of (4.5.17), but with the coefficient K having a different value.

When the coefficient K is known, as well as the frequency dependence of the refractive index at some distance from the absorption band being considered, the quantity $\omega\delta_{123}/c$ (Sect. 4.1.1) can be estimated. Its value, in particular, determines the width of the frequency region, adjacent to the resonance, where three waves may appear (Sect. 4.1).

When we compare (4.1.1) and (4.5.16), we find that in cubic crystals, for example,

$$\frac{\omega_L}{c}\delta_{123}=\frac{\gamma}{4\pi\Lambda^{(0)}}\,. \tag{4.6.1}$$

On the other hand, as follows from (4.5.15 b), outside an absorption band

$$n^2(\omega)=\varepsilon_0(\omega)=\varepsilon_{00}+\frac{4\pi\Lambda^{(0)}/\rho}{\omega_L^2-\omega^2}\equiv\varepsilon_{00}+\frac{K_1\lambda^2}{\lambda^2-\lambda_L^2}\,, \tag{4.6.2}$$

where

$$\lambda=\frac{2\pi c}{\omega}\,,\quad \lambda_L=\frac{2\pi c}{\omega_L}\,,\quad K_1=\frac{\Lambda^{(0)}}{\rho}\,\frac{\lambda_L^2}{\pi c^2}\,.$$

Therefore

$$\frac{\omega_L}{c}\delta_{123}=\frac{K_2}{\pi\lambda K_1^2}\,, \tag{4.6.3}$$

where K_2 is the coefficient in the equation

$$\varphi(\lambda_0)=\frac{K_2\lambda^2}{(\lambda^2-\lambda_L^2)^2}\,, \tag{4.6.4}$$

and it is obvious, see (4.5.17), that $K_2=K\lambda_L^4/4\pi^2c^2$. According to [4.76], the absorption band at $\lambda_L\approx 4930$ Å of a cinnabar crystal corresponds to $K_1\approx 0.56$ and $K_2\approx 1.06\,\pi\times 10^{-10}$ m. Hence, we obtain $\omega_L\delta_{123}/c = 2\pi\delta_{123}/\lambda_L\approx 0.8\times 10^{-3}$. For the organic crystal of benzyl, the band $\lambda_L\approx 2400$ Å corresponds to $K_1\approx 0.4$ and $K_2\approx 0.35\,\pi\times 10^{-10}$ m. Therefore, $\omega_L\delta_{123}/c\approx 10^{-3}$. According to data presented by the same author, the band at $\lambda_L\approx 900$ Å of the cubic crystal of sodium chlorate corresponds to $K_1=1.18$ and $K_2=0.06\,\pi\times 10^{-10}$ m and, consequently, $\omega_L\delta_{123}/c\approx 4\times 10^{-5}$. The band at $\lambda_L=1850$ Å corresponds to $K_1=0.08$ and $K_2=0.07\,\pi\times 10^{-1}$ m, so that $\omega_L\delta_{123}/c\approx -0.5\times 10^{-4}$.

Cholesteric liquid crystals possess a very high rotatory power [4.126]. It may be as high as 200 revolutions per mm of thickness so that the value of the quantity $\omega_L\delta_{123}/c$ can be substantially higher than for the crystals considered above.

Of especial interest is a study of gyrotropy at low temperatures since the damping of the exciton states may be small and one can expect fine effects of spatial dispersion, associated with the exciton-band structure, etc., to be manifested. Circular dichroism in the vicinity of certain absorption lines of a sodium uranyl acetate crystal at low temperatures have been investigated in [4.127, 128].

For molecules in solution, the magnitude of circular dichroism, i.e., the ratio of the difference between the absorption coefficients of right- and left-handed circularly polarized light to their sum, is small, in general, and of the order of a/λ, a being the size of the molecule and λ the wavelength of light. Thus $a/\lambda \approx 10^{-2}$ to 10^{-3}. In sodium uranyl acetate, however, the magnitude of circular dichroism for certain weak lines approaches unity; this corresponds to relatively strong absorption of only one of the circularly polarized waves. Since sodium uranyl acetate crystals consist of nongyrotropic molecules, the observed phenomenon is typical only of the crystalline state. It is due to differences in the structure of the exciton bands corresponding to right- and left-handed circular polarization of the dipole moment vector [4.129, 130]. Typical of the sodium uranyl acetate crystal is the absence of circular dichroism in the vicinity of intensive absorption bands, where the broadening caused by strong intramolecular interaction evidently "masks" the fine effects of the band structure, due to weak intermolecular interaction. The observation of a new wave may prove to be difficult (owing to the low oscillator strength) in the vicinity of certain weak absorption lines that are typical of only the crystalline state. An investigation of the rotatory power of the crystal as a function of the frequency is feasible, however, and essential in order to elucidate the nature of the "crystalline" absorption lines under discussion (see also [4.131]).

To date [4.132] reports on the only investigation in which a new wave was observed in a gyrotropic crystal. Specifically, the research concerned Raman scattering of light in quartz. The exciting light of a laser was directed perpendicular to the quartz plate, along its optical axis, and the spectral composition was measured for the backscattered light (the scattering angle was $\theta = 180°$). If $\boldsymbol{k}_l$ is the wave vector of the light from the laser and $\boldsymbol{k}_s$ is the wave vector of the scattered light then, in the case under discussion, $\boldsymbol{k}_s \approx -\boldsymbol{k}_l$. Thus the momentum transmitted to the crystal is $\hbar\boldsymbol{k} = \hbar(\boldsymbol{k}_s - \boldsymbol{k}_l) \approx -2\hbar\boldsymbol{k}_l$. In [4.132] the radiation of helium-neon and argon lasers was employed. In both cases, $k_l \gg \omega_L/c$. This means that $k \gg \omega_L/c$ as well. Consequently, only elementary excitations for which it is unnecessary to take retardation into account could be excited in the quartz. On the other hand, the vector $\boldsymbol{k} \parallel \boldsymbol{k}_l$ and is therefore directed along the optical axis. Hence, polaritons in the frequency region of the dipole-active optical vibrations of the lattice, as shown in Sect. 4.1.2, should follow the linear dispersion law, see also (4.5.20),

$$\omega_{\pm} = \omega_L \pm \alpha k .$$

This means (Fig. 4.1) that in the spectra of backscattered light, at a given $k = 2k_l$, there should be two $\omega_{\pm}$ lines, which is exactly what was observed in [4.132].

Besides confirming the linear dispersion law, the application of lasers with different frequencies also enabled the value of α to be determined. In the case under consideration ($\omega_L = 128\ \mathrm{cm}^{-1}$), the value of α was found approximately to equal 8×10^6 m/s. Further research in the scattering of light in gyrotropic crystals is certainly of great interest. However, experimental investigations of other optical effects, associated with new waves in gyrotropic crystals (Sect. 4.5), would be no less interesting.

The existence of three waves in a crystal having the same frequency but different refractive indices, absorption coefficients and polarization, should, under favorable conditions (at low temperatures), as shown in Sects. 4.5.6 and 9, lead to a number of phenomena. Among others, it was shown in Sect. 4.5.6 that to the right of the turning point[14] (Fig. 4.2) a region of total reflection ($R = 1$) begins for "left"-polarized waves when absorption is neglected. If spatial dispersion is ignored, the region of strong reflection in dielectrics is located to the right (on a frequency scale) of the resonance. The observation of strong reflection to the left of the resonance and outside the absorption line would therefore confirm the shape of the dispersion curves, shown in Fig. 4.2, for a gyrotropic crystal in the vicinity of a dipole absorption line.

An observation of the states of polarization of linearly polarized light transmitted through a plane-parallel plate (Sect. 4.5.9) seems to be of no less interest, since this light, after passing through the plate, is converted into a sum of a linearly polarized wave and a circularly polarized wave which interfere with each other. For frequencies ω less than the frequency of the turning point (Fig. 4.2), the amplitude of the circularly polarized wave is small, but it becomes of the same order of magnitude as the amplitude of the linearly polarized wave as the frequency approaches the turning point (Sect. 4.5.9). We also point out that when circularly polarized light with a frequency in this region is allowed to pass through the plate and its polarization coincides with that of the third (new) wave, then the light emerging from the plate has the same polarization, but its intensity oscillates as the plate thickness is varied, see (4.5.57) with $a_2 = 0$. If, however, the frequency of the linearly polarized light incident on the plate is somewhat to the right of the turning point, then the emerging light is found to be circularly polarized: for the case dealt with in Sect. 4.5.9 we have

$$e^{i\omega n_1 l/c} = e^{i\omega n_3 l/c} \approx 0\,, \qquad A_{1x}^- = A_{3x}^- = 0\,, \qquad A_{2x}^+ \neq 0\,.$$

In analyzing experimental data, all of these possibilities should be kept in mind [4.133]. In order to realize them it is, of course, necessary that the width

[14] As is evident from Fig. 4.4, the turning point vanishes when damping of the waves is taken into account. Nevertheless, we shall continue to use this term because in the case when absorption is ignored or is sufficiently weak, the frequency corresponding to the turning point divides the whole frequency region into two parts in which normal waves propagate in quite different ways.

of the absorption line is sufficiently narrow because this is the only case in which a new (third) wave can be propagated in a gyrotropic crystal (Sect. 4.1).

In concluding we mention the crystals of symmetry groups C_{3v}, C_{4v} and C_{6v} (Sect. 3.2.1). In such crystals only the exraordinary waves are polarized elliptically and the plane of the polarization ellipse coincides with that containing the $\boldsymbol{c}$ axis and the wave vector $\boldsymbol{k}$ [4.133]. This particular case of optical activity, called "weak" optical activity, has been demonstrated recently for CdS and AgI crystals in the region of the excition lines in a series of reflectivity experiments [4.73a, 133a]. The effect of weak optical activity on photoluminescence spectra was investigated in [4.133b].

4.6.2 Nongyrotropic Crystals

The optical effects associated with spatial dispersion of the dielectric constant tensor in nongyrotropic crystals were studied from the very beginning for the nonorganic crystals Cu_2O [4.45–48], CdS [4.61–63], Si [4.134] and GaAs [4.135], as well as for certain organic crystals (anthracene, stilbene, etc. [4.136–138] for more recent work see below). Investigations of anisotropy in the vicinity of a quadrupole absorption line and of the influence of external effects on the optical properties of Cu_2O and CdS have already been mentioned (Chap. 1 and Sect. 4.3.1).

First we shall discuss in greater detail investigations of the optical anisotropy of cuprous oxide in the neighborhood of quadrupole lines. A complex structure in the form of two hydrogen-like absorption-line series, which are associated, using the language of perturbation theory, with the excitation of Wannier-Mott excitons in the Cu_2O lattice, occur on the long-wavelength fundamental absorption edge of the Cu_2O crystal. At the long-wavelength side of the "yellow" series, which, when $n \geqslant 2$, obeys the formula $\nu_n = (17460 - 785/n^2)\ \mathrm{cm}^{-1}$, a line is observed at $\lambda = 6125$ Å, which is much weaker and narrower than the other lines of this series. It has been assigned the number $n = 1$. In investigating the absorption spectra of Cu_2O with single-crystal specimens, anisotropic light absorption was observed at $T = 77$ K, precisely at this line with $n = 1$. This effect was never observed previously and is therefore unusual for cubic crystals. It was found that in the spectrum of light that has passed through a Cu_2O plate, the intensity of this absorption line, in two perpendicular states of polarization, is not the same, in general, and depends strongly on the direction of light propagation. If the light is propagated along one of the fourfold symmetry axes or along any of the body diagonals of the cube, the absorption is independent of polarization. If, however, the wave vector of the light is directed along a face diagonal of the cube, the absorption of light at the frequency of this line can be observed only when the electric field strength vector of the light wave is perpendicular to the plane of the face.

The results of these investigations enable us to draw an unambiguous conclusion about the quadrupole nature of light absorption in Cu_2O at

$\lambda = 6125$ Å. As a matter of fact (Sect. 4.3.1) light absorption in cubic crystals near dipole exciton lines differs, when spatial dispersion is taken into account, in accordance with the polarization and direction of light propagation. For any direction of light propagation, however, absorption is nonzero regardless of the direction of the electric vector, i.e., of the polarization of the light. It follows that light absorption at $\lambda = 6125$ Å is not of dipole character (i.e., it is not associated with the excitation of exciton states which, at $\boldsymbol{k} = 0$, correspond to the representation F_1' of group O_h; see Table 4.2) and in order to analyze its nature it is necessary to consider the quadrupolar exciton states.

A comparison of the results of experimental investigations discussed above with the results obtained in examining the anisotropy of absorption and dispersion of light in the vicinity of quadrupolar exciton lines in cubic crystals of the class O_h (Sect. 4.3) enables us to conclude that absorption in Cu_2O crystals at $\lambda = 6125$ Å is of quadrupole nature. Such a comparison also indicates that we are dealing with the excitation of excitons whose wave functions at $\boldsymbol{k} = 0$ are transformed according to an irreducible representation F_2 of the group O_h (Table 4.2). Note that this conclusion, which, like the results of Sect. 4.3, is not associated with the use of any specific model of the exciton, is confirmed by investigations of the angular distribution of absorption. This distribution, as shown in [4.35], is described by (4.3.46). The fact that the line at $\lambda = 6125$ Å is associated with the quadrupole transition $A_1 \to F_2$ is also confirmed by experiments conducted to study the splitting of the line $\lambda = 6125$ Å due to a deformation of the crystals [4.46] or a magnetic field [4.48].

Birefringence

The topic discussed above was the anisotropy of light absorption in cubic crystals. The optical anisotropy of cubic crystals in the transparent region, also due to spatial dispersion, leads to birefringence. Birefringence was first observed in cubic crystals of silicon Si [4.134] and gallium arsenide (GaAs) [4.135] only in 1971.

As shown in Sect. 4.3, birefringence (i.e., the dependence of the refractive index on light polarization) occurs, in particular, for a vector $\boldsymbol{k}$ directed along a face diagonal (see Sect. 4.3.1; for the transparent region, to be discussed in the following, results obtained by using the expansions of the direct and inverse dielectric-constant tensors coincide. Thus we make use of the relations given in Sect. 4.3.1). In this case, as is evident from (4.3.7, 8), the difference of the refractive indices $\Delta n = n_1 - n_2$ [where the n_1 satisfies (4.3.8) and n_2 satisfies (4.3.7)] is determined by the equation

$$\Delta n = \frac{1}{2}\left[\beta_{xyxy} + \frac{1}{2}(\beta_{xxyy} - \beta_{xxxx})\right] n^5 \frac{\omega^2}{c^2} . \tag{4.6.5}$$

In obtaining this relation from (4.3.7, 8), it was assumed that the condition $4\varepsilon_0^2 \beta\omega^2/c^2 \ll 1$ is satisfied. Since $\varepsilon_0 \approx n^2$ and $\beta \sim a^2$, where a is a length of the order of the lattice constant, the preceding condition can also be written in the form $4n^4(2\pi a/\lambda)^2 \ll 1$. In [4.134] birefringence in silicon (single crystals of silicon having a length of about 2 cm and a low dislocation concentration were used) was investigated in the wavelength range $\lambda = 1.1$ to 1.5 μm. In this spectral region silicon has a refractive index $n \approx 3.5$ and the quantity $4\varepsilon_0^2\beta\omega^2/c^2$ is small compared to unity. Details concerning the experimental procedures can be found in [4.134]. Here we note only that owing to the weakness of dispersion – the dependence of n on ω – the relation $\Delta n \sim \omega^2$, see (4.6.5), can be used in the indicated spectral region. For light propagating along the [110] direction, the maximum value of Δn was found to be equal to $\Delta n = (5.04 \pm 0.12) \times 10^{-6}$. This yields a value of approximately 5×10^{-18} cm^2 for the combination of components of tensor β_{ijlm} within the square brackets of (4.7.5).

Employing the procedure of [4.134], birefringence in gallium arsenide was investigated [4.135] for the wavelength range from 0.95 to 1.8 μm. In this research, a strengthening of the phenomenon was observed in the region of the absorption band edge (in GaAs the minimum straight gap $\Delta \approx 1.43$ eV). To explain this effect, the dielectric-constant tensor of the GaAs crystal was calculated in [4.135], taking spatial dispersion into account within the framework of the four-zone semiconductor model taken from [4.139]. The birefringence in CuBr was observed recently [4.140].

New Waves in Nongyrotropic Crystals

Brodin and *Pekar* [4.136] proposed a method for experimentally proving the existence of new (additional) light waves in a nongyrotropic crystal. The method consists of measuring the intensity of monochromatic light, passed through a plane-parallel plate of the crystal, as a function of the plate thickness. The observation of the interference of the light waves emerging from the crystal is to be taken as an indication that two waves exist in the crystal. Such interference should lead to oscillations in intensity as a function of thickness. It was supposed that oscillations, due to reflection and triple passage of one of the waves, are either negligible or can be taken into account in an appropriate way.

Experimental data indicate that the intensity of light absorption in anthracene plates, in the region of the self-absorption band with a maximum at $\omega_L = 25\,200$ cm^{-1} at $T = 20$ K, actually does oscillate as the thickness is varied [4.137]. Assuming that I is the intensity of the transmitted light and I_0 that of the incident light, in the thickness range from 0.05 to 0.3 μm, $\ln(I/I_0)$ was found to be an oscillating function with the distance (difference in plate thickness) between the abscissa of the maxima (with a period), $\Delta d \approx 0.06$ μm. The most distinct oscillations were observed for the frequency $\omega = 25\,108$ cm^{-1}. These oscillations were considered proof of the existence of two waves of the same polarization, with a difference in refractive indices equal to 6.9 [4.136].

It should be pointed out, however, that this interpretation of the oscillations contradicts the results of the calculations for absorption and dispersion curves given in Sect. 4.2. As a matter of fact, according to the data given in [4.137, 138], the value $A \approx 0.1$, see (4.1.13), corresponds to the transition being considered in anthracene. At $\xi = (\omega - \omega_l)/\omega_l = -4\times 10^{-3}$ this value of A yields $n^2 = -A/\xi = 25$, which agrees with [4.20, 137]. At $\delta = 10^{-3}$ and $\xi = -4\times 10^{-3}$, the calculations for the new wave (Figs. 4.6–13) yield $n_2 \approx 35$ and $\varkappa_2 \approx 9$. Consequently, even at $d \approx 0.1\ \mu m$, the new wave practically cannot "reach" the other surface of the film because, at $\lambda_0 = 0.4\ \mu m$, the factor $\exp(-2\pi\varkappa d/\lambda_0) \approx 10^{-6} \ll 1$. These estimates correspond to the value $|\beta'| = 10^{-5}$. Actually, however, the quantity $|\beta'|$ may be much smaller than this value, so that the role played by absorption becomes even greater. We mention also that in analyzing experiments with very thin films an approach based on the use of the tensor $\varepsilon_{ij}(\omega,\boldsymbol{k})$ may prove unsuitable when the depth of the subsurface layer, whose properties are due to the presence of the boundary, is comparable to the film thickness. In this connection, see also [4.141–143].

The experiments conducted in [4.136] were not repeated and thus the origin of these oscillations is not clear. They may possibly be explained by the interference of the ray that passed once through the plate with a ray that passed through it three times. In this case, however (when, moreover, absorption is neglected), the value of the refractive index of anthracene would have to be $n = \lambda_0/2\Delta d \approx 3.45$. This disagrees with the value $n \gtrsim 5$ obtained in [4.20, 137], but one cannot draw conclusions from such a discrepancy since the relation $n = \lambda_0/2\Delta d$ is inaccurate in the case of strong absorption.

Similar experiments, but with a Cu_2O crystal [4.144], differ from those reported in [4.136] in that the oscillations in cuprous oxide were observed in the region of exciton absorption at the quadrupole line $\lambda = 6125$ Å. As previously mentioned, this line pertains to the "yellow" exciton series of Cu_2O, and is ascribed the principal quantum number $n = 1$. The measurements in [4.144] were made at $T = 93$ K and the distance between maxima of oscillations amounted to about 0.2 mm, thereby exceeding the oscillation period that would be observed in the interference of repeatedly reflected waves ($\Delta d = \lambda_0/2n \approx 10^{-4}$ mm) by three orders of magnitude. If we assume, however, that the oscillations observed in this case are caused by the appearance of a new wave, we must also assume that the distance between the maxima is $\Delta d = \lambda_0/2(n_1 - n_2)$, where n_1 and n_2 are the refractive indices of the ordinary and new waves. This, obviously, is a question of adding the two oscillations $A \cos(\omega t - \omega n_1 d/c)$ and $B \cos(\omega t - \omega n_2 d/c)$. The time-averaged square of the amplitude of the resultant oscillation equals $I = (A^2 + B^2)/2 + AB \cos[2\pi(n_1 - n_2)d/\lambda_0]$. If the plate thickness d is changed by the amount Δd, the value of I changes so that for neighboring minima and maxima $\Delta d = \lambda_0/2(n_1 - n_2)$. This result varies only slightly if absorption is taken into account, provided that absorption is sufficiently weak. Hence, at $\Delta d \approx 2\times 10^{-4}$ m and $\lambda_0 \approx 6\times 10^{-7}$ m, we obtain $n_1 - n_2 \approx 10^{-3}$.

At $\xi = 0$ (vicinity of a resonance) we obtain from (4.3.53) that $\tilde{n}_2^2 \approx 0$ and $\tilde{n}_1^2 = \varepsilon_0 - \nu/\mu$. Therefore, when the quantity ε_0 is approximately equal to the ratio ν/μ, the difference $|n_1 - n_2|$ may be very small. Since, however, both the quantity ε_0 and the ratio ν/μ can, in general, assume different values independently of each other for various quadrupole lines and also in various crystals, the occurrence of such a small value of the difference $|n_1 - n_2|$ is extremely unlikely. On the other hand, such small differences of the refractive indices may be observed in the vicinity of quadrupole lines for waves of mutually perpendicular polarizations when quadrupole absorption occurs only for the waves of one of the polarizations (in the vicinity of dipole lines this difference is of the order of unity)[15]. Hence, in the case of insufficiently accurate alignment of the crystal, the intensity oscillations mentioned above may occur. These oscillations have nothing whatsoever to do with the existence of new waves. We could mention other possible causes of similar oscillations. It seems to us, therefore, premature to draw conclusions in connection with the effects observed in Cu_2O until these oscillations have been analyzed in terms of further experimental investigations.

Hitherto we discussed the effects of spatial dispersion in crystals, studied by passing light through them. As mentioned before, the strong absorption of new waves is of prime importance in these experiments. In this connection, it is of interest first of all to consider some experiments on the reflection of light from the surface of a CdS crystal at 1.6 to 2.4 K. The most complete results for the investigation of the frequency dependence of the reflection coefficient of light in the neighborhood of the exciton line $\hbar\omega_L(0) \approx 2.5528$ eV ("A"-exciton line in the notation of [4.145]) are found in [4.82]. Since the "A" excitons are excited by light in the dipole approximation, the dielectric-constant tensor in the corresponding frequency region may be represented by

$$\begin{aligned} &\varepsilon_{011}(\omega) = \varepsilon_{022}(\omega) = \varepsilon_{00}^{\perp} + \frac{2A\,\omega_L^2(0)}{\omega_L^2(0) - \omega^2}\,, \\ &\varepsilon_{033}(\omega) = \varepsilon_{00}^{\parallel}\,, \quad \varepsilon_{0ij}(\omega) = 0\,, \quad i \neq j\,, \end{aligned} \tag{4.6.6}$$

where the meaning of A is clear from (4.1.13). The quantities $\varepsilon_{00}^{\perp}$ and $\varepsilon_{00}^{\parallel}$ are determined by the contributions of the other resonances; they may be assumed constant in the frequency region under consideration. In (4.6.6) use was made of the fact that a mechanical "A" exciton is polarized in a plane perpendicular to the optical axis (the axis $\boldsymbol{c}$, taken as the z axis) and is doubly degenerate. When the wave vector $\boldsymbol{k}$ of the exciton makes a certain angle with the optical axis, degeneracy is removed if the Coulomb long-range interaction is taken into account. Hence, instead of a doubly degenerate mechanical exciton, we

[15] In the vicinity of the exciton transition F_2, when light is propagated along the face diagonal of the cube, only a wave whose electric vector is perpendicular to the plane of the face (Table 4.3) is subject to quadrupolar absorption.

have a transverse exciton (a polarization wave, see Sect. 2.2.2) whose frequency as $\boldsymbol{k} \to 0$ remains equal to $\omega_\perp = \omega_L(0)$, and a longitudinal exciton (a fictitious longitudinal wave, see Sect. 2.2.2) whose limiting frequency depends on the angle θ. This frequency is determined from (2.2.38) combined with (4.6.6). It is equal, consequently, to

$$\omega'_\parallel(0) = \omega_\perp \sqrt{1 + 2A\sin^2\theta/(\varepsilon_{00}^\perp + \Delta\varepsilon_{00}\cos^2\theta)}\ ,$$

where $\Delta\varepsilon_{00} = \varepsilon_{00}^\parallel - \varepsilon_{00}^\perp$.

Moreover, it can readily be seen that the frequency $\omega = \omega_\perp(0) \equiv \omega_L(0)$ determines the position of the absorption line for light whose electric vector is perpendicular to the optical axis and the direction of $\boldsymbol{k}$, whereas the frequency $\omega = \omega'_\parallel(0)$ determines the position of the absorption line of light whose electric vector lies in the same plane as the optical axis and the vector $\boldsymbol{k}$. When the vector $\boldsymbol{k}$ is perpendicular to the optical axis, we have a special situation. In this case, the Coulomb exciton with the frequency $\omega'_\parallel(0)$ becomes longitudinal, remaining so when retardation is taken into account, and does not manifest itself in absorption. Here the refractive index for light, polarized perpendicular to the optical axis $\boldsymbol{c}$ and vector $\boldsymbol{k}$, is determined, in the absence of absorption and spatial dispersion, by

$$n^2(\omega) = \varepsilon_{00}^\perp + \frac{2A\,\omega_L^2(0)}{\omega_\perp^2(0) - \omega^2}\,. \tag{4.6.7}$$

The refractive index for light propagating along the optical axis has the same value. In the frequency region $\omega_\perp(0) < \omega < \omega_\parallel(0)$, the quantity $n^2(\omega) < 0$. Consequently, the reflection coefficient $R = 1$ should correspond to this region. However, as was experimentally established in [4.82], the maximum value of R at $T = 4.2$ K does not exceed 0.5 and, moreover, the function $R(\omega)$ differs somewhat for the two cases in which $\boldsymbol{k} \perp \boldsymbol{c}$ and $\boldsymbol{k} \parallel \boldsymbol{c}$ (Fig. 4.17 [4.82]). This difference is associated with the different intensities of the additional reflection peak in the vicinity of the frequency $\omega = \omega_\parallel(0)$ [in Fig. 4.17b, the frequency $\omega_0 \equiv \omega_\perp(0)$].

When absorption and spatial dispersion are neglected and $n^2(\omega)$ is determined by (4.6.7), the function $R(\omega)$ should be the same in the two cases: $\boldsymbol{k} \perp \boldsymbol{c}$ (and $\boldsymbol{E} \perp \boldsymbol{c}$) and $\boldsymbol{k} \parallel \boldsymbol{c}$.

If absorption is taken into account and (4.6.6) is replaced by

$$\varepsilon_{011}(\omega) = \varepsilon_{022}(\omega) = \varepsilon_{00}^\perp + \frac{2A\,\omega_L^2(0)}{\omega_L^2(0) - \omega^2 - \mathrm{i}\nu\omega}\,, \tag{4.6.8}$$

the value of the reflection coefficient decreases in the frequency region under consideration (Fig. 4.17b). The agreement between calculation and experi-

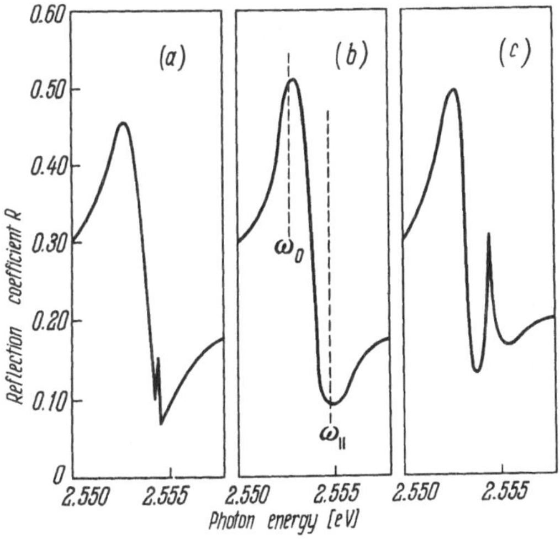

Fig. 4.17a–c. Relation of the light reflection coefficient for a CdS crystal in the vicinity of an "*A*"-exciton absorption line [25]; (**a**) with the wave vector perpendicular to the optical axis (obtained experimentally); (**b**) results of calculations for $\delta = 2\times10^{-4}$, $A = 0.0047$ and $\varepsilon_{00}^{\perp} = 8.1$; (**c**) with the wave vector parallel to the optical axis (obtained experimentally)

ment is best, as established in [4.82], when the following values are chosen: $\hbar\gamma = 10^{-3}$ eV and $A = 0.0047$. In this connection it was noted that the half-width of the line with the energy $\hbar\omega_L(0) = 2.5528$ eV at $T \approx 4$ K does not exceed 10^{-4} eV, and the reduced reflection must therefore be due to spatial dispersion. The reflection coefficient $R(\omega)$ was calculated, accordingly, for $\nu \approx 0$ by the expression

$$\varepsilon_{011}(\omega, \boldsymbol{k}) = \varepsilon_{022}(\omega, \boldsymbol{k}) = \varepsilon_{00}^{\perp} + \frac{2A\,\omega_L^2(0)}{\omega_L^2(0) - \omega^2 + \hbar k^2 \omega_L(0)/m_{\text{exc}}}, \qquad (4.6.9)$$

where m_{exc} is the effective mass of the exciton (Sect. 4.2.2). The calculated results are given in Figs. 4.18, 19. To explain the additional reflection peak it proved essential to assume the existence of a certain subsurface layer of thickness l, where the exciton polarization is supposed to vanish (Sect. 4.5.3, 6), and to take the value $A = 0.0625$. The function $R(\omega)$ was calculated for various l by means of (4.5.45), and the quantity $\varepsilon_{00}^{\perp} = n_0^2$ in (4.6.9), equal to the dielectric constant at $\omega \gg \omega_L(0)$, was taken to be 8. The quantities Γ_t and $1/T_t$, N, appearing in (4.5.33) – for comparison, see (4.5.46) – were taken equal to zero. With these assumptions agreement with experiments is reached (Fig. 4.17; with respect to another model of the "dead" layer, see [4.112a, 146]).

We note in this connection that values of $\gamma(\omega)$, see (4.6.8), as small as those mentioned above, had to be postulated to interpret experimental data obtained in studying the spectra of light reflection from thin films. We shall discuss these investigations in more detail below. Here we point out only that in ideal crystals the quantity $\gamma(\omega)$ is determined by the scattering of polaritons with the frequency ω by phonons. At $\omega \lesssim \omega_\perp$, see (4.6.7), the same process determines the amount of light absorption in the region of the long-wavelength edge of the exciton bands. As shown in Sect. 7.3.3, the damping of

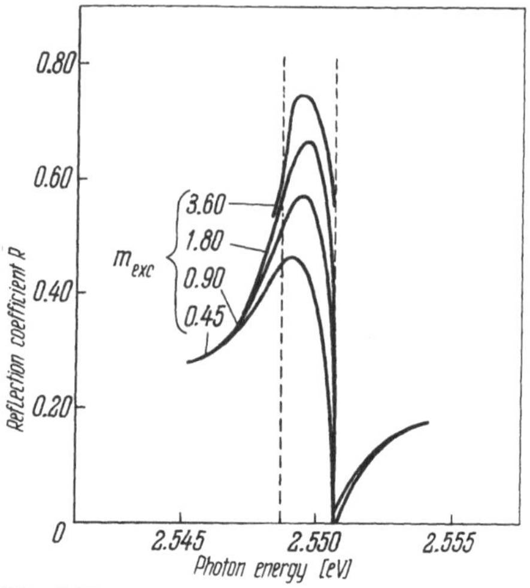

Fig. 4.18

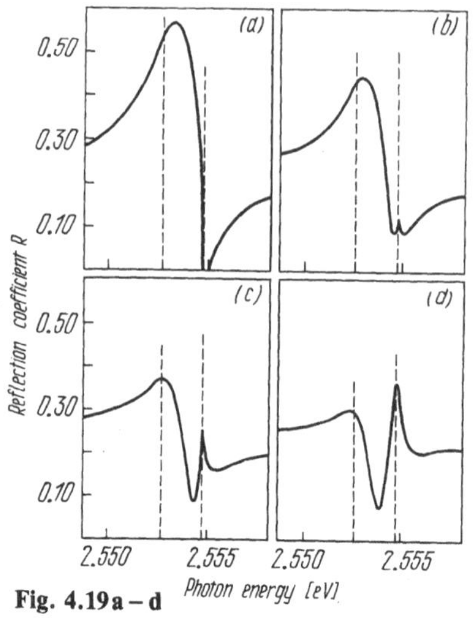

Fig. 4.19a – d

Fig. 4.18. Light reflection coefficient as a function of the effective mass m_{exc} of the exciton in the vicinity of an excition absorption line, with spatial dispersion taken into account and absorption neglected [4.82]. Calculations with $A = 0.0625$ and $\varepsilon_{00}^{\perp} = 8$

Fig. 4.19a – d. Light reflection coefficient as a function of the thickness l of the subsurface layer in the vicinity of an exciton absorption line, with spatial dispersion taken into account and absorption neglected [4.82]. Calculations with $a = 0.0625$, $m_{\text{exc}} = 0.9\,m$ and $\varepsilon_{00}^{\perp} = 8$; **(a)** $l = 0$, **(b)** $l = 77$ Å, **(c)** $l = 116$ Å and **(d)** $l = 154$ Å

polaritons in this spectral region is due to Raman scattering at the frequency $\omega \lesssim \omega_L$ (such Raman scattering is said to be resonance Raman scattering) by phonons. Thus, in principle, independent estimates of the quantity $\gamma(\omega)$ are feasible.

The first successful experimental observations of the interference structure of reflection and transmission spectra for thin crystals ($l \approx 0.1$ to $2\ \mu$m) at low temperatures ($T = 1.6$ to 4.2 K) and in the vicinity of exciton absorption bands, which led to the detection of the additional light wave, were described in [4.147]. The experiments were conducted on specimens of CdS and CdSe crystals. Along with the ordinary interference structure (Sect. 4.5.8) in the spectra of reflected light, structures corresponding to the interference of additional waves were observed. The spectrum of reflection $R(\omega)$ of a CdSe crystal with thickness $l = 0.24\ \mu$m for the region of the exciton line $A\,(n = 1)$ is shown in Fig. 4.20. The light was normally incident, and its polarization was $\boldsymbol{E} \perp \boldsymbol{c}$, $\boldsymbol{c}$ being the direction of the optical axis of the crystal. Also shown in Fig. 4.20 is the dispersion curve for polaritons; the order N of interference is indicated. On the long-wavelength side of the frequency $\omega_{\parallel} = 14730\ \text{cm}^{-1}$, the reflection maxima when $\omega \to \omega_{\parallel}$ become more crowded owing to the increase in $n(\omega)$ as $\omega \to \omega_{\parallel}$. Since there is only a single normal wave 1 (Fig.

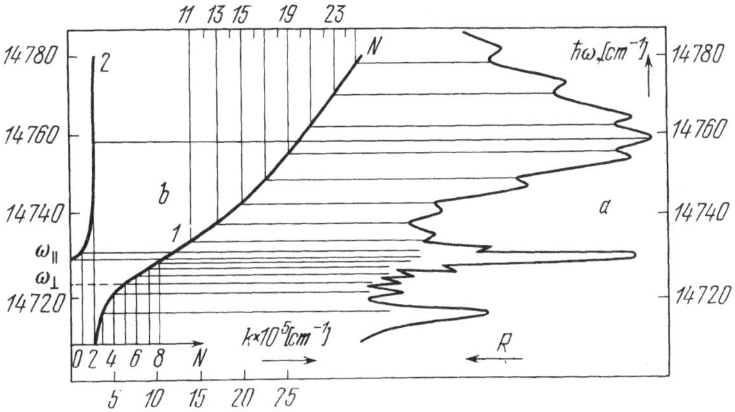

Fig. 4.20. Microphotometric curve *a* of the reflection spectrum (*R*) for a thin CdSe crystal in the vicinity of the exciton line $A(\upsilon = 1)$; *b* – dispersion curves for polaritons (*1* and *2*), plotted from experimental values of the frequencies and the determined values of the order *N* of interference

4.20) in this region of the spectrum, the observed interference corresponds to the ordinary case.

A change in the nature of the interference is observed on the short-wavelength side of the frequency $\omega_{\parallel}$, i.e., at $\omega > \omega_{\parallel}$. Here, the interference of the waves 1 and 2, as they emerge from the plate, is superimposed on the ordinary interference pattern for wave 1. Since the refractive index of wave 2 is much greater than that of wave 1, it is precisely the interference of the waves 1 and 2 that leads to the fine structure of the reflection spectrum $R(\omega)$ at $\omega > \omega_{\parallel}$. The interpretation of the reflection spectra advanced in [4.147] enables one to estimate the values for a number of parameters characterizing excitons in CdS and CdSe, such as the effective mass of the exciton, the damping constant γ, the kind of ABC required, etc. [Ref. 4.52a, p. 141].

The most direct method for the determination of the dispersion relation for refractive indices of a given material is, of course, the observation of the refraction of light by a prism. In the region near the absorption line, however, the transmission, in general, is very small. Therefore, this kind of spectroscopy was not performed until 1981. Using very thin prismatic crystals with small prism angles, *Broser* et al. [4.148] have partially overcome these difficulties. They carried out measurements with prismatic CdS platelets having a thickness from 10 to 100 μm and a prism angle from 0.5° to 2°. The *c* axis was parallel to the two major planes. A beam of monochromatic radiation was deflected by the prism due to refraction; this deflection was then measured with high precision.

The simplest relation for the dielectric-constant tensor was used to compare experimental data with theory. For $E \parallel c$ only the B excition is active, i.e.,

$$\varepsilon^{\|} = \varepsilon_0^{\|} + \frac{4\pi\alpha_B^{\|}\omega_B^2}{\omega_B^2 - \omega^2 + \dfrac{\hbar\omega_B\omega^2 n_{\|}^2}{c^2 m_B^{\perp}} - \mathrm{i}\Gamma_B\omega} . \tag{4.6.10}$$

The contributions of both the A and B excitons were taken into consideration for $E \perp c$, and the following expression, more general than (4.6.9), was used in the region $\omega < \omega_B$:

$$\varepsilon_{\perp} = \varepsilon_0^{\perp} + \frac{4\pi\alpha_A^{\perp}\omega_A^2}{\omega_A^2 - \omega^2 + \dfrac{\hbar\omega_A\omega^2 n_{\perp}^2}{c^2 m_A^{\perp}} - \mathrm{i}\Gamma_A\omega} + \frac{4\pi\alpha_B^{\perp}\omega_B^2}{\omega_B^2 - \omega^2} . \tag{4.6.11}$$

Excellent agreement for refraction and transmission measurements at $T \approx 1.6$ K was obtained if the following values for the exciton parameters are assumed:

$$\hbar\omega_A = 2.5528 \pm 0.0002 \text{ eV} , \quad \hbar\omega_B = 2.5680 \pm 0.0002 \text{ eV} ,$$

$$4\pi\alpha_A^{\perp} = 0.0130 \pm 0.0001 , \quad 4\pi\alpha_B^{\|} = 0.0100 \pm 0.0001 ,$$
$$4\pi\alpha_B^{\perp} = 0.0080 \pm 0.0001 ,$$

$$\varepsilon_0^{\|} = 7.4 \pm 0.1 , \quad \varepsilon_0^{\perp} = 7.4 \pm 0.1 ,$$

$$m_A^{\perp} = (0.9 \pm 0.05)\, m_0 , \quad m_B^{\perp} = (1.3 \pm 0.2)\, m_0 ,$$

$$\Gamma_A = (3.0 \pm 0.3) \times 10^{-5} \text{ eV} , \quad \Gamma_B = (10 \pm 1) \times 10^{-5} \text{ eV} .$$

In picosecond spectroscopy measurements of polariton propagation times have been made at various polariton energies for many crystals. The polariton is expected to propagate in the crystal at the group velocity determined from the dispersion curve. Hence, the data for the polariton group velocity re-established dispersion in spectral regions in which energy dissipation is negligible and $v_g = d\omega/dk$. Up to the present time, time-of-flight measurements have been carried out for CuCl by *Segawa* and co-workers [4.149a, b], for GaAs by *Ulbrich* and *Fehrenbach* [4.150], for CdS by *Aoyagi* et al. [4.151], and for CdSe by *Itoh* and co-workers [4.152a, b].

In the majority of cases the information gained from experimental research agrees well with the results obtained when employing other methods. But when sufficiently thin crystals are employed in time-of-flight measurements, the transmission of light can be observed near the exciton resonance frequency $\omega_{\perp}$, i.e., in the region where energy dissipation becomes appreciable and where the ordinary expression for v_g is no longer valid. This necessitates the analysis of pulse propagation in a highly dispersive and absorptive medium (such media were discussed in Sect. 2.3.2). In particular, the results of time-of-flight measurements in CdSe were discussed on the basis of *Loudon*'s theory [4.153] by *Duong* et al. [4.152b].

The above discussion refers to the observation of spatial dispersion in the spectral region of exciton transitions by the method of linear optics. Light scattering and nonlinear optical methods provide for many new and interesting possibilities. To illustrate the aforesaid, we mention the results of Mandelstam-Brillouin resonant scattering in GaAs [4.154] and in CdS [4.155]. The intensity of this scattering, like that of Raman scattering, increases drastically as the frequency of the exciting light approaches the resonance of the dielectric constant. If this resonance is due to the contribution of the exciton state, the special features of the exciton-polariton spectrum should appear in the scattered light spectra. The exciting laser radiation at frequency ω, close to that of a dipole-active exciton with the main quantum number $n=1$, was used for the investigation in [4.154]. The radiation was perpendicularly incident onto a plate made of a GaAs crystal. Polaritons with the frequency ω and wave vector $\boldsymbol{k}$, perpendicular to the surface, were excited in the crystal. The spectrum of scattered radiation emerging from the crystal along a normal to its surface was measured [4.154]. The photons of the scattered light are formed on the surface of the crystal when polaritons with the wave vector $\boldsymbol{k}'$, directed along the outer normal of the crystal, are incident on the surface from inside the crystal. If $\hbar\Omega(\boldsymbol{q})$ is the energy of a phonon (in the case under study, the main contribution to the scattering process is due to longitudinal acoustic phonons [4.154], then the frequency of the scattered polariton $\omega'=\omega\pm\Omega(\boldsymbol{q})$, with $\boldsymbol{q}=\boldsymbol{k}-\boldsymbol{k}'$. If the additional wave is ignored, a doublet should be observed in the backscattered radiation spectrum. If, on the other hand, the additional wave is taken into account, then an octet should be formed at $\omega>\omega_{\parallel}$ in the scattered-radiation spectrum, as previously pointed out in [4.156].

This is evident from Fig. 4.21 which presents the dispersion curve for polaritons at $k>0$ (for polaritons with the energy $\hbar\omega$) and at $k<0$ (for polaritons with the energy $\hbar\omega'$); only scattering processes in which a phonon is emitted are considered. In this figure k_1 and k_2 are wave vectors of polaritons with energy $\hbar\omega$, which are formed in the crystal by the effect of laser radiation with frequency ω. It is clear that the frequencies in the octet of the backscattered light depend upon the frequency of the exciting radiation. This fact opens the way for studying the dispersion of polaritons [4.154]. As a result, the value of longitudinal-transverse splitting (0.8 ± 0.02) meV was determined for GaAs, and the translational mass of the exciton $m_{\text{exc}}=(0.6\pm0.1)$ m (for heavy excitons in the [100] direction). In calculating the intensity for the process under consideration, it is necessary to apply ABC twice: in finding the amplitudes of the polaritons $(\omega,\boldsymbol{k})$ and also in determining the probability that they will be converted into photons in vacuum [4.156]. According to [4.157], the best agreement with experimental results is reached when using the condition $dP/dz=0$ (for the results of further investigations

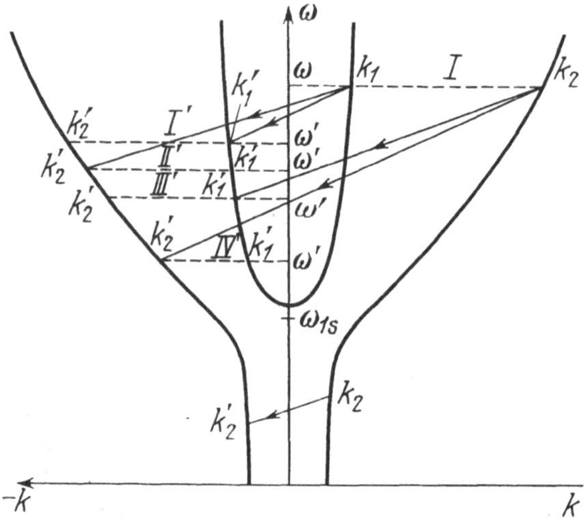

Fig. 4.21. Dispersion curves of polaritons inside the crystal for the $\boldsymbol{k}$- and $-\boldsymbol{k}$-directions (shown schematically). The figure illustrates the kinematics of backward Mandelstam-Brillouin scattering; I denotes the initial polariton state, I' through IV' denotes four possible final states. Arrows indicate the transitions that proceed with the emission of a low-frequency phonon, for which the laws of conservation of energy and momentum are complied with (see [4.157])

the reader is referred to [4.157, 158], to references quoted therein and to [Ref. 452a, pp. 27, 83]). We point out that 18 lines should be observed in similar spectra of scattered radiation at $\omega < \omega_\perp$ in gyrotropic cubic crystals. Experiments have not yet been conducted with such crystals.

It has already been mentioned that in interpreting the dependence of the reflection coefficient R on frequency for the region of exciton resonances, it is necessary to postulate the existence of "dead" layer. The nature of this layer has frequently been questioned, but in many cases it has not been resolved so far. It is natural to reason that a layer of this type should always exist if the subsurface region of the crystal has been sufficiently strongly impaired by impurities, structural defects and other kinds of imperfections. It follows, however, from experimental data that even for crystals whose surface can be assumed "ideal", it becomes necessary sometimes to assume l to be of the order of hundreds of angstroms. In these cases, the presence of a "dead" layer can evidently be related to the diameter of the exciton localized at the surface.

If the distance from the center of the exciton to the surface of the crystal is large compared to the radius of the exciton, the effect of the boundary on its spectrum is negligible. Otherwise, the presence of a boundary leads to a shift in the level of the exciton, and its contribution to the polarization of the medium in the region of a "bulk" resonance becomes small. Such a phenomenon (discussed in Sect. 4.5.3, 6) certainly appears when the exciton touches the boundary. In this case, polarization for the bulk resonance region becomes small for the layer thickness $l \approx 2r_{\rm exc}$, $r_{\rm exc}$ being the exciton radius at a point far away from the boundary. Such statements, of course, are qualitative. It is all the more remarkable, then, that they are quite well confirmed, in some cases by experimental data. Thus, the work [4.159], devoted to precisely the question under discussion, investigates spectra of light reflected from the

Table 4.6

CdS	Thickness of "dead layer" [Å]	Size of exciton [Å]
$n=1$	≈ 70	95
$n=2$	≈ 300	380
$n=3$	≈ 980	855

surface of a CdS crystal at $T=2$ K. It was shown that the thickness of the "dead" layer can be varied by employing various kinds of surface treatment. If the surface of the crystal is properly prepared and can be considered to be sufficiently "good", there is an unambiguous correspondence between the thickness of the "dead" layer and the size of the exciton. Results of such a comparison are listed in Table 4.6. Here the thicknesses of the "dead" layer are given for the region of excitonic transitions in CdS that correspond to the quantum numbers $n = 1, 2, 3$, ($r_{\text{exc}} = a_{\text{exc}} n^2$, a_{exc} being the Bohr radius of the exciton; see [4.160, 161].

Further investigations deal with reflection spectra for oblique incidence [4.162, 163]. These spectra have a rather rich structure, enabling appreciably more information to be gained from them. The first results for the Rayleigh resonant scattering of polaritons (in CdS) were obtained in [4.168].

Finally, we mention that attempts were made to verify experimentally the dispersion relation (2.2.70) for light frequencies ω in the vicinity of exciton absorption lines [4.138, 164]. With certain crystals and at low temperatures this relation was found to be invalid, but the deviations gradually vanish as the temperature is raised. It is quite clear that such investigations are of interest and require further studies. Among others, *Brodin* and *Strashnikova* [4.164] discovered deviations from the dispersion relation for the "A"-exciton band in CdS. The CdS crystal is uniaxial. Hence, if spatial dispersion is neglected, the function $\tilde{n}_l^2(\omega, s)$ has no singularities at $\omega = \omega' + \mathrm{i}\omega''$, where $\omega'' > 0$, for each of the normal waves [there are no essential multiple roots of $\tilde{n}_l^2(\omega, s)$; see Sect. 2.2.5] and the dispersion relation (2.2.70) must hold for each of these functions. It is natural, therefore, to attempt to associate the deviations observed in [4.164] with the effect of spatial dispersion whose presence, as already noted in Sect. 2.2.5, can sometimes lead to singularities of $\tilde{n}_l^2(\omega, s)$ in the upper half-plane for the complex ω. This, consequently, can lead to deviations from the (2.2.70) for the normal wave under consideration [a related question concerns the validity of (4.2.12), see Sect. 4.2.2]. Under the specific conditions of the experiment described in [4.164], where the ordinary and new waves are not separated, the violation of the dispersion relation could be attributed to the excitation of the new wave and the fact that $n_l(\omega)$ and $\varkappa_l(\omega)$ were not measured for a single normal wave; only certain effective quantities were determined. If the values $A = 0.0047$, $\varepsilon_{00} = 8$ and $\delta = 10^{-3}$ are taken for the above-mentioned exciton line, the quantity $|\beta'| = 4 \times 10^{-4}$ (this value corresponds to an effective exciton mass equal to the mass of the electron) and (4.2.8) are used to determine $\tilde{n}_{1,2}$, then, at $\xi=0$, $\varepsilon_0 \approx 8+5\mathrm{i}$, $|n_1| \approx 10$ and $|n_2| \approx 15$, both waves can appear even at $\delta = 10^{-3}$, if the quantity $|\beta'| = 4\times 10^{-4}$ being used has not been overestimated by too much. Further investigations of the optical properties of the CdS crystal in the region of the exciton line under consideration seem to be exceptionally promising [4.165–167].

In concluding our review of the principal features of spatial dispersion in crystals, we point out that significant advances have been achieved in this field in the last decade. This primarily concerns, nongyrotropic crystals, although in gyrotropic crystals the effects of spatial dispersion should, in general, be considerably stronger.

Further progress will evidently require an even more extensive application of the various methods, including not only those of linear optics, but also those employing processes of inelastic scattering of light in crystals, as well as techniques of nonlinear optics.

5. Surface Excitons and Polaritons

This chapter describes the main properties of surface polaritons in isotropic and anisotropic media, along with their excitation and propagation techniques. Surface-polariton dispersion is determined not only by the dielectric permeabilities of the adjacent media, but also reveals some information on the properties of the transition layers (thin films). Recent trends in the spectroscopy of surfaces, interfaces, transition layers and thin films, based on the use of surface polaritons, are discussed.

5.1 Polaritons at the Interface of Isotropic Media

The effect of crystal boundaries on the exciton states may be of a dual nature. First of all, the boundary affects the behavior, near the surface of the crystal of the wave function of "bulk" excitons and polaritons, i.e., excitations whose existence is in no way associated with the presence of a boundary. This effect has been discussed in Sects. 4.5.5 – 9. Moreover, in certain cases the presence of a boundary leads to the appearance of special surface excitons (polaritons), which are analogous to Rayleigh surface waves from elasticity theory and Tamm surface levels of electrons in a bounded crystal. Thus, although there is, in principle, nothing unexpected in the appearance of surface excitons or polaritons, their wave functions and energies have not been investigated to any appreciable extent and, in general, require the application of microtheoretical methods.

An exception to the aforesaid is the case when the thickness of the layer in which the surface excitations are localized is substantially greater than the lattice constant of the crystal. Such surface states (surface polaritons) can be investigated within the framework of phenomenological electrodynamics (see, e.g. [Ref. 5.1, § 68] as well as [5.2 – 4, 4a]).

As a matter of fact, we can presume that the electric and magnetic field-strength vectors, $\boldsymbol{E}$ and $\boldsymbol{H}$, of a surface electromagnetic wave of frequency ω satisfy Maxwell's macroscopic equations in both media (in vacuum and in the crystal); the crystal is assumed to be nonmagnetic[1]. Accordingly, $\boldsymbol{H}=\boldsymbol{B}$; see (2.1.1):

$$\operatorname{rot}\boldsymbol{E}=\mathrm{i}\frac{\omega}{c}\boldsymbol{H}, \quad \operatorname{rot}\boldsymbol{H}=-\mathrm{i}\frac{\omega}{c}\boldsymbol{D}, \qquad \operatorname{div}\boldsymbol{D}=0, \quad \operatorname{div}\boldsymbol{H}=0, \quad D_i=\varepsilon_{ij}(\omega,\boldsymbol{k})E_j. \tag{5.1.1}$$

We shall assume that the surface of the crystal coincides with the xy plane so that the half-space $z>0$ corresponds to vacuum and the half-space $z<0$ to the crystal.

Since we are interested in solutions of Maxwell's equations which decrease when $z\to\pm\infty$, we shall seek them in the following form[2]:

$$\begin{aligned}\boldsymbol{E}^{(1)}(\boldsymbol{r})&=\boldsymbol{E}_0^{(1)}\exp[\mathrm{i}(k_1x+k_2y+\mathrm{i}\varkappa_1z)]\equiv\boldsymbol{E}_0^{(1)}\mathrm{e}^{\mathrm{i}\boldsymbol{K}^{(1)}\cdot\boldsymbol{r}}, \quad z>0, \quad \mathrm{Re}\{\varkappa_1\}>0,\\ \boldsymbol{E}^{(2)}(\boldsymbol{r})&=\boldsymbol{E}_0^{(2)}\exp[\mathrm{i}(k_1x+k_2y-\mathrm{i}\varkappa_2z)]\equiv\boldsymbol{E}_0^{(2)}\mathrm{e}^{\mathrm{i}\boldsymbol{K}^{(2)}\cdot\boldsymbol{r}}, \quad z<0, \quad \mathrm{Re}\{\varkappa_2\}>0,\end{aligned} \tag{5.1.2}$$

where the vectors

$$\begin{aligned}\boldsymbol{K}^{(1)}&=k_1\boldsymbol{e}_1+k_2\boldsymbol{e}_2+\mathrm{i}\varkappa_1\boldsymbol{e}_3,\\ \boldsymbol{K}^{(2)}&=k_1\boldsymbol{e}_1+k_2\boldsymbol{e}_2-\mathrm{i}\varkappa_2\boldsymbol{e}_3 \quad \text{and} \quad \boldsymbol{e}_i\,(i=1,2,3)\end{aligned} \tag{5.1.2a}$$

are unit vectors directed along the coordinate axes. The expressions for $\boldsymbol{H}^{(1)}$ and $\boldsymbol{H}^{(2)}$ are to be written in the same form. The fields, (5.1.2), and the essentially analogous fields have the form of plane waves, (2.2.2), but with complex $\boldsymbol{k}$, i.e., they are a special case of nonuniform plane waves. It is therefore evident that the application of the field equations (2.2.1) or (5.1.1) leads to the dispersion equation (2.2.10) for waves with $\boldsymbol{E}\neq0$. In the notation of (5.1.2), this equation is of the form

$$\left|\frac{\omega^2}{c^2}\varepsilon_{ij}(\omega,\boldsymbol{K})-\boldsymbol{K}^2\delta_{ij}+K_iK_j\right|=0. \tag{5.1.3}$$

In vacuum $\varepsilon_{ij}=\delta_{ij}$ and $\boldsymbol{K}=\boldsymbol{K}^{(1)}$, and (5.1.3) reduces to

$$\omega^2=c^2\boldsymbol{K}^2=c^2(\boldsymbol{K}^{(1)})^2,$$

---

[1] The theory of surface waves in magnetic media was discussed in [5.2, 5]. For surface spin waves and surface polaritons, see [5.6–9]. We also point out that the waves under consideration should be distinguished from two-dimensional excitations that appear in layers with two-dimensional motion of charge carriers [5.10, 11].

[2] Another method for finding surface waves has been developed in [5.12] on the basis of the Ewald-Oseen extinction theorem.

i.e., to

$$\varkappa_1 = \sqrt{k_1^2 + k_2^2 - \omega^2/c^2}\,. \tag{5.1.4}$$

At $\boldsymbol{K} = \boldsymbol{K}^{(2)}$, (5.1.3) can evidently be applied only if the tensor $\varepsilon_{ij}(\omega, \boldsymbol{k})$ retains its meaning where the values of $\boldsymbol{k} = \boldsymbol{K}^{(2)}$ are complex. Since in the present case the quantity $\varkappa_2$ is assumed to be small ($\varkappa_2 a \ll 1$; this being the only case in which the phenomenological approach is applicable), the assumption that this condition is fulfilled seems to be quite natural and agrees with the assumption that it is possible to use the tensor $\varepsilon_{ij}(\omega, \boldsymbol{k})$ in the presence of damping.

To determine the region of frequencies ω that corresponds to surface excitons, it is necessary to require that the boundary conditions are satisfied at $z = 0$, i.e.,

$$E_{01}^{(1)} = E_{01}^{(2)}, \quad E_{02}^{(1)} = E_{02}^{(2)}, \quad E_{03}^{(1)} = D_{03}^{(2)}, \tag{5.1.5a}$$

$$\boldsymbol{H}_0^{(1)} = \boldsymbol{H}_0^{(2)}. \tag{5.1.5b}$$

It is evident that if we comply with (5.1.5b), the boundary condition $E_{03}^{(1)} = D_{03}^{(2)}$ is automatically fulfilled. The other two conditions can be written in the form

$$\boldsymbol{e}_3 \times (\boldsymbol{K}^{(1)} \times \boldsymbol{H}_0^{(1)}) = \boldsymbol{e}_3 \times \boldsymbol{c}, \quad c_i = \varepsilon_{ij}^{-1} (\boldsymbol{K}^{(2)} \times \boldsymbol{H}_0^{(1)})_j. \tag{5.1.6}$$

If, in the case of an isotropic medium, we neglect small terms of the order of $(a/\lambda)^2$ and set

$$\varepsilon_{ij}^{-1}(\omega, \boldsymbol{k}) = \frac{1}{\varepsilon(\omega, \boldsymbol{k})}\,\delta_{ij},$$

(5.1.6) assumes the form

$$H_{03}^{(1)} \boldsymbol{K}^{(1)} - \boldsymbol{H}_0^{(1)} K_3^{(1)} = \frac{1}{\varepsilon(\omega, \boldsymbol{K}^{(2)})} (\boldsymbol{K}^{(2)} H_{03}^{(1)} - \boldsymbol{H}_0^{(1)} K_3^{(2)})\,. \tag{5.1.7}$$

According to (5.1.1), $\boldsymbol{H}^{(1)}\boldsymbol{K}^{(1)} = 0$ and $\boldsymbol{H}^{(2)}\boldsymbol{K}^{(2)} = 0$. This, if we take (5.1.2a, 5b) into account, is satisfied only at $H_{03} = 0$. It then follows from (5.1.7) that $K_3^{(1)} = K_3^{(2)}/\varepsilon$, i.e.,

$$\varkappa_1 = -\varkappa_2/\varepsilon(\omega, \boldsymbol{K}^{(2)})\,. \tag{5.1.7a}$$

Making use of (5.1.4), as well as (5.1.3) which, for an isotropic medium, can be reduced to the form

$$\varkappa_2 = \sqrt{k_1^2 + k_2^2 - \frac{\omega^2}{c^2}\,\varepsilon(\omega, \boldsymbol{K})}\,, \tag{5.1.8}$$

and combining with (5.1.7a), we find

$$k_1^2+k_2^2 \equiv k_{\|}^2=\frac{\omega^2}{c^2}\,\frac{\varepsilon(\omega,\boldsymbol{K})}{\varepsilon(\omega,\boldsymbol{K})+1}\,. \tag{5.1.9}$$

Equations (5.1.7a, 8 and 9) are also valid for complex ε. The frequency ω of the surface wave and the quantity $\varkappa_2$ can thus be expressed in terms of the sum $k_1^2+k_2^2=k_{\|}^2$, to find the dispersion law. Let us first consider the dispersion of surface waves under the assumption that there is no dissipation in the medium.

5.1.1 Dispersion of Surface Waves for Lossless Media

For a real value of $\varepsilon(\omega,\boldsymbol{K})$, and real and positive values of $\varkappa_1$ and $\varkappa_2$, (5.1.7a) can be satisfied only in the frequency region where $\varepsilon(\omega,\boldsymbol{K})<0$. It follows, in this case from the condition $k_1^2+k_2^2>0$, see (5.1.9), that

$$|\varepsilon(\omega,\boldsymbol{K})|>1\,. \tag{5.1.10}$$

From (5.1.9, 10) we see that $k_1^2+k_2^2>\omega^2/c^2$ and, consequently, $\varkappa_1>0$, see (5.1.4), as was assumed. Hence the frequency range in which the surface states under consideration may exist is limited by the condition, see (5.1.10)[3],

$$\varepsilon(\omega,\boldsymbol{K})<-1\,. \tag{5.1.11}$$

The band of these surface excitons is narrower, at least in the absence of spatial dispersion, than the difference $\omega_{\|}-\omega_{\perp}$ [recall that $\varepsilon(\omega_{\|})=0$ and $\varepsilon(\omega_{\perp})=\pm\infty$]. Since we take retardation into account and, in general, consider the complete system of equations of the electromagnetic field, it is clear that the surface excitons under investigation can also be called surface polaritons. The medium on whose surface these polaritons exist is sometimes said to be surface-active.

5.1.2 General Case

In the more general case, when a medium with the dielectric constant $\varepsilon(\omega,k)\equiv\varepsilon_2(\omega,k)$ borders along the plane $z=0$, not with the vacuum, but with a medium having a real dielectric constant $\varepsilon_1>0$ in the frequency region

[3] Dissipation will be taken into account below. In [5.13], Poynting's vector has been calculated, as well as the energy density distribution in a surface polariton and the velocity of polariton propagation when energy dissipation is taken into account. In the absence of dissipation this velocity equals the group velocity $v_{\mathrm{gr},i}=(\partial/\partial k_i)\,[\omega(k_i,0,\varkappa)]$, $i=x,y$; for the energy density and its flux see also Sect. 6.4.2b. Quantization of the field of surface waves was also suggested in [5.13].

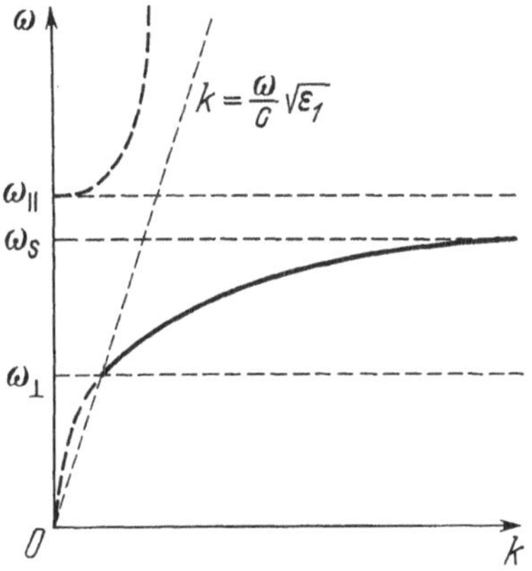

Fig. 5.1. Dispersion of a surface polariton at the interface between isotropic media (*full line curve*). The heavy dashed lines indicate parts of the dispersion curve of waves having a radiation width. (These waves are unstable with respect to transformation into bulk polaritons)

under consideration, the dispersion law for surface polaritons assumes, instead of (5.1.9), the form

$$k_{\|}^{2}=\frac{\omega^{2}}{c^{2}}\,\frac{\varepsilon_{1}\varepsilon}{\varepsilon_{1}+\varepsilon}\,. \tag{5.1.9a}$$

The existence of surface polaritons is determined by the inequality

$$\varepsilon(\omega,k)<-\varepsilon_{1}\,.$$

When (4.1.13) is applied, (5.1.9a) yields the dispersion curve for the surface polariton, illustrated schematically in Fig.5.1 by a heavy line.

5.1.3 Surface Polaritons for Layered Structures

The spectrum of surface polaritons becomes more complicated for layered structures. This can be illustrated by investigating a three-layer structure: $\varepsilon_1>0$, $\varepsilon_3>0$ and $\varepsilon_2<0$ (Fig. 5.2) [5.14]. The results enable us to observe the effect of thickness d of the surface-active layer on the spectrum of surface polaritons. According to [5.14] this spectrum is described by the relation

$$\left(1+\frac{\varkappa_3}{\varkappa_2}\,\frac{\varepsilon_2}{\varepsilon_3}\right)\left(1+\frac{\varkappa_1}{\varkappa_2}\,\frac{\varepsilon_2}{\varepsilon_1}\right) - \exp(-2\varkappa_2 d)\left(1-\frac{\varkappa_1}{\varkappa_2}\,\frac{\varepsilon_2}{\varepsilon_1}\right)\left(1-\frac{\varkappa_3}{\varkappa_2}\,\frac{\varepsilon_2}{\varepsilon_3}\right)=0\,, \tag{5.1.9b}$$

where

$$\varkappa_i=\left(k_{\|}^{2}-\frac{\varepsilon_i\omega^2}{c^2}\right)^{1/2},\quad i=1,2,3\,, \tag{5.1.9c}$$

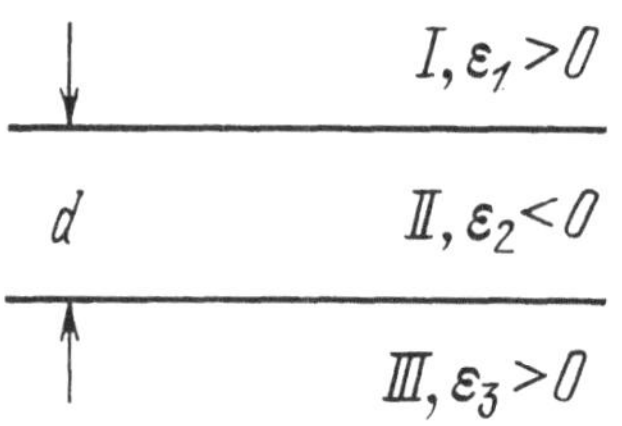

Fig. 5.2

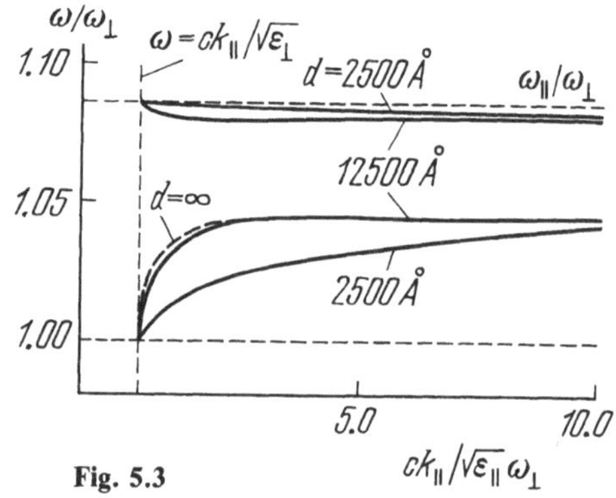

Fig. 5.3

Fig. 5.2. A three-layer system

Fig. 5.3. Results of calculations on the dispersion relation for surface polaritons in a GaAs film located on a sapphire substrate [5.14]

and $\boldsymbol{k}_{\|}(k_1, k_2)$ is the wave vector of the polariton. Phenomenological theories for surface waves in plates were reviewed in [5.15], including spectra for small crystals. Spectra of surface waves on spherical and cylindrical surfaces were investigated in [5.16, 17].

At large values of d, (5.1.9b) is transformed into independent equations of the form of (5.1.9a) for each of the interfaces, I – II and II – III (Fig. 5.2). The nature of the dispersion relations for finite values of d and for $d = \infty$ is illustrated in Fig. 5.3.

Relations (5.1.9, 9a and b) establish the dependence of surface-exciton frequency on $k_1^2 + k_2^2$ with retardation. The transition to the Coulomb surface exciton corresponds to the neglect of retardation, i.e., to the neglect of the displacement current. This means that in (5.1.1) [and, consequently, in (5.1.4 and 8)] it is necessary to discard the term with ω^2/c^2. Therefore, $\varkappa_1 = \varkappa_2 = \sqrt{k_1^2 + k_2^2}$, whereas, by virtue of (5.1.7a), $\varkappa_1 = -\varkappa_2/\varepsilon(\omega, \boldsymbol{K}^{(2)})$. Hence, in the present case, the frequency of a Coulomb surface exciton is determined by

$$\varepsilon(\omega, \boldsymbol{K}^{(2)}) = -1 . \tag{5.1.12}$$

Formally, the same result is obtained from (5.1.9) if the transition $c \to \infty$ is performed with fixed values of k_1, k_2 and ω. In the more general case of (5.1.9a), the condition (5.1.12) is of the form

$$\varepsilon(\omega, \boldsymbol{K}^{(2)}) = -\varepsilon_1 .$$

Neglecting spatial dispersion we set, in accordance with (4.1.13),

$$\varepsilon(\omega, \boldsymbol{k}) = \varepsilon_{00} - \frac{2A\,\omega_{\perp}^2(0)}{\omega^2 - \omega_{\perp}^2(0)} , \tag{5.1.13}$$

then (5.1.12) leads to the following surface-exciton frequency:

$$\omega_s^2(0) = \omega_\perp^2(0) + \frac{2A\,\omega_\perp^2(0)}{\varepsilon_{00}+1}\,. \tag{5.1.14}$$

Hence, the frequency of a Coulomb surface exciton is determined by the relation

$$\omega_s(0) = \omega_\perp(0)\sqrt{1+\frac{2A}{\varepsilon_{00}+1}} < \omega_\|(0)\,, \tag{5.1.15}$$

because $\omega_\|(0) = \omega_\perp(0)\sqrt{1+2A/\varepsilon_{00}}$ in the model under consideration. Since resonances in light absorption correspond to frequencies of nonlongitudinal Coulomb excitons, the position of exciton-absorption lines, associated with surface excitons, correspond, approximately to the frequency $\omega_s(0)$ as determined by (5.1.15).

The fact that the surface polaritons considered above can be excited by charged particles and light is important. In the latter case, however, this excitation can take place only with the participation of phonons, crystal lattice defects or defects at the interface (irregularities). In the absorption of a bulk photon by a surface polariton ($\omega_{\rm ph} = c\sqrt{k_1^2+k_2^2+k_3^2}$), not only the frequency of the wave, but also the tangential components k_1 and k_2 of the wave vector should be conserved. Substituting $\omega = \omega_{\rm ph}$ into (5.1.4) and neglecting defects and phonons yields an imaginary value for $\varkappa_1$ [compare with (5.1.2)].

For the same reason (violation of the laws of conservation of energy and momentum), a surface polariton cannot spontaneously become a bulk polariton (or, as is sometimes said, a surface polariton on an ideal boundary has a radiation width equal to zero). This is in no way surprising because both bulk electromagnetic waves and surface waves are independent normal solutions of Maxwell's equations, and can be transformed one into the other only if we take into account perturbing factors, such as the roughness of the interface (discussed above) or optical nonlinearity which is beyond the scope of Maxwell's linear equations.

For surface polaritons with $\boldsymbol{E} = 0$, whereas $\boldsymbol{D} \neq 0$ by analogy with the case of bulk "polarization waves", see Sect. 2.2.2 and (2.2.41), it is evident that

$$|\varepsilon_{ij}^{-1}(\omega,\boldsymbol{K}^{(2)})| = 0 \tag{5.1.16}$$

is to be satisfied instead of (5.1.3), together with the additional condition $\boldsymbol{K}^{(2)}\boldsymbol{D}^{(2)} = 0$. If $\varkappa_2 a \ll 1$ surface waves with $\boldsymbol{E} = 0$ and $\boldsymbol{D} \neq 0$ (surface "polarization waves") can also be considered within the framework of macrotheory. In this case, however, it is necessary to apply ABC [5.3, 18]. Note that the states of long-wavelength surface Coulomb excitons were treated also within the framework of microtheory in [5.19–23]. However, it is clear that in investigating the properties of surface dipole exciton states at small values of $k_{\|}$ and $\varkappa$, microtheory, with regard to a functional dependence, does not yield anything new, as compared to the phenomenological approach[4]. Moreover, within the range of applicability of equations of the form of (5.1.11 – 13), a microtheoretical treatment of surface excitons (having $\boldsymbol{E} \neq 0$ or $\boldsymbol{D} \neq 0$ when $|k_{\|}|$ and $\varkappa \ll 1/a$) is not necessary because these equations contain only parameters that determine a bulk quantity – the dielectric constant, (4.1.13), and the ABC.

a) The Brewster Waves

The above discussion was subject to a certain idealization because the dielectric constant of one of the media (the surface-active medium) was assumed real (negative) in the spectral region under consideration. For such conditions, the existence of surface-wave solutions of Maxwell's equations was first noted by *Fano* in [5.25]. He also pointed out the interconnection between the spectrum of surface waves and the dispersion of the Brewster angle, i.e., the dependence of this angle on frequency.

In the case of light polarized in the plane of incidence and impinging on the surface of a medium with $\varepsilon(\omega)>0$ at the Brewster angle θ (i.e., for Brewster's waves; here, as previously, we assume a boundary with vacuum for simplicity), the reflected wave disappears [5.1]. Since $\tan\theta = \sqrt{\varepsilon(\omega)}$, we obtain for the equal tangential components of the wave vector of both incident and refracted waves $[k_{\|} = (\omega/c)\sin\theta]$:

$$k_{\|}^2 \equiv k_1^2 + k_2^2 = \frac{\varepsilon(\omega)}{\varepsilon(\omega)+1}\,\frac{\omega^2}{c^2}, \tag{5.1.9d}$$

i.e., exactly the same relation that describes the dispersion of surface polaritons (in Fig. 5.1 the heavy dashed line corresponds to the region of the Brewster waves). This circumstance becomes clear if we note that the dispersion laws, (5.1.9, 9a), do not contain the z components of the wave vector, and that for light incident on a crystal at the Brewster angle, the structure of the fields (Brewster's waves) has the same form, (5.1.2), on both sides of the interface due to the absence of the reflected wave.

The difference in the solutions obtained for Brewster's waves and for surface waves is due to the fact that it $\varepsilon(\omega)>0$ (the Brewster-angle case), the quantity $k_{\|}^2$ turns out to be less than the quantities $(\boldsymbol{K}^{(1)})^2$ and $(\boldsymbol{K}^{(2)})^2$, so that the value of k_3 is real in both media. If, however, $\varepsilon(\omega)<0$, the values of k_3 in both media, see (5.1.4, 8), are found to be imaginary, and the waves are surface waves.

---

[4] Effects of the electrostatic image forces on the spectra of surface Frenkel excitons and Wanier-Mott excitons have especially been treated in [5.20 – 22] (see also [5.24]).

b) Propagation of Surface Waves with Damping

Let us introduce the complex refractive index $k_{\|}^2 c^2/\omega^2 = \tilde{n}^2 = (n+\mathrm{i}\varkappa)^2$ for a surface wave of the given frequency ω. According to (5.1.9), with $\varepsilon = \varepsilon' + \mathrm{i}\varepsilon''$ and weak damping ($\varkappa \ll n$), we have

$$n^2(\omega) = 1 - \frac{\varepsilon'(\omega)+1}{[\varepsilon'(\omega)+1]^2 + [\varepsilon''(\omega)]^2}, \quad 2n\varkappa = \frac{\varepsilon''(\omega)}{[\varepsilon'(\omega)+1]^2 + [\varepsilon''(\omega)]^2}. \tag{5.1.17}$$

In particular, for metals in the region of the normal skin effect (for a more general treatment, see [5.25a]), we can write

$$\varepsilon(\omega) = 1 - \frac{\omega_p^2}{\omega(\omega+\mathrm{i}\Gamma)}, \tag{5.1.18}$$

and, if $\Gamma < \omega_p$, $\omega \ll \omega_p$, the refraction index $n(\omega) \approx 1$, whereas $\varkappa(\omega) = \Gamma\omega/2\omega_p^2$. Therefore, the propagation length $L = c/2\omega\varkappa$ of the polariton is determined by the relation $L = (c/\Gamma)(\omega_p/\omega)^2$. For example, if $\omega_p/\omega = 50$ and $\Gamma = 10^{14}$ Hz, the length $L \approx 1$ cm, as observed experimentally for many metals (Sect. 5.5.4).

For strong dissipation we must use, in discussing the structure of the surface wave, the previously derived relation

$$K_3^{(2)}/K_3^{(1)} = \varepsilon(\omega) = \varepsilon' + \mathrm{i}\varepsilon''. \tag{5.1.19}$$

Setting $K_3^{(1)} = a + \mathrm{i}\varkappa_1$, $K_3^{(2)} = b - \mathrm{i}\varkappa_2$, a, b, $\varkappa_{1,2}$ are real values, $\varkappa_1 > 0$ and $\varkappa_2 > 0$, we obtain

$$ab = \varkappa_1\varkappa_2 + \varepsilon'(\omega)(a^2+\varkappa_1^2), \quad \text{and} \tag{5.1.20a}$$

$$a\varkappa_2 + b\varkappa_1 = -\varepsilon''(\omega)(a^2+\varkappa_1^2). \tag{5.1.20b}$$

In the frequency region in which $\varepsilon'(\omega) > 0$ or else $\varepsilon'(\omega) < 0$, but $|\varepsilon'(\omega)|$ is sufficiently small compared to $|\varepsilon''(\omega)|$ (as is the case, for instance, for metals if $\omega \ll \Gamma$), the product $ab > 0$, so that a and b are quantities of the same sign. Since $\varepsilon''(\omega) > 0$ in a medium at equilibrium, it follows from (5.1.20b) that $a < 0$ and $b < 0$. This case of surface waves was investigated as far back as the beginning of this century [5.26] (the possibility of surface-wave propagation on the interface between two conductors was discussed in [5.27]).

For analogous reasons, in taking dissipation into account, (5.1.9d) permits solutions, of the form of the surface waves previously considered, also for the frequency region where Brewster's waves may exist at $\mathrm{Im}\{\varepsilon\} = 0$ (heavy dashed curve in Fig. 5.1). If damping is assumed to approach zero for these surface solutions, the region of wave localization at the surface increases without restriction, whereas the structure of the field is transformed into the structure considered in deriving (5.1.9d) for real values of $\varepsilon(\omega)$. The waves discussed above correspond to the case when $\mathrm{Im}\{\omega\}=0$ and $\mathrm{Im}\{k_\parallel\}>0$, i.e., to waves with a fixed (real) value of the frequency ω, that are damped in propagation along the interface. However, for complex values of $\varepsilon(\omega)\equiv\varepsilon_2$ or $\varepsilon_1(\omega)$, or both at the same time, the dispersion equation (5.1.9a) predicts surface waves of a more general form, too. As a matter of fact, for complex values of ε and ε_1, (5.1.9a) corresponds to two equations for the real and imaginary parts. Since, in general, ω and k can be complex, these two equations relate four variables (ω', ω'', k', k'', $\omega=\omega'+\mathrm{i}\omega''$, $k=k'+\mathrm{i}k''$). Consequently, a third relation is required, which would allow any three of these variables to be expressed as a function of the fourth. An example of one such additional relation ($\mathrm{Im}\{\omega\}=0$, $\mathrm{Im}\{k_\parallel\}>0$) has already been used above. Waves with damping only in time ($\mathrm{Im}\{k_\parallel\}=0$, $\mathrm{Im}\{\omega\}<0$) can be dealt with in an analogous way. The form of the additional condition is generally determined by exciting and detecting surface waves. Thus, in applying the attenuated total reflection (ATR) method (Sect. 5.5) and scanning with respect to the frequency of the exciting light, the additional relation is of the form

$$\frac{\mathrm{Re}\{\omega\}+\mathrm{i}\,\mathrm{Im}\{\omega\}}{\mathrm{Re}\{k_\parallel\}+\mathrm{i}\,\mathrm{Im}\{k_\parallel\}}\,n_p\sin\varphi=\mathrm{const}\,,$$

where n_p is the refractive index total internal reflection, and φ is the angle of incidence.

c) Reflection, Diffraction and Refraction of Surface Waves at Surface Boundaries

Let us assume that medium I [$\varepsilon_1(\omega)>0$, ω denoting the surface-wave frequency] borders along the plane interface (plane xy) on medium II [$\varepsilon_2(\omega)<0$], where $x<0$ and on medium III [$\varepsilon_3(\omega)<0$], where $x>0$ (Fig. 5.4a). Here the boundary line – the y axis – divides the surface into two parts. In each of these parts the region of existence and the dispersion law of the surface polaritons differs from those in the other part. This naturally poses the question: according to what laws do reflection and refraction of surface waves take place at the boundary line? Since translational symmetry is re-

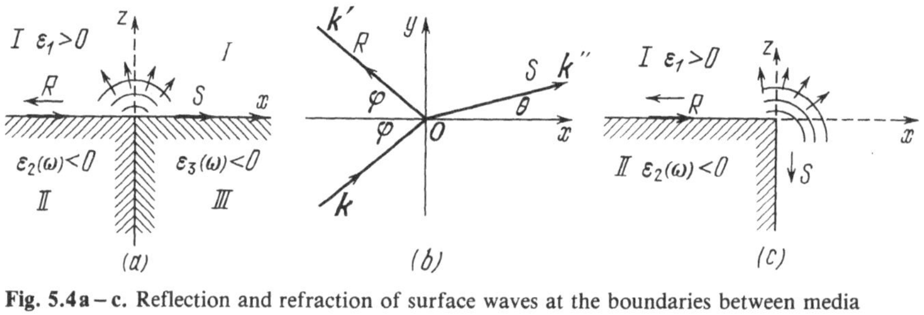

Fig. 5.4a–c. Reflection and refraction of surface waves at the boundaries between media

tained in this situation only with respect to displacement along the y axis, the y projection of the wave vector should be conserved in reflection and refraction. It immediately follows that the angle of incidence is equal to the angle of reflection (Fig. 5.4b). The angle of refraction θ should be determined from the condition

$$n_{\mathrm{I,II}} \sin\varphi = n_{\mathrm{I,III}} \sin\theta \quad \text{or} \quad \sin\theta/\sin\varphi = n_{\mathrm{I,II}}/n_{\mathrm{I,III}} ,$$

where n_{AB} is the refractive index of the surface wave at the interface between media A and B. We assume here that the frequency ω is confined to the region of existence of surface waves for the interface between media I and III too. In the case under discussion, incidence of a surface wave on the boundary line, no surface wave propagates along the boundary between media II and III because, owing to the assumed negative values of ε_2 and ε_3, there simply are no such waves. In anisotropic media, however, it is possible that two refracted waves are formed simultaneously. Of interest is the fact that if $n_{\mathrm{I,II}}/n_{\mathrm{I,III}} > 1$ then, at a certain value $\varphi = \varphi_0$ of the angle of incidence, the angle $\theta = \pi/2\,(\sin\varphi_0 = n_{\mathrm{I,III}}/n_{\mathrm{I,II}})$. This means that at $\varphi > \varphi_0$, total internal reflection of the surface wave is realized. Nevertheless, this does not mean that the energy of the reflected wave is equal to that of the incident wave. Previously we only referred to refracted surface waves. In the case under consideration, however, a bulk wave can be formed in medium I upon reflection from the boundary line. This circumstance is associated with a violation of translational invariance along the x axis, and can readily be understood on the basis of the following simple reasoning. Let us assume, for simplicity, that the angle of incidence $\varphi = 0$. In this case the only conserved component of the wave vector, i.e., the quantity $k_y = 0$. Hence, by proper selection of the values of q_x and q_z (q is the wave vector of the bulk wave, and $q_y = 0$) for the bulk wave in medium I, we can always satisfy the resonance condition

$$\frac{1}{\sqrt{\varepsilon_1}} c \sqrt{q_x^2 + q_z^2} = \omega_s(k_x) .$$

There may also be a similar situation at values of $\varphi \neq 0$ if, for such angles, values of q_z and q_x exist for which there is a real value of k_y that complies with the equation

$$\frac{c}{\sqrt{\varepsilon_1}} \sqrt{q_x^2 + q_z^+ + k_y^2} = \omega_s\left(\frac{k_y}{\sin\varphi}\right) .$$

Hence it follows that the transformation of a surface wave into a bulk wave is possible at the boundary line. Analogous phenomena also occur when a medium with $\varepsilon(\omega) < 0$ forms a wedge of the type illustrated in Fig. 5.4c. Then, if a surface wave propagates in the plane $z = 0$ from infinity to the y

axis, a refracted wave S, propagating in the plane $x = 0$, appears, in addition to the reflected wave R. Here, as in the preceding case, surface waves are transformed into bulk waves along the y axis.

Intensity and angular-distribution calculations for the induced bulk wave, as well as calculations of the amplitude of the reflected wave for the situations illustrated in Fig. 5.4, require a solution of the problem of wave diffraction by a wedge (Fig. 5.4c) or by two contiguous wedges (Fig. 5.4a). These problems, usually arising in connection with problems of radiowave propagation, are among the principal problems of the mathematical theory of diffraction [5.28]. They have been solved only in the impedance approximation [5.29]. When solving an electromagnetic problem outside the wedge (or wedges) in this approximation, we may disregard the field within the wedge, using the approximate relation [5.1] as a boundary condition:

$$E_t = Z(H_t \times n)\,,$$

where E_t and H_t are the tangential components of the electric and magnetic field strengths, n is a normal to the face of the wedge, and Z is the impedance of the face. The application of impedance boundary conditions is justified only if $|\varepsilon| \gg 1$. However, in the region of surface waves this inequality is always satisfied for frequencies $\omega \gtrsim \omega_\perp$ (Fig. 5.1) and continues to be satisfied for all branches or surface-wave frequencies provided that the wedge ($\varepsilon_2(\omega) < 0$) borders on a medium for which $\varepsilon_1 \gg 1$ (Fig. 5.4) rather than with vacuum. Assuming that the condition $|\varepsilon(\omega)| \gg 1$ is fulfilled, let us assess the intensity of bulk waves appearing upon diffraction of a surface wave by an impedance wedge (Fig. 5.4c). According to [5.29] the modulus of the amplitude of the reflected surface wave (taking the amplitude of the incident wave equal to unity) is

$$R = \sqrt{\frac{1+\cos(\pi^2/\alpha)}{\cosh(2\pi\xi/\alpha) - \cos^2(\pi^2/\alpha)}}\,\sinh\left(\frac{\pi\xi}{\alpha}\right),$$

whereas the modulus of the amplitude of the refracted wave (Fig. 5.4) is

$$S = \sqrt{\frac{1-\cos(\pi^2/\alpha)}{\cosh(2\pi\xi/\alpha) + \cos^2(\pi^2/\alpha)}}\,\sinh\left(\frac{\pi\xi}{\alpha}\right),$$

where α is the exterior angle of the wedge ($\alpha = 3\pi/2$ in Fig. 5.4c). The quantity ξ is determined from the relation $Z = -\mathrm{i}\sinh\xi$. If the wedge borders on vacuum, then $\sinh\xi = 1/\sqrt{|\varepsilon|}$ and, if $|\varepsilon| \gg 1$, then $\xi \approx 1/\sqrt{|\varepsilon|}$. It follows from the expressions for R and S that if $|\varepsilon| \gg 1$, then $R \ll 1$ and $S \ll 1$. Hence, practically all the energy of the incident surface wave is transformed into the energy of the generated bulk waves.

This conclusion was confirmed in the experimental work [5.30], which also included a directional diagram for the radiation of surface waves breaking away from the edge of a metal. In addition to the principal maximum with a halfwidth of approx 2°, a number of less intense secondary maxima were also observed.

The effects of spatial dispersion prove to be important when applying thin films to the surface of a metal. For this case, diffraction theory requires more general formulation. The results obtained by this theory concerning the diffraction of surface polaritons at the impedance step are discussed below in Sect. (5.4).

5.2 Spectra of Surface Polaritons in Anisotropic Crystals

The spectra of surface polaritions in anisotropic crystals have a number of specific features [5.31 – 39]. Of primary interest among them is the fact that in anisotropic crystals surface electromagnetic waves of some frequency regions cannot exist at arbitrary orientations of the surface of the crystal with respect to its crystallographic axes. No less important is the strong dependence of the dispersion of a surface polariton on the direction of its two-dimensional wave vector (nonanalyticity with respect to $\boldsymbol{k}$). These features have been experimentally confirmed. Calculations have been carried out for a number of simple orientations in [5.12, 33 – 39]. In particular, in [5.35, 36, 39], as well as in [5.31, 32], the properties of surface waves in uniaxial crystals were investigated, whereas [5.36 – 38] are devoted to properties of surface polaritons in biaxial crystals[5]. The basic though purely formal difficulty, which makes dispersion calculations for surface polaritons somewhat cumbersome, emerges when the normal to the surface of or interface between media does not coincide with the direction of one of the principal axes of the tensor $\varepsilon_{ij}(\omega)$. In other respects these calculations are quite elementary even though they are carried out in the above-mentioned papers only for a number of special situations. In this connection, in accordance with [5.40], we shall, in the following, first find the basic relations that determine the dispersion and polarization of surface H waves (sometimes called TH waves) in arbitrary anisotropic crystals. For this purpose we shall apply the inverse tensor $\varepsilon_{ij}^{-1}(\omega) = A_{ij}$ rather than the tensor $\varepsilon_{ij}(\omega)$, thereby making it substantially simpler to obtain most of the aforementioned results. Then we shall deal with the question of finding surface waves of more general form and, neglecting retardation, we shall examine surface waves in anisotropic crystals of arbitrary symmetry, for any directions of propagation.

As in discussing the properties of surface polaritons in an isotropic medium, we shall assume that the interface corresponds to the xy plane and that

[5] There may be situations in which surface polaritons can propagate only in certain directions [5.31].

the medium is isotropic for $z>0$ [with dielectric constant $\varepsilon_{ij}=\varepsilon_1\delta_{ij}$, $\varepsilon_1>0$] and anisotropic for $z<0$ [with dielectric constant $\varepsilon_{ij}=\varepsilon_{ij}(\omega)$, $A_{ij}(\omega)\equiv\varepsilon_{ij}^{-1}(\omega)$].

It follows from Maxwell's equations that in the regions where $z>0$ (superscript I) and where $z<0$ (subscript II), the amplitudes of the required fields satisfy the system of equations

$$H=\frac{c}{\omega}(K^{\mathrm{I}}\times E^{\mathrm{I}}),\quad D^{\mathrm{I}}=-\frac{c}{\omega}(K^{\mathrm{I}}\times H),\quad E_i^{\mathrm{I}}=\varepsilon_1^{-1}D_i^{\mathrm{I}}\quad(z>0),$$

$$H=\frac{c}{\omega}(K^{\mathrm{II}}\times E^{\mathrm{II}}),\quad D^{\mathrm{II}}=-\frac{c}{\omega}(K^{\mathrm{II}}\times H),\quad E_i^{\mathrm{II}}=A_{ij}D_j^{\mathrm{II}}\quad(z<0),\tag{5.2.1}$$

where K is a vector with the components k_1, k_2 and $\mathrm{i}\varkappa_1$ ($\mathrm{Re}\{\varkappa_1\}>0$) for fields where $z>0$ ($K=K^{\mathrm{I}}$), and with the components k_1, k_2 and $-\mathrm{i}\varkappa_2$ ($\mathrm{Re}\{\varkappa_2\}>0$) for fields where $z<0$ ($K=K^{\mathrm{II}}$). Also taken into account in (5.2.1) is the fact that all three components of H are continuous at the interface ($H^{\mathrm{I}}=H^{\mathrm{II}}=H$). It follows from the orthogonality conditions $K^{\mathrm{I}}\cdot H=0$ and $K^{\mathrm{II}}\cdot H=0$ that $H_3=0$ in waves with $H\neq 0$. Further simplification is achieved if the coordinate system is selected so that the x axis is directed along $k\,(k_1,k_2)$ (with $k_2=0$), where we have omitted the sign $\|$ in denoting the two-dimensional vector $k\equiv(k_1,k_2)$. From the orthogonality of H and K, (5.2.1) and the continuity of E_2 at $z=0$ in this case, it follows that $E_2^{\mathrm{I}}=E_2^{\mathrm{II}}=0$ and $H_1^{\mathrm{I}}=H_1^{\mathrm{II}}=0$, whereas $H_2^{\mathrm{I}}=H_2^{\mathrm{II}}=H\neq 0$. The waves being considered are H waves [5.1]. Substituting $E_i=-(c/\omega)A_{ij}(K\times H)_j$ into (5.2.1), we obtain the relation

$$A_{11}K_3^2+A_{33}k^2-2A_{13}kK_3=\omega^2/c^2,\tag{5.2.2}$$

which enables the quantities $\varkappa_1$ and $\varkappa_2$ to be expressed in terms of k and ω for each of the regions ($z>0$ and $z<0$). In particular, for the region with $z>0$, we have $A_{11}=A_{33}=1/\varepsilon_1$, $A_{13}=0$ and $K_3=\mathrm{i}\varkappa_1$, so that for $\varkappa_1$ we obtain

$$\varkappa_1=\sqrt{k^2-\frac{\omega^2}{c^2}\varepsilon_1}\,.\tag{5.2.3}$$

For the region where $z<0$, we have $K_3=-\mathrm{i}\varkappa_2$, and for the quantity $\varkappa_2$ (where $\mathrm{Re}\{\varkappa_2\}>0$), we find the value

$$\varkappa_2=\mathrm{i}kA_{13}A_{11}^{-1}+\sqrt{(A_{33}A_{11}^{-1}-A_{13}^2A_{11}^{-2})k^2-\frac{\omega^2}{c^2}A_{11}^{-1}}\,.\tag{5.2.4}$$

If $\mathrm{Im}\{\varkappa_2\} \neq 0$ oscillatory damping of the fields occurs when $z \to -\infty$.

Also following from the condition $E_2^{\mathrm{II}} = 0$ is the relation

$$A_{21}D_1 + A_{23}D_3 = 0 ,$$

from which we find that for the frequencies of the surface waves under consideration it is necessary to satisfy the condition

$$\mathrm{i}A_{21}\varkappa_2 + A_{23}k = 0 . \tag{5.2.4a}$$

This relation is extremely restrictive and results in strong limits on the region of existence of surface H waves.

The last of the boundary conditions which we still have to satisfy is the continuity condition for E_1 at the interface. Making use of (5.2.1), we find that the condition $E_1^{\mathrm{I}} = E_1^{\mathrm{II}}$ is satisfied if

$$E_1^{\mathrm{II}} = -\frac{Hc}{\omega}(A_{13}k + i\varkappa_2 A_{11}) = E_1^{\mathrm{I}} = \mathrm{i}\frac{Hc}{\omega\varepsilon_1}\varkappa_1 ,$$

i.e., if

$$A_{13}k + \mathrm{i}A_{11}\varkappa_2 = -\mathrm{i}\varkappa_1/\varepsilon_1 .$$

Combining this with (5.2.4) and cancelling $\varkappa_2$, the preceding relation can also be written in the form

$$\frac{\varkappa_1}{\varepsilon_1} = -A_{11}\sqrt{\left(\frac{A_{33}}{A_{11}} - \frac{A_{13}^2}{A_{11}^2}\right)k^2 - \frac{\omega^2}{c^2}A_{11}^{-1}} = -A_{11}\mathrm{Re}\{\varkappa_2\} , \tag{5.2.5}$$

from which it follows that, by virtue of the fact that $\mathrm{Re}\{\varkappa_2\} > 0$ in the frequency region of surface polaritons, $A_{11}(\omega) < 0$. The quantity ε_1 is assumed to be positive, as has been indicated. It is clear that the condition $\mathrm{Re}\{\varkappa_2\} > 0$, in general, restricts not only the frequency region of surface H waves, but the region of possible directions of their propagation as well. With respect to further restrictions, see (5.2.3a, 4b and 6), which are given below.

In view of the aforesaid, it follows from (5.2.3, 4) that the conditions $\mathrm{Re}\{\varkappa_1\} > 0$ and $\mathrm{Re}\{\varkappa_2\} > 0$ can be fulfilled in the region of frequencies ω such that

$$k^2 > \frac{\omega^2}{c^2}\varepsilon_1 \tag{5.2.3a}$$

and where, in satisfying the inequality $A_{11} < 0$, we also have

$$\frac{A_{33}}{A_{11}} - \left(\frac{A_{13}}{A_{11}}\right)^2 \geqslant 0 . \tag{5.2.4b}$$

This last inequality indicates that the quantities A_{33} and A_{11} should have the same sign (negative). If, however, A_{11} and A_{33} differ in sign ($A_{33} > 0$), then the condition $\mathrm{Re}\{\varkappa_2\} > 0$ can also be satisfied if

$$k^2 < \frac{\omega^2}{c^2} |A_{11}^{-1}| \left| \frac{A_{33}}{A_{11}} - \left(\frac{A_{13}}{A_{11}}\right)^2 \right|^{-1} . \tag{5.2.4c}$$

This means that if $A_{11}(\omega) < 0$, but $A_{33}(\omega) > 0$, the possible region of surface polariton existence is determined by the inequality [here we took (5.2.3a) into account]:

$$|A_{11}^{-1}| \left| \frac{A_{33}}{A_{11}} - \left(\frac{A_{13}}{A_{11}}\right)^2 \right|^{-1} > \varepsilon_1 . \tag{5.2.6}$$

Substituting the explicit expression for $\varkappa_1$ into (5.2.5) and determining the quantity k^2 from the relation thus obtained, we finally find the dispersion law for a surface wave:

$$k^2 = \frac{\omega^2}{c^2} \frac{A_{11} - \varepsilon_1^{-1}}{A_{11}A_{33} - A_{13}^2 - \varepsilon_1^{-2}} . \tag{5.2.7}$$

To illustrate this relation we shall consider several of the most interesting examples. In particular, if the x and z axes coincide with the principal axes of the tensor A_{ij}, i.e., if $A_{ij} = \delta_{ij}/\varepsilon_{ii}$, (5.2.4) is satisfied, and (5.2.7) can be written in the form

$$k^2 = \frac{\omega^2}{c^2} \varepsilon_1 \varepsilon_{33} \frac{\varepsilon_{11} - \varepsilon_1}{\varepsilon_{11}\varepsilon_{33} - \varepsilon_1^2} , \tag{5.2.8}$$

as obtained in [5.36]. For $\varepsilon_{11} = \varepsilon_{33} = \varepsilon$, it can be transformed into (5.1.9a) for the case of isotropic media. For uniaxial crystals (5.2.4a) can also be fulfilled if the optical axis lies in the xz plane and makes an arbitrary angle θ with the z axis (with such orientation of the axis, $A_{21} = A_{23} = 0$). Since, in this case,

$$\begin{aligned} A_{11} &= \frac{\cos^2\theta}{\varepsilon^{\perp}} + \frac{\sin^2\theta}{\varepsilon^{\parallel}} , \quad A_{33} = \frac{\cos^2\theta}{\varepsilon^{\parallel}} + \frac{\sin^2\theta}{\varepsilon^{\perp}} , \\ A_{13} &= \sin\theta\cos\theta \left(\frac{1}{\varepsilon^{\parallel}} - \frac{1}{\varepsilon^{\perp}}\right) , \end{aligned} \tag{5.2.9}$$

we obtain $A_{11}A_{33} = 1/\varepsilon^{\parallel}\varepsilon^{\perp}$, by virtue of which, see (5.2.7),

$$k^2 = \frac{\omega^2}{c^2} \frac{(\varepsilon^{\parallel}\cos^2\theta + \varepsilon^{\perp}\sin^2\theta) - \varepsilon^{\perp}\varepsilon^{\parallel}/\varepsilon_1}{\varepsilon_1^2 - \varepsilon^{\perp}\varepsilon^{\parallel}} \varepsilon_1^2 . \tag{5.2.10}$$

It follows from this relation that resonances of the refraction index $n^2 = k^2c^2/\omega^2$ are brought about at values of the frequency ω that comply with the condition $\varepsilon^{\perp}\varepsilon^{\parallel} = \varepsilon_1^2$ and, consequently, their position is independent of the angle θ (only the intensity of the resonance depends upon angle θ).

It has been pointed out above that (5.2.7) describes the dispersion law only for the simplest surface H waves. In anisotropic crystals the structure of surface waves may be substantially more complicated. The point is that two generally different values of $\varkappa_2$, with $\mathrm{Re}\{\varkappa_2\} > 0$, can be found from the Fresnel equation for an anisotropic crystal at given values of ω and $\boldsymbol{k}(k_1, k_2)$. This means that when seeking surface-wave solutions in the general form, it is necessary to join not just a single wave, but a superposition of two solutions with the field in medium I at the interface, as is usually done for bulk waves in the presence of birefringence[6]. This kind of general approach turns out to be extremely cumbersome, even for uniaxial crystals, and has not yet provided any significant results that could not be obtained within the framework of H-wave theory. At the same time, H waves are realized, as has been indicated, only for special directions of propagation of surface waves, thereby providing only a very incomplete idea of their spectrum. In connection with the aforesaid, we shall consider the spectrum of surface polaritons, within the scope of the same geometry as in the previous treatment, in the region of sufficiently large values of $k \gg \omega/c$, where retardation can be neglected (a similar treatment for uniaxial crystals can be found in [5.31]). Here Maxwell's equations, when $c \to \infty$, are reduced to the form

$$\mathrm{rot}\,\boldsymbol{E} = 0, \;\; \mathrm{div}\,\boldsymbol{D} = 0, \;\; D_i = \varepsilon_{ij}E_j, \qquad \mathrm{rot}\,\boldsymbol{H} = 0, \;\; \mathrm{div}\,\boldsymbol{H} = 0 .$$

We shall examine a field with $\boldsymbol{H} = 0$, because for fields with $\boldsymbol{E} = 0$, but $\boldsymbol{H} \neq 0$, it can be readily seen that no surface waves appear in nonmagnetic media.

For a plane wave with the wave vector $\boldsymbol{K}(k, 0, K_3)$, $\boldsymbol{D} \cdot \boldsymbol{K} = 0$ and $\boldsymbol{E} = C\boldsymbol{K}$, C being a scalar. Hence

$$\varepsilon_{ij}(\omega)K_iK_j = 0 . \tag{5.2.11}$$

This relation enables us to express the quantity K_3 in terms of ω and k. In particular, in medium I, $K_3 = \mathrm{i}\varkappa_1$, where $\varkappa_1 = k$, and in medium II, $K_3 = -\mathrm{i}\varkappa_2$, where

$$\varkappa_2 = -\mathrm{i}k\frac{\varepsilon_{13}}{\varepsilon_{33}} + k\sqrt{\frac{\varepsilon_{11}}{\varepsilon_{33}} - \left(\frac{\varepsilon_{13}}{\varepsilon_{33}}\right)^2} . \tag{5.2.12}$$

---

[6] In [5.41] such a treatment is used to analyze surface plasmons in semiconductors located in a magnetic field (see also [5.42]).

It follows from the requirement $\mathrm{Re}\{\varkappa_2\} > 0$ that for the frequencies of surface waves the quantities ε_{11} and ε_{33} should have the same sign. The continuity of $\boldsymbol{E}_t$ at the interface is provided for by the choice $C^{\mathrm{I}} = C^{\mathrm{II}}$. As for the continuity of D_3, i.e., $D_3^{\mathrm{I}} = D_3^{\mathrm{II}}$, it is the case if $\varepsilon_1 K_3^{\mathrm{I}} = \varepsilon_{31} k + \varepsilon_{33} K_3^{\mathrm{II}}$, i.e., if

$$\varepsilon_1 = -\varepsilon_{33} \sqrt{\frac{\varepsilon_{11}}{\varepsilon_{33}} - \left(\frac{\varepsilon_{13}}{\varepsilon_{33}}\right)^2} . \tag{5.2.13}$$

For real values of $\varepsilon_{ij}(\omega)$ and $\varepsilon_1 > 0$, (5.2.13) can be fulfilled only if $\varepsilon_{33} < 0$ (and, consequently, $\varepsilon_{11}(\omega) < 0$ as well, see above). Then (5.2.13) assumes the form

$$\varepsilon_1 = \sqrt{\varepsilon_{11}(\omega)\,\varepsilon_{33}(\omega) - \varepsilon_{13}^2(\omega)} . \tag{5.2.14}$$

Relation (5.2.14) completely determines the values of the limiting frequencies of surface polaritons in arbitrary anisotropic crystals when $k \gg \omega/c$ (neglecting spatial dispersion) and it is precisely the values of these frequencies $\omega = \omega_s$, satisfying (5.2.14), which determine the resonances of the refractive indices $\tilde{n}^2(\omega)$ of possible surface waves. For the previously discussed H waves, this follows directly from (5.2.7) provided we take into account that in fulfilling (5.2.4a) in the region of large values of k, i.e., at $A_{12} = 0$ and $A_{23} = 0$ (consequently, $\varepsilon_{12} = \varepsilon_{23} = 0$ as well),

$$A_{11} A_{33} - A_{13}^2 = (\varepsilon_{11}\varepsilon_{33} - \varepsilon_{13}^2)^{-1} .$$

The aforesaid is also valid for the dispersion of surface waves which are not H waves and for which the dispersion law has not yet been found in the explicit form.

For the coordinate system employed here (the z axis $\|$ to the normal to the surface, and $\boldsymbol{k} \|$ to the x axis), the values of the components of tensor ε_{ij} contained in (5.2.14) are found to be dependent on the orientation of the principal axes of ε_{ij}. This circumstance leads to the previously mentioned strong dependence of the frequencies ω_s, satisfying (5.2.14), on the direction of propagation of the surface wave (i.e., to the nonanalyticity of the dependence of ω_s on $\boldsymbol{k}$ when $\boldsymbol{k} \to 0$).

In conclusion, we point out (see [5.33], as well as a detailed analysis in [5.37]) that along those branches in the spectrum of surface waves $\omega = \omega(\boldsymbol{k})$ that extend into the region of large values of $\boldsymbol{k}$, surface polaritons may, in general, exist in regions of k values that are limited, not only below, but above as well.

For H waves, this conclusion follows from a comparison of inequalities (5.2.3a and 4c), and is valid for those directions of wave propagation and those frequency regions where $A_{11}(\omega)/A_{33}(\omega) < 0$. Such a solution may be obtained only at $k \sim \omega/c$, whereas these states do not exist if retardation is ignored. The observation of such states was mentioned in [5.15, 37, 39]. Sometimes [5.37] they have been called virtual surface polaritons.

Under conditions in which the dispersion law (5.2.7) is valid, the solutions under discussion are realized in the frequency region where $\varepsilon_{11}(\omega) = A_{11}^{-1}(\omega) < 0$ and $\varepsilon_{33} = A_{33}^{-1}(\omega) > \varepsilon_1$ [the last inequality follows from (5.2.6)].

To illustrate the aforesaid, we shall present here, in accordance with [5.37], results obtained in examining the simplest case, in which an optically isotropic medium ($\varepsilon^{\mathrm{I}} = \varepsilon_1 > 0$) borders along the plane $z = 0$ with an anisotropic medium for which the principal axes of the dielectric-constant tensor are directed along the x, y and z axes. We assume in this medium that

$$\varepsilon_{11}^{\mathrm{II}}(\omega) = 1 + \frac{S_1 \omega_{\perp x}^2}{\omega_{\perp x}^2 - \omega^2}, \qquad \varepsilon_{33}^{\mathrm{II}}(\omega) = 1 + \frac{S_3 \omega_{\perp z}^2}{\omega_{\perp z}^2 - \omega^2}, \tag{5.2.15}$$

and that the wave vector $\boldsymbol{k}$ of the surface wave is directed along the x axis. If $\omega_{\perp z} > \omega_{\perp x}$, then, in the frequency range $(\omega_{\perp x}, \omega_{\|x})$, $\varepsilon_{11}^{\mathrm{II}}(\omega_{\|x}) = 0$, the quantity $\varepsilon_{11}^{\mathrm{II}}(\omega) < 0$ and the inequality $\varepsilon_{33}^{\mathrm{II}} > \varepsilon_1$ can also be satisfied for a sufficiently large value of the coefficient S_3. If such a situation actually exists, then, using (5.2.15), the dispersion law (5.2.8) leads to the result illustrated schematically in Fig. 5.5.

5.3 Transition-Layer Effects in Surface Polariton Spectra

Previously, in discussing the properties of surface polaritons within the framework of Maxwell's equations, boundary conditions were applied at a sharp interface, and in the absence of surface currents and charges. In this simplest case the properties of surface waves are completely determined by the dielectric-constant tensors of the two media. Hence information on the dispersion

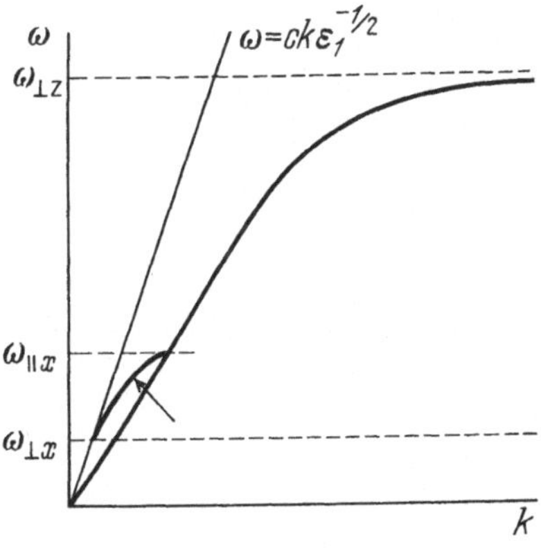

Fig. 5.5. The "virtual" excitation surface polariton dispersion curve (indicated by the arrow). Also shown (at $\omega < \omega_{\perp z}$) is the dispersion relation for bulk polaritons having a wave vector $\boldsymbol{k} \parallel \boldsymbol{x}$ and the electric field strength $\boldsymbol{E} \parallel z$

of polaritons, obtained from their experimental investigation, can be and is being used to find the above-mentioned tensors.

If surface currents and charges are present on the interfaces between media or, more generally, if there is a certain subsurface transition layer, the characteristics of the surface polaritons are found to be dependent on the properties of this layer as well. Evidently, there are only two nontrivial situations which can lead to qualitative effects if the transition layer is taken into account for surface polaritons. One situation corresponds to the presence of a thin layer of metal on the surface of a dielectric or a semiconductor, which leads to "metallic" quenching of the surface polariton and a similar phenomenon can occur even when the surface conductivity is sufficiently high. This quenching is accompanied by substantial broadening, and a shift of the surface polariton lines [5.43]. Their measurement, as is shown below and in [5.43], can be used, for instance, to determine the electrical conductivity of thin metal films in the region of the polariton spectrum.

The other nontrivial case corresponds to the presence of a dielectric transition layer under the condition that one of the natural frequencies of the dipole vibrations is within the frequency range of the surface polariton. The resulting resonance situation leads to the appearance of a gap in the surface polariton spectrum [5.44]. It is common practice to assume that the presence of a transition layer leads to optical effects of the order of d/λ, d being the thickness of the layer and λ the wavelength of the light. This actually holds true in the absence of a resonance with the surface polariton. But if there is such a resonance, the gap that appears in the surface polariton spectrum has a magnitude only of the order of $\sqrt{d/\lambda}$. Both of these situations are discussed in greater detail in the following (see also [5.40] and *Agranovich* in [Ref. 5.4a, p. 187]).

5.3.1 Transition Layers with High Electric Conductivity

Consider a semi-infinite isotropic crystal coated by a thin film of metal having electrical conductivity σ. If the film thickness a is much less than the depth of the skin layer, the presence of the film can be taken into consideration by introducing surface currents. Then, in view of the fact that the discontinuity in the tangential component of the magnetic-field strength due to the surface current $\sigma a E_t$ is $H_y^{(1)} - H_y^{(2)} = (4\pi/c)\sigma a E_1$ (the equation $E_1^{(1)} = E_1^{(2)}$ remains as the second boundary condition), it can readily be seen that the frequency of the surface polariton satisfies the equation[7]

$$\frac{\varepsilon}{\varkappa_2} + \frac{1}{\varkappa_1} = \mathrm{i}\frac{4\pi\sigma a}{\omega}, \tag{5.3.1}$$

[7] We shall assume from now on that the quantity $\sigma(\omega)$ is real; this holds only when $\omega \ll 1/\tau$, τ being the mean collision time of the electron in a metal (for a more detailed discussion, see below).

where $\varepsilon(\omega)$ is the dielectric constant of the crystal,

$$\varkappa_1 = \sqrt{k^2 - \omega^2/c^2}\,, \qquad \varkappa_2 = \sqrt{k^2 - \varepsilon\omega^2/c^2}\,,$$

k is the wave vector of the surface polariton and ω is its frequency. In particular, it follows from (5.3.1) that in the nonrelativistic region ($k \gg \omega/c$ and $k \gg (\omega/c)\,|\varepsilon|$), the width of the surface polariton is determined by

$$\Gamma = \Gamma_0 + \frac{2\pi\sigma(\varepsilon_0 - \varepsilon_\infty)}{(\varepsilon_0 + 1)(\varepsilon_\infty + 1)}\,ka\,. \tag{5.3.2}$$

The occurrence of an additional term in (5.3.2) is evidently due to the Joule energy losses of the polariton in the metallic film. The first observations of "metallic" quenching of surface polaritons were published for Au on SiO_2 in [5.45] and for Ag and Bi on SiO_2 in [5.46], as well as in [5.47].

A distinctive feature of (5.3.2) is the linear increase in Γ with increasing k. Since the quantities Γ and Γ_0 can, in principle, be measured experimentally for the system under consideration, it becomes feasible to find the electrical conductivity σ of thin metallic films and its dependence on various factors (magnetic field, temperature, etc.) for the frequencies of surface polaritons (i.e., at $\omega \approx 10^{11} - 10^{15}$ Hz). It is unnecessary to have a continuous film for this purpose. The only essential restriction is the requirement that the quantity Γ be small in comparison to the width of the gap between the frequencies of the longitudinal and transverse bulk optical phonons (or excitons). Other parameters being equal, this restriction actually provides for the very existence of the surface polariton. However, this requirement may not be met even for thin layers of suitable (high-conductivity) metals. For example, if we have a gold film of thickness $a = 10$ Å on the surface of quartz, $\Delta\Gamma = \Gamma - \Gamma_0 \gtrsim 4 \times 10^{13}$ Hz for a polariton in the frequency region $\omega \approx \omega_1 = 1072\ \mathrm{cm}^{-1}$ ($\varepsilon_0 = 3.03$ and $\varepsilon_\infty = 2.36$), whereas the gap width is approximately 10^{13} Hz. If in this problem we assume a layer of dielectric with finite thickness D, having a metallic layer on only one of its surfaces, the quenching of a surface polariton, localized at the other surface layer, can be substantially weakened [5.43].

Along with the broadening of the lines of surface polaritons, the presence of a metallic film or a metallic substrate also has a substantial influence on the surface-wave dispersion law itself. To discuss various possible situations that may exist here, we shall consider a four-layer plane-parallel system (Fig. 5.6) in which medium II [$\varepsilon_2 = \varepsilon(\omega) < 0$] is surface active (with the layer thickness d); medium IV is a metal (with the film thickness a and with $\varepsilon_4 \equiv \varepsilon_M(\omega)$; the normal skin effect is assumed); whereas media I and III ($\varepsilon_1 > 0$ and $\varepsilon_3 > 0$) are unbound. It can be shown [5.47] that if we take the layers to be macroscopic (i.e., assuming that d and a are large compared to the interatomic distance), the dispersion law of surface polaritons in such a system is determined by

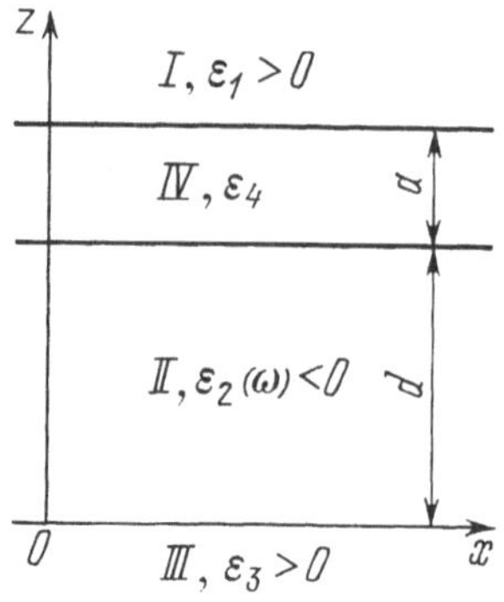

Fig. 5.6. A four-layer system

$$\frac{(Z_2+\zeta)(Z_2+Z_3)}{(Z_2-\zeta)(Z_2-Z_3)}=\mathrm{e}^{-2\varkappa_2 d}, \tag{5.3.3}$$

where (5.1.9c)

$$Z_l=\frac{\mathrm{i}c}{\omega\varepsilon_l}\sqrt{k^2-\frac{\omega^2}{c^2}\varepsilon_l}\equiv\frac{\mathrm{i}c}{\omega\varepsilon_l}\varkappa_l, \tag{5.3.4a}$$

$$\zeta=Z_4\frac{Z_4+Z_1\coth a\varkappa_4}{Z_1+Z_4\coth a\varkappa_4}. \tag{5.3.4b}$$

If $a\to 0$, then $\zeta\to Z_1$ and (5.3.3) is transformed into (5.1.9b). For arbitrary values of a and $d\to\infty$, the dispersion equation for surface waves at the boundary with a metal is, according to (5.3.3), of the form

$$Z_2+\zeta=0 \tag{5.3.5}$$

and, taking (5.3.4b) into account, coincides with (5.1.9b) for a three-layer medium.

It can readily be seen that this equation has a solution in the frequency region where $\varepsilon(\omega)<0$ only if the penetration of the polariton field into the metal is accounted for.

If the metal-film thickness a is small compared to the depth of the skin layer ($\delta\sim 1/|\varkappa_4|$), then, to first order in a and making use of (5.3.4b), we obtain (for frequencies $\omega\ll\omega_p$, and with $|\varepsilon_4|\gg 1$)

$$\zeta\approx Z_1+\mathrm{i}\frac{\omega Z_1^2}{c}\varepsilon_4(\omega)a.$$

The substitution of this relation into (5.3.5) yields (5.3.1) provided we take into account that in the frequency region under consideration the dielectric constant of the metal

$$\varepsilon_4(\omega) = 1 + \mathrm{i}\frac{4\pi\sigma}{\omega} \approx \mathrm{i}\frac{4\pi\sigma}{\omega}\,.$$

If the metal-film thickness a is large compared to the depth of the skin layer, then $\zeta = Z_4$ [see (5.3.4b); assuming that $Z_4 + Z_1 \neq 0$]. In this case (5.3.3) determines two branches of surface polaritons. In the case of sufficiently large values of d, however, when $\exp(-2kd) \ll 1$, polaritons that are localized along the interfaces $z = 0$ and $z = d$ are "split" and, as a first approximation, can be investigated independently. In particular, in the system under investigation, a polariton localized along plane $z = d$ is strongly damped and is evidently difficult to observe. Hence, we shall consider the properties of a polariton localized along the plane $z = 0$ in more detail in the following. The dispersion law for this polariton is determined by the relation

$$Z_2 + Z_3 = \mathrm{e}^{-2d\varkappa_2}\frac{2Z_3(Z_3+Z_4)}{Z_4 - Z_3}\,.$$

Since $|Z_4| \ll |Z_3|$, this relation can also be written in the form

$$\frac{\varkappa_2}{\varepsilon_2} + \frac{\varkappa_3}{\varepsilon_3} = -2\frac{\varkappa_3}{\varepsilon_3}\mathrm{e}^{-2\varkappa_2 d}\left(1 + 2\frac{\varkappa_4\varepsilon_3}{\varepsilon_4\varkappa_3}\right). \tag{5.3.6}$$

For the values of the wave vector $\boldsymbol{k}$ in the nonrelativistic region where

$$k \gg (\omega/c)\sqrt{|\varepsilon_2|}\,, \quad k \gg (\omega/c)\sqrt{|\varepsilon_3|}\,, \quad \text{but}$$

$$k \ll (\omega/c)|\varepsilon_4|^{1/2}\,, \quad \varkappa_2 = \varkappa_3 = k \quad \text{and} \quad \varkappa_4 = (\omega/c)\sqrt{-\varepsilon_4(\omega)}\,.$$

In this region of the surface-wave spectrum, (5.3.6) assumes (taking the inequality $\exp(-2kd) \ll 1$ into account) the simpler form

$$\frac{1}{\varepsilon_3} + \frac{1}{\varepsilon_2(\omega)} \approx 4\mathrm{e}^{-2kd}\frac{\varkappa_4}{k\varepsilon_4}\,.$$

When there is no "metallic" quenching (i.e., if $|\varepsilon_4| \to \infty$), the surface polariton frequency $\omega_s + \mathrm{i}\Gamma_0$ is determined from the condition $\varepsilon_2(\omega) = -\varepsilon_3$. The presence of "metallic" quenching leads to additional broadening, in which case

$$\Gamma = \Gamma_0 + \frac{4\omega_s}{ck}\varepsilon_3^2\left(\frac{d\varepsilon_2}{d\omega}\right)^{-1}_{\omega_s}\mathrm{Re}\left\{\sqrt{\frac{1}{\varepsilon_4(\omega)}}\right\}.$$

Making use of (4.1.13) for $\varepsilon_2(\omega)$, we find that in the frequency region of the normal skin effect, where $\mathrm{Re}\{\sqrt{1/\varepsilon_4}\} = \mathrm{Re}\{\sqrt{\omega/8\pi\sigma}\}$,

$$\Delta\Gamma = \Gamma - \Gamma_0 = \frac{2\omega_\perp^2\varepsilon_3^2(\varepsilon_0-\varepsilon_\infty)\,e^{-2kd}}{ck(\varepsilon_3+\varepsilon_\infty)^2}\,\mathrm{Re}\left\{\sqrt{\frac{\omega}{8\pi\sigma}}\right\}.$$

Thus, in contrast to the case in which the quenching of the surface polariton is due to the presence of thin metallic films, see (5.3.2), contact with a thick layer of metal leads to quenching that decreases with increasing σ. This is evidently due to a reduction in the depth to which the field penetrates into the metal as σ increases.

5.3.2 Transition Layers in the Presence of a Resonance with the Surface Polariton

a) The Effective Boundary Conditions

The boundary conditions used in deriving (5.3.1) are generally insufficient for examining the properties of polaritons in the presence of a transition layer. This is associated with the fact that in deriving (5.3.1), we disregarded the polarizability of the transition layer in the direction perpendicular to the plane separating the media. There may be cases, of course, for which resonance with the surface polariton is reached precisely because there are vibrations in the transition layer that are at least partly polarized across the layer. To allow for such vibrations in the transition layer it is necessary to apply more general boundary conditions. In order to discuss these conditions, we note, first of all, that the question of the effect of macroscopic transition layers (i.e, layers with a thickness $d \gg a$ and $d \ll \lambda$, where a is the crystal lattice constant or the size of the molecule, and λ is the wavelength of light) on the optical properties of condensed media (though not in connection with their effect on the properties of surface waves) has a very long history and has been examined in great detail (see, e.g. [5.48 – 50]). The known simplicity of this case is based on the following: if we have a macroscopic transition layer, the boundary conditions for fields outside the layer can be derived directly within the framework of Maxwell's equations. Assuming, for instance, that the dielectric constant within the layer $(0 < z < d)$ is constant $(\tilde{\varepsilon}_{ij} = \tilde{\varepsilon}_i\delta_{ij})$, we integrate the equation $\mathrm{div}\boldsymbol{D} = 0$ with respect to z between 0 and d.

For fields of the form $D = D(z)\exp(\mathrm{i}k_1x + \mathrm{i}k_2y)$, such integration evidently yields the condition

$$D_3(d) - D_3(0) = -\mathrm{i}d\tilde{\varepsilon}_1\boldsymbol{k}_t\boldsymbol{E}_t(0)\,,$$

where $\boldsymbol{k}_t \equiv (k_1, k_2)$. Recalling next that $D_3(d) = D_3(1)$ and $D_3(0) = D_3(2)$, where the indices 1 and 2 correspond to the media with $\varepsilon \equiv \varepsilon_1$ when $z > d$ and with $\varepsilon \equiv \varepsilon_2$ when $z < 0$, respectively, we obtain one of the boundary conditions:

$$D_3(2) - D_3(1) = \mathrm{i}\gamma\boldsymbol{k}_t\boldsymbol{E}_t(1)\,, \tag{5.3.6a}$$

where $\gamma = d(\tilde{\varepsilon} - \varepsilon_1)$, and in which the presence of the transition layer is effectively accounted for to first order in $kd \ll 1$. The remaining boundary conditions can be obtained in a similar way (see, e.g., [5.40b]):

$$E_t(2) - E_t(1) = -i\mu D_n(1)k_t, \quad (5.3.6b)$$

$$H_t(2) - H_t(1) = -ik_0\gamma n \times E_t(1), \quad (5.3.6c)$$

$$H_n(2) - H_n(1) = 0, \quad (5.3.6d)$$

where $k_0 = \omega/c$ and $\mu = d/(1/\tilde{\varepsilon}_3 - 1/\varepsilon_1)$. Of these four conditions only two are independent. Equations (5.3.6b, c) are applied in the following. If in these boundary conditions we pass over to a metallic transition layer ($|\tilde{\varepsilon}_i| \gg 1$), retaining only terms of the order of σd, σ being the electrical conductivity of the metal, we obtain precisely the boundary conditions that have already been applied previously (for a metal the relation $\tilde{\varepsilon}_1 = 1 + 4\pi\sigma/i\omega$ should be taken into account).

The general form of the boundary conditions (5.3.6) is retained for microscopic transition layers as well [5.51]. Only the relations expressing the quantities μ and γ in terms of the polarizability of the molecules of the transition layer are changed. We shall not consider these relations here in detail because they were found to depend strongly on the model of the transition layer (this problem was dealt with in [5.11] for the case of semiconductor transition layers, taking quantization of transverse motion of quasiparticles into account. If the layer thickness is small compared to the Bohr radius of the exciton in a bulk specimen, the quantities γ and μ become nonlinear functions of the layer thickness d. Closely related is the problem of two-dimensional plasmons [5.52].

b) Polariton Spectrum Splitting for TH Surface Waves

We make use of the fact that since γ and μ depend upon the polarizability of the molecules along and across the layer, their resonance frequencies should, in general, differ. We shall consider, for example, the frequency region $\omega \approx \omega_0$, where ω_0 is the resonance frequency for $\gamma(\omega)$ [where $\gamma(\omega_0) = \infty$; the role of excitation damping is dealt with later]. In this case, only terms proportional to γ need be retained in the right-hand side of (5.3.6). Hence (5.3.6b, c) assume the form

$$E_t(2) - E_t(1) = 0, \quad H_t(2) - H_t(1) = -ik_0\gamma n \times E_t(1). \quad (5.3.7)$$

We shall assume that the wave vector k_t of the surface field is directed along the x axis, and shall denote the dielectric constants of media I and II by ε_1 (where $\varepsilon_1 > 0$) and $\varepsilon(\omega)$ [where $\varepsilon(\omega) < 0$]. Then, making use of (5.1.1), we find that the quantities $H_y(1)$ and $H_y(2)$ satisfy the system of equations

$$\frac{\varkappa}{\varepsilon}H_y(2)+\frac{\varkappa_1}{\varepsilon_1}H_y(1)=0\,,\qquad H_y(2)-H_y(1)=\gamma\frac{\varkappa_1}{\varepsilon_1}H_y(1)\,,$$

which has a nontrivial solution only under the condition that

$$\frac{\varepsilon_1}{\varkappa_1}+\frac{\varepsilon}{\varkappa}=-\gamma\,. \tag{5.3.8}$$

In the resonance region $\gamma(\omega)\approx A\,\omega_0/(\omega_0-\omega)$, $A>0$. If $\omega_s(k)$ is the dispersion law of the surface polariton that corresponds to the solution of (5.3.8) at $\gamma=0$, then at $\omega\approx\omega_s(k)$

$$\frac{\varepsilon_1}{\varkappa_1}+\frac{\varepsilon}{\varkappa}=\frac{\omega-\omega_s(k)}{C(k)}\,,\qquad \text{where} \tag{5.3.9}$$

$$C^{-1}(k)=\frac{\partial}{\partial\omega}\left(\frac{\varepsilon_1}{\varkappa_1}+\frac{\varepsilon}{\varkappa}\right)_{\omega=\omega_s(k)}>0\,.$$

Substituting (5.3.9) into (5.3.8), making use of the explicit expression for $\gamma(\omega)$ and solving (5.3.8) for ω, we obtain the dispersion law of the surface polariton:

$$\omega_{1,2}(k)=\tfrac{1}{2}[\omega_0+\omega_s(k)]\pm\tfrac{1}{2}\sqrt{[\omega_s(k)-\omega_0]^2+4A\,\omega_0C(k)}\,. \tag{5.3.10}$$

If ω_0 is found to be within the surface-polariton zone $\omega_s(k)$, i.e., if at a certain $k=k_0$ the frequency $\omega_s(k_0)=\omega_0$, then, in the presence of a transition layer, the gap $\Delta=2\sqrt{A\,\omega_0C(k_0)}$ (Fig. 5.7) appears in the polariton spectrum at $k=k_0$. If the magnitude of Δ exceeds the surface-polariton linewidth due to damping, the gap in the polariton spectrum can be observed experimentally. To obtain a rough estimate of the gap magnitude we can assume that $A\approx d$

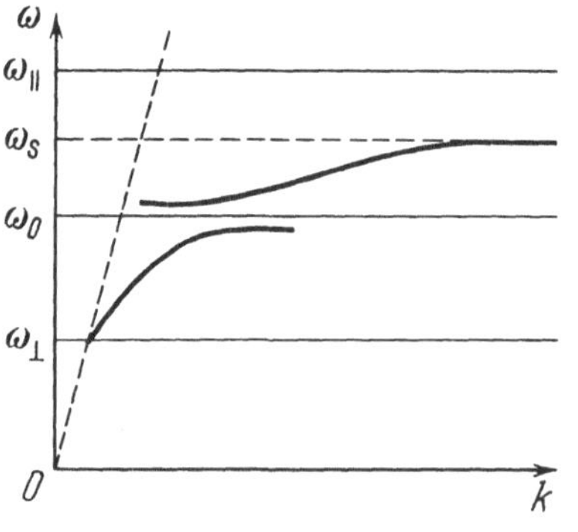

Fig. 5.7. Dispersion of a surface polariton in the case of resonance with the vibrations in the transition layer

and $C(k_0) \approx (\varepsilon_0 - \varepsilon_\infty)\,\omega_0 k_0$, so that $\Delta \approx 2\,\omega_0 \sqrt{dk_0(\varepsilon_0 - \varepsilon_\infty)}$. Hence, at $d \approx 10\,\text{Å}$, $k_0 \approx 5\times 10^3\,\text{cm}^{-1}$ and $\varepsilon_0 - \varepsilon_\infty = 4$, the ratio $\Delta/\omega_0 \approx 0.08$. This gap magnitude may exceed or be of the order of the polariton linewidth ($\Gamma \sim 2\times 10^{-2}\omega_0$), thereby allowing the gap to be observed experimentally. For example, in applying the ATR method (see [5.53, 54, 54a] and below in Sect. 5.5.1 for experimental investigations of this effect), two reflection minima, close to each other in position and intensity, should appear in the region of the gap. If, however, we turn to experiments investigating the propagation of surface waves of a given frequency, the presence of the transition layer leads to the appearance of resonances at $\omega = \omega_0$ of the absorption and refractive indices of the surface polariton. This conclusion follows even from Fig. 5.7. Moreover, owing to the identity

$$\varepsilon_1/\varkappa_1 + \varepsilon/\varkappa \equiv B(\omega)[n^2 - n_0^2(\omega)]\,, \tag{5.3.11}$$

where $n_0^2(\omega)$ is the refractive index of the surface polariton at $\gamma = 0$ and $B(\omega) > 0$, we actually do find, making use of (5.3.8, 11), that at $\gamma \neq 0$

$$n^2(\omega) = n_0^2(\omega) + \gamma(\omega)/B(\omega)\,.$$

When damping is taken into account, the quantity n becomes complex: $n \to \tilde{n} = n + \mathrm{i}\varkappa$. Hence, in the region of resonance with the transition layer, where

$$\gamma(\omega) \approx A\,\omega_0(\omega_0 - \omega - \mathrm{i}\nu)^{-1}\,,$$

and with sufficiently weak damping ($\varkappa \ll n$),

$$2n\varkappa \approx 2n_0(\omega)\varkappa_0(\omega) + \frac{1}{B(\omega_0)}\,\frac{A\,\omega_0\nu}{(\omega_0 - \omega)^2 + \nu^2}\,. \tag{5.3.11a}$$

If the film (transition-layer) thickness is so small that the gap size $\Delta \lesssim \nu$, the effect of the film on the real part of the refractive index n can be neglected. In this case the resonance at the frequency $\omega = \omega_0$ corresponds to the resonance $\varkappa$. Consequently, in investigating the dependence of the propagation length of surface waves on the frequency we directly find the frequency of vibrations in the transition layer.

The surface-wave method was employed for high-resolution observations of the vibrational spectrum of a hydrogen monolayer adsorbed on the surface (100) of tungsten [5.55].

We shall now pass over to the case of a macroscopic transition layer (Fig. 5.2), ignoring its anisotropy ($\tilde{\varepsilon}_{11} = \tilde{\varepsilon}_{22} = \tilde{\varepsilon}_{33} = \tilde{\varepsilon}$; distinctive features of the spectrum of surface waves in the case of an anisotropic transition layer have been discussed in [5.56]). It is readily evident, in this case, that the equation for the frequency of surface polaritons has the form [5.40b, 44]

$$\frac{\varepsilon}{\varkappa}+\frac{\varepsilon_1}{\varkappa_1}=\frac{d(\varepsilon_1-\tilde{\varepsilon})}{\tilde{\varepsilon}\varkappa_1^2}\left[k^2(\varepsilon_1+\tilde{\varepsilon})-\frac{\omega^2}{c^2}\tilde{\varepsilon}\varepsilon_1\right]. \tag{5.3.12}$$

Naturally, this equation also follows from the dispersion equation (5.1.9b) for surface polaritons in a three-layer medium, if this equation is expanded in terms of the small parameter $\varkappa_2 d \ll 1$ and only terms linear in d are retained (it follows that using (5.3.12) to analyze the dispersion of surface polaritons is justified only for waves with $k<1/d$). The right-hand side of (5.3.12) vanishes at $d=0$, or at $\varepsilon_1=\tilde{\varepsilon}$ or $\tilde{\varepsilon}=\varepsilon$. In the last two cases the transition layer also vanishes and the interface is displaced by the amount d. Setting

$$\tilde{\varepsilon}(\omega)=\tilde{\varepsilon}_\infty\frac{\omega^2-\tilde{\omega}_\|^2}{\omega^2-\tilde{\omega}_\perp^2},$$

we find that the resonances of the right-hand side of (5.3.12) occur at $\omega=\tilde{\omega}_\perp$ and $\omega=\tilde{\omega}_\|$. If in each of these cases we retain only the resonance terms, we obtain a relation of the form of (5.3.8) and

$$A^\perp=d\tilde{\varepsilon}_\infty\frac{\tilde{\omega}_\|^2-\tilde{\omega}_\perp^2}{2\tilde{\omega}_\perp^2},\qquad A^\|=\frac{d}{\tilde{\varepsilon}_\infty}\,\frac{\tilde{\omega}_\|^2-\tilde{\omega}_\perp^2}{2\tilde{\omega}_\|^2}\,\frac{k_0^2}{\varkappa_0^2},\qquad \varkappa_0^2\equiv k_0^2-\frac{\tilde{\omega}_\|^2}{c^2}$$

(here we assumed $\varepsilon_1=1$). Thus, in the case under discussion, both longitudinal and transverse vibrations of the transition layer may resonate with the surface polariton provided that $\tilde{\varepsilon}(\omega)$ has been suitably chosen. In some cases this leads to the appearance of two gaps (the gaps may overlap) in the surface polariton spectrum [5.54a].

c) TE Surface Waves in the Transition Layers' Resonance Region

All of the preceding discussion concerned the effect of the transition layer on the TH surface wave spectrum. But, in some cases, the presence of even an isotropic transition layer can lead to the formation of TE-type surface waves. For these waves the vector $\boldsymbol{E}=(0, E_2, 0)$, whereas the vector $\boldsymbol{H}=(H_1, 0, H_3)$ (as before the x axis is direct along vector $\boldsymbol{k}(k_1, k_2)$ and this means that $k_2=0$). At $Z>0$ and $Z<0$ it follows from $\operatorname{div}\boldsymbol{H}=0$ that

$$kH_1(1)+\mathrm{i}\varkappa_1 H_3(1)=0,$$
$$kH_1(2)-\mathrm{i}\varkappa_2 H_3(2)=0,$$

or, taking (5.3.6c) into consideration,

$$k[H_1(1)-H_1(2)]+\mathrm{i}H_3(1)(\varkappa_1+\varkappa_2)=0. \tag{5.3.13}$$

It follows from (5.3.6c) that

$$H_1(1)-H_1(2) = -\mathrm{i}k_0\gamma E_2(1)\,.$$

Substituting this into (5.3.13) we find

$$-\mathrm{i}k_0 k\gamma E_2(1)+\mathrm{i}(\varkappa_1+\varkappa_2)H_3(1)=0\,. \tag{5.3.14}$$

But it follows from (5.1.1) that

$$\mathrm{i}k_0 H_3(1)=\mathrm{i}kE_2(1)\,, \tag{5.3.15}$$

so that (5.3.14) can be written in the form

$$H_3(1)(\varkappa_1+\varkappa_2-k_0^2\gamma)=0\,.$$

It follows from this equation that $H_3(1)\neq 0$ and consequently $E_2(1)\neq 0$ [see (5.3.15)] if [5.40b]

$$\varkappa_1+\varkappa_2-k_0^2\gamma=0\,. \tag{5.3.16}$$

As was shown by *Kravtsov* [5.62a] the solution of this equation corresponds to TE surface waves with the dispersion law

$$k^2=\frac{\omega^2}{c^2}\left[1+\frac{1}{4}\left(\frac{1-\varepsilon(\omega)}{\delta(\omega)}-\delta(\omega)\right)^2\right], \tag{5.3.17}$$

$$\delta(\omega)=\frac{\omega}{c}d(\tilde{\varepsilon}(\omega)-1)\,.$$

In the region where TE waves exist, the inequality

$$\delta(\omega)\geqslant\sqrt{1-\varepsilon(\omega)}$$

should be complied with. It follows from this inequality that at small values of $(\omega/c)d$, the region where TE surface waves exist correponds to large positive values of $\tilde{\varepsilon}(\omega)$ [i.e., the region located below the resonance frequency of $\tilde{\varepsilon}(\omega)$]. This conclusion can be readily understood it we recall that in a plate with a sufficiently high dielectric constant, the formation of both TH- and TE-waveguide modes is feasible. The solution reached above genetically corresponds to the TE-waveguide mode of zero order, whose field varies only slightly along the film thickness. Of interest is the fact that the dispersion branch of the TE-type surface wave is located exactly within the gap formed in the TH surfrace wave spectrum, as has been observed in the ATR spectrum ([5.62a], a LiF film on a CaF_2 substrate; frequency region $\omega\approx 307.5\ \mathrm{cm}^{-1}$).

In conclusion, we point out that the transition layer is not necessarily associated with the presence of foreign molecules on the surface of the crystal. In cases in which the existence of a crystal boundary, retardation being neglected, leads to the formation of microscopic surface excitons, i.e. excitons with a penetration depth d of the order of the lattice constant, the boundary conditions (5.3.6a – d) retain their form even though the quantities μ and γ may differ considerably for different models of excitons. Hence the results given above remain qualitatively valid in this case as well. This enables us to understand how the presence of microscopic surface (Coulomb) excitons affects the spectrum of the surface polaritons. Evidently, the most interesting case here is one in which the frequency of the microscopic surface exciton is within the frequency region of surface polaritons. Experimental investigations of the gap that appears in this case could, in principle, yield information on the characteristics of the surface exciton (as such we make no distinction here between excitons and optical phonons).

Of no less interest, however, is the case in which the region of frequency change of the surface polariton includes the range of two-particle states, corresponding, for instance, to the overtone or summation tone of optical surface phonons. In this case (i.e., with a Fermi resonance with the surface polariton), the surface polariton resonates with a whole continuum of states. Hence the surface polariton that appears as a result of the dispersion law is found to depend strongly both on the density-of-states distribution in the range of two-particle states and on the surface biphonon or biexciton spectrum (for details concerning this theory see [5.57]); in this case the quantity γ in (5.3.8) is found from the relation

$$\gamma(\omega) = \frac{A\,\omega_0}{\omega_0 - \omega} + \int \frac{f(\omega')\,d\omega'}{\omega' - \omega},$$

where ω_0 is the biphonon frequency, and integration is carried out over the range of two-particle states. Experimental observations on this phenomenon have been presented in [5.58].

We also point out that resonance with a surface plasmon, similar to the phenomenon under discussion, is found in semiconductors the surface of which is covered with a layer of low concentration of charge carriers. Here the plasma vibrations at the surface (their frequency is small compared to the plasma frequency in the bulk of the semiconductor) resonate with the "ideal" surface plasmon (plasmopolariton) [5.59], altering its spectrum. The results of such an examination depend strongly upon the structure of the transition layer (see [5.60 – 62] and references therein).

5.3.3 Effect of Surface Roughness (Irregularities) on the Path Length of Surface Polaritons

So far in the discussion of the properties of surface polaritons we assume the boundary or interface between media to be an ideally smooth plane surface. In this case the path length of the surface polariton (by definition, its amplitude decreases by a factor $\sqrt{e}$ over this length) is completely determined by the energy dissipation in the layer. The specimens that are actually used in experiments always have some surface roughness (irregularities), leading to various processes that reduce the path length of the surface polariton, as does the dissipation mentioned above. We should bear in mind, in estimating this effect, that notwithstanding the care taken in preparing specimens for optical investigations in the visible or infrared regions of the spectrum, their surface irregularities are of a size of the order of hundreds or even thousands of angstroms. The theory of surface-wave scattering by irregularities on the interface between isotropic media was developed in [5.63]. It was shown that surface roughness stimulates two processes. One of them corresponds to the process of elastic scattering in which a surface polariton is transformed into a bulk polariton (radiative decay of the surface polariton; this process must be taken into account in considering the possible contribution of surface polaritons to exciton luminescence spectra [5.64, 65]). The second process corresponds to scattering into another state of the surface polariton. We shall not go into details of the calculations given in [5.63]; after several remarks, we shall only cite the results obtained. In [5.63], the shape of the surface was specified by the function $z=\zeta(x,y)$, the average value of ζ being zero, i.e., $\langle\zeta(x,y)\rangle=0$, whereas the quantity $\langle\zeta(x,y)\,\zeta(0)\rangle$ is assumed to be nonzero. The scattered waves were calculated, in a first approximation with respect to ζ, actually making use thereby of an analogue of the first Born approximation of the quantum-mechanical scattering theory.

In the surface-wave frequency region where $|\varepsilon(\omega)|\gg 1$, the scattering length at the boundary with vacuum is determined mainly by the "breaking away" of surface waves into vacuum[8]. Here the path length is found to be equal to

$$l=\frac{3\pi c^5|\varepsilon|^{1/2}}{4\omega^5}F_\zeta^{-1}(0)\,, \tag{5.3.18}$$

where $F_\zeta(\boldsymbol{k})$ is the Fourier component of the correlation function:

$$F_\zeta(\boldsymbol{k})=\int dx\,dy\,\mathrm{e}^{\mathrm{i}(k_1x+k_2y)}\langle\zeta(x,y)\,\zeta(0)\rangle\,. \tag{5.3.19}$$

[8] Such a process has probably been observed in a ZnO crystal [5.66]. The surface polaritons were excited in the process of nonlinear absorption of the radiation from two lasers with frequencies ω' and ω'', where $\omega'+\omega''=\omega_s(k)$. For calculations of the intensity, see [5.67].

In particular, if the function $\langle \zeta(x,y)\,\zeta(0)\rangle$ has the Gaussian form

$$\langle \zeta(r)\,\zeta(0)\rangle = \delta^2 \mathrm{e}^{-r^2/a^2},$$

then

$$l = \frac{3c^5}{4\delta^2 a^2 \omega^5}|\varepsilon|^{1/2}.$$

In the frequency region where $|\varepsilon| \lesssim 3$, the principal process caused by roughness is, according to [5.63], the process of scattering the surface polariton into some other surface polariton state. In particular, at $|\varepsilon| \approx 1$, i.e., at $\omega \approx \omega_s$, where $\varepsilon(\omega_s) = -1$, the path length is

$$l \approx \frac{2c^5(|\varepsilon| - 1)^{9/2}}{3\omega^5 \delta^2 a^2},$$

which is very small. For surface polaritons propagating along the surfaces of metals and having a frequency $\omega \ll \omega_p$ (and with $|\varepsilon| \gg 1$), the equation for l (at $\delta \sim a \approx 0.3$ µm, $\lambda = 10$ µm and $|\varepsilon| = 4000$) yields the value $l \approx 1500$ $\lambda \approx 1.5$ cm, which is of the order of the experimentally observed values. This signifies that for the specified wavelengths and for $a \sim \delta \approx 0.3$ µm, the effect of roughness is not appreciable. At the same time, the role played by irregularities drastically increases for surface polaritons of shorter wavelengths and in some cases can lead even to their localization, i.e., to the appearance of so-called local (or shape) resonances. The problem of local resonances has attracted more and more attention in recent years, especially in connection with the investigation of surface enhanced Raman scattering [5.67a]. Its investigation is also important for the preparation of surfaces with preset optical properties. For the theory of local resonances on rough surfaces of bulk specimens, see [5.67a], for thin rough films, [5.67b].

a) Surface Polariton Scattering in the Vicinity of Phase-Transition Points

The propagation length of the surface polariton calculated above can be employed to discuss its properties in the vicinity of the point $T = T_c$ of a phase transition in a thin dielectric film (with the film thickness D being much less than the polariton wavelength) applied on a metallic substrate [5.68].

The point is that the presence of irregularities is equivalent to the presence at $z = 0$ of a film whose polarizability per unit area

$$\chi = \frac{\varepsilon - 1}{4\pi}\zeta(x,y) \tag{5.3.20}$$

is a function of x and y (ε is the dielectric constant of the metal). In the vicinity of the phase-transition point, the fluctuating part of the polarizability of the film (because of the fluctuations of the order parameter η) undergoing a phase transition can be written in the form

$$\chi - \chi_0 = \frac{D}{4\pi}\left(\frac{d\varepsilon_0}{d\eta}\right)_0 \delta\eta(x,y)\,, \tag{5.3.21}$$

where χ_0 is the average polarizability of the film, $\delta\eta = \eta - \eta_0$, $\langle\eta\rangle = \eta_0$, whereas $\varepsilon_0(\eta)$ is the dielectric constant of the film material, corresponding to the value of the order parameter η. This relation indicates that we are concerned with a phase transition for which, at small values of η, the part of $\varepsilon_0(\eta)$ depending on η is proportional to η.

Making use of the expression given above for the propagation length l, we find that the value of l determined by the fluctuations η satisfies the relation

$$\frac{1}{l} = \frac{4\omega^2 D^2 (d\varepsilon_0/d\eta)_0^2 F_\eta(0)}{3\pi^2 c^2 |\varepsilon|^{5/2}}\,, \tag{5.3.22}$$

so that at $T \approx T_c$ the character of the temperature dependence of the scattering intensity $I \approx l^{-1}$ coincides with that for the quantity $F_\eta(0)$.

In crystals in which $\chi - \chi_0 \sim \eta^2$, at $T < T_c$, $\eta^2 = \eta_0^2 + 2\eta_0\delta\eta$, and the scattering intensity $I \sim l^{-1} \sim \eta_0^2 F_\eta(0)$. Here the role of the fluctuations of the order parameter are found to be weakened [5.69].

In the Landau theory of second-order phase transitions, $\eta_0^2 \sim T_c - T$ and $F_\eta(0) \sim |T - T_c|^{-1}$. Therefore, with a linear relation between $\varepsilon_0(\eta)$ and η, the scattering intensity $I(T)$ increases without limit as $T \to T_c$. If, however, $\Delta\varepsilon_0(\eta) \sim \eta^2$, the application of the Landau phase-transition theory enables us to expect only a finite increase in $I(T)$ as $T \to T_c$.

The scattering of surface polaritons by fluctuations of the order parameter in massive specimens was discussed in [5.70] and more recently in [5.71]. This effect can be investigated experimentally by the procedure developed in [5.66].

5.4 Effect of Spatial Dispersion on the Spectra of Surface Polaritons, and Additional ("New") Surface Waves

We shall first discuss in more detail the frequency dispersion of the complex refractive index of surface polaritons in the resonance region.

According to (5.3.9a),

$$\tilde{n}^2 = (n + \mathrm{i}\varkappa)^2 = \varepsilon_1\varepsilon_2/(\varepsilon_1 + \varepsilon_2)\,. \tag{5.4.1}$$

Using this equation we shall examine the relations $n = n(\omega)$ and $\varkappa = \varkappa(\omega)$ for the case of sufficiently weak damping ($\varkappa \ll n$), assuming $\varepsilon_1 = \text{const}$ and using the following expression for $\varepsilon_2(\omega)$:

$$\varepsilon_2(\omega) = \varepsilon(\omega) = \varepsilon_\infty + \frac{(\varepsilon_0 - \varepsilon_\infty)\,\omega_\perp^2}{\omega_\perp^2 - \omega^2 - \mathrm{i}\,\omega\Gamma(\omega)}\,. \tag{5.4.2}$$

Substituting (5.4.2) into (5.4.1), we obtain

$$(n + \mathrm{i}\varkappa)^2 = \frac{\varepsilon_1 \varepsilon_\infty}{\varepsilon_1 + \varepsilon_\infty}\,\frac{\omega^2 - \omega_\parallel^2 + \mathrm{i}\,\omega\Gamma}{\omega^2 - \omega_s^2 + \mathrm{i}\,\omega\Gamma}$$

by virtue of which

$$\begin{aligned} n^2(\omega) &= \frac{\varepsilon_1 \varepsilon_\infty}{\varepsilon_1 + \varepsilon_\infty}\,\frac{(\omega_\parallel^2 - \omega^2)(\omega_s^2 - \omega^2) + \omega^2\Gamma^2}{(\omega^2 - \omega_s^2)^2 + \omega^2\Gamma^2}\,, \\ 2n\varkappa &= \frac{\varepsilon_1 \varepsilon_\infty}{\varepsilon_1 + \varepsilon_\infty}\,\frac{\omega\Gamma(\omega_\parallel^2 - \omega_s^2)}{(\omega^2 - \omega_s^2)^2 + \omega^2\Gamma^2}\,, \end{aligned} \tag{5.4.3}$$

where the frequency ω_s of the Coulomb surface exciton satisfies the equation

$$\varepsilon_1 + \varepsilon_2(\omega_s) = 0 \tag{5.4.4}$$

and, consequently, in the absence of damping (at $\Gamma = 0$), is equal to

$$\omega = \omega_s = \omega_\perp \sqrt{(\varepsilon_1 + \varepsilon_0)/(\varepsilon_1 + \varepsilon_\infty)}\,. \tag{5.4.5}$$

It follows from (5.4.3) that with weak damping (i.e, for sufficiently small values of Γ) the refractive index $n(\omega)$ of the surface polariton may become anomalously large as $\omega \to \omega_s$. This naturally poses the question of taking into account the dependence of the frequency ω_s of the surface Coulomb exciton on the two-dimensional wave vector $\boldsymbol{k}(k_x, k_y)$.

If spatial dispersion is taken into consideration in (5.4.4), then $\varepsilon = \varepsilon(\omega, \boldsymbol{K})$, where $\boldsymbol{K} = (\boldsymbol{k}, -\mathrm{i}\varkappa)$, and in the nonrelativistic limit $\varkappa = k$. In an isotropic medium and with weak spatial dispersion, $\varepsilon(\omega, \boldsymbol{K}) = \varepsilon(\omega) + \alpha K^2 = \varepsilon(\omega)$ because for surface waves $\boldsymbol{K}^2 \equiv \boldsymbol{K} \cdot \boldsymbol{K} = k^2 - \varkappa^2 = 0$. This means that for the model of the medium under consideration, such an elementary approach does not allow even the first term in the expansion of ω_s with respect to k to be found.

The interaction of zones of surface waves with zones of bulk excitations, which becomes possible due to the overlapping of their spectra when spatial dispersion is taken into account, has been discussed in [5.23] in connection with an investigation of the dispersion of surface vibrations for a model of an ionic crystal[9] (the results of this investigation may also be found in [5.15]), and in a more general formulation within the framework of the phenomenological description in [5.73]. This interaction damps the surface waves even when anharmonicity is neglected and, moreover, leads to the appearance of terms linear in k in the dispersion law (except for special cases of intersection with the van Hove point). Further on we shall return to these results. Here we show, on a phenomenological level [5.40], that it is precisely the transition layer in bounded media, which always appears when some kind of boundary condition is taken into account, which leads to a linear dependence of ω_s on k. As a matter of fact, in the Coulomb limit $\omega/c \to 0$ the boundary conditions for $\boldsymbol{E}$ and $\boldsymbol{D}$ assume the form, see (5.3.6),

$$D_{\mathrm{n}}(2) - D_{\mathrm{n}}(1) = \mathrm{i}\gamma \boldsymbol{k}_{\mathrm{t}} \boldsymbol{E}_{\mathrm{t}}(1), \quad \boldsymbol{E}_{\mathrm{t}}(2) - \boldsymbol{E}_{\mathrm{t}}(1) = -\mathrm{i}\mu D_{\mathrm{n}}(1) \boldsymbol{k}_{\mathrm{t}}. \tag{5.4.6}$$

If retardation is neglected $\boldsymbol{E} = A\boldsymbol{K}$, A being a scalar. Hence in media 1 and 2: $\boldsymbol{E}(1) = A_1 \boldsymbol{K}_1$, $\boldsymbol{E}(2) = A_2 \boldsymbol{K}_2$, $\boldsymbol{K}_1 = (k, 0, \mathrm{i}k)$ and $\boldsymbol{K}_2 = (k, 0, -\mathrm{i}k)$. Substituting these relations into (5.4.6) we obtain two linear homogeneous equations in A_1 and A_2. The condition that the determinant of this system of equations becomes zero leads to the required dispersion equation

$$\varepsilon(\omega) + \varepsilon_1 = -(\gamma - \varepsilon_1^2 \mu) k. \tag{5.4.7}$$

Making use now of (5.4.2) we find that when the transition layer is taken into consideration, the frequency of the surface polariton

$$\omega_s(k) = \omega_s(0) - \frac{\alpha'}{2\,\omega_s(0)} k, \tag{5.4.8}$$

where the quantity $\omega_s(0)$ is determined by (5.4.5), and where

$$\alpha' = \frac{(\gamma - \varepsilon_1^2 \mu)(\varepsilon_0 - \varepsilon_\infty)}{(\varepsilon_\infty + \varepsilon_1)^2} \omega_\perp^2,$$

with the quantities μ and γ, along with ε_1, assumed to be independent of ω in the frequency region $\omega \approx \omega_s$. It becomes necessary to take the dependence of μ or γ on ω into consideration if the resonance frequency of μ or γ is close to $\omega_s(0)$. In particular, when these frequencies coincide, ω_s is found to depend on $\sqrt{k}$ rather than on k for small values of k. For microscopic transition layers, the factor α' can, of course, be either positive or negative. It is noteworthy that the sign of α' is difficult to check in this case. The situation is different for macroscopic transition layers (in this connection, see also [5.74]). If the transition layer is macroscopic and homogeneous, then $\gamma - \varepsilon_1^2 \mu =$

[9] The distinctive features of surface waves in polar semiconductors were dealt with in [5.72].

$d(\tilde{\varepsilon}-\varepsilon_1^2/\tilde{\varepsilon})$, d being the thickness of the layer and $\tilde{\varepsilon}$ its dielectric constant. Hence, if $\tilde{\varepsilon} > \varepsilon_1$, then $\alpha' > 0$.

We also point out that in the case of a transition layer whose thickness d is large compared to the length characterizing the effects of spatial dispersion in the bulk of the medium (i.e., compared to the lattice constant, the Debye length, the free-path length or the length v_F/ω, v_F being the Fermi velocity, etc.), the term in (5.4.8) that is linear in k should evidently be the principal one and should exceed (in modulus) other corrections to the dispersion law.

The dispersion law for a surface polariton in the presence of a macroscopic transition layer can also be obtained from (5.1.9b). Proceeding to the limit $\omega/c \to 0$ and carrying out an expansion into a series with respect to the small parameter $kd \ll 1$, we obtain (5.4.7). The application of boundary conditions of the type of (5.4.6), which are a natural generalization of the conditions at a sharp boundary, enabled us to express the dispersion law for a surface wave in terms of γ and μ, which characterize the transition layer. However, the calculation of the phenomenological parameters γ and μ can be carried out only within the framework of a consistent microtheory of surface waves with spatial dispersion taken into account. It is clear that such a theory must depend substantially on the model, spectral region, and conditions at the boundary. This, precisely, is the reason why the development of such a theory is still far from complete. One of the first results was obtained in [5.75, 76]. It was shown that in the mirror reflection of electrons from the boundary of a metal, the dispersion of the surface plasmons $\omega_s(k_x, k_y=0)$ with spatial dispersion taken into account is determined, in the nonrelativistic limit ($\omega/c \to 0$), by the relation

$$\frac{2k_x}{\pi}\int_0^\infty \frac{dk_z}{(k_x^2+k_z^2)\,\varepsilon(\omega,k_x,k_z)} = -1\,.$$

This relation (also valid for the exciton region of the spectrum in the mirror reflection of excitons) was generalized for a metal in a constant magnetic field where the dielectric-constant tensor becomes anisotropic [5.77]. For surface excitons in a ZnO crystal, see [5.78].

Surface waves with spatial dispersion in an isotropic medium for the exciton region of the spectrum were also investigated [5.73]. The dielectric constant $\varepsilon(\omega,\boldsymbol{K})$, where $\boldsymbol{K} \equiv (k,0,\mathrm{i}\varkappa)$ and $K^2 = k^2-\varkappa^2$, for the resonance region was used in the form:

$$\varepsilon(\omega,\boldsymbol{K}) = \varepsilon_\infty + \frac{\omega_p^2}{\omega_\perp^2-\omega^2+D(k^2-\varkappa^2)}\,.$$

The substitution into solutions for fields in the form of nonuniform plane waves leads to the equation

$$\varepsilon(\omega,K)\,[c^2K^2/\omega^2-\varepsilon(\omega,K)]=0\,,$$

which provides for three values of $\varkappa_i$ (where $i=1,2,3$), expressible in terms of ω and k.

This means that the field of the surface polariton in the medium is the sum of three nonuniform plane waves. Therefore, ordinary boundary conditions are insufficient to determine the values of all the amplitudes. Hence ABC of the form

$$\left.\frac{dP}{dz}\right|_{z=0+}+\mathrm{i}\Gamma(k)P|_{z=0+}=0$$

were used, where $\Gamma(k)$ is a certain parameter of the ABC (Sect. 4.5.4). As a result, when terms linear in k are taken into account, the equation determining the frequency of the surface wave assumes the form

$$\varepsilon(\omega_s)+1=\frac{k}{\mathrm{i}\Gamma\varkappa}\left[\varkappa\left(1+\frac{\omega_p^2}{2D\Gamma^2}\right)+\mathrm{i}\Gamma\right],$$

$$\varkappa=\left(\frac{\omega_\parallel^2-\omega_s^2(0)}{D}\right)^{1/2},\qquad \varepsilon(\omega_\parallel)=0 \quad \text{at} \quad k=0\,.$$

At small values of k

$$\varepsilon(\omega_s)+1=\left(\frac{d\varepsilon}{d\omega}\right)_{\omega=\omega_s(0)}[\omega_s(k)-\omega_s(0)]\,,$$

where $\omega_s(0)$ satisfies the condition $\varepsilon(\omega_s(0))+1=0$, so that the required dependence of the surface polariton on k is of the form

$$\omega_s(k)=\omega_s(0)+\left(\frac{d\varepsilon}{d\omega}\right)^{-1}_{\omega=\omega_s(0)}\left[\varkappa\left(1+\frac{\omega_p^2}{2D\Gamma^2}\right)+\mathrm{i}\Gamma\right]\frac{k}{\mathrm{i}\Gamma\varkappa}\,.$$

Consequently, as in the case when the transition layer is phenomenologically taken into account, see (5.4.8), $\omega_s(k)$ is a linear function of k. A comparison of the result obtained for $\omega_s(k)$, corresponding to the use of a certain macroscopic model of the transition layer, with (5.4.8) enables us to express the phenomenological quantity α' in terms of the parameters determining the dielectric constant and of the phenomenological parameter Γ appearing in the ABC.

In [5.73] the quantity Γ was determined within the framework of an extremely crude model [5.79], mentioned in Sect. 4.5. For this reason, we shall not rewrite the relation derived above using the expression for Γ obtained in [5.73].

The more general theory for the surface wave in the excition-resonance region and for different types of the ABC has been discussed in [5.65, 80, 81], for experiment see [5.65]. We point out that when $\alpha' > 0$ an additional ("new") surface wave [5.40] appears. In this case, in fact, $\omega_s^2 = \omega_s^2(0) - \alpha' k$. Substituting this relation into (5.4.3), we obtain the following equations for determining $n(\omega)$, neglecting damping:

$$n^2 = \frac{A}{\zeta - \alpha n}, \tag{5.4.9}$$

where

$$\alpha = \frac{\alpha'}{2c\omega_s(0)}, \quad A = \frac{\varepsilon_1 \varepsilon_\infty}{\varepsilon_1 + \varepsilon_\infty} \frac{\omega_\parallel^2 - \omega_s^2(0)}{2\omega_s^2(0)}, \quad \xi = \frac{\omega_s(0) - \omega}{\omega_s(0)}.$$

Let us assess the values of the parameters using, as an example, data for NaCl ($\omega_\perp = 3.1 \times 10^{13}$ Hz, $\varepsilon_0 = 5.6$ and $\varepsilon_\infty = 2.25$) and setting $\varepsilon_1 = 1$. Then $A \approx 0.2$ and $\alpha \approx d(\tilde{\varepsilon} - 1/\tilde{\varepsilon}) \times 10^2\,\mathrm{cm}^{-1}$. At $\tilde{\varepsilon} = 5$ and $d = 5 \times 10^2$ Å, the quantity $\alpha \approx 2 \times 10^{-3}$. The dependence $n(\xi)$ is shown in Fig. 5.8. It follows from this figure that at $\xi < \xi_0$, each value of the frequency corresponds to two values of $n_{1,2}$. The quantity $\xi = \xi_0$, where $n_1 = n_2$, is determined by the relation $\xi_0 = (3/\sqrt[3]{4})\alpha^{2/3} A^{1/3}$. Substituting the parameters for NaCl we obtain $\xi_0 \approx 10^{-2}$. The linewidth of the surface polaritons is $\approx 10^{-2}\,\omega_\perp$. This signifies that at $\xi = \xi_0$ damping should be taken into consideration. However, for frequencies $\xi < \xi_0$ damping does not have such a strong effect on the dispersion law of the waves. The dispersion law (5.4.8) holds for the additional wave ($n = n_2$) in the spectral region $\xi \ll \xi_0$, whereas the results of (5.4.3) can be used to find n_1. The dispersion law (5.4.8) could be verified in studying the Raman scattering of light by surface polaritons at small scattering angles. When, however, waves of a given frequency propagate along a surface, the additional wave may cause interference effects of the type dealt with in Sect. 4.6 for bulk waves with spatial dispersion being accounted for.

We shall explain the aforesaid by considering waves on a surface formed by two contiguous wedges (Fig. 5.4a). Assume that a surface wave is excited in some way in the interface between media I and II, that this wave propagates in a direction perpendicular to the boundary line between the surfaces (i.e., perpendicular to the y axis) and has a frequency ω that corresponds, in the interface between media I and III, to two surface waves rather than one. Since the component of the wave vector (i.e., the quantity k_x) perpendicular to the boundary line is not conserved in this case, the effect of the initial surface wave in the interface between media I and III leads to the excitation of two surface waves, in general, with the same frequency ω and also propagating perpendicular to the y axis, but having different values of k_x.

To determine the amplitudes of these waves it is necessary, within the framework of the theory of surface-wave diffraction by a wedge, to take into account spatial dispersion and to obtain a relation for the wave amplitudes having the sense of additional boundary conditions (ABC). This problem, as formulated here will be discussed below. Here, we shall only mention possibilities of experimental observation of interference effects.

If, for example, lines are ruled on the interface between the media, perpendicular to the propagation direction of the "ordinary" and "additional" surface waves, the waves are converted into bulk electromagnetic waves of the same frequency ω (the role of ruled lines is dealt with in Sect. 5.5). The total intensity of the bulk waves should oscillate in accordance with the path travelled by the surface waves up to ruled lines, and these oscillations could indicate the role of spatial dispersion. It is clear, of course, that other imperfections of the surface could, in principle, be utilized instead of ruled lines.

It is also possible to find interesting effects of spatial dispersion in the region of resonance of the surface polariton with the vibrations in the transition layer. A more detailed analysis of the dispersion equation (5.3.12), carried out in [5.82], indicates that along with the splitting of the dispersion curve (Fig. 5.7), the appearance of an additional surface polariton is also possible in this case. With an increase in k, as shown in [4.82], the frequency of the surface waves of the lower branch ($\omega < \omega_0$, Fig. 5.7) is reduced (according to a linear law) so that for a certain frequency region $\omega < \omega_0$, a single value of ω corresponds to two surface waves with different values of their wave vectors (similar to the situation with $\omega < \omega_s$ illustrated in Fig. 5.8). Additional surface waves have not yet been observed. It is apparent that attempts could be made to detect and investigate their properties by utilizing the interference and diffraction phenomenon.

The appearance of an additional surface wave may be considered, in a certain sense, an effect of spatial dispersion due to the parameter kd, k being

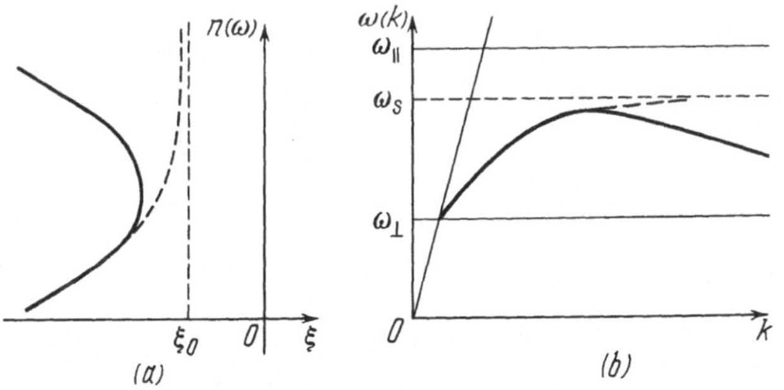

Fig. 5.8a, b. Dispersion of a surface polariton when spatial dispersion is taken into account. **(a)** Dependence of the refractive index of the surface wave on the frequency (neglecting damping); **(b)** dependence of the surface wave frequency on the wave vector. The heavy dashed line corresponds to the dispersion law when spatial dispersion is neglected

the surface polariton wave vector and d the thickness of the film. As in bulk crystaloptics, where the inclusion of additional waves results in a considerable modification of Fresnel formulas, a similar situation also occurs when the transformation of surface waves is considered at the impedance step. In connection with the results obtained in [5.82], a theory was developed by *Agranovich* et al. [5.83] for the diffraction of surface polaritons at the impedance step formed on depositing on a metal a film of thickness $d \ll \lambda = 2\pi c/\omega$ (ω being the surface polariton frequency) in the region of existence of the additional wave. The amplitudes of the reflected and refracted surface waves, as well as the angular distribution of intensity of the bulk radiation that is produced, were found by applying the factorization technique. One of the most important results of this study is the prediction of an anomalously high efficiency of transformation of the surface-wave energy into the energy of bulk radiation near resonance with oscillations in the film. This means that at the incidence of white surface radiation (with a broad frequency spectrum) on the impedance step, the spectrum of the bulk radiation must contain frequencies of the film resonances. This phenomenon may prove useful for experimental investigations of the optical properties of surfaces and thin films.

Similar problems for the region of exciton resonances in thin semiconductor films applied to a metal were discussed by *Agranovich* et al. [5.84]. It is also necessary, in this case, to take into account the spatial dispersion in the film. Consequently, the number of additional waves increases. Besides, the theory predicts the possible existence of edge modes, one-dimensional waves capable of propagating along the edge of the film.

It was shown in [5.83, 84] that if additional surface waves are taken into account, the role of additional boundary conditions (ABC) is reduced to the necessity of introducing certain additional terms into the boundary conditions for the fields in the impedance-step region. These terms have the meaning of linear current. In this sense, the situation here is similar to that in bulk crystaloptics, where taking spatial dispersion into account may lead to the appearance of surface currents in the boundary conditions (Chap. 2).

The first observations of interference phenomena involving the participation of surface polaritons was reported in [5.85]. *Schlesinger* and *Sievers* [5.85] observed the interference of a surface polariton with bulk radiation appearing on an impedance step. The impedance step was produced by applying a dielectric film on the surface of a metal. The region of the metal coated by the film was prepared in the shape of a wedge. In the experiments, one could then observe oscillations of the radiation intensity, which depended upon the length of the path travelled by the surface polariton in the film.

Similar experiments for the resonance region of the surface polariton with the vibrations in the film could be used to observe surface waves in this region of the spectrum.

We note in conclusion that the surface polaritons discussed above can exist at interfaces between media that differ in the sign (for some frequency range) of their dielectric constants (Sect. 5.1, as well as Sect. 5.2, which deal with the conditions for the existence of surface waves in anisotropy crystals). Although, as shown above, spatial dispersion may lead to the appearance of additional solutions for these waves, its role in the case under consideration is not, as a rule, decisive.

5.4.1 Surface Electromagnetic Waves at the Right-Left Gyrotropic Crystal Interface

The search for media and situations in which intrinsic surface electromagnetic waves are formed due to the fact that spatial dispersion of the dielectric-constant tensor is taken into consideration is of some interest. Evidently, a discussion of the conditions for the existence of surface polaritons at the interface between media [5.86] whose dielectric-constant tensors differ only if spatial dispersion is accounted for is of particular value. In particular, such an interface may be the boundary separating the "right" and "left" modifications of a gyrotropic crystal (for instance, quartz) or a gyrotropic liquid crystal that has "right" and "left" phases under certain conditions of existence [5.87]. Following [5.86], we shall limit ourselves to a discussion of the case of gyrotropic media and, in the expansion of the dielectric-constant tensor, take into account only terms linear with respect to the wave vector $\boldsymbol{k}$.

In calculating the surface waves, we shall find the fields from the conditions at the interface indicated in Sect. 2.3.2, see (2.3.51 c, d) and (2.1.2). We assume that the "right" ($\alpha = \mathrm{I}$ and $z<0$) and "left" ($\alpha = \mathrm{II}$ and $z>0$) non-magnetic media are separated by the plane $z=0$. In accordance with the discussion in Sect. 2.3.3, the equation relating the electric induction $\boldsymbol{D}$ and field strength $\boldsymbol{E}$ to the frequency ω in each of the media $\alpha = \mathrm{I}, \mathrm{II}$ is presented in the form

$$D_i^\alpha(\boldsymbol{r}) = \varepsilon_{ij}(\omega)E_j^\alpha(\boldsymbol{r}) - \frac{1}{2}e_{ijm}g_{ml}^\alpha(\boldsymbol{r})\frac{\partial}{\partial x_l}E_j^\alpha(\boldsymbol{r}) - \frac{1}{2}e_{ijm}\frac{\partial}{\partial x_l}[g_{ml}^\alpha(\boldsymbol{r})E_j^\alpha(\boldsymbol{r})] , \tag{5.4.10}$$

where $\varepsilon_{ij}(\omega)$ is the dielectric-constant tensor when spatial dispersion is neglected, $g_{ij}(\boldsymbol{r})$ is the gyration pseudotensor, $g_{ij}^{\mathrm{I}} = -g_{ij}^{\mathrm{II}}$, and e_{ijm} is the completely antisymmetric unit pseudotensor. We shall seek surface solutions for fields, which decrease when $z \to \pm\infty$ and extend along the x axis, in the usual form

$$\boldsymbol{E}^\alpha(\boldsymbol{r}) = \boldsymbol{E}^\alpha \exp(\mathrm{i}\boldsymbol{K}^\alpha \cdot \boldsymbol{r}) , \quad \boldsymbol{H}^\alpha(\boldsymbol{r}) = \boldsymbol{H}^\alpha \exp(\mathrm{i}\boldsymbol{K}^\alpha \cdot \boldsymbol{r}) , \quad \alpha = \mathrm{I}\ \text{and}\ \mathrm{II} ,$$
$$\boldsymbol{K}^\alpha = (k, 0, k_3^\alpha) , \quad \mathrm{Im}\{k_3^{\mathrm{I}}\} < 0 , \quad \mathrm{Im}\{k_3^{\mathrm{II}}\} > 0 . \tag{5.4.11}$$

At some distance from the boundary the dielectric-constant tensor is given by

$$\varepsilon_{ij}^{\alpha}(\omega,\boldsymbol{K}^{\alpha}) = \varepsilon_{ij}(\omega) - \mathrm{i}e_{ijm}g_{ml}^{\alpha}K_l^{\alpha}\,, \qquad D_i^{\alpha} = \varepsilon_{ij}^{\alpha}(\omega,\boldsymbol{K}^{\alpha})E_j^{\alpha}\,. \tag{5.4.12}$$

We can then obtain boundary conditions for fields, (5.4.11), that satisfy Maxwell's equations in the same manner as we did in Sect. 2.3.2 for bulk waves:

$$\operatorname{rot}\boldsymbol{E}^{\alpha}(\boldsymbol{r}) = \mathrm{i}\frac{\omega}{c}\boldsymbol{H}^{\alpha}(\boldsymbol{r})\,, \qquad \operatorname{rot}\boldsymbol{H}^{\alpha}(\boldsymbol{r}) = -\mathrm{i}\frac{\omega}{c}\boldsymbol{D}^{\alpha}(\boldsymbol{r})\,. \tag{5.4.13}$$

As an example [5.86], we can consider crystals of class C_{2v} in the rhombic system and uniaxial crystals of class D_{2d} in the tetragonal system, where (with the crystallographic axes fixed in the same way as in Sect. 3.2 so that twofold symmetry axes are normal to the interface) the tensors ε_{ij} are diagonal and the gyration tensors are determined by a single parameter:

$$\varepsilon_{ij}(\omega) = \varepsilon_i(\omega)\delta_{ij}\,, \qquad g_{ij}^{\alpha} = (e_{ij3})^2 g^{\alpha}\,, \qquad g^{\mathrm{II}} = -g^{\mathrm{I}} = g > 0\,; \tag{5.4.14}$$

for uniaxial crystals, moreover, $\varepsilon_1 = \varepsilon_2 = \varepsilon_{\perp}$ and $\varepsilon_3 = \varepsilon_{\parallel}$. It can be readily seen that the boundary conditions at $z = 0$ for the fields, (5.4.11), then have the form:

$$\varepsilon_3 E_3^{\mathrm{I}} - \mathrm{i}gkE_1^{\mathrm{I}} = \varepsilon_3 E_3^{\mathrm{II}} + \mathrm{i}gkE_1^{\mathrm{II}}\,, \qquad \boldsymbol{H}^{\mathrm{I}} = \boldsymbol{H}^{\mathrm{II}}\,, \qquad E_{1,2}^{\mathrm{I}} = E_{1,2}^{\mathrm{II}}\,. \tag{5.4.15}$$

The use of Maxwell's equations (5.4.13) leads to linear relations between the amplitudes of the fields, for each of the media I and II. An elementary analysis of these relations indicates that they, as well as (5.4.15), are satisfied by solutions of the type (5.4.11) for which only the quantities E_3 and H_2 are nonzero in both media, and

$$\begin{aligned}
&E_3^{\mathrm{I}} = E_3^{\mathrm{II}} = E_3 \neq 0\,, && H_2^{\mathrm{I}} = H_2^{\mathrm{II}} = -\frac{kc}{\omega}E_3\,,\\
&D_1^{\mathrm{I}} = -D_1^{\mathrm{II}} = \mathrm{i}gkE_3\,, && D_2^{\mathrm{I}} = D_2^{\mathrm{II}} = 0\,,\\
&D_3^{\mathrm{I}} = D_3^{\mathrm{II}} = \varepsilon_3 E_3\,, && k_3^{\mathrm{II}} = -k_3^{\mathrm{I}} = \mathrm{i}g\frac{\omega^2}{c^2}\,.
\end{aligned}$$

It is also found that the dispersion law for surface waves has the form

$$k^2 = \frac{\omega^2}{c^2}\varepsilon_3(\omega)\,, \tag{5.4.16}$$

and, consequently, these waves exist in the transparent region, i.e., in the frequency region where $\varepsilon_3(\omega) > 0$.

Hence the surface waves are transverse and their dispersion law (5.4.16) is independent of the spatial dispersion parameter g. At the same time, this parameter appears in the relations between the field amplitudes, and also determines the rate of their decrease, as $|z|$ increases (see the expression for k_3; as $g \to 0$, $k_3 \to 0$, i.e., the wave becomes a "bulk" one). It is of interest that as $c \to \infty$, $k^{\mathrm{I,II}} \to 0$ again. This means that the solution exists only when retardation is accounted for.

The depth of penetration of the field into media I and II equals $l = |k_3|^{-1} = c^2/\omega^2 g$, i.e., it is $c/\omega g$ times greater than the wavelength in vacuum. In the transparent region of the spectrum for ordinary crystals, $\omega g/c \approx 10^{-2}$ to 10^{-3}; hence $l \approx$ (100 to 1000) $\lambda_0 \approx 5 \times 10^{-2}$ cm.

Note that if the processes of energy dissipation are accounted for, $k^2 = (\omega^2/c^2)(n + i\varkappa)^2$, see (5.4.16). Thus, with weak damping, when $\varepsilon_3(\omega) = \varepsilon_3' + i\varepsilon_3''$, where $\varepsilon_3'' \ll \varepsilon_3'$, the quantity $\varkappa$ is determined by the relation $\varkappa = \varepsilon_3''/2\sqrt{\varepsilon_3'(\omega)}$. In this way, the length L over which the intensity of the propagating surface wave is decreased e-fold, is equal to $L_0 = [(\omega/c)(\varepsilon_3''/\sqrt{\varepsilon_3'})]^{-1}$, i.e., it is found to be equal to the corresponding length for bulk waves of the same polarization when gyrotropy is neglected. It is known that for many crystals L_0 may be of the order of 10 to 100 cm^{-1} in certain spectral regions (in this case $\mathrm{Im}\{k\} \equiv \omega\varkappa/c \approx (10^{-1}$ to $10^{-2})$ cm^{-1}; this corresponds, at $\omega/c \approx 10^5$ cm^{-1}, to the value $\varepsilon_3'' \approx 10^{-6}$ to 10^{-7}). The smallness of $\mathrm{Im}\{k\}$ is extremely important because at large values of $\mathrm{Im}\{k\}$ a surface wave of the type under consideration acquires a radiation width, and the process (in a linear approximation) of its conversion into a bulk wave becomes feasible. Generally speaking, bulk electromagnetic waves also exist in the frequency region where the quantity $\varepsilon_3(\omega) > 0$. In comparing the spectra of surface waves and the above-mentioned bulk waves, we shall confine ourselves, for the sake of simplicity, to the case of uniaxial crystals. Moreover, since the tangential component of the wave vector should be conserved in the transformation processes just mentioned, we shall consider the dispersion of bulk waves with the wave vector $\boldsymbol{K} = (k, 0, k_3)$, k_3 being real. For "ordinary" bulk waves ($E_1 = E_3 = 0$ and $E_2 \neq 0$), gyrotropy does not appear in (5.4.12). Here the dispersion law has the ordinary form $(\omega^2/c^2)\varepsilon_\perp^0(\omega) = k^2 + k_3^2$. The spectrum of these waves overlaps that of surface waves only at $\varepsilon_\perp(\omega) > \varepsilon_\parallel(\omega)$, but the process of surface polariton decay, even in the presence of weak damping, is forbidden owing to the orthogonality of the fields. At the same time, the spectrum of "extraordinary" bulk waves ($E_2 = 0$, $E_1 \neq 0$ and $E_3 \neq 0$) is determined by the relation

$$\frac{\omega^2}{c^2}\varepsilon_\perp(\omega)\varepsilon_\parallel(\omega) - \varepsilon_\parallel(\omega)k_3^2 - \varepsilon_\perp(\omega)k^2 - \frac{\omega^2}{c^2}g^2k^2 = 0\,. \tag{5.4.17}$$

It can readily be seen that the relation $\omega = \omega(k)$ found from this equation is not compatible with the dispersion law for surface waves $\omega_s = \omega_s(k)$, obtained from (5.4.16), for any real value, whatsoever, of k_3. At $k_3 = 0$, however, it follows from (5.4.17) that the quantity k for a bulk wave equals

$$k = \sqrt{\frac{(\omega^2/c^2)\,\varepsilon_\perp(\omega)\,\varepsilon_\parallel(\omega)}{\varepsilon_\perp(\omega) + (\omega^2/c^2)\,g^2}}\,,$$

i.e., it differs, in general, from the corresponding value at the same frequency for a surface wave $k_s = (\omega/c)\sqrt{\varepsilon_\parallel(\omega)}$, see (5.4.16), only slightly:

$$\Delta k = k - k_s \simeq -\frac{k_s}{2}\left(\frac{\omega g}{c}\right)^2 \frac{1}{\varepsilon_\perp} = -\frac{1}{2}\,\frac{\omega^3}{c^3}\,\frac{\sqrt{\varepsilon_\parallel(\omega)}}{\varepsilon_\perp(\omega)}\,g^2\,. \tag{5.4.18}$$

This means that the surface waves under consideration show no radiation width, only under the condition that

$$\mathrm{Im}\{k_3\} \ll |\Delta k| \qquad \text{or} \tag{5.4.19}$$

$$\mathrm{Im}\{\varepsilon_\parallel(\omega)\} \ll \frac{\varepsilon'_\parallel(\omega)}{\varepsilon_\perp(\omega)}\left(\frac{\omega}{c}\,g\right)^2 . \tag{5.4.20}$$

For ordinary gyrotropic crystals in the visible region of the spectrum, the right-hand side of inequality (5.4.20) is a quantity of the order of 10^{-5} to 10^{-6}. Hence, for such crystals (5.4.20) can be satisfied only in the region of high transparency, where $\mathrm{Im}\{\varepsilon_\parallel(\omega)\} \approx 10^{-6}$ to 10^{-7}. For liquid crystals, however, the existence region for the stable surface waves under consideration seems considerably more extensive.

Experimental observation of surface waves at the interface between "right" and "left" gyrotropic media would reveal new opportunities to study gyrotropy, including that of crystals with a magnetic structure.

5.5 Experimental Investigations of Surface Polaritons

In recent years a definite breakthrough has occurred in the development of experimental investigations of surface polaritons in metals, semiconductors and dielectrics. These elementary excitations can be studied with great success by the attenuated total reflection (ATR) method. Research has been initiated, utilizing low-energy electron diffraction (LEED) [5.88] Raman scattering of light (RSL) [5.89], and methods of active laser spectroscopy [5.90]. These methods are based, as expressed by their names, on the various physical processes and have different degrees of accuracy. Nevertheless, they are all important because they supplement one another, enabling surface wave spectra to be investigated in a wide range of wave vector. This will be elucidated in the following discussion of the distinctive features of the various methods for studying surface polaritons, and the results these methods can yield. We note that ATR, LEED and RSL can also be used to excite surface polaritons, and that this fact and the opportunities is provides should be taken into consideration in discussing the conceivable sources and detectors of surface waves that are required for the development of surface crystal optics [5.4a, 40, 91].

5.5.1 The Attenuated Total Reflection (ATR) Method

The well-known ATR method [5.92] for investigating surface plasmons in metals was first applied in 1968 [5.93]. Subsequently, *Ruppin* [5.94] drew attention to the feasibility of employing this method to study surface phonons as well[10]. The first experimental investigations of surface polariton spectra in the infrared region of the spectrum were carried out for cubic crystals NaCl, HBr, NaF, LiF, CdF_2, GaP and many others in [5.99 – 102]. The spectra of surface polaritons in the uniaxial crystals MgF_2, TiO_2 [5.35, 103], quartz, sapphire and lithium niobate [5.39, 104] were investigated only for certain orientations of the crystal surface and directions of polariton propagation. Surface polaritons were observed in the exciton region of the spectrum by the ATR method in crystals of ZnO [5.105], CuBr [5.106] and anthracene [5.107]. Observations of electronic surface polaritons at room temperature have been carried out for a number of crystals of various kinds [5.108].

In ATR the reflection spectrum of electromagnetic radiation incident from an optically denser medium upon the plane boundary of two media, which is the plane of total internal reflection (Fig. 5.9), is measured. The presence of an absorbing medium (i.e., medium III, Fig. 5.9, which in our case is the crystal in which the spectrum of surface waves is being investigated) leads to a reduction in the intensity of reflected light. Penetration of the electromagnetic field beyond the plane of total internal reflection and the possibility of its dissipation in an optically less dense medium was experimentally studied as far back as the turn of the century [5.109], see also [5.97]. Nevertheless, the meaning of the comments made in the literature [5.93, 94] is not trivial because the possibility, in principle, of absorption does not yet signify that this absorption can be due to the excitation of a surface polariton. This can be elucidated by returning to the spectra of polaritons (Fig. 5.1). It follows from the appearance of these spectra that the frequency $\omega_s(\boldsymbol{k})$ of the surface polariton satisfies the condition $\omega_s(\boldsymbol{k}) < cK/\sqrt{\varepsilon_\infty} < ck$, where $\boldsymbol{k}(k_x, k_y)$ is the wave vector of

[10] The possibility of using ATR for exciting electromagnetic waves in waveguides [5.95] was pointed out earlier in [5.96]. The related phenomenon of beam displacement at total reflection, the so-called Goos-Haenchen effect, was comprehensively studied by *Lotsch* [5.97], and its significance in integrated optics has been discussed in [5.98].

the surface polariton. Since in the interaction processes of a photon incident, for instance, from vacuum on the surface of a crystal, the projection of the wave vector on the crystal surface should be conserved, the inequality $\omega = c\sqrt{k^2+k_z^2} > \omega_s(k)$, ω being the frequency of the photon in vacuum, makes it impossible (neglecting the role of phonons and similar factors) to excite surface polaritons by ordinary bulk photons. If, however, we turn our attention to plane waves having an imaginary k_z, i.e., to fields which decrease exponentially with the distance from plane $z=0$ (as is known, such a field exists below the plane of total internal reflection; Fig. 5.9), the law of conservation of energy $\hbar\omega = \hbar\omega_s(\boldsymbol{k})$ can be satisfied. This forms the basis for applying the ATR method in investigating surface polaritons. The following should be kept in mind. One might expect that the excitation of surface polaritons should be especially intense at very small values of $d \to 0$ (Fig. 5.9). Actually, at $d=0$ these waves are not excited at all because the region of space where the exciting waves have an imaginary value of k_z vanishes (the spectrum of surface waves is also deformed to some extent; the role of the substrate is discussed below). Therefore, an optimal spacing d is chosen, i.e., one large enough so that the presence of the prism has practically no effect on the spectrum of surface polaritons, but the reduction in the reflection coefficient can, nevertheless be detected experimentally. Calculations of the reflection coefficient in ATR have often been carried out [5.100, 110]. Hence, we shall only present the formula for the reflection coefficient in a situation that corresponds to Fig. 5.9 and to a sufficiently large value of d. (In experiments, the minimum value of d for which the frequency of the minimum $R(\omega)$ is independent of d is selected for each value of k.) If ε_1 and ε_2 are the dielectric constants, respectively, of the material of the prism and of the substance in the gap, and $\varepsilon_3(\omega)$ is the dielectric constant of the crystal [$\varepsilon_3(\omega) < 0$; ε_1, ε_2 and ε_3 are assumed to be real], then the reflection coefficient is

$$R(\omega) \approx 1 - 16\pi \frac{\varepsilon_3^2 \varkappa_1}{\varepsilon_1 \varkappa_3^2} \delta \left(\frac{\varepsilon_3(\omega)}{\varkappa_3} + \frac{\varepsilon_2}{\varkappa_2} \right) \mathrm{e}^{-\varkappa_2 d}, \tag{5.5.1}$$

where $\varkappa_i = \sqrt{k^2 - \omega^2 \varepsilon_i / c^2}$. It follows from this equation that the reflection

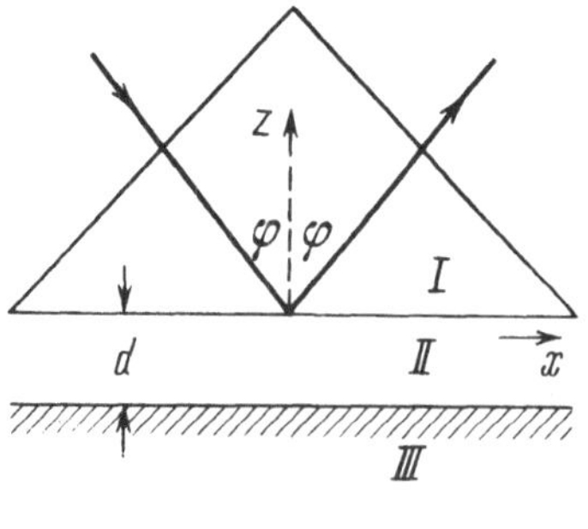

Fig. 5.9. Radiation-path diagram for the ATR method

coefficient has a value less than unity only for the frequencies ω and the angles of incidence φ for which the quantities ω and $k \equiv (\omega/c)\sqrt{\varepsilon_1}\sin\varphi$ satisfy the equation

$$\varepsilon_3(\omega)/\varkappa_3 + \varepsilon_2/\varkappa_2 = 0 . \tag{5.5.2}$$

This is exactly what determines the dispersion of the surface polariton at the boundary between media II and III when $d \to \infty$. The presence of the δ function in (5.5.1) is due to the fact that dissipation processes are neglected. The integral quantity of reduction in the reflection coefficient, within definite limits independent of dissipation and proportional to the total probability of exciting a surface polariton with the wave vector $\boldsymbol{k}$ in ATR, is determined by the expression

$$W(k) = 16\pi \frac{\varepsilon_3^2 \varkappa_1}{\varepsilon_1 \varkappa_3^2} C(k) \mathrm{e}^{-\varkappa_2 d} , \tag{5.5.3}$$

where the frequency ω should be assumed equal to $\omega_s(\boldsymbol{k})$ of the surface polariton with the wave vector $\boldsymbol{k}$. The quantity $C(k)$ is determined by the expression

$$\frac{d}{d\omega}\left(\frac{\varepsilon_3}{\varkappa_3} + \frac{\varepsilon_2}{\varkappa_2}\right)\Bigg|_{\omega=\omega_s(k)} \equiv \frac{1}{C(k)} . \tag{5.5.4}$$

If $\Gamma(k)$ is the linewidth of the surface polariton, then, in the region of the line center, the reduction $R(\omega)$ is approximately equal to $W(k)/\Gamma(k)$. We estimate the quantity $W(k)$ approximately, neglecting relativistic terms in $\varkappa_i$. In this approximation $\varkappa_i = k$. Hence, making use of (4.1.13) for ε_3, we obtain

$$C(k) = \frac{k(\varepsilon_0 - \varepsilon_\infty)\,\omega_\perp^2}{2\,\omega_s(k)(\varepsilon_2 + \varepsilon_\infty)} . \tag{5.5.5}$$

Consequently, in order of magnitude, $C(k) \sim k\omega_\perp/10$. Thus $W/\Gamma \sim 20(\omega_\perp/\Gamma)\exp(-kd)$ and, if $\omega_\perp/\Gamma \approx 10$ and $d \approx \lambda$, the ratio $W/\Gamma \approx 0.5$. The large values of the ratio W/Γ are precisely what has permitted the successful application of the ATR method. Typical experimental curves, given in Fig. 5.10, represent the dependence of the frequency ω [which, at the given k, corresponds to min $R(\omega)$] on the angle φ. This figure also exhibits curves of the relation $\omega_s = \omega_s(k)$ [5.104].

A review of both experimental research (using the ATR method) and theoretical investigations of surface polaritons for the exciton region of the spectrum can be found in [5.65, 80, 111].

The ATR method has also been applied to study the effects of a transition layer (thin film of LiF) on the dispersion of a surface polariton in sapphire

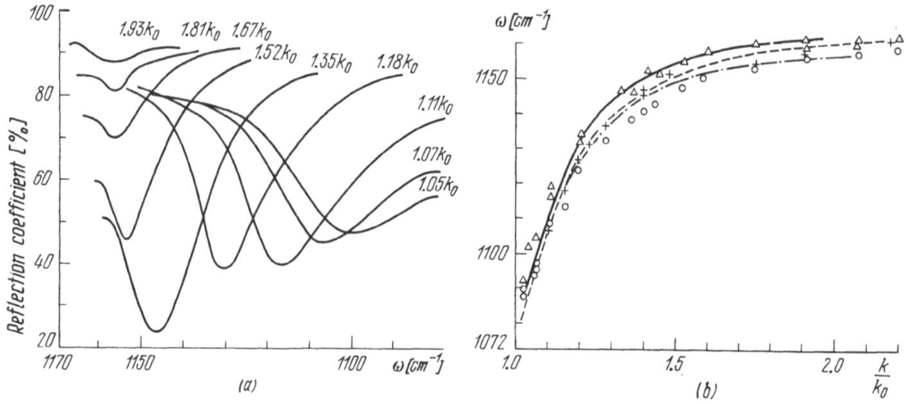

Fig. 5.10a, b. Experimental data from [5.104]. **(a)** Dependence of the reflection coefficient R on the frequency ω for various values of q_x, the projections of the bulk photon wave vector on the plane of total internal reflection ($q_x = k_0 n_0 \sin\varphi$, where φ is the angle of incidence, n_0 is the refractive index of the prism material and $k_0 = \omega/c$; (**b** dependence of the surface polariton frequency on k (various curves correspond to different orientations of the surface [5.104])

and rutile crystals. Since, in this case, resonance of vibrations in the transition layer with the surface polariton is reached, a gap was observed in the polariton spectrum. The magnitude of this gap, in agreement with theory (Sect. 5.4), is proportional to $\sqrt{d}$, d denoting the thickness of the LiF film (at $d = 100$ Å, the gap magnitude $\Delta \approx 20\,\mathrm{cm}^{-1}$ [5.53]).

As indicated in [5.54], the gap size can increase substantially in the visible region of the spectrum. In this work, the splitting effect was observed for surface waves propagating along the surface of aluminum coated with silver films ($d \approx 20$ to 60 Å). The frequency of plasma vibrations in the silver are in the frequency-change region for surface polaritons on aluminum, and the gap size Δ at $d \approx 26$ Å was found to be approximately 0.4 eV.

The same effect was observed for surface wave propagating along the surface of silver coated with dye monolayer in the frequency-change region for surface polaritons on silver [5.54a] ($\Delta \approx 0.1$ eV).

Research on surface polaritons in the metallic film is also reported in [5.110]. If different dielectrics make contact with the metal at the two sides of the film, then the branches of the surface polaritons are split (Sect. 5.1). In this case, as pointed out in [5.112], the analysis of the conditions at each film surface is simplified (on the basis of the ATR spectra).

We note that the ATR method can be used to observe the surface polariton spectrum only up to values of the wave vector $k \lesssim (\omega/c)\sqrt{\varepsilon_1}$. Consequently, this method does not allow surface-wave dispersion to be studied in the region of large values of k. RSL and LEED may prove especially effective in this spectral region.

5.5.2 The Ruled Diffraction-Grating Method

Suppose lines parallel, for instance, to the y axis are ruled at equal distances d from one another on the surface $z=0$ of a crystal. Then, upon interaction between light waves and a surface polariton, the projection k_x of the wave vector is conserved with an accuracy only to order $2\pi/d$. The conversion of bulk photons with frequency $\omega=\omega_s(k_x^m)$ into surface polaritons becomes possible [with a simultaneous reduction of the reflection coefficient $R(\omega)$] because at certain values of $k_x=k_x^m$, notwithstanding the fact that the inequality

$$\frac{c}{\sqrt{\varepsilon_\infty}}\sqrt{k_x^2+k_z^2}>\omega_s(k_x)\,,$$

is satisfied, the condition

$$\frac{c}{\sqrt{\varepsilon_\infty}}\sqrt{\left(k_x^m-m\frac{2\pi}{d}\right)^2+k_z^2}=\omega_s(k_x^m) \tag{5.5.6}$$

can be satisfied by selecting integral values of $m=\pm1,\ \pm2,\ \dots$ Under the conditions under consideration the last equation expresses the laws of conservation of energy and of the wave vector k_x. It is clear that such a reduction of $R(\omega)$ should occur at angles of incidence φ determined by the relation $(\omega/c)\sin\varphi=k_x-2\pi m/d$, as was observed in experiments [5.113–117]. In most experiments the surface periodicity (grating on the surface) was produced by a machining process. This, of course, violated the structure of the subsurface layer in ways which are difficult to detect and led to additional damping of the waves. These unfavorable factors were avoided in [5.118] by using surface acoustic (Rayleigh) waves [5.91] to provide the surface periodicity (for a theoretical analysis of the problems posed here, see [5.119]). As a whole, however, the ruled diffraction-grating method is found to be less convenient than the ATR method, because the calculations of the reflectivity of photons from a system of ruled lines are extremely crude. More accurate calculations require an exact knowledge of the profiles of the ruled lines [5.120–123]. In the ATR method, on the other hand, the minima $R(\omega)$ quite accurately correspond, as previously shown, to the frequencies of surface polaritons if sufficiently large clearances are provided between the prism and the surface of the crystal.

5.5.3 Raman Scattering of Light by Surface Polaritons

The feasibility of using RSL for studying the dispersion of surface polaritons has been studied repeatedly [5.124, 125], descriptions of respective experi-

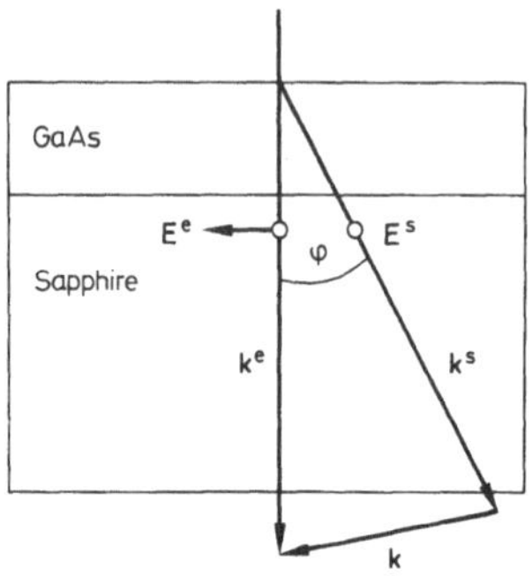

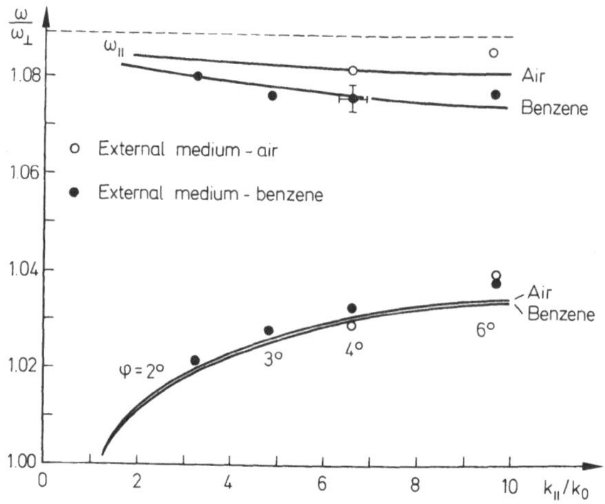

Fig. 5.11

Fig. 5.12

Fig. 5.11. Experimental arrangement for the investigation of Raman scattering of light by a surface polariton in gallium arsenide [5.127]

Fig. 5.12. Experimental data on the dispersion of a surface polariton at the GaAs-sapphire boundary [5.127, 128]. The full line curves represent the theoretical values [5.14]

ments can be found in [5.126–128]. The early experiments consisted of measuring the spectra of RSL of a thin film (approx. 2500 Å) of GaAs, epitaxially grown on a sapphire substrate. The source of radiation was the line 4880 Å of a continuous argon ion laser with a power of 400 MW. The exciting light was polarized perpendicular to the plane of scattering, and was normally incident on the film (Fig. 5.11). Although gallium arsenide very strongly absorbs laser light (penetration depth ≈900 Å), the use of a thin film enabled measurements to be made, not only for backward, but for forward scattering geometries as well. If ω_s is the measured frequency of a photon scattered at the angle φ to the direction of incidence of the laser photons ($\omega = \omega_l$) on the film, then the frequency of the surface polariton $\omega = \omega_l - \omega_s$ corresponds to the wave vector $k = q \sin\varphi$, q being the wave vector of the light from the laser: $q = \omega_l n_l / c$, and n_l the refractive index of light in GaAs at the frequency ω_l.

The experimental data given in [5.127, 128] are presented in Fig. 5.12. The solid curves have been plotted on the basis of the calculated results for the dispersion of a surface polariton in the three-layer structure: vacuum-GaAs-sapphire [5.14]. As is evident from the data given, there is good agreement between theory and experiment.

In the geometry illustrated in Fig. 5.11, dispersion of the surface polaritons is determined by an equation obtained in [5.14], and in which anisotropy of the substrate (sapphire) is taken into account. This equation is of the form

$$\left[1+\frac{\varkappa_0}{\varkappa_1}\varepsilon(\omega)\right]\left(1+\frac{\varkappa_2}{\varkappa_1}\frac{\varepsilon(\omega)}{\varepsilon_\perp}\right)$$
$$-\exp(-2\varkappa_1 d)\left[1-\frac{\varkappa_0}{\varkappa_1}\varepsilon(\omega)\right]\left(1-\frac{\varkappa_2}{\varkappa_1}\frac{\varepsilon(\omega)}{\varepsilon_\perp}\right)=0\,, \qquad (5.5.7)$$

where d is the thickness of the GaAs film,

$$\varkappa_0=\sqrt{k^2-\omega^2/c^2}\,,\qquad \varkappa_1=\sqrt{k^2-\varepsilon(\omega)\,\omega^2/c^2}\,,$$
$$\varkappa_2=\sqrt{(\varepsilon_\perp/\varepsilon_\parallel)k^2-\varepsilon_\parallel\omega^2/c^2}\,,$$

$\varepsilon(\omega)$ is the dielectric constant of gallium arsenide, $\omega_\perp \approx 270\ \mathrm{cm}^{-1}$, $\varepsilon_0 = 13.1$, $\varepsilon_\infty = 11.1$, and $\varepsilon(\omega)$, $\varepsilon_\parallel$ and $\varepsilon_\perp$ are the dielectric constants of sapphire (a uniaxial crystal; in the experiments [5.127], its optical axis was directed perpendicular to the plane interface between the media).

Equation (5.5.7), as could be expected, determines two branches of the surface polaritons. As $d \to \infty$, these branches convert into branches of a surface polariton at the vacuum-GaAs boundary and a surface polariton at the GaAs-sapphire boundary. Owing, however, to the small value of the ratio $(\varepsilon_0-\varepsilon_\infty)/(\varepsilon_0+\varepsilon_\infty)$ for GaAs, the branch of the first of these polaritons is located in the immediate vicinity of the longitudinal wave frequency in GaAs. It had not been detected in [5.127] due to line broadening in Raman scattering by bulk longitudinal waves. The upper branch was observed only in the experiments by *Prieur* and *Ushioda* [5.128]. They utilized an argument made in [5.129], according to which the Raman scattering of light by surface polaritons should be suppressed in a backward-scattering geometry (the cause of such an effect for GaAs films is elucidated in Sect. 6.4.2b). For this reason, the measured spectra of RSL in the backward-scattering geometry were subtracted from the measured spectra of RSL in the forward geometry (i.e. the background produced by RSL by bulk longitudinal and transverse polaritons was removed). As a result, the part of the spectrum that was due only to the excitation of surface waves (Fig. 5.13) was obtained. The second (lower) branch of surface polaritons, converting (as $d \to \infty$) into polaritons at the GaAs-sapphire boundary, was found to be substantially displaced, due to the high values of $\varepsilon_\parallel$ and $\varepsilon_\perp$ for sapphire ($\varepsilon_\parallel = 11.6$ and $\varepsilon_\perp = 9.35$), from the frequencies of bulk longitudinal and transverse waves, thereby enabling the first experiments to be successfully performed [5.128]. In order to explain the "mechanism" caused by the substrate, we note, first of all, that for sufficiently large values of k, values at which the inequality $k \gg 1/d$, and not only $k^2 \gg (\omega^2/c^2)\,|\varepsilon|$, is satisfied, the polariton dispersion law, as is evident from (5.4.11), is independent of d. It coincides with the dispersion law in a nonrelativistic region at $d=\infty$. Hence, in layered structures of the type dealt with in [5.127], the limiting value of the surface-wave frequency ω_s can be

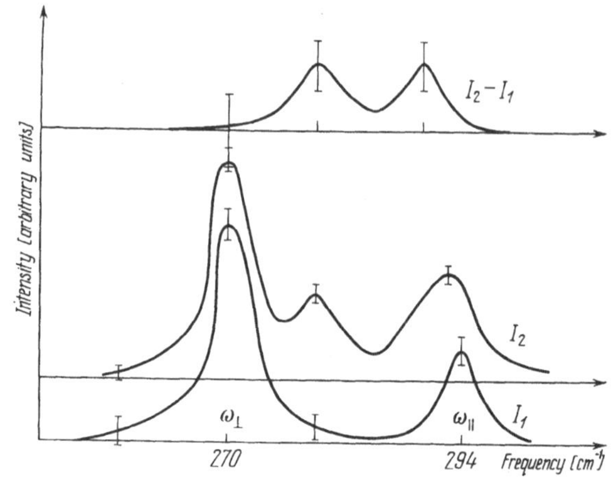

Fig. 5.13. Spectra of Raman scattering of light in a layer system (Fig. 5.11) obtained for backward (I_1) and forward (I_2) scattering geometries; the difference of these spectra (I_2-I_1) is shown above [5.128]

found from simpler relations. Since the frequency of a bulk longitudinal wave $\omega_\parallel = \omega_\perp\sqrt{\varepsilon_0/\varepsilon_\infty}$, the difference between frequencies $\omega_\parallel$ and ω_s in this case is

$$\Delta\omega = \omega_\parallel - \omega_s = \omega_\perp\left(-\sqrt{\frac{\varepsilon_0+\varepsilon_1}{\varepsilon_\infty+\varepsilon_1}} + \sqrt{\frac{\varepsilon_0}{\varepsilon_\infty}}\right). \tag{5.5.8}$$

Thus, at the vacuum-GaAs boundary ($\varepsilon_1 = 1$, ε_0 and $\varepsilon_\infty \gg \varepsilon_1$), $\Delta\omega \approx \omega_\parallel \times (\varepsilon_\infty^{-1} - \varepsilon_0^{-1})/2 \approx 10^{-2}\ \omega_\parallel < \delta$, δ being the linewidth of RSL in GaAs: $\delta \sim \omega_\parallel/30$. However, at the GaAs-sapphire boundary, $\Delta\omega > \delta$; moreover, the curve $\omega_s(k)$ does not become asymptotic at $d = 2500$ Å as readily as for the case of the vacuum-GaAs boundary (Fig. 5.3).

It was evidently not an accident that the first experiments of RSL by surface polaritons were conducted with GaAs. Crystals of GaAs have an extremely high value of nonlinear polarizability χ_{ijl}, which determines the intensity of the process. This high nonlinear polarizability is not the only factor that determined the candidate materials. The point is that the intensity of RSL by bulk phonons and polaritons increases with the thickness of the crystal, whereas the intensity of RSL by surface polaritons depends upon the thickness of the crystal only if $d \lesssim \varkappa$, where $\varkappa = \sqrt{k^2 - (\omega^2/c^2)\varepsilon(\omega)}$. Hence, at large values of d, RSL by surface waves is "drowned" in the background of RSL by bulk polaritons.

We note in conclusion that even though the RSL method is one of the chief ones used in investigating bulk polaritons, it cannot be applied to media with an inversion center because in such media the nonlinear susceptibility tensor, determining the intensity of the process, becomes identically zero. We have a quite different situation, however, if we resort to RSL by surface polaritons under conditions in which the medium under investigation is bound along its surface by a medium having no inversion center. Since the electromagnetic field in the surface polariton is nonzero at distances of the order of the wavelength ($\lambda \approx 10\ \mu m$) on each side of the plane interface and, consequently, is also nonzero in the region where $\chi_{ijl} \neq 0$, the intensity of RSL by a surface polariton is different from zero and sufficient for observations (by the violated symmetry method; [5.130] and Sect. 6.4.3 b).

5.5.4 Surface Polariton Propagation Over Long Distances. Crystal Optics of Surfaces

The propagation of a surface polariton along a copper-air boundary was investigated in [5.131]; the bulk light waves were coupled to the surface waves by means of two NaCl prisms (Fig. 5.14). The surface waves were pumped by a CO_2 laser ($\lambda_0 = 10.6\ \mu m$; power level: about 250 MW). The distance L, corresponding to an e-fold decrease in the surface wave intensity, was found to be 1.6 cm, in agreement with the estimated value (Sect. 5.1). At the frequency of the CO_2 laser, the depth of penetration of the field into the copper is approximately 250 Å. For this reason, a copper film with a thickness of 3000 Å was used in the experiments. It was applied by evaporation on a glass substrate. The clearance d between the prism and the metal surface was chosen to be 15 μm.

The work of *Schoenwald* et al. [5.131] stimulated similar investigations for many other metals [5.132, 133]. In particular, the first successful experiments were performed [5.134, 135] which permitted an estimate of the effect of thin dielectric coating (i.e., the effect of the transition layer; Sect. 5.3) on the propagation length of surface polaritons as a function of their frequency. If the frequency of the polariton is within the region of light absorption in the film, its propagation length should decrease, as observed in [5.134, 135]. In connection with these experiments we note that in crystal optics the propagation in crystals is usually investigated by bulk plane monochromatic light waves, specified by definite values of the frequency ω and wave vector $\boldsymbol{k}$. An experimental determination of the dispersion law for these waves (i.e., the relation $\omega = \omega(\boldsymbol{k})$ or the equivalent dependence of the complex refractive index $\tilde{n}$ of light on ω) and a comparison of this law with that obtained within the framework of Maxwell's equations (making use of the dielectric-constant tensor)

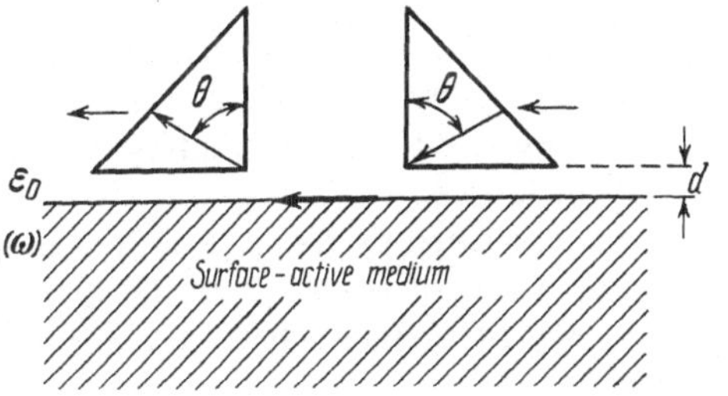

Fig. 5.14. Experimental arrangement for investigating the propagation of surface polaritons over macroscopic distances [5.131]

yield information mainly on the spectrum of bulk excitations of the medium or, for instance, on its conductivity (metal optics). It is clear that in analyzing the surface properties of a medium the maximum amount of information could be obtained by investigating the propagation of surface waves in crystals, rather than bulk electromagnetic waves, because the energy of the former is propagated only along the interfaces. A study of the damping of such surface waves, as well as of their reflection and refraction by interfaces (or boundary lines) or, so to speak, surface crystal optics, could play as important a role in the development of the physics of surfaces as ordinary (bulk) crystal optics plays in spectral investigations of the bulk properties of crystals.

In connection with the aforesaid, it is clear that further experiments can undoubtedly yield much valuable information on the structure and properties of excited states at the contacting material regions. The importance of such experiments for the development of the physics of surfaces cannot be too strongly emphasized.

We also note that further research in the field of nonlinear optical processes with the participation of surface waves is very promising. One of the problems posed here, namely, the generation of the second harmonic by a surface polariton, was discussed in [5.136]. Further problems are touched on in [5.67, 97, 137].

5.5.5 The Inelastic Low-Energy Electron Scattering (ILES) Method

The principle of inelastic low-energy electron scattering (ILES) [5.88] provides a well-known method for analyzing the surface vibrations of crystals. Since the depth of penetration of low-energy electrons (electrons with an energy from 1 to 100 eV are usually employed) into a crystal is extremely small (several lattice constants), the spectra analyzed are for electrons that do not pass through the crystal, but are reflected from its surface (an ultrahigh vacuum of about 10^{-10} torr is required, of course, as is an atomically pure surface). The main features of spectrometers in use today were discussed by *Roy* and *Carette* in [Ref. 5.88, Chap. 2]. Here we note only that the energy resolution attainable at the present time for low-energy electrons is never better than 10^{-3} eV. This does not allow the structure of the zone of surface vibrations to be studied by the ILES method. However the frequencies of the surface vibrations of the atoms, determined from the loss spectra, for example, by mirror reflection of electrons, can be compared with the limiting values of the frequencies of optical vibrations of the lattice.

The first successful experiments were described by *Ibach* [5.138]. Among others, the frequencies found for various faces of the ZnO crystal [5.138a] were in good agreement with the frequency of surface vibration determined from the condition

$$\varepsilon(\omega_s) = -1\,, \qquad \omega_s \approx (68 \pm 0.5)10^{-3}\,\text{eV}\,. \tag{5.5.9}$$

The probability of exciting surface vibrations, determined on the basis of a phenomenological theory [5.139],

$$W(\hbar\omega)\,d(\hbar\omega) = \frac{4}{a_0 k_0}\,\text{Im}\left\{\left(-\frac{1}{\varepsilon+1}\right)\right\}\frac{d(\hbar\omega)}{\hbar\omega}\,, \tag{5.5.10}$$

where $a_0 = 5.29 \times 10^{-11}$ m is the Bohr radius and k_0 is the wave vector of the electrons, also agrees well with experiment [5.138a]. The application of a phenomenological theory is justified if sufficiently long-wavelength phonons are excited ($k \ll \pi/a$, a being the lattice constant). This occurs when the inequality $v \gg \omega a$ is satisfied, v being the velocity of the exciting electrons. If $\omega \approx 10^{13}$ Hz, then the preceding inequality is satisfied even for electrons with an energy of 1 eV. Hence, the region of applicability is indeed very wide. Note also that an estimate for the wavelengths of the excited phonons, carried out in [5.138a] on the basis of an analysis of the angular distributions of the electrons, yielded the value: $\lambda \approx 20$ Å. Thus the ILES method enables frequency values to be obtained for surface phonons at values of the wave vectors about two orders of magnitude greater than the ones that can be used when the RSL method is applied.

Besides investigating the surface vibrations of atomically pure surfaces, the ILES method can also be employed to determine whether local vibrations, due to the presence of adsorbed impurities, appear on the surface of crystals [5.139]. *Barsukov* and *Naryshkina* [5.140] developed the theory of the Cherenkov radiation for surface waves; and *Bishop* and *Maradudin* [5.141] proposed that transition radiation be used in investigating surface polaritons (see also [5.142]).

5.5.6 Nonlinear Surface Electromagnetic Waves

Until now theoretical studies of surface waves (surface polaritons) were carried out based on linear approximation, i.e., the approximation in which the dielectric tensor of none of the contacting media depends on the electromagnetic fields in the media. Recently, however, interest has turned to the study of nonlinear waves which require for their existence at least one medium in the layered system characterized by a dielectric tensor

$$\varepsilon_{ij}(\omega, |\boldsymbol{E}|^2)\,,$$

that does depend on the electromagnetic field in it [5.143 – 150]. In this case the nonlinearities are not regarded as small and give rise to new types of surface electromagnetic waves that in some cases have no counterpart in the absence of the nonlinearity (i.e., in linear theory). *Chen* and *Carter* [5.151], using ATR, observed for the first time the dependence of the dispersion of a surface polariton on the intensity of the electromagnetic field in it. The method used in [5.151] is useful for the experimental determination of third-order nonlinear susceptibilities, and in particular of their frequency dependence. The system studied in [5.151] was a single interface system consisting of a metal ($\varepsilon_0(\omega) < 0$) 1 μm thick and a nonlinear material [$\varepsilon(\omega) > 0$] $\frac{1}{4}$ mm thick, which was a semiconductor. The metal – semiconductor interface is sinusoidally corrugated (a grating interface), with a period $d = 2600$ Å, and a peak-to-peak amplitude of ~500 Å. In the experiment a p-polarized laser beam derived from a high-power pulsed dye laser (100 mJ/pulse, 10 pulses/s with a pulse width ≈5 ns) was incident on the semiconductor – air interface with an angle of incidence θ. The plane of incidence was perpendicular to the grooves of the grating. The incident radiation was tunable in the range 1.0 – 1.2 μm. The maximum infrared output in the laser beam was ≈1 mJ/pulse and the collimation of the beam was better than 0.2°.

The corrugation of the metal/semiconductor interface was to couple the radiative incident and reflected radiation to the nonradiative surface polariton that propagates along the interface, by relaxing the condition that the component of the wave vector of the incident (and reflected) light must be continuous across the interface. In fact, the surface polariton is resonantly excited when the parallel component of the wave vector equals (Sect. 5.5.2)

$$k = \left| \frac{\omega}{c} \sin\theta + \frac{2\pi}{d} m \right| , \tag{5.5.11}$$

where m is an integer, and the frequency ω of the incident light equals the frequency of the surface polariton for this value of k.

Two different semiconductors, GaAs and Si, were used as the nonlinear dielectric media. For both the GaAs – Ag and Si – Ag cases the surface polaritons were coupled by the $m = -1$ order. The surface polariton coupling angles were determined by measuring the reflected intensity of the laser beam as a function of the angle of incidence θ. A dip in this intensity is observed at a value of θ for which k given by (5.5.11) with $m = -1$ is a point on the surface polariton dispersion curve at the frequency of the laser, since energy is then drained away from the incident laser beam.

The low-intensity reflection curve was obtained with an input laser energy of 35 μJ/pulse and the high intensity reflection curve was obtained with an input laser energy of 350 μJ/pulse. Since GaAs is transparent at this wavelength, the magnitude and width of reflection dip (25% and 2°, respectively) are due primarily to the dielectric loss of the silver film and to the diffraction (radiative) loss of the grating.

It was shown in [5.151] that with an increase in the input laser energy the position of the reflection dip shifts toward smaller angles which, from

(5.5.10), indicates that the wave vector k of the surface polariton increases with increasing laser intensity.

To interpret the results from *Chen* and *Carter* in [5.146] a perturbation theory was developed for weakly nonlinear p-polarized surface polaritions. The dielectric tensor for nonlinear medium was taken in the form

$$\varepsilon_{ij} = \varepsilon\delta_{ij} + \rho_{ij}(|E|)E_j, \quad \rho_{ij} = \delta_{ij}\rho_i, \quad \rho_i = (a-b)|E_i|^2 + b|E|^2$$

and $\varepsilon_{ij} = \varepsilon_1\delta_{ij}$ for linear. In [5.146] it was shown that the dispersion relation for weakly nonlinear p-polarized surface polaritons assumes the form

$$\frac{\varepsilon_1}{\varkappa_1} + \frac{\varepsilon_{\mathrm{NL}}}{\varkappa_{\mathrm{NL}}} = 0\,, \tag{5.5.12}$$

$$\varkappa_1 = \sqrt{k^2 - \frac{\omega^2}{c^2}\varepsilon_1}\,, \quad \varkappa_{\mathrm{NL}} = \sqrt{k^2 - \frac{\omega^2}{c^2}\varepsilon_{\mathrm{NL}}}\,,$$

$$\varepsilon_{\mathrm{NL}} = \frac{a(\varkappa^2 + k^4/\varkappa^2) + 2bk^2}{2(k^2+\varkappa^2)} + \varepsilon\,, \quad \varkappa = \sqrt{k^2 - \frac{\omega^2}{c^2}\varepsilon}\,.$$

In the case $a = b = 4\pi\chi^{(3)}$ ($\chi^{(3)}$ of third-order nonlinear susceptibility) the formula (5.5.12) yields:

$$\varepsilon_{\mathrm{NL}} = \varepsilon + \frac{a}{2}(|E_1(0)|^2 + |E_3(0)|^2)\,.$$

In conclusion we draw attention to the discussion of the possibilities of self-induced transparency (SIT) of the short pulses of surface polaritons and wave pulses in waveguides [5.152–154]. The SIT waves both in the bulk and on the surface, as linear and nonlinear surface polaritons, are the solutions of the macroscopic Maxwell equations. However, the SIT waves are specific nonlinear waves which in certain conditions may have the features of stationary (stable) pulses – solitons. In [5.152, 153] such pulses were obtained on surfaces and in flat waveguides covered with a thin film of a resonance substance.

6. Microscopic Theory
Calculation of the Tensor $\varepsilon_{ij}(\omega, \boldsymbol{k})$

This chapter deals with the dielectric-constant tensor $\varepsilon_{ij}(\omega, \boldsymbol{k})$, as well as the transverse dielectric-constant tensor $\varepsilon_{\perp,ij}(\omega, \boldsymbol{k})$. These tensors are expressed for the region of exciton resonances. This expression is used as the basis for examining the dispersion of the refractive index for normal waves. The procedure for calculating tensor $\varepsilon_{ij}(\omega, \boldsymbol{k})$ for molecular crystals is described in detail, in particular, by the local field method. Special features of the shape of the exciton absorption bands in the region of their long-wavelength edge are discussed. An account is given of the theory of Raman scattering of light and x rays with the production of excitons, and bulk and surface polaritons. Also treated is the question of Cherenkov and transition radiation in media displaying spatial dispersion.

6.1 General Expressions for $\varepsilon_{ij}(\omega, \boldsymbol{k})$

6.1.1 Quantum-Mechanical Derivation of $\varepsilon_{ij}(\omega, \boldsymbol{k})$

The calculation of the tensor $\varepsilon_{ij}(\omega, \boldsymbol{k})$ for a crystal, or for any model of a crystal, is a microtheoretical problem. The microtheory of ionic crystals, applicable to the infrared part of the spectrum, was developed long ago [6.1, 2]. Using ionic crystals as an example, the frequency dependence of normal oscillations (waves) on the direction of the wave vector $\boldsymbol{k}$ was discussed for $\boldsymbol{k} \to 0$. It was shown, for instance, that if the Coulomb interaction is taken fully into account but retardation is neglected, the frequencies of normal oscillations may be nonanalytic functions of $\boldsymbol{k}$ (as $\boldsymbol{k} \to 0$). In this connection, we point out that in a general investigation, a number of features require no microscopic analysis because the corresponding conclusions follow directly from the field equations and the existence of the tensor $\varepsilon_{ij}(\omega, \boldsymbol{k})$. As an example, we mention the relation (2.2.37 or 38) which reveals the dependency of the frequencies "fictitious" longitudinal waves $\omega'_{\parallel}(\boldsymbol{k})$ on the direction of $\boldsymbol{k}$ if $\boldsymbol{k} \to 0$.

An important concept, also established in the theory of ionic crystals, is the Ewald procedure, allowing the longitudinal part of the macroscopic field strength to be separated. This, in turn, makes it possible to determine the frequency and the nature of the oscillations of mechanical excitons for ionic crystals (for mechanical excitons, see Sect. 2.2.2). Using these results, the dielectric-constant tensor can be determined [Ref. 6.2, §44]. In the following, we shall show how the knowledge about the mechanical exciton states can be used to find the dielectric-constant tensor in the general quantum-mechanical treatment. The relevant procedure is well known and consists of finding the current density induced by the classical external field.

It proves convenient for this purpose to gauge the potentials so that the scalar potential of the perturbing field equals zero and thus $\boldsymbol{E} = -(1/c)\,\partial \boldsymbol{A}/\partial t$, $\boldsymbol{A}$ being the vector potential of the field induced in the medium by external sources. For the sake of simplicity we shall assume that all the rest of the field, with the exception of the perturbing field $\boldsymbol{E}$, is described by a scalar potential. Such a description would be insufficient, for example, in the case of appreciable magnetic interaction between the particles, where it would be necessary to introduce the vector potential $\boldsymbol{A}_\mu$ of the microfield.

Let us denote by Ψ_{n0} the wave functions of the medium when the field $\boldsymbol{E}$ is equal to zero or neglected, and by $\Psi_n = \Psi_{n0} + \delta\Psi_n$ the corresponding functions in the presence of the field $\boldsymbol{E}$. In a linear approximation, the operator describing the interaction between the charges and external field is equal to

$$\hat{U} = -\sum_\alpha \frac{e_\alpha}{2m_\alpha c}[\boldsymbol{p}^\alpha \cdot \boldsymbol{A}(\boldsymbol{r}_\alpha, t) + \boldsymbol{A}(\boldsymbol{r}_\alpha, t) \cdot \hat{\boldsymbol{p}}^\alpha]\,, \tag{6.1.1}$$

where $\boldsymbol{r}_\alpha$ is the radius vector of the α-th particle having the charge e_α and mass m_α, and $\sum_\alpha$ indicates summation over all the particles of the crystal and

$$\hat{\boldsymbol{p}}^\alpha = -\mathrm{i}\hbar \frac{\partial}{\partial \boldsymbol{r}_\alpha}\,.$$

Assuming

$$\boldsymbol{A} = -\frac{\mathrm{i}c}{2\omega}[\boldsymbol{E}_0 \mathrm{e}^{\mathrm{i}(\boldsymbol{k}\cdot\boldsymbol{r} - \omega t)} - \boldsymbol{E}_0^* \mathrm{e}^{-\mathrm{i}(\boldsymbol{k}\cdot\boldsymbol{r} - \omega t)}]\,,$$

we find that

$$\hat{U} = \hat{F}\mathrm{e}^{-\mathrm{i}\omega t} + \hat{G}\mathrm{e}^{\mathrm{i}\omega t}\,, \quad \hat{F}(\boldsymbol{k}) = \hat{G}^*(\boldsymbol{k}) = -\frac{\mathrm{i}c}{2\omega}\hat{\boldsymbol{M}}(\boldsymbol{k}) \cdot \boldsymbol{E}_0\,,$$
$$\hat{\boldsymbol{M}}(\boldsymbol{k}) = -\sum_\alpha \frac{e_\alpha}{2m_\alpha c}[\hat{\boldsymbol{p}}^\alpha \mathrm{e}^{\mathrm{i}\boldsymbol{k}\cdot\boldsymbol{r}_\alpha} + \mathrm{e}^{\mathrm{i}\boldsymbol{k}\cdot\boldsymbol{r}_\alpha}\hat{\boldsymbol{p}}^\alpha]\,. \tag{6.1.1a}$$

Hence, applying perturbation theory, we obtain to a first approximation [Ref. 6.3, §40],

$$\Psi_n = \Psi_{n0} - \sum_{m \neq n} \left[\frac{F_{mn} e^{-i\omega t}}{\hbar(\omega_{mn} - \omega)} + \frac{F^*_{mn} e^{i\omega t}}{\hbar(\omega_{mn} + \omega)} \right] \Psi_{m0} = \Psi_{n0} + \delta\Psi_n , \quad (6.1.2)$$

where

$$F_{mn} = \int \Psi^*_{m0} \hat{F} \Psi_{n0} d\tau , \qquad \hbar\omega_{mn} = W_m - W_n \equiv \hbar(\omega_m - \omega_n) ,$$

$$d\tau = dr_1 \ldots dr_N$$

(W_m is the energy eigenvalue, corresponding to the function Ψ_{m0}). The average value of the current-density operator in the state Ψ_n equals

$$j^{(n)}(r,t) = \langle \Psi_n | \sum_\alpha \frac{e_\alpha}{2m_\alpha} \left\{ \left[-i\hbar \frac{\partial}{\partial r_\alpha} - \frac{e_\alpha}{c} A(r_\alpha, t) \right] \delta(r - r_\alpha) \right.$$

$$\left. + \delta(r - r_\alpha) \left[-i\hbar \frac{\partial}{\partial r_\alpha} - \frac{e_\alpha}{c} A(r_\alpha, t) \right] \right\} | \Psi_n \rangle ,$$

where

$$\langle \Psi_n | \hat{Q} | \Psi_n \rangle = \int \Psi^*_n \hat{Q} \Psi_n d\tau$$

($\hat{Q}$ being an arbitrary operator). In a linear approximation, the expression for the average current density $\delta j\,(n)$ induced by the field E is of the form

$$\delta j^{(n)} = \left\langle \delta\Psi_n \left| \sum_\alpha \frac{e_\alpha}{2m_\alpha} [p^\alpha \delta(r - r_\alpha) + \delta(r - r_\alpha) \hat{p}^\alpha] \right| \Psi_{n0} \right\rangle$$

$$+ \left\langle \Psi_{n0} \left| \sum_\alpha \frac{e_\alpha}{2m_\alpha} [\hat{p}^\alpha \delta(r - r_\alpha) + \delta(r - r_\alpha) \hat{p}^\alpha] \right| \delta\Psi_n \right\rangle$$

$$- \left\langle \Psi_{n0} \left| \sum_\alpha \frac{e^2_\alpha}{m_\alpha c} \delta(r - r_\alpha) A(r_\alpha, t) \right| \Psi_{n0} \right\rangle . \quad (6.1.3)$$

This relation determines the magnitude of the induced current density at each point in space and time (r, t). Consequently, in order to obtain the dielectric-constant tensor it is necessary to pass over to the Fourier components of the quantity $\delta_j^{(n)}$. Before we perform this transition, however, we shall deduce the selection rule for the matrix elements F_{mn}.

Since the wave functions Ψ_{m0} have the symmetry of the irreducible representations of the space group of the crystal, the action of operator $\hat{T}_a$ (i.e., the operator of the translation of all the particles of the crystal over the integral-valued lattice vector a) on the wave function Ψ_{m0} leads to the following result:

$$\hat{T}_a \Psi_{m0} = e^{-iq_m \cdot a} \Psi_{m0} , \quad (6.1.4)$$

where the vector $\boldsymbol{q}_m$ determines the irreducible representation of the translation subgroup to which the state Ψ_{m0} refers (see, e.g., [6.4] and Appendix B). According to (6.1.1 a)

$$\hat{T}_a \hat{M}(\boldsymbol{k}) = \mathrm{e}^{\mathrm{i}\boldsymbol{k}\cdot\boldsymbol{a}} \hat{M}(\boldsymbol{k}) . \tag{6.1.4a}$$

Thus the integrand in the formula for the matrix element $M_{mn}(\boldsymbol{k}) = \langle \Psi_{m0} | \hat{M}(\boldsymbol{k}) | \Psi_{n0} \rangle$ remains unchanged, upon a translation of the coordinates (with respect to which we integrate) by an arbitrary integral-valued lattice vector, only if $\boldsymbol{q}_m = \boldsymbol{q}_n - \boldsymbol{k} + 2\pi\boldsymbol{b}$. Obviously, this condition must be fulfilled for the matrix element $F_{mn}(\boldsymbol{k})$ to be nonzero as well. In a similar way we arrive at the conclusion that the matrix elements $M_{nm}(\boldsymbol{k})$ and $F_{nm}(\boldsymbol{k})$ are nonzero if $\boldsymbol{q}_m = \boldsymbol{q}_n + \boldsymbol{k} + 2\pi\boldsymbol{b}$, $\boldsymbol{b}$ being an arbitrary integral-valued vector of the reciprocal lattice.

Taking the aforesaid into account and making use of the expression

$$\delta(\boldsymbol{r} - \boldsymbol{r}_\alpha) = \frac{1}{V} \sum_{\boldsymbol{k}'} \mathrm{e}^{\mathrm{i}\boldsymbol{k}'\cdot(\boldsymbol{r} - \boldsymbol{r}_\alpha)} ,$$

where V is the volume of the crystal, we find that

$$\delta j_i^{(n)} = j_i^{(n)}(\boldsymbol{k} + 2\pi\boldsymbol{b}, \omega) = \sigma_{ij}^{(n),\boldsymbol{b}}(\omega, \boldsymbol{k}) E_j(\omega, \boldsymbol{k}) , \tag{6.1.5}$$

where the tensor

$$\begin{aligned}
\sigma_{ij}^{(n),\boldsymbol{b}}(\omega, \boldsymbol{k}) = {} & \frac{\mathrm{i}\delta_{ij}}{V\omega} \left(\sum_\alpha \frac{e_\alpha^2}{m_\alpha} \mathrm{e}^{\mathrm{i}2\pi\boldsymbol{b}\cdot\boldsymbol{r}_\alpha} \right)_{nn} \delta_{\boldsymbol{k}-\boldsymbol{k}'} \\
& + \frac{\mathrm{i}}{V\hbar\omega} \sum_{\substack{m \neq n \\ \alpha, \beta}} \frac{e_\alpha e_\beta}{4 m_\alpha m_\beta} \left[\frac{(p_i^\alpha \mathrm{e}^{-\mathrm{i}\boldsymbol{k}'\cdot\boldsymbol{r}_\alpha} + \mathrm{e}^{-\mathrm{i}\boldsymbol{k}'\cdot\boldsymbol{r}_\alpha} p_i^\alpha)_{nm}}{\omega - \omega_m - \omega_n} \right. \\
& \times (p_j^\beta \mathrm{e}^{\mathrm{i}\boldsymbol{k}\cdot\boldsymbol{r}_\beta} + \mathrm{e}^{\mathrm{i}\boldsymbol{k}\cdot\boldsymbol{r}_\beta} p_j^\beta)_{mn} \\
& \left. - \frac{(p_i^\alpha \mathrm{e}^{-\mathrm{i}\boldsymbol{k}'\cdot\boldsymbol{r}_\alpha} + \mathrm{e}^{-\mathrm{i}\boldsymbol{k}'\cdot\boldsymbol{r}_\alpha} p_i^\alpha)_{mn}}{\omega + \omega_m - \omega_n} (p_j^\beta \mathrm{e}^{\mathrm{i}\boldsymbol{k}\cdot\boldsymbol{r}_\beta} + \mathrm{e}^{\mathrm{i}\boldsymbol{k}\cdot\boldsymbol{r}_\beta} p_j^\beta)_{mn} \right] .
\end{aligned} \tag{6.1.6}$$

Here $\boldsymbol{k}' = \boldsymbol{k} + 2\pi\boldsymbol{b}$; $\delta_{\boldsymbol{k}} = 0$ if $\boldsymbol{k} \neq 0$ and $\delta_{\boldsymbol{k}} = 1$ if $\boldsymbol{k} = 0$.

The relation (6.1.6) enables us to arrive at the conclusion that with an increase in $|\boldsymbol{b}|$ the quantities $\sigma_{ij}^{(n),\boldsymbol{b}}$ decrease like the Fourier components of the smoothly varying function $\Psi_{n0}^* \Psi_{m0}$. Since the interaction energy in the crystal is a continuous function of the coordinates, the function Ψ_{n0} has at least a continuous second derivative. Under similar conditions, the quantities $\sigma_{ij}^{(n),\boldsymbol{b}}$

decrease faster than $1/|\boldsymbol{b}|^3$ for large values of $|\boldsymbol{b}|$. Actually, under real conditions, it is probable that all the derivatives of the function Ψ_{m0} are continuous. At the same time, for large values of $|\boldsymbol{b}|$ the quantities $\sigma_{ij}^{(n),\boldsymbol{b}}$ decrease exponentially with an increase in $|\boldsymbol{b}|$.

It follows from (6.1.5) that in a crystal subjected to an electric field of amplitude $\boldsymbol{E}(\boldsymbol{k},\omega)$ an induced electric current appears, whose nonzero Fourier components have the value of the wave vector $\boldsymbol{k}'=\boldsymbol{k}+2\pi\boldsymbol{b}$, $\boldsymbol{b}$ being an arbitrary vector of the reciprocal lattice. If $\boldsymbol{b}=0$ we obtain

$$j_i^{(n)}(\boldsymbol{k},\omega)=\sigma_{ij}^{(n)}(\omega,\boldsymbol{k})E_j(\boldsymbol{k},\omega)\,, \tag{6.1.7a}$$

where

$$\sigma_{ij}^{(n)}(\omega,\boldsymbol{k})=\sigma_{ij}^{(n),\boldsymbol{b}=0}(\omega,\boldsymbol{k}) \tag{6.1.7b}$$

is the complex conductivity tensor. If this tensor is known, we can also find the dielectric-constant tensor by making use of the relation

$$\varepsilon_{ij}^{(n)}(\omega,\boldsymbol{k})=\delta_{ij}+\mathrm{i}\frac{4\pi}{\omega}\sigma_{ij}^{(n)}(\omega,\boldsymbol{k})\,. \tag{6.1.7c}$$

Hence, with (6.1.6) and (6.1.7b and c), we obtain the final expression:

$$\varepsilon_{ij}^{(n)}(\omega,\boldsymbol{k})=\left(1-\frac{4\pi}{\omega^2 V}\sum_\alpha\frac{e_\alpha^2}{m_\alpha}\right)\delta_{ij}$$
$$-\frac{4\pi c^2}{\hbar\omega^2 V}\sum_{m\neq n}\left[\frac{M_{nm}^i(-\boldsymbol{k})M_{mn}^j(\boldsymbol{k})}{\omega-\omega_m+\omega_n}-\frac{M_{mn}^i(-\boldsymbol{k})M_{nm}^j(\boldsymbol{k})}{\omega+\omega_m-\omega_n}\right], \tag{6.1.8}$$

where

$$\boldsymbol{M}_{nm}(\boldsymbol{k})=\langle\Psi_{n0}|\boldsymbol{M}(\boldsymbol{k})|\Psi_{m0}\rangle\,.$$

This expression plays the role of the tensor ε_{ij} in (3.1.6). This is because the wave functions Ψ_{n0} used above were obtained with the entire short-wave field taken into account (see also Sect. 3.1.1).

If we are not interested in the infrared region of the spectrum, the contribution to the dielectric-constant tensor made by the direct interaction between the ions and the electromagnetic field need not be considered (this means that in (6.1.6) the mass m_α of the ions is assumed to be infinite). In this approximation

$$\frac{4\pi}{V}\sum_\alpha\frac{e_\alpha^2}{m_\alpha}=\frac{4\pi e^2 N_0}{m}=\omega_0^2\,,$$

where N_0 is the total electron concentration, and ω_0 is the 'plasma' frequency corresponding to the electrons. If the crystal is in a state of thermodynamic equilibrium, the tensor $\varepsilon_{ij}(\omega,\boldsymbol{k})$ is obtained by using a Gibbs distribution to average (6.1.8). Hence, in a state of thermodynamic equilibrium, the current density $\boldsymbol{j} = \partial(\boldsymbol{D}-\boldsymbol{E})/4\pi\partial t$, appearing in the field equations (2.1.1), is given by

$$\boldsymbol{j} = \sum_n \exp\left(-\frac{\mathrm{F}-\mathrm{W}}{kT}\right)\boldsymbol{j}^{(n)},$$

where the free energy $F = -kT \ln \sum \exp(-W_n/kT)$. A more general expression, corresponding to averaging by means of a statistical matrix, is of the form $\boldsymbol{j} = \mathrm{Tr}\{\exp[(F-\hat{H})/kT]\hat{\boldsymbol{j}}\}$, where $\hat{H}$ is the total Hamiltonian of the system, including the field, and $\hat{\boldsymbol{j}}$ is the current density operator (see above). The fields $\boldsymbol{E}$ and $\boldsymbol{B}$ in (2.1.1) correspond to the operators $\hat{\boldsymbol{E}}$ and $\hat{\boldsymbol{B}}$, averaged by means of a statistical matrix.

As $T\to 0$, the tensor $\varepsilon_{ij}(\omega,\boldsymbol{k}) \approx \varepsilon_{ij}^0(\omega,\boldsymbol{k})$ takes the following form (the superscript $n=0$ refers to the ground state of the crystal):

$$\varepsilon_{ij}(\omega,\boldsymbol{k}) = \left(1-\frac{4\pi e^2 N_0}{m\omega^2}\right)\delta_{ij} - \frac{4\pi c^2}{\hbar\omega^2 V}\sum_{m\neq 0}\left[\frac{M_{0m}^i(-\boldsymbol{k})M_{m0}^j(\boldsymbol{k})}{\omega-\omega_m} - \frac{M_{m0}^i(-\boldsymbol{k})M_{0m}^j(\boldsymbol{k})}{\omega+\omega_m}\right]. \tag{6.1.9}$$

Before we discuss the properties of $\varepsilon_{ij}(\omega,\boldsymbol{k})$ which are due to the presence of exciton states, it is appropriate to point out that a procedure, similar to the one described above, can be used to determine the tensor $\varepsilon_{\perp,ij}(\omega,\boldsymbol{k})$, see (2.2.63). For normal waves the latter vector enables us to express the polarization vector $\boldsymbol{P} = (1/4\pi)(\boldsymbol{D}-\boldsymbol{E})$ in terms of the transverse part of the electric field strength alone. For this purpose we should choose as the "unperturbed" states the states that correspond to the Coulomb excitons obtained when the Coulomb interaction is fully taken into account, but retardation is neglected (vortex field). Then the perturbation is due only to the transverse part of the electric field strength. Consequently, for $\varepsilon_{\perp,ij}(\omega,\boldsymbol{k})$ we obtain

$$\varepsilon_{\perp,ij}^{(n)}(\omega,\boldsymbol{k}) = \delta_{ij} + 4\pi\chi_{\perp,ij}^{(n)}, \tag{6.1.8a}$$

where the tensor

$$\chi_{\perp,ij}^{(n)} \equiv b_{ij}^{(n)}\eta_{ij}, \tag{6.1.8b}$$

and

$$b_{ij}^{(n)}(\omega,\boldsymbol{k}) = -\frac{1}{\omega^2 V}\sum_\alpha \frac{e_\alpha^2}{m_\alpha}\delta_{ij} - \frac{c^2}{\hbar\omega^2 V}$$

$$\times \sum_{m \neq n}\left[\frac{M_{nm}^i(-\boldsymbol{k})M_{mn}^j(\boldsymbol{k})}{\omega-\Omega_m+\Omega_n} - \frac{M_{mn}^i(-\boldsymbol{k})M_{nm}^j(\boldsymbol{k})}{\omega+\Omega_m-\Omega_n}\right]. \quad (6.1.8\mathrm{c})$$

In this expression Ω_m are the frequencies of the Coulomb excitons and the matrix elements $\boldsymbol{M}_{mn}(\pm\boldsymbol{k})$ are constructed on the wave functions of these excitons.

It was shown in Sect. 2.2.4 how the tensor $\varepsilon_{\perp,ij}(\omega,\boldsymbol{k})$ can be determined if the tensor $\varepsilon_{ij}(\omega,\boldsymbol{k})$ is known, see (2.2.58 and 59). This procedure is unambiguous. Also of interest, in this connection, is whether it is possible to reconstruct the tensor $\varepsilon_{ij}(\omega,\boldsymbol{k})$ if the tensor $\varepsilon_{\perp,ij}(\omega,\boldsymbol{k})$ is known. In [6.5] the tensor ε_{ij} is reconstructed in the following way. Since for normal waves, where $D_i s_i = 0$,

$$\eta_{il}E_l = E_i - E_{\parallel,i} = E_i + 4\pi s_i(\boldsymbol{P}\cdot\boldsymbol{s}),$$

we can make use of (6.1.8a) to obtain the relation

$$D_i \equiv E_i + 4\pi P_i = E_i + 4\pi b_{il}^{(n)}[E_l + 4\pi s_l(\boldsymbol{P}\cdot\boldsymbol{s})],$$

from which

$$4\pi P_i = 4\pi d_{il}^{-1} b_{lj}^{(n)} E_j, \qquad d_{ij} = \delta_{ij} - 4\pi b_{im}^{(n)} s_m s_j .$$

If we postulate that this relation is also valid at $D_i s_i \neq 0$, we find that

$$\varepsilon_{ij}^{(n)}(\omega,\boldsymbol{k}) = \delta_{ij} + 4\pi d_{il}^{-1} b_{lj}^{(n)} . \quad (6.1.8\mathrm{d})$$

This procedure, however, demands proof, since, as will be shown in the following, there are many tensors of the form (6.1.8c) to which the same tensor $\varepsilon_{\perp,ij}^{(n)}$ corresponds. Actually, since the product of the tensor $s_i s_j$ and tensor η_{ij} equals zero, i.e.

$$s_i s_l \eta_{lj} = s_i s_l(\delta_{lj} - s_l s_j) = s_i s_j - s_i s_j = 0,$$

the tensor (6.1.8a) remains unchanged if $b_{ij}^{(n)}$ is replaced by $b_{ij}^{(n)} + \alpha_{il} s_l s_j$, where α_{il} is an arbitrary vector. At the same time, the tensor of type (6.1.8d) assumes the following form:

$$\begin{aligned} \varepsilon_{ij}^{(n)} &= \delta_{ij} + 4\pi \tilde{d}_{il}^{-1}(b_{lj}^{(n)} + a_{lm}s_m s_j), \\ \tilde{d}_{ij} &= \delta_{ij} - 4\pi b_{im}^{(n)} s_m s_j - 4\pi\alpha_{im} s_m s_j \end{aligned} \quad (6.1.8\mathrm{e})$$

and is no longer equal to the tensor (6.1.8 d), although they both correspond, insofar as their origin is concerned, to the tensor (6.1.8 a). To confirm the validity of this statement it is sufficient, for instance, to put $b_{ij}^{(n)} = 0$ after which we can readily see that the tensor $\varepsilon_{ij}^{(n)} - \delta_{ij}$, obtained from (6.1.8 e), is not identically equal to zero. Hence, the same tensor (6.1.8 a) corresponds to all tensors of the form of (6.1.8 d and e). Consequently, the problem of deciding which of the tensors [of (5.1.8 d or e)] is the "true" one cannot be solved within the framework of the approach described in [6.5].

For an isotropic medium, as well as for an isolated nondegenerate transition, we were able to show (using the results of [6.6], based on the application of Green's functions) that (6.1.8 d) is correct to terms of the order of k^2 if we assume the absence of absorption. In the more general case, however, the connection between (6.1.8 d) and the true tensor $\varepsilon_{ij}(\omega, \boldsymbol{k})$ has not yet been clarified.

Perturbation theory was applied in deriving the expression for the tensor $\varepsilon_{ij}^{(n)}(\omega, \boldsymbol{k})$. But in some cases this technique requires generalization (in its application, for example, the effect of the long-wavelength absorption edge, see Sect. 6.3.2, is not accounted for). Moreover, for a number of systems, methods have been developed for finding the wave functions and the energies of Coulomb excitons rather than those of mechanical excitons used in deriving (6.1.9). Therefore, a calculation that would enable us to determine the dielectric-constant tensor on the basis of a knowledge of the Coulomb excitons is undoubtedly of interest.

As shown in [6.6], a method based on Green's functions [6.7 – 9] can be efficiently applied for this purpose. Here, we have neither the space for nor the intention of discussing this method in any detail. However, we should dwell, at least briefly, on this question[1].

The main function to be determined is the retarded Green function $D_{ij}^E(\boldsymbol{r}, t; \boldsymbol{r}', t')$ whose Fourier components, given by

$$D_{ij}^E(\boldsymbol{r}, t; \boldsymbol{r}', t') = \frac{V}{(2\pi)^4} \int D_{ij}^E(\boldsymbol{k}, \omega) \mathrm{e}^{\mathrm{i}\boldsymbol{k}\cdot(\boldsymbol{r}-\boldsymbol{r}') - \mathrm{i}\omega(t-t')} d\omega\, d\boldsymbol{k} ,$$

satisfy the equation

$$\left[\varepsilon_{il}(\omega, \boldsymbol{k}) - \frac{c^2 k^2}{\omega^2} \delta_{il} + \frac{c^2 k_i k_l}{\omega^2}\right] D_{lj}^E(\boldsymbol{k}, \omega) = 4\pi \delta_{ij} .$$

From this it is clear that the dielectric-constant tensor

$$\varepsilon_{ij}(\omega, \boldsymbol{k}) = \frac{c^2 k^2}{\omega^2} \eta_{ij} + 4\pi [D^E(\boldsymbol{k}, \omega)]_{ij}^{-1} .$$

[1] We point out that the method involving Green's functions was also used to calculate the dielectric-constant tensor in [6.10, 11].

Thus the calculation of $\varepsilon_{ij}(\omega, \boldsymbol{k})$ is reduced to the calculation of $D^E_{ij}(\boldsymbol{k}, \omega)$. As shown in [6.8], the function $D^E_{ij}(\boldsymbol{r}, t)$ is determined by

$$D^E_{ij}(\boldsymbol{r}, t; \boldsymbol{r}', t') = \mathrm{i}\langle[\hat{E}_i(\boldsymbol{r}, t)\hat{E}_j(\boldsymbol{r}', t')]\rangle\theta(t-t') + 4\pi\delta_{ij}\delta(\boldsymbol{r}-\boldsymbol{r}')\delta(t-t'),$$

where $\hat{E}(\boldsymbol{r}, t)$ is the electric field strength operator at the point $\boldsymbol{r}$ and at the time t; $\theta(t) = 0$ if $t < 0$ and $\theta(t) = 1$ if $t > 0$. The expression $\langle[\hat{A}\hat{B}]\rangle$ indicates statistical averaging of the commutator of two operators:

$$[\hat{A}\hat{B}] \equiv \hat{A}\hat{B} - \hat{B}\hat{A}, \qquad \langle[AB]\rangle \equiv \mathrm{Tr}\{\mathrm{e}^{(F-\hat{H})/kT}[\hat{A}\hat{B}]\},$$

where $\hat{H}$ is the total Hamiltonian of the system, including, among others, the retardation interaction between the charges along with the Coulomb interaction, F is the free energy of the system and T is its temperature.

Functions of the form $\chi\Psi_{n0}$, where Ψ_{n0} is the eigenfunction of the crystal ($H_{\mathrm{Coul}}\Psi_{n0} = W_{n0}\Psi_{n0}$) and χ is the wave function of the photon field, can be used as the complete system of wave functions required to find the sum of diagonal elements of a certain operator $\hat{C}$, i.e., its trace $\mathrm{Tr}\{\hat{C}\}$. If a Coulomb-gauged potential is used, the total Hamiltonian of the system is $\hat{H} = \hat{H}_{\mathrm{Coul}} + \hat{H}_{\perp} + \hat{H}_{\mathrm{int}}$, where $\hat{H}_{\mathrm{Coul}}$ takes account of the instantaneous Coulomb interaction between the particles, $\hat{H}_{\perp}$ is the Hamiltonian corresponding to the free field of the transverse photons, and $\hat{H}_{\mathrm{int}}$ describes the interaction between the charges and the transverse photon field. Here the functions Ψ_{n0} and the eigenvalues of W_{n0} in the exciton region of the spectrum correspond to Coulomb excitons, whereas the functions χ are eigenfunctions of the operator $\hat{H}_{\perp}$. In operating with functions of the type $\chi\Psi_{n0}$, the trace $\mathrm{Tr}\{\exp[(F-\hat{H})/kT][\hat{E}_i(\boldsymbol{r}, t), \hat{E}_j(\boldsymbol{r}', t')]\}$ evidently consists of an infinite number of terms, even if $T = 0$, because the eigenfunctions of the operator $\hat{H}$ do not coincide with $\chi\Psi_{n0}$. Hence, in determining $D^E_{ij}(\boldsymbol{r}, t; \boldsymbol{r}', t')$, it proves much more convenient to use the eigenfunctions of the crystal in taking retardation into account, i.e., the eigenfunctions of the operator $\hat{H}$. This, specifically, is the case in which the method applied in [6.6] is found to be effective. We shall return to this subject in Sect. 6.3.2, in connection with the problem of damping.

Here we restrict ourselves to one note. In all the above expressions for $\varepsilon_{ij}(\omega, k)$ [e.g., (6.1.7c, 8, 9)] a nonphysical pole at $\omega = 0$ is present. In Sect. 6.1.2 it will be shown how this pole can be eliminated (in dipole approximation). If the Goepert-Mayer [6.12a] transformation is used for the perturbing electromagnetic fields, the form of perturbation (6.1.1) changes due to a transition from the vector potential to the fields and in the expression for $\varepsilon_{ij}(\omega, k)$ no pole arises at $\omega = 0$ [6.12b].

6.1.2 Contribution of the Exciton States to the Tensor $\varepsilon_{ij}(\omega, \boldsymbol{k})$

When a dielectric is in the ground state, the quasimomentum equals zero because the wave function of this state is invariant with respect to all symmetry operations of the crystal, including operations of the translation subgroup. The exciton states are, as is known, characterized by a single continuous quantum number, the quasimomentum $\boldsymbol{q}$ and furthermore by a set of discrete quantum numbers s. Hence, for the exciton states, we have

$$\Psi_{m0}(\boldsymbol{r}_1, \boldsymbol{r}_2, \ldots) \equiv \Psi_{qs}(\boldsymbol{r}_1, \boldsymbol{r}_2, \ldots),$$

where $T_a \Psi_{qs} = \exp(-\mathrm{i}\boldsymbol{q}\cdot\boldsymbol{a})\, \Psi_{qs}$ according to (6.1.4). As already indicated in Sect. 6.1.1, the property of the operator $\hat{\boldsymbol{M}}(\boldsymbol{k})$ expressed by (6.1.4a) leads us to the conclusion that the matrix element $M_{mn}(\boldsymbol{k}) \neq 0$, provided that $\boldsymbol{q}_m = \boldsymbol{q}_n - \boldsymbol{k} + 2\pi \boldsymbol{b}$ and, analogously, $M_{nm}(\boldsymbol{k}) \neq 0$ if $\boldsymbol{q}_n = \boldsymbol{q}_m - \boldsymbol{k} + 2\pi\boldsymbol{b}$, $\boldsymbol{b}$ being an integral-valued vector of the reciprocal lattice. Therefore, since the ground state of the crystal $n = 0$ corresponds to $\boldsymbol{q}_n = 0$, it is clear that with the vector $\boldsymbol{q}_n$ selected within a basic cell of the reciprocal lattice, we obtain for small $\boldsymbol{k}$

$$M^i_{m0}(\boldsymbol{k}) \equiv M^i_{qs;0}(\boldsymbol{k}) = M^i_{-ks;0}(\boldsymbol{k})\, \delta_{q+k}\,, \tag{6.1.10}$$

$$M^i_{0m}(\boldsymbol{k}) \equiv M^i_{0;qs}(\boldsymbol{k}) = M^i_{0;ks}(\boldsymbol{k})\, \delta_{k-q}\,. \tag{6.1.11}$$

In a perfect dielectric crystal, the states over which summation is carried out in (6.1.9) may correspond, in general, to the excitation of several mechanical excitons (among which there may be phonons, etc.) rather than to the excitation of a single one. In this case, the state m is characterized by several continuous quantum numbers, the quasimomenta of the various excitons, and (6.1.10, 11) are valid here as well, but $\boldsymbol{q}$ should be construed as the total quasimomentum in the state m.

The existence of an exciton-phonon coupling or any other interactions that change the states of the mechanical excitons, whose energies lie in the region of the spectrum that interests us, leads to the fact (for a more detailed discussion, see Sect. 6.3) that even for a crystal in the ground state the tensor $\varepsilon_{ij}(\omega, \boldsymbol{k})$, (6.1.9), becomes complex even when both ω and $\boldsymbol{k}$ are real. Since problems associated with the damping of electromagnetic waves are dealt with in more detail in Sect. 6.3, we shall assume here that absorption is absent, and the states m, over which summation is carried out in (6.1.9), are to be understood as states of the crystal in which only a single mechanical exciton is excited, $m \equiv (\boldsymbol{k}, s)$. Then, making use of (6.1.9, 11), we obtain

$$\varepsilon_{ij}(\omega, \boldsymbol{k}) = \delta_{ij}\left(1 - \frac{4\pi e^2 N_0}{m\omega^2}\right)$$

$$-\frac{4\pi c^2}{V\hbar\omega^2}\sum_s\left[\frac{M^i_{0;-ks}(-k)M^j_{-ks;0}(k)}{\omega-\omega_s(-k)}-\frac{M^i_{ks;0}(-k)M^j_{0;ks}(k)}{\omega+\omega_s(k)}\right]. \tag{6.1.12}$$

In the neighborhood of the isolated s-th absorption line, the variation of the tensor $\varepsilon_{ij}(\omega, \boldsymbol{k})$ as a function of the frequency ω is determined mainly by one of the resonance terms in the dispersion equation (6.1.12). In this case, with some approximation,

$$\varepsilon_{ij}(\omega,\boldsymbol{k})=\varepsilon^{(0)}_{ij}(\omega)$$

$$-\frac{4\pi c^2}{V\hbar\omega^2}\left[\frac{M^i_{0;-ks}(-k)M^j_{-ks;0}(k)}{\omega-\omega_s(-k)}-\frac{M^i_{ks;0}(-k)M^j_{0;ks}(k)}{\omega+\omega_s(k)}\right], \tag{6.1.12a}$$

where $\varepsilon^{(0)}_{ij}(\omega)$ is a smoothly varying function of ω in the frequency range under consideration. However, if the state of the mechanical exciton is degenerate at $\boldsymbol{k}=0$, the number of resonance terms increases accordingly.

Note that for exciton states $(\boldsymbol{k}, s)$, whose wave functions are transformed at $\boldsymbol{k}=0$ like the components of a polar vector, the matrix elements $M^i_{0s;0}(0)$, where $i=1,2,3$, are nonzero (at least some of them are). It is clear that only these exciton states (which we previously called dipole exciton states) contribute to the dielectric-constant tensor when spatial dispersion is neglected. Since, in this case,

$$M^j_{0s;0}(0)=\mathrm{i}\frac{\omega_s(0)}{c}\sqrt{N}D^j_{0s;0}, \tag{6.1.12b}$$

where $\boldsymbol{D}_{0s;0}$ is the matrix element of the operator of the dipole moment of a unit cell in the crystal, and N is the number of unit cells in the basic volume of the crystal,

$$\varepsilon_{ij}(\omega,0)=\left(1-\frac{4\pi e^2N_0}{m\omega^2}\right)\delta_{ij}-\frac{8\pi}{v\hbar\omega^2}\sum_s\frac{D^i_{0;0s}(0)D^j_{0s;0}(0)}{\omega^2-\omega_s^2(0)}\,\omega_s^3(0), \tag{6.1.13}$$

with v denoting the volume of a unit cell of the crystal. Accordingly, instead of (6.1.12a), we have in the case under consideration

$$\varepsilon_{ij}(\omega,0)=\varepsilon^{(0)}_{ij}(\omega)-\frac{8\pi}{v\hbar\omega^2}\,\frac{D^i_{0;0s}(0)D^j_{0s;0}(0)}{\omega^2-\omega_s^2(0)}\,\omega_s^3(0). \tag{6.1.13a}$$

The pole at $\omega=0$ in (6.1.13) is not actual, so can easily be excluded. In fact, taking into consideration that

$$\frac{\omega_s^2(0)}{\omega^2[\omega^2-\omega_s^2(0)]}=\frac{1}{\omega^2-\omega_s^2(0)}-\frac{1}{\omega^2}$$

and that [6.15]

$$\frac{4\pi e^2 N_0}{m}\delta_{ij}=\frac{8\pi}{v\hbar}\sum_s D^i_{0;0s}(\boldsymbol{k})D^j_{0s;0}(\boldsymbol{k})\,\omega_s(\boldsymbol{k})\,, \tag{6.1.14a}$$

we obtain

$$\varepsilon_{ij}(\omega,0)=\delta_{ij}-\frac{8\pi}{v\hbar}\sum_s\frac{D^i_{0;0s}(0)\,D^j_{0s;0}(0)}{\omega^2-\omega_s^2(0)}\,\omega_s(0) \tag{6.1.14b}$$

instead of (6.1.13). In cubic crystals all states $(0s)$ for which $\boldsymbol{D}_{0s;0}(0)\neq 0$ are triply degenerate $(s=s_1,s_2,s_3)$ and, if the wave functions of these states are chosen so that each one is transformed like one of the components of a polar vector (x or y or z), then

$$D^i_{0;0s_\alpha}(0)=D^1_{0;0s_1}(0)\,\delta_{i\alpha}\,.$$

Consequently

$$\sum_\alpha D^i_{0;0s_\alpha}(0)\,D^j_{0s_\alpha;0}(0)=|D^1_{0;0s_1}(0)|^2\delta_{ij}$$

and, taking into account that $\omega_{s_1}(0)=\omega_{s_2}(0)=\omega_{s_3}(0)$, we obtain for a cubic crystal

$$\varepsilon_{ij}(\omega,0)=\left[1-\frac{8\pi}{3v\hbar}\sum_s\frac{|D^1_{0;0s}(0)|^2\,\omega_s(0)}{\omega^2-\omega_s^2(0)}\right]\delta_{ij} \tag{6.1.14c}$$

instead of the general equation (6.1.14b). Equation (6.1.14b) can be transformed in a similar way for crystals of lower symmetry.

Along with the dipole exciton states, it proves convenient to separate out of the exciton states the quadrupolar excitons, whose wave functions at $\boldsymbol{q}=0$ are transformed like the product of the components of two polar vectors. It can be shown in a way similar to the one used in [6.13, 14] that for such states, in a first approximation with respect to $\boldsymbol{k}$, we obtain

$$M^i_{ks;0}(\boldsymbol{k})=-\mathrm{i}\frac{e}{2mc}\sum_{l=1}^{3}\langle 0,s|(p_i^\alpha r_l^\alpha+r_l^\alpha p_i^\alpha)|0\rangle k_l\,. \tag{6.1.15}$$

The difference between (6.1.15) and the analogous equation obtained in [6.13, 14] is that (6.1.15) was written for the wave functions of mechanical excitons, not of Coulomb excitons. In general, the wave functions of the latter depend nonanalytically on the quasimomentum at small quasimomenta, a fact that was neglected in [6.13, 14].

Using the notation of (4.3.24), we can see that

$$M^{i}_{0;\,-ks}(-\boldsymbol{k})M^{j}_{-ks;0}(\boldsymbol{k})$$

$$= -\frac{e^2}{4m^2c^2}\sum_{l,m=1}^{3}\langle 0|\hat{T}_{il}|0,s\rangle\langle 0,s|\hat{T}_{jm}|0\rangle k_l k_m$$

$$\equiv -\frac{e^2}{8m^2c^2}\sum_{l,m=1}^{3}(\langle 0|\hat{T}_{il}|0,s\rangle\langle 0,s|\hat{T}_{jm}|0\rangle$$

$$+\langle 0|\hat{T}_{im}|0,s\rangle\langle 0,s|\hat{T}_{jl}|0\rangle)k_l k_m\,.$$

If the frequency $\omega\approx\omega_s(0)$, the main contribution to the summation of (6.1.12) is due to the term proportional to $[\omega-\omega_s(0)]^{-1}$. Since the exciton term may be degenerate ($s=s_1,s_2,\dots$), we obtain (4.3.23), which has been previously used.

a) Refractive Index Near Exciton Resonances of Anisotropic Crystals

In conclusion, we point out that if we neglect spatial dispersion and damping, the tensor $\varepsilon_{\perp,ij}(\omega,\boldsymbol{k})$, see (6.1.8a), can be written for $T=0$, similar to (6.1.14b), in the following form

$$\varepsilon_{\perp,ij}(\omega,0)=\delta_{ij}-\frac{8\pi}{v\hbar}\sum_s\frac{\Omega_s(0)D^{i}_{0;0s}(0)D^{l}_{0s;0}(0)}{\omega^2-\Omega_s^2(0)}\eta_{lj}\,. \tag{6.1.16}$$

Here $(0s)$ denotes the state of the s-th Coulomb (not mechanical) exciton at $\boldsymbol{k}=0$, and $\Omega_s(0)$ is the frequency of this Coulomb exciton. With the aid of (6.1.16) we can readily obtain a convenient formula for the refractive index of light waves in a crystal. As a matter of fact, for normal electromagnetic waves

$$D_i=\eta_{ij}D_j=\eta_{ij}\varepsilon_{\perp,jl}(\omega,0)E_l=n^2\eta_{ij}E_j\,. \tag{6.1.17}$$

Hence, on the basis of (6.1.16) and since $\eta_{ij}=\eta_{il}\eta_{lj}$, we find that (6.1.17) can be written in the form

$$\eta_{il}\varepsilon_{\perp,lm}(\omega,0)\,\eta_{mn}\eta_{nj}E_j=n^2\eta_{ij}E_j\,. \tag{6.1.18}$$

Next we direct the z axis along the vector $\boldsymbol{s}$, and the x and y axes along the unit vectors $\boldsymbol{e}^{\alpha}$ (where $\alpha=1,2$), $\boldsymbol{s}\cdot\boldsymbol{e}^{\alpha}=0$ and $\boldsymbol{e}^{(1)}\cdot\boldsymbol{e}^{(2)}=0$. Then, evidently,

$$\eta_{3l}\varepsilon_{\perp,lm}(\omega,0)\,\eta_{mi} = \eta_{il}\varepsilon_{\perp,lm}(\omega,0)\,\eta_{m3} = 0\,, \qquad i = 1,2,3\,,$$

$$\eta_{\alpha l}\varepsilon_{\perp,lm}(\omega,0)\,\eta_{m\beta} = \delta_{\alpha\beta} - \frac{8\pi}{v\hbar}\sum_s \frac{\Omega_s(0)(\boldsymbol{e}^\alpha\cdot \boldsymbol{D}_{0;0s}(0))(\boldsymbol{D}_{0s;0}(0)\cdot \boldsymbol{e}^\beta)}{\omega^2 - \Omega_s^2(0)}\,, \tag{6.1.19}$$

$$\alpha,\ \beta = 1,2\,.$$

It follows from (6.1.18) that the equation for determining $n^2(\omega,s)$ has the following form

$$\begin{vmatrix} n^2 - \eta_{1l}\varepsilon_{\perp,lm}\eta_{m1}\,, & -\eta_{1l}\varepsilon_{\perp,lm}\eta_{m2} \\ -\eta_{2l}\varepsilon_{\perp,lm}\eta_{m1}\,, & n^2 - \eta_{2l}\varepsilon_{\perp,lm}\eta_{m2} \end{vmatrix} = 0\,,$$

from which [6.15]

$$\begin{aligned} n_{1,2}^2(\omega,s) = 1 &- \frac{1}{2}\sum_s \frac{\omega_0^2 f_s(s)\sin^2\varphi(s,\boldsymbol{k})}{\omega^2 - \Omega_s^2(0)} \\ &\pm \frac{1}{2}\left[\left(\sum_s \frac{\omega_0^2 f_s(s)[\cos^2\varphi_1(s,\boldsymbol{k}) - \cos^2\varphi_2(s,\boldsymbol{k})]}{\omega^2 - \Omega_s^2(0)}\right)^2\right. \\ &\left. + 4\left(\sum_s \frac{\omega_0^2 f_s(s)\cos\varphi_1(s,\boldsymbol{k})\cos\varphi_2(s,\boldsymbol{k})}{\omega^2 - \Omega_s^2(0)}\right)^2\right]^{1/2}. \end{aligned} \tag{6.1.20}$$

This expression uses the following notation: ω_0^2 is the square of the plasma frequency, f_s is the oscillator strength corresponding to the s-th exciton band, $\varphi(s,\boldsymbol{k})$ is the angle between the vector $\boldsymbol{k}$ and vector $\boldsymbol{D}_{0s;0}(0)$, and $\varphi_\alpha(s,\boldsymbol{k})$ (with $\alpha = 1,2$) is the angle between the unit vector $\boldsymbol{e}^\alpha$ and vector $\boldsymbol{D}_{0s;0}(0)$. By definition

$$f_s(s) = \frac{1}{\omega_0^2}\frac{8\pi}{v\hbar}|\boldsymbol{D}_{0s;s}(0)|^2\,\Omega_s(0)\,.$$

Equation (6.1.20) enables us to analyze the dependence of the quantities $n_{1,2}(\omega,s)$ on the direction of light propagation in the crystal and may be used to determine the direction of the transition vector $\boldsymbol{D}_{0;0s}$ in an experimental investigation of light dispersion outside of the absorption bands.

6.2 Mechanical Excitons and the Tensor $\varepsilon_{ij}(\omega, \boldsymbol{k})$ in Molecular Crystals and in the Classical Oscillator Model

6.2.1 Molecular Crystals. Mechanical Excitons

In order to calculate the tensor $\varepsilon_{ij}(\omega, \boldsymbol{k})$ by the methods described in Sect. 6.1, it is necessary that the wave functions Ψ_{m0} and the energy eigenvalues W_m for mechanical excitons be known. According to the discussion in Sect. 2.2.2, the excited states, called mechanical excitons, correspond to a solution of the problem when the action of the macroscopic (long-wavelength) field is neglected or there is no such field.

Let us expound the theory of excitons for the case of molecular crystals in the approximation of fixed molecules. Note that the Heitler-London approximation can be used for this purpose [6.16] only if the molecular interaction is sufficiently weak. Actually, a number of molecular crystals exist in which the oscillator strengths are large and the molecular interaction in many spectral regions cannot be said to be weak (though this interaction does not violate the neutrality of the individual molecules, it causes considerable mixing of the electronic configurations). A relevant theory, not based on the Heitler-London approximation, was developed in [6.17]. Here, for the sake of simplicity, we shall apply the Heitler-London method because the results obtained in the following can be refined in a way similar to that described in [6.17].

Assume that a unit cell of the crystal contains σ molecules. The Hamiltonian of the complete Coulomb problem is

$$\hat{H} = \sum_{n\alpha} \hat{H}_{n\alpha} + \frac{1}{2} \sum_{n\alpha \neq m\beta} \hat{V}_{n\alpha, m\beta}, \tag{6.2.1}$$

where $\boldsymbol{n}$ and $\boldsymbol{m}$ are integral-valued vectors of the crystal lattice; $\alpha, \beta = 1, 2, 3, \ldots, \sigma$; $\hat{H}_{n\alpha}$ is the Hamiltonian of the molecule $(\boldsymbol{n}\alpha)$; and $V_{n\alpha, m\beta}$ is the Coulomb interaction operator between the molecules $\boldsymbol{n}\alpha$ and $\boldsymbol{m}\beta$. In deriving the wave functions of the ground state and the excited states of the crystal, we shall not account for the intermolecular exchange of electrons because this exchange does not play any appreciable role in the region of the lower excited states. Let us introduce the normalized functions $\varphi^l_{n\alpha}$, which are antisymmetric with respect to all coordinates of the electrons:

$$\hat{H}_{n\alpha} \varphi^l_{n\alpha} = W^0_l \varphi^l_{n\alpha} \tag{6.2.2}$$

(here $l = 0$ corresponds to the ground state of the molecule; in the case when the molecular term is degenerate, the superscript l should be regarded as being complex: $l \to (l, r)$, $r = 1, 2, \ldots, t$, where t is the degeneracy).

In the Heitler-London approximation, the wave function of the ground state of the crystal is

$$\Phi_0 = \prod_{n\alpha} \varphi^0_{n\alpha}, \tag{6.2.3}$$

and that of an excited state is

$$\Phi^l = \sum_{n\alpha, r} a^{lr}_{n\alpha} \chi^{lr}_{n\alpha}, \qquad \text{where} \tag{6.2.4}$$

$$\chi^{lr}_{n\alpha} = \varphi^{lr}_{n\alpha} \prod_{m\beta \neq n\alpha} \varphi^0_{m\beta}. \tag{6.2.5}$$

The energy of the ground state is $W_0 = \langle \Phi_0^* | \hat{H} | \Phi_0 \rangle$. To determine the energy of the excited states, it is necessary to find the minimum of $H = \langle \Phi^l | \hat{H} | \Phi^l \rangle$ with respect to the set $\{a^{lr}_{n\alpha}\}$ under the additional condition

$$\sum_{n\alpha, r_{lr}} |a^{lr}_{n\alpha}|^2 = 1 .$$

In minimizing we find that the quantities $a^{lr}_{n\alpha}$ satisfy the system of equations

$$\sum_{\substack{m\beta \neq n\alpha \\ r'}} M^{l;rr'}_{m\beta, n\alpha} a^{lr'}_{m\beta} - \mu^l a^{lr}_{n\alpha} = 0 , \qquad \mu^l = W - \mathscr{D}_l - \Delta_l , \tag{6.2.6}$$

where W is the energy of the excited state of the crystal, $\Delta_l \equiv W^0_l - W^0_0$ is the excitation energy of an isolated molecule,

$$M^{l;rr'}_{m\beta, n\alpha} = \int \varphi^{*\,lr}_{n\alpha} \varphi^{*\,0}_{m\beta} \hat{V}_{n\alpha, m\beta} \varphi^0_{n\alpha} \varphi^{lr'}_{m\beta} d\tau , \tag{6.2.7}$$

and the quantity $\mathscr{D}_l$ is equal to the change in the interaction energy between the molecule and its surroundings due to transition of the molecule from the ground to the excited state (see also footnote [3] on page 349).

It is clear from considerations of translational symmetry that

$$a^{lr}_{n\alpha} = \frac{1}{\sqrt{N}} a^{lr}_{\alpha}(\boldsymbol{k}) \, \mathrm{e}^{\mathrm{i}\boldsymbol{k} \cdot \boldsymbol{r}_{n\alpha}} , \tag{6.2.8}$$

where N is the total number of unit cells in the basic volume of the crystal. Hence, instead of (6.2.6), we obtain

$$\sum_{\substack{\beta \neq \alpha \\ r'}}{}' \Gamma^{l;rr'}_{\beta\alpha}(\boldsymbol{k}) a^{lr'}_{\beta}(\boldsymbol{k}) - \mu^l a^{lr}_{\alpha} = 0 , \tag{6.2.9}$$

where the prime of the summation sign indicates that the term $\boldsymbol{n} = \boldsymbol{m}$, $\alpha = \beta$ is omitted,

$$\Gamma^{l;rr'}_{\beta\alpha}(\boldsymbol{k}) = \sum_{m}{}' M^{l;rr'}_{m\beta, n\alpha} \exp[\mathrm{i}\boldsymbol{k} \cdot (\boldsymbol{r}_{m\beta} - \boldsymbol{r}_{n\alpha})] . \tag{6.2.10}$$

It follows from (6.2.9) that the excitation-energy values in the complete Coulomb problem $W = W_0 + \Delta_l + \mu^l$ are expressed in terms of μ^l, the eigenvalues of the Hermitian matrix $\hat{\Gamma}(\boldsymbol{k})$, and the quantities $a_\alpha^{lr}(\boldsymbol{k})$, where $\alpha = 1, 2, \ldots, \sigma$ and $r = 1, 2, \ldots, t$, are the components of its eigenvectors.

Considering now the problem of the mechanical exciton, we separate out of the interaction matrix $\hat{\Gamma}(\boldsymbol{k})$ the interaction that is due to the long-wavelength field (only the region of small values of $\boldsymbol{k}$ is actually of interest to us, because the long-wavelength field is separated out by using Ewald's procedure for these values of the wave vector). Moreover, for the sake of simplicity we restrict ourselves, in dealing with molecular interactions, to the dipole-dipole interaction which dominates in some interesting cases. In this approximation

$$\hat{V}_{n\alpha,m\beta} = \frac{1}{|\boldsymbol{r}_{n\alpha} - \boldsymbol{r}_{m\beta}|^3} \left[\hat{\boldsymbol{R}}_{n\alpha} \cdot \hat{\boldsymbol{R}}_{m\beta} - \frac{3}{|\boldsymbol{r}_{n\alpha} - \boldsymbol{r}_{m\beta}|^2} (\hat{\boldsymbol{R}}_{n\alpha}, \boldsymbol{r}_{n\alpha} - \boldsymbol{r}_{m\beta}) \times (\hat{\boldsymbol{R}}_{m\beta}, \boldsymbol{r}_{n\alpha} - \boldsymbol{r}_{m\beta}) \right], \tag{6.2.11}$$

where $\boldsymbol{R}_{n\alpha}$ is the operator of the dipole moment of the molecule $n\alpha$. Therefore, making use of (6.2.7, 10) we find that

$$\Gamma_{\beta,\alpha}^{l;rr'}(\boldsymbol{k}) = -\boldsymbol{R}_\alpha^{0;lr} \cdot \boldsymbol{E}_{n\alpha,\beta}^{lr';0}(\boldsymbol{k})\, \mathrm{e}^{-\mathrm{i}\boldsymbol{k}\cdot\boldsymbol{r}_{n\alpha}}, \tag{6.2.12}$$

where

$$\boldsymbol{R}_\alpha^{0;lr} = \int \varphi_{n\alpha}^{*lr} \hat{\boldsymbol{R}}_{n\alpha} \varphi_{n\alpha}^0 d\tau, \tag{6.2.13}$$

$$\boldsymbol{E}_{n\alpha,\beta}^{lr';0}(\boldsymbol{k}) = -\sum_m{}' \frac{\mathrm{e}^{\mathrm{i}\boldsymbol{k}\cdot\boldsymbol{r}_{m\beta}}}{|\boldsymbol{r}_{n\alpha} - \boldsymbol{r}_{m\beta}|^3} \times \left[\boldsymbol{R}_\beta^{0;lr'} - 3 \frac{\boldsymbol{r}_{n\alpha} - \boldsymbol{r}_{m\beta}}{|\boldsymbol{r}_{n\alpha} - \boldsymbol{r}_{m\beta}|^2} (\boldsymbol{R}_\beta^{0;lr'}, \boldsymbol{r}_{n\alpha} - \boldsymbol{r}_{m\beta}) \right]. \tag{6.2.14}$$

From the form of (6.2.14) it is clear that $\boldsymbol{E}_{n\alpha,\beta}^{lr';0}$ is the electric field at the point $\boldsymbol{r} = \boldsymbol{r}_{n\alpha}$, set up by the dipoles located at lattice sites of the β kind, and the magnitude of the dipoles varies from site to site according to the law

$$\boldsymbol{R}(m\beta) = \boldsymbol{R}_\beta^{lr';0} \mathrm{e}^{\mathrm{i}\boldsymbol{k}\cdot\boldsymbol{r}_{m\beta}}. \tag{6.2.15}$$

The total field (6.2.14) actually consists of two parts: the long-wavelength (macroscopic) field and the short-wavelength field. The method of dividing the total field in a dipole lattice into the macroscopic field and a short-wavelength field is what is known as the Ewald method, or procedure.

The operation of separating the macroscopic part from the total field (6.2.14) is described in detail in [Ref. 6.2, § 30]. Therefore we give here only the final result:

$$[E^{lr';0}_{n\alpha,\beta}(\boldsymbol{k})]_j = E_j \mathrm{e}^{\mathrm{i}\boldsymbol{k}\cdot\boldsymbol{r}_{n\alpha}} + \mathrm{e}^{\mathrm{i}\boldsymbol{k}\cdot\boldsymbol{r}_{n\alpha}} \sum_{j'} Q_{jj'}\begin{pmatrix}\boldsymbol{k}\\ \alpha\beta\end{pmatrix} R^{lr';0}_{\beta,j'}, \tag{6.2.16}$$

where E is the amplitude of the macroscopic part of the field:

$$E_j = -\frac{4\pi}{v}\frac{k_j}{|\boldsymbol{k}|}\left(\frac{\boldsymbol{k}}{|\boldsymbol{k}|}\sum_\alpha \boldsymbol{R}^{lr';0}_\alpha\right), \tag{6.2.17}$$

and the coefficients $Q_{jj'}\begin{pmatrix}\boldsymbol{k}\\ \alpha\beta\end{pmatrix}$ depend only on the cell structure for a given value of the wave vector $\boldsymbol{k}$ (their specific expression, not needed here, is determined by [Ref. 6.3, Eq. (30.31)]. It is essential that, as $\boldsymbol{k} \to 0$, the coefficients $Q_{jj'}\begin{pmatrix}\boldsymbol{k}\\ \alpha\beta\end{pmatrix}$ are analytic functions of the wave vector. Hence, the entire non-analyticity of the summation in (6.2.14) is concentrated in (6.2.17). The coefficients $Q_{jj'}\begin{pmatrix}\boldsymbol{k}\\ \alpha\beta\end{pmatrix}$ satisfy the following relations:

$$\begin{aligned}
Q_{jj'}\begin{pmatrix}\boldsymbol{k}\\ \alpha\beta\end{pmatrix} &= Q_{j'j}\begin{pmatrix}\boldsymbol{k}\\ \alpha\beta\end{pmatrix},\\
Q_{jj'}\begin{pmatrix}-\boldsymbol{k}\\ \alpha\beta\end{pmatrix} &= Q^*_{jj'}\begin{pmatrix}\boldsymbol{k}\\ \alpha\beta\end{pmatrix},\\
Q_{jj'}\begin{pmatrix}\boldsymbol{k}\\ \alpha\beta\end{pmatrix} &= Q^*_{jj'}\begin{pmatrix}\boldsymbol{k}\\ \beta\alpha\end{pmatrix}.
\end{aligned} \tag{6.2.18}$$

Omitting in (6.2.16) the first term at the right-hand side of the equation, we see that the interaction matrix for the mechanical exciton has the form:

$$\tilde{\Gamma}^{l;rr'}_{\beta\alpha}(\boldsymbol{k}) = -\sum_{jj'} Q_{jj'}\begin{pmatrix}\boldsymbol{k}\\ \alpha\beta\end{pmatrix} R^{0;lr}_{\alpha j} R^{lr';0}_{\beta j'}. \tag{6.2.19}$$

Making use of (6.2.18, 19), we readily find that the matrix $\tilde{\Gamma}(\boldsymbol{k})$, in the same way as matrix $\Gamma(\boldsymbol{k})$, is Hermitian:

$$\tilde{\Gamma}^{l;r'r}_{\beta\alpha}(\boldsymbol{k}) = \tilde{\Gamma}^{*l;rr'}_{\alpha\beta}(\boldsymbol{k}). \tag{6.2.19a}$$

It is clear that for mechanical exciton a relation of the form of (6.2.8) also exists, but, in contrast to (6.2.9), the amplitudes $\tilde{a}^{lr}_\alpha(\boldsymbol{k})$ satisfy the system of equations

$$\sum_{\substack{\beta \neq \alpha \\ r'}} \tilde{\Gamma}_{\beta\alpha}^{l';r'r}(\boldsymbol{k})\tilde{a}_{\beta}^{lr'}(\boldsymbol{k}) - \tilde{\mu}^{l}\tilde{a}_{\alpha}^{lr}(\boldsymbol{k}) = 0 \tag{6.2.20}$$
$$(\tilde{\mu}_l = \tilde{W} - \mathscr{D}_l - \Delta_l),$$

where $\tilde{W}$ is the energy of the mechanical exciton. For each value of $\boldsymbol{k}$, the operator $\tilde{\Gamma}(\boldsymbol{k})$ has σt eigenvalues $\tilde{\mu}_{\rho}^{l}(\boldsymbol{k})$, $\rho = 1, 2, \ldots, \sigma t$. The corresponding orthonormal eigenvectors will be denoted by $\{\tilde{a}_{\alpha}^{lr,\rho}\}$. It is clear that in the approximation under consideration the discrete quantum number determining the exciton state, like the continuous quantum number $\boldsymbol{k}$, is the complex quantity $s \equiv (l, \rho)$. Assume that at $\boldsymbol{k} = 0$ the quantity $\tilde{\mu}^{l}(0)$ corresponds to the nondegenerate state of the mechanical exciton. Then, owing to the analyticity of the matrix $\tilde{\Gamma}(\boldsymbol{k})$ as $\boldsymbol{k} \to 0$, the functions $\tilde{\mu}^{l}(\boldsymbol{k})$, as well as $\tilde{a}_{\infty}^{lr}(\boldsymbol{k})$, will be analytic functions of $\boldsymbol{k}$ as $\boldsymbol{k} \to 0$.

If the state of the mechanical exciton is degenerate at $\boldsymbol{k} = 0$, the function $\tilde{\mu}^{l}(\boldsymbol{k})$ can have a singular point (branch point) at $\boldsymbol{k} = 0$.

In crystals containing one molecule per unit cell, only a single exciton band corresponds to the l-th nondegenerate molecular term ($\sigma = 1$, $t = 1$). In this case

$$\tilde{\mu}^{l}(\boldsymbol{k}) = \tilde{\Gamma}_{11}^{l}(\boldsymbol{k}), \qquad \tilde{a}^{l}(\boldsymbol{k}) = 1 .$$

In more complex crystals, for instance those consisting of two molecules per unit cell, the quantities $\tilde{\mu}_{\rho}^{l}(\boldsymbol{k})$ are determined as the roots of the equation (assuming the molecular term to be nondegenerate)

$$[\tilde{\Gamma}_{11}^{l}(\boldsymbol{k}) - \tilde{\mu}^{l}][\tilde{\Gamma}_{22}^{l}(\boldsymbol{k}) - \tilde{\mu}^{l}] - \tilde{\Gamma}_{12}^{l}(\boldsymbol{k})\tilde{\Gamma}_{21}^{l}(\boldsymbol{k}) = 0 . \tag{6.2.21}$$

The quantity $\Gamma_{11}^{l}(\boldsymbol{k})$ depends only on the interaction of molecules with $\alpha = 1$, and quantity $\Gamma_{22}^{l}(\boldsymbol{k})$ on the interaction of molecules with $\alpha = 2$. Since at $\boldsymbol{k} = 0$ these quantities are independent of the direction $\boldsymbol{k}/k$, it is readily apparent that $\Gamma_{11}^{l}(0) = \Gamma_{22}^{l}(0)$ (in contrast to $\Gamma_{11}^{l}(\boldsymbol{k})|_{k\to 0} \neq \Gamma_{22}^{l}(\boldsymbol{k})|_{k\to 0}$). Hence, two states of mechanical excitons correspond to the molecular term at $\boldsymbol{k} = 0$ [6.16]. For these states

$$\begin{aligned}
\tilde{\mu}_{\mathrm{I}}^{l}(0) &= \tilde{\Gamma}_{11}(0) + |\tilde{\Gamma}_{12}(0)|, \\
\tilde{a}_{1}^{l}(0,\mathrm{I}) &= \frac{\tilde{\Gamma}_{12}(0)}{|\tilde{\Gamma}_{12}(0)|}\tilde{a}_{2}^{l}(0,\mathrm{I}) = \frac{1}{\sqrt{2}}, \\
\tilde{\mu}_{\mathrm{II}}^{l}(0) &= \tilde{\Gamma}_{11}(0) - |\tilde{\Gamma}_{12}(0)|, \\
\tilde{a}_{1}^{l}(0,\mathrm{II}) &= -\frac{\tilde{\Gamma}_{12}(0)}{|\tilde{\Gamma}_{12}(0)|}\tilde{a}_{2}^{l}(0,\mathrm{II}) = \frac{1}{\sqrt{2}} .
\end{aligned} \tag{6.2.22}$$

A knowledge of the wave functions of the mechanical excitons enables us to present directly the quantities appearing in the expression for the dielectric-constant tensor in terms of the characteristics of the individual molecules. Using (6.2.3, 4), as well as (6.1.1a), we find that the quantities appearing in (6.1.12) for the tensor $\varepsilon_{ij}(\omega,\boldsymbol{k})$ are determined in the following manner for molecular crystals:

$$\begin{aligned} \frac{1}{\sqrt{V}} M^{i}_{-ks;0}(\boldsymbol{k}) &= \frac{e/2mc}{\sqrt{v}} \sum_{\alpha,r} a^{*lr;\rho}_{\alpha}(-\boldsymbol{k}) \\ &\quad\times \int \varphi^{*lr}_{n\alpha} \sum_{\nu\in n\alpha} (p^{\nu}_{i} \mathrm{e}^{i\boldsymbol{k}\cdot\boldsymbol{\rho}_{\nu}} + \mathrm{e}^{i\boldsymbol{k}\cdot\boldsymbol{\rho}_{\nu}} p^{\nu}_{i})\, \varphi^{0}_{n\alpha} d\tau\,, \\ \frac{1}{\sqrt{V}} M^{i}_{0;-ks}(-\boldsymbol{k}) &= \frac{e/2mc}{\sqrt{v}} \sum_{\alpha,r} a^{lr;\rho}_{\alpha}(-\boldsymbol{k}) \\ &\quad\times \int \varphi^{*0}_{n\alpha} \sum_{\nu\in n\alpha} (p^{\nu}_{i} \mathrm{e}^{-i\boldsymbol{k}\cdot\boldsymbol{\rho}_{\nu}} + \mathrm{e}^{-i\boldsymbol{k}\cdot\boldsymbol{\rho}_{\nu}} p^{\nu}_{i})\, \varphi^{lr}_{n\alpha} d\tau\,, \end{aligned} \tag{6.2.23}$$

where ν labels the electron pertaining to the molecule $n\alpha$, and $\boldsymbol{\rho}_{\nu}$ is the radius vector of this electron with respect to the lattice site $\boldsymbol{r}_{n\alpha}$. The frequencies $\omega_s(\boldsymbol{k})$ appearing in (6.1.12) are determined in the Heitler-London approximation, according to (6.2.20), by the relations ($s \equiv l, \rho$)

$$\omega_s(\boldsymbol{k}) \equiv \omega^{l}_{\rho}(\boldsymbol{k}) = \frac{1}{\hbar}[\mathscr{D}_l + \Delta_l + \tilde{\mu}^{l}_{\rho}(\boldsymbol{k})]\,. \tag{6.2.24}$$

Thus, for example, the frequencies ω_s (where $s = 1, 2$) of the exciton transition for exciton states that correspond to (6.2.22) are determined by

$$\omega_1(0) = \frac{1}{\hbar}[\Delta_l + \mathscr{D}_l + \tilde{\Gamma}_{11}(0) + |\tilde{\Gamma}_{12}(0)|]\,, \tag{6.2.24a}$$

$$\omega_2(0) = \frac{1}{\hbar}[\Delta_l + \mathscr{D}_l + \tilde{\Gamma}_{11}(0) - |\tilde{\Gamma}_{12}(0)|]\,.$$

For the same states, the matrix elements of the cell dipole moment operator, appearing in (6.1.14b), i.e., the quantities $\boldsymbol{D}_{0s;0}$ ($\boldsymbol{D}_{0s;0} = \sum_{\alpha} a^{l}_{\alpha} \boldsymbol{R}^{0l}_{\alpha}$), are of the form

$$\boldsymbol{D}_{01;0} = \frac{1}{\sqrt{2}}(\boldsymbol{R}^{0l}_{1} + \boldsymbol{R}^{0l}_{2})\,, \quad \boldsymbol{D}_{02;0} = \frac{1}{\sqrt{2}}(\boldsymbol{R}^{0l}_{1} - \boldsymbol{R}^{0l}_{2}) \tag{6.2.24b}$$

and consequently (since the modulus $|\boldsymbol{R}^{0l}_{\alpha}|$ is independent of α, i.e., of the position of the molecule in the cell) are orthogonal to each other.

6.2.2 Calculating the Dielectric-Constant Tensor of Molecular Crystals by the Local-Field Method

We have already discussed methods for calculating the dielectric-constant tensor $\varepsilon_{ij}(\omega,\boldsymbol{k})$ for a crystal, which are based on the use of the exciton states of the crystal for the zeroth-approximation states. As was indicated, the choice of the zeroth-approximation states uniquely determines the form of the perturbation operator that leads to the appearance of induced currents. For example, in applying mechanical excitons the perturbing field is the total average field $\boldsymbol{E}$, which satisfies Maxwell's equations. If, however, Coulomb excitons are used, the perturbation is due, not to the total field $\boldsymbol{E}$, but only to its transverse part $\boldsymbol{E}_{\perp}$. In particular, it was shown in Sect. 6.2.1 how the states of the mechanical excitons can be found for molecular crystals and how these states can be used as a basis for determining the tensor $\varepsilon_{ij}(\omega,\boldsymbol{k})$. However, it is precisely for molecular crystals that it is not always necessary to use exciton states for calculating the tensor $\varepsilon_{ij}(\omega,\boldsymbol{k})$ because in such crystals the states of the individual molecules can be chosen as the zeroth-approximation states. In this case, the so-called effective (local or internal) field, which polarizes one or another molecule of the crystal, serves as the perturbation. This field is found to be different for different molecules inside the unit cell and also differs substantially from the average electric field $\boldsymbol{E}$, appearing in the field equations (2.1.1) and dealt with throughout this book.

It will be shown below how a knowledge of the molecular polarizabilities, as well as of the relations expressing the field $\boldsymbol{E}_{\text{eff}}$, acting on the molecules of the crystal, in terms of the average field $\boldsymbol{E}$, enables us to calculate the polarization in the crystal and the tensor $\varepsilon_{ij}(\omega,\boldsymbol{k})$. Here we point out that the feasibility of applying the above-mentioned procedure for calculating the tensor $\varepsilon_{ij}(\omega,\boldsymbol{k})$ in molecular crystals is due to the fact that in the spectral region of the lowest singlet electronic (or vibrational) intramolecular excitations of such crystals, the interaction of the molecules with one another leads to practically no violation of their neutrality. This means that the intermolecular interaction is of a purely classical nature. This is what makes it possible to take this interaction into account by the local-field method, which goes back to *Lorentz*, and it led him to the well-known formula for the refractive index of light in isotropic media (the Lorenz-Lorentz equation).

Let us recall the derivation of this equation which, in its explicit form, is based on the fact that in condensed media the electric field $\boldsymbol{E}_{\text{eff}}$, acting on a given molecule and causing its polarization, is not equal to the average field $\boldsymbol{E}$ that satisfies Maxwell's phenomenological equations. For example, for models of point dipoles in a cubic lattice $\boldsymbol{E}_{\text{loc}} = \boldsymbol{E} + 4\pi\boldsymbol{P}/3 = (\varepsilon+2)\boldsymbol{E}/3$, $\boldsymbol{P}$ being the polarization per unit volume, $\boldsymbol{P} = (\varepsilon-1)\boldsymbol{E}/4\pi$ and ε is the dielectric constant of the medium. On the other hand, the polarization per unit volume $\boldsymbol{P} = N_0 a \boldsymbol{E}_{\text{loc}}$ (a being the polarizability of a molecule and N_0 the number of molecules in the unit volume) and the induction vector $\boldsymbol{D}$ is determined by the relation $\boldsymbol{D} = \boldsymbol{E} + 4\pi\boldsymbol{P} = [1 + 4\pi N_0(\varepsilon+2)/3]\boldsymbol{E} = \varepsilon\boldsymbol{E}$. This leads to the Lorenz-Lorentz equation

$$\frac{\varepsilon-1}{\varepsilon+2} = \frac{4\pi}{3} N_0 a \,. \tag{6.2.25}$$

However, even for cubic crystals, which consist of isotropic molecules interacting with one another by means of Van der Waals forces, this formula, expressing the dielectric constant of the medium in terms of the polarizability of an individual molecule, is extremely approximate. Specifically, it does not take spatial dispersion into account at all and, moreover, neglects the contribution (displayed even at $\boldsymbol{k} = 0$) of the higher multipoles to the energy of intermolecular interaction, i.e., the interaction that leads to the distinction between the optical properties of a crystal and the optical properties of molecules in dilute gases.

As the result of a generalization of the effective field method, it was shown in [6.18] how the above-mentioned factors can be taken into account. Here we shall show how the dielectric-constant tensor of an anisotropic molecular crystal can be found, with spatial dispersion, within the framework of the approach being considered.

Assume that a unit cell of the crystal contains σ identical molecules, differently oriented with respect to the crystallographic axes. When a plane electromagnetic wave of amplitude $\boldsymbol{E}(\omega, \boldsymbol{k})$, propagating in the crystal, acts on the molecule α, the electric field $\boldsymbol{E}^{\alpha}$ (we omit the subscript "loc" in the following) is not equal to the average field $\boldsymbol{E}$ and is determined by the relation[2] [Ref. 6.2, § 30]

$$E_i^{\alpha} = E_i + \sum_{\beta j} Q_{ij}^{\alpha\beta}(\boldsymbol{k}) p_j^{\beta} \,, \tag{6.2.26}$$

where $\boldsymbol{p}^{\alpha}$ is the amplitude of the dipole moment induced in molecules of type α, and the coefficients $Q_{ij}^{\alpha\beta}(\boldsymbol{k})$ (coefficients of the internal field) are determined only by the structure of the lattice. If $a_{ij}^{\alpha}(\omega)$ is the polarizability of a molecule in its orientation α (here and in all of the following we assume, for the sake of simplicity, that the molecules have no static dipole moments[3]), then

---

[2] Here, for simplicity, we take only the dipole-dipole interaction into account. How higher multipoles are taken into consideration is shown in [6.18] and with the use of Goepert-Mayer canonical transformation [6.12b]. The methods used to calculate lattice sums for multipole interactions have been reviewed in [6.19] (see also [6.19a]).

[3] At the same time, molecules can, of course, have static moments of higher multipole order both in the ground and excited states. These moments differ, in general, for the ground state and the excited state. Also differing in these states are the energies of interaction of a molecule with its surroundings, leading to a certain frequency shift of intramolecular transitions with respect to the frequencies of transitions in vacuum. Henceforth we assume that the tensor $a_{ij}(\omega, \boldsymbol{k})$ differs from the corresponding tensor for a molecule in vacuum only as the result of this shift. Above, see (6.2.6), the energy corresponding to this shift was denoted by $\mathscr{D}_l$.

$$p_i^\alpha = a_{ij}^\alpha(\omega) E_j^\alpha . \tag{6.2.27}$$

Substituting (6.2.27) into (6.2.26), we obtain

$$E_j^\alpha = E_j + \sum_{\beta j_1} Q_{ij_1}^{\alpha\beta}(\boldsymbol{k}) a_{j_1 j}^\beta(\omega) E_j^\beta , \tag{6.2.28}$$

and the magnitude of the local field $\boldsymbol{E}^\alpha$ expressed in terms of $\boldsymbol{E}$:

$$E_i^\alpha(\omega,\boldsymbol{k}) = A_{ij}^\alpha(\omega,\boldsymbol{k}) E_j(\omega,\boldsymbol{k}) .$$

If the tensor A_{ij}^α is known, we readily find an expression for the dielectric-constant tensor of the crystal. In fact, if the polarization per unit volume is

$$P_i = \frac{1}{v} \sum_{\alpha j_1} a_{ij_1}^\alpha E_{j_1}^\alpha = \frac{1}{v} \sum_{\alpha j_1} a_{ij_1}^\alpha A_{j_1}^\alpha E_j , \tag{6.2.29}$$

where $v \equiv 1/N_0$ is the volume of the unit cell, we obtain

$$\varepsilon_{ij}(\omega,\boldsymbol{k}) = \delta_{ij} + \frac{4\pi}{v} \sum_{\alpha j_1} a_{ij_1}^\alpha A_{j_1 j}^\alpha .$$

To specify this relation still further, it is necessary to determine the frequency dependence of the polarizability of the molecule.

In cubic crystals, consisting of isotropic molecules, for the vicinity of an isolated resonance $\omega = \omega_1$ we have

$$a(\omega) = a_0 + a_1(\omega) , \tag{6.2.30}$$

where $a_1 = f_1/(\omega_1^2 - \omega^2)$, $f_1 = 2R_1^2\omega_1/\hbar$, R_1 is the dipole moment of the transition, whereas the quantity a_0 is determined by the contribution of distant resonances to the polarizability of a molecule. In the frequency region $\omega \approx \omega_1$, a_0 can be considered to be a constant quantity, independent of ω. Substituting (6.2.30) into the Lorenz-Lorentz equation (6.2.25), we obtain the following expression for the quantity $\varepsilon(\omega)$

$$\varepsilon(\omega) = \varepsilon_0 + \frac{\frac{4\pi}{v} f_1 \left(\frac{\varepsilon_0 + 2}{3}\right)^2}{\omega_\perp^2 - \omega^2} , \tag{6.2.31}$$

where

$$\varepsilon_0 = 1 + \frac{4\pi}{v} a_0 \left(1 - \frac{4\pi}{3v} a_0\right)^{-1} \tag{6.2.31 a}$$

is the "background" dielectric constant, i.e., a quantity that would be equal to the dielectric constant of a crystal under the condition that there is no resonance term at the frequency ω_1 in the polarizability of its constituent molecules [i.e., that $a_1(\omega) \equiv 0$]. The value of the resonance frequency $\omega_\perp$, appearing in (6.2.31), is determined by

$$\omega_\perp^2 = \omega_1^2 - \frac{4\pi}{3v} f_1 \left(\frac{\varepsilon_0 + 2}{3}\right). \tag{6.2.32}$$

Relations (6.2.31, 32) indicate in what way the intermolecular interaction in crystals of the type under consideration renormalizes the oscillator strength of the transition and its frequency.

a) Davydov Splitting

We shall now apply an analogous treatment [6.18] for an anisotropic crystal as well, for example, in the vicinity of the frequency of a nondegenerate dipole intramolecular vibration. Neglecting, for simplicity, the contribution of distant resonances to the polarizability of an isolated molecule, we assume that

$$a_{ij}(\omega) = f_1 l_i l_j (\omega_1^2 - \omega^2)^{-1},$$

where f_1 is a quantity proportional to the oscillator strength of the $0 \to 1$-th transition, see (6.2.30) and $\boldsymbol{l}$ is a unit vector directed along the transition dipole moment vector. Since the various molecules in the unit cell differ from one another by their orientation, their polarizability tensor is evidently

$$a_{ij}^{\alpha}(\omega) = f_1 l_i^{\alpha} l_j^{\alpha} (\omega_1^2 - \omega^2)^{-1}, \qquad \alpha = 1, 2, \ldots, \sigma.$$

Substituting this relation into (6.2.28) and subjecting the equation to scalar multiplication by $\boldsymbol{l}^\alpha$, we obtain a system of σ equations for the quantities $\boldsymbol{E}^\alpha \boldsymbol{l}^\alpha$ (where $\alpha = 1, 2, \ldots$):

$$\boldsymbol{E}^{\alpha} \cdot \boldsymbol{l}^{\alpha} - \sum_{\beta} M_{\alpha\beta}(\omega, \boldsymbol{k})(\boldsymbol{E}^{\beta} \cdot \boldsymbol{l}^{\beta}) = \boldsymbol{E} \cdot \boldsymbol{l}^{\alpha}, \tag{6.2.33}$$

where

$$M_{\alpha\beta}(\omega, \boldsymbol{k}) = \frac{f_1}{\omega_1^2 - \omega^2} \sum_{ij} Q_{ij}^{\alpha\beta}(\boldsymbol{k}) l_i^{\alpha} l_j^{\beta}.$$

We shall consider crystals with two molecules per unit cell in more detail. The optical properties of such molecular crystals have especially been investigated. This group of crystals includes the crystals of anthracene, naphthalene and many other crystals of the aromatic series.

Crystals of the anthracene type have the elements of symmetry that convert molecules $\alpha = 1$ into molecules $\alpha = 2$. Therefore, at $\boldsymbol{k} = 0$, or for vectors $\boldsymbol{k} \neq 0$ directed along or perpendicular to the monoclinic axis, the relations $M_{11}(\omega, \boldsymbol{k}) = M_{22}(\omega, \boldsymbol{k})$ and $M_{12}(\omega, \boldsymbol{k}) = M_{21}(\omega, \boldsymbol{k})$ are valid. For such $\boldsymbol{k}$ the solution of the system of equations (6.2.33) is not too cumbersome. We can readily see that in this case

$$\boldsymbol{E}^{\alpha} \cdot \boldsymbol{l}^{\alpha} = \frac{1}{2}\left[\frac{L_j^{(1)}}{1 - M_{11}(\omega, \boldsymbol{k}) - M_{12}(\omega, \boldsymbol{k})} - \frac{(-1)^{\alpha} L_j^{(2)}}{1 - M_{11}(\omega, \boldsymbol{k}) + M_{12}(\omega, \boldsymbol{k})}\right] E_j ,$$

$$\alpha = 1, 2 ,$$

where $\boldsymbol{L}^{(1)} = \boldsymbol{l}^{(1)} + \boldsymbol{l}^{(2)}$ and $\boldsymbol{L}^{(2)} = \boldsymbol{l}^{(1)} - \boldsymbol{l}^{(2)}$. Substituting these relations into the expression for the polarization (6.2.29), we obtain the expression for the dielectric-constant tensor:

$$\varepsilon_{ij}(\omega, \boldsymbol{k}) = \delta_{ij} + \frac{2\pi}{v} f_1 \left[\frac{L_i^{(1)} L_j^{(1)}}{\Omega_1^2(\boldsymbol{k}) - \omega^2} + \frac{L_i^{(2)} L_j^{(2)}}{\Omega_2^2(\boldsymbol{k}) - \omega^2}\right], \tag{6.2.34}$$

where

$$\Omega_1^2(\boldsymbol{k}) = \omega_1^2 - f_1 \sum_{ij} Q_{ij}^{11}(\boldsymbol{k})\, l_i^{(1)} l_j^{(1)} - f_1 \sum_{ij} Q_{ij}^{12}(\boldsymbol{k})\, l_i^{(1)} l_j^{(2)} ,$$

$$\Omega_2^2(\boldsymbol{k}) = \omega_1^2 - f_1 \sum_{ij} Q_{ij}^{11}(\boldsymbol{k})\, l_i^{(1)} l_j^{(1)} + f_1 \sum_{ij} Q_{ij}^{12}(\boldsymbol{k})\, l_i^{(1)} l_j^{(2)} .$$

It should be readily clear that at $\boldsymbol{k} = 0$ expression (6.2.34) coincides with (6.1.14b) if we take the relations (6.2.24a, b) into consideration.

The vectors $\boldsymbol{L}^{(1)}$ and $\boldsymbol{L}^{(2)}$ are orthogonal. Hence, if the coordinate axes x and y are directed along the directions of $\boldsymbol{L}^{(1)}$ and $\boldsymbol{L}^{(2)}$, for example, we find that the tensor ε_{ij} has been reduced to the diagonal form

$$\varepsilon_{xx}(\omega, \boldsymbol{k}) = 1 + \frac{2\pi}{v}\, \frac{F_1}{\Omega_1^2(\boldsymbol{k}) - \omega^2}, \quad F_1 = f_1 |\boldsymbol{L}^{(1)}|^2$$

$$\varepsilon_{yy}(\omega, \boldsymbol{k}) = 1 + \frac{2\pi}{v}\, \frac{F_2}{\Omega_2^2(\boldsymbol{k}) - \omega^2}, \quad F_2 = f_1 |\boldsymbol{L}^{(2)}|^2$$

$$\varepsilon_{zz}(\omega, \boldsymbol{k}) = 1 .$$

It follows from these relations that for light propagating along the z axis, with the direction of the electric vector $\boldsymbol{E} \| \boldsymbol{L}^{(1)}$, absorption should occur at the frequency $\Omega_1(\boldsymbol{k})$. If, however, the direction of the electric vector $\boldsymbol{E} \| \boldsymbol{L}^{(2)}$, the light is absorbed at $\omega = \Omega_2(\boldsymbol{k})$. Thus, even though transition in an isolated molecule at the frequency ω_1 was assumed to be nondegenerate, two absorption lines for differently polarized light should appear in the absorption spectrum for a crystal with two molecules per unit cell. When the tensor ε_{ij} (6.2.34) is known, we can, of course, find the position of the absorption lines for arbitrary polarizations and direction of light propagation. We can also account for the effects of spatial dispersion.

Up to the present time this phenomenon (Davydov splitting) has been studied for a great many different specimens. Of interest is the fact that even though this phenomenon was first revealed in generalizing the Frenkel exciton theory for the case of crystals with several molecules per unit cell [6.20], the theory of small-radius excitons was not required in investigating the phenomenon. As has been shown above, it is sufficient, for this purpose, to generalize the Lorenz-Lorentz equation for the case of anisotropic crystals [6.19b]. This is not true for exciton spectra in semiconductors. As is known, their explanation required the theory of large-radius excitons. This is not to be implied, of course, as depreciating the theory of small-radius excitons. Only within the framework of the small-radius theory can we calculate the exciton energies over the whole range of allowable values of the wave vectors and consistently study the processes of exciton-phonon and exciton-exciton interaction. This is exactly what makes it possible, within the scope of small-radius exciton theory, to understand, more or less, such phenomena as energy transfer of electronic excitation in crystals, the optical properties of molecular crystals with high excitation levels, nonlinear optical effects, fine details in the structure of light absorption and luminescence spectra in molecular crystals [6.16, 20–22]. The theory of the local field effect on a surface is developed in [6.22a].

6.2.3 The Oscillator Model

The model of molecular crystals dealt with in Sect. 6.2.1, includes, as a limiting case, the classical crystal model with classical dipoles located at the lattice sites. The transition to this model is especially simple within the scope of the method described in Sect. 6.2.2, and the discussion given here enables the essence of the assumptions that were made to be more fully revealed.

To prove that this remark is valid, we should take into account the fact that in a crystal, along with the lattice constant a, a whole set of parameters a_i, equal to the effective dimensions of atoms and molecules in various states, as well as the effective lengths that determine the matrix elements of the transitions between the atomic and molecular states exist. Among other things, these values a_i determine the magnitudes of the matrix elements for various multipole (dipole, quadrupole, etc.) transitions. Instead of the parameters a and a_i we can therefore use a and the values of certain effective multipole moments (or, more exactly, their matrix elements), assuming these multipoles to be of the point type and located at the lattice sites. In the neighborhood of intense dipole lines we can restrict ourselves to dipole moments alone and thus arrive at a crystal model consisting of point dipoles. Here the only kind of interaction is the dipole-dipole type and, instead of (6.2.23), we have

$$\frac{1}{\sqrt{V}} M^{j}_{0;\,-ks}(\pm k) = \frac{\mathrm{i}\,\omega^{l}_{\rho}(-k)}{c\sqrt{v}} \sum_{\alpha r} a^{lr;\rho}_{\alpha}(-k) R^{lr;0}_{\alpha j}\,,$$

$$\frac{1}{\sqrt{V}} M^{j}_{-ks;0}(\pm k) = -\frac{\mathrm{i}\,\omega^{l}_{\rho}(-k)}{c\sqrt{v}} \sum_{\alpha r} a^{*lr;\rho}_{\alpha}(-k) R^{0;lr}_{\alpha j}\,,$$

where $R^{lr;0}_{\alpha j}$ is the j-th component of the dipole matrix element of the transition from the ground to the lr-th excited state of the molecule located at an α-type site, see (6.2.23 and 13).

Let us consider, for example, that the Bravais lattice of the crystal is of the cubic type, and that a unit cell has one isotropic molecule. The excited states of such a molecule, in which transitions are allowed in the dipole approximation, are triply degenerate. Let us choose the corresponding wave functions so that the vectors $\boldsymbol{R}^{lr}$, where $r = 1, 2, 3$, are directed along the fourfold symmetry axes of the cube. It is evident, in this case, that if the coordinate axes are also directed along these symmetry axes, then $R^{lr;0}_{j} = R^{l,0}\delta_{rj}$ and

$$\frac{1}{\sqrt{V}} M^{j}_{\pm ks;0}(k) = -\frac{\mathrm{i}\,\omega^{l}_{\rho}(\pm k)}{c\sqrt{v}} \overset{*}{a}{}^{lj;\rho}(\pm k) R^{l,0} = \frac{1}{\sqrt{V}} \overset{*}{M}{}^{j}_{0;\,\pm ks}(k)\,.$$

Consequently, recalling that the subscript $s \equiv (l, \rho)$,

$$\frac{1}{V} M^{i}_{0;\,-ks}(-k) M^{j}_{-ks;0}(-k) = \frac{[\omega^{l}_{\rho}(k)]^2}{c^2 v} a^{li;\rho}(-k) \overset{*}{a}{}^{lj;\rho}(-k) |R^{l0}|^2\,.$$

If we neglect the anisotropy of the effective mass of the mechanical excitons, then, for small values of k, to terms of the order of k^2, we obtain

$$\omega^{l}_{1}(k) = \omega^{l}_{2}(k) = \omega^{l}_{3}(k) \equiv \omega_l(k)\,.$$

Using the relation

$$\sum_{\rho} \overset{*}{a}{}^{li;\rho}(k) a^{lj;\rho}(k) = \delta_{ij}$$

with $\sigma = 1$ and $t = 3$, we obtain the dielectric-constant tensor in the above approximation. It is reduced to the scalar

$$\begin{aligned} \varepsilon_{ij}(\omega, k) &= \varepsilon(\omega, k)\,\delta_{ij} \\ &= \left[1 - \frac{4\pi e^2 N_0}{m\omega^2} - \frac{8\pi}{\hbar v \omega^2} \sum_{l} \frac{|R^{l,0}|^2 \omega^3_l(k)}{\omega^2 - \omega^2_l(k)}\right] \delta_{ij}\,. \end{aligned} \tag{6.2.35}$$

If we make use of the summation rule, (6.1.14a), the tensor (6.2.35) can be represented in the form

$$\varepsilon_{ij}(\omega,\boldsymbol{k}) = \varepsilon(\omega,\boldsymbol{k})\,\delta_{ij} \equiv \left[1 - \sum_{l} \frac{\omega_0^2 f_l(\boldsymbol{k})}{\omega^2 - \omega_l^2(\boldsymbol{k})}\right]\delta_{ij}\,, \tag{6.2.36}$$

where $\omega_0 = \sqrt{4\pi e^2 N_0/m}$ is the "plasma" frequency and $f_l(\boldsymbol{k})$ is the oscillator strength

$$f_l(\boldsymbol{k}) = \frac{2|R^{l,0}|^2 \omega_l(\boldsymbol{k})\, m}{3\hbar v e^2 N_0}\,, \qquad \sum_l f_l(\boldsymbol{k}) = 1\,. \tag{6.2.36a}$$

Equation (6.2.36) still remains a general quantum-mechanical expression. The model requires further specification in order to pass to the classical limit. We shall assume that each site of the unit cell accommodates isotropic identical harmonic oscillators, one at each site, corresponding to the effective masses and charges, m and e, respectively. The frequency of the oscillators is triply degenerate. When the interaction between the various cells of the crystal is taken into account, degeneracy is removed so that instead of a single frequency there are three energy bands of mechanical excitons $\hbar\omega_\rho(\boldsymbol{k})$, where $\rho = 1$, 2, 3. At small values of $\boldsymbol{k}$, however, the frequency $\omega_1(\boldsymbol{k}) = \omega_2(\boldsymbol{k}) = \omega_3(\boldsymbol{k}) \equiv \omega(\boldsymbol{k})$, i.e., the degeneracy can be assumed to be conserved. For this model we have a single resonance term in (6.2.36). Moreover, in the case under consideration, the number of electrons per unit volume is $N_0 = 1/v$, and in the approximation being applied (Heitler-London method, small values of $|\boldsymbol{k}|$), $|R^{l,0}|^2 = e^2\hbar/2m\omega(\boldsymbol{k})$. Hence, in accordance with (6.2.36a), $f(\boldsymbol{k}) = 1$ and, consequently,

$$\varepsilon_{ij}(\omega,\boldsymbol{k}) = \varepsilon(\omega,\boldsymbol{k})\,\delta_{ij} = \left[1 - \frac{\omega_0^2}{\omega^2 - \omega^2(\boldsymbol{k})}\right]\delta_{ij}\,.$$

In this classical model the refractive index for transverse normal waves is evidently equal to $n(\omega) = \sqrt{\varepsilon(\omega, \omega n s/c)}$ [6.23]. Normal waves in any direction are either longitudinal or transverse. The frequencies $\omega(\boldsymbol{k})$ of polarization waves (Sect. 2.2.2) correspond to the poles $\varepsilon(\omega,\boldsymbol{k})$, whereas the quantity $\varepsilon(\omega,\boldsymbol{k})$ vanishes at the frequencies of longitudinal waves. Fictitious longitudinal waves do not exist in the model under investigation.

It is evident that less symmetrical models of dipole lattices can be dealt with an a similar manner.

6.3 Absorption

6.3.1 The Absorption Mechanism. Absorption in a First Approximation

The damping of electromagnetic waves in crystals is due primarily to the possibility of irreversible conversion of their energy into the energy of other degrees of freedom. In nonmetals the absorption of electromagnetic waves in the visible and ultraviolet spectral ranges is due mainly to the conversion of electron-excitation energy into nuclear-vibration energy of the crystal lattice. The existence of this process leads to a situation in which the photon states in the medium (real excitons) become quasi-steady, and the tensor $\varepsilon_{ij}(\omega,\boldsymbol{k})$ becomes non-Hermitian even when ω and $\boldsymbol{k}$ are real. The determination of this tensor, in a first approximation, is a procedure entirely analogous to that described in Sect. 6.1. The main difference is that in taking lattice vibration into account the state of the crystal is characterized, not only by the electron states, but by nuclear motion or, in quantum-mechanical language, by the states of the phonon subsystem.

The necessity for taking the motion of the nuclei in the crystal into account along with the motion of the electrons substantially complicates the problem of determining the steady states in the unperturbed problem (i.e., the problem in which the macroscopic electric field is neglected). All aspects of the theory of these states have not yet been fully investigated and the whole problem undoubtedly requires further discussion. In the present book, we do not deal with these matters and assume that the steady states in the unperturbed problem are known. A discussion of related problems can be found, for example, in [6.16, 21, 22, 24–28].

According to the statement of the problem these states are characterized by approximate eigenfunctions of the energy operator of the system. There is always an interaction (for instance, anharmonic terms in the lattice vibration equations) that leads to transitions between the above-mentioned approximate states of the system even in the absence of external perturbation. Owing to these transitions, the lifetimes of the excited states of the crystal and, in particular, those of normal electromagnetic waves, are finite. We refer here to normal electromagnetic waves in which the wave vector $\boldsymbol{k}$ is assumed to be real and the frequency ω is determined by the dispersion equation (2.2.10, 22); in the case of absorption, the frequency ω is complex.) In order to take the above-mentioned transitions into account in calculating the induced current, due to the action of external periodic perturbance, one must consider not only this periodic perturbation, but also the time-independent perturbation mentioned above.

Let $\hat{V}$ be an operator corresponding to the time-independent perturbation. In determining the "perturbed" wave function we should then use the operator

$$\hat{U} = \hat{F}e^{-i\omega t} + \hat{G}e^{i\omega t} + \hat{V} \tag{6.3.1}$$

instead of (6.1.1). As usual [6.3], we shall seek the "perturbed" wave function of the crystal in the state n as the sum

$$\Psi_n(t) = \exp\left(-i\frac{W_n}{\hbar}t\right)\Psi_{n0} + \sum_{m \neq n} a_{mn}(t)\,\Psi_{m0}, \quad |a_{mn}| \ll 1. \tag{6.3.2}$$

Here, as in Sect. 6.1.1, we assume that the function Ψ_{m0} is time-independent. Moreover, it proves convenient to choose the reference point for the energy in the system so that $W_n = 0$. Then, substituting (6.3.2) into the Schrödinger time equation and taking the perturbation (6.3.1) into consideration, we obtain

$$i\hbar\dot{a}_{mn} = W_m a_{mn} + \sum_k u_{mk}(t)\,a_{kn}(t), \tag{6.3.3}$$

where

$$u_{mk}(t) = \int \Psi^*_{m0}\,\hat{U}(t)\,\Psi_{k0}\,d\tau$$

are the matrix elements of the operator (6.3.1). Turning our attention to the initial conditions for $a_{kn}(t)$, we divide all the excited states of the crystal, corresponding to the presence of mechanical excitons and phonons in the crystal, into two groups[4]. In the first group we include a comparatively small number of states to which the system may make a transition from state n due to the effect of light. These are states for which the matrix element F_{nm}, see (6.1.2), is fairly large. Belonging to the second group are the states for which F_{nm} is negligibly small and to which, consequently, a light-induced transition from the state n is forbidden. The system can make a transition to these states from the states of the first group only as a result of the perturbation $\hat{V}$, see (6.3.1). States of the first group will be denoted by the subscript m, as before, whereas for the states of the second group we shall introduce the subscript q. Actually, for the cases that we are interested in, the states n and m differ only with respect to the state of the electrons. For example, $n = 0$ is the ground state and $m = 1$ is the state with one mechanical exciton with $F_{nm} \neq 0$, and the state of the phonons is the same in states n and m. In this example, state q is obtained from the state $m = 1$, e.g., as the result of the emission or absorption of a phonon by an exciton, which is thereby converted into a mechanical exciton with different energy and momentum.

Of interest to us are the solutions of (6.3.3) under the conditions that the states of the first group are steady, so that $a_{mn}(t)$ can be sought in the following form, cf. (6.1.2):

---

[4] Certain steps in our calculation of the wave function (6.3.2) follow [6.5]. The more particular case, in which the operator $\hat{V}$ appearing in (6.3.1) is the exciton-phonon interaction operator, was discussed in [6.21, 29, 30].

$$a_{mn}(t) = A_m(-\omega)\mathrm{e}^{-\mathrm{i}\omega t} + A_m(\omega)\mathrm{e}^{\mathrm{i}\omega t}, \tag{6.3.4}$$

where the amplitudes A_m are independent of time. The coefficients $a_{qn}(t)$ can be determined by making use of the initial condition

$$a_{qn}(t=0) = 0\,. \tag{6.3.5}$$

This condition can be expediently used if there are very many states q (states of the second group) and, in connection with this circumstance, all the states of the q type are weakly excited so that at all values of time t they can be considered as not having yet reached steady values. In other words, the states of the second group form a dissipative system. Using this terminology, it can be said that the states of the first group form a dynamic system.

Next we shall seek a solution of the system of equations (6.3.3) that satisfies the formulated conditions. We select the basic functions Ψ_{m0} in such a way that the matrix $V_{mn'}$ is diagonal in the subspace of functions of the first group. In this case (6.3.3) assumes the form

$$\mathrm{i}\hbar\dot{a}_{mn} = W_m a_{mn} + \sum_q V_{mq} a_{qn} + F_{mn}\mathrm{e}^{-\mathrm{i}\omega t} + G_{mn}\mathrm{e}^{\mathrm{i}\omega t}, \tag{6.3.6}$$

$$\mathrm{i}\hbar\dot{a}_{qn} = W_q a_{qn} + \sum_m V_{qm} a_{mn} + \sum_{q'} V_{qq'} a_{q'n}\,. \tag{6.3.7}$$

It follows from (6.3.6) that the coefficients a_{qn} are quantities, with respect to the matrix elements of the perturbation, of a higher order of smallness compared to a_{mn}. Therefore, in a first approximation, we can omit the sum $\sum_{q'} V_{qq'} a_{q'n}$ in (6.3.7). Then, substituting (6.3.4) into (6.3.7), and taking (6.3.5) into consideration, we obtain

$$a_{qn} = a_{qn}^{\omega} + a_{qn}^{-\omega}, \tag{6.3.8}$$

where

$$a_{qn}^{\pm\omega}(t) = (\mathrm{e}^{\mp\mathrm{i}\omega t} - \mathrm{e}^{\mp\mathrm{i}\omega_q t}) \sum_m \frac{V_{qm} A_m(\mp\omega)}{(\pm\hbar\omega - W_q)}, \qquad W_q = \hbar\omega_q\,.$$

Substituting (6.3.4, 8) into (6.3.6), we obtain a system of equations for determining the coefficients $A_m(\pm\omega)$:

$$\begin{aligned}
&(W_m - \hbar\omega)A_m(-\omega) + \sum_{m'} V^{\mathrm{a}}_{mm'}(-\omega)A_{m'}(-\omega) + F_{mn} = 0\,,\\
&(W_m + \hbar\omega)A_m(\omega) + \sum_{m'} V^{\mathrm{a}}_{mm'}(\omega)A_{m'}(\omega) + G_{mn} = 0\,.
\end{aligned} \tag{6.3.9}$$

Here the quantity

$$V^{\mathrm{a}}_{mm'}(-\omega) = \sum_q V_{mq} V_{qm'} \frac{1-\exp[-\mathrm{i}(\omega_q - \omega)t]}{\hbar\omega - W_q} \tag{6.3.10}$$

is actually independent of time for sufficiently large values of t [Ref. 6.31, § 16], since

$$\frac{1-\exp[-\mathrm{i}(\omega_q-\omega)t]}{\hbar(\omega-\omega_q)} = \frac{\mathscr{P}}{\hbar\omega - W_q} - \mathrm{i}\pi\delta(\hbar\omega - W_q)$$

as $t \to \infty$. Therefore

$$V^{\mathrm{a}}_{mm'}(-\omega) = \mathscr{P}\sum_q \frac{V_{mq}V_{qm'}}{\hbar\omega - W_q} - \mathrm{i}\pi \sum_q V_{mq}V_{qm'}\delta(\hbar\omega - W_q)\,. \tag{6.3.11}$$

The symbol $\mathscr{P}$ in (6.3.11) indicates that the corresponding integral (the summation over the states q) is to be taken as the principal value. When the off-diagonal elements $V^{\mathrm{a}}_{mm'}(\pm\omega)$ of (6.3.9) are omitted in a first approximation, we have

$$\begin{aligned} A_m(-\omega) &= -\frac{F_{mn}}{W_m + V^{\mathrm{a}}_{mm}(-\omega) - \hbar\omega}\,,\\ A_m(\omega) &= -\frac{G_{mn}}{W_m + V^{\mathrm{a}}_{mm}(\omega) + \hbar\omega}\,. \end{aligned} \tag{6.3.12}$$

Let us consider in greater detail the case in which the state n of the crystal, see (6.3.2), is the ground state $n = 0$. Since, in this case, $W_q - W_0 > 0$ for all other states, it is clear that the imaginary part of the function $V^{\mathrm{a}}_{mm}(\omega)$ vanishes at $\omega > 0$ in accordance with (6.3.11). Making use of the derivation of the expression for the tensor $\varepsilon_{ij}(\omega, \boldsymbol{k})$, given in Sect. 6.1.1, and taking into account that $\hat{V} \neq 0$ and, consequently, the finite lifetimes of the crystal states to which a light-induced transition from the ground state is allowed, we find that the dielectric-constant tensor of the crystal at low temperatures is determined by

$$\begin{aligned} \varepsilon_{ij}(\omega,\boldsymbol{k}) = {} & \left(1 - \frac{4\pi}{\omega^2 V}\sum_\alpha \frac{e_\alpha^2}{m_\alpha}\right)\delta_{ij} - \frac{4\pi c^2}{\omega^2 V}\\ & \times \sum_{m\neq 0}\left[\frac{M^i_{0m}(-\boldsymbol{k})M^j_{m0}(\boldsymbol{k})}{\hbar\omega - \hbar\omega_m(-\boldsymbol{k}) - V^{\mathrm{a}}_{mm}(-\omega)} - \frac{M^i_{m0}(-\boldsymbol{k})M^j_{0m}(\boldsymbol{k})}{\hbar\omega + \hbar\omega_m(\boldsymbol{k}) + V^{\mathrm{a}}_{mm}(\omega)}\right], \end{aligned} \tag{6.3.13}$$

in which we used (6.3.11) and

$$V^{a}_{mm}(-\omega) = \Delta W_m(\omega) - \frac{i\hbar\nu_m(\omega)}{2}, \qquad \nu_m > 0,$$

$$V^{a}_{mm}(\omega) = \Delta W_m(-\omega).$$

In this case

$$\hbar\nu_m(\omega) = 2\pi \sum_q |V_{qm}|^2 \delta(\hbar\omega - W_q),$$

$$\Delta W_m(\omega) = \mathscr{P} \sum_q \frac{|V_{qm}|^2}{\hbar\omega - W_q}. \tag{6.3.14}$$

Since $\nu_m \neq 0$, the tensor (6.3.13) is non-Hermitian:

$$\varepsilon_{ij}(\omega, \boldsymbol{k}) \neq \varepsilon^*_{ji}(\omega, \boldsymbol{k}).$$

Moreover, it follows from (6.3.13) that the poles of this tensor are located, as was to be expected (Sect. 2.1.1), in the lower half-plane of the complex variable ω.

When ε_{ij} is to be calculated for the case with absorption, it would seem natural to replace the frequencies ω_m of undamped mechanical excitons in (6.1.9) without absorption by frequencies for damping accounted for. In other words, the point is to deal with mechanical excitons, taking into consideration the damping due to their interaction, for instance, with phonons. We find, however, that this approach is wrong because the perturbation frequency (i.e., the frequency of the field) differs from ω_m. From a formal aspect, the result (6.3.14) obtained for $\nu_m(\omega)$ and $\Delta W_m(\omega)$ differs from the values of $\nu_m(\omega_m)$ and $\Delta W(\omega_m)$ for mechanical excitons only in that ω_m is replaced by ω.

In the neighborhood of a separate resonance $\omega \approx \omega_m$ and, on the basis of (6.3.13), we can write

$$\varepsilon_{ij}(\omega, \boldsymbol{k}) = \varepsilon^0_{ij} - \frac{4\pi c^2}{\omega_m^2 V} \frac{M^i_{0m}(-\boldsymbol{k}) M^j_{0m}(\boldsymbol{k})}{\hbar\omega - \hbar\omega_m - \Delta W_m + i\hbar\nu_m(\omega)/2}. \tag{6.3.15}$$

Equations (6.3.13–15) confirm the previously used expression for $\varepsilon(\omega)$ in the case of damping, see (4.1.13a), and, moreover, enable us to understand the reasons for the dependence of the "effective" collision frequency $\nu_m(\omega)$ on ω.

As we see from (6.3.13), the tensor $\varepsilon_{ij}(\omega, \boldsymbol{k})$ has no singularities with real values of $\boldsymbol{k}$ in the presence of absorption. Furthermore, complex values of $\boldsymbol{k}$, corresponding to the poles of the tensor $\varepsilon_{ij}(\omega, \boldsymbol{k})$ (assuming that the frequency ω is real), move farther and farther away from the real axis, as absorption increases. In the integral relation (2.1.4), the kernel $\hat{\varepsilon}_{ij}(\tau, \boldsymbol{R})$ is determined, by virtue of (2.1.6), by the expression

$$\mathcal{E}_{ij}(\tau,\boldsymbol{R}) = \frac{1}{(2\pi)^4} \int_{-\infty}^{+\infty} d\omega \int d\boldsymbol{k}\, \varepsilon_{ij}(\omega,\boldsymbol{k})\, \mathrm{e}^{\mathrm{i}(\boldsymbol{k}\cdot\boldsymbol{R}-\omega\tau)}\,.$$

For large values of R, when in the integral with respect to $\boldsymbol{k}$ the main contribution is made by small values of $\boldsymbol{k}$ and (6.3.13) is applicable, the kernel $\mathcal{E}_{ij}(\tau,\boldsymbol{R})$ decreases more rapidly with an increase in R than any power of $1/R$. This can be illustrated by an example of the simplest function $\varepsilon(k) = 1/(k^2+a^2)$, where $a^2>0$, which, like $\varepsilon_{ij}(\omega,\boldsymbol{k})$, has no singularities if k is real. In this case

$$\mathcal{E}(R) = \int_{-\infty}^{+\infty} \mathrm{e}^{\mathrm{i}kR}\varepsilon(k)\,dk = \frac{\pi}{a}\mathrm{e}^{-a|R|}$$

which thereby decreases exponentially with increasing $|R|$.

Note that in a similar way we can determine the tensor $\varepsilon_{\perp,ij}$ (Sect. 2.2.4) which, for normal waves, links the induction vector with the transverse part of the electric-field strength. In the absence of damping, this is a problem which has already been dealt with in Sect. 6.1.1. In the presence of damping processes, (6.1.8a) remains valid, but (6.1.8c) for crystals in the ground state should be replaced by

$$\begin{aligned} b_{ij}^{(n)} = &-\frac{4\pi}{\omega^2 V}\sum_\alpha \frac{e_\alpha^2}{m_\alpha}\delta_{ij} \\ &-\frac{4\pi c^2}{\omega^2 V}\sum_{m\neq 0}\left[\frac{M_{0m}^i(-\boldsymbol{k})M_{m0}^j(\boldsymbol{k})}{\hbar\omega-\hbar\Omega_m-V_{mm}^{\mathrm{a}}(-\omega)}-\frac{M_{m0}^i(-\boldsymbol{k})M_{0m}^j(\boldsymbol{k})}{\hbar\omega+\hbar\omega_m+V_{mm}^{\mathrm{a}}(\omega)}\right]; \end{aligned} \tag{6.3.16}$$

where, as previously, the quantities V_{mm}^{a} are determined by (6.3.14). The states of the unperturbed problem are now understood to be states corresponding to a situation in which Coulomb interaction is fully taken into account.

Calculations of the frequency dependence of the quantities $\nu_m(\omega)$, see (6.3.15), which can be carried out for specific models [6.21, 22, 29, 30], enable the shape of the absorption band to be considered, as well as its dependence on the temperature and other factors.

6.3.2 Absorption of Normal Electromagnetic Waves in the Vicinity of an Exciton Transition Frequency

To illustrate the nature of the frequency dependence of the quantity $\nu_m(\omega)$, which determines the shape of the exciton absorption line, we shall consider an important case in more detail. Here the basic mechanism leading to the damping of normal electromagnetic waves in a dielectric crystal is the exciton-phonon interaction. Since the exciton absorption lines are located in the vicinity of Coulomb-exciton frequencies, it proves more convenient to deal with the problem of the shape of an exciton absorption line by considering the states of Coulomb, rather than mechanical, excitons.

For a crystal in which the vector potential is taken in the Coulomb gauge and the retarded interaction of the charges is neglected (corresponding to neglecting the interaction of the charges with the field of the transverse photons), the Hamiltonian operator can be represented in the form

$$\hat{H}_{\text{Coul}} = \hat{H}_1 + \hat{H}_2 + \hat{V}, \tag{6.3.17}$$

where $\hat{H}_1$ and $\hat{H}_2$ are the Hamiltonians of the exciton and phonon subsystems, and $\hat{V}$ is the exciton-phonon interaction operator (here and in the following, excitons are understood to be Coulomb excitons).

Let us assume that the exciton-phonon interaction is weak. Then the approximately steady states of the crystal in the exciton region of the spectrum are determined by the simultaneously given states $(\boldsymbol{k},s)$ of the electron subsystem (this consists of specifying the wave vector $\boldsymbol{k}$ of the Coulomb exciton and the discrete number s of the exciton band), as well as the states of the phonon subsystem, i.e., the set of numbers $\{\dots,N_l(\boldsymbol{p}),\dots\}$, where $N_l(\boldsymbol{p})$ is the quantum number ($N_l(\boldsymbol{p}) = 0,1,2,\dots$) corresponding to the state of the phonon of the lth branch of crystal lattice vibrations with quasimomentum $\boldsymbol{p}$.

For simplicity, we assume that the crystal has temperature $T=0$. Consequently, the crystal is in the ground state and there are no excitons or phonons. Now, if only electrons are taken into account in the operator $\hat{\boldsymbol{M}}(\boldsymbol{k})$ (see, in (6.1.1 a), the sum with respect to α), and small terms corresponding to interaction of the nuclei with the external electric field are neglected, the matrix elements $\boldsymbol{M}_{0m}(\boldsymbol{k})$ are nonzero only for the states of the crystal in which, as in the ground state, there are no phonons and the electron subsystem is in the exciton state with the wave vector $\boldsymbol{k}$ (Sect. 6.1). If, in this case, $\boldsymbol{M}_{0m}(0) \neq 0$, we evidently have a dipole transition; but if $\boldsymbol{M}_{0m}(0) = 0$ we have either a quadrupole or some higher-multipole transitions.

Let us consider in greater detail the neighborhood of the isolated nondegenerate dipole band s, sufficiently remote from other exciton bands so that interband transitions, caused by weak exciton-phonon interaction, can be neglected. In this case we can assume that the first group of states, discussed in Sect. 6.3.1, consists of the single state $m = (\boldsymbol{k},s,\dots,N_l(\boldsymbol{p})=0,\dots)$, whereas, for the states of the second group, $q = (\boldsymbol{k},s,\dots,N_l(\boldsymbol{p}),\dots)$ and, in general, $N_l(\boldsymbol{p}) \neq 0$. Since the crystal was assumed to be perfect, the exciton-phonon interaction operator, leading to a transition from state m to state q and, consequently, playing the part of the operator $\hat{V}$, see (6.3.1), can be represented to a first approximation in the form

$$\hat{V} = \sum_{\boldsymbol{k};\boldsymbol{p},l} F(\boldsymbol{k}+\boldsymbol{p};\boldsymbol{k};\boldsymbol{p},l)(B^+_{\boldsymbol{k}+\boldsymbol{p}}B_{\boldsymbol{k}} + B_{\boldsymbol{k}}B^+_{\boldsymbol{k}+\boldsymbol{p}})(b_{\boldsymbol{p},l} + b^+_{-\boldsymbol{p},l}), \tag{6.3.18}$$

where B_k^+ and B_k are the creation and destruction (Bose) operators of the exciton $(\boldsymbol{k}, s)$, and $b_{p,l}^+$ and $b_{p,l}$ are the creation and destruction operators of the phonon $(\boldsymbol{p}, l)$. Here, since $\hat{V}$ is a self-adjoint operator,

$$F(\boldsymbol{k}+\boldsymbol{p};\boldsymbol{k};\boldsymbol{p},l) = F^*(\boldsymbol{k};\boldsymbol{k}+\boldsymbol{p};-\boldsymbol{p},l)\,. \tag{6.3.19}$$

By virtue of (6.3.18)

$$|V_{qm}|^2 = |F(\boldsymbol{k}+\boldsymbol{p};\boldsymbol{k};\boldsymbol{p},l)|^2$$

and, in accordance with (6.3.14),

$$\begin{aligned}\nu_m(\omega) &\equiv \nu_s(\omega,\boldsymbol{k})\\ &= \frac{2\pi}{\hbar}\sum_{p,l}|F(\boldsymbol{k}+\boldsymbol{p};\boldsymbol{k};\boldsymbol{p},l)|^2\delta[\hbar\omega - W_s(\boldsymbol{k}-\boldsymbol{p}) - \hbar\omega_l(\boldsymbol{p})]\,,\end{aligned} \tag{6.3.20}$$

where $W_s(\boldsymbol{k})$ is the energy of the Coulomb exciton $(\boldsymbol{k}, s)$. For dipole excitons we can put $\boldsymbol{k}=0$ in (6.3.20). For quadrupole excitons $\nu_m(\omega) \sim k^2$, etc. The explicit form of the function $\nu_m(\omega)$ can be determined only if specific values are given for the coefficients F in (6.3.19) [6.26]. We note here only one special feature of (6.3.20). If follows from this equation that the quantity $\nu_m(\omega)$ is zero in the approximation under consideration if the frequency $\omega < W_s^{\min}/\hbar$, $W_s^{\min}$ being the minimum energy of the Coulomb exciton in the s-th band. If the minimum in the exciton band corresponds to the quasi-momentum $\boldsymbol{k}=0$, then $W_s^{\min} = W_s(0) \equiv W_s(\boldsymbol{k}=0)$ and $\nu_m(\omega) = 0$ when $\hbar\omega < W_s(0)$. This conclusion remains valid if in (6.3.18) we also take into account the subsequent terms in the expansion of the operator $\hat{V}$ of nuclei displacements, i.e., if along with the terms linear with respect to b and b^+, we also take into consideration terms of the form of $b \cdot b$, $b \cdot b^+$, etc.

Actually, however, the value of $\omega = W_s^{\min}/\hbar$ is not the long-wavelength edge of the exciton absorption band, because a more accurate consideration of the exciton-photon interaction leads to damping of the light even for $\omega < W_s^{\min}/\hbar$.

6.3.3 The Long-Wavelength Edge of the Exciton Absorption Bands: Raman and Brillouin-Mandelstam Scattering of Polaritons

In order to elucidate the nature of long-wavelength absorption in the region of frequencies $\omega < W_s^{\min}/\hbar$, we shall consider the question: what kinds of elementary excitations are there in the crystal if we take into account the retarded interaction between the charges, as well as the instantaneous Coulomb interaction?

When the Coulomb gauge is used, the total Hamiltonian operator for the crystal (including also the energy operator of the transverse photon field) is equal to

$$\hat{H} = \hat{H}_{\mathrm{Coul}} + \hat{H}_{\perp} + \hat{H}_{\mathrm{int}} ,$$

where the operator $\hat{H}_{\mathrm{Coul}}$ is given by (6.3.17), $\hat{H}_{\perp}$ is the Hamiltonian of the transverse photon field and $\hat{H}_{\mathrm{int}}$ describes the interaction of charges with the field of these photons.

When a photon with frequency ω and wave vector $\boldsymbol{k}$ (which, obviously, can only be a transverse photon) is incident on a crystal from vacuum, at $\hat{H}_{\mathrm{int}} = 0$ the photon passes through the crystal without interacting with the excitons. If, however, the operator $\hat{H}_{\mathrm{int}}$ is regarded as a weak perturbation, the state of the transverse photons becomes damped. Let us set the temperature $T = 0$ so that phonons and excitons can only be emitted, but not absorbed. Then, in the vicinity of an isolated Coulomb exciton band, the decay of a photon with frequency ω and wave vector $\boldsymbol{k}$ can occur either with the formation of a Coulomb exciton and phonons [which are exactly the kind of processes referred to in (6.3.20)] or with the formation of some other photon $(\omega', \boldsymbol{k}')$ and phonons (we ignore the highly improbable process in which a photon of energy $\hbar\omega$ is converted only into phonons). If we take into consideration the process of the production of a single phonon, this takes place when the following conservation laws are satisfied:

$$\omega = \omega' + \omega_l(\boldsymbol{p}) , \quad \boldsymbol{k} = \boldsymbol{k}' + \boldsymbol{p} , \tag{6.3.21}$$

where $\omega_l(\boldsymbol{p})$ is the phonon frequency of the lth branch of vibrations with the wave vector $\boldsymbol{p}$. Practically, this process consists of Raman or Brillouin-Mandelstam scattering of a transverse photon with the formation of a phonon.

Relations of the type (6.3.21) impose no restrictions on the frequency ω, but at $\omega > W_s^{\min}/\hbar$ the intensity of this process of Raman scattering is usually small compared to that of processes of photon decay into a Coulomb exciton and phonons. Consequently, in this frequency region the last-mentioned decay processes completely determine the intensity of damping of the transverse photon states in the crystal (the existence of defects is ignored). The cross section for Raman or Brillouin-Mandelstam scattering of transverse photons with phonon production can be calculated by means of the operator (6.3.1), in second-order perturbation theory (Sect. 6.4.2). Here, the exciton appears in the intermediate state; in this case it is a Coulomb exciton. Such calculations are unsuitable, however, if the photon frequency ω lies in the vicinity of the frequency $W_s(0)/\hbar$ of the Coulomb exciton. If the exciton band under consideration corresponds to high oscillator strengths, the frequency region in which perturbation theory is inapplicable becomes quite extensive.

Let us examine this problem in greater detail. At low values of the quasimomentum $\boldsymbol{k}$ the energy $W_s(\boldsymbol{k})$ of the Coulomb exciton can be represented as follows:

$$W_s(\boldsymbol{k}) = W_s(0) + \alpha_i k_i + \beta_{ij} k_i k_j + \dots ,$$

where $W_s(0)$, α_i and β_i depend, in general, on $\boldsymbol{s} = \boldsymbol{k}/k$. The spectrum of the transverse photons is determined by the equation

$$W_\perp(\boldsymbol{k}) = \hbar |\boldsymbol{k}| c ,$$

and in quasimomentum space, at low values of $|\boldsymbol{k}| \sim W_s(0)/\hbar c \ll 1/a$, the energy branches of the Coulomb excitons and the transverse photons intersect. Hence, if even a weak interaction between these excitons and transverse photons (i.e., retarded interaction) is taken into account, the shape of the spectrum of elementary excitations is altered in the vicinity of the intersection. Such a situation is sometimes said to be a "mixing" of the states of Coulomb excitons and transverse photons. Thus, for example, in cubic crystals (Fig. 6.1) the transverse Coulomb excitons, like the transverse optical vibrations of ionic lattice, "mix" with the states of the transverse photons and form observable optical waves, i.e., normal electromagnetic waves – polaritons – or real excitons (under the conditions illustrated in Fig. 6.1 absorption is neglected). Polariton spectra in the region of overtones and combination tones of vibration were dealt with in [6.32].

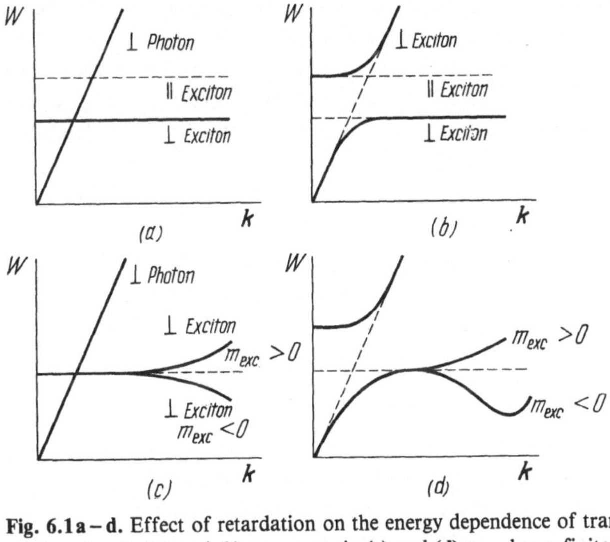

Fig. 6.1a–d. Effect of retardation on the energy dependence of transverse (⊥) excitons on the wave vector. In **(a)** and **(b)** $m_{exc} = \infty$, in **(c)** and **(d)** m_{exc} has a finite value; the branch of longitudinal excitons is not marked; retardation was taken into account in **(b)** and **(d)**

It is evident from Fig. 6.1 that when we take retardation into account, the group velocity of the normal waves is changed in the vicinity of the intersection of the Coulomb exciton and photon branches. This proves to be of particular importance in considering the interaction of these waves with phonons. The "mixing" of the states, considered here, is an indication of the obvious fact that normal electromagnetic waves in a medium (polaritons), especially when they are close to resonances, differ essentially from normal waves in vacuum.

The effect of retarded interaction between electrons on exciton-phonon interaction can be most simply treated in the following way [6.21, 22, 24, 33, 34].

In the zeroth approximation we take both Coulomb and retarded interaction between the charges into account, but neglect the exciton-phonon interaction. We obtain, in this case, undamped polaritons [of energy $W(\boldsymbol{k})$]. We shall assume the interaction with phonons to be weak so that when it is taken into consideration in the subsequent approximation a small imaginary contribution $\mathrm{i}\delta W(\boldsymbol{k})$ is added to the energy $W(\boldsymbol{k})$ of the polariton in the zeroth approximation, thus damping these states. If the quantity $\delta W(\boldsymbol{k})$ is known, we can calculate, to a first approximation, the damping factor $\varkappa(\omega,\boldsymbol{s})$ for normal waves propagating in the direction $\boldsymbol{s}$ and having the frequency ω. As a matter of fact, denoting $\mathrm{Im}\{\boldsymbol{k}\}$ by $\delta\boldsymbol{k}$, we obtain

$$\mathrm{i}\frac{\omega}{c}\varkappa(\omega,\boldsymbol{s})\boldsymbol{s} \equiv \delta\boldsymbol{k} = \frac{d\boldsymbol{k}}{d\omega(\boldsymbol{k})}\mathrm{i}\delta\omega(\boldsymbol{k}), \qquad \delta\omega(\boldsymbol{k}) = \frac{\delta W(\boldsymbol{k})}{\hbar}. \tag{6.3.22}$$

The quantity $v_{\mathrm{gr}} = d\omega(\boldsymbol{k})/d\boldsymbol{k}$ is the group velocity of waves to lowest order. It is essential that at $\omega \approx W_s(0)/\hbar$ (Fig. 6.1), $v_{\mathrm{gr}} \ll c$. If the frequency ω becomes appreciably less than the limiting frequency $W(0)/\hbar$, then $v_{\mathrm{gr}} \sim c$ (for instance, as $\omega \to 0$, the velocity $v_{\mathrm{gr}} = c/\sqrt{\varepsilon_0}$, where $\varepsilon_0 = \mathrm{const}$ is the static dielectric constant) and the value of the absorption coefficient $\varkappa(\omega,\boldsymbol{s})$ decreases drastically. Moreover, the decrease in the value of $\varkappa(\omega,\boldsymbol{s})$ is also associated with the fact that when $\omega \ll W_s(0)/\hbar$ and when, at the same time, ω is far away from other resonances, the normal waves in the medium differ comparatively little from the states of the transverse photons and interact weakly with phonons. Note that in calculating $\varkappa(\omega,\boldsymbol{s})$ this is precisely the frequency region in which we can apply the previously mentioned second-order perturbation theory. This approximation determines here the probability of polariton scattering with phonon production (Sect. 4.6.2).

Estimates indicate [6.34–36] that the frequency range in which it is essential to take retardation into account in calculating $\varkappa(\omega,\boldsymbol{s})$ is wider, when the oscillator strength of the transition is greater. It may sometimes reach values of the order of $100\ \mathrm{cm}^{-1}$. Hence the method described above, which makes use of (6.3.22), actually enables us to take the effect of the long-wavelength edge into account, to a first approximation. At the same time, however, the effect of damping on the group velocity and, in general, on the function $n(\omega)$ remains unaccounted for. Besides, the complex tensor $\varepsilon_{ij}(\omega, \boldsymbol{k})$ remains unknown in this approximation. This tensor was calculated in [6.6] with the aid of a method which uses Green's functions. We shall not give an account of this here, but shall restrict ourselves to indicating the changes in the expression for $\tilde{n}^2(\omega, \boldsymbol{s})$ that appear in comparison to the expression for $\tilde{n}^2(\omega, \boldsymbol{s})$ obtained by using perturbation theory (Sects. 6.1 and 6.3.1). As has been mentioned in Sect. 6.1, to determine the quantities $\tilde{n}^2(\omega, \boldsymbol{s})$ it is sufficient to know the tensor $\varepsilon_{\perp, ij}(\omega, \boldsymbol{k})$. According to [6.6], accounting for retardation leads to a situation in which the tensor $\varepsilon_{\perp, ij}(\omega, \boldsymbol{k})$ retains the form $\varepsilon_{\perp, ij} = \delta_{ij} + 4\pi b_{il}\eta_{lj}$, with the tensor b_{ij} being determined by (6.3.16). In this equation, however, the term $\mathrm{Im}\{V^a_{mm}(-\omega)\}$ is replaced by the imaginary additional term to the energy $\hbar\omega = W(\boldsymbol{k})$ which, in a first approximation, can be determined by applying an approximation procedure [6.34] and appears in (6.3.22). This relation holds with an accuracy to terms linear with respect to $\delta W(\boldsymbol{k})$ [6.6]. An experimental investigation of the long-wavelength edge may prove convenient in studying the shape of the exciton band for small values of $\boldsymbol{k}$ simply because light is considerably more weakly absorbed in the region of the long-wavelength edge than in the fundamental absorption band.

6.4 Raman Scattering of Light and X Rays Accompanied by Exciton Production. Influence of Spatial Dispersion on Energy Losses, and on Cherenkov and Transition Radiation of Charged Particles

By measuring the refractive index $n_l(\omega, \boldsymbol{s})$ for a given normal wave of type l, we can find the function $\omega_l(\boldsymbol{k})$ because $\boldsymbol{k} = (\omega/c)\tilde{n}_l(\omega, \boldsymbol{s})\boldsymbol{s}$. At the same time, the energy of the real exciton, corresponding to the normal wave, is equal to $\hbar\omega_l(\boldsymbol{k})$ and the function $\omega_l(\boldsymbol{k})$ can evidently be found both by the Raman scattering method and by nonoptical methods, such as neutron scattering or x-ray diffraction, etc. The application of nonoptical methods in exciton research is of particularly great interest due to the difficulties in investigating various effects of spatial dispersion encountered in crystal optics in some cases.

If by its nature an exciton is essentially coupled to ion vibrations, as occurs in ionic crystals for the so-called optical vibrations, an effective research method is offered by the investigation of inelastic neutron scattering [6.37, 38] and the Mössbauer effect [6.39, 40]. In the case of "electron-type" excitons, which are only weakly coupled to the lattice vibrations, other methods are available. These include, besides linear and nonlinear optical methods (Sect. 4.6), the measurement of the discrete energy losses when electrons pass through thin films [6.41] and the study of inelastic scattering of neutrons [6.42] and x rays in crystals. We shall begin with a discussion of this latter phenomenon [6.43].

6.4.1 Raman Scattering of X Rays Accompanied by Exciton Production

Assume that x rays of frequency ω_r are incident on a crystal and that the scattered radiation with the frequency ω_r' is observed at an angle Φ (Φ is the angle between the wave vectors $\boldsymbol{k}_r$ and $\boldsymbol{k}_r'$ of the incident and scattered waves). We presume, in this case, that a real exciton (polariton) is produced during the scattering process. From the conservation laws (we neglect exciton absorption here because it leads to distortion of the picture) it follows that

$$\omega = \omega_r - \omega_r' , \quad \boldsymbol{k} = \boldsymbol{k}_r - \boldsymbol{k}_r' , \tag{6.4.1}$$

where $\omega = \omega_l$ and $\boldsymbol{k}$ are the frequency and wave vector of the exciton. For long-wavelength excitons ($k = 2\pi/\Lambda \ll 2\pi/a$, a being the lattice constant) we obtain from (6.4.1) that

$$2\Lambda \sin\frac{\Phi}{2} = \lambda_r = \frac{2\pi c}{\omega_r} . \tag{6.4.2}$$

Here, the fact that for x rays $n(\omega_r) \approx n(\omega_r') \approx 1$ is taken into consideration. Owing to anisotropy at the given angle Φ and a definite orientation of the crystal, the frequency ω depends, in general, on the azimuthal angle as well (i.e., on the direction of vector $\boldsymbol{k}$ for a given angle Φ). Moreover, the frequencies $\omega(\boldsymbol{k})$ differ, of course, for excitons of different types, for instance, those that differ in the value of the discrete quantum number s (Sect. 6.1.2).

For $\lambda_r \approx 1$ to 3 Å and the smallest attainable angles $\Phi \approx 10'$, the exciton wavelength is $\Lambda \approx \lambda_r/\Phi \sim (3$ to $10) \times 10^{-6}$ cm, whereas at $\Phi \approx 1°$ to $3°$, $\Lambda \approx 2\times10^{-7}$ to 2×10^{-6} cm or $k = 2\pi/\Lambda \approx 3\times10^{6}$ to 3×10^{7} cm^{-1}. Energies typical of "electronic" excitons are in the range $\hbar\omega \approx 1$ to 5 eV, and the widths of the corresponding energy bands range from 0.01 to 1 eV.[5] At the same time the energy resolution of the x-ray spectral apparatus amounts to only 0.1 eV (at $\lambda_r \approx 3$ Å), thereby appreciably restricting the possibilities of the method. Therefore, it is only for excitons with $\hbar\omega \gtrsim 1$ eV and a bandwidth Δ of $\Delta \gtrsim 0.5$ eV that reliable information can be obtained on the function $\omega(\boldsymbol{k})$, provided that the intensity of the Raman scattering is sufficiently high.

The emission conditions (6.4.1) can be written in a form different from (6.4.2). For short-wavelength radiation (x rays) the medium can be regarded as isotropic and characterized by the refractive index $n_0(\omega_r)$. For the long-wavelength exciton under consideration $n = n_l(\omega, \boldsymbol{s})$, where $\boldsymbol{s} = \boldsymbol{k}/k$. Next, as always,

$$\boldsymbol{k}_r = \frac{\omega_r}{c} n_0(\omega_r)\boldsymbol{s}_r , \quad \boldsymbol{k}_r' = \frac{\omega_r'}{c} n_0(\omega_r')\boldsymbol{s}_r' , \quad \boldsymbol{k} = \frac{\omega}{c} n_l(\omega, \boldsymbol{s})\boldsymbol{s} .$$

---

[5] By way of comparison we point out that in scattering by free electrons the change in the energy of the x-ray photon, $\hbar\omega \approx (\hbar\omega_r/2mc^2)\Phi^2\hbar\omega_r$ at $\Phi \approx 1°$ and $\lambda_r \approx 3$ Å is $\hbar\omega \approx 4\times10^{-4}$ eV.

Then, for the angle θ_l between $\boldsymbol{k}$ and $\boldsymbol{k}_r$ we obtain from (6.4.1)

$$\cos\theta_l = \frac{\omega_r n_0(\omega_r)}{\omega n_l(\omega,s)} - \frac{\omega_r' n_0(\omega_r')}{\omega n_l(\omega,s)} \sqrt{1 - \left(\frac{\omega n_l(\omega,s)}{\omega_r' n_0(\omega_r')}\right)^2 \sin^2\theta_l}$$
$$= \frac{\omega_r^2 n_0^2(\omega_r) - \omega_r'^2 n_0^2(\omega_r') + \omega^2 n_l^2}{2\,\omega_r n_0(\omega_r)\,\omega n_l}\,. \qquad (6.4.3)$$

If $\omega \ll \omega_r$, then

$$\cos\theta_l \approx \frac{d\omega_r n_0(\omega_r)/d\omega_r}{n_l(\omega,s)} = \frac{c}{n_l(\omega,s)\,v_{\mathrm{gr}}(\omega_r)}\,. \qquad (6.4.4)$$

This condition coincides with the condition for Cherenkov radiation, see (6.4.45), of particles travelling with the velocity $v = v_{\mathrm{gr}}(\omega_r)$. In the present case we can also speak of Cherenkov radiation (emission of an exciton, i.e., a photon with the frequency ω in the medium), which is produced by short-wavelength radiation (a photon) with the frequency ω_r rather than by a particle. The fact that the group velocity $v_{\mathrm{gr}}(\omega_r)$ is involved in this process, instead of the particle velocity v, has a simple physical meaning. As a matter of fact, the relations (6.4.2 – 4) represent the conditions for satisfying the conservation laws or, in classical terms, the condition for interference (the classical picture adequately represents the problem here because Planck's constant $\hbar$ does not appear in the equation). These conditions are evidently the same for any emitter: charge, dipole, wave packet (pulse), etc. The intensity of radiation, on the contrary, depends essentially on the nature of the emitter. In particular, the principle of superposition is complied with within the framework of linear electrodynamics and the electromagnetic field, propagating in the medium, cannot produce Cherenkov radiation, i.e., the intensity of this radiation equals zero. When nonlinearity is taken into consideration, the wave packet is evidently equivalent to a certain emitter, travelling in the medium at a velocity equal to that of the packet, i.e., at the group velocity v_{gr} [6.44].

It is thus possible to regard this emission of excitons by x-ray photons as a form of Cherenkov radiation (it is clear from what follows that the intensity of this radiation is nonzero only on nonlinear grounds).

It is especially simple to calculate the intensity or the effective cross section for the process under consideration for the limiting case in which the frequency of the x radiation is high compared to all the natural frequencies of the medium. The effective coherent scattering cross section of nonpolarized radiation is determined by [Ref. 6.45, § 99]

$$d\sigma = r_0^2 \frac{1+\cos^2\Phi}{2} \left|\int N_0(\boldsymbol{r})\, \mathrm{e}^{i\boldsymbol{k}\boldsymbol{r}} d\boldsymbol{r}\right|^2 d\Omega\,, \qquad (6.4.5)$$

where $N_0(\boldsymbol{r})$ is the time-average electron concentration and $r_0 = e^2/mc^2$ and $d\Omega = \sin\Phi\, d\Phi\, d\varphi$.

The quantum-mechanical approach yields the same result if the expression $(e^2/2mc^2)N_0(\boldsymbol{r})A^2(\boldsymbol{r})$, where $\boldsymbol{A}$ is the vector potential of the radiation field, is used as the interaction energy in the Hamiltonian.

In considering Raman scattering of an electromagnetic wave in a medium (exciton), we can no longer assume that the concentration $N(\boldsymbol{r})$ is given and we can write:

$$\begin{aligned} N(\boldsymbol{r}) &= N_0(\boldsymbol{r}) + N'(\boldsymbol{r}), \\ 4\pi e N'(\boldsymbol{r}) &= \operatorname{div} \boldsymbol{E}_l = \mathrm{i}(\boldsymbol{k}\cdot\boldsymbol{e})E_l^0 \exp[\mathrm{i}(\boldsymbol{k}\cdot\boldsymbol{r} - \omega_l t)] . \end{aligned} \tag{6.4.6}$$

Here

$$\boldsymbol{E}_l(\boldsymbol{r}) = \boldsymbol{e}E_l^0 \exp[\mathrm{i}(\boldsymbol{k}\cdot\boldsymbol{r} - \omega_l t)]$$

is the electric field in the normal wave (exciton) of type l (the unit vector $\boldsymbol{e}$ characterizes the polarization). The energy associated with the excitation of this wave can be represented in the form of the excitation energy of an oscillator:

$$a_l(\boldsymbol{k})\boldsymbol{E}_l^*\cdot\boldsymbol{E}_l V = p_l^* p_l + \omega_l^2 q_l^* q_l = 2\omega_l^2 q_l^* q_l = \hbar\omega_l n , \tag{6.4.7}$$

where the last term represents the quantization ($n = 0, 1, 2, \dots$), V is the volume of the crystal and the coefficient $a_l(\boldsymbol{k})$ can be calculated in considering the energy of the system.

The energy of the interaction with x rays is then equal to, see (6.4.6, 7),

$$\begin{aligned} &\frac{e^2}{2mc^2}[N_0(\boldsymbol{r}) + N'(\boldsymbol{r})]A^2(\boldsymbol{r}) \\ &= \frac{e^2}{2mc^2}\left\{N_0(\boldsymbol{r}) + \frac{\mathrm{i}(\boldsymbol{k}\cdot\boldsymbol{e})q_l\omega_l\sqrt{2}}{4\pi e\sqrt{Va_l}} \exp[\mathrm{i}(\boldsymbol{k}\cdot\boldsymbol{r} - \omega_l t)]\right\}A^2(\boldsymbol{r}) . \end{aligned} \tag{6.4.8}$$

When the quantum number $n \gg 1$, the problem is a classical one and is reduced to the scattering of transverse electromagnetic waves (x rays) by waves (excitons) having a longitudinal field component [6]. Since there are practically no excitons in crystals in the initial equilibrium state, in scattering we

---

[6] This type of scattering is not restricted, of course, to the kinds of waves under consideration here. In the case of an isotropic plasma, for instance, we could study the scattering of transverse waves by longitudinal ones, or in a magnetically active plasma, the scattering of one kind of normal waves by another kind, provided that the latter are not strictly transverse [Ref. 6.46, § 39].

should consider the transition, in the exciton subsystem, from the state $n = 0$ to the state $n = 1$, see (6.4.7). The matrix element for this transition is

$$\langle 0|q_l \mathrm{e}^{-\mathrm{i}\omega_l t}|1\rangle = \sqrt{\hbar/2\omega_l}\,.$$

In this case the problem of x-ray scattering need not be solved again because we can obtain (6.4.5) from (6.4.8) with $q_l = 0$, and for $q_l \neq 0$ the calculations remain essentially unchanged. Hence, from (6.4.5, 8) we immediately arrive at the Raman scattering cross section:

$$\begin{aligned} d\sigma_k &= r_0^2 \frac{1+\cos^2\Phi}{2}(\boldsymbol{k}\cdot\boldsymbol{e})^2 \frac{V\hbar\omega_l}{(4\pi e)^2 a_l}\,d\Omega \\ &= r_0^2 \frac{1+\cos^2\Phi}{2}(\boldsymbol{k}\cdot\boldsymbol{e})^2 \frac{\hbar\omega_l N_0 V}{4\pi a_l m\omega_0^2}\,d\Omega\,, \end{aligned} \tag{6.4.9}$$

where $\omega_0^2 = (4\pi e^2/m)N_0$ is the plasma frequency and N_0 is the average electron concentration.

In incoherent scattering by free electrons, we have in (6.4.5)

$$|\int N_0(\boldsymbol{r})\,\mathrm{e}^{\mathrm{i}\boldsymbol{k}\boldsymbol{r}}d\boldsymbol{r}|^2 = N_0 V\,,$$

because $N_0(\boldsymbol{r}) = \sum_i \delta(\boldsymbol{r}-\boldsymbol{r}_i)$. Therefore, the ratio of the cross section, (6.4.9), to the cross section $d\sigma_0$ for scattering by free electrons is equal to

$$\frac{d\sigma_k}{d\sigma_0} = \frac{(\boldsymbol{k}\cdot\boldsymbol{e})^2\hbar\omega_l}{4\pi a_l m\omega_0^2} = \frac{\hbar\omega_r}{mc^2}\,\frac{\omega_r\omega_l}{\omega_0^2}\,\frac{\sin^2(\Phi/2)\cos^2\varphi}{\pi a_l}\,, \tag{6.4.10}$$

where φ is the angle between $\boldsymbol{k}$ and $\boldsymbol{e}$; moreover, the relation $k^2 = (2\omega_r/c)^2 \times \sin^2(\Phi/2)$, evident from (6.4.2), was used.

If the vector $\boldsymbol{k}$ is oriented along the corresponding symmetry axes or is arbitrarily directed in cubic crystals (neglecting spatial dispersion), the normal waves $(\boldsymbol{k}, l)$ can be classified as transverse and longitudinal waves. In the transverse waves $\boldsymbol{k}\cdot\boldsymbol{e} = 0$, and, in the approximation used, there is no Raman scattering, see (6.4.10), whereas for longitudinal waves $\cos\varphi = 1$ and $d\sigma_k \neq 0$. For longitudinal (plasma) waves in a gas of free electrons (plasma), the frequency $\omega_l = \omega_0$ and $a_l = 1/2\pi$, see (6.4.11), when spatial dispersion is neglected (this means that $\omega_0^2 \gg v_0^2 k^2$, where v_0 is a certain average velocity of the electrons). In the case of an arbitrary optically isotropic medium (metals, dielectrics, etc.) with sufficiently weak absorption, in the frequency region under investigation we have for longitudinal waves [see (2.3.16); consideration should be given to the fact that in (6.4.6) the field E is complex]

$$\varepsilon(\omega_l) = 0\,, \qquad a_l = \frac{\omega_l}{4\pi}\left(\frac{d\varepsilon}{d\omega}\right)_{\omega=\omega_l}, \tag{6.4.11}$$

where $\varepsilon(\omega)$ is the dielectric constant of the medium (for free electrons $\varepsilon = 1-\omega_0^2/\omega^2$ and $a_l = 1/2\pi$).

For a plasma of free electrons (i.e., for the Raman effect on plasmons) we have

$$\frac{d\sigma_{k,0}}{d\sigma_0} = \frac{\hbar k^2}{2m\omega_0} = 2\frac{\hbar\omega_r}{mc^2}\frac{\omega_r}{\omega_0}\sin^2\frac{\Phi}{2}. \tag{6.4.10a}$$

This equation is equivalent to that found in the classical calculation of scattering by a plasma wave whose charge density varies according to the law $N'\exp[\mathrm{i}(\boldsymbol{k}\cdot\boldsymbol{r}-\omega_0 t)]$, where

$$|N'| = \sqrt{\hbar k^2 N_0/2m\omega_0 V}.$$

This value is obtained for the amplitude when (6.4.6) is applied; the value of the electric field strength is deduced from the relation $E^*EV/2\pi = \hbar\omega_0$. At $\Phi\approx 3°$, $\hbar\omega_r \approx 5\times 10^3$ eV ($\lambda_r\approx 3$ Å) and $\hbar\omega_0\approx 10$ eV, the ratio $d\sigma_{k,0}/d\sigma_0 \approx 10^{-2}$.

Next we estimate the value of $d\sigma_k/d\sigma_0$, see (6.4.10), for the case of bound electrons in cubic crystals. Making use of (6.1.14a) we obtain

$$\varepsilon(\omega) = \varepsilon_\infty - \frac{8\pi|\boldsymbol{D}_{0,0l}|^2\omega_\perp}{v\hbar(\omega^2-\omega_\perp^2)},$$

where ε_∞ is determined by the contribution of the rest of the resonances. According to (6.4.11) we find that

$$a_l = \frac{\omega_\parallel}{4\pi}\left(\frac{d\varepsilon}{d\omega}\right)_{\omega=\omega_\parallel} = \frac{\varepsilon_\infty^2\omega_\parallel^2 v\hbar}{16\pi^2|\boldsymbol{D}_{0;0l}|^2\omega_\perp},$$

where $\omega_\parallel$ is a root of the equation $\varepsilon(\omega)=0$. On substituting into (6.4.10) we obtain

$$\frac{d\sigma_k}{d\sigma_0} = 16\pi\frac{\omega_r^2}{\omega_0^2}\frac{|\boldsymbol{D}_{0,0l}|^2}{mc^2 v}\sin^2\frac{\Phi}{2}. \tag{6.4.12}$$

This equation was obtained for $\varphi = 0$ (longitudinal waves) and $\omega_\perp/\omega_\parallel \approx 1$, which is valid when $d\sigma_k/d\sigma_0$ is determined as a function of the oscillator strength (in the linear approximation). Equation (6.4.12) can be obtained, of course, within the framework of a systematic microtheory [6.43]. For inelastic x ray scattering in an anisotropic crystal, the natural generalization of (6.4.12) is valid [6.43]:

$$\frac{d\sigma_k}{d\sigma_0} = 16\pi\frac{\omega_r^2}{\omega_0^2}\frac{|\boldsymbol{D}_{0;0l}|^2}{mc^2 v}\sin^2\frac{\Phi}{2}\cos^2\varphi.$$

Hence, in Raman scattering of x rays involving the production of dipole excitons, the cross section is proportional to the oscillator strength of the transition. For example, the most intensive dipole transitions in molecular crystals are found in anthracene (the second electron transition [6.47]), for which $|\boldsymbol{D}_{0;0l}|/e \approx 3\times 10^{-8}$ cm. Assuming $v \approx 125\times 10^{-24}$ cm^3, $\Phi \approx 3°$, $\cos^2\varphi \approx 1/3$ and $\omega_r/\omega_0 \approx 5\times 10^2$, we obtain: $d\sigma_k/d\sigma_0 \approx 4\times 10^{-3}$. Our estimates indicate that the intensity of Raman scattering is high enough for reliable measurement, provided that the corresponding oscillator strengths are not too small [6.48].

It was assumed in the foregoing that the exciton frequency ω is very low compared to the frequency ω_r of the incident radiation. We can, of course, consider a more general case of Raman scattering in which the inequality $\omega \ll \omega_r$ is not well satisfied. This may be the case, for instance, in Raman scattering of ultraviolet or visible light with the production of two photons having comparable frequencies. In this case the general equation (6.4.1) and, within limits, (6.4.3) as well, remain valid, but the radiation conditions (6.4.2, 4) are no longer applicable (this means that it is impermissible to neglect recoil, i.e., the change in the momentum of the incident photon). On the other hand, if we consider the scattering of visible light of frequency $\omega \gg \omega_l(\boldsymbol{k})$ by lattice vibrations, the situation resembles the above-mentioned one. Hence, (6.4.9) can be applied to estimate the intensity for scattering by dipole vibrations, provided that the frequency ω is within the transparency range of the crystal (for a more rigorous treatment, see Sect. 6.4.2). In the same case, but when the above inequality $\omega \gg \omega_l$ is not satisfied or when the light frequency is not within the transparency range, the situation is more complicated and its analysis requires a more detailed consideration [6.49 – 58]. It is also worth noting that the problem of taking spatial dispersion into account arises the study of light scattering close to points of second-order phase transitions [6.59 – 61].

6.4.2 Raman Scattering of Light by Polaritons

a) Raman Scattering of Light by Bulk Polaritons

Initially it seemed that Raman scattering with the production of excitons could be more readily observed in the x-ray range. Attempts at such observations have been made [6.48], but, by and large, Raman scattering at optical frequencies was found to be considerably more important for exciton research. This is undoubtedly due to the possibility of employing highly perfected lasers and interference apparatus. In Sect. 4.5.1 we mentioned Raman scattering of light with the production of polaritons (excitons) or, as they say, scattering by polaritons. Now we shall consider this process in more detail.

We repeat again that in the Raman scattering of light (RSL), comparatively high-frequency radiation [or, in quantum-mechanical language, a photon in a medium (polariton)] with frequency ω^{e} and wave vector $\boldsymbol{k}^{\mathrm{e}}$ produces or absorbs a relatively low-frequency polariton (exciton) wave with frequency ω and wave vector $\boldsymbol{k}$. In this process, the following relation holds

$$\omega^{\mathrm{e}} = \omega^{\mathrm{s}} \pm \omega \,, \qquad \boldsymbol{k}^{\mathrm{e}} = \boldsymbol{k}^{\mathrm{s}} \pm \boldsymbol{k} \,,$$

where ω^{s} and $\boldsymbol{k}^{\mathrm{s}}$ are the frequency and wave vector of the scattered wave (scattered photon in the medium); to avoid confusion, high-frequency polaritons e and s will be called, in the following, photons in a medium or simply photons. If these equations are multiplied by Planck's constant $\hbar$, they become the laws of conservation of energy and momentum (or, more exactly, the quasi-momentum [6.62]). Both photons in a medium and polaritons are quasi-particles (elementary excitations) of a crystal.

In the optical range, the wavelength of all the waves employed is large compared to the lattice constant. Therefore, the interaction energy of the waves (anharmonicity operator) that causes scattering can be expressed in terms of the macroscopic characteristics of the medium. In fact, the nonlinear part of the polarization due to lattice vibrations is a macroscopic quantity and the above-mentioned anharmonicity operator is

$$\hat{\mathscr{H}}' = -\tfrac{1}{2}\int \hat{\boldsymbol{E}}(\boldsymbol{r})\,\delta\hat{\boldsymbol{P}}(\boldsymbol{r})\,dv \,, \tag{6.4.13}$$

in which the nonlinear part of the polarization is

$$\delta\hat{P}_i(\boldsymbol{r}) = \chi_{ijl}\hat{E}_j(\boldsymbol{r})\hat{E}_l(\boldsymbol{r}) \,. \tag{6.4.14}$$

Here χ_{ijl} is the nonlinear polarizability tensor (in centrosymmetric crystals, $\chi_{ijl} = 0$).

Below we shall discuss only nonresonant RSL, i.e., we shall assume that the frequencies ω^{e} and ω^{s} lie well away from the resonances of the medium[7]. At the same time, the polariton frequency ω can lie within the region of lattice resonances, so that the dependence of the quantity χ_{ijl} on ω can, in general, be quite strong. In order to determine this dependence it is necessary to discuss the nature of nonlinear polarizability in greater detail.

The polarizability of a crystal, in the region of the frequencies ω^{e} and ω^{s}, is almost completely determined by the contribution of the electron transitions. The relative displacement of the charges $\boldsymbol{u}$ and the low-frequency electric field $\boldsymbol{E}^{\mathrm{p}}$ of the polariton they set up have only a weak effect on the electronic excitations of the crystal, leading to a certain change in the electronic polarizability. To a first approximation, this change can be written in the form of the sum of the deformation and electrooptic contributions:

---

[7] If the photon frequency ω^{e} is within the region of an exciton resonance, a number of special features appear in the processes of photon scattering (enhancement of RSL, etc.). See Sect. 6.3.3 and [6.33–36, 49–58].

$$\frac{1}{4\pi}\delta\varepsilon_{ij} = a_{ijl}u_l + b_{ijl}E_l^{\mathrm{p}}, \tag{6.4.15}$$

where the derivatives of $\varepsilon_{ij}(\omega, \boldsymbol{u}, \boldsymbol{E}^{\mathrm{p}})$

$$a_{ijl} = \frac{1}{4\pi}\frac{\partial\varepsilon_{ij}}{\partial u_l}, \qquad b_{ijl} = \frac{1}{4\pi}\frac{\partial\varepsilon_{ij}}{\partial E_l^{\mathrm{p}}} \tag{6.4.16}$$

should be taken at $\boldsymbol{u} = 0$ and $\boldsymbol{E}^{\mathrm{p}} = 0$. Consequently, like the dielectric-constant tensor of a crystal in an undistorted lattice, these derivatives can be regarded as constant quantities for the frequency region ω^{e} and ω^{s}. In cubic crystals belonging to the crystal classes T and T_d (we shall restrict ourselves here to a discussion of these)

$$a_{ijl} = a\,|e_{ijl}|, \qquad b_{ijl} = b\,|e_{ijl}|,$$

where e_{ijl} is a completely antisymmetric pseudotensor of rank three, denotes a modulus. A treatment on more general grounds can be found in [6.63 – 66].

In the region of an isolated dipole-active resonance $\omega \approx \omega_r$ in such crystals, the displacement $\boldsymbol{u}$ and field $\boldsymbol{E}^{\mathrm{p}}$ are connected by the relation (to take damping into account, see below)

$$u_i = \beta(\omega)E_i^{\mathrm{p}}, \qquad \beta(\omega) = \frac{e_r^*/M_r}{\omega_r^2 - \omega^2}, \tag{6.4.17}$$

where e_r^* and M_r are the effective charge and reduced mass, corresponding to the rth optical mode. Therefore, (6.4.15) can also be written in the form

$$\frac{1}{4\pi}\delta\varepsilon_{ij} = [\beta(\omega)a_{ijl} + b_{ijl}]E_l^{\mathrm{p}} \equiv \chi_{ijl}E_l^{\mathrm{p}},$$

where

$$\chi_{ijl} = [a\beta(\omega) + b]\,|e_{ijl}| \tag{6.4.18}$$

is the nonlinear susceptibility tensor, whose frequency dependence (having in mind the polariton frequency ω) has already been written explicitly.

Since the tensor χ_{ijl} is determined by two terms, their contributions are compensated by each other in many crystals for certain frequencies of the polariton branches of the spectrum. When damping is taken into consideration, (6.4.17) corresponds to the complex dielectric constant of the crystal

$$\varepsilon(\omega) = \varepsilon_\infty + \frac{\omega_r^2 F}{\omega_r^2 - \omega^2 - \mathrm{i}\omega\gamma_r}, \qquad F = \frac{4\pi N(e_r^*)^2}{VM_r\omega_r^2} \equiv \varepsilon_0 - \varepsilon_\infty, \tag{6.4.19}$$

ε_0 and ε_∞ being the low- and high-frequency values of $\varepsilon(\omega)$, which will be used below. At first, however, we shall neglect energy dissipation, assuming that $\gamma_r = 0$, as indicated above.

The electric field strength operator is determined by the relation [Ref. 6.21, Chap. III]

$$\hat{E}(r) = \sum_{\rho,k} [E^\rho(k)\,\xi_\rho(k)\,e^{ik\cdot r} + \overset{*}{E}{}^\rho(k)\,\xi_\rho^+(k)\,e^{-ik\cdot r}]\,, \tag{6.4.20}$$

where $\xi_\rho(k)$ and $\xi_\rho^+(k)$ are polariton (ρ, k), annihilation and creation operators, $E^\rho(k)$ is the amplitude of the electric field strength in the polariton (ρ, k), ρ being the spectral branch number and k the wave vector of the polariton; we call polaritons $(\rho \equiv e, k^e)$ and $(\rho \equiv s, k^s)$ photons in the medium.

Making use of the expression for the operator $\hat{\mathscr{H}}'$ given in (6.4.13, 14), we can readily find its matrix element corresponding to the RSL process [involving, for instance, the production of the polariton (p, k), which agrees with the change in the quantum number (occupation number) $n_p \to n_p + 1$]. The matrix element is determined by the relation

$$|\hat{\mathscr{H}}'|^{if} = (e_i^e e_j^s e_m^p \chi_{ijm})\,|E^e||E^s||E^p(k)|(1+n_p)^{1/2}\,V\Delta(k^e - k^s - k)\,,$$

where $\Delta(k)$ is the Kronecker symbol: $\Delta(k) = 0$, $k \neq 0$ and $\Delta(0) = 1$; e^e, e^s and e^p are unit vectors that determine the direction of the strength of the electric field of quasi-particles that participate in the RSL process. Therefore, the total probability of the photon-scattering process per unit time is

$$\begin{aligned}\mathscr{P} &= \frac{2\pi}{\hbar}\sum_f |\hat{\mathscr{H}}'|^{if}\delta(E^e - E^s - \hbar\omega)\\ &= \frac{2\pi}{\hbar^2}V^2\sum_{k^s} |e_i^e e_j^s e_m^p \chi_{ijm}|^2 |E^e|^2 |E^s|^2 |E^p|^2 (1+n_p)\\ &\quad \times \Delta(k^e - k^s - k)\,\delta(\omega^e - \omega^s - \omega)\,.\end{aligned}$$

It follows from this expression that the differential of the scattering probability is

$$d\mathscr{P} = \frac{2\pi}{\hbar^2}|\chi|^2|E^e|^2|E^s|^2|E^p|^2(1+n_p)\,\delta(\omega^e - \omega^s - \omega)\,V^2\frac{V(k^s)^2 d\Omega}{(2\pi)^3 v_{gr}^s}\,d\omega^s\,,$$

where $\chi \equiv e_i^e e_j^s e_m^p \chi_{ijm}$ and $v_{gr}^s = d\omega^s/dk^s$ is the group velocity of the scattered photon.

The quantity $d\mathscr{P}$ is the probability that the photon frequency after scattering falls within the frequency range $d\omega^s$, and the direction k^s, within an element of the solid angle $d\Omega$. If the quantity $d\mathscr{P}$ is divided by the initial photon flux, i.e., by the quantity $I = v^{\mathrm{e}}_{\mathrm{gr}} S/V$, where S is the total cross section of the specimen, we obtain the spectral differential cross section of the process:

$$\frac{\partial^2 \sigma}{\partial \Omega \partial \omega^{\mathrm{s}}} = \frac{2\pi}{\hbar^2} |\chi|^2 |E^{\mathrm{e}}|^2 |E^{\mathrm{s}}|^2 |E^{\mathrm{p}}|^2 \frac{L V^3 (k^{\mathrm{s}})^2 (1+n_p)}{(2\pi)^3 v^{\mathrm{e}}_{\mathrm{gr}} v^{\mathrm{s}}_{\mathrm{gr}}} \delta(\omega^{\mathrm{e}} - \omega^{\mathrm{s}} - \omega) , \tag{6.4.21}$$

where $\chi = (a\beta + b)(|e_{ijm}| e^{\mathrm{e}}_i e^{\mathrm{s}}_j e^{\mathrm{p}}_m)$ and, in the state of thermal equilibrium

$$n_p = [\exp(\hbar\omega/T) - 1]^{-1} .$$

Since the frequencies ω^{e} and ω^{s} lie far from the resonances in the medium, and the difference $|\omega^{\mathrm{e}} - \omega^{\mathrm{s}}| \ll \omega^{\mathrm{e}}$ or ω^{s}, we can ignore the difference between $|E^{\mathrm{e}}|$ and $|E^{\mathrm{s}}|$.

The value of the amplitude $E^\rho(k)$ for an arbitrary transverse polariton (ρ, k) is determined from the normalization condition, i.e., from the condition of equality of the energy of the electromagnetic field in a polariton to the magnitude $\hbar\omega$ of the quantum.

When spatial dispersion is neglected, the energy density of the electromagnetic field in a nonmagnetic dispersive medium is determined by (2.3.16). Hence, the amplitude E^ρ of the polariton [taking into account the fact that the factor 1/2 appearing in (2.3.5), is not explicitly present in (6.4.20)] is determined from

$$\frac{\hbar\omega_\rho}{V} = \frac{1}{4\pi} \left\{ \varepsilon(\omega_\rho) + \frac{\partial}{\partial \omega_\rho} [\omega_\rho \varepsilon(\omega_\rho)] \right\} |E^\rho|^2 ,$$

so that

$$|E^\rho|^2 = \frac{2\pi\hbar\omega_\rho}{V\left[\varepsilon(\omega_\rho) + \frac{1}{2}\omega_\rho \frac{d\varepsilon(\omega_\rho)}{d\omega_\rho}\right]} . \tag{6.4.22}$$

Hence, for high-frequency polaritons (photons) e and s

$$|E^{\mathrm{e}}|^2 \approx |E^{\mathrm{s}}|^2 = \frac{2\pi\hbar\omega^{\mathrm{s}}}{\varepsilon_\infty V} , \tag{6.4.23 a}$$

whereas, see (6.4.19),

$$|E^{\mathrm{p}}|^2 = \frac{2\pi\hbar\omega}{V\varepsilon_\infty} \left[1 + \frac{F/\varepsilon_\infty}{(\omega_r^2 - \omega^2)^2} \omega_r^4 \right]^{-1} . \tag{6.4.23 b}$$

Substituting the above values for the amplitudes into (6.4.21), and taking (6.4.17, 18) into account, we obtain the relation derived in [6.67] which has the form

$$\frac{\partial^2\sigma}{\partial\Omega\,\partial\omega^{\mathrm{s}}}=\frac{\Delta(\omega^{\mathrm{e}})^4}{c^4}\left|a+b\frac{M_r}{e_r^*}(\omega_r^2-\omega^2)\right|^2(|e_{ijm}|e_i^{\mathrm{e}}e_j^{\mathrm{s}}e_m^{\mathrm{p}})^2L$$

$$\times(1+n_p)\frac{\hbar}{2M_r\omega_r}S(\omega)\,\delta(\omega^{\mathrm{e}}-\omega^{\mathrm{s}}-\omega)\,, \tag{6.4.24}$$

where $\Delta=V/N$ is the volume of the unit cell, whereas

$$S(\omega)=\frac{F\omega\omega_r^3/\varepsilon_\infty}{(\omega_r^2-\omega^2)^2+F\omega_r^4/\varepsilon_\infty}\,. \tag{6.4.25}$$

The function $S(\omega)$ is commonly called the phonon-strength function [6.68]. It determines the ratio of the squared amplitude of ion displacement in a polariton with frequency ω to the analogous quantity for the corresponding optical phonon with frequency $\omega=\omega_r$. Since, as $\omega\to\omega_r$, the polariton is "converted" into an optical phonon, whose properties can be determined without accounting for retardation, it is natural that the function $S(\omega\to\omega_r)\to 1$. If, however, $\omega\ll\omega_r$ and, in addition, $|\omega-\omega_r|\gg F\omega_r^4/\varepsilon_\infty$, then

$$S(\omega)\approx\frac{F}{\varepsilon_\infty}\,\frac{\omega}{\omega_r}\ll 1\,.$$

An expression similar to (6.4.24) is also obtained for the scattering processes that involve the absorption of a polariton (for this purpose it is necessary to replace $1+n_p$ by n_p in (6.4.24) and to change the sign of ω in the δ function).

b) General Expression for the Scattering Intensity

The fact that (6.4.24) contains a δ function leads to zero width of the scattered radiation line. In order to obtain the correct result, it is necessary to account for the possibility of energy dissipation in the polariton. This requires a more general approach than that advanced above. Let us now discuss this approach.

It is known that light-scattering processes in a medium, including RSL by polaritons, are obtained when we take into account fluctuations of the tensor of the dielectric constant of the medium with respect to its average value appearing in Maxwell's equations. Owing to the fluctuation part of the dielectric-constant tensor, denoted below by $\delta\varepsilon_{\mu\nu}(\boldsymbol{r},t)$, a fluctuation part of the induction appears in Maxwell's equations:

$$\tilde{D}_i(\boldsymbol{r},t) = \delta\varepsilon_{ij}(\boldsymbol{r},t)E_j(\boldsymbol{r},t)\,, \tag{6.4.26}$$

so that the electric field strength equation assumes the form

$$\text{rot rot}\,\boldsymbol{E} + \frac{1}{c^2}\frac{\partial^2 \boldsymbol{D}}{\partial t^2} = -\frac{1}{c^2}\frac{\partial^2 \tilde{\boldsymbol{D}}}{\partial t^2} \approx -\frac{1}{c^2}\delta\,\varepsilon\frac{\partial^2 \boldsymbol{E}}{\partial t^2}$$

(the fluctuations are assumed to be slow, i.e., $|\partial\delta\varepsilon_{ij}/\partial t| \ll \omega|\delta\varepsilon_{ij}|$). To linear terms in $\delta\varepsilon$ the scattered wave field $\boldsymbol{E}^{\rm s}(\boldsymbol{r},t)$, satisfying the preceding equation, can be written in terms of its Green function $M_{ij}(\boldsymbol{r},\boldsymbol{r}',t,t')$ as follows:

$$E_i^{\rm s}(\boldsymbol{r},t) = \frac{1}{c^2}\int M_{ij}(\boldsymbol{r},\boldsymbol{r}';t,t')\frac{\partial^2 \tilde{D}_j(\boldsymbol{r}',t')}{\partial^2 t'}\,d\boldsymbol{r}'\,dt'\ .$$

The Green's function M_{ij} applied here satisfies the equation

$$\left(\text{rot}_{im}\,\text{rot}_{ml} + \frac{1}{c^2}\,\varepsilon_{il}\frac{\partial^2}{\partial t^2}\right)M_{lj}(\boldsymbol{r},\boldsymbol{r}';t,t') = -\delta_{ij}\delta(\boldsymbol{r}-\boldsymbol{r}')\delta(t-t')\,.$$

It is evident that for a homogeneous isotropic medium, the function M_{ij} is determined by

$$\begin{aligned} M_{ij}(\boldsymbol{r},\boldsymbol{r}';t,t') &\equiv \frac{1}{2\pi}\int d\omega M_{ij}(\boldsymbol{r}-\boldsymbol{r}',\omega)\,{\rm e}^{-{\rm i}\omega(t-t')} \\ &= \frac{1}{(2\pi)^4}\int d\omega\,d\boldsymbol{k}\,d_{ij}(\omega,\boldsymbol{k})\exp\{{\rm i}[\boldsymbol{k}\cdot(\boldsymbol{r}-\boldsymbol{r}')-\omega(t-t')]\}\,, \end{aligned}$$

where the tensor d_{ij} is given by [Ref. 6.7, § 28]

$$d_{ij}(\omega,\boldsymbol{k}) = \frac{\delta_{ij}-c^2k_ik_j/\omega^2\varepsilon(\omega,\boldsymbol{k})}{\varepsilon(\omega,\boldsymbol{k})\,\omega^2/c^2-k^2}\,.$$

In (6.4.26) the field $\boldsymbol{E}(\boldsymbol{r},t)$ can be regarded as given and as equal to the field $\boldsymbol{E}^{\rm e}$ of the exciting radiation (for instance, the radiation field of a laser). Setting

$$\boldsymbol{E}^{\rm e}(\boldsymbol{r},t) = \boldsymbol{E}^{\rm e}(\boldsymbol{k}^{\rm e},\omega^{\rm e})\exp[{\rm i}(\boldsymbol{k}^{\rm e}\cdot\boldsymbol{r}-\omega^{\rm e}t)]\,,$$

we find that the electric field strength of the radiation scattered by the medium is determined by

$$\boldsymbol{E}^{\rm s}(\boldsymbol{r},t) = \int \boldsymbol{E}^{\rm s}(\boldsymbol{k},\omega)\exp[{\rm i}(\boldsymbol{k}\cdot\boldsymbol{r}-\omega t)]\,d\boldsymbol{k}\,d\omega\,,$$

where

$$E^{\mathrm{s}}_{\alpha}(\boldsymbol{k}^{\mathrm{s}},\omega^{\mathrm{s}}) = \frac{(\omega^{\mathrm{e}})^2}{c^2}\int\frac{d\boldsymbol{r}'\,dt'}{(2\pi)^4}\,d_{\alpha\beta}(\boldsymbol{k}^{\mathrm{s}},\omega^{\mathrm{s}}) \times \delta\varepsilon_{\beta\gamma}(\boldsymbol{r}',t')\,E^{\mathrm{e}}_{\gamma}(\boldsymbol{k}^{\mathrm{e}},\omega^{\mathrm{e}})\exp[\mathrm{i}(\boldsymbol{k}\cdot\boldsymbol{r}'-\omega_p t')]$$

and

$$\boldsymbol{k}=\boldsymbol{k}^{\mathrm{e}}-\boldsymbol{k}^{\mathrm{s}}\,,\qquad \omega_p=\omega^{\mathrm{e}}-\omega^{\mathrm{s}}\,.$$

Next we shall write the expression for the average value with respect to fluctuations of Poynting's vector for scattered radiation:

$$\boldsymbol{S}=\frac{c}{4\pi}\,\langle\boldsymbol{E}^{\mathrm{s}}(\boldsymbol{r},t)\times\boldsymbol{H}^{\mathrm{s}}(\boldsymbol{r},t)\rangle\,.$$

Since the correlation function $\langle\delta\varepsilon_{\beta\gamma}(\boldsymbol{r}',t)\,\delta\varepsilon_{\beta'\gamma'}(\boldsymbol{r}'',t'')\rangle$ depends only on $\boldsymbol{r}'-\boldsymbol{r}''$ and $t'-t''$,

$$\boldsymbol{S}=\int d\omega^{\mathrm{s}}dk\,d\Omega\,S(\omega^{\mathrm{s}},\boldsymbol{k})\,,$$

where

$$S(\omega^{\mathrm{s}},\boldsymbol{k})=\boldsymbol{k}\,\frac{k^2c^2}{4\pi\omega^{\mathrm{s}}}\left(\frac{\omega^{\mathrm{e}}}{c}\right)^4[d_{ij}(\omega^{\mathrm{s}},\boldsymbol{k})\,\overset{*}{d}_{il}(\omega^{\mathrm{s}},\boldsymbol{k}) \times \varphi_{jmlm'}(\omega^{\mathrm{e}}-\omega^{\mathrm{s}},\boldsymbol{k}^{\mathrm{e}}-\boldsymbol{k}^{\mathrm{s}})\,E^{\mathrm{e}}_{m}(\boldsymbol{k}^{\mathrm{e}},\omega^{\mathrm{e}})\,\overset{*}{E}{}^{\mathrm{e}}_{m'}(\boldsymbol{k}^{\mathrm{e}},\omega^{\mathrm{e}})+\mathrm{c.c.}] \quad (6.4.27)$$

and

$$\varphi_{jmlm'}(\omega,\boldsymbol{k})=\int\frac{d\boldsymbol{r}\,dt}{(2\pi)^4}\langle\delta\varepsilon_{jm}(\boldsymbol{r},t)\,\delta\overset{*}{\varepsilon}_{lm'}(0,0)\rangle\exp\{\mathrm{i}(\boldsymbol{k}\cdot\boldsymbol{r}-\omega t)\}\,.$$

c) The Scattering Cross Section by Polariton. Linewidth

As has been pointed out, (6.4.15) can be used for finding the quantity $\delta\varepsilon_{\alpha\beta}$. On the other hand, since the displacement $\boldsymbol{u}(r,t)$ appearing in (6.4.15) determines the dipole moment $\boldsymbol{P}$ per unit volume, due to lattice vibrations,

$$\boldsymbol{P}=\frac{e^*}{\Delta}\boldsymbol{u}\,,$$

and the strength $\boldsymbol{E}^{\mathrm{p}}$ of the electric field corresponding to this polarization is determined by the Maxwell equation

$$\nabla\times\nabla\times\boldsymbol{E}^{\mathrm{p}}(\boldsymbol{r},\omega)-\varepsilon_{\infty}\frac{\omega^2}{c^2}\boldsymbol{E}^{\mathrm{p}}=4\pi\frac{\omega^2}{c^2}\boldsymbol{P}(\boldsymbol{r},\omega)\,,$$

for the polarization $\boldsymbol{P}$ and the field $\boldsymbol{E}^{\mathrm{p}} \sim \exp(-\mathrm{i}\omega t + \mathrm{i}\boldsymbol{k}\cdot\boldsymbol{r})$, the displacement $\boldsymbol{u}$ can be eliminated from (6.4.15). As a result, this relation, for example, corresponding to the excitation of a transverse polariton, can be presented for cubic crystals of the type of zincblende (ZnS) in the form

$$\delta\varepsilon_{\alpha\beta}(\boldsymbol{r},t) = 4\pi\left[a\frac{\Delta}{4\pi \overset{*}{e}}\left(\frac{k^2c^2}{\omega^2} - \varepsilon_\infty\right) + b\right]|e_{\alpha\beta\mu}|E_\mu^{\mathrm{p}}(\boldsymbol{r},t)\,,$$

where $E_\mu^{\mathrm{p}}(\boldsymbol{r},t) = e_\mu^{\mathrm{p}}|E^{\mathrm{p}}(\boldsymbol{r},t)|$, and $\boldsymbol{e}^{\mathrm{p}}$ is the polarization vector of the polariton. Since, according to our assumption, the frequency of the incident and the scattered radiation lies in the transparency region of the crystal, the aforementioned waves are transverse [for them: $k^2c^2 = \omega^2\varepsilon'(\omega)$, $\varepsilon = \varepsilon' + \mathrm{i}\varepsilon''$, $\varepsilon''(\omega) \ll \varepsilon'(\omega)$, $n(\omega^{\mathrm{e}}) \simeq n(\omega^{\mathrm{s}})$ and $\omega_{\mathrm{e}}/\omega_{\mathrm{s}} \approx 1$]. For this frequency region, the expression above for $d_{ij}(\omega,\boldsymbol{k})$ reduces to

$$d_{ij}(\omega,\boldsymbol{k}) = \frac{\delta_{ij} - k_ik_j/k^2}{\varepsilon(\omega)\,\omega^2/c^2 - k^2} = [\varepsilon(\omega)\,\omega^2/c^2 - k^2]^{-1}\sum_{\alpha=1,2} e_i^\alpha e_j^\alpha\,,$$

where $\boldsymbol{e}^\alpha$ (with $\alpha = 1,2$) are unit vectors of the transverse polarization, so that integration in (6.4.27) with respect to k can be performed if we take into account the fact that

$$\frac{1}{|\varepsilon(\omega)\,\omega^2/c^2 - k^2|^2} = \left(\frac{c}{\omega}\right)^4\frac{1}{(\varepsilon' - k^2c^2/\omega^2)^2 + (\varepsilon'')^2} \approx \frac{\pi L}{2k_s^2}\,\delta(k - k_s)\,,$$

where $k_s = \omega n(\omega)/c$, $n(\omega) = \sqrt{\varepsilon'(\omega)}$, $L \equiv c/2\omega\varkappa(\omega)$ and $\varkappa(\omega)$ is the imaginary part of the index of refraction at the frequency ω. To obtain the final result, it is necessary to know the fluctuation intensity.

The spectral function of electric-field fluctuation density is determined by [Ref. 6.7, §28]

$$\int d\boldsymbol{r}\,dt\langle E_\mu(\boldsymbol{r},t)E_{\mu'}(0,0)\rangle\exp(-\mathrm{i}\boldsymbol{k}\cdot\boldsymbol{r} + \mathrm{i}\omega t)$$
$$= \frac{4\pi\mathrm{i}\omega^2\hbar}{c^2}(1 + n_p)[d_{\mu\mu'}(\boldsymbol{k},\omega + \mathrm{i}\eta) - d_{\mu\mu'}(\boldsymbol{k},\omega - \mathrm{i}\eta)]\,, \qquad \eta \to +0\,,$$

where n_p is the Planck distribution.

Thus the spectral differential scattering cross section (per unit frequency interval and unit solid angle), for example, for transverse polaritons is

$$\frac{\partial^2\sigma}{\partial\omega^{\mathrm{s}}\partial\Omega} = \frac{\hbar(\omega^{\mathrm{e}})^4\omega^{\mathrm{s}}\Delta L}{\pi M_r c^4}|(e_\alpha^{\mathrm{s}}e_\beta^{\mathrm{e}}e_\gamma^{\mathrm{p}}|e_{\alpha\beta\gamma}|)|^2\left|a + b\frac{4\pi e^*}{\Delta}\,\frac{\omega_p^2}{k^2c^2 - \varepsilon_\infty\omega_p^2}\right|^2$$
$$\times\frac{(k^2c^2/\omega_p^2 - \varepsilon_\infty)^2}{F\omega_r^2}\,\mathrm{Im}\left\{\frac{1}{k^2c^2/\omega_p^2 - \varepsilon(\omega_p)}\right\}(1 + n_p)\,. \quad (6.4.28)$$

If the damping of the incident radiation is taken into account,

$$L \to L_{\text{eff}} \qquad \text{where}$$

$$L_{\text{eff}} = \left(\frac{1}{L(\omega^{\text{e}})} + \frac{1}{L(\omega^{\text{s}})} \right)^{-1} \approx \frac{L(\omega^{\text{e}})}{2} .$$

The scattering cross section σ has been determined as usual by the expression

$$\sigma = \frac{S_{\text{s}}(\omega^{\text{s}})}{\hbar\omega^{\text{s}}} \frac{\hbar\omega^{\text{e}}}{S_{\text{e}}(\omega^{\text{e}})} ,$$

$$S_{\text{e}}(\omega^{\text{e}}) = \frac{c}{2\pi} n(\omega^{\text{e}}) |E^{\text{e}}(\omega^{\text{e}}, k^{\text{e}})|^2 .$$

If, for $\varepsilon(\omega)$, we make use of (6.4.19), we obtain

$$\operatorname{Im}\left\{ \frac{1}{k^2c^2/\omega_p^2 - \varepsilon(\omega_p)} \right\}$$

$$= \frac{F\omega_r^2\gamma_r\omega_p^5}{[(k^2c^2 - \varepsilon_\infty\omega_p^2)(\omega_p^2 - \omega_r^2) - F\omega_r^2\omega_p^2]^2 + \omega_p^2\gamma_r^2(k^2c^2 - \varepsilon_\infty\omega_p^2)^2} .$$

The substitution of this relation into (6.4.28) yields the following expression for the spectral differential scattering cross section [6.56, 69 – 71]

$$\frac{\partial^2\sigma}{\partial\omega^{\text{s}}\partial\Omega} = \frac{(\omega^{\text{e}})^4 \Delta L_{\text{eff}}\hbar}{\pi c^4 M_r} (e_\alpha^{\text{e}} e_\beta^{\text{s}} e_\gamma^{\text{p}} |e_{\alpha\beta\gamma}|)^2 (1 + n_p) \left| a + b \frac{4\pi \overset{*}{e}_r \omega_p^2/\Delta}{c^2k^2 - \omega_p^2\varepsilon_\infty} \right|^2$$

$$\times \frac{\gamma_r\omega_p(\omega_p^2 - c^2k^2/\varepsilon_\infty)^2}{(\omega_p^2 - \omega_+^2)(\omega_p^2 - \omega_-^2)^2 + \gamma_r\omega_p^2(\omega_p^2 - c^2k^2/\varepsilon_\infty)^2} , \qquad (6.4.28\text{a})$$

where $\omega_\pm^2(k)$ are the larger and smaller roots of the dispersion equation at $\gamma_r = 0$, i.e., the equation

$$\frac{c^2k^2}{\omega^2} = \varepsilon_\infty + \frac{(\varepsilon_0 - \varepsilon_\infty)\omega_r^2}{\omega_r^2 - \omega^2} \equiv \varepsilon(\omega) .$$

Let us assume, for example, that the frequency $\omega = \omega^{\text{e}} - \omega^{\text{s}}$ is close to the frequency $\omega_-(k)$. In this case, the last of the terms in (6.4.28a) can be written in the form

$$\frac{\omega\gamma_r(\omega^2 - c^2k^2/\varepsilon_\infty)^2}{(\omega^2 - \omega_+^2)^2(\omega^2 - \omega_-^2)^2 + \gamma_r^2\omega^2(\omega^2 - c^2k^2/\varepsilon_\infty)^2}$$

$$\equiv \frac{1}{2\omega_r} S(\omega) \frac{\Gamma(\omega_-)/2}{(\omega-\omega_-)^2+[\Gamma(\omega_-)/2]^2},$$

where

$$\Gamma(\omega) = \gamma_r \frac{\omega}{\omega_r} S(\omega) = \frac{\omega^2 \omega_r^2 F}{F\omega_r^4 + \varepsilon_\infty(\omega_r^2-\omega^2)^2} \gamma_r . \tag{6.4.29}$$

Thus, the scattering line has a Lorentz shape with half-width $\Gamma(\omega)$. As the scattering angle is reduced, the polariton frequency decreases, as does the product $\omega_-(\boldsymbol{k})S[\omega_-(\boldsymbol{k})]$. This means that in reducing the scattering angle by the polaritons of the lower branch ω_-, the Stokes' frequency shift should decrease (this effect was first mentioned in [6.49]) and the line width of the scattered light should simultaneously become narrower.

The theory of light scattering by polaritons has been confirmed by a great number of experimental investigations. The law of polariton dispersion in the infrared region of the spectrum could determined for many kinds of crystals and, at the same time, the relevant dependence of the complex dielectric constant on the frequency has been found. Experimental investigations of the narrowing of Raman scattering lines by polaritons when the scattering angle is decreased provide the dependence of the quantity γ_r on ω according to (6.4.29) [6.72–75]. A wealth of information can be found in proceedings of numerous conferences devoted to the subject [6.70, 76–79]. For a comprehensive review, see [6.66].

We note that the application of the above method is especially convenient in investigating light scattering when the interface between media is taken into account. In this case the Green's function of the field-strength equation should be found by accounting for the pertinent boundary conditions. In addition, this provides information on light-scattering processes that involve the excitation of surface polaritons [6.80].

Equation (6.4.27), for the scattering intensity, is quite general and, in particular, can be applied to study light scattering close to phase-transition points. Here the quantity $\delta\varepsilon$ can be represented as an expansion with respect to the order parameter η. Hence the scattering intensity $I(\omega^s)$ is found to be proportional to the correlation functions $\langle\eta(\boldsymbol{r},t)\eta(0,0)\rangle$ if $\delta\varepsilon \sim \eta$, and to $\langle\eta^2(\boldsymbol{r},t)\eta^2(0,0)\rangle$ if $\delta\varepsilon \sim \eta^2$ (for a more detailed treatment, see [6.53]). Since the quantity $\int\langle\eta(\boldsymbol{r},t)\eta(0,0)\rangle \exp(-\mathrm{i}\boldsymbol{k}\cdot\boldsymbol{r}+\mathrm{i}\omega t)d\boldsymbol{r}\,dt$ increases close to the point $T = T_c$ of a phase transition of the second kind when we have small values of $\boldsymbol{k}$ and ω, the scattering cross section should also increase as $T \to T_c$. To date, a substantial number of papers have been devoted to the theory of light scattering in the vicinity of phase transitions [6.53, 78, 81].

d) Raman Scattering of Light by Surface Polaritons

Next we shall consider some special features of light scattering by surface polaritons. To begin with we note that the presence of an interface leads to a situation in which only the tangential component $\boldsymbol{k}_{\|}(k_1, k_2)$ of the momentum is conserved (so that $\boldsymbol{k}_{\|}^{\mathrm{e}} = \boldsymbol{k}_{\|}^{\mathrm{s}} \pm \boldsymbol{k}_{\|}$) along with the energy $\hbar\omega^{\mathrm{e}} = \hbar\omega^{\mathrm{s}} \pm \hbar\omega$ (ω being the surface polariton frequency) in RSL processes involving surface polaritons. At the present time RSL by surface polaritons forms the basis, along with attenuated total reflection of light (Chap. 5), of effective methods of studying the properties of surface waves. Since the characteristics of surface waves are determined by the properties of both media, the Raman scattering of light by surface polaritons offers new possibilities for studying the dispersion of media in the optical region of the spectrum (these possibilities are not available when light scattering by bulk polaritons is applied). We have already mentioned that Raman scattering of light by bulk polaritons does not occur in media having a center of symmetry, where the tensor χ_{ijl} of nonlinear polarizability, governing the intensity of the process, vanishes identically. However, the situation is radically changed when we investigate RSL by surface polaritons under conditions in which the medium being investigated has a common boundary with a medium lacking a center of symmetry. If the latter medium has no strong dispersion in the frequency region of the surface polariton and is transparent, its presence has only a slight effect on the dispersion law of the surface polariton and can be easily accounted for, [see (5.1.9a); this implies that $\varepsilon_1 = \mathrm{const}$]. Since the electromagnetic field of the surface polariton is nonzero at distances of the order of $1/k$ (k being the wave vector of the polariton) on both sides of the interface and, consequently, in the region of space (the substrate) where $\chi_{ijl} \neq 0$, the intensity of RSL by the surface polariton is nonzero and, as will be shown below, large enough to be observed. In some cases, the nonzero field of the surface polariton on both sides of the interface and the dependence of RSL intensity on the values of the tensor χ_{ijl} in the two contacting media may produce effects of opposite sign. The sum of such effects may reduce the intensity of RSL by a surface polariton for certain scattering angles (compensation effect). The possibility of this occurring (see below) is based on the fact that in the surface polariton the normal components of the electric-field strength differ not only in magnitude, but in sign as well (this follows directly from the boundary condition $D_{\mathrm{n}}^{\mathrm{I}} = D_{\mathrm{n}}^{\mathrm{II}}$).

In order to elucidate the aforesaid we shall first calculate the cross section for RSL by a surface polariton, assuming, for generality, that the tensor χ_{ijl} is nonzero in the two adjoining media. Subsequently, we shall consider more specific cases.

We assume that the interface between media I and II is in the plane $z = 0$ and that in medium I ($z > 0$) the tensor $\chi_{ijl} \equiv \chi_{ijl}^{\mathrm{I}}$, and in medium II ($z < 0$) the tensor $\chi_{ijl} \equiv \chi_{ijl}^{\mathrm{II}}$. Neglecting possible anisotropy of the media we assume the dielectric constant in region $z > 0$ to be equal to $\varepsilon_1 > 0$, and that in the region $z < 0$ to be negative ($\varepsilon_2 \equiv \varepsilon(\omega) < 0$). The dispersion of the surface polariton is determined by (5.1.9a). For the perturbation operator leading to RSL, we should consider an operator of the form

$$\hat{\mathscr{H}} = -\int \chi_{ijl}(z) \hat{E}_i^{\mathrm{e}}(\boldsymbol{r}) \hat{E}_j^{\mathrm{s}}(\boldsymbol{r}) \hat{E}_l^{\mathrm{p}}(\boldsymbol{r}) d\boldsymbol{r}\,, \tag{6.4.30}$$

where the dependence of the tensor χ_{ijl} on z is given explicitly, and where $\hat{E}^{e}$, $\hat{E}^{s}$ and $\hat{E}^{p}$ are the field-strength operators of the laser-radiation electric-field, the scattered-wave electric field and the electric field of the surface polariton. We shall neglect the difference in the refractive indices of laser radiation in media I and II, and thus not take into account the possibility (negligible in assessing the magnitude of the cross section for the process) of the reflection of the high-frequency field from the interface. Moreover, we assume that the laser beam propagates along the z axis (wave vector $\boldsymbol{k}^{e} = (0, 0, k^{e})$) and is polarized parallel to the x axis ($\boldsymbol{E}^{e} \| \boldsymbol{x}$), whereas the scattered light (wave vector $\boldsymbol{k}^{s} \equiv (k_1^{s}, 0, k_3^{s})$ is polarized along the y axis ($\boldsymbol{E}^{e} \| \boldsymbol{y}$) (Fig. 5.11). Then, in the surface polariton with the wave vector $k = k_1^{s} - k_1^{e}$, parallel to the x axis, the nonzero amplitudes of the electromagnetic field (Chap. 5) are of the form

$$
\begin{aligned}
E_1^{p} &= \mathrm{i}\frac{\varkappa_1}{k} E_3^{p} = \mathrm{i}\frac{c\varkappa_1}{\omega\varepsilon_1} H_2^{p}, \qquad \varkappa_1 = \sqrt{k^2 - \frac{\omega^2}{c^2}\varepsilon_1} \quad \text{at} \quad z > 0, \\
E_1^{p} &= -\mathrm{i}\frac{\varkappa_2}{k} E_3^{p} = -\mathrm{i}\frac{c\varkappa_2}{\omega\varepsilon} H_2^{p}, \qquad \varkappa_2 = \sqrt{k^2 - \frac{\omega^2}{c^2}\varepsilon} \quad \text{at} \quad z < 0 .
\end{aligned}
\tag{6.4.31}
$$

The absolute values of these amplitudes are determined from the condition of the equality of the energy of the surface-polariton electromagnetic field and the quantity $\hbar\omega$. Taking into account the nature of the dependence of the amplitudes on x, y and z (Chap. 5), as well as (6.4.31), we find that (with two isotropic media in contact) the energy densities are

$$
\begin{aligned}
W(\boldsymbol{r}) &= \frac{|E_1^{p}|^2}{4\pi}\left[\tilde{\varepsilon}_1\left(1 + \frac{k^2}{\varkappa_1^2}\right) + \frac{\varepsilon_1^2\omega^2}{c^2\varkappa_1^2}\right] \mathrm{e}^{-2\varkappa_1 z}, \quad z > 0; \\
W(\boldsymbol{r}) &= \frac{|E_1^{p}|^2}{4\pi}\left[\tilde{\varepsilon}\left(1 + \frac{k^2}{\varkappa_2^2}\right) + \frac{\varepsilon^2\omega^2}{c^2\varkappa_2^2}\right] \mathrm{e}^{2\varkappa_2 z}, \quad z < 0;
\end{aligned}
\tag{6.4.32}
$$

where the quantity $\tilde{\varepsilon} = \varepsilon + \omega d\varepsilon/d\omega$.

Consequently, the energy density (per 1 cm²) is

$$
\begin{aligned}
\rho &= \int_{-\infty}^{+\infty} W(\boldsymbol{r})\, dz = \frac{1}{4\pi}\frac{|E_1^{p}|^2}{\Phi} \\
\Phi(k) &= \left[k^2\left(\frac{\varepsilon_1}{\varkappa_1^3} + \frac{\varepsilon}{\varkappa_2^3}\right) + \frac{\omega}{2\varkappa_1}\left(1 + \frac{k^2}{\varkappa_1^2}\right)\frac{d\varepsilon_1}{d\omega} \right. \\
&\quad \left. + \frac{\omega}{2\varkappa_2}\left(1 + \frac{k^2}{\varkappa_2^2}\right)\frac{d\varepsilon}{d\omega}\right]^{-1}
\end{aligned}
\tag{6.4.33}
$$

and the energy flux is $I = \rho v_{gr}$, where $v_{gr} = d\omega/dk$.

From the condition $\hbar\omega = \int W dr$, see (2.3.16) and using (6.4.32), we obtain

$$|E_1^p|^2 = \frac{4\pi\hbar\omega}{S}\,\Phi(k)\,, \tag{6.4.34}$$

where S is the area of the interface.

If we take (6.4.31) and (5.1.9a) into consideration, the expression for $\Phi(k)$ can be written as

$$\Phi = \frac{\varkappa_2}{\varepsilon}(\varepsilon-\varepsilon_1)^{-1}\left(\frac{\omega}{2\varepsilon^2}\frac{d\varepsilon}{d\omega}+\frac{\omega}{2\varepsilon_1^2}\frac{d\varepsilon_1}{d\omega}+\frac{\varepsilon+\varepsilon_1}{\varepsilon\varepsilon_1}\right)^{-1}. \tag{6.4.33a}$$

We choose the initial phase so that the quantity E_1^p is both real and positive. In this case

$$E_3^p = -\mathrm{i}\frac{k}{\varkappa_1}\sqrt{\frac{4\pi\hbar\omega}{S}}\sqrt{\Phi} \quad \text{at} \quad z>0\,,$$

$$E_3^p = \mathrm{i}\frac{k}{\varkappa_2}\sqrt{\frac{4\pi\hbar\omega}{S}}\sqrt{\Phi} \qquad \text{at} \quad z<0\,.$$

The fact that the quantity E_3^p appearing in (6.4.30) (with the polarizations of the incident and scattered light chosen beforehand) differs in media I and II, not only in absolute value, but in sign as well, leads in some cases, owing to the compensation of the contributions, to the suppression of RSL by a surface polariton. This will be shown in the following.

As before, we set $\chi_{ijl} = \chi\,|e_{ijm}|$, e_{ijm} being a completely antisymmetric unit tensor of rank three. The matrix element of the operator, (6.4.30), corresponding to the RSL process with the production of a polariton, is equal, when we take the chosen polarizations of the incident and scattered light into account, to

$$(\hat{\mathscr{H}}')^{if} = \mathrm{i}k\sqrt{4\pi\hbar\omega S\Phi}\sqrt{1+n(\omega)}\left[\frac{\chi^{(1)}}{\varkappa_1(\varkappa_1-\mathrm{i}\sigma)}-\frac{\chi^{(2)}}{\varkappa_2(\varkappa_2+\mathrm{i}\sigma)}\right]|E_1^e||E_2^s|, \tag{6.4.35}$$

where E_1^e and E_2^s are the electric-field amplitudes in the exciting and the scattered radiation, $\chi^{(1)}$ and $\chi^{(2)}$ are the nonlinear polarizabilities in the spatial regions $z>0$ and $z<0$, respectively, $\sigma = k_3^e - k_3^s$, and $n(\omega) = [\exp(\hbar\omega/T)-1]^{-1}$.

The probability for the process under consideration is

$$\mathscr{P}^{if} = \frac{2\pi}{\hbar^2}\sum_f |\mathscr{H}^{if}|^2 V\frac{(k^s)^2 d\Omega d\omega^s}{(2\pi)^3 v_{gr}^s}\Delta(k_1^e - k_1^s - k)$$
$$\times[1+n(\omega)]\,\delta(\omega^e-\omega^s-\omega)\,,$$

where $\Delta(k)$ is the Kronecker symbol, and $v_{\rm gr}^{\rm s} = d\omega^{\rm s}/dk^{\rm s}$ is the group velocity of the scattered photon. Considering the spectral differential efficiency $\partial^2 I/\partial\Omega\,\partial\omega^{\rm s}$ of the process, we find that

$$\frac{\partial^2 I}{\partial\Omega\,\partial\omega^{\rm s}} = \frac{2\pi}{\hbar^2}|\mathcal{H}^{if}|^2 \frac{V^2(k^{\rm s})^2}{S(2\pi)^3 v_{\rm gr}^{\rm e} v_{\rm gr}^{\rm s}}\,\delta(\omega^{\rm e}-\omega^{\rm s}-\omega)\,.$$

Taking (6.4.35) into account, this can be written in the form

$$\frac{\partial^2 I}{\partial\Omega\,\partial\omega^{\rm s}} = \frac{4\pi\hbar(\omega^{\rm e})^2(k^{\rm s})^2\omega\,\Phi[1+n(\omega)]}{v_{\rm gr}^{\rm e} v_{\rm gr}^{\rm s}\varepsilon_2(\omega^{l})} \times\left|\frac{\chi^{(1)}}{\varkappa_1(\varkappa_1-{\rm i}\sigma)} - \frac{\chi^{(2)}}{\varkappa_2(\varkappa_2+{\rm i}\sigma)}\right|^2 \delta(\omega^{\rm e}-\omega^{\rm s}-\omega)\,. \qquad (6.4.36)$$

e) Compensation Effect

Before calculating numerical estimates of the process efficiency, let us consider certain special features of (6.4.36). In the region of small scattering angles θ, the quantity $\sigma \sim k^{\rm e}\theta^2/2$, whereas $\varkappa_2$ and $\varkappa_1 \sim k^{\rm e}\theta$. Hence, we have approximately

$$\left|\frac{\chi^{(1)}}{\varkappa_1(\varkappa_1-{\rm i}\sigma)} - \frac{\chi^{(2)}}{\varkappa_2(\varkappa_2+{\rm i}\sigma)}\right|^2 \approx \left(\frac{\chi^{(1)}}{\varkappa_1^2} - \frac{\chi^{(2)}}{\varkappa_2^2}\right)^2,$$

so that if for some k the equality $\varkappa_2^2/\varkappa_1^2 = \chi^{(2)}/\chi^{(1)}$ is satisfied, the probability for RSL by a surface polariton becomes zero ([6.82a]; for a more accurate treatment see [6.82b]). Obviously, such compensation of the contributions to the RSL cross section does not always occur. It is only possible in media for which the quantities $\chi^{(2)}$ and $\chi^{(1)}$ are simultaneously both positive or both negative. This is not the only condition; in fact, the ratio $\varkappa_2^2/\varkappa_1^2$ becomes infinite at $\omega = \omega_\perp$, and tends to unity at large values of k, see (5.1.4, 8): $\varepsilon_2(\omega) < 0$. Therefore, the compensation effect occurs only when the inequality $\chi^{(2)}/\chi^{(1)} > 1$ is satisfied. The compensation effect enables the values of χ to be compared for two media in contact, and if the value of χ is known for one medium, the value of χ can be found for the other.

Shown in Fig. 6.2 are curves illustrating the dependence of the intensity of Raman scattering by a surface polariton on the scattering angle under conditions when the compensation effect is manifested. This figure shows the results of calculations carried out in [6.82b] to determine the quantity $I(k_1)$ (the scattering angle θ is determined from the condition $\theta \sim (ck_1/\omega^{l})/\sqrt{\varepsilon_\infty})$ for surface polaritons propagating along the interface between two isotropic media. The linear and nonlinear dielectric constants for a GaP crystal in medium I and for a GaAs crystal in medium II were used in these calculations.

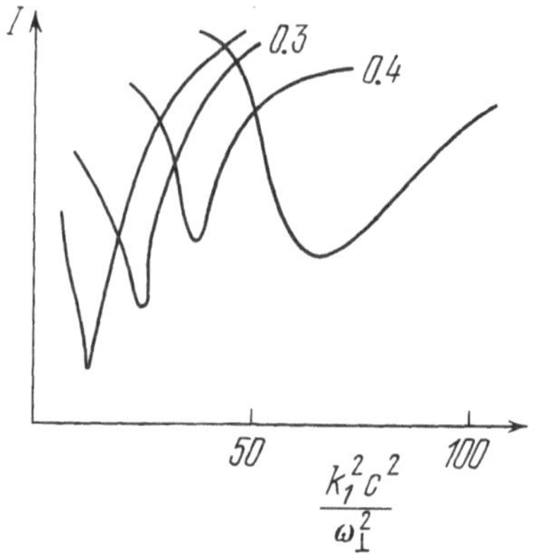

Fig. 6.2. The intensity of Raman-scattering of light as a function of k_x under conditions in which the compensation effect is manifested at various ratios of the nonlinear polarizabilities of the contacting media. For GaP and GaAs in contact this ratio is $\chi_1/\chi_2 = 0.3$

We now turn our attention to certain special cases and estimates of the RSL cross section. We assume at first that medium I is the vacuum. Then $\chi^{(1)} = 0$, $\varepsilon_1 = 1$, $\chi^{(2)} \equiv \chi$, and

$$\Phi(k) = \left[k^2\left(\frac{1}{\varkappa_1^3} + \frac{\varepsilon}{\varkappa_2^3}\right) + \frac{\omega}{2\varkappa_2}\left(1 + \frac{k^2}{\varkappa_2^2}\right)\frac{d\varepsilon}{d\omega}\right]^{-1} .$$

In the "nonrelativistic" limit, when $k \gg (\omega/c)\,|\,\varepsilon(\omega)\,|$,

$$\Phi(k) \approx k\frac{\omega_\perp^2}{\omega^2}\,\frac{(\varepsilon_0 - \varepsilon_\infty)}{(\varepsilon_\infty + 1)} .$$

In this case the integrated intensity with respect to the polariton linewidth

$$\frac{dI}{d\Omega} \approx \frac{4\pi\hbar(\omega^{\mathrm{e}})^4\omega_\perp^2\chi^2(\varepsilon_0 - \varepsilon_\infty)^2}{c^4 k\,\omega(\varepsilon_\infty + 1)}$$

decreases with increasing k. At $\omega^{\mathrm{e}}/c = 10^5$, $k = 10^3$, $\omega_\perp = 10^{13}$, $1 + \varepsilon_\infty = 10$, $\varepsilon_0 - \varepsilon_\infty = 3$ and $\chi = 10^{-6}$ (for the sake of definiteness we use the data for GaP in the radiation region of a He – Ne laser; all quantities are in cgs units), we obtain $dI/d\Omega \approx 4 \times 10^{11}$, which is quite sufficient to observe RSL by a surface polariton (primarily, of course, this is indicated by the fact that RSL has been observed [6.83, 84, 84a]). It should be kept in mind that bulk polaritons are also excited in RSL processes in the situations under discussion, together with the surface polaritons. Therefore, when thick crystals are used, processes involving the excitation of surface waves will be "lost" in the background of bulk processes. This presents difficulties in observing RSL by surface polaritons and requires the use of thin layers [6.83, 84].

f) Broken Symmetry Method

It follows from (6.4.36) that even in the case when medium II has a center of symmetry (i.e., when $\chi^{(2)} = 0$), the cross section for RSL by a surface polariton in the spectral region where $\varepsilon(\omega) < 0$ is nevertheless nonzero, provided that $\chi^{(1)} \neq 0$ for medium I (i.e., the substrate). As mentioned in [6.85], this permits the spectra of surface polaritons for centrally symmetrical media to be investigated by means of RSL and, consequently, the determination of the dielectric constant of these media.

This method has many features worth mentioning. For example, bulk polaritons of a centrosymmetric medium are not excited at all in the situation under consideration. It is important, therefore, to keep any kind of overtones or combination tones of the nonlinear substrate, active in RSL spectra, out of the surface-wave spectral region. If this cannot be avoided, the role of such overtones can be diminished, other things being equal, if the nonlinear substrate is sufficiently thin. An estimate of the RSL cross section by means of the impaired-symmetry method indicates that for the previously selected values of the parameters this cross section is found to be about the same as for a medium with $\chi \neq 0$, adjoining a vacuum.

Next we shall discuss [6.86] still another possibility, in principle, of exciting polaritons in a centrosymmetric crystal. It does not involve the artificial choice of a nonlinear substrate (i.e., a substrate with $\chi_{ijl} \neq 0$). This possibility makes use of ionic crystals. When these crystals are in a state of thermodynamic equilibrium, the concentrations of anion and cation vacancies are equal to each other, in general, only at some distance from the interface. In the subsurface region, owing to the different effects of the crystal boundary on the energy required to form vacanies of opposite signs, the local neutrality is violated and, as was shown in [6.87], an electric double layer is formed with a thickness of the order of hundreds of angstroms. The equilibrium distribution of vacancies within this layer was determined in [6.88]. Obviously, the layer region behaves as if it were pyroelectric with variable spontaneous polarization along the layer thickness. This polarization is perpendicular to the surface of the crystal. More important to us here is the fact that the layer region is also piezoelectric, because deformation of the layer should be accompanied by additional polarization in the layer, proportional to the amount of deformation. Such surface piezoelectricity may be the cause of a number of phenomena and, in particular, it may influence the spectra of Rayleigh surface waves and their damping. Moreover, certain nonlinear optical processes become allowed in the region of a surface piezoelectric layer. Such processes include RSL by polaritons or frequency doubling, which are forbidden at some distance from the boundary of the crystal due to the presence of a center of symmetry. The problems posed in this connection, in particular because of the feasibility of their experimental investigation, require special consideration.

Let us now discuss the linewidth of RSL by a surface polariton.

Line broadening, due to dissipation processes, can be found from the condition [6.85] giving the pole of the Green's function for an electromagnetic field in the presence of an interface. This condition determines the dispersion law for surface polaritons and, consequently, see (5.1.9a), is of the form [the vector $\boldsymbol{k} \equiv (k_1, k_2, 0)$ is two-dimensional]

$$A(\omega) \equiv \varepsilon(\omega) + \frac{\varepsilon_1 k^2}{k^2 - \varepsilon_1 \omega^2/c^2} = 0\,, \tag{6.4.37}$$

where, when damping processes are taken into account, $\varepsilon(\omega)$ is given by (6.4.19). Since the polariton frequency is several hundred times less than the frequency of the laser radiation, the difference in the lengths of the vectors $\boldsymbol{k}^{\mathrm{e}}$ and $\boldsymbol{k}^{\mathrm{s}}$ can be ignored in determining the polariton wave vector $\boldsymbol{k} = \boldsymbol{k}_{\mathrm{t}}^{\mathrm{e}} - \boldsymbol{k}_{\mathrm{t}}^{\mathrm{s}}$. This means that in RSL processes in which the wave vectors $\boldsymbol{k}^{\mathrm{e}}$ and $\boldsymbol{k}^{\mathrm{s}}$ form the specified angle θ, polaritons are created in the crystal with the same value of the wave vector $\boldsymbol{k}$, a value given by the conditions of the experiment. However, owing to dissipation processes, this value of $\boldsymbol{k}$ corresponds to a whole set of polariton frequencies, rather than to a single frequency value. To determine the width of this set, the real and imaginary parts of the expression for $A(\omega)$ should be separated out, assuming that ω and k are real, writing the following equation up to the unessential common factor $C(k)$:

$$A(\omega) = C[\omega - \omega_s(k) + \mathrm{i}\Gamma(k)]\,,$$

where $\omega_s(k)$ is the dispersion of the surface polariton ($\mathrm{Re}\{A(\omega_s)\} = 0$) and $\Gamma(k)$ is the required width. Since, as a rule, $\omega_s \gg \Gamma(k)$ for the waves under consideration, the quantity $\Gamma(k)$ can be determined by

$$\Gamma(\omega_s) = \mathrm{Im}\{A(\omega_s)\} \left(\frac{d}{d\omega} \mathrm{Re}\{A\} \right)^{-1}_{\omega = \omega_s} .$$

Up to quadratic terms with respect to γ and taking (6.4.19) into consideration, we obtain in the above approximation

$$A(\omega) = \varepsilon_\infty + \frac{\omega_\perp^2(\varepsilon_0 - \varepsilon_\infty)}{\omega_\perp^2 - \omega^2} + \frac{\varepsilon_1 k^2}{k^2 - \varepsilon_1 \omega^2/c^2} - \frac{2i\omega\gamma\omega_\perp^2(\varepsilon_0 - \varepsilon_\infty)}{(\omega_\perp^2 - \omega^2)^2} .$$

Thus, the linewidth for a surface polariton with the frequency $\omega = \omega_s(k)$ is determined by

$$\Gamma(\omega) = \frac{\varepsilon_1 \omega^2 \omega_\perp^2 (\varepsilon_0 - \varepsilon_\infty)}{\varepsilon_\infty(\varepsilon_1 + \varepsilon_\infty)(\omega_s^2 - \omega^2)^2 + \varepsilon_1 \omega_s^2 \omega_\perp^2 (\varepsilon_0 - \varepsilon_\infty)} \gamma\,,$$

where ω_s is the root of equation $\varepsilon_1 + \varepsilon(\omega) = 0$, i.e., the quantity $\omega = \omega_s(k)$ at large values of k.

Let us consider the limiting cases $k^2 \gg \varepsilon_1 \omega^2/c^2$ and $k^2 \approx \varepsilon_1 \omega_\perp^2/c^2$. It can readily be seen that in the first case $\Gamma(k) \approx \gamma$. In the second case

$$\Gamma(k) = \gamma k^2 [k^2 + (\varepsilon_0 - \varepsilon_\infty)\,\omega_\perp^2/c^2]^{-1}\,.$$

Since at large values of k this expression tends to the correct limit ($\Gamma(k) \to \gamma$)., it can be used as an interpolation expression for the whole region of allowed values of $k (k \geqslant (\omega_\perp/c)\,\varepsilon_1^{1/2})$. In particular, for $k^2 = \varepsilon_1 \omega_\perp^2/c^2$ the quantity $\Gamma(k) = \gamma \varepsilon_1 (\varepsilon_1 + \varepsilon_0 - \varepsilon_\infty)^{-1}$. Hence, at $\varepsilon_0 - \varepsilon_\infty > \varepsilon_1$ there is a substantial narrowing of the RSL line as the scattering angle is decreased. If, on the contrary, $\varepsilon_0 - \varepsilon_\infty \ll \varepsilon_1$, the quantity Γ is practically independent of k.

We note that to a linear approximation in γ, the quantity $\omega_s(k)$ is equal to the surface polariton frequency when damping processes are neglected (at $\gamma = 0$; it is a simple matter to carry out the corresponding, more general calculations). The dependence of $\Gamma(k)$ for multilayer media can also be dealt with in a similar way. It should be emphasized, however, that the expressions given above for the linewidth of RSL by a surface polariton are meaningful and can be applied only in cases when, taking dissipation into account, the spectra of RSL by bulk and surface polaritons do not overlap. Otherwise, interference occurs between the various scattering processes and the expression for the intensity becomes considerably more complicated. The theory required for these conditions has been developed in [6.80 – 82, 89], in the last two papers the method of calculation is based on the application of the fluctuation-dissipation theorem; the Green's function for an electromagnetic field was used in [6.80] (for a general review, see [6.84a]).

We note here that in the calculation of the RSL intensity due to surface polaritons the reflection and refraction of the exciting and scattered light at the interface between the media was ignored. It was therefore assumed that both the exciting radiation (for instance, laser light) and the scattered wave have the form of the same type of plane waves, rather than their superposition, throughout the whole space, notwithstanding the presence of an interface between the media. Since in the initial and final states, corresponding (within the framework of perturbation theory) to the states of the zeroth approximation with respect to anharmonicity, the field should satisfy linear Maxwell's equations, it follows that these fields should also satisfy correct boundary conditions. In the general case, however, the states satisfying the correct boundary conditions are degenerate; this poses the problem of finding the correct linear combinations.

The aforesaid can be clarified by the previously discussed simplest example of RSL from a semi-infinite medium ($z > 0$). Assume that photons are incident on the boundary of this medium from the side of negative values of z. At the boundary some these photons are reflected and some are transmitted,

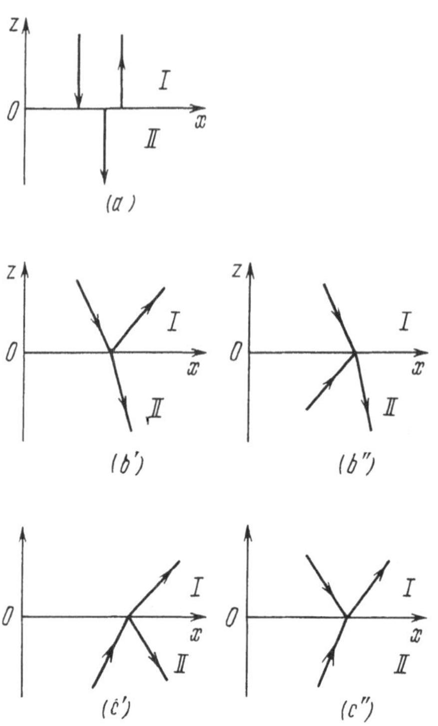

Fig. 6.3a – c. Direction of the wave vector of exciting and scattering radiation with the reflection and refraction of the waves at the interface between the media taken into account.

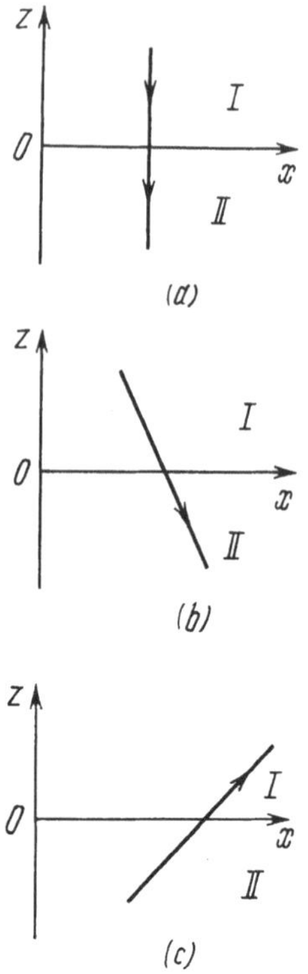

Fig. 6.4a – c. Similar to Fig. 6.3, but with reflection and refraction neglected at the interface between media.

as is illustrated for normal incidence in Fig. 6.3a. Here the initial state is not degenerate.

However, as a result of RSL by a surface polariton (with the frequency ω_p and wave vector k_x) a scattered photon appears with a projection of its wave vector on the x axis equal to k_x. Shown schematically in Fig. 6.4b, c are plane waves, the same throughout the whole space and corresponding to the scattered photons (in a forward scattering geometry in Fig. 6.4b and in a backward scattering geometry in Fig. 6.4c). These plane waves were previously considered in calculating the intensity of RSL by surface polaritons, there the reflection and refraction of the high-frequency field at the interface was neglected. Two linearly independent states are shown in Fig. 6.3b′, c′ (with the same frequency ω and with the same the projections of their wave vectors on the x axis equal to k_x), in which the transformation of the waves at the boundary has exactly been taken into account. Two other linearly independent states are shown in Fig. 6.3b″, c″, each being a linear superposition of the first two. This poses the question: which of the above-mentioned states with given (k_x, ω) should be chosen as the final ones in calculating the probability of RSL by surface polaritons according to perturbation theory? The answer to this question can be obtained [6.82b] by applying the results of a more general theory, namely, the theory of scattering with energy dissipation in the medium, if, in the final expression for the RSL intensity, we pass to the limit of weak dissipation. It was found that the correct final states are the states presented in Fig. 6.3b″ (forward scattering geometry) and in Fig. 6.3c″ (backward scattering geometry). We also note that the problem under discussion is quite general and an analogous choice of states must be made in the theory for any nonlinear processes that take place in the boundary region of the specimen (see, e.g., [6.55] which deals with resonance Mandelstam-Brillouin scattering in semiconductors).

In conclusion, we give the expressions for the intensity of RSL by surface polaritons when the reflection and refraction of the high-frequency field by the interface are taken into account. As shown in [6.82b], these expressions are of the form (I_t is the intensity of RSL by transmission and I_r is the reflected intensity)

$$\frac{\partial^2 I_t}{\partial\omega^s\,\partial\Omega} = \frac{4\pi\hbar(\omega^e\omega^s)^2\omega\,\Phi[n_p(\omega)+1]\,n_{II}}{c^4 n_I}\,t_e^2|\chi|^2 \times\frac{k^2}{\varkappa^2}\left|\frac{1}{|k_z^e|-|k_z^s|+\mathrm{i}\varkappa}-\frac{r_s}{|k_z^e|+|k_z^s|+\mathrm{i}\varkappa}\right|^2\delta(\omega^e-\omega^s-\omega)\,,$$

$$\frac{\partial^2 I_r}{\partial\omega^s\,\partial\Omega} = \frac{4\pi\hbar(\omega^e\omega^s)^2\omega\,\Phi[n_p(\omega)+1]}{c^4}\,t_e^2 t_s^2|\chi|^2 \times\frac{k^2}{\varkappa^2}\left|\frac{1}{|k_z^e|+|k_z^s|+\mathrm{i}\varkappa}\right|^2\delta(\omega^e-\omega^s-\omega)\,,$$

where

$$t_e = 2n_I(n_I+n_{II})^{-1}\,,$$
$$t_s = 2n_I\cos\theta_1(n_I\cos\theta_1+n_{II}\cos\theta_2)^{-1}\,,$$
$$r_s = (n_I\cos\theta_1-n_{II}\cos\theta_2)(n_I\cos\theta_1+n_{II}\cos\theta_2)^{-1}\,,$$

n_I and n_{II} are refractive indices at the frequency $\omega^e\approx\omega^s$ in the media I and II, and θ_1 and θ_2 are the angles of incidence and refraction of scattered radiation. In the case of RSL with forward geometry, $|k_z^e|\approx|k_z^s|$ and the inequality $I_t \gg I_r$, discussed in Chap. 5, follows from the above relations.

g) Coherent Anti-Stokes Raman Spectroscopy (CARS)

The previous discussion concerned spontaneous, or thermal, Raman scattering of light by polaritons. As has been pointed out, this process is due to the fluctuations of the low-frequency electromagnetic field in the medium, which lead to fluctuations of the dielectric constant. In recent years, along with investigations of the processes of spontaneous scattering, there have been significant developments in coherent anti-Stokes Raman spectroscopy (CARS) [6.90]. In CARS, Raman scattering of light by the excitations of the medium (polaritons, phonons, etc.), which are specially produced by means of some auxiliary process that proceeds independently is observed. It is assumed that the scattering intensity is not yet strong enough to lead to stimulated RSL. Typical of scattering by such excitations is a drastic growth in intensity (compared to scattering by thermal fluctuations). In many cases this offers new opportunities for investigating the spectra of elementary excitations in condensed media. Among other factors, the method of coherent active spectroscopy of Raman scattering by polaritons is based on the use of their excitation in the field of two sufficiently intense laser beams. These beams have the frequencies $\omega_1(\boldsymbol{k}_1)$ and $\omega_2(\boldsymbol{k}_2)$, whose difference is equal or close to the polariton frequency $\omega_p(\boldsymbol{k})$, where $\boldsymbol{k} = \boldsymbol{k}_1 - \boldsymbol{k}_2$.

Since polaritons are coherent elementary excitations, i.e., characterized not only by their frequency, but by their wave vector as well[8], the intensity of the forced vibrations of the field and, consequently, the intensity of scattered light (the signal) depends not only on the difference in the frequencies of the exciting fields, but also on the difference of their wave vectors. The investigation of this dependence (each time at a fixed difference $\omega_1 - \omega_2$), i.e., so to speak, spectroscopy in $\boldsymbol{k}$ space [6.94], is an independent method for studying the dispersion relation of polaritons. Let us consider the theoretical problems that are posed [6.95–97]. If ω_p is the frequency of the probe wave, then, in the process under consideration, the frequency of the scattered wave $\omega^{\mathrm{s}} = \omega_p \pm \Omega$, where $\Omega = \omega_1 - \omega_2$. In quantum-mechanical language we can say that the process is initiated by the collision of a photon of the probe wave with a photon in the radiation from one of the lasers. This collision, however, must be one in which two photons are produced, one being a quantum of scattered radiation with the frequency ω^{s}, whereas the second has the frequency of the other laser radiation. Thus the scattering process is a four-photon process. Therefore, its intensity in a centrosymmetric crystal is completely determined by the nonlinear polarizability $\chi^{(3)}_{ijlm}$ (in this case the four-photon process is said to be direct). However, in acentric crystals, in which the tensor $\chi^{(2)}_{ijl}$ is also nonzero, the intensity of scattering of the probe wave is determined by the contribution of the so-called cascade three-photon processes, and not only that of the direct four-photon processes.

The possibility that these cascade processes can occur becomes clear when considering the polarization in a medium due to the effect of waves with frequencies ω_1 and ω_2, i.e.,

$$P_i(\boldsymbol{r},t) = \chi^{(2)}_{ijl} E^{(1)}_j \overset{*}{E}{}^{(2)}_l \exp\{\mathrm{i}(\boldsymbol{k}_1 - \boldsymbol{k}_2) \cdot \boldsymbol{r} + (\omega_1 - \omega_2)t]\}.$$

[8] CARS with scattering by incoherent excitations has been discussed in [6.91–93].

This polarization leads to the appearance of an electric field (for the sake of simplicity we shall assume that the medium has cubic symmetry), whose strength is determined by the relation

$$E_i(\boldsymbol{r},t) = \frac{4\pi P_i(\boldsymbol{r},t)}{c^2Q^2/\Omega^2 - \varepsilon(\Omega)},$$

where $\Omega = \omega_1 - \omega_2$, and $\boldsymbol{Q} = \boldsymbol{k}_1 - \boldsymbol{k}_2$. The interaction of this field with the probe wave (assumed to have the frequency ω_1) leads to polarization at the Stokes ($\omega^{\mathrm{s}} = \omega_1 - \Omega$) and anti-Stokes ($\omega^{\mathrm{a}} = \omega_1 + \Omega$) frequencies. In particular, this polarization in the latter case is

$$P_i^{\mathrm{a}} = \chi_{ijl}^{(2)} E_j^{(1)} E_l \exp\{\mathrm{i}[(\boldsymbol{Q}+\boldsymbol{k}_1)\cdot \boldsymbol{r} + (\Omega + \omega_1)t]\}.$$

For the same case, when the direct four-photon processes are also taken into account, we arrive at the following expression for the total polarization:

$$P_i^{\mathrm{tot}}(\boldsymbol{r},t) = \chi_{ijlm} E_j^{(1)} E_l^{(1)} \overset{*}{E}{}_m^{(2)} \exp\{\mathrm{i}[(2\boldsymbol{k}_1 - \boldsymbol{k}_2)\cdot \boldsymbol{r} + (2\omega_1 - \omega_2)t]\},$$

where

$$\chi_{ijlm} = \chi_{ijlm}^{(3)} + \frac{4\pi}{c^2Q^2/\Omega^2 - \varepsilon(\Omega)} \chi_{ijn}^{(2)} \chi_{nlm}^{(2)}.$$

It can be shown [6.95] that this expression corresponds to the following intensity of anti-Stokes scattering:

$$I(\omega^{\mathrm{a}}) \sim I_1^2 I_2 |\chi|^2 L^2 \left(\frac{\sin(\Delta k_{\mathrm{a}} L/2)}{\Delta k_{\mathrm{a}} L/2}\right)^2,$$

where $\chi = \chi_{ijlm} e_i^{\mathrm{a}} e_j^{(1)} e_l^{(1)} e_m^{(2)}$, and $\Delta \boldsymbol{k}_{\mathrm{a}} = 2\boldsymbol{k}_1 - \boldsymbol{k}_2 - \boldsymbol{k}_{\mathrm{a}}$, and $\boldsymbol{k}_{\mathrm{a}}$ is the wave vector of a photon with the frequency $\omega^{\mathrm{a}} = 2\omega_1 - \omega_2$ in the medium. The vanishing of $\Delta \boldsymbol{k}_{\mathrm{a}}$ (the so-called synchronism condition) corresponds to the maximum of the scattering intensity[9]. At the same time, the relative contribution of the cascade processes, as follows from the expression for χ at fixed values of the frequencies ω_1 and ω_2, has a maximum for the orientations of vectors $\boldsymbol{k}_1$ and $\boldsymbol{k}_2$ at which the following condition, for the excitation of polaritons $(\boldsymbol{Q}, \Omega)$ in the medium, is met $|\boldsymbol{k}_1 - \boldsymbol{k}_2|^2 \approx [(\omega_1 - \omega_2)^2/c^2]\,\varepsilon(\omega_1 - \omega_2)$, $\varepsilon(\omega_1 - \omega_2)$ being the dielectric constant. The observation of this relative maximum is precisely what enables the dispersion of polaritons to be studied by spectroscopy in $\boldsymbol{k}$ space [6.94]. The fact that polaritons with frequency Ω and wave vector $\boldsymbol{Q}$ are

9 The dependence of the scattering intensity in a crystal of $LiNbO_3$ on the frequency difference $\omega_1 - \omega_2$ was discussed in [6.98]. Special features of the interference of the direct and the cascade processes were investigated in [6.79, 99].

produced under the conditions of the above-mentioned resonance can be detected, as shown in [6.100], by the radiation at the polariton frequency emerging from the crystal (provided that the crystal is sufficiently transparent in this frequency region).

If the frequency difference $\Omega = \omega_1 - \omega_2$ is within the region of negative values of $\varepsilon(\Omega)$, the maximum of the CARS signal due to direct processes occurs, as before, when the synchronism condition $\Delta \boldsymbol{k}_\mathrm{a} = 0$ is satisfied. At the same time, the relative contribution of the cascade processes at $\varepsilon(\Omega) < 0$ is maximum under the condition that

$$Q_\| = \frac{\Omega}{c}\sqrt{\frac{\varepsilon(\Omega)}{1+\varepsilon(\Omega)}} \, ,$$

where $Q_\|$ is the tangential component (parallel to the surface) of the difference $\boldsymbol{Q} = \boldsymbol{k}_1 - \boldsymbol{k}_2$. Since, in this case, the relation between Ω and $Q_\|$ is that for the dispersion of surface polaritons, it becomes possible to investigate this relation by the CARS technique [6.101–103]. The observation of this phenomenon was treated in [6.103].

6.4.3 Energy Losses and Cherenkov Radiation of a Charge Travelling with Uniform Motion Through a Medium Displaying Spatial Dispersion. Transition Radiation

One of the methods for the observation of new waves that appear when spatial dispersion is accounted for can consist of exciting these waves by means of the Vavilov-Cherenkov effect. In this connection, the more general problem of theoretically investigating the energy losses of a charge travelling through a medium with spatial dispersion arises [6.104].

Assume that an electron travels at the velocity v through an anisotropic nonmagnetic medium having arbitrary spatial dispersion. The system of field equations then is of the form:

$$\begin{aligned} &\operatorname{div} \boldsymbol{B} = 0 \, , && \operatorname{rot} \boldsymbol{E} = -\frac{1}{c}\frac{\partial \boldsymbol{B}}{\partial t} \, , \\ &\operatorname{div} \boldsymbol{D} = 4\pi e \delta(\boldsymbol{r} - \boldsymbol{v}t) \, , && \operatorname{rot} \boldsymbol{B} = \frac{1}{c}\frac{\partial \boldsymbol{D}}{\partial t} + \frac{4\pi}{c} e \boldsymbol{v} \delta(\boldsymbol{r} - \boldsymbol{v}t) \, , \qquad (6.4.38) \\ &D_i(\boldsymbol{r},t) = \int \varepsilon_{ij}(\omega, \boldsymbol{k}) E_j(\boldsymbol{k}, \omega) \exp[\mathrm{i}(\boldsymbol{k}\cdot\boldsymbol{r} - \omega t)]\, d\boldsymbol{k} \, . \end{aligned}$$

A transition to the Fourier components of the quantities $\boldsymbol{E}$, $\boldsymbol{D}$ and $\boldsymbol{B}$, and the elimination of the vectors $\boldsymbol{D}$ and $\boldsymbol{B}$, yield

$$E_i(\boldsymbol{r},t) = \frac{\mathrm{i}e^2}{2\pi^2c^2}\int(\boldsymbol{k}\cdot\boldsymbol{v})A_{ij}^{-1}v_j\exp[\mathrm{i}\boldsymbol{k}\cdot(\boldsymbol{r}-\boldsymbol{v}t)]\,d\boldsymbol{k}\,,$$
$$A_{ij} = k^2\eta_{ij} - \frac{(\boldsymbol{k}\cdot\boldsymbol{v})^2}{c^2}\varepsilon_{ij}(\boldsymbol{k}\cdot\boldsymbol{v},\boldsymbol{k})\,, \tag{6.4.39}$$

where $\eta_{ij} = \delta_{ij} - k_ik_j/k^2$ is the projection tensor. The energy losses of the travelling charge, of interest to us, are determined by the work done by the retarding force acting on the particle from the side of the field it sets up [Ref. 6.45, Chap. 12). Taking the magnitude of the field at the point where the electron is at the instant t, i.e., at $\boldsymbol{r} = \boldsymbol{v}t$, and making use of the identity

$$A_{ij}^{-1}v_j \equiv B_{ij}^{-1}v_j + \frac{(B_{lm}^{-1}k_lv_m)B_{ij}^{-1}k_j}{1-B_{st}^{-1}k_sk_t}\,,$$
$$B_{ij} = k^2\delta_{ij} - \frac{(\boldsymbol{k}\cdot\boldsymbol{v})^2}{c^2}\varepsilon_{ij}(\boldsymbol{k}\cdot\boldsymbol{v},\boldsymbol{k})\,, \tag{6.4.40}$$

we obtain the following expression for the total losses per unit length of path over which the charge travels [6.104]

$$R = -\frac{\mathrm{i}e^2}{2\pi^2vc^2}\int(\boldsymbol{k}\cdot\boldsymbol{v})$$
$$\times\frac{|B_{ij}^{-1}v_ik_j|^2-(B_{lm}^{-1}v_lv_m)(B_{rs}^{-1}k_rk_s)+B_{nk}^{-1}v_nv_k}{1-B_{sv}^{-1}k_sk_v}\,d\boldsymbol{k}\,. \tag{6.4.41}$$

In the nonrelativistic approximation (6.4.41) can be simplified to

$$R = -\frac{\mathrm{i}e^2}{2\pi^2v}\int\frac{\boldsymbol{k}\cdot\boldsymbol{v}}{\varepsilon_{lm}k_lk_m}\,d\boldsymbol{k}\,.$$

We shall discuss the quantity R of (6.4.41) in greater detail for the case of an isotropic nongyrotropic medium. According to (3.2.1), see also (2.1.48), we assume that

$$\varepsilon_{ij}(\omega,\boldsymbol{k}) = \varepsilon_\perp(\omega,k)\left(\delta_{ij}-\frac{k_ik_j}{k^2}\right) + \varepsilon_\parallel(\omega,k)\frac{k_ik_j}{k^2}\,,\qquad \tilde{\varepsilon} = \varepsilon_\parallel - \varepsilon_\perp\,,$$

and then obtain

$$R = -\frac{\mathrm{i}e^2}{\pi}\int_{-\infty}^{+\infty}\omega\,d\omega\int_0^\infty k\,dk$$

$$\times \frac{v^{-2} - \varepsilon_{\|}(\omega, \sqrt{k^2+\omega^2/v^2})/c^2 + \dfrac{\omega^2}{\omega^2+k^2v^2}\tilde{\varepsilon}(\omega, \sqrt{k^2+\omega^2/v^2})}{\varepsilon_{\|}(\omega, \sqrt{k^2+\omega^2/v^2})\{k^2+\omega^2[v^{-2}-c^{-2}\varepsilon_{\perp}(\omega, \sqrt{k^2+\omega^2/v^2})]\}} . \tag{6.4.42}$$

It follows from (6.4.42) that the zero terms within the curly brackets in the denominator (and, consequently, the losses incurred by the excitation of transverse waves [Ref. 6.45, § 86]) can appear when the condition

$$v > \frac{c}{n_l(\omega)} \tag{6.4.43}$$

is satisfied. The refractive index $n_l(\omega)$ of the normal waves, see (6.4.43), can be determined by

$$n^2 = \varepsilon_{\perp}\left(\omega, \frac{\omega}{c}n\right) . \tag{6.4.44}$$

If spatial dispersion is neglected, (6.4.44) for waves with any polarization determines a single function $n(\omega)$. When spatial dispersion is accounted for, this equation has two solutions, $n_l = n_l(\omega)$, with $l = 1, 2$, in the vicinity of the resonance $\varepsilon_{\perp}(\omega, 0)$ and the general situation concerning losses from Cherenkov radiation, generally speaking, changes [6.104, 105].

Let θ_l be the angle between the direction of particle motion and the direction of radiation. Then, as known [6.45, 75, 106],

$$\cos\theta_l = \frac{c}{v n_l(\omega)} . \tag{6.4.45}$$

We arrive at the conclusion that Cherenkov radiation with the frequency ω propagates in a medium whose spatial dispersion is taken into account along the surface of two cones with the apex angles θ_l, with $l = 1, 2$, determined by (6.4.45).

Next we consider the intensity distribution of Cherenkov radiation with respect to these cones.

Integrating in (6.4.42) with an appropriate choice of path around the poles in the complex $\boldsymbol{k}$ plane [Ref. 6.45, § 86], we find that the energy losses of Vavilov-Cherenkov radiation in the frequency range from ω to $\omega + d\omega$ are determined by

$$dR = \frac{e^2}{c^2}\sum_l \left(1 - \frac{c^2}{v^2 n_l^2(\omega)}\right)\left|1 - \frac{1}{2n_l(\omega)}\frac{\partial}{\partial n_l}\varepsilon_{\perp}\left(\omega, \frac{\omega}{c}n_l\right)\right|^{-1}\omega\, d\omega . \tag{6.4.46}$$

This equation gives the total intensity by a sum of the intensities distributed over the individual cones, (6.4.45).

If the frequency of the Cherenkov radiation is far away from the natural frequencies of the medium, the solution of (6.4.44) can be sought by making use of the relation, see (3.2.1), $\varepsilon_\perp(\omega,\boldsymbol{k}) = \varepsilon_0(\omega) - \alpha_\perp(\omega)n^2$. In this case we find, on the basis of (6.4.44), a unique solution to this equation: $n^2(\omega) = \varepsilon_0(\omega)[1+\alpha_\perp(\omega)]^{-1}$, and, therefore, the Cherenkov radiation is distributed over the surface of a single cone. Since $|\alpha_\perp(\omega)| \ll 1$ in the spectral region under consideration, the fact that the spatial dispersion of the medium is taken into account has practically no effect on the Cherenkov radiation of the electron. It can be quite another matter if the Cherenkov-radiation frequencies are in the vicinity of an exciton absorption band, because (6.4.44) has two roots in this case. If we apply (3.2.3) then, in an isotropic medium,

$$1 - \frac{1}{2n}\frac{\partial}{\partial n}\varepsilon_\perp\left(\omega, \frac{\omega}{c}n\right) = 1 + \beta_\perp(\omega)n^4 .$$

Consequently, the total intensity of Cherenkov radiation in the frequency range from ω to $\omega + d\omega$ is determined by

$$dR = \frac{e^2}{c^2}\sum_{l=1,2}\frac{1 - c^2/v^2 n_l^2(\omega)}{|1 + \beta_\perp(\omega)n_l^4(\omega)|}\,\omega\, d\omega . \qquad (6.4.47)$$

In the frequency range in which the refractive indices n_1 and n_2 are close to each other, it is essential to take spatial dispersion into consideration. In this case, the radiation is distributed over two cones and the corresponding intensities are found to be of the same order of magnitude. As the frequency is taken farther and farther away from the resonance region, one of the roots of (6.4.44) increases and the contribution of radiation corresponding to the anomalous cone decreases drastically. A similiar situation arises for gyrotropic cubic crystals as well. Here the intensity of Cherenkov radiation in the neighborhood of dipole exciton lines is distributed over three radiation cones.

As is clear from symmetry considerations, the group velocity in the isotropic medium can be directed only parallel or antiparallel to the wave vector $\boldsymbol{k}$. It was shown in Sect. 4.2.2 that the group velocity vector is always parallel to $\boldsymbol{k}$ for ordinary waves outside the absorption band, i.e., $\boldsymbol{v}_{\mathrm{gr}}\boldsymbol{k} > 0$, whereas for a new wave it is possible that $\boldsymbol{v}_{\mathrm{gr}}\boldsymbol{k} < 0$. This circumstance is especially important for the emergence of new light waves from the crystal into vacuum, such waves being excited by the charge travelling in the crystal.

Let us deal with this problem in more detail [6.107, 108]. Assume that a fast charged particle travels along the z axis from vacuum ($z < 0$) and passes through a plane-parallel plate bounded by the planes $z = 0$ and $z = d$ (Fig. 6.5). By virtue of (6.4.45), $\cos\theta_l = c/v n_l(\omega)$ for a Cherenkov wave of frequency ω and refractive index $n_l(\omega)$. Moreover, from the condition of refrac-

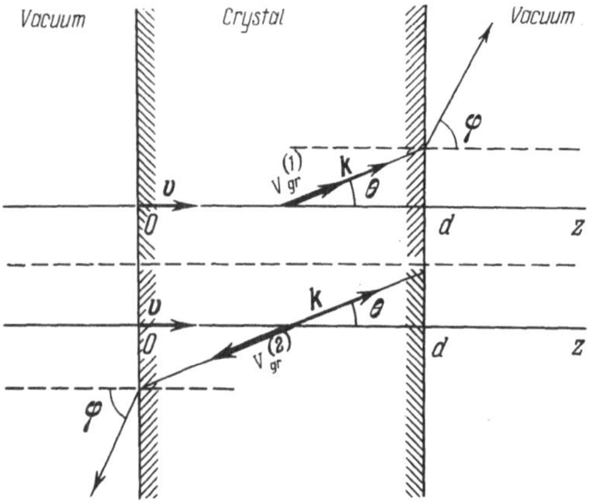

Fig. 6.5. Direction of wave propagation for Cherenkov radiation, with spatial dispersion taken into account, for the ordinary wave (superscript 1) and the new wave (superscript 2)

tion of a wave on the crystal surface, $n_l(\omega) = \sin\varphi_l/\sin\theta_l$, φ_l being the angle between the direction of wave propagation in vacuum and the z axis. Since $\sin^2\varphi_l \leqslant 1$ and, at the same time,

$$\begin{aligned}\sin^2\varphi_l &= n_l^2(\omega)\sin^2\theta_l = n_l^2(\omega)[1-\cos^2\theta_l]\\ &= n_l^2(\omega)[1-c^2/n_l^2(\omega)v^2] = n_l^2(\omega)-c^2/v^2\,, \qquad (6.4.48)\end{aligned}$$

we arrive at the conclusion that only Cherenkov waves for which

$$n_l^2(\omega) \leqslant 1+c^2/v^2\,, \qquad (6.4.49)$$

can emerge from the crystal into vacuum. Together with the condition $n_l^2(\omega) \geqslant c^2/v^2$, see (6.4.45), for the generation of Cherenkov radiation, the inequality (6.4.49) determines the frequency of the Cherenkov waves that emerge from the crystal and, under favorable conditions (sufficiently weak absorption), can be observed in vacuum. It is clear that for these waves

$$c^2/v^2 \leqslant n_l^2(\omega) \leqslant 1+c^2/v^2\,. \qquad (6.4.50)$$

In order to explain in what way the spectrum of the Cherenkov waves emerging from the crystal changes when spatial dispersion is taken into account, in accordance with the velocity of the charge, we shall consider the case of an isotropic nongyrotropic medium with $\beta' < 0$ (Sect. 4.2.2), neglecting absorption. If the velocity v of the charge is high, so that $c^2/v^2+1 < n_0^2$ (where n_0 is the value of n corresponding to the multiple root, see Fig. 6.6), then the condition (6.4.50) is satisfied only for the ordinary wave. Since $\boldsymbol{v}_{\mathrm{gr}}\boldsymbol{k} > 0$ for this wave, a comparatively wide spectrum of Cherenkov radiation is observed at the side ($z > d$) where the charge emerges from the crystal. The width of this spectrum decreases with the velocity of the charge and, at

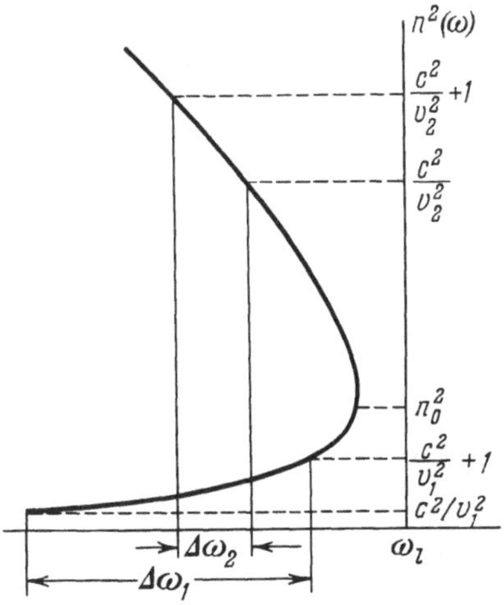

Fig. 6.6. Dependence of the Cherenkov radiation spectrum on the charge velocity when spatial dispersion is taken into account. Two values of the velocity have been chosen as an example: for $v = v_1$ the inequality $n_0^2 > 1 + c^2/v_1^2$ is complied with, whereas for $v = v_2$, on the contrary, $n_0^2 < c^2/v_2^2 < 1 + c^2/v_2^2$

$n_0^2 < 1 + c^2/v^2$ the condition (6.4.50) is satisfied for the new wave in a certain frequency region. As has been indicated, $\boldsymbol{v}_{gr}\boldsymbol{k}$ may assume negative values for this wave. In this case, Cherenkov radiation appears in the form of a comparatively thin line of radiation in front of the plate in the region $z < 0$. Since this line is associated with Cherenkov excitation of a new wave, its experimental observation would be direct proof of the existence of such a wave. Absorption reduces the intensity of the radiation emerging from the crystal. Therefore, and for other reasons as well, the vicinity of dipole exciton absorption lines in gyrotropic crystals at low temperatures offers conditions that are highly promising for the observation of new waves.

Even for a bounded medium, only Cherenkov radiation was dealt with in the above. When a charge travelling at uniform velocity passes through a boundary, transition radiation also appears [6.44, 75, 109] (we do not concern ourselves here with 'bremsstrahlung' and other kinds of radiation associated with changes in the motion of the charge)[10]. The problem of spatial dispersion in the theory of transition radiation was discussed in [6.44, 105, 108, 110–112]. In [6.108] it was shown that in the neighborhood of an exciton line the contribution of transition radiation is small compared to that of Cherenkov radiation and has practically no effect on our general understanding of the phenomenon.

---

[10] Surface excitons and polaritons, such as are considered in Chap. 5, may also be excited when a charge passes through the interface between two media. For the theory for molecular crystals see [6.113], for thin films [6.114] and references therein.

7. Conclusion

As a rule, only the absorption is measured in the first stage of investigating excitons by optical techniques, without analyzing the problem of the shape of the absorption lines. This is a natural procedure, provided that the subject of research is the detection of more or less distinct excited levels in crystals or the elucidation of the corresponding series laws, etc. If the crystal can be regarded as being optically isotropic, the situation here is quite similar to that for determining the atomic energy levels in gases. This analogy can also be extended to an anisotropic medium if we replace the ordinary gas by an "oriented gas", i.e., an assembly of independent molecules with fixed directions of certain axes (e.g., the directions of normal vibrations of an anisotropic oscillator). Since, in the case of weak absorption, the position of the absorption lines determines the frequencies $\Omega(\boldsymbol{k})$ of the Coulomb excitons, these are precisely the frequencies that are measured in the experiments we referred to.

It is obvious, however, that in a detailed investigation of the energy spectrum of crystals in the region of optical frequencies we should deal with a more general statement of the problem. On the one hand, the shape of the absorption lines is to be analyzed; on the other hand, in addition to absorption, we can and should examine the dispersion, i.e., measure the refractive index. It has been found in recent years that these problems can be especially effectively solved by using the picosecond spectroscopy method, the methods of nonlinear optics and Raman scattering of light with the excitation (or absorption) of excitons or polaritons (i.e., normal electromagnetic waves or, employing a different terminology, excitons dealt with taking into account the retarded electromagnetic interaction). We cannot restrict ourselves, in this case, to a consideration of an optically isotropic medium because even cubic crystals are optically anisotropic when spatial dispersion is taken into consideration. Thus, an investigation of excitons is found to be intimately linked to both classical crystal optics and to crystal optics with spatial dispersion.

However, until a relatively short time ago, crystal optics was being developed only in an approximation in which spatial dispersion was ignored (somewhat conditionally we refer to such an approximation as classical crystal optics). But it is impossible, generally speaking, to analyze many basic problems of condensed matter and plasma optics [7.1] without taking spatial dispersion into account.

It was the specific aim of the first edition of this book, published in 1965, to expound crystal optics with spatial dispersion in sufficient detail and from a unified point of view. In this aspect, of course, crystal optics contains all of classical crystal optics as a limiting (or special) case. At that time the role of spatial dispersion in crystal optics was not yet sufficiently known and, consequently, some topics were discussed in great detail. Although we made no claim to a complete treatment, most of the known literature was taken into consideration. Today the situation has changed and, on the one hand, the place of spatial dispersion in crystal optics has been fully established, while on the other hand, numerous papers, dealing with many special problems, have been published. In this connection, notwithstanding the additional material included in this edition which has almost doubled the volume of the book, the purpose of the book has changed. We regard the book in its present form to be an introduction to the problem, especially with respect to microtheory.

Notwithstanding the advances in this field, the idea that all the interesting problems have already been investigated within the framework of crystal optics with spatial dispersion is not true. Suffice it to mention that even in classical crystal optics, whose development extends over many decades, interesting problems and insufficiently investigated cases are still being encountered (for instance, the problem of singular optical axes). A great many problems in crystal optics could be solved by taking spatial dispersion into consideration. But the solution of many such problems is not justified from the point of view of the practicable requirements, which are determined by the experimental possibilities and the value of specific items of information in crystal theory. It is our opinion, therefore, that any further development of theory should be linked, primarily, to an analysis of the data obtained from and the possibilities offered by a definite line of experimental research. Even in the absence of a direct connection with experiments it is worthwhile to investigate the effects of weak spatial dispersion on the propagation of electromagnetic waves in crystals of various classes, close to the optical axes, for crystal plates of various orientations, in the vicinity of surfaces (with various assumptions as to their properties), etc.

Among other things, the following topics also deserve further study: new waves in gyrotropic and nongyrotropic crystals, dispersion and absorption close to quadrupole absorption lines in crystals, and the effects of external electric and magnetic fields, as well as stresses and deformation, on the optical properties of crystals.

No less interesting are many of the problems mentioned in this book. This refers, particularly, to nonlinear effects in crystal optics, which are already attracting close attention and will evidently play an even more prominent role in optics in the future. In this book, as a whole, we have restricted ourselves to the treatment of linear crystal optics, though we did discuss certain questions that are closely related to nonlinear crystal optics (the scattering of light and x rays, and the effects of external electromagnetic fields on the propagation of light).

The development of the optics and spectroscopy of surfaces and interfaces led to the establishment of new trends in research in this field [7.2]. For instance, the investigation of resonance phenomena due to the presence on the surfaces of one or another kind of transition layers (for example, thin metal, dielectric or semiconductor films) required the development of diffraction theory with spatial dispersion, in particular, with additional surface polaritons taken into account [7.3]. Investigations in the field of SERS (Surface Enhanced Raman Scattering) [7.4] stimulated the development of the electrodynamics of rough surfaces, in particular, when spatial dispersion is taken into account [7.5]. Unfortunately, we do not have the opportunity here to discuss to a sufficient extent the results obtained in this research.

We also cannot discuss the properties of nonlinear electromagnetic waves (both of the bulk and surface types) nor the problem of self-induced transparency in the exciton region of the spectrum [7.6, 7], as well as a number of allied problems, whose analysis requires that spatial dispersion be taken into account [7.8, 9].

An important feature, pointed out repeatedly, is the complete groundlessness of neglecting macroscopic (phenomenological) crystal optics, which makes use of the tensor $\varepsilon_{ij}(\omega,\boldsymbol{k})$, or contrasting it with microscopic theory. There is no doubt that microscopic calculations (e.g., calculations of $\varepsilon_{ij}(\omega,\boldsymbol{k})$ or $\tilde{n}_l(\omega,\boldsymbol{s})$ for some specific model) can in a sense yield more information than macroscopic theory (of crystal optics, in this case). Nevertheless, the results of correct calculations for a given reasonable model not only include all the consequences deduced from macrotheory applied to this model, but should enable a number of relationships, for example, the frequency dependence of $\varepsilon_{ij}(\omega,\boldsymbol{k})$, to be specifically defined. This does not, however, clear up the question: which results depend upon the given model and which can be obtained with any model (i.e., are entirely independent of the model)? It is of course clear that the use of some model or approximation and the comparison of results of the corresponding calculations with experimental data is of value primarily when we are dealing with the consequences and features due specifically to the application of the given model, rather than those that are general and independent of the chosen model. Hence, the application of macrotheory is, in general, not only expedient, but is indispensable in solving the problem of the value of definite model or approximation. Moreover, this statement of the question eliminates the need for microcalculations of quantities that are, in essence, derived from more fundamental quantities [it is, for example, more expedient, in the general case, to calculate $\varepsilon_{ij}(\omega,\boldsymbol{k})$ rather than $\tilde{n}_l(\omega,\boldsymbol{s})$].

The aforesaid is, of course, not only a specific aspect of the macro- and microtheory of the optical properties of crystals, but is of a general (and well-known) nature. As we have previously pointed out, a harmonic combination of the macroscopic and microscopic approaches was not attained, for various reasons, in the first stage of development of exciton theory. One of the principal aims of the two editions of the present book was and is to reveal and underline the nature of the interrelations between the microscopic and macroscopic approaches in crystal optics.

Owing to the extensive application of crystal optics with spatial dispersion and its proper combination with the microtheory of excitons, research in the field of the optical and certain other properties of crystals has been facilitated to a great extent. We believe that further developments will be along the same lines.

Appendix

A.1 Crystal-Symmetry Notation

In applying the dielectric-constant tensor $\varepsilon_{ij}(\omega, \boldsymbol{k})$, the crystal is considered to be an essentially spatially homogeneous (continuous) medium. The properties that characterize such a medium are often said to be macroscopic because they depend only on direction (for instance, the direction of the wave vector of light), but not on position. Crystal symmetry leads to the equality of a number of directions; along these equivalent directions all the macroscopic properties of the crystals are identical (see also Appendix A.2).

Crystals may have the following macroscopic symmetry elements: a center of symmetry, planes of mirror symmetry, axes of one-, two-, three-, four- and sixfold symmetry, and rotation-reflection or inversion axes of the same orders.

A combination of these macroscopic symmetry elements for a given body forms a group. It is said to be a point group because all the axes and planes of symmetry have a common point of intersection.

All crystals are subdivided into 32 crystallographic classes to each of which its point group corresponds. On the other hand, crystals are divided into crystallographic, or simply crystal, systems, in accordance with the symmetry of their space lattices. International symbols for the symmetry elements are listed in Table A.1 together with other frequently applied notations.

If a crystal has a symmetry center, the properties of the crystal in a given direction are identical to those in the opposite direction. The existence of a plane of symmetry leads to the equivalence of crystals obtained upon mirror reflection by the given plane. The existence of an n-fold symmetry axis indicates equivalence of the crystal when it is turned about this axis through an angle of $360°/n$. In the case of an inversion axis, reflection (inversion) at the origin is required, in addition to rotation. Rotation-reflection axes are frequently employed in place of inversion axes. In this case, instead of inversion, reflection is performed in a plane perpendicular to the axis. An n-fold rotation-reflection axis is designated by the symbol $L_n^{n/2}$ (there are L_2^1, L_4^2 and L_6^3 axes).

Table A.1. Symbols for elements for symmetry

Elements of symmetry	International symbol	Other notations	
Center of symmetry	$\bar{1}$	C	I
Plane of symmetry	m	P	σ
Rotational axes			
Onefold symmetry	1	L^1	C_1
Twofold symmetry	2	L^2	C_2
Threefold symmetry	3	L^3	C_3
Fourfold symmetry	4	L^4	C_4
Sixfold symmetry	6	L^6	C_6
Inversion axes			
Onefold symmetry axis ≡ center of symmetry	$\bar{1}$	$L_i^1 \equiv C$	I
Twofold symmetry axis ≡ plane of symmetry perpendicular to the axis	$\bar{2}$ ($\equiv m$)	$L_i^2 \equiv P$	σ_h
Threefold symmetry axis ≡ threefold rotational symmetry axis + center of symmetry	$\bar{3}$	L_i^3	
Fourfold symmetry axis (includes a twofold rotational symmetry axis)	$\bar{4}$	L_i^4	
Sixfold symmetry axis ≡ threefold rotational symmetry axis + a plane of symmetry perpendicular to this axis	6 ($\equiv 3/m$)	L_i^6	
Rotation-reflection axes			
Twofold symmetry axis ≡ center of symmetry		$L_2^1 \equiv C$	$S_2 = I$
Fourfold symmetry axis		L_4^2	S_4
Sixfold symmetry axis		L_6^3	S_6

The division of crystal classes into systems and their designation are given in Table A.2.

The cubic system is sometimes called the higher system, the hexagonal, trigonal and tetragonal systems are said to be intermediate, and rhombic, monoclinic and triclinic systems are called the lower ones.

The Schönflies crystal symbols are based on the following principle. The letter C (from the Latin word *cyclus*, meaning cycle) indicates the existence of a single symmetry axis, with the subscript specifying the kind of axis (for instance, crystals of class C_2 have a single-axis twofold symmetry). The existence in the crystal of a plane of symmetry is indicated by a subscript lower-case letter. The subscript h (horizontal) specifies a plane perpendicular to the axis, whereas the subscript v (vertical) is used for a plane parallel to the axis. Hence, $C_{3h} = L^3P$ and $C_{3v} = L^3 3P$. The capital letter D denotes the

Table A.2. Crystal classes

System	Class		
	Symmetry element	International symbols	Schönflies symbols
Triclinic	L^1	1	C_1
	C	$\bar{1}$	$C_i = S_2$
Monoclinic	L^2	2	C_2
	P	m	$C_{1h} = C_s$
	L^2PC	$2/m$	C_{2h}
Rhombic	$3L^2$	222	$D_2 = V$
	$L^2 2P$	$mm2$	C_{2v}
	$3L^2 3PC$	mmm	$D_{2h} = V_h$
Tetragonal	L^4	4	C_4
	L_4^2	$\bar{4}$	S_4
	L^4PC	$4/m$	C_{4h}
	$L^4 4L^2$	422	D_4
	$L^4 4P$	$4mm$	C_{4v}
	$L_4^2 2L^2 2P$	$\bar{4}2m$	$D_{2d} = V_d$
	$L^4 4L^2 5PC$	$4/mmm$	D_{4h}
Trigonal (rhombohedral)	L^3	3	C_3
	$L_6^3 C$	$\bar{3}$	$C_{3i} = S_6$
	$L^3 3L^2$	32	D_3
	$L^3 3P$	$3m$	C_{3v}
	$L_6^3 3L^2 3PC$	$\bar{3}m$	D_{3d}
Hexagonal	L^6	6	C_6
	L^3P	$\bar{6}$	C_{3h}
	L^6PC	$6/m$	C_{6h}
	$L^6 6L^2$	622	D_6
	$L^6 6P$	$6mm$	C_{6v}
	$L^3 3L^2 4P$	$\bar{6}m2$	D_{3h}
	$L^6 6L^2 7PC$	$6/mmm$	D_{6h}
Cubic	$3L^2 4L^3$	23	T
	$3L^2 4L_6^3 3PC$	$m3$	T_h
	$3L^4 4L^3 6L^2$	432	O
	$3L_4^2 4L^3 6P$	$\bar{4}3m$	T_d
	$3L^4 4L_6^3 6L^2 9PC$	$m3m$	O_h

classes that have only symmetry axes, i.e., $D_2 = 3L^2$ and $D_3 = L^3 3L^2$ (the figure preceding L or P indicates the number of axes or planes). Classes with rotation-reflection axes are denoted by the capital letter S (from 'Spiegelachse', which is German for mirror axis). The remaining notation is clear from Table A.2.

The symmetry elements of crystal classes can be visually represented by means of a stereographic projection [Ref. A.1, Table 30]. It also proves convenient in such projections to indicate the directions of the commonly chosen coordinate axes. This has been done in Table 3.1.

A.2 Information from Space Group Theory [1]

Two points in a crystal or two directions in a crystal are said to be equivalent if all the physical and geometrical properties of the crystal are identical at these points or in these directions (this holds for the average statistical properties of the crystal in a state of equilibrium). The set of symmetry elements of the crystal, i.e., the set of operations that transfer each point of the crystal to its equivalent point and each direction to its equivalent direction, constitute the symmetry group G of the crystal. All in all, nonmagnetic media can be classed into 230 different crystal groups, which are commonly called space, or Fedorov, groups.

The combination of all the rotations and rotoflections that transfer each direction into the equivalent direction constitutes the point group of crystal class F (whose elements are indicated by the lower-case letter r), which specifies directional symmetry in the crystal. Each element g of the crystal group G can be represented in the form

$$g = T_a r \,, \tag{A.1}$$

where T_a is the translation operation corresponding to a displacement by the vector $\boldsymbol{a}$, and the elements r, corresponding to the elements g, constitute a point group as mentioned above: the point group of the crystal class F. The translational vector $\boldsymbol{a}$ appearing in (A.1) is equal to

$$\begin{aligned} &\boldsymbol{a} = \boldsymbol{m} + \boldsymbol{\alpha} \,, \qquad \boldsymbol{m} = m_1\boldsymbol{a}_1 + m_2\boldsymbol{a}_2 + m_3\boldsymbol{a}_3 \,, \\ &\boldsymbol{\alpha} = \alpha_1\boldsymbol{a}_1 + \alpha_2\boldsymbol{a}_2 + \alpha_3\boldsymbol{a}_3 \,, \end{aligned} \tag{A.2}$$

where $\boldsymbol{a}_1$, $\boldsymbol{a}_2$ and $\boldsymbol{a}_3$ are the vectors of the unit cell, m_1, m_2 and m_3 are integers equal to $0, \pm 1, \pm 2, \ldots$, and α_1, α_2 and α_3 are positive fractions of a value less than unity.

In the given space group G each element r of the group F corresponds to a definite vector $\boldsymbol{\alpha}$, whereas the values of m_1, m_2 and m_3 may be arbitrary. It follows that any element of the crystal space group can be represented in the form of the product of a translation operation, corresponding to translation

[1] Proofs of a number of the statements made here, as well as a more detailed and comprehensive description of the theory of space groups can be found in [A.2 – 5].

over an integral-valued lattice vector $\boldsymbol{m} = m_1\boldsymbol{a}_1 + m_2\boldsymbol{a}_2 + m_3\boldsymbol{a}_3$, with one of the "rotation elements" $T_{\alpha_1}r_1$, $T_{\alpha_2}r_2, \ldots, T_{\alpha_N}r_N$, where N is the number of elements in the point group of crystal class F. If the space group of the crystal contains no essential screw axes or slip planes (this means that all $\boldsymbol{\alpha}_i = 0$), the rotation elements[2] $g = T_\alpha r$ of the symmetry group G of the crystal coincide with the elements r of group F, thereby constituting a point group. If, however, not all $\boldsymbol{\alpha}_i = 0$, then group F is evidently not a subgroup of group G.

Since the product of the translation operations over an integral-valued lattice vector is also a translation operation over an integral-valued lattice vector (where $T_n \times T_m = T_{n+m}$), it is clear that translations over an integral-valued lattice vector forms a subgroup of the space group G of the crystal.

The representation A of a group G is said to be specified in a certain linear space L when an operator $A(g)$ (which may be specified by a matrix) of the space L corresponds to each element g of G. Thus, the product of elements of a group corresponds to the product of the operators A, i.e.,

$$A(g_1)A(g_2) = A(g_1 g_2)\,. \tag{A.3}$$

The dimensionality of the space L (the number of linearly independent elements) is called the dimensionality of the representation.

Since the Hamiltonian of the crystal is invariant with respect to the symmetry elements of the space group of the crystal, the set of all solutions Ψ_W of the Schrödinger equation for the energy W is a linear space L_W, whose elements transform under operations by the space group elements such that they remain in the same space. Here a matrix corresponds to each symmetry element of the crystal. It is a transformation matrix of a dimensionality equal to the degeneracy of level W, i.e., the dimensionality of the space L_W. The set of all such matrices forms the representation of the space group of the crystal, corresponding to the level W.

By virtue of the translational symmetry of the crystal, the solutions of the Schrödinger equation can be chosen such that they are the eigenfunctions of the translation operator for an integral-valued lattice vector. If Ψ_{ks} is such a solution, then

$$T_m \Psi_{ks} = \mathrm{e}^{-\mathrm{i}\boldsymbol{k}\cdot\boldsymbol{m}}\,\Psi_{ks}\,. \tag{A.4}$$

Each of the functions Ψ_{ks} is an eigenfunction of T_m, i.e., it is transformed, by the action of the operators T_m, according to (A.4), which does not contain any other solutions of the Schrödinger equation. If follows that each of the functions Ψ_{ks} is transformed according to a one-dimensional representation of the translation subgroup. Moreover, as is evident from (A.4), wave functions that

---

[2] We note that even in the presence of essential screw axes or slip planes, vectors $\boldsymbol{\alpha}$ exist which are equal to zero (for example, the vector $\boldsymbol{\alpha}$ corresponding to an identity operation can always be taken to be zero).

correspond to the value $\boldsymbol{k} = 0$ are not changed by the translation. This fact enables us to conclude that wave functions corresponding to $\boldsymbol{k} = 0$ provide a representation of the factor group of the translation subgroup.

Since the concept of a factor group is relatively seldom employed in physics, we shall discuss this question in greater detail. In this discussion our main goal is to prove that the factor group of the translation subgroup T is isomorphic to a group of the crystal class F.

We begin by introducing the concept of conjugate elements, classes and an invariant subgroup. If g is an arbitrary element of a group, then gg_1g^{-1} is said to be an element conjugate to g_1 (g^{-1} is the inverse of g, i.e., $gg^{-1} = E$). It can readily be shown that if two elements g_1 and g_2 are conjugate to the element g_3, they are also conjugate to each other. The elements of a group that are conjugate to one another form a class. It is clear that a class is determined by the assignment of one of its elements g_1 and that it unites the set of elements of the type gg_1g^{-1}, where the role of g is played by all the elements of the group. All the elements of a group can be divided into classes because each element of the group is found in one and only in one class. The identity (unit) element E of the group forms a class of its own because it is not conjugate to any other element: $gEg^{-1} = E$ for all values of g. With the exception of this class, consisting of a single element, no other class forms a subgroup because no other class contains the unit element E.

In certain groups a subgroup can be formed which consists exclusively of entire classes of the initial group. Such a subgroup is said to be invariant. Groups that have no invariant subgroups are said to be simple. Note that no such space group exists because the translation subgroup is an invariant subgroup. In fact, since a group element of the form gT_mg^{-1}, g being an arbitrary element of the group, is a translation, the translation subgroup includes, in addition to the unit element, all elements that are conjugate to any translation operation.

The invariant subgroup permits the introduction of the concept of an adjacent class corresponding to the given invariant subgroup R. If g is a certain element of the group, then the set of elements of the form gR_1, gR_2, gR_3, ..., where R_1, R_2, etc., are elements of a subgroup R, forms the class adjacent to g. It can be shown that the set of elements of the form gR coincides with the set of elements Rg. If the elements of one adjacent class Rg_1 are multiplied by all the elements of another adjacent class Rg_2, the product yields the elements of the adjacent class Rg_1g_2. Hence, if we regard the adjacent classes corresponding to a certain invariant subgroup R as new objects and define the product of two adjacent classes as an adjacent class with elements obtained by multiplying the elements of two adjacent classes by one another, than the adjacent classes themselves form a group. This group is called the factor group of an invariant subgroup. The unit element of the factor group is this invariant subgroup. The order, i.e., the number of elements in the factor group, is equal to the number of adjacent classes for the invariant subgroup (in the following we consider the invariant translation subgroup T of the group G). The factor group should not be confused with a subgroup of G. Elements of the subgroup are elements of G, whereas the elements of the factor group are the adjacent classes.

Let us return now to the space group of a crystal. As mentioned previously, this group has an invariant subgroup: the translation subgroup T (the operations T_m being its elements). At the same time, each adjacent class in T is a set consisting of elements of the form $T_m T_\alpha r$, where the operation $r \neq E$ is fixed and T_m comprises all the elements of the translation subgroup T. If $r = E$, the corresponding adjacent class coincides with the translation subgroup. It is clear that in the case of the space group G under consideration, the number of elements in the factor group is equal to the number of elements of the point group of the crystal class because the number of rotation elements $T_\alpha r$ is equal to the number of elements of the group F. Moreover, there is a one-to-one correspondence between the elements of the factor group and those of the group F. Here the product of the elements of the factor group corresponds to the product of the respective elements of the group of the crystal class. Thus, in the given case, the factor group is isomorphic to F.

We point out that since the elements of the space group, included in a certain adjacent class corresponding to invariant subgroup T, are of the form $T_m T_\alpha r$, i.e., they differ from one another only by a translation operation, they all yield the same result when applied to the function Ψ_{ks} at $\boldsymbol{k} = 0$. As a matter of fact

$$T_m T_\alpha r \Psi_{0s} = T_\alpha r T_m \Psi_{0s} = T_\alpha r \Psi_{0s} . \tag{A.5}$$

Let us define the result of the application of an element of the factor group Φ_r to the function Ψ_{0s} as

$$\Phi_r \Psi_{0s} = T_\alpha r \Psi_{0s} .$$

Since the operator $T_\alpha r$ belongs to the space group in the crystal, the function $T_\alpha r \Psi_{0s}$ is a linear combination of the functions $\Psi_{0s'}$, which, like Ψ_{0s}, correspond to the same energy W of the crystal. It is clear that we thereby define the transformation matrix for each element of the factor group Φ_r. These matrices constitute the representation of the factor group in the linear space of the eigenfunctions Ψ_{0s} of the Hamilton operator, corresponding to the same energy W.

The linear space of all functions Ψ_{0s} can, in general, be divided into so-called invariant subspaces. Functions Ψ^i_{0s}, where $i = 1, 2, \ldots, t$, included in the given invariant subspace are transformed when subject to the effect of elements of a factor group, but remain in the same space; i.e., for any element of the factor group Φ_r, the function $\Phi_r \Psi^i_{0s}$ remains in the invariant subspace. Consequently, a representation of the group Φ_r of dimensionality t corresponds to each invariant subspace. If it is impossible in the invariant subspace to separate out an invariant subspace of lower dimensionality, then the corresponding representation is said to be irreducible. Otherwise, the representation is said to be reducible.

Isomorphic groups have the same irreducible representations. This is an important fact because it means that to classify the states of mechanical excitons, corresponding to $\boldsymbol{k} = 0$, we can make use of the irreducible representations of a point group of the crystal class F. This class, as we have seen above, is isomorphic to the factor group of the invariant translation subgroup. Thus, it is possible to use the irreducible representations of F, notwithstanding the fact that F, in general, contains elements that are not symmetry elements of the crystal (this is the case, as previously indicated, in the presence of essential screw axes and glide planes).

Group representations are frequently investigated with the aid of the so-called characters of representations. The character $\chi(g)$ of a representation A is the sum of the diagonal elements of the matrix $A(g)$, corresponding to element g of some group. The scalar product of the characters $\chi^{(1)}(g)$ and $\chi^{(2)}(g)$ for two different representations $A^{(1)}$ and $A^{(2)}$ is understood to be the expression

$$(\chi^{(1)}, \chi^{(2)}) = \frac{1}{N} \sum_g \chi^{(1)}(g)\chi^{(2)}(g)\,, \tag{A.6}$$

where N is the number of elements g in the group (the order of the group) and $\sum_g$ indicates that summation is to be over all the elements of the group.

For irreducible representations

$$(\chi^{(i)}, \chi^{(k)}) = \delta_{ik}\,. \tag{A.7}$$

For any reducible representation $A(g)$, its character $\chi(g)$ is equal to the sum of the characters of all the irreducible representations $A_\nu(g)$ contained in $A(g)$ [each representation is counted as many times as it is contained in $A(g)$]. If m_ν denotes the number indicating how many times the irreducible representation A_ν is contained in the representation A, then

$$\chi(g) = \sum_\nu m_\nu \chi^\nu(g) \quad \text{and} \quad m_\nu = (\chi, \chi_\nu)\,.$$

Therefore, by virtue of (A.7),

$$(\chi, \chi) = \sum_\nu m_\nu^2\,.$$

If the representation is reducible, the scalar square (χ, χ) of its character is greater than unity. This statement contains a convenient criterion of the reducibility of the representation.

We note also that since a unit matrix corresponds to the unit element $g = E$ of the group in any representation, the character $\chi(E)$ is always a positive integer equal to the order of the representation matrix, i.e., to its dimension.

A.2.1 Classification of the States of Mechanical Excitons

Since long-range interaction is either absent or ignored in determining the states of mechanical excitons, the energy $W_s(\boldsymbol{k})$ of a mechanical exciton at $\boldsymbol{k} \to 0$ is independent of $\boldsymbol{s} = \boldsymbol{k}/k$ (in contrast to the energy of Coulomb excitons, see Sect. 6.2). Hence, in the classification of the states of mechanical excitons, the direction $\boldsymbol{s}$ at $\boldsymbol{k} = 0$ is unimportant. This classification turns out to be entirely analogous to the classification of the excited states of atoms and molecules.

As an example let us consider a crystal belonging to the crystal class of the point group O_h. The characters of this group are listed in Table 4.2. We can conclude, on the basis of this table, that in the crystal being considered there are ten types of exciton bands for mechanical excitons (according to the number of irreducible representations)[3], four of these types of bands are triply degenerate as $\boldsymbol{k} \to 0$ (the wave functions Ψ_{0s} corresponding to these bands are transformed according to the irreducible representations F_1, F_2, F_1' and F_2'), two types of bands are doubly degenerate at $\boldsymbol{k} \to 0$ (representations E and E'), and the rest of the zones are nondegenerate at $\boldsymbol{k} \to 0$.

If this crystal is subjected to the action of an external constant electric field of a direction along the axis of fourfold symmetry (the z axis), certain of the representations that are irreducible in group O_h become reducible in the new group of symmetry (group C_{4v}). Let us consider, for example, the representation E (not to be confused with the unit element E) of group O_h. Instead of three axes C_4^2 in the new group we have only one; instead of six axes of fourfold symmetry C_4, only two such axes are left; instead of three operations IC_4^2 and six symmetry operations IC_4, only two symmetry operations remain in each case. It is clear that the same representation matrix $A(g)$ and, consequently, the same value of the character χ correspond to each element of symmetry that is retained in the new group. Therefore, the characters of the representation E of group O_h can be represented in the group C_{4v} in accordance with the following table.

	E	C_4^2	$2C_4$	$2IC_4^2$	$2IC_4$
χ	2	2	0	2	0

Since, according to (A.6), $(\chi, \chi) = (4+4+2\times0+2\times4+2\times0)/8 = 2 > 1$, the representation under consideration is reducible. Making use of the characters of the irreducible representations A_1, A_2, B_1, B_2 and E of group C_{4v} (Table 4.4), we find, see (A.6), that $(\chi, \chi^{A_1}) = 1$, $(\chi, \chi^{A_2}) = 0$, $(\chi, \chi^{B_1}) = 1$ and $(\chi, \chi^{B_2}) = (\chi\chi^E) = 0$. Here χ are the characters of the matrices of the reducible representations listed in the table. Thus, the representation E of the

[3] If states with $\boldsymbol{k} \neq 0$ are taken into consideration, each term is evidently converted into a band.

group O_h falls into two one-dimensional representations: $E \to A_1 + B_1$. If the wave functions of mechanical excitons are transformed at $\boldsymbol{k} = 0$ according to the irreducible representation E of group O_h, this means that they all correspond to the same energy and that the number of such functions is equal to two (doubly degenerate level). Since in the group C_{4v} the representation E of group O_h falls into two one-dimensional representations (see above), it is clear that the doubly degenerate symmetry term of representation E in group O_h splits, in general, into two nondegenerate terms whose wave functions are transformed according to the one-dimensional irreducible representations of the group C_{4v}.

Other two- and three-dimensional representations of group O_h can also be considered in a similar way.

Notation[1]

$\varepsilon_{ij}(\omega, \boldsymbol{k}) = \delta_{ij} + 4\pi\chi_{ij}(\omega, \boldsymbol{k})$ $= \varepsilon'_{ij}(\omega, \boldsymbol{k}) + \mathrm{i}\varepsilon''_{ij}(\omega, \boldsymbol{k})$ $= Re\{\varepsilon_{ij}(\omega, \boldsymbol{k})\} + \mathrm{i}\ \mathrm{Im}\{\varepsilon_{ij}(\omega, \boldsymbol{k})\}$	Complex dielectric-constant tensor
$\varepsilon_{ij}^{-1}(\omega, \boldsymbol{k}) \equiv \nu_{ij}$	Inverse tensor of $\varepsilon_{ij}(\omega, k)$
$\chi_{ij}(\omega, \boldsymbol{k})$	Complex susceptibility tensor
$\delta_{ij} = 1$ if $i = j$, and $\delta_{ij} = 0$ if $i \neq j$ ε'_{ijh} ε''_{ij}	Hermitian tensors
$\mathrm{Re}\{\varepsilon_{ij}\}$ and $\mathrm{i}\ \mathrm{Im}\{\varepsilon_{ij}\}$	Real and imaginary parts parts of ε_{ij}
$\varepsilon_{ij}(\omega)$	Complex dielectric-constant tensor if there is no spatial dispersion (classical crystal optics)
$\varepsilon_{ij}(t, t', \boldsymbol{r}, \boldsymbol{r}')$, $\hat{\varepsilon}_{ij}(t-t', \boldsymbol{r}-\boldsymbol{r}')$ $\hat{\varepsilon}_{ij}(\tau, \boldsymbol{R})$	Kernels of integral relations, see (2.1.3-6)
$\varepsilon_{\alpha\beta}$, $\varepsilon_{\alpha\beta}^{-1} = \nu_{\alpha\beta}$	Two-dimensional tensors ($\alpha, \beta = 1, 2$)
$\omega = \omega' + \mathrm{i}\omega'' = \omega' + \mathrm{i}\gamma$	Frequency, $\omega' = \mathrm{Re}\{\omega\}$
$\omega'' = \mathrm{Im}(\omega)$	In the great majority of cases dealt with in this book, the frequency ω is assumed to be real, i.e., $\omega = \omega'$
$\boldsymbol{k} = \boldsymbol{k}' + \mathrm{i}\boldsymbol{k}''$	Wave vector (the vectors $\boldsymbol{k}'$ and $\boldsymbol{k}''$ are real)

[1] Rarely used symbols are not listed; the same applies to common ones such as e, $\hbar = h/2\pi$ and c, which are the elementary charge, Planck's constant and the velocity of light. If different quantities are denoted by the same latter, they appear, as a rule, in different sections.

$\boldsymbol{k} = (k' + \mathrm{i}k'')\boldsymbol{s}$	Wave vector of uniform plane monochromatic waves, which, with a few exceptions, are the only ones dealt with in this book
$\boldsymbol{k} = (\omega/c)\tilde{n}(\omega, \boldsymbol{s})\boldsymbol{s}$	Wave vector of uniform normal waves
$\boldsymbol{s}$	Real unit vector characterizing the direction of propagation of uniform plane waves
$\boldsymbol{S} = \boldsymbol{S}^{(0)} + \boldsymbol{S}^{(1)}$	Energy flux, see (2.3.11, 12)
$\tilde{n}(\omega, s) = \mathrm{n} + \mathrm{i}\varkappa$	Complex refractive index where $n = \mathrm{Re}\{\tilde{n}\}$ and $\varkappa = \mathrm{Im}\{n\}$
n	Refractive index
$\varkappa$	Absorption index
$\mu = 2\omega\varkappa/c$	Absorption coefficient
$\boldsymbol{E}$	Electric field strength
$\boldsymbol{D} = \boldsymbol{E} + 4\pi\boldsymbol{P}$	Electrical induction
$\boldsymbol{P}$	Polarization
$\boldsymbol{B}$	Magnetic induction
$\boldsymbol{j}_{\mathrm{ext}}$	Current density of external sources
ρ_{ext}	Charge density of external sources
$\boldsymbol{B}_{\mathrm{ext}}$	Induction of a time-independent magnetic field (for brevity such a field is called the external field)
$\lambda = \lambda_0/n$	Wavelength in a medium
$\lambda_0 = 2\pi c/\omega$	Wavelength in vacuum
a or a_{s}	Lattice constant or a length on an atomic scale
ω_s	Natural frequency
$\omega_m(\boldsymbol{k})$	Frequency of mechanical excitons
$\omega_{\|\|}$	Frequency of longitudinal waves
$\omega'_{\|\|}$	Frequency of "fictitious" longitudinal waves
ω_p	Frequency of "polarization waves"
$\lvert a_{ij} \rvert$	Determinant of matrix a_{ij}
$\boldsymbol{E}_{\perp}$	Strength of a transverse field, see (2.2.56)
$\boldsymbol{E}_{\|\|}$	Strength of a longitudinal field, see (2.2.56)
$\varepsilon_{\perp, ij}$	Tensor, see (2.2.58)
$\chi_{\perp, ij}$	Tensor, see (2.2.59)
$\eta_{ij} = \delta_{ij} - s_i s_j$	Projection tensor
$\boldsymbol{v}_{\mathrm{gr}} = d\omega/d\boldsymbol{k}$	Group-velocity vector
Ψ	Angle between $\boldsymbol{v}_{\mathrm{gr}}$ and $\boldsymbol{k}$

References

Chapter 1

1.1 L.D. Landau, E.M. Lifshitz: *Electrodynamics of Continuous Media* (Gostekhizdat, Moscow 1957) [English transl.: Pergamon, Oxford 1960]
1.2 M. Born: *Optik* (Springer, Berlin 1933); M. Born, E. Wolf: *Principles of Optics* (Pergamon, Oxford 1964)
1.3 S. Szivessy: "Kristalloptik", in *Handbuch der Physik*, Vol.20 (J. Springer, Berlin 1928) p.635
1.4 G.N. Ramachandran, S. Ramaseshan: "Crystal Optics," in *Encyclopedia of Physics*, Vol.25/1 (Springer, Berlin, Göttingen, Heidelberg 1961)
1.5 V.L. Ginzburg: Zh. Eksp. Teor. Fiz. *34*, 1953 (1958) [English transl.: Sov. Phys. JETP *7*, 1096 (1958)]
1.6 V.L. Ginzburg: *The Propagation of Electromagnetic Waves in Plasma* (Nauka, Moscow 1967) [English transl.: Pergamon, Oxford 1970]
1.7 V.L. Ginzburg, A.A. Rukhadze: *Waves in Magnetoactive Plasma* (Nauka, Moscow 1975) [in Russian]
1.8 V.L. Ginzburg: *Theoretical Physics and Astrophysics* (Nauka, Moscow 1976) [English transl.: Pergamon, Oxford 1979]
1.9 V.P. Silin, A.A. Rukhadze: *Electromagnetic Properties of Plasma and Plasma-Like Media* (Gosatomizdat, Moscow 1961) [in Russian]
1.10 V.L. Ginzburg, G.P. Motulevich: Usp. Fiz. Nauk *55*, 469 (1955) [German transl.: Fortschr. Phys. *3*, 309 (1955)]
1.11 A.A. Abrikosov, L.P. Gorkov, I.E. Dzyaloshinsky: *Methods of Quantum Field Theory in Statistical Physics* (Fizmatgiz, Moscow 1962) [in Russian]
1.12 H.A. Lorentz: *Collected Papers*, Vol.2 (1936) p.79; Vol.3 (1936) p.314 (M. Nijhoff, The Hague)
1.13 K.H. Hellwege: Z. Phys. *129*, 626 (1951)
1.14 E.F. Gross, A.A. Kaplyansky: Dokl. Akad. Nauk SSSR *132*, 93 (1960); and *139*, 75 (1961)
1.15 B.N. Gershman, V.L. Ginzburg: Radiofizika Izv. Vuzov *5*, 31 (1962)
1.16 M. Born: *Dynamik der Kristallgitter* (B.G. Teubner, Leipzig 1915) Chap.3
1.17 S.I. Pekar: Zh. Eksp. Teor. Fiz. *33*, 1022 (1957) [English transl.: Sov. Phys. JETP *6*, 785 (1958)]
1.18 V.M. Agranovich, V.L. Ginzburg: Usp. Fiz. Nauk *76*, 643 (1962); *77*, 663 (1962); Fortschr. Phys. *11*, 163 (1963) [English transl.: Sov. Phys. Usp. *5*, 323 and 675 (1962/63)]
1.19 B.E. Tsekvava: Fiz. Tverd. Tela *3*, 1164 (1961) [English transl.: Sov. Phys. Solid State *3*, 847 (1961)]
1.20 I.S. Gorban, V.B. Timofeyev: Dokl. Akad. Nauk SSSR *141*, 791 (1961)
1.21 J.J. Hopfield, D.G. Thomas: Phys. Rev. *132*, 561 (1963)
1.22 V.M. Agranovich, V.L. Ginzburg: Zh. Eksp. Teor. Fiz. *40*, 913 (1961) [English transl.: Sov. Phys. JETP *13*, 638 (1961)]
1.23 V.M. Agranovich, A.A. Rukhadze: Zh. Eksp. Teor. Fiz. *35*, 1171 (1958) [English transl.: Sov. Phys. JETP *8*, 819 (1959)]
1.24 C.H. Henry, J.J. Hopfield: Phys. Rev. Lett. *15*, 964 (1965)

1.25 M. Cardona (ed.): *Light Scattering in Solids I*, 2nd ed., Topics Appl. Phys., Vol.8 (Springer, Berlin, Heidelberg, New York 1982)
M. Cardona, G. Güntherodt (eds.): *Light Scattering in Solids II-IV*, Topics Appl. Phys., Vols.50,51,54 (Springer, Berlin, Heidelberg, New York 1982/83)
1.26 D. Fröhlich, E. Mohler, P. Wiesner: Phys. Rev. Lett. *28*, 554 (1971)
1.27 B. Honerlage, A. Bivas, V.D. Phach: Phys. Rev. Lett. *41*, 49 (1978)
1.28 W. Brenig, R. Zeyher, J.L. Birman: Phys. Rev. B*6*, 4617 (1972)
1.29 E.I. Rashba, M.D. Sturge (eds.): *Excitons* (North-Holland, Amsterdam 1982)
1.30 V.A. Kiselev, B.S. Razbizin, I.N. Uraltsev: Phys. Status Solidi B*72*, 161 (1975)
1.31 Y. Segawa, Y. Aoyagi, K. Azuma, S. Namba: Solid State Commun. *28*, 853 (1978)
1.32 I. Broser, M. Rosenzweig, R. Broser, E. Beckman, E. Birkicht: J. Phys. Soc. Jpn. A*49*, 401 (1980)
1.33 Yu.S. Barash, V.L. Ginzburg: Usp. Fiz. Nauk *116*, 5 (1976); Radiofizika Izv. Vuzov *21*, 1637 (1978)
1.34 A. Bardasis, F.K. Schrieffer: Phys. Rev. *121*, 1050 (1961)
1.35 V.L. Ginzburg: Zh. Eksp. Teor. Fiz. *41*, 828 (1961) [English transl.: Sov. Phys. JETP *14*, 594 (1962)]
1.36 S.V. Vonsovsky, Y.A. Izyumov, E.Z. Kurmaev: *Superconductivity of Transition Metals*, Springer Ser. Solid-State Sci., Vol.27 (Springer, Berlin, Heidelberg, New York 1982)
1.37 V.L. Ginzburg: Usp. Fiz. Nauk *69*, 537 (1959) [English transl.: Sov. Phys. Usp. *2*, 874 (1960)]
1.38 R. Loudon: *The Quantum Theory of Light* (Clarendon, Oxford 1973)
1.39 D.L. Mills, E. Burstein: Rep. Prog. Phys. *37*, 817 (1974)
1.40 V.M. Fain: *Quantum Radiophysics. Photons and Nonlinear Media* (Soviet Radio Publishers, Moscow 1972) [in Russian]
1.41 M. Born, K. Huang: *Dynamical Theory of Crystal Lattices* (Clarendon, Oxford 1954)
1.42 H. Haken: Fortschr. Phys. *6*, 271 (1958)
1.43 R.S. Knox: Theory of Excitons, in *Solid State Physics*, Vol.5 (Academic, London 1963)
1.44 K. Cho (ed.): *Excitons*, Topics Curr. Phys., Vol.14 (Springer, Berlin, Heidelberg, New York 1979)
1.45 S. Nakajima, Y. Toyozawa, R. Abe: *The Physics of Elementary Excitation*, Springer Ser. Solid-State Phys., Vol.12 (Springer, Berlin, Heidelberg, New York 1980)
1.46 V.M. Agranovich: *Theory of Excitons* (Nauka, Moscow 1968) [in Russian]
1.47 E. Burstein, F. De Martini (eds.): *Polaritons* (Pergamon, Oxford 1974)
1.48 R.M. White: *Quantum Theory of Magnetism*, Springer Ser. Solid-State Sci., Vol.32 (Springer, Berlin, Heidelberg, New York 1982)
1.49 K. Huang: Proc. Roy. Soc. London A*208*, 352 (1951)
1.50 K.B. Tolpygo: Zh. Eksp. Teor. Fiz. *20*, 497 (1950)
1.51 U. Fano: Phys. Rev. *103*, 1202 (1956)
1.52 J.J. Hopfield: Phys. Rev. *112*, 1555 (1958)
1.53 V.M. Agranovich: Zh. Eksp. Teor. Fiz. *37*, 430 (1959)
1.54 V.M. Agranovich, I.I. Lalov: Phys. Lett. A*53*, 169 (1975); Zh. Eksp. Teor. Fiz. *69*, 647 (1975); Fiz. Tverd. Tela *18*, 515, 621 (1976)
1.55 F. Saccetti. A Selloni: Phys. Rev. B*13*, 2286 (1976)
1.56 L.N. Ovander: Fiz. Tverd. Tela *3*, 2394 (1961); *4*, 157 and 294 (1962) [English transl.: Sov. Phys. Solid State *3*, 1737 (1961); *4*, 112 and 212 (1962)]; Usp. Fiz. Nauk *86*, 3 (1965)
1.57 J.J. Hopfield: J. Phys. Soc. Jpn. Suppl. *21*, 77 (1966)
1.58 V.M. Agranovich: Usp. Fiz. Nauk *115*, 199 (1975)
1.59 V.M. Agranovich, D.L. Mills (eds.): *Surface Polaritons* (North-Holland, Amsterdam 1982)

Chapter 2

2.1 V.P. Silin, A.A. Rukhadze: *Electromagnetic Properties of Plasma and Plasma-Like Media* (Gosatomizdat, Moscow 1961) [in Russian]
2.2 V.L. Ginzburg: *The Propagation of Electromagnetic Waves in Plasma* (Nauka, Moscow 1967) [English transl.: Pergamon, Oxford 1970]
2.3 V.M. Fain: *Quantum Radiophysics. Photons and Nonlinear Media* (Soviet Radio Publishers, Moscow 1972) [in Russian]
2.4 N. Bloembergen: *Nonlinear Optics* (Benjamin, New York 1965)
2.5 D.C. Hanna, M.A. Yuratich, D. Cotter: *Nonlinear Optics of Free Atoms and Molecules*, Springer Ser. Opt. Sci., Vol.17 (Springer, Berlin, Heidelberg, New York 1979)
2.6 I.E. Tamm: *Fundamentals of the Theory of Electricity* (Nauka, Moscow 1976)
2.7 L.D. Landau, E.M. Lifshitz: *Electrodynamics of Continuous Media* (Gostekhizdat, Moscow 1957) [English transl.: Pergamon, Oxford 1960]
2.8 G.S. Krinchik, M.V. Chetkin: Zh. Eksp. Teor. Fiz. *41*, 673 (1961) [English transl.: Sov. Phys. JETP *14*, 485 (1962)]
2.9 D.A. Krizhnits: Usp. Fiz. Nauk *119*, 357 (1976)
2.10 V.L. Ginzburg: Zh. Eksp. Teor. Fiz. *29*, 748 (1955) [English transl.: Sov. Phys. JETP *2*, 589 (1956)]
2.11 V.L. Ginzburg, N.N. Meiman: Zh. Eksp. Teor. Fiz. *46*, 243 (1964) [English transl.: Sov. Phys. *19*, 169 (1964)]
2.12 M.A. Leontovich: Zh. Eksp. Teor. Fiz. *40*, 907 (1961) [English transl.: Sov. Phys. JETP *13*, 634 (1961)]
2.13 V.L. Ginzburg: Dokl. Akad. Nauk SSSR *36*, 9 (1942)
2.14 A.A. Andronov: Radiofizika Izv. Vuzov *3*, 645 (1960)
2.15 V.L. Ginzburg: Akust. Zh. *1*, 31 (1955)
2.16 E.M. Hornreich, S. Shtriman: Phys. Rev. *171*, 1065 (1968)
2.17 B.K. Vainshtein: *Modern Crystallography I*, Springer Ser. Solid-State Sci., Vol.15 (Springer, Berlin, Heidelberg, New York 1981)
2.18 J.F. Nye: *Physical Properties of Crystals* (Clarendon, Oxford 1964)
2.19 S. Nakajima, Y. Toyozawa, R. Abe: *The Physics of Elementary Excitation*, Springer Ser. Solid-State Sci., Vol.12 (Springer, Berlin, Heidelberg, New York 1980)
2.20 K. Cho (ed.): *Excitons*, Topics Curr. Phys., Vol.14 (Springer, Berlin, Heidelberg, New York 1979) Chap.4
2.21 V.M. Agranovich: In *Surface Excitations*, ed. by V.M. Agranovich, R. Loudon (North-Holland, Amsterdam 1984)
2.22 G.N. Ramachandran, S. Ramaseshan: "Crystal Optics", in *Encyclopedia of Physics*, Vol.25/1 (Springer, Berlin, Göttingen, Heidelberg 1961)
2.23 F.I. Fedorov: *Optics of Anisotropic Media* (BSSR Akad. Nauk Publishers, Minsk 1958) [in Russian]
2.24 A.P. Khapalyuk: Kristallografiya *7*, 724 (1962)
2.25 A.F. Alexandrov, L.S. Bogdankevich, A.A. Rukhadze: *Principles of Plasma Electrodynamics*, Springer Ser. Electrophys., Vol.9 (Springer, Berlin, Heidelberg, New York 1983)
2.26 W. Voigt: Gött. Nachr. *5*, 269 (1902); Ann. Phys. *9*, 367 (1902); *27*, 1002 (1908)
2.27 P.S. Petrov, F.I. Fedorov: Opt. Spektrosk. *15*, 792 (1963)
2.28 S.I. Pekar: Usp. Fiz. Nauk *77*, 309 (1962) [English transl.: Sov. Phys. Usp. *5*, 515 (1962)]
2.29 V.N. Piskovoi: Fiz. Tverd. Tela *5*, 3, 168, 701 (1963) [English transl.: Sov. Phys. Solid State *5*, 1, 115, 511 (1963)]
2.30 S.I. Pekar: Zh. Eksp. Teor. Fiz. *33*, 1022 (1957) [English transl.: Sov. Phys. JETP *6*, 785 (1958)]
2.31 V.S. Barash, V.I. Karpman: Zh. Eksp. Teor. Fiz. *85*, 1962 (1983)

2.32 A.S. Davydov: Zh. Eksp. Teor. Fiz. *43*, 1832 (1962)
2.33 B.N. Gershman, V.L. Ginzburg: Radiofizika Izv. Vuzov *5*, 31 (1962)
2.34 A. Einstein: Ann. Phys. *23*, 371 (1907)
2.35 W. Pauli: *Theory of Relativity* (Pergamon, Oxford 1964)
2.36 V.L. Ginzburg: Radiofizika Izv. Vuzov *5*, 473 (1962)
2.37 V.L. Ginzburg: Radiofizika Izv. Vuzov *4*, 74 (1961)
2.38 J. Neufeld: Phys. Lett. *6*, 246 (1963)
2.39 M.E. Gertsenshtein: Zh. Eksp. Teor. Fiz. *26*, 680 (1954)
2.40 Yu.S. Barash, V.L. Ginzburg: Usp. Fiz. Nauk *118*, 523 (1976); Zh. Eksp. Teor. Fiz. *69*, 1179 (1975); Fiz. Tverd. Tela *19*, 946 (1977)
2.41 R. Loudon: J. Phys. A*3*, 233 (1970)
2.42 S.M. Rytov: Zh. Eksp. Teor. Fiz. *17*, 930 (1947)
2.43 L.A. Vainshtein: Usp. Fiz. Nauk *118*, 337 (1976)
2.44 J.L. Birman, M.J. Frankel: Opt. Commun. *13*, 303 (1975); Phys. Rev. A*15*, 2000 (1977)
2.45 A. Sommerfeld, L. Brillouin: Ann. Phys. *44*, 177 (1914); L. Brillouin: *Wave Propagation and Group Velocity* (Academic, London 1960)
2.46 K.A. Barsukov, V.L. Ginzburg: Radiofizika Izv. Vuzov *7*, 1187 (1964)
2.47 P. Agarwal, J.L. Birman, D.N. Pattanayak, A. Puri: Phys. Rev. B*25*, 2715 (1982)
2.48 J.L. Birman, D.N. Pattanayak: In *Light Scattering in Solids*, ed. by J.L. Birman, H.Z. Cummins, K.K. Rebane (Plenum, New York 1979) p.131
2.49 V.L. Ginzburg, V.A. Ugarov: Usp. Fiz. Nauk *118*, 175 (1976); Usp. Fiz. Nauk *110*, 309 (1973); *122*, 325 (1977)
2.50 L.P. Pitayevsky: Zh. Eksp. Teor. Fiz. *39*, 1450 (1960)
2.51 Yu.S. Barash, V.L. Ginzburg: Usp. Fiz. Nauk *116*, 5 (1976); Radiofizika Izv. Vuzov *21*, 1637 (1978)
2.52 B.V. Bokut, A.N. Serdyukov: Zh. Eksp. Teor. Fiz. *16*, 1808 (1971)
2.53 V.M. Agranovich, V.L. Ginzburg: Zh. Eksp. Teor. Fiz. *63*, 838 (1972)
2.54 V.M. Agranovich, V.I. Yudson: Opt. Commun. *9*, 58 (1973)
2.55 B.V. Bokut, A.N. Serdyukov: Zh. Priklad. Spektrosk. *20*, 677 (1974)
2.56 U. Schlagheck: Opt. Commun. *13*, 273 (1975)
2.57 V.M. Agranovich, V.I. Yudson: Opt. Commun. *5*, 422 (1972)
2.58 N. Bloembergen, R.K. Chang, S.S. Iha, C.H. Lee: Phys. Rev. *174*, 813 (1968)
2.59 V.M. Agranovich, S.A. Darmanyan: Zh. Eksp. Teor. Fiz. Pis'ma *35*, 68 (1982)
2.60 V.L. Ginzburg: Usp. Fiz. Nauk *69*, 537 (1959) [English transl.: Sov. Phys. Usp. *2*, 874 (1960)]
2.61 L.A. Vainshtein: *Electromagnetic Waves* (Soviet Radio Publishers, Moscow 1957) [in Russian]
2.62 E.L. Feinberg: *Propagation of Radiowaves Along the Earth's Surface* (USSR Akad. Nauk Publishers, Moscow 1961) [in Russian]
2.63 V.L. Ginzburg: Zh. Eksp. Teor. Fiz. *34*, 1953 (1958) [English transl.: Sov. Phys. JETP *7*, 1096 (1958)]
2.64 F. Goos, H. Hänchen: Ann. Phys. Leipzig *1*, 333 (1947)
2.65 H.K.V. Lotsch: Optik *32*, 116, 190, 550 (1970)
2.66 B.R. Horowitz, T. Tamir: J. Opt. Sci. Am. *61*, 586 (1971)
2.67 A. Mazer, C. Imbert, S. Huard: C.R. Acad. Sci. Ser. B*273*, 592 (1971)
2.68 O. Costa de Beauregard, C. Imbert: Phys. Rev. D*7*, 3555 (1973)
2.69 J.L. Birman, D.N. Pattanayak, A. Puri: Phys. Rev. Lett. *50*, 1664 (1983)

Chapter 3

3.1 V.L. Ginzburg: *The Propagation of Electromagnetic Waves in Plasma* (Nauka, Moscow 1967) [English transl.: Pergamon, Oxford 1970]
3.2 M. Born: *Dynamik der Kristallgitter* (Leipzig, 1915) Chap.3
3.3 M. Born, K. Huang: *Dynamical Theory of Crystal Lattices* (Clarendon, Oxford 1954)
3.4 V.M. Agranovich, V.L. Ginzburg: Usp. Fiz. Nauk *76*, 643 (1962); *77*, 663 (1962); Fortschr. Phys. *11*, 163 (1963) [English transl.: Sov. Phys. Usp. *5*, 323 and 675 (1962/63)]
3.5 V.L. Ginzburg, A.A. Rukhadze, V.P. Silin: Fiz. Tverd. Tela *3*, 1835, 2890 (1961) [English transl.: Sov. Phys. Solid State *3*, 1337,2110 (1961/62)]; J. Phys. Chem. Sol. *23*, 85 (1962)
3.6 I.E. Tamm: *Fundamentals of the Theory of Electricity* (Nauka, Moscow 1976)
3.7 V.L. Ginzburg: Zh. Eksp. Teor. Fiz. *34*, 1953 (1958) [English transl.: Sov. Phys. JETP *7*, 1096 (1958)]
3.8 V.M. Agranovich, Yu.V. Konobeyev: Fiz. Tverd. Tela *5*, 2544 (1963) [English transl.: Sov. Phys. Solid State *5*, 1858 (1963)]
3.9 N.G. Van Kampen: Math. Rev. *20*, 1227 (1959)
3.10 S.I. Pekar: Zh. Eksp. Teor. Fiz. *38*, 1787 (1960) [English transl.: Sov. Phys. JETP *11*, 1286 (1960)]
3.11 V.M. Agranovich, A.A. Rukhadze: Zh. Eksp. Teor. Fiz. *35*, 982 (1958) [English transl.: Sov. Phys. JETP *35*, 685 (1959)]
3.12 S.I. Pekar: Fiz. Tverd. Tela *4*, 1301 (1962) [English transl.: Sov. Phys. Solid State *4*, 953 (1962)]
3.13 L.D. Landau, E.M. Lifshitz: *Electrodynamics of Continuous Media* (Gostekhizdat, Moscow 1957) [English transl.: Pergamon, Oxford 1960]
3.14 S. Szivessy: "Kristalloptik", in *Handbuch der Physik*, Vol.20 (J. Springer, Berlin 1928) p.635
3.15 G.N. Ramachandran, S. Ramaseshan: "Crystal Optics", in *Encyclopedia of Physics*, Vol.25/1 (Springer, Berlin, Göttingen, Heidelberg 1961)
3.16 J.F. Nye: *Physical Properties of Crystals* (Clarendon, Oxford 1964)
3.17 B.K. Vainshtein: *Modern Crystallography I*, Springer Ser. Solid State Sci., Vol.15 (Springer, Berlin, Heidelberg, New York 1981)
3.18 F.I. Fedorov: Opt. Spektrosk. *6*, 85, 377 (1959)
3.19 W.F. Brown, S. Shtrikman, D. Treves: J. Appl. Phys. *34*, 1233 (1965)
3.20 R.R. Birss: Philos. Mag. *15*, 687 (1967)
3.21 V.M. Agranovich, V.L. Ginzburg: In *Progress in Optics*, Vol.9, ed. by E. Wolf (North-Holland, Amsterdam 1971) p.372
3.22 M.I. Kaganov, R.R. Yankelevich: Fiz. Tverd. Tela *10*, 2771 (1968)
3.23 O.S. Yeritsyan: Fiz. Tverd. Tela *19*, 1545 (1977)
3.24 L.D. Landau, E.M. Lifshitz: *Mechanics of Continuous Media* (Gostekhizdat, Moscow 1953) Part II. §10 [English transl.: Pergamon, Oxford 1960]

Chapter 4

4.1 M. Born: *Optik* (Springer, Berlin 1933); M. Born, E. Wolf: *Principles of Optics* (Pergamon, Oxford 1964)
4.2 S. Szivessy: "Kristalloptik", in *Handbuch der Physik*, Vol.20 (J. Springer, Berlin 1928) p.635
4.3 G.N. Ramachandran, S. Ramaseshan: "Crystal Optics", in *Encyclopedia of Physics*, Vol.25/1 (Springer, Berlin, Göttingen, Heidelberg 1961)
4.4 V.M. Agranovich, A.A. Rukhadze: Zh. Eksp. Teor. Fiz. *35*. 982 (1958) [English transl.: Sov. Phys. JETP *35*, 685 (1959)]

4.5 V.M. Agranovich: Dokl. Akad. Nauk SSSR *97*, 797 (1954); Opt. Spektrosk. *1*, 338 (1956)
4.6 Yu.A. Tsvirko: Zh. Eksp. Teor. Fiz. *38*, 1615 (1960) [English transl.: Sov. Phys. JETP *11*, 1163 (1960)]
4.7 A.A. Kaminskii: *Laser Crystals*, Springer Ser. Opt. Sci., Vol.14 (Springer, Berlin, Heidelberg, New York 1981)
4.8 B.V. Bokut, A.N. Serdyukov: Zh. Eksp. Teor. Fiz. *16*, 1808 (1971)
4.9 V.M. Agranovich, V.L. Ginzburg: Zh. Eksp. Teor. Fiz. *63*, 838 (1972)
4.10 V.M. Agranovich, V.I. Yudson: Opt. Commun. *9*, 58 (1973)
4.11 V.L. Ginzburg: Zh. Eksp. Teor. Fiz. *34*, 1953 (1958) [English transl.: Sov. Phys. JETP *7*, 1096 (1958)]
4.12 S.I. Pekar: Zh. Eksp. Teor. Fiz. *33*, 1022 (1957) [English transl.: Sov. Phys. JETP *6*, 785 (1958)]
4.13 F.I. Fedorov: *Optics of Anisotropic Media* (BSSR, Akad. Nauk Publishers, Minsk 1958) [in Russian]
4.14 A.P. Khapalyuk: Kristallografiya *7*, 724 (1962)
4.15 V.M. Agranovich, V.L. Ginzburg: Fortschr. Phys. *13*, 175 (1965)
4.16 A.G. Molchanov: Fiz. Tverd. Tela *8*, 1156 (1966)
4.17 A.A. Andronov: Radiofizika Izv. Vuzov *3*, 645 (1960)
4.18 A.S. Pine: Phys. Rev. B*2*, 2049 (1970)
4.19 M.F. Bryzhina, S.Kh. Yesayan, V.V. Lemanov: Zh. Eksp. Teor. Fiz. Pis'ma *25*, 513 (1977)
4.20 I.V. Obreimov, A.F. Prikhotko: In *Collection Pamyati S.I. Vavilova* (USSR Akad. Nauk, Moscow 1952) p.197 [in Russian]
4.21 V.M. Agranovich, V.L. Ginzburg: In *Progress in Optics*, Vol.9, ed. by E. Wolf (North-Holland, Amsterdam 1971) p.237
4.22 P.G. de Gennes: *The Physics of Liquid Crystals* (Clarendon, Oxford 1974)
4.23 W. Helfrich, G. Heppke (eds.): *Liquid Crystals of One- and Two-Dimensional Order*, Springer Ser. Chem. Sci., Vol.11 (Springer, Berlin, Heidelberg, New York 1980)
4.24 V.G. Kamensky, E.I. Katz: Opt. Spektrosk. *45*, 1106 (1978)
4.25 W.R. Heller, A. Markus: Phys. Rev. *84*, 809 (1951)
4.26 A. Sommerfeld: *Optics* (Academic, London 1954)
4.27 E.E. Lysenko: Zh. Eksp. Teor. Fiz. *6*, 787 (1936)
4.28 A.S. Davydov: Zh. Eksp. Teor. Fiz. *19*, 181 (1949); Tr. Inst. Fiz. Akad. Nauk UKSSR *3*, 36 (1952);
A.S. Davydov, E.I. Rashba: Ukr. Fiz. Zh. *2*, 226 (1957)
4.29 M.S. Brodin, A.F. Prikhotko, M.S. Soskin: Opt. Spektrosk. *6*, 28 (1959)
4.30 A. Bosacchi, B. Bosacchi, S. Franchi: Phys. Rev. Lett. *36*, 1086 (1976)
4.31 P.P. Feofilov, A.A. Kaplyansky: Usp. Fiz. Nauk *76*, 201 (1962)
4.32 H.A. Lorentz: *Collected Papers*, Vol.2 (1936) p.79; Vol.3 (1936) p.314 (M. Nijhoff, The Hague)
4.33 K.H. Hellwege: Z. Phys. *129*, 626 (1951)
4.34 J.F. Nye: *Physical Properties of Crystals* (Clarendon, Oxford 1964)
4.35 E.F. Gross, A.A. Kaplyansky: Dokl. Akad. Nauk SSSR *132*, 93 (1960) and *139*, 75 (1961)
4.36 S.I. Pekar, B.E. Tsekvava: Fiz. Tverd. Tela *2*, 261 (1960) [English transl.: Sov. Phys. Solid State *2*, 242 (1960)]
4.37 A.S. Davydov: *Theory of Light Absorption in Molecular Crystals* (UKSSR Akad. Nauk, Kiev 1951) [in Russian]
4.38 H. Winston: J. Chem. Phys. *19*, 156 (1951)
4.39 G.Ya. Lyubarsky: *Group Theory and Its Application in Physics* (Fizmatgiz, Moscow 1958) [in Russian]
4.40 B.K. Vainshtein: *Modern Crystallography I*, Springer Ser. Solid-State Sci., Vol.15 (Springer, Berlin, Heidelberg, New York 1981)

4.41 J.L. Birman: "Theory of Crystal Space Groups and Infrared and Raman Lattice Processes of Insulating Crystals", in *Encyclopedia of Physics*, Vol.25/2b (Springer, Berlin, Heidelberg, New York 1974)
4.42 L.D. Landau, E.M. Lifshitz: *Quantum Mechanics - Nonrelativistic Theory* (Nauka, Moscow 1974) [in Russian]
4.43 V.I. Cherepanov, V.S. Galishev: Fiz. Tverd. Tela *3*, 1085 (1961)
4.44 V.E. Tsekvava: Fiz. Tverd. Tela *3*, 1164 (1961) [English transl.: Sov. Phys. Solid State *3*, 847 (1961)]
4.45 E.F. Gross: Usp. Fiz. Nauk *63*, 575 (1957) [English transl.: Sov. Phys. Usp. Nauk *63*, 576 (1957)]
4.46 E.F. Gross, A.A. Kaplyansky: Fiz. Tverd. Tela *2*, 1676, 2963 (1960) [English transl.: Sov. Phys. Solid State *2*, 1518, 2637 (1960)]
4.47 E.F. Gross, B.P. Zakharchenya, L.M. Kanskaya: Fiz. Tverd. Tela *3*, 972 (1961) [English transl.: Sov. Phys. Solid State *3*, 706 (1961)]
4.48 E.F. Gross, A.G. Zhilich, B.P. Zakharchenya, A.V. Varfolomeyev: Fiz. Tverd. Tela *3*, 1445 (1961) [English transl.: Sov. Phys. Solid State *3*, 1048 (1961)]
4.49 V.I. Cherepanov: Fiz. Tverd. Tela *3*, 1493 (1961) [English transl.: Sov. Phys. Solid State *3*, 1082 (1961)]
4.50 A.G. Zhilich: Fiz. Tverd. Tela *3*, 204 (1961) [English transl.: Sov. Phys. Solid State *3*, 1483 (1961)]
4.51 V.I. Cherepanov: Fiz. Tverd. Tela *3*, 2183 (1961) [English transl.: Sov. Phys. Solid State *3*, 1583 (1961)]
4.52 S.A. Moskalenko: Fiz. Tverd. Tela *2*, 1755 (1960) [English transl.: Sov. Phys. Solid State *2*, 1587 (1960)]
4.52a E.I. Rashba, M.D. Sturge (eds.): *Excitons* (North-Holland, Amsterdam 1982)
4.53 N.N. Akhmediyev, A.K. Zvezdin: Zh. Eksp. Teor. Fiz. *38*, 167 (1983)
4.54 N.B. Baranova, Yu.V. Bogdanov, B.Ya. Zel'dovich: Usp. Fiz. Nauk *123*, 349 (1977)
4.55 T.S. Moos, G.J. Burrell, A. Hetherington: Proc. Roy. Soc. A*308*, 125 (1968)
4.56 L.A. Almazov, F.T. Vasko, I.M. Dykman: Zh. Eksp. Teor. Fiz. Pis'ma *16*, 305 (1972);
F.T. Vasko: Fiz. Tverd. Tela *14*, 3680 (1972);
L.A. Almazov: Fiz. Tekh. Poluprovodnikov *9*, 657 (1975)
4.57 L.E. Vorobyov, E.L. Ivchenko, G.E. Pikus, I.I. Farbshtein, V.A. Shaligan, A.V. Shturbin: Zh. Eksp. Teor. Fiz. Pis'ma *29*, 485 (1979)
4.58 L.I. Korovina, A.E. Shabad: Zh. Eksp. Teor. Fiz. *67*, 1032 (1974)
4.59 I.S. Zheludov: Usp. Fiz. Nauk *120*, 702 (1976)
4.60 T. Koda, T. Murahashi, T. Mutani, S. Sakoda, Y. Onodera: Phys. Rev. B*5*, 705 (1972)
4.61 E.F. Gross, B.P. Zakharchenya, O.V. Konstantinov: Fiz. Tverd. Tela *3*, 305 (1961) [English transl.: Sov. Phys. Solid State *3*, 221 (1961)]
4.62 D.G. Thomas, J.J. Hopfield: Phys. Rev. Lett. *4*, 357 (1960); Phys. Rev. *124*, 657 (1961)
4.63 D.G. Thomas, J.J. Hopfield: Phys. Rev. Lett. *5*, 505 (1960)
4.63a H. Venghaus, S. Suga, K. Cho: Phys. Rev. B*16*, 4419 (1977)
4.63b K. Cho: Phys. Rev. B*14*, 4463 (1976)
4.63c B.S. Razbirin, A.N. Starukhin, E.M. Gamarts, M.I. Karaman, V.P. Mushinskii: Zh. Eksp. Teor. Fiz. Pis'ma *27*, 341 (1978)
4.64 A.G. Zhilich: Fiz. Tverd. Tela *6*, 2058 (1964) [English transl.: Sov. Phys. Solid State *6*, 1624 (1965)]; Fiz. Tverd. Tela *7*, 670 (1965)
4.65 V.P. Silin, E.P. Fetisov: Zh. Eksp. Teor. Fiz. *41*, 159 (1961) [English transl.: Sov. Phys. JETP *14*, 115 (1962)]
4.65a K.L. Kliewer: Surf. Sci. *101*, 57 (1980)
4.65b F. Forstmann, R.R. Gerhardts: In *Festkörperprobleme*, Adv. Solid State Phys., Vol.22 (Vieweg, Braunschweig 1982) p.291

4.66 V.M. Agranovich, V.E. Pafomov, A.A. Rukhadze: Zh. Eksp. Teor. Fiz. *36*, 238 (1959)
4.67 B.L. Zhelnov: Zh. Eksp. Teor. Fiz. *40*, 170 (1961) [English transl.: Sov. Phys. JETP *13*, 117 (1961)]
4.68 V.M. Agranovich, V.L. Ginzburg: Usp. Fiz. Nauk *76*, 643 (1962); *77*, 663 (1962); Fortschr. Phys. *11*, 163 (1963) [English transl.: Sov. Phys. Usp. *5*, 323 and 675 (1962/63)]
4.69 Y. Osaka, Y. Imai, Y. Takeuti: J. Phys. Soc. Jpn. *24*, 236 (1968)
4.70 V.I. Sugakov: Fiz. Tverd. Tela *5*, 2207, 2682 (1963); *6*, 1361 (1964) [English transl.: Sov. Phys. Solid State *5*, 1607, 1959 (1963); *6*, 1064 (1964)]
4.71 V.S. Mashkevich: Zh. Eksp. Teor. Fiz. *38*, 906 (1960); *40*, 1803 (1961); *42*, 135 (1962) [English transl.: Sov. Phys. JETP *11*, 653 (1960); *13*, 1267 (1961)]
4.72 K. Huang: Proc. Roy. Soc. A*208*, 352 (1951)
4.73a E.L. Ivchenko, S.A. Permogorov, A.V. Selkin: Solid State Commun. *28*, 345 (1978);
4.73b G. Fishman: Solid State Commun. *27*, 1097 (1978)
4.74 M. Born, K. Huang: *Dynamical Theory of Crystal Lattices* (Clarendon, Oxford 1954)
4.75 V.M. Agranovich, M.I. Kaganov: Fiz. Tverd. Tela *4*, 1681 (1962) [English transl.: Sov. Phys. Solid State *4*, 1236 (1962)]
4.76 S. Chandrasekhar: Proc. Indian Acad. Sci. A*36*, 103 (1952); A*37*, 468 (1953); A*39*, 243 (1954)
4.77 V.M. Agranovich: Opt. Spektrosk. *2*, 738 (1957)
4.78 G. Harbeke: "Optical Properties of Semiconductors", in *Optical Properties of Solids*, ed. by F. Abelès (North-Holland, Amsterdam 1972) p.21
4.79 E.L. Ivchenko, G.E. Pikus: Fiz. Tverd. Tela *16*, 967 (1974)
4.80 K. Natori: J. Phys. Soc. Jpn. *39*, 1013 (1975)
4.81 A.G. Aronov, A.S. Ioselevich: Fiz. Tverd. Tela *20*, 2615 (1978)
4.81a E.L. Ivchenko: In *Excitons*, ed. by E.I. Rashba, M.D. Sturge (North-Holland, Amsterdam 1982) p.141
4.82 J.J. Hopfield, D.G. Thomas: Phys. Rev. *132*, 561 (1963)
4.83 V.L. Ginzburg, G.P. Motulevich: Usp. Fiz. Nauk *55*, 469 (1955) [German transl.: Fortschr. Phys. *3*, 309 (1955)]
4.84 G.E.H. Reuter, E.H. Sondheimer: Proc. Roy. Soc. London A*195*, 336 (1948)
4.85 R. Zeyher, J.L. Birman, W. Brenig: Phys. Rev. B*6*, 4613, 4617 (1972)
4.86 V.M. Agranovich, V.I. Yudson: In *Up-to-Date Problems of Optics and Nuclear Physics* (Naukova Dumka, Kiew 1974) p.78 [in Russian]
4.87 D.L. Mills: "Spatial Dispersion and Its Effect on the Properties of Dielectrics", in *Polaritons*, Taormina 1972, ed. by E. Burstein, F. de Martini (Pergamon, Oxford 1974) p.147;
V.V. Hyzhnyakov, A.A. Maradudin, D.L. Mills: Phys. Rev. B*11*, 3149 (1945)
4.88 D.P. Craig, P.G. Hobbins: J. Chem. Soc. *539*, 2309 (1955)
4.89 C.S. Ting, M.J. Frankel, J.L. Birman: Solid State Commun. *17*, 1285 (1975)
4.90 M.F. Bishop: Solid State Commun. *20*, 779 (1976)
4.91 N.N. Akhmediyev, V.V. Yatsyshen: Fiz. Tverd. Tela *18*, 1679 (1976)
4.92 A. Stahl: Phys. Status Solidi B*94*, 221 (1979); B*106*, 575 (1981)
4.93 S. Sakoda: J. Phys. Soc. Jpn. *40*, 152 (1976)
4.93a I. Balslev: Phys. Status Solidi B*88*, 155 (1978)
4.94 A. D'Andrea, R. Del Sole: Phys. Rev. B*25*, 3714 (1982)
4.95 V.M. Agranovich, V.I. Yudson: Opt. Commun. *7*, 121 (1973)
4.96 O.V. Konstantinov, M.M. Panakhov, Sh.R. Saifulayev: Fiz. Tverd. Tela *17*, 3551 (1975)
4.97 G.S. Agarwal, D.N. Pattanayak, E. Wolf: Phys. Rev. B*10*, 1447 (1974); B*11*, 1342 (1975);
J.L. Birman, J.J. Sein: Phys. Rev. B*6*, 2482 (1972);
A.A. Maradudin, D.L. Mills: Phys. Rev. B*7*, 2787 (1973)
4.98 J.T. Foley, A.J. Devaney: Phys. Rev. B*12*, 3104 (1975)
4.99 M.F. Bishop, A.A. Maradudin: Phys. Rev. B*14*, 3384 (1976)

4.100 L.D. Johnson, P.R. Rimbey: Phys. Rev. B*14*, 2398 (1976)
4.101 F. Flores, F. Garcia-Moliner, R. Monreal: Phys. Rev. B*15*, 5085 (1977)
4.102 D.V. Sivukhin: Zh. Eksp. Teor. Fiz. *18*, 976 (1948); *21*, 94, 367 (1952); *30*, 374 (1956)
4.103 N.N. Akhmediyev, V.V. Yatsyshen: Solid State Commun. *27*, 357 (1978)
4.104 C.W. Deutche, C.A. Mead: Phys. Rev. A*138*, 63 (1965); G.D. Mahan, G. Obermair: Phys. Rev. *183*, 834 (1969); J.E. Sipe, J. Van Kranendonk: Can. J. Phys. *53*, 2095 (1975); C.A. Mead: Phys. Rev. B*15*, 519 (1977); B*17*, 4644 (1978)
4.105 M.R. Philpott: Phys. Rev. B*14*, 3471 (1976); J. Chem. Phys. *60*, 1410, 2520 (1974)
4.106 N.N. Akhmediyev: Zh. Eksp. Teor. Fiz. *78*, 1615 (1980)
4.107 V.A. Kizel: *Reflection of Light* (Nauka, Moscow 1973) [in Russian]
4.108 V.L. Ginzburg: *The Propagation of Electromagnetic Waves in Plasma* (Nauka, Moscow 1967) [English transl.: Pergamon, Oxford 1970]
4.109 A.V. Alexandrov, L.S. Bogdankevich, A.A. Rukhadze: *Principles of Plasma Electrodynamics*, Springer Ser. Electrophys., Vol.9 (Springer, Berlin, Heidelberg, New York 1983)
4.110 A. Selkin: Phys. Status Solidi B*83*, 47 (1977)
4.111 J.L. Birman, D.N. Pattanayak: In *Light Scattering in Solids*, ed. by J.L. Birman, H.Z. Cummins, K.K. Rebane (Plenum, New York 1979) p.131
4.112 Yu.A. Tsvirko: Fiz. Tverd. Tela *5*, 1498 (1963) [English transl.: Sov. Phys. Solid State *5*, 1089 (1963)]
4.112a P. Halevi, G. Hernandez-Cocoletzi: Phys. Rev. Lett *48*, 1500 (1982)
4.112b V.M. Agranovich, V.E. Kravtsov, T.A. Leskova, A.G. Mal'shukov, G. Hernandez-Cocoletzi, A.A. Maradudin: Phys. Rev. B (1984) (in press)
4.113 M.I. Nathan, A.B. Fowler, G. Burns: Phys. Rev. Lett. *11*, 152 (1963)
4.114 G.K. Vlasov, A.V. Kritsky, Yu.A. Kupchenko: Zh. Eksp. Teor. Fiz. *72*, 2064 (1977)
4.115 N.N. Akhmediyev, V.V. Yatsychen: Fiz. Tverd. Tela *21*, 3529 (1979)
4.116 N.N. Akhmediyev: Zh. Eksp. Teor. Fiz. *79*, 1534 (1980)
4.117 J. Voigt, F. Spiegelberg, M. Senoner: Phys. Status Solidi B*91*, 189 (1979)
4.118 L.I. Mandelstam: Zh. Eksp. Teor. Fiz. *15*, 475 (1945); *Collected Papers* (USSR Akad. Nauk Publishers, Moscow 1947) Vol.2, p.334 [in Russian]
4.119 M. Born: Phys. Zs. *16*, 251, 437 (1915); Ann. Phys. (Leipzig) *55*, 177 (1917)
4.120 C. Oseen: Ann. Phys. (Leipzig) *48*, 1 (1915)
4.121 K. Nomura: Phys. Rev. Lett. *5*, 500 (1960)
4.122 N. Henrion, T. Eckart: Z. Naturforsch. *19*a, 1024 (1964)
4.123 V.I. Burkov, V.A. Kizel, M.F. Kostenko, Z.B. Perikalina: Dokl. Akad. Nauk. SSSR, *199*, 806 (1968)
4.124 Yu.V. Denisov, V.A. Kizel, V.V. Mnev, E.P. Sukhenko, V.G. Tishchenko: Zh. Eksp. Teor. Fiz. Pisma. *22*, 242 (1975)
4.125 V.A. Kizel, Yu.I. Krasilov, V.I. Burkov: Usp. Fiz. Nauk. *114*, 295 (1974)
4.126 I.G. Chistyakov: Kristallografia, *5*, 962 (1961)
4.127 B.N. Samoilov: Zh. Eksp. Teor. Fiz. *18*, 1030 (1948)
4.128 M.S. Brodin, Ya.O. Dovgy: Opt. Spektrosk. *12*, 285 (1962)
4.129 V.M. Agranovich: Usp. Fiz. Nauk. *71*, 141 (1960) [English transl.: Sov. Phys.-Usp. *3*, 427 (1961)]
4.130 V.M. Agranovich: Fiz. Tverd. Tela, *2*, 1197 (1960)
4.131 A.G. Molchanov: Fiz. Tverd. Tela, *11*, 1184 (1969)
4.132 A.S. Pine, G. Dresselhaus: Phys. Rev. *188*, 1489 (1969)
4.133 F.I. Fedorov: Opt. Spektrosk. *6*, 85, 377 (1959)
4.133a E.L. Ivchenko, A.V. Selkin: Zh. Eksp. Teor. Fiz. *76*, 1837 (1979)
4.133b E.L. Ivchenko, A.B. Pevtsov, A.V. Selkin: Solid State Commun. *39*, 453 (1980)
4.134 J. Pastnak, K. Vedam: Phys. Rev. B*3*, 2567 (1971)
4.135 P.Y. Yu, M. Cardona: Solid State Commun. *9*, 1421 (1971)
4.136 M.B. Brodin, S.I. Pekar: Zh. Eksp. Teor. Fiz. *38*, 71 1910 (1960) [English transl.: Sov. Phys.-JETP *11*, 1373 (1960)]

4.137 M.S. Brodin, A.F. Prikhotko: Opt. Spectrosk. *7*, 132 (1959)
4.138 M.S. Brodin, A.F. Prikhotko, M.S. Soskin: Opt. Spektrosk. *6*, 28 (1959)
4.139 C.W. Higginbotham, M. Cardona, F.H. Pollak: Phys. Rev. *184*, 821 (1969)
4.140 O.V. Gogolin, E.G. Tsitsishvily, J.L. Dais, K. Klingshirn, V.E. Solomko: Zh. Eksp. Teor. Fiz. Pis'ma *34*, 328 (1981)
4.141 A.S. Davydov: Zh. Eksp. Teor. Fiz. *45*, 723 (1963) [English transl.: Sov. Phys.-JETP *18*, 496 (1964)]
4.142 V.M. Agranovich, A.S. Davydov: Zh. Eksp. Teor. Fiz. *21*, 667 (1951)
4.143 A.S. Davydov: Usp. Fiz. Nauk, *82*, 393 (1964) [English transl.: Sov. Phys.-Usp. *7*, 145 (1964)]
4.144 I.S. Gorban, V.B. Timofeyev: Dokl. Akad. Nauk SSSR *141*, 791 (1961)
4.145 J.J. Hopfield, D.G. Thomas: Phys. Rev. *116*, 512 (1959)
4.146 V.A. Kiselov: Fiz. Tverd. Tela, *20*, 2173 (1978)
4.147 V.A. Kiselov, B.S. Razbirin, I.N. Uraltsev: Phys. Status Solidi B*72*, 161 (1975);
V.A. Kiselov, I.V. Makarenko, B.S. Razbirin, I.N. Uraltsev: Fiz. Tverd. Tela, *19*, 1348 (1977);
I.V. Makarenko, I.N. Uraltsev, V.A.K. Kiselev: Phys. Status Solidi B*98*, 773 (1980)
4.148 I. Broser, R. Broser, E. Beckmann, E. Birkicht: Solid State Commun *39*, 1209 (1981)
4.149 Y. Segawa, Y. Aoyagi, K. Azuma, S. Namba: Solid State Commun. *28*, 853 (1978)
Y. Segawa, Y. Aoyagi, S. Namba: Solid State Commun. *32*, 299 (1979)
4.150 R.G. Ulbrich, G.W. Fehrenbach: Phys. Rev. Lett. *43*, 963 (1979)
4.151 Y. Aoyagi, Y. Segawa, T. Baba, S. Namba: In *Picosecond Phenomena II*, ed. by R.M. Hochstrasser, W. Kaiser and C.V. Shank, Springer Ser. Chem. Phys. Vol. 14 (Springer, Berlin, Heidelberg, New York 1980) p. 298
4.152 T. Itoh, P. Lavallard, J. Raydellet, C. Benoit â la Guillaume: Solid State Commun. *37*, 925 (1981)
P.H. Duong, T. Itoh, P. Lavallard: Solid State Commun. *43*, 879 (1982)
4.153 R. Loudon: J. Phys. A: Gen. Phys. *3*, 233 (1970)
4.154 R.G. Ulbrich, G. Weisbuch: Phys. Rev. *38*, 865 (1977)
4.155 G. Winterling, E. Koteles: Solid State Commun. *23*, 95 (1977)
4.156 W. Brenig, R. Zeyher, J.L. Birman: Phys. Rev. B*6*, 4617 (1972)
4.157 E.S. Koteles, G. Winterling: Phys. Rev. B*20*, 628 (1979)
4.158 P.Y. Yu: Solid State Commun. *32*, 29 (1979)
4.159 F. Evangelisti, A. Frova, F. Patella: Phys. Rev. B*10*, 4253 (1974); Solid State Commun. *20*, 23 (1976)
4.160 R.S. Knox: Theory of Excitons, in *Solid State Physics*, Vol. 5 (Academic, London 1963)
4.161 V.M. Agranovich: *Theory of Excitons* (Nauka, Moscow 1968) in Russian
4.162 S.A. Permogorov, V.V. Travnikov, A.V. Selkin: Fiz. Tverd. Tela *14*, 3642 (1972); *15*, 1822 (1973); Izv. Akad. Nauk SSSR, Ser. Fiz. *37*, 711 (1973);
S.A. Permogorov, A.V. Selkin: Fiz. Tverd. Tela *15*, 3025 (1973)
4.163 E. Tosatti, G. Harbeke: Nuovo Cimento, B*22*, 87 (1974)
4.164 M.S. Brodin, I.N. Strashnikova: Fiz. Tverd. Tela *4*, 2454 (1962) [English transl.: Sov. Phys.-Solid State, *4*, 1798 (1962)]
4.165 M.S. Brodin, N.A. Davydova, M.I. Strashnikova: Phys. Status Solidi B*70*, 365 (1975)
4.166 M.S. Brodin, N.A. Davydova, M.I. Strashnikova: Zh. Eksp. Teor. Fiz. Pisma *19*, 567 (1974)
4.167 M.I. Strashnikova: Fiz. Tverd. Tela *17*, 729 (1975): Zh. Eksp. Teor. Fiz. *74*, 2206 (1978)
4.168 A.B. Pevtsov, S.A. Permogorov, A.V. Selkin: Zh. Eksp. Teor. Fiz. Pis'ma *33*, 419 (1981)

Chapter 5

5.1 L.D. Landau, E.M. Lifshitz: *Electrodynamics of Continuous Media* (Gostekhizdat, Moscow 1957) [English transl.: Pergamon, Oxford 1960]
5.2 D.L. Mills, E. Burstein: Rep. Prog. Phys. *37*, 817 (1974)
5.3 V.M. Agranovich, Yu.V. Konobeyev: Fiz. Tverd. Tela *7*, 111 (1965)
5.4 V.M. Agranovich, V.L. Ginzburg: Fortschr. Phys. *13*, 175 (1965)
5.4a V.M. Agranovich, D.L. Mills (eds.): *Surface Polaritons* (North-Holland, Amsterdam 1982)
5.5 F.A. Olivera, A.F. Khater, E.F. Sarmento, D.R. Tilley: J. Phys. C*12*, 4021 (1979)
5.6 V.V. Bryksin, A.V. Raitsev, Yu.A. Firsov: Fiz. Tverd. Tela *17*, 685 (1975)
5.7 C. Shu, A. Caillé: Solid State Commun. *42*, 233 (1982)
5.8 R. Damon, J. Eshbach: J. Phys. Chem. Solids *19*, 308 (1961)
5.9 D.L. Mills: In *Surface Excitations*, ed. by V.M. Agranovich, R. Loudon (North-Holland, Amsterdam 1984) (in press)
5.10 A.V. Chaplik, M.V. Krasheninnikov: Solid State Commun. *27*, 1297 (1978)
5.11 L.V. Keldysh: Zh. Eksp. Teor. Fiz. Pis'ma *29*, 717 (1979); *30*, 244 (1979)
5.11a R.R. Guseinov: Fiz. Tverd. Tela *25*, 2120 (1983)
5.12 G.S. Agarwal: Phys. Rev. B*8*, 4768 (1973)
5.13 J. Nkoma, R. Loudon, D.R. Tilleye: J. Phys. C*7*, 3547 (1974)
5.14 D.L. Mills, A.A. Maradudin: Phys. Rev. Lett. *31*, 372 (1973)
5.15 V.V. Bryksin, D.N. Mirlin, Yu.A. Firsov: Usp. Fiz. Nauk *113*, 29 (1974)
5.16 R. Englman, R. Ruppin: J. Phys. C*1*, 614, 630, 1515 (1968); Rep. Progr. Phys. *33*, 149 (1970)
5.17 C.A. Pfeiffer, E.N. Economou, K.L, Ngai: Phys. Rev. B*10*, 3038 (1974)
5.18 V.A. Kiselov: Fiz. Tverd. Tela *20*, 2173 (1978)
5.19 G.F. Kventsel, S.I. Pekar: Fiz. Tverd. Tela *6*, 811 (1964) [English transl.: Sov. Phys. Solid State *6*, 630 (1964)]
5.20 V.M. Agranovich, A.G. Malshukov, M.A. Mekhtiyev: Zh. Eksp. Teor. Fiz. *63*, 2274 (1972)
5.21 V.L. Ginzburg, V.V. Kelle: Zh. Eksp. Teor. Fiz. Pis'ma *17*, 428 (1973)
5.22 Yu.E. Lozovik, V.N. Nishanov: Fiz. Tverd. Tela *18*, 3267 (1976)
5.23 V.M. Agranovich, M.D. Galanin: *Electron Excitation Energy Transfer in Condensed Media* (Nauka, Moscow 1978) [in Russian]
5.24 V.V. Bryksin, Yu.A. Firsov: Fiz. Tverd. Tela *11*, 2167 (1969); *14*, 1148 (1972)
5.25 U. Fano: J. Opt. Soc. Am. *31*, 213 (1941)
5.25a Alig R. Casanova: Solid State Commun. *13*, 1603 (1973)
5.26 J. Zenneck: Ann. Phys. *23*, 5846 (1907)
5.27 P. Halevi: Phys. Rev. B*12*, 4032 (1975); B*18*, 590 (1978)
5.28 E.L. Feinberg: *Radiowave Propagation Along the Earth's Surface* (USSR Akad. Nauk Publishers, Moscow 1961) [in Russian];
L.A. Vainshtein: *Theory of Diffraction and the Factorization Method* (Soviet Radio Publishers, Moscow 1966) [in Russian];
V.Yu. Zavadsky: *Calculating Wave Fields in Open Regions and Waveguides* (Nauka, Moscow 1972) [in Russian];
V.M. Babič, N.Y. Kirpičnikova: *The Boundary-Layer Method in Diffraction Problems*, Springer Ser. Eletrophys., Vol.3 (Springer, Berlin, Heidelberg, New York 1979)
5.29 G.D. Malyuzhinetz: Dokl. Akad. Nauk SSSR *121*, 436 (1959);
N.N. Lebedyev, I.P. Skalskaya: Zh. Eksp. Teor. Fiz. *32*, 1174 (1962);
R.P. Starovoitova, M.S. Bobrovnikov, V.N. Kislitsyna: Radiotekh. Elektron. *7*, 250 (1962)
5.30 G.N. Zhizhin, M.A. Moskalova, E.V. Shomina, V.A. Yakovlev: Zh. Eksp. Teor. Fiz. Pis'ma *29*, 533 (1979)
5.31 V.M. Agranovich: *Theory of Excitons* (Nauka, Moscow 1968) [in Russian]
5.32 V.M. Agranovich, O.A. Dubovsky: Fiz. Tverd. Tela *7*, 2885 (1965)
5.33 O.A. Dubovsky: Fiz. Tverd. Tela *12*, 3054 (1970)

5.34 V.N. Lyubimov, D.G. Sannikov: Fiz. Tverd. Tela *14*, 675 (1972)
5.35 V.V. Bryksin, D.N. Mirlin, I.I. Reshina: Fiz. Tverd. Tela *15*, 1118 (1973)
5.36 Yu.A. Romanov: Radiofizika Izv. Vuzov *8*, 1203 (1965);
A. Hartstein, E. Burstein, J.J. Brion, R.F. Wallis: Surface Sci. *34*, 81 (1973)
5.37 A. Hartstein, E. Burstein, J.J. Brion, R.F. Wallis: Solid State Commun. *12*, 1083 (1973);
E. Burstein, F. De Martini (eds.): *Polaritons* (Pergamon, Oxford 1974) p. 111
5.38 G. Borstel: Phys. Status Solidi B*60*, 427 (1973); *65*, 123 (1974)
5.39 G. Borstel, H.J. Falge: Phys. Status Solidi B*83*, 11 (1977)
5.40 V.M. Agranovich: Usp. Fiz. Nauk *115*, 199 (1975);
V.M. Agranovich: In *Surface Polaritons*, ed. by V.M. Agranovich, D.L. Mills (North-Holland, Amsterdam 1982) p.187
5.41 R.F. Wallis, J.J. Brion, E. Burstein, A. Hartstein: Phys. Rev. B*9*, 3424 (1974)
5.42 G.A. Puchkovskaya, V.L. Strizhevsky, S.P. Yashir: Phys. Status Solidi B*89*, 27 (1978)
5.43 V.M. Agranovich, T.A. Leskova: Fiz. Tverd. Tela *16*, 1796 (1974)
5.44 V.M. Agranovich, A.G. Malshukov: Opt. Commun. *11*, 169 (1974)
5.45 G.N. Zhizhin, M.A. Moskalova, V.G. Nazin, V.A. Yakovlev: Fiz. Tverd. Tela *16*, 1402 (1974)
5.46 D.N. Mirlin, I.I. Reshina: Fiz. Tverd. Tela *16*, 1141 (1974)
5.47 E.A. Vinogradov, G.N. Zhizhin, A.G. Malshukov, V.I. Yudson: Solid State Commun. *23*, 915 (1977)
5.48 M. Born: *Optik* (Springer, Berlin 1933);
M. Born, E. Wolf: *Principles of Optics* (Pergamon, Oxford 1964)
5.49 P. Drude: *Lehrbuch der Optik* (Hirzel, Leipzig 1912)
5.50 V.M. Agranovich, V.I. Yudson: Opt. Commun. *5*, 422 (1972)
5.51 D.V. Sivukhin: Zh. Eksp. Teor. Fiz. *18*, 976 (1948); *21*, 94, 367 (1952); *30*, 374 (1956)
5.52 A.V. Chaplik: Zh. Eksp. Teor. Fiz. *60*, 1845 (1971) [English transl.: Sov. Phys. JETP *33*, 997 (1971)]; Zh. Eksp. Teor. Fiz. *62*, 746 (1972) [English transl.: Sov. Phys. JETP *35*, 395 (1972)]
5.53 V.A. Yakovlev, V.G. Nasin, G.N. Zhizhin: Opt. Commun. *15*, 293 (1975);
G.N. Zhizhin, M.A. Moskalova, V.G. Nasin, V.A. Yakovlev: Zh. Eksp. Teor. Fiz. *72*, 687 (1977)
5.54 T. Lopez-Rios, F. Abelès, G. Vuye: J. Phys. Paris *39*, 645 (1978)
5.54a I. Pockrand, A. Brillante, D. Möbius: J. Chem. Phys. *77*, 6289 (1982)
5.55 Y.J. Chabal, A.J. Sievers: Phys. Rev. Lett. *44*, 944 (1980)
5.56 V.M. Agranovich, S.A. Darmanyan, A.G. Malshukov: Opt. Commun. *33*, 234 (1980)
5.57 V.M. Agranovich, I.I. Lalov: Opt. Commun. *16*, 239 (1976)
5.58 B. Fischer, W.J. Buckel, D. Bauerle: Solid State Commun. *15*, 1801 (1974); *14*, 291 (1974)
5.59 M. Cardona: In *Light Scattering in Solids II*, Topics Appl. Phys., Vol.50, ed. by M. Cardona, G. Güntherodt (Springer, Berlin, Heidelberg, New York (1982) Chap.2
5.60 D. Guidotti, S.A. Rice, H.L. Lemberg: Solid State Commun. *15*, 113 (1974)
5.61 S.L. Cunningham, A.A. Maradudin, R.F. Wallis: Phys. Rev. B*10*, 3342 (1974)
5.62 E.M. Conwell: Phys. Rev. B*11*, 1508 (1975);
E.M. Conwell, C.C. Kao: Solid State Commun. *18*, 1123 (1976)
5.62a V.E. Kravtsov, E.I. Firsov, V.A. Yakovlev, G.N. Zhizhin: Solid State Commun. (1984) (in press)
5.63 D.L. Mills: Phys. Rev. B*12*, 4036 (1975); B*14*, 5539 (1976);
E. Kretschmann: Opt. Commun. *5*, 331 (1972); *6*, 185 (1972); *10*, 336 (1974);
A.A. Maradudin: In [5.4a], p.405;
H. Rather: In [5.4a], p.331

5.64 V.M. Agranovich, T.A. Leskova: Zh. Eksp. Teor. Fiz. Pis'ma *29*, 151 (1979)
5.65 B. Fischer, J. Lagois: In *Excitons*, Topics Curr. Phys., Vol.14, ed. by K. Cho (Springer, Berlin, Heidelberg, New York 1979) p.59
5.66 F. De Martini, S.E. Kohn, Y.R. Shen: Phys. Rev. Lett. *38*, 1223 (1977)
5.67 M. Fukui, V.C.Y. So, G.L. Stegeman: Solid State Commun. *30*, 683 (1979);
5.67a R.K. Chang, T.E. Furtak (eds.): *Surface Enhanced Raman Scattering* (Plenum, New York 1981);
D.W. Berreman: Phys. Rev. *163*, 855 (1967);
R. Ruppin: Solid State Commun. *39*, 908 (1981);
P.C. Das, J.I. Gersten: Phys. Rev. B*25*, 6281 (1982);
A.G. Mal'shukov, Sh.A. Shekhmametiev: Fiz. Tverd. Tela (1983)
5.67b V.M. Agranovich, V.E. Kravtsov, T.A. Leskova: Solid State Commun. *47*, 925 (1983)
5.68 V.M. Agranovich: Zh. Eksp. Teor. Fiz. Pis'ma *24*, 602 (1976)
5.69 V.L. Ginzburg, A.P. Levanyuk, A.A. Sobyanin: Usp. Fiz. Nauk *130*, 615 (1980) [English transl.: Phys. Rep. *57(3)*, 151 (1980)]
5.70 V.M. Agranovich, T.A. Leskova: Solid State Commun. *21*, 1065 (1977)
5.71 V.M. Agranovich, V.E. Kravtsov, T.A. Leskova: In *Surface Polaritons*, ed. by V.M. Agranovich, D.L. Mills (North-Holland, Amsterdam 1982) p.511
5.72 L.E. Gurevich, R.G. Tarkhanyan: Fiz. Tverd. Tela *17*, 1944 (1975);
R.G. Tarkhanyan: Phys. Status Solidi B*72*, 111 (1975)
5.73 A.A. Maradudin, D.L. Mills: Phys. Rev. B*7*, 2787 (1973)
5.74 A.Ya. Blank, V.L. Berezinsky: Zh. Eksp. Teor. Fiz. *75*, 2317 (1978)
5.75 K.L. Kleiwer, R. Fuchs: Phys. Rev. *172*, 607 (1968)
5.76 R.H. Ritchie, A.L. Marusak: Surface Sci. *4*, 234 (1966)
5.77 Y. Omura, M. Tsuji: J. Phys. Soc. Jpn. *40*, 107 (1976)
5.78 P.R. Rimbey: Phys. Rev. B*15*, 1215 (1977)
5.79 G.S. Agarwal, D.N. Pattanayak, E. Wolf: Phys. Rev. B*10*, 1447 (1974); B*11*, 1342 (1975);
J.L. Birman, J.J. Sein: Phys. Rev. B*6*, 2482 (1972);
A.A. Maradudin, D.L. Mills: Phys. Rev. B*7*, 2787 (1973)
5.80 V.M. Agranovich: In *Surface Excitations*, ed. by V.M. Agranovich, R. Loudon (North-Holland, Amsterdam 1984) (in press)
5.81 P.R. Rimbey: Phys. Rev. B*15*, 1215 (1977)
5.82 V.M. Agranovich: Zh. Eksp. Teor. Fiz. *77*, 1125 (1979); in *Light Scattering in Solids*, ed. by J.L. Birman, H.Z. Cummins, K.K. Rebane (Plenum, New York 1979) p.113
5.83 V.M. Agranovich, V.E. Kravtsov, T.A. Leskova: Zh. Eksp. Teor. Fiz. *81*, 1828 (1981)
V.M. Agranovich, V.E. Kravtsov, T.A. Leskova: Solid State Commun. *40*, 687 (1981)
5.84 V.M. Agranovich, V.E. Kravtsov, T.A. Leskova: Zh. Eksp. Teor. Fiz. *84*, 103 (1983)
5.85 Z. Schlesinger, A.J. Sievers: Report Nr.4177, Materials Science Center, Cornell University, Ithaca (1979)
5.86 V.M. Agranovich, O.A. Dubovsky: Zh. Eksp. Teor. Fiz. Pis'ma *26*, 641 (1977)
5.87 Ya.B. Zeldovich: Zh. Eksp. Teor. Fiz. *67*, 2357 (1974)
5.88 H. Ibach (ed.): *Electron Spectroscopy for Surface Analysis*, Topics Curr. Phys., Vol.4 (Springer, Berlin, Heidelberg, New York 1977)
5.89 M. Cardona (ed.): *Light Scattering in Solids*, Topics Appl. Phys., Vol.8 (Springer, Berlin, Heidelberg, New York 1975);
M. Cardona, G. Güntherodt (eds.): *Light Scattering in Solids II* and *Light Scattering in Solids III*, Topics Appl. Phys., Vols.50,51 (Springer, Berlin, Heidelberg, New York 1982)
5.90 W. Demtröder: *Laser Spectroscopy*, Springer Ser. Chem. Phys., Vol.5 (Springer, Berlin, Heidelberg, New York 1981)
5.91 A.A. Onliner: *Acoustic Surface Waves*, Topics Appl. Phys., Vol.24 (Springer, Berlin, Heidelberg, New York 1978)

5.92 H. Wolter: "Optik dünner Schichten," in *Handbuch der Physik*, Vol.24, (Springer, Berlin, Göttingen, Heidelberg 1956) p.461;
N.I. Herrick: *Internal Reflection Spectroscopy* (Interscience, New York 1967)
5.93 A. Otto: Z. Phys. *216*, 398 (1968)
5.94 R. Ruppin: Solid State Commun. *8*, 1129 (1970)
5.95 H.K.V. Lotsch: Optik *27*, 239 (1968)
5.96 L.V. Iogansen: Bull. Invention *19* (1962); J. Opt. Soc. *66*, 972 (1976)
5.97 H.K.V. Lotsch: Optik *32*, 116,189,299,553 (1970/71)
5.98 T. Tamir: *Integrated Optics*, Topics Appl. Phys., Vol.7, 2nd ed. (Springer, Berlin, Heidelberg, New York 1979);
R.G. Hunsperger: *Integrated Optics: Theory and Technology*, Springer Ser. Opt. Sci., Vol.33 (Springer, Berlin, Heidelberg, New York 1982)
5.99 V.V. Bryksin, Yu.M. Gerbshtein, D.N. Mirlin: Fiz. Tverd. Tela *13*, 2125 (1971); *14*, 543 (1972)
5.100 V.V. Bryksin, Yu.M. Gerbshtein, D.N. Mirlin: Phys. Status Solidi B*51*, 901 (1971)
5.101 N. Marschall, B. Fischer: Phys. Rev. Lett. *28*, 811 (1972)
5.102 P. Halevi: In *Electromagnetic Surface Modes*, ed. by A.D. Boardman (Wiley, New York 1982) p.249
5.103 V.V. Bryksin, Yu.M. Gerbshtein, D.N. Mirlin: Zh. Eksp. Teor. Fiz. Pis'ma *15*, 445 (1972)
5.104 V.A. Yakovlev, G.N. Zhizhin: Zh. Eksp. Teor. Fiz. Pis'ma *19*, 333 (1974)
5.105 J. Lagois, B. Fischer: Phys. Rev. Lett. *36*, 680 (1976); Solid State Commun. *18*, 1519 (1976)
5.106 I. Hirabayashi, T. Koda, Y. Tokura, J. Murata, Y. Kaneko: J. Phys. Soc. Jpn. *40*, 1215 (1976)
5.107 K. Tomioka, S.A. Rice, M.G. Sceats: Bull. Am. Phys. Soc. Ser.II *21*, 355 (1976)
5.108 M.R. Philpott, A. Brillante, I.R. Pokrand, J.D. Swalen: Mol. Cryst. Liq. Cryst. *50*, 139 (1979)
5.109 L.I. Mandelstam: Phys. Z. *15*, 220 (1914)
5.110 A. Otto: *Spectroscopy of Surface Polaritons by Attenuated Total Reflection. Advances in Solid State Physics* (Pergamon, New York 1974)
5.111 B. Fischer, J. Lagois: In *Excitons*, Topics Curr. Phys., Vol.14, ed. by K. Cho (Springer, Berlin, Heidelberg, New York 1979) p.183
5.112 F. Abelès, T. Lopez-Rios: Opt. Commun. *11*, 89 (1974)
5.113 Y.Y. Teng, E.A. Stern: Opt. Commun. *19*, 511 (1967)
5.114 R.H. Ritchie, E.T. Arakawa, J.J. Cowan, R.N. Hamm: Opt. Commun. *21*, 1530 (1968)
5.115 D. Beaglehole: Opt. Commun. *22*, 708 (1969)
5.116 N. Marschall, B. Fischer, H.J. Queisser: Opt. Commun. *27*, 95 (1971)
5.117 B. Fischer, N. Marschall, H.J. Queisser: Surface Sci. *34*, 50 (1973)
5.118 J. Schoenwald, E. Burstein, R.F. Wallis: Bull. Am. Phys. Soc. Ser.III *16*, 1409 (1971)
5.119 D.L. Mills: Phys. Rev. B*15*, 3097 (1977)
5.120 U. Fano: Ann. Phys. NY *32*, 393 (1938);
A. Hessel, A.A. Onliner: Appl. Opt. *4*, 1275 (1965)
5.121 J. Hügglung, F. Sellberg. J. Opt. Soc. Am. *56*, 1031 (1966)
5.122 C.E. Wheeler, E.T. Arakawa, R.H. Ritchie: Phys. Rev. B*13*, 2372 (1976)
5.123 R. Petit (ed.): *Electromagnetic Theory of Gratings*, Topics Curr. Phys., Vol.22 (Springer, Berlin, Heidelberg, New York 1980)
5.124 R. Ruppin, R. Engelman: In *Light Scattering Spectra of Solids*, ed. by G.B. Wright (Springer, New York 1969) p.157
5.125 V.M. Agranovich, V.L. Ginzburg: *Light Scattering in Solids*, ed. by M. Balkanski (Flammarion Sci., Paris 1971) p.226; Zh. Eksp. Teor. Fiz. *61*, 1243 (1971)
5.126 S. Ushioda: In *Progress in Optics*, Vol.19, ed. by E. Wolf (North-Holland, Amsterdam 1981) p.139

5.126a S. Ushioda, R. Loudon: In *Surface Polaritons*, ed. by V.M. Agranovich, D.L. Mills (North-Holland, Amsterdam 1982) p.535
5.127 D.I. Evans, S. Ushioda, J.D. McMullen: Phys. Rev. Lett. *31*, 369 (1973)
5.128 J.Y. Prieur, S. Ushioda: Phys. Rev. Lett. *34*, 1012 (1975)
5.129 Y.J. Chen, E. Burstein, D.L. Mills: Phys. Rev. Lett. *34*, 1516 (1975)
5.130 V.M. Agranovich: Zh. Eksp. Teor. Fiz. Pis'ma *19*, 28 (1974)
5.131 J. Schoenwald, E. Burstein, J.M. Elson: Solid State Commun. *12*, 185 (1973)
5.132 J.D. McMullen: Solid State Commun. *17*, 331 (1975)
5.133 C.A. Ward, R.W. Alexander, R.J. Bell: Phys. Rev. B*12*, 3293 (1975)
5.134 D.A. Bryan, D.L. Begley, K. Bhasin, R.W. Alexander: Surface Sci. *57*, 53 (1976)
5.135 G.N. Zhizhin, M.A. Moskalova, E.V. Shomina, V.A. Yakovlev: Zh. Eksp. Teor. Fiz. Pis'ma *24*, 221 (1976)
5.136 D.L. Mills: Solid State Commun. *24*, 669 (1977)
5.137 Y.R. Shen, F. de Martini: In *Surface Polaritons*, ed. by V.M. Agranovich, D.L. Mills (North-Holland, Amsterdam 1982) p.629
5.138 H. Ibach: Phys. Rev. Lett. *24*, 1416 (1970); *27*, 253 (1971)
5.139 R.H. Ritchie: Phys. Rev. *106*, 874 (1957);
H. Boersch, T. Geiger, W. Stickel: Z. Phys. *222*, 130 (1968)
5.140 K.A. Barsukov, L.G. Naryshkina: Zh. Tekh. Fiz. *36*, 800 (1966)
5.141 M.F. Bishop, A.A. Maradudin: Solid State Commun. *12*, 1225 (1973)
5.142 V.L. Ginzburg, V.N. Tsytovich: Usp. Fiz. Nauk *126*, 553 (1978) [English transl.: Phys. Rep. *49(1)*, 1 (1979)];
5.143 W.J. Tomlinson: Opt. Lett. *5*, 323 (1980)
5.144 V.M. Agranovich, V.S. Babichenko, V.Ya. Chernyak: Zh. Eksp. Teor. Fiz. Pis'ma *32*, 352 (1980) [English transl.: Sov. Phys. JETP Lett. *32*, 512 (1980)]
5.145 A.A. Maradudin: Z. Phys. B*41*, 341 (1981);
A.A. Maradudin: In *Optical and Acoustic Waves in Solids - Modern Topics* (World Scientific, Singapore 1983) p.72
5.146 V.M. Agranovich, V.Ya. Chernyak: Solid State Commun. *44*, 1309 (1982)
5.147 N.N. Akhmediev: Zh. Eksp. Teor. Fiz. *83*, 545 (1982)
5.148 V.K. Fedyanin, D. Mihalache: Z. Phys. B*47*, 167 (1982)
5.149 M.Y. Yu: Phys. Rev. A*27*, N2 (1983)
5.150 V.M. Agranovich, V.Ya. Chernyak: Phys. Lett. A*88*, 423 (1982)
5.151 Y.C. Chen, G.M. Carter: Bull. Am. Phys. Soc. *27*, 343 (1982)
5.152 V.M. Agranovich, V.I. Rupasov, V.Ya. Chernyak: Zh. Eksp. Teor. Fiz. Pis'ma *33*, 196 (1981); Opt. Commun. *37*, 363 (1981); Fiz. Tverd. Tela *24*, 2992 (1982)
5.153 V.M. Agranovich: In *Optical and Acoustic Waves in Solids - Modern Topics* (World Scientific, Singapore 1983) p.51
5.154 G.T. Adamashvili, V.A. Monakov: Solid State Commun. *48*, 381 (1983)

Chapter 6

6.1 M. Born: *Dynamik der Kristallgitter* (Leipzig 1915) Chap.3
6.2 M. Born, K. Huang: *Dynamical Theory of Crystal Lattices* (Clarendon, Oxford 1954)
6.3 L.D. Landau, E.M. Lifshitz: *Quantum Mechanics - Nonrelativistic Theory* (Nauka, Moscow 1974) [in Russian]
6.4 G. Ya. Lyubarsky: *Group Theory and Its Application in Physics* (Fizmatgiz, Moscow 1958) [in Russian]
6.5 S.I. Pekar: Zh. Eksp. Teor. Fiz. *36*, 451 (1959) [English transl.: Sov. Phys. JETP *9*, 314 (1959)]
6.6 V.M. Agranovich, Yu.V. Konobeyev: Fiz. Tverd. Tela *5*, 2544 (1963) [English transl.: Sov. Phys. Solid State *5*, 1858 (1963)]
6.7 A.A. Abrikosov, L.P. Gorkov, I.E. Dzyaloshinsky: *Methods of Quantum Field Theory in Statistical Physics* (Fizmatgiz, Moscow 1962) [in Russian]

6.8 I.E. Dzyaloshinksky, L.P. Pitayevsky: Zh. Eksp. Teor. Fiz. *36*, 1797 (1959) [English transl.: Sov. Phys. JETP *9*, 1282 (1959)]
6.9 E.N. Economou: *Green's Functions in Quantum Physics*, Springer Ser. Solid-State Sci., Vol.7, 2nd ed. (Springer, Berlin, Heidelberg, New York, Tokyo 1983)
6.10 G.M. Gandelman, V.M. Yermachenko: Zh. Eksp. Teor. Fiz. *45*, 522 (1963) [English transl.: Sov. Phys. JETP *18*, 358 (1964)]
6.11 B.Sh. Zinger: Zh. Eksp. Teor. Fiz. *72*, 1858 (1977)
6.12a M. Goepert-Mayer: Ann. Phys. *9*, 273 (1931)
6.12b V.N. Piskovoi, B.E. Tsekvava: Fiz. Tverd. Tela *25*, 1938 (1983)
6.13 V.I. Cherepanov, V.S. Galishev: Fiz. Tverd. Tela *3*, 1085 (1961)
6.14 A.G. Zhilich, V.I. Cherepanov, Yu.A. Kargapolov: Fiz. Tverd. Tela *3*, 1317 (1961) [English transl.: Sov. Phys. Solid State *3*, 1317 (1961)]
6.15 V.M. Agranovich: Zh. Eksp. Teor. Fiz. *37*, 430 (1959)
6.16 A.S. Davydov: *Theory of Light Absorption in Molecular Crystals* (UkSSR Akad. Nauk, Kiev 1951) [in Russian]
6.17 V.M. Agranovich: Fiz. Tverd. Tela *3*, 811 (1961) [English transl.: Sov. Phys. Solid State *3*, 592 (1961)]
6.18 V.M. Agranovich: Usp. Fiz. Nauk *112*, 132 (1974)
6.19 P.G. Cummins, D.A. Dunmur, R.W. Munn, R.J. Newham: Acta Cryst. A*32*, 847 (1976)
6.19a C.K. Purvis, P.L. Taylor: Phys. Rev. B*26*, 4547 (1982)
6.19b L.N. Ovander, Yu.G. Pashkevich, N.S. Tyu: Opt. Spektrosk. *50*, 1129 (1981);
Yu.G. Pashkevich, A.D. Petrenko: Opt. Spektr. *50*, 554 (1981)
6.20 A.S. Davydov: *The Theory of Molecular Excitons* (Nauka, Moscow 1968) [in Russian]
6.21 V.M. Agranovich: *Theory of Excitons* (Nauka, Moscow 1968) [in Russian]
6.22 V.M. Agranovich, M.D. Galanin: *Electron Excitation Energy Transfer in Condensed Media* (North-Holland, Amsterdam 1982)
6.23 U. Fano: Phys. Rev. *118*, 451 (1960)
6.24 W. Egler, H. Haken: Z. Phys. *28*, 51 (1977)
6.25 I. Toyozawa: J. Phys. Chem. Solids *25*, 59 (1964)
6.26 V.M. Agranovich, Yu.V. Konobeyev: Fiz. Tverd. Tela *6*, 644 (1964) [English transl.: Sov. Phys. Solid State *6*, 644 (1964)]
6.27 H. Haken: Fortschr. Phys. *6*, 271 (1958)
6.28 R.S. Knox: "Theory of Excitons," in *Solid State Physics*, Vol.5 (Academic, London 1963)
6.29 E.E. Lysenko: Zh. Eksp. Teor. Fiz. *6*, 787 (1936)
6.30 A.S. Davydov: Zh. Eksp. Teor. Fiz. *19*, 181 (1949); Tr. Inst. Fiz. Nauk UkSSR *3*, 36 (1952);
A.S. Davydov, E.I. Rashba: Ukr. Fiz. Zh. *2*, 226 (1957)
6.31 W. Heitler: *Quantum Theory of Radiation* (Clarendon, Oxford 1954)
6.32 V.M. Agranovich: In *Spectroscopy and Excitation - Dynamics of Condensed Molecular Systems*, ed. by V.M. Agranovich, R.M. Hochstrasser (North-Holland, Amsterdam 1983) p.83
6.33 V.M. Agranovich: Usp. Fiz. Nauk *71*, 141 (1960) [English transl.: Sov. Phys. Usp. *3*, 427 (1961)]
6.34 V.M. Agranovich, Yu.V. Konobeyev: Fiz. Tverd. Tela *3*, 360 (1961) [English transl.: Sov. Phys. Solid State *3*, 260 (1961)]
6.35 A.A. Demidenko: Fiz. Tverd. Tela *5*, 489 (1963)
6.36 W.C. Tait, R.L. Weicher: Phys. Rev. *166*, 769 (1968)
6.37 L.S. Kothari, K.S. Singwi: Solid State Phys. *8*, 110 (1959)
6.38 B. Dorner: *Coherent Inelastic Neutron Scattering in Lattice Dynamics*, Springer Tracts Mod. Phys., Vol.93 (Springer, Berlin, Heidelberg, New York 1982)
6.39 Yu. Kagan (ed.): *Mössbauer Effect*, Contributed Volume (Inostrannaya Literatura, Moscow 1962)

6.40 U. Gonser (ed.): *Mössbauer Spectroscopy*, Topics Appl. Phys., Vol.5 (Springer, Berlin, Heidelberg, New York 1975)
6.41 O. Klemperer, J.P.G. Shepherd: Adv. Phys. *12*, 355 (1963)
6.42 V.M. Agranovich, I.I. Lalov: Phys. Lett. A*53*, 169 (1975); Zh. Eksp. Teor. Fiz. *69*, 647 (1975); Fiz. Tverd. Tela *18*, 515, 621 (1976)
6.43 V.M. Agranovich, V.L. Ginzburg: Zh. Eksp. Teor. Fiz. *40*, 913 (1961) [English transl.: Sov. Phys. JETP *13*, 638 (1961)]
6.44 V.L. Ginzburg, V.N. Tsytovich: Usp. Fiz. Nauk *126*, 553 (1978) [English transl.: Phys. Rep. *49(1)*, 1 (1979)]
6.45 L.D. Landau, E.M. Lifshitz: *Electrodynamics of Continuous Media* (Gostekhizdat, Moscow 1957) [English transl.: Pergamon, Oxford 1960]
6.46 V.L. Ginzburg: *The Propagation of Electromagnetic Waves in Plasma* (Nauka, Moscow 1967) [English transl.: Pergamon, Oxford 1970]
6.47 D.P. Craig, P.G. Hobbins: J. Chem. Soc. *539*, 2309 (1955)
6.48 V.A. Bushuyev, R.N. Kuzmin: Usp. Fiz. Nauk *122*, 81 (1977)
6.49 L.N. Ovander: Fiz. Tverd. Tela *3*, 2394 (1961); *4*, 157 and 294 (1962) [English transl.: Sov. Phys. Solid State *3*, 1737 (1961); *4*, 112 and 212 (1962)]; Usp. Fiz. Nauk *86*, 3 (1965)
6.50 R.M. Martin, L.M. Falicov: In *Light Scattering in Solids*, Topics Appl. Phys., Vol.8, 2nd ed., ed. by M. Cardona (Springer, Berlin, Heidelberg, New York 1983) p.79
6.51 R. Zeyher, J.L. Birman, W. Brenig: Phys. Rev. B*6*, 4613 and 4617 (1972)
6.52 J.J. Hopfield: Phys. Rev. *182*, 945 (1969)
6.53 D.L. Mills, E. Burstein: Phys. Rev. *188*, 1465 (1969)
6.54 B. Bendow, J.L. Birman: Phys. Rev. B*1*, 1678 (1970)
6.55 B. Bendow: Phys. Rev. B*2*, 5051 (1970)
6.56 A.S. Barker, Jr., R. Loudon: Rev. Mod. Phys. *44*, 18 (1972)
6.57 B. Bendow: In *Electronic Structure of Noble Metals and Polariton-Mediated Ligh Scattering*, Springer Tracts Mod. Phys., Vol.82, (Springer, Berlin, Heidelberg, New York 1978) p.69
6.58 P.Y. Yu: In *Excitons*, Topics Curr. Phys., Vol.14, ed. by K. Cho (Springer, Berlin, Heidelberg, New York 1979) p.211
6.59 V.L. Ginzburg, A.P. Levanyuk: J. Phys. Chem. Solids *6*, 51 (1958); Zh. Eksp. Teor. Fiz. *39*, 192 (1960) [English transl.: Sov. Phys. JETP *12*, 138 (1961)]
6.60 V.L. Ginzburg: Usp. Fiz. Nauk *77*, 621 (1962) [English transl.: Sov. Phys. Usp. *5*, 649 (1963)]
6.61 V.L. Ginzburg, A.P. Levanyuk, A.A. Sobyanin: Usp. Fiz. Nauk *130*, 615 (1980) [English transl.: Phys. Rep. *57(3)*, 151 (1980)]
6.62 V.L. Ginzburg, V.A. Ugarov: Usp. Fiz. Nauk *118*, 175 (1976); Usp. Fiz. Nauk *110*, 309 (1973); *122*, 325 (1977)
6.63 R. Claus, L. Merten, J. Brandmüller: *Light Scattering by Phonon-Polaritons* (Springer, Berlin, Heidelberg, New York 1975)
6.64 J.F. Scott: Am. J. Phys. *39*, 1360 (1971)
6.65 R. Loudon: In [6.70] p.25; E. Burstein, S. Ushioda, A. Pinczukj, J.F. Scott: in [6.70] p.43
6.66 M. Cardona (ed.): *Light Scattering in Solids*, Topics Appl. Phys., Vol.8, 2nd ed. (Springer, Berlin, Heidelberg, New York 1983); M. Cardona, G. Güntherodt (eds.): *Light Scattering in Solids II* and *Light Scattering in Solids III*, Topics Appl. Phys., Vols.50,51 (Springer, Berlin, Heidelberg, New York 1982) M. Cardona, G. Güntherodt (eds.): *Light Scattering in Solids IV* Topics Appl. Phys., Vol.54 (Springer, Berlin, Heidelberg, New York 1983)
6.67 R. Loudon: Proc. Phys. Soc. *82*, 393 (1963)
6.68 E. Burstein: Comments Solid State Phys. *1*, 202 (1969); E. Burstein, D.L. Mills: Comments Solid State Phys. *2*, 111 (1969); Comments Solid State Phys. *3*, 12 (1970)

6.69 H.J. Benson, D.L. Mills: Phys. Rev. B*1*, 4835 (1970)
6.70 G.B. Wright (ed.): *Light Scattering Spectra of Solids* (Springer, Berlin, Heidelberg, New York 1969)
6.71 V.I. Emelyanov, Yu.L. Klimontovich: Zh. Eksp. Teor. Fiz. *62*, 778 (1972)
6.72 V.M. Agranovich, B.N. Mavrin, Kh.E. Sterin: Usp. Fiz. Nauk *113*, 710 (1974)
6.73 V.M. Agranovich, V.L. Ginzburg: In *Light Scattering in Solids*, 2nd Intern. Conference, ed. by M. Balkansi (Flammarion Sci., Paris 1971) p.226; Zh. Eksp. Teor. Fiz. *61*, 1243 (1971)
6.74 V.L. Ginzburg: Usp. Fiz. Nauk *106*, 151 (1972)
6.75 V.L. Ginzburg: *Theoretical Physics and Astrophysics* (Nauka, Moscow 1976) [English transl.: Pergamon, Oxford 1979]
6.76 B. Balkanski (ed.): *Light Scattering in Solids*, 2nd Intern. Conference Paris 1971 (Flammarion Sci., Paris 1971)
6.77 M. Balkansi, R.C.C. Leite, S.P.S. Porto (eds.): *Light Scattering in Solids*, 3rd Intern. Conference (Flammarion Sci., Paris 1975)
6.78 B. Bendow, J.L. Birman, V.M. Abranovich (eds.): *Theory of Light Scattering in Condensed Matter* (Plenum, New York 1976)
6.79 Yu.N. Polivanov: Usp. Fiz. Nauk *126*, 185 (1978)
6.80a D.L. Mills, Y.J. Chen, E. Burstein: Phys. Rev. B*13*, 4419 (1976)
6.80b N.N. Akhmediyev: Kvantovaya Elektron. *3*, 1354 (1976)
6.81 H.Z. Cummins, A.P. Levanyuk (eds.): *Light Scattering Near Phase Transitions* (North-Holland, Amsterdam 1983)
6.82a V.M. Agranovich: Opt. Commun. *11*, 389 (1974)
6.82b V.M. Agranovich, T.A. Leskova: Fiz. Tverd. Tela *17*, 1367 (1975); Fiz. Tverd. Tela *19*, 804 (1977)
6.83 D.I. Evans, S. Ushioda, J.D. McMullen: Phys. Rev. Lett. *31*, 369 (1973)
6.84 J.Y. Prieur, S. Ushioda: Phys. Rev. Lett. *34*, 1012 (1975)
6.84a S. Ushioda, R. Loudon: In *Surface Polaritons*, ed. by V.M. Agranovich, D.L. Mills (North-Holland, Amsterdam 1982) p.535
6.85 V.M. Agranovich: Zh. Eksp. Teor. Fiz. Pis'ma *19*, 28 (1974)
6.86 V.M. Agranovich: Usp. Fiz. Nauk *115*, 199 (1975)
6.87 Ya.I. Frenkel: *Kinetic Theory of Liquids* (USSR Akad. Nauk, Moscow 1945) Chap.2 [in Russian]
6.88 I.M. Lifshitz, Ya.B. Geguzin: Fiz. Tverd. Tela *7*, 62 (1965)
6.89 J.S. Nkoma, R. Loudon: J. Phys. C*8*, 1950 (1975)
6.90 A. Weber (ed.): *Raman Spectroscopy of Gases and Liquids*, Topics Curr. Phys., Vol.11 (Springer, Berlin, Heidelberg, New York 1979) Chaps.6,7
6.91 R.W. Terhune, P.D. Marker: Phys. Rev. A*137*, 801 (1965)
6.92 J.A. Giordmaine, W. Kaiser: Phys. Rev. *144*, 676 (1966)
6.93 S.A. Akhmanov, N.I. Koroteyev: Zh. Eksp. Teor. Fiz. *67*, 1306 (1974); Usp. Fiz. Nauk *123*, 405 (1977)
6.94 J.P. Coffinet, F. De Martini: Phys. Lett. *22*, 60 (1969)
6.95 J.J. Wynne: Comments Solid State Phys. *7*, 7 (1975)
6.96 D.N. Klyshko: Kvantovaya Elektron. *2*, 265 (1975)
6.97 V.L. Strizhevsky, Yu.N. Yashkir: Kvantovaya Elektron. *2*, 995 (1975)
6.98 J.J. Wynne: Phys. Rev. Lett. *29*, 650 (1972)
6.99 Yu.N. Polivanov, R.Sh. Sayanov, A.T. Sukhodolsky: Kratk. Soobshch. Fiz. *12*, 16 (1976); Yu.N. Polivanov, A.T. Sikhodolsky: Zh. Eksp. Teor. Fiz. Pis'ma *25*, 240 (1977)
6.100 F. De Martini: Phys. Rev. B*4*, 4556 (1971)
6.101 N.N. Akhmediyev: Opt. Spektrosk. *39*, 779 (1975)
6.102 V.L. Strizhevsky, Yu.N. Yashkir: Kvantovaya Elektron. *2*, 2602 (1975)
6.103 F. De Martini, G. Giuliana, P. Martini, E. Palange, Y.R. Shen: Phys. Rev. Lett. *37*, 440 (1976)
6.104 V.M. Agranovich, A.A. Rukhadze: Zh. Eksp. Teor. Fiz. *35*, 1171 (1958) [English transl.: Sov. Phys. JETP *8*, 819 (1959)]

6.105 B.J. Hoenders, D.N. Pattanayak: Phys. Rev. D*13*, 282, 291 (1976)
6.106 V.L. Ginzburg: Usp. Fiz. Nauk *69*, 537 (1959) [English transl.: Sov. Phys. Usp. *2*, 874 (1960)]
6.107 V.M. Agranovich, V.E. Pafomov, A.A. Rukhadze: Zh. Eksp. Teor. Fiz. *36*, 238 (1959)
6.108 F.G. Bass, M.I. Kaganov, V.M. Yakovenko: Fiz. Tverd. Tela *4*, 3260 (1962) [English transl.: Sov. Phys. Solid State *5*, 2386 (1962)]
6.109 V.L. Ginzburg, I.M. Frank: Zh. Eksp. Teor. Fiz. *16*, 15 (1946)
6.110 B.L. Zhelnov: Zh. Eksp. Teor. Fiz. *40*, 170 (1961) [English transl.: Sov. Phys. JETP *13*, 117 (1961)]
6.111 V.Ya. Eidman: Radiofizika Izv. Vuzov *5*, 478 (1962)
6.112 F.G. Bass, V.M. Yakovenko: Usp. Fiz. Nauk *86*, 189 (1965)
6.113 Ya.M. Strelniker: Fiz. Tverd. Tela *25*, 2230 (1983)
6.114 W. Ekardt: Phys. Rev. B*23*, 3723 (1981)

Chapter 7

7.1 V.L. Ginzburg: *The Propagation of Electromagnetic Waves in Plasmas*, 2nd ed. (Pergamon, Oxford 1970);
V.L. Ginzburg, A.A. Rukhadze: "Waves and Resonances in Magneto-Active Plasma," in *Handbuch der Physik*, Vol.49/4 (Springer, Berlin, Heidelberg, New York 1972) p.395
7.2 V.M. Agranovich, D.L. Mills (eds.): *Surface Polaritons* (North-Holland, Amsterdam 1982)
7.3 V.M. Agranovich, V.E. Kravtsov, T.A. Leskova: Zh. Eksp. Teor. Fiz. *84*, 103 (1983)
7.4 R.K. Chang, T.E. Furtak (eds.): *Surface Enhanced Raman Scattering* (Plenum, New York 1982)
7.5 V.M. Agranovich, V.E. Kravtsov, T.A. Leskova: Solid State Commun. *47*, 925 (1983)
7.6 H. Haken, A. Schenzle: Z. Phys. *258*, 231 (1973)
7.7 V.M. Agranovich, V.I. Rupasov: Fiz. Tverd. Tela *18*, 801 (1976); Chem. Phys. Lett. (1984, in press)
7.8 V.I. Karpman: *Nonlinear Waves in Dispersive Media* (Pergamon, Oxford 1975)
7.9 B.B. Kadomtsev: *Collective Phenomena in Plasma* (Nauka, Moscow 1976)

Appendix

A.1 J.F. Nye: *Physical Properties of Crystals* (Clarendon, Oxford 1964)
A.2 G.Ya. Lyubarsky: *Group Theory and Its Application in Physics* (Fizmatgiz, Moscow 1958) [in Russian]
A.3 E.P. Wigner: *Group Theory and Its Application to Quantum Mechanics of Atomic Spectra* (Academic, New York 1959)
A.4 J.L. Birman: "Theory of Crystal Space Groups and Infrared and Raman Lattice Processes of Insulating Crystals", in *Handbuch der Physik*, Vol.25/2b (Springer, Berlin, Heidelberg, New York 1974)
A.5 B.K. Vainshtein: *Modern Crystallography I*, Springer Ser. Solid-State Sci., Vol.15 (Springer, Berlin, Heidelberg, New York 1981)

Subject Index

Additional boundary condition (ABC) 206, 207
Absorption
  integral exciton absorption 164, 165, 247
  in the vicinity of an exciton resonance 361
  linewidth 163, 164
  long-wavelength edge 363, 366
  quadrupole 7, 171
Additional waves 6, 15
Anisotropy, optical 6, 173
  in the presence of external influences 168 – 190
Approximation of classical crystal optics 37 – 40
Attenuated total reflection (ATR) method 315 – 317
Average heat evolved per unit time 73
Average of Poynting's vector 76

Boundary conditions, Maxwell 20, 205
  for a gyrotropic medium 96
  for optically nonlinear medium with a center of inversion 98
  in the case of spatial dispersion near a separate resonance 219
  with "dead" layer 224
  with normal component of surface current 21
Brewster waves 278
Brillouin-Mandelstam scattering of polaritons 363
Broken symmetry method (in Raman scattering by surface polaritons) 38

Causality principle 22, 26
Cherenkov radiation, intensity for new waves 399 – 401
Coherent anti-Stokes Raman spectroscopy (CARS) by polaritons 393 – 395
Conductivity, complex non-Hermitian 28
Conservation of energy for a field 83
Conservation of momentum for a field 93
Coulomb excitons 51
Crystal optics of surfaces 323
Crystal symmetry (notations) 406

Davydov splitting 352
"Dead" layer problem 224
  theory 225 – 226
Denisty of an electromagnetic pulse 93
Dichroism, circular 251, 253
Dielectric constant
  longitudinal 40 – 41
  transverse 40 – 41
Dispersion equation 44
Dispersion relations 31, 32, 68
  for a complex refractive index 66

Edge modes (one-dimensional waves) at the edge of a film on substrate 310
Effect or retardation on the excitons 13 – 15, 365
Electrical analog of the Faraday effect 194
Electromagnetic waves
  inhomogeneous plane 45
  longitudinal 14
  new 6, 15, 145, 150
  normal 1, 42, 149
Energy conservation law 76, 237
Energy density of the electric field 77
Energy flow velocity 87
Equations of an electromagnetic field 18
Ewald method 344
Excitons
  Frenkel's 9
  real, Coulomb, mechanical 10
  Wannier-Mott 9

Experimental investigation of spatial dispersion in gyrotropic crystal 253
- anisotropy of light absorption in cubic crystal 257
- birefringence 258
- circular dichroism 255
- frequency dependence of the rotation 254
- in nongyrotropic crystals 255
- new wave 255–256

Experimental investigations of surface polaritons
- ATR method 318
- diffraction-grating method 319
- ILES method 325
- propagation of a surface polariton along a copper-air boundary 323
- Raman scattering in layer system 322

Experimental new waves observations
- interference phenomena 259, 264–265
- Mandelstam-Brillouin resonant scattering 267
- reflection spectra 263, 264, 269
- refraction of light by a prism 265
- time-of-flight experiments 266
- verification of dispersion relations 269

Fermi resonance with surface polariton 300
"Fictitious" longitudinal waves 53
Frequency dependence of tensor 23
Frequency dispersion 2
Fresnel's equation 47

Goos-Hänchen shift 91, 92, 315
Group velocity 6, 78, 84, 165
- direction 6

Gyrotropy 3, 5
- angle of rotation of the polarization plane 142
- caused by terms of the order of k^3 142–143
- for regions of semiconductors with interband transitions 214
- gyromagnetic birefringence 129
- near a separated resonance of dielectric 213
- near magnetic phase transition points 129
- weak 128

Index of refraction 2
Interference conditions 245

Kramers-Kronig (dispersion) relations 31–33, 269

Local-field method 348–350

Magnetic-field-induced spatial dispersion 194
Magnetic-field inversion 196
Magnetic permeability and magnetic susceptibility 38–39
Magnetoactive medium 59
Magneto-Stark effect 202
Mechanical excitons 54, 56
- in molecular crystals 343–344

Mechanical stress, influence of 188
Microtheory
- afrionic crystals 328
- based on knowledge of the Coulomb excitons 335
- contribution of the dipole exciton states 338–339
- contribution of the quadrupolar exciton states 339–340
- for molecular crystals 342, 348
- for temperature $T=0$ 333
- quantum-mechanical treatment 329–332
- theory of 334
- with absorption 355

Normal waves in gyrotropic crystals 138
- non-orthogonality 140
- polarization 140

Orthogonality of normal waves 99
- in gyrotropic crystal 140

Oscillator model 353

Plasmon 10
Plasmopolariton 300
Polariton (real exciton) 14, 51
- microscopic theory
- surface

Polariton dispersion in gyrotropic cubic crystals 144
Polarization waves 97
Poynting theorem 94
- applications

Principle of growth of entropy 34
Principle of symmetry of the kinetic coefficients 29, 97
- for nongyrotropic medium 30
- in the presence of a magnetic field 102, 193

Quadrupole transitions
- new waves 185
- selection rules 185

Quantum mechanical derivation of $\varepsilon_{ij}(\omega, \boldsymbol{k})$ and $\varepsilon_{ij}^{(n)}(\omega, \boldsymbol{k})$ 329–334
near isolated exciton line 338–339
on the application of Green's function 335
Raman scattering by bulk polaritons, cross section 377
general expression 381
linewidth 383
Raman scattering by surface polaritons, cross section 387
compensation effect 388
Raman scattering of polaritons 363, 364
Raman scattering of x rays, cross section 369–371
Reciprocity theorem 100
generalized 102
Reflection of a finite-duration pulse 87
Refracted wave in a medium with spatial dispersion 252
Refractive index near exciton resonances 341
Representations, irreducible 175, 187, 190
Rotation of the polarization plane 141, 142

Scattering of neutrons by laser photons 16
Scattering of polaritons by phonons 15
Selection of the roots of the dispersion equation 252
Self-induced transparency of surface polariton 327
Singular optical axes 59, 61
Space group theory (information) 409
Spatial dispersion 3
Spatial dispersion, weak 117
Stress tensor 93
Subsurface layer effects 243
Surface current in ABC 228–233
Surface excitons (polaritons) 271
at the interface of isotropic media 272–273
dispersion relations 274, 275
for four-layered structure 292
for three-layered structure 275
transformation at surface boundaries (reflection, refraction and diffraction) 280–283
with damping 279
Surface nonlinear polaritons 325–327
Surface polariton linewidth 390
Surface polariton scattering in the vicinity of phase-transition points 303
Surface polariton with spatial dispersion
additional (new) surface waves 309, 310
at the right-left gyrotropic crystal interface 311
diffraction of surface polaritons at the impedance step 310
in Coulomb limit ($\omega/c \to 0$) 305
in general case 306–308
interference phenomena 310
Surface polaritons in anisotropic crystals
dispersion law 286
"virtual" 289
without retardation (Coulomb) 288
Surface roughness effects 301
path length of surface polariton 301
Surface TE polariton 298
Surface TH polariton spectra at an interface with transition layer (TL) 289
for high electric conductivity TL 291, 293
in the presence of a resonance 295–296
resonance dependence of the propagation length 297

Tensor 23, 63
energy-momentum 93
general properties 27–32
in gyrotropic crystals 122–128
introduction 108
inverse 24
symmetry properties 33
with absorption 359
Total internal reflection with spatial dispersion 107
for gyrotropic cubic crystal 211
for isotropic nongyrotropic media 216
near an isolated resonance 210
Transitions 167
dipole 167, 238
quadrupole 167, 174, 185, 233
selection rules 174

Van der Waals forces 349

Topics in Applied Physics

Founded by **H. K. V. Lotsch**

Volume 8

Light Scattering in Solids I

Introductory Concepts

Editor: **M. Cardona**

2nd corrected and updated edition. 1983.
111 figures. XV, 363 pages
ISBN 3-540-11913-2

Contents: *M. Cardona:* Introduction. – *A. Pinczuk, E. Burstein:* Fundamentals of Inelastic Light Scattering in Semiconductors and Insulators. – *R. M. Martin, L. M. Falicov:* Resonant Raman Scattering. – *M. V. Klein:* Electronic Raman Scattering. – *M. H. Brodsky:* Raman Scattering in Amorphous Semiconductors. – *A. S. Pine:* Brillouin Scattering in Semiconductors. – *Y.-R. Shen:* Stimulated Raman Scattering. – Overview. – Additional References with Titles. – Subject Index. – Contents of Light Scattering in Solid II, III and IV.

Volume 50

Light Scattering in Solids II

Basic Concepts and Instrumentation

Editors: **M. Cardona, G. Güntherodt**

1982. 88 figures. XIII, 251 pages
ISBN 3-540-11380-0

Contents: *M. Cardona, G. Güntherodt:* Introduction. – *M. Cardona:* Resonance Phenomena. – *R. K. Chang, M. B. Long:* Optical Multichannel Detection. – *H. Vogt:* Coherent and Hyper-Raman Techniques. – Subject Index.

Volume 51

Light Scattering in Solids III

Recent Results

Editors: **M. Cardona, G. Güntherodt**

1982. 128 figures. XI, 281 pages
ISBN 3-540-11513-7

Contents: *M. Cardona, G. Güntherodt:* Introduction. – *M. S. Dresselhaus, G. Dresselhaus:* Light Scattering in Graphite Intercalation Compounds. – *D. J. Lockwood:* Light Scattering from Electronic and Magnetic Excitations in Transition-Metal Halides. – *W. Hayes:* Light Scattering by Superionic Conductors. – *M. V. Klein:* Raman Studies of Phonon Anomalies in Transition-Metal Compounds. – *J. R. Sandercock:* Trends in Brillouin Scattering: Studies of Opaque Materials, Supported Films, and Central Modes. – *C. Weisbuch, R. G. Ulbrich:* Resonant Light Scattering Mediated by Excitonic Polaritons in Semiconductors. – Subject Index.

Volume 54

Light Scattering in Solids IV

Editors: **M. Cardona, G. Güntherordt**

ISBN 3-540-11942-6
In preparation

Contents: Introduction. – *G. Abstreiter, M. Cardona, A. Pinczuk:* Light Scattering by Free Carrier Excitations of Semiconductors. – *S. Geschwind, R. Romestain:* Spin Flip Raman Scattering in CdS. – *G. Güntherodt, R. Zeyher:* Spin-dependent Raman Scattering in Magnetic Semiconductors. – *G. Güntherodt, R. Merlin:* Raman Scattering in Rare Earth Chalcogenides. – *A. Otto:* Surface Enhanced Raman Scattering (Experiment). – *R. Zeyher, K. Arya:* Surface Enhanced Raman Scattering (Theory). – *B. A. Weinstein, R. Zallen:* Pressure-Raman Effects in Covalent and Molecular Solids.

Springer Series in Solid-State Sciences

Editors: M. Cardona P. Fulde H.-J. Queisser

1 **Principles of Magnetic Resonance** 2nd Edition 2nd Printing By C. P. Slichter

2 **Introduction to Solid-State Theory** 2nd Printing. By O. Madelung

3 **Dynamical Scattering of X-Rays in Crystals** By Z. G. Pinsker

4 **Inelastic Electron Tunneling Spectroscopy** Editor: T. Wolfram

5 **Fundamentals of Crystal Growth I** Macroscopic Equilibrium and Transport Concepts. 2nd Printing By F. Rosenberger

6 **Magnetic Flux Structures in Superconductors** By R. P. Huebener

7 **Green's Functions in Quantum Physics** By E. N. Economou 2nd Edition

8 **Solitons and Condensed Matter Physics** 2nd Printing Editors: A. R. Bishop and T. Schneider

9 **Photoferroelectrics** By V. M. Fridkin

10 **Phonon Dispersion Relations in Insulators** By H. Bilz and W. Kress

11 **Electron Transport in Compound Semiconductors** By B. R. Nag

12 **The Physics of Elementary Excitations** By S. Nakajima, Y. Toyozawa, and R. Abe

13 **The Physics of Selenium and Tellurium** Editors: E. Gerlach and P. Grosse

14 **Magnetic Bubble Technology** 2nd Edition By A. H. Eschenfelder

15 **Modern Crystallography I** Symmetry of Crystals, Methods of Structural Crystallography By B. K. Vainshtein

16 **Organic Molecular Crystals** Their Electronic States. By E.A. Silinsh

17 **The Theory of Magnetism I** Statics and Dynamics. By D. C. Mattis

18 **Relaxation of Elementary Excitations** Editors: R. Kubo and E. Hanamura

19 **Solitons,** Mathematical Methods for Physicists. 2nd Printing By G. Eilenberger

20 **Theory of Nonlinear Lattices** By M. Toda

21 **Modern Crystallography II** Structure of Crystals. By B. K. Vainshtein, V. M. Fridkin, and V. L. Indenbom

22 **Point Defects in Semiconductors I** Theoretical Aspects By M. Lannoo and J. Bourgoin

23 **Physics in One Dimension** Editors: J. Bernasconi, T. Schneider

24 **Physics in High Magnetic Fields** Editors: S. Chikazumi and N. Miura

25 **Fundamental Physics of Amorphous Semiconductors** Editor: F. Yonezawa

26 **Elastic Media with Microstructure I** One-Dimensional Models. By I.A. Kunin

27 **Superconductivity of Transition Metals** Their Alloys and Compounds. By S.V. Vonsovsky, Yu.A. Izyumov, and E.Z. Kurmaev

28 **The Structure and Properties of Matter** Editor: T.Matsubara

29 **Electron Correlation and Magnetism in Narrow-Band Systems** Editor: T. Moriya

30 **Statistical Physics I** By M. Toda, R. Kubo, N. Saito

31 **Statistical Physics II** By R. Kubo, M. Toda, N. Hashitsume

32 **Quantum Theory of Magnetism** By R. M. White

33 **Mixed Crystals** By A. I. Kitaigorodsky

34 **Phonons: Theory and Experiments I** Lattice Dynamics and Models of Interatomic Forces. By P. Brüesch

35 **Point Defects in Semiconductors II** Experimental Aspects By J. Bourgoin and M. Lannoo

36 **Modern Crystallography III** Crystal Growth By A. A. Chernov et al.

37 **Modern Crystallography IV** Physical Properties of Crystals By L. A. Shuvalov et al.

38 **Physics of Intercalation Compounds** Editors: L. Pietronero and E. Tosatti

39 **Anderson Localization** Editors: Y. Nagaoka and H. Fukuyama